ELECTRONIC STRUCTURE AND PROPERTIES OF TRANSITION METAL COMPOUNDS

ELECTRONIC STRUCTURE AND PROPERTIES OF TRANSITION METAL COMPOUNDS

Introduction to the Theory

Isaac B. Bersuker
The University of Texas at Austin

A WILEY-INTERSCIENCE PUBLICATION

JOHN WILEY & SONS, INC.

New York · Chichester · Brisbane · Toronto · Singapore

Library of Congress Cataloging in Publication Data:

Bersuker, I. B. (Isaak Borisovich)
    Electronic structure and properties of transition metal compounds
    introduction to the theory / Isaac B. Bersuker.
        p.   cm.
    Includes bibliographical references and indexes.
    ISBN 0-471-13079-6 (cloth : acid-free paper)
    1. Transition metal compounds. I. Title.
    QD172.T6B48   1995   95-461
    546'.6—dc20

To my grandsons Eugene and Kirill

FOREWORD

While several major treatises have been published in the English language which purport to cover the topic of this volume, none has appeared in more than a decade. Advances in computing during this time have put the tools of the theoretical chemist in the hands of the experimentalist. Use of these tools requires a working knowledge of the basis for understanding the electronic structure and properties of the transition elements and their compounds. Therefore, theoretical features of the electronic structure of transition metal compounds are an important component of the training of both the experimental and theoretical chemist contributing solutions to problems in this area. Since this chemistry permeates industrial and biological chemistry as well as catalysis, solutions to problems in this field take on considerable importance. The field of transition metal chemistry is, indeed, fortunate that Isaac Bersuker, the leading contributor to the theory of transition metal electronic structure in the late period of the former Soviet Union, has translated and edited his numerous contributions published originally in Russian and combined them with new work into this modern English language text.

Isaac Bersuker became recognized as an authority in the Soviet Union on transition metal chemistry theory with publication of his 1962 book, which in many ways covered the same material that was published in English as *An Introduction to Transition-Metal Chemistry* by L. E. Orgel (Frome, Somerset, England: Butler and Tanner Ltd., 1960). While Orgel's book was "must" reading for transition metal chemists in the West in the 1960s, Bersuker's book served the same role in the Soviet Union. To our knowledge, Bersuker's Russian language book was never fully translated into English, although one of us (JPF) had a Russian-speaking student translate much of it for him in the late 1960s. As Bersuker's interaction with Western scientists increased during the 1970s and 1980s, his contribution to the understanding of the electronic structure of metal systems became increasingly known and appreciated. While in the former Soviet Union, Bersuker directed a powerful Academy of Sciences program in Kishinev, Moldova which focused on problems involving coupling between electronic and vibrational phenomena. In 1993 he moved to the United States (Austin, Texas), where he continues to be active. He recently authored a major review on "The Concept of Vibronic Interactions in Crystal Stereochemistry of Transition Metal Compounds," *J. Coord. Chem.*, 289–338 (1995).

Many readers of this Foreword will be familiar with Isaac Bersuker's contributions to the understanding of the Jahn–Teller effect. His book *The*

Jahn–Teller Effect and Vibronic Interactions in Modern Chemistry (New York: Plenum Press, 1984) is an important contribution used by persons interested in evaluating those systems in which vibrational and electronic states have similar energy separations which interact with each other. The useful concept of *orbital vibronic constants* is first presented in text form in the Jahn–Teller volume. This topic is brought into focus for the reader of this book in Chapter 7, following along after several introductory chapters that build the basic theory of transition metal chemistry. Incidentally, relativistic effects for transition metal compounds are treated for what is perhaps the first time in a book directed toward inorganic chemistry. Molecular orbital methods are introduced and compared, various bonding types are classified, electronic band shapes are interpreted, and magnetic properties are discussed. Stereochemistry is a fundamental and important part of this book, with symmetry and group theory widely used throughout, including the classification of terms, selection rules, crystal structures, and vibronic effects (vibronic stereochemistry). The book concludes with an investigation of the electronic properties of transition metal complexes, including the use of modern techniques such as EXAFS and the development of the concepts associated with electron transfer, a topic of fundamental importance to biological catalysis. Chemical activation and the direct calculation of energy barriers in chemical reactions complete the coverage in the book. Thus the student of this text is armed with an important, even essential arsenal of theory with which to handle virtually any electronic structure problem in transition metal chemistry.

We trust that this book will be a suitable primer for every serious practitioner of modern transition metal chemistry.

F. A. COTTON
J. P. FACKLER, JR.

PREFACE

The work on this book has a long 35-year history. In the late 1950s, after I completed my thesis at Leningrad University in the group of V. A. Fock (one of the authors of the Hartree–Fock method), A. V. Ablov invited me to join him at the Academy of Sciences of the USSR, Moldavian branch, in investigations of optical and magnetic properties of transition metal compounds. At that time there were no books in this field, and we decided to prepare a general presentation of the subject that would be suitable for chemists. That book, with the title *Chemical Bonding in Complex Compounds*, was published in 1962 simultaneously with the well-known book by C. J. Ballhausen on the same topic and shortly after a similar publication by L. E. Orgel. Since then interest in the structure and properties of transition metal compounds has grown very rapidly; I reviewed this field periodically and published another three books on this subject in 1971, 1976, and 1986. The experience with those four books (which sold over 18,000 copies) along with published reviews, critical comments, and suggestions served as a basis for the preparation of this book. However, the reader will find that this newest book is significantly different from its predecessors with respect to both topics and perspectives.

Presently, transition metal compounds are being investigated widely, with vast applications ranging from a variety of magnetic, ferroelectric, and superconducting materials to all kinds of catalysts and metallobiochemical systems of vital importance. The main goal of this book is to provide a comprehensive discussion of the latest developments in the study of electronic structure and related properties of transition metal coordination systems, and to present the subject in a form suitable for chemists and physicists—students, researchers, and teachers.

Most attention is paid to a better understanding of the basic principles, general features, and specifics of electronic properties affecting ligand bonding, stereochemistry and crystal chemistry, chemical reactivity, electron transfer and redox phenomena, as well as spectroscopic, magnetic, and electronic density distribution properties. The discussion of relativistic effects in bonding, presented in a book context for the first time, elucidates the origin of important properties, including, for example, the "nobleness" and the yellow color of gold.

Quite novel are the implications of vibronic effects in chemical and physical phenomena presented in this book. The concept of vibronic interactions developed during the last two decades as a perturbational approach to the coupling between the electronic motion and nuclear configuration contributes significantly to the solution of a number of problems. These include, for instance, band shapes of electronic and photoelectron spectra, local stereo-chemistry and structural phase transitions in crystals, plasticity, distortion isomers, and temperature-dependent conformers, molecular pseudorotations, chemical activation by coordination, and electron-conformational effects in biological systems.

One of the special features of this book is that it includes both the theory of electronic structure and its applications to various problems. Significant efforts were made to present the entire topic in a unified fashion with indications of direct and indirect links between its numerous more specific aspects and to make the presentation understandable without oversimplification. Many examples are provided that will assist the reader in understanding how theoretical concepts might be applied to laboratory problems.

During the preparation of this book I benefited from the help and cooperation of my students and co-workers as well as from suggestions of many colleagues throughout the world. I am grateful to all of them. I especially thank J. P. Fackler, Jr., who, in addition to publishing two of my previous books in his series "Modern Inorganic Chemistry," contributed significantly to the publication of this book. Thanks are also due to S. Kirschner and A. B. P. Lever for helpful suggestions, Roald Hoffmann for critical comments, and to A. V. Ablov, F. Basolo, H. Bill, J. E. Boggs, F. A. Cotton, V. A. Fock, J. Gazo, O. Kahn, A. D. Liehr, W. J. A. Maaskant, K. A. Müller, R. S. Pearlman, Yu. E. Perlin, D. Reinen, Ch. Simmons, M. Verdaguer, and M. G. Veselov for their cooperation in relation to various aspects of this book. My co-workers G. I. Bersuker, S. A. Borshch, L. F. Chibotaru, A. S. Dimoglo, N. N. Gorinchoy, M. D. Kaplan, I. Ya. Ogurtsov, V. Z. Polinger, A. O. Solonenko, S. S. Stavrov, B. S. Tsukerblat, and B. G. Vekhter are also especially acknowledged. I am grateful to D. S. Mason and M. K. Leong for help in preparation of the manuscript, and to my wife, L. B. Bersuker, for her continued patience over the years.

My sincere apologies are due to many authors who contributed significantly to the development of the main topics of this vast field but, for various reasons, may not have been cited. The many contributors who *are* cited certainly do not reflect the author's preferences; they have been selected as illustrative examples that may allow the reader to find further developments of the corresponding topics.

ISAAC B. BERSUKER

Austin, Texas

CONTENTS

MATHEMATICAL SYMBOLS

A	electron affinity
	hyperfine constants
$[A]$	atomic core
A, B	nondegenerate terms
A, B, C	Racah parameters
B	magnetic induction
C	redox capacitance
$C^{(i)}$	anharmononicity correction coefficients
c	speed of light
c_{ij}	LCAO coefficients
D	barrier height
D_s, D_t	low-symmetry crystal field corrections
d	atomic state
	barrier width
E	energy
E, e	double degenerate representation (term)
E_{JT}	Jahn–Teller stabilization energy
e	electronic charge
e_σ, e_π	AOM parameters
F	linear vibronic constant
F_k	radial components of crystal field matrix elements
$F^{(k)}$	Slater–Condon parameter
$F(S)$	structure factor in x-ray scattering
$F(\Omega)$	form function for spectroscopic band shapes
f	angular dependence function in ESR spectra
	atomic state
	linear orbital vibronic constant
	multiplicity of degeneracy
	oscillator strength
G	quadratic vibronic constants
	symmetry operator
G_{ij}	group overlap integral
$G^{(k)}$	Slater–Condon parameter
g	g-factor
	quadratic orbital vibronic constant
H	Hamiltonian
$\mathscr{H}$	magnetic field intensity

$\mathcal{H}_n$	Hermite polinomial
I	ionization potential
	light intensity
	nuclear spin
I_{ij}	Coulomb interaction
i, j	atomic and molecular states
	multiply used index
J	exchange coupling constant
	quantum number
	total momentum
K_{ij}	exchange interaction
$K_{\Gamma*}^{\Gamma}$	force (elastic) constant (AP curvature)
$K_\Gamma(\Gamma*)$	vibronic reduction factor
$K(\Omega)$	coefficient of light absorption
k	multiply used index
k^i	force constant coefficients
L	orbital momentum
L_k^i	Laguerre polynomial
l, m, n	direction cosines
	quantum numbers of:
	atomic states
	vibronic energy levels
M	magnetic (electric) dipole moment
	magnetization
	nucleus mass
$[M]$	molecular core electronic configuration
$M(S)$	magnetic structure factor
M_{12}	transition moment
m	electron mass
n	vibrational quantum number
P	electron transfer probability
P_a	amplification coefficient
P_{ik}	operator of permutation of i and k
P_l^m	associated Legendre polinomial
$P_{\mu\nu}$	bond orders
	density matrix element
$p = K_E(A_2)$	vibronic reduction factor in the $E-e$ problem
Q	normal (symmetrized) nuclear coordinates
Q_{xx}	gradient of electric field
$q = K_E(E)$	vibronic reduction factor in the $E-e$ problem
q_i	MO occupation numbers (MO population)
q_μ	atomic effective charge
Δq	charge transfer
Δq_i	orbital charge transfer
R	interatomic distance
	nuclear coordinate

R_{nl}	atomic radial wavefunction
r_i	electron coordinate
S	electron spin
S, S_{ij}	overlap integral
$S, P, D, F, \ldots$	atomic terms
s, p, d, f	atomic one-electron states
T	temperature
	period of vibrations
$T_{1,2}$	triple degenerate representations (terms)
U	effective potential
	electron–nuclear interaction
V	electron–nuclear plus nuclear–nuclear interaction
	crystal field potential
v	electron speed
W	ionic interaction potential
	vibronic interaction terms
w	intramolecular electron transfer constant
X	character of representation
x, y, z	Cartesian coordinates
Z	nuclear charge
Z_{eff}	effective charge

Greek Symbols

α	radial wavefunction parameter
	angle of rotation of polarized light
α_i	Coulomb integrals
α, β	multiple used index
β	anharmonicity correction
	Bohr magneton
	nephelauxetic ratio
Γ	irreducible representation (term)
γ	anharmonicity coefficient
	covalency parameter
	core screening parameter
	line of degenerate irreducible representations
γ_{AB}	electron repulsion integral
Δ	crystal field parameter
	electron energy gap
ΔX	change of X
δ	chemical shift
	extrastabilization energy
	Kroneker index (δ symbol)
	tunneling splitting
$\delta(x)$	delta function

δX	variation of X
$\delta \Omega$	optical bandwith
ε	one-electron orbital energy
ε, ϑ	two components of the doubly degenerate E representation
$\varepsilon(\Omega)$	extinction coefficients
ξ, η, ζ	three components of the triply-degenerate T_2 representation
ζ	one-electron spin–orbital coupling constant
η	asymmetry parameter
	chemical hardness
	relativistic contraction
λ	covalency parameter
	spin–orbital interaction constant
λ_Γ	dimensionless vibronic constant
μ	chemical potential
	ligand dipole moment
	magnetic moment
ν	vibrational frequency, vibrational quantum number
Π	pairing energy
	term of linear molecule
π	π MO, π bonds
ρ, ϑ, ϕ	nuclear polar coordinates
ρ	electron charge density
$\Delta\rho$	deformation density
$\displaystyle\sum_i$	summation over i
$\displaystyle\sum_{i=m}^{n}$	summation over i from m to n
$\displaystyle{\sum_{ij}}'$	summation over i and j except $i = j$
σ	σ MO, σ bond
	Pauli matrix
τ	lifetime, relaxation time
Φ	ligand MO wavefunction
	multielectron wave function
φ	one-electron atomic or MO wavefunction
χ	nuclear wavefunction
	magnetic susceptibility
Ψ	total wavefunction
ψ	one-electron MO wavefunction
Ω	electron transition frequency
	incident irradiation frequency
ω	vibrational frequency

ABBREVIATIONS

acac	acethylacetone
AES	Auger electron spectroscopy
AO	atomic orbitals
AOM	angular overlap model
AP	adiabatic potential
BA	border atom
bipy	bipyridyl
bpym	2,2′-bipyrimidine
bzp	bromazepan
CA	central atom
CFT	crystal field theory
CI	configuration interaction
CNDO	complete neglect of differential overlap
CoTPP	cobalt–tetraphenylporphyrin
Cp*	pentamethylcyclopentadienide
CT	Creutz–Taube
DD	deformation density
dto	ditiooxalato
ECP	effective core potential
ECE	electron-conformational effect
EIVC	energy of ionization of the valence state
emu	electromagnetic units
En, en	ethylenediamine
Eq.	equation
Eqs.	equations
Et	ethyl
EXAFS	extended x-ray absorption fine structure
HDVV	Heisenberg–Dirac–Van Vleck (model)
HF	Hartree–Fock
hfac	hexafluoroacetylacetonate
HFR	Hartree–Fock–Roothaan
HOMO	highest occupied MO
ImH	imidazol
IT	intervalence transition
INDO	intermediate neglect of differential overlap
JT	Jahn–Teller
LUMO	lowest unoccupied MO

M	metal atom
MCD	magnetic circular dichroism
MCZDO	multicenter zero differential overlap
Me	methyl
MINDO	modified INDO
MM	molecular mechanics
MO	molecular orbitals
MO LCAO	MO–linear combinations of AO
MP	metal porphyrin
MPc	metal phtalocyanine
MV	mixed valence
NDDO	neglect of diatomic differential overlap
NITEt	ethylnitronylnitroxide
NR	nonrelativistic
obbz	oxamidobis(benzoato)
OVC	orbital vibronic constant
Ox	oxalate
pba	1,3-propylene bis(oxamato)
Ph	phenyl
pyz	pyrazine
QR	quasi-relativistic
QRA	quasi-relativistic approach
QSAR	quantitative SAR
RE	relativistic effect
SAR	structure–activity relationship
SCCC	self-consistent charge and configuration
SCF	self-consistent field
SOMO	single occupied MO
SP	square planar
TCNE	tetracyanooethylene
TCNQ	7,7,8,8-tetracyano-p-quinodimethane
TDETMC	three-dimensional ETMC
UPS	ultraviolet photoelectron spectroscopy
VSEPR	valence-shell electron pair repulsion
VMO	vibronic MO
XES	x-ray emission spectroscopy
XPS	x-ray photoelectron spectcroscopy

ELECTRONIC STRUCTURE
AND PROPERTIES
OF TRANSITION
METAL COMPOUNDS

1 Introduction: Subject and Methods

The electronic theory of transition metal coordination systems pioneers a way of thinking in chemistry.

This chapter is to introduce the reader to the objectives of the book and its main purpose, and to define the subject, the methods of its exploration, and its "ecological niche" in the fast development of science and increasing demands for generalized information. After a brief discussion of the role and place of this book among the others available (Section 1.1), chemical bonding and coordination systems are defined (Section 1.2), followed by a short outline of the main ideas of quantum chemistry employed in the subsequent presentation (Section 1.3).

1.1. OBJECTIVES

Molecular Engineering and Intuitive Guesswork

The approaching threshold of the next century (even the next millennium of human civilization) inclines us to sum up the achievements of this century and to relate to what is expected in the coming new age. In the twentieth century the theory of structure and properties of transition metal coordination compounds, as well as polyatomic systems in general, advanced tremendously and reached impressive results. A basic understanding of the nature of chemical bonding and chemical transformations was provided and purposeful synthesis of new coordination compounds with required properties was promoted significantly. As a result of the rapid development of this trend in science, especially in the second half of the century, the solution of the problem of *molecular engineering*, which includes design and consequent synthesis of newly designed compounds, is approaching fast. *In the twenty-first century the majority of new chemical compounds will be obtained based on molecular design, and we should be prepared, from both the practical and psychological points of view, to meet this challenge.*

Molecular engineering is based primarily on knowledge of the molecular structure, including electronic structure. To design a compound with specific properties, the laws that control the formation and structure of molecular systems, as well the correlation between structure and properties, must be known in detail. Therefore, *the study of electronic structure and properties of polyatomic systems is one of the most important tasks of modern chemistry* in view of its trends and developments in the near future.

1

However, so far the majority of chemical compounds with required properties have been obtained based on intuitive knowledge without molecular engineering. For the most part, the electronic structure and methods of control of the properties of molecular systems remain unknown. The preparation of new compounds in these cases depends on the art of the researcher — on his or her intuition. On the other hand, intuition, or intuitive knowledge, does not emerge from a vacuum: Implicitly, it is based on real knowledge or, more precisely, on *understanding* of the phenomena that underlie the processes under consideration. Intuitive guesswork is also a kind of "engineering." Deep (implicit) understanding of phenomena allows us to ignore the lack of detailed information about laws controlling a process under study and to come to a correct result which from the outside appears "unexpected." The better the visual images and comparisons, the more fruitful the intuitive thinking.

It is clear that the smaller the region of lack of knowledge, the easier it is to jump over it. The less knowledge there is, the more the results of intuitive guesswork are unique and accidental in nature; they become more frequent and more purposeful as knowledge increases. Hence the preparation of new compounds based on intuitive thinking is ultimately also dependent on a deep understanding of the phenomena — understanding based on knowledge of the laws controlling the formation of new compounds and their properties.

Preparation of new compounds with specific properties based on either molecular engineering or intuitive conjecture requires (*in both cases*) *knowledge of the structure and properties of such compounds.*

The term *understanding* used above is not trivial and needs clarification. We use this term in the following sense: *To understand the origin of a new phenomenon means to be able to reduce it to more simple* (*usual or normal*) *conventional images.* To deepen or extend understanding means to introduce more complicated basic images to which the phenomena should be reduced. In the 1950s the basic images in an understanding of the origin of properties of transition metal compounds were created by *crystal field theory* (Chapter 4), which arose instead of and in addition to the image of a two-electron valence bond. Subsequently, deeper understanding was reached by introducing more complicated images of *molecular orbitals*, which continue to serve as basic images (Chapters 5 and 6). In the last decades, new basic images have developed based on *vibronic coupling* (Chapter 7). Note that the new images of understanding, being more complicated, do not completely negate the old ones but complement them with new content. New images are produced by the theory. With progress in science the images become more complicated, approaching reality.

Lack of understanding means that it is impossible to reduce the phenomenon to well-established conventional images. This requires creation of new images. Sometimes the latter differ drastically from the usual ones. In the history of science the most dramatic new images have been introduced by quantum mechanics. *Wave–particle duality* — the fact that a micro-object (e.g., a electron) exhibits properties of both particles (i.e., it can be localized at a single point of space) and plane waves (i.e., it is delocalized over the entire space) — cannot be understood within the existing images; it must be taken as such in a conventional manner until it becomes normal.

In view of what has been said above about understanding, to make the book intelligible means to reduce the properties of transition metal compounds to basic images. Hence we should describe the newest basic images that provide understanding most appropriate to the real phenomena. The main images in the theory of electronic structure of coordination compounds mentioned above (crystal fields, molecular orbitals, vibronic coupling) should be presented such as to become normal ways of thinking in chemistry (in fact, molecular orbitals are now such elements). This, in turn, requires *simplicity* and *visualization* to the greatest extent possible. Simplicity in this aspect means less abstracted presentation with more concrete examples, avoiding as much as possible bulky mathematical deductions. As pointed out by Heisenberg [1.1], "even for the physicist, description in plain language will be a criterion of the degree of understanding that has been reached."

However, aspiring to simplicity involves the danger of *oversimplification*. The latter takes place when the phenomenon under consideration is presented by a "smoothed" picture in which angles are cut off and important details are omitted. For instance, in many books and papers it is stated that as a result of the Jahn–Teller effect, distorted molecular configurations should be observed. This statement is definitively an oversimplification, since, in fact, Jahn–Teller distortions can be observed only under certain important additional conditions (Sections 7.3 and 9.2). Besides misunderstanding, oversimplification may create illusions of "easy access to science," whereas, in fact, much stronger efforts are needed; this may have a negative influence on scientific thinking.

Comparison with Other Books

Finally, the relation of this book to others with similar goals should be notified. There are many books and review articles devoted to the electronic structure and properties of transition metal coordination compounds or, more often, to particular aspects of this problem (see, e.g. [1.2–1.16]). The present book differs significantly from these sources in many aspects.

First, this book attempts to give *a generalized view of the modern state of art of the entire topic*, beginning from the main ideas of quantum chemistry and atomic states through theories of electronic structure and vibronic coupling to physical methods of investigation and applications to various chemical and physical problems. The advantages of this presentation, compared with many publications devoted to a more narrow aspect of the problem, is that the latter give a generalized view of what is going on in that narrow field, whereas this book generalizes the trend as a whole, including the main particular aspects of the problem.

We emphasize that *the whole trend is not equal to the sum of particular trends*. A general view of the topic as a whole, given as an entire subject with direct interrelations between its different more particular aspects, provides a significantly higher level of understanding of both the particular problems and the whole trend. Presented by the same author in a unified way and on the same level, different problems are expected to be better understood by the reader.

Many aspects of the problem treated in this book are novel; they have not been fully considered before in books on coordination compounds. This is, first, *the concept of vibronic interactions* considered in Chapter 7 and used to solve various problems of coordination compounds (Chapters 9 to 11). The treatment of *electronic structure, relativistic effects in bonding, optical band shapes, electronic and vibronic origin of stereochemistry, electron transfer in mixed-valence compounds, chemical activation by coordination*, and others is also novel. Even for those parts of the problem that were solved long before and considered repeatedly in books and review publications, renewal of their presentation updated in accordance with the novel achievements of the theory is required periodically. Previous books on electronic structure of coordination compounds with goals similar to those of this book have long been published [1.3–1.11]; more attention has been paid recently to methods of numerical computation [1.17–1.19, 1.21, 1.22]. The majority of the present book's sections include novel, original treatments for these "classical" problems, too (see, e.g., the definition of the coordination bond given in Sections 1.2 and 6.1).

Applications of theory to various chemical problems (Chapters 9 to 11), together with the theory itself in the same book, is, in general, seldom presented. Meanwhile, this presentation allows readers, interested in solutions to applied problems to consult the theoretical background of these solutions directly and to consider their transferrability to other problems. The treatment of different chemical properties from the same point of view also has the advantage of stimulating the search for new effects, rules, and laws that emerge from these direct comparisons.

Note, however, that applications of the theory considered in Chapters 9 to 11 cannot claim completeness; they are just diversified examples of such applications, which stimulate further developments in corresponding fields. Direct applications of the theory of electronic structure are expanding continuously. For reasons of limited space, several related issues (e.g., clusters and multicenter transition metal systems with unusual coordinations [1.23], extended and solid state systems [1.14]) are underrepresented in this book; also, some of the physical methods of investigation in widespread use in this field (e.g., Mossbauer spectroscopy) are not included, while others are considered very briefly (vibrational spectroscopy, x-ray diffraction). To summarize: *The main objectives of this book are to give a general view on the theory of electronic structure and properties of transition metal coordination compounds in a form intelligible to chemists and physicists — students, researchers, and teachers in these and related fields.*

Some comments are worthwhile about the phrase "introduction to the theory" given in the book's subtitle. This means that the book is addressed also to those who have not studied any special theory of electronic structure of transition metal coordination compounds (but who have a background of quantum mechanics or quantum chemistry in the volume of a regular course for chemists). This also implies that the book is not devoted to advances in the theory itself, or its sophisticated formulations and methodologies. Instead, the latest achievements of the theory are presented together with explanations of how they have been obtained (but without bulky mathematical deductions) and how they can be used to solve the chemical (physicochemical) problems. The problems of the theory itself form a special field of quantum chemistry well presented in literature [1.16–1.22].

An important question concerning the theory is the real meaning implied by this term. The theory of electronic structure forms one of the principal parts of modern quantum chemistry (others being nuclear dynamics, intermolecular interactions, molecular transformations, influence of external perturbations, etc.). Its particular emphasis — numerical computation of the electronic structure for fixed nuclei — is at present most advanced. Modern computers and supercomputers allow us, in principle, to compute the electronic structure of any coordination system of reasonable size and to get relatively accurate figures for its energies and wavefunctions, energy barriers of chemical reactions, and so on. Note that until recently, metal-containing systems with active d and f orbitals have been a challenge to quantum chemistry [1.17, 1.21]. With the development of computers and advanced computer algorithms and programs, these calculations tend to become routine.

However, the numerical data of the computed electronic structure themselves cannot be regarded as a theory. Indeed, these data characterize a single compound (for which the computation has been carried out), and in general, they cannot be transferred directly to other compounds. From this point of view computer data look similar to many other characteristics of the compounds obtained by different experimental facilities. *In fact, numerical results on electronic structure computation are some outputs of a computer experiment*, the computer thus being similar to a numerical spectrometer that yields the energy spectrum of the system.

To transform experimental data into a theory, they should be properly accumulated and generalized. The latter means to put the data in correspondence with *analytical models* obtained by simplifications and reasonable assumptions introduced in the general principles. In this way the experimental data can be rationalized and shown to express some laws, rules, trends, and characteristic orders of magnitude. The same is true for computer numerical data. The latter are thus most appreciated when they are obtained for series of compounds with similar structures and/or similar properties that can be directly generalized. In particular, this is true for different nuclear configurations of the same system: potential energy surfaces and chemical reaction energy barriers [1.17, 1.21, 1.22] (Chapters 6 and 11).

Thus the numerical data of quantum-chemical computations only do not form the theory of electronic structure — they should be processed through analytical models. Note also that the *ab initio* calculated wavefunctions of coordination compounds are given in thousands of determinants, which in general can be neither read nor understood without specific rationalizations by means of physically grounded simplified schemes. Nevertheless, *the results of numerical calculation are of inestimable value to the theory of electronic structure: Together with other (experimental) data they form the informational basis of the theory and allow us to discriminate the best theoretical models among the many possible.*

1.2. DEFINITIONS OF CHEMICAL BONDING AND TRANSITION METAL COORDINATION SYSTEM

Chemical Bonding as an Electronic Phenomenon

Chemical bonding is usually defined as an interaction between two or several atoms which causes the formation of a chemically stable polyatomic system (molecule, radical, molecular ion, complex, crystal, chemisorbed formation on surfaces, etc.). However, this formulation is not sufficiently rigorous, since without additional explanation it is not clear when the system should be considered as chemically stable. In fact, in this definition, admitting that chemical bonding is a kind of interaction, we introduce for the characterization of the latter a new term, *chemically stable system*, which is no more clear than the initial one, the chemical bond.

One may try to discriminate chemical bonding from other (say, intermolecular) interactions by the bonding energy. However, the latter, as is well known from experimental data, is not sufficiently informative: For chemical bonds the bonding energy varies from several to several hundred kcal per mole, thus being both smaller and larger than intermolecular bonding (which reaches about 20 kcal/mol) and the hydrogen bond (1 to 8 kcal/mol) (cf. the energies of the bond UBr_5—Br, equal to 13 kcal/mol, or the reaction $ClO_2 \rightarrow Cl + O_2$, equal to 4 kcal/mol [1.24]). It can be shown that bond lengths are also not always sufficiently informative with respect to the nature of the bonding.

A more rigorous discrimination of the chemical bonding can be made based on the differences in the electronic structure. *The main feature of chemical interaction is that it results in a significant reorganization (restructuring) of the electronic shells of the bonding atoms.* This reorganization is characterized by *collectivization* of the valence electrons and *charge transfer* (in case of different atoms). The electron collectivization is a more general characteristic of the bond since it can take place without charge transfer, whereas charge transfer cannot be realized without collectivization; the limiting case of pure (100%) ionic compounds does not exist.

We define the chemical bond as an interaction between atoms associated with a collectivization of the valence electronic orbitals [1.25]. This definition is sufficiently rigorous and allows one to distinguish the chemical bonding from, say, intermolecular interaction or physical adsorption on surfaces (from the point of view of this definition the hydrogen bond, which is associated with electron collectivization and charge transfer, is a type of chemical bonding).

Any rigorous definition of a physical quantity should contain, explicitly or implicitly, an indication of the means of its observation. In the definition of the chemical bonding given above, the means of its observation are implied: The collectivization of the electrons affects all the main physical and chemical properties of the system, and therefore the set of all these properties forms an experimental criterion of chemical bonding. In this set, such an important characteristic of the bond as its energy, which is an integral feature of the bond, may be less sensitive to the electronic structure than, for instance, the electronic spectra. In the example above, the bond UBr_5—Br, with a bonding energy of ~ 13 kcal/mol (which is less than the intermolecular limit of 20 kcal/mol), could be attributed to intermolecular bonding, but the electronic spectra testify to the chemical bonding. Besides bonding energy and electronic spectra, the chemical bonding affects essentially all the other spectra in whole-range spectroscopy, magnetic and electric properties, and so on.

The electronic nature of chemical bonding leads directly to the conclusion of its quantum origin. The motions of the electrons in atomic systems can be described correctly only by means of quantum mechanics. The nature of the bonding between two neutral atoms in the hydrogen molecule was first revealed by Heitler and London in 1927 by means of a quantum-mechanical description [1.26]. It was shown that the bonding results from the exchange part of the energy, which is negative and occurs due to the *indistinguishability* of the electrons and the *Pauli principle*; the exchange energy is a quantum effect and has no classical (nonquantum) analog. The Heitler–London work lies at the base of the quantum electronic theory of chemical bonding and quantum chemistry.

However, it is incorrect to state that the chemical bonding is due to exchange forces that keep the neutral atoms together. Analysis of the Heitler–London approximation for the H_2 molecule clearly shows that the only forces which lead to the formation of the chemical bond are the electrostatic interaction forces between four particles, two protons and two electrons. In fact, the bonding is caused by the quantum wave properties of the electrons. The interference of the wavefunctions of the two electrons from the two bonding hydrogen atoms, under certain conditions, results in extra electronic charge concentration in the region between the two nuclei, thus keeping them bonded. In many cases a significant part of the bonding energy is due to the reduction of the kinetic energy of the collectivized electrons. The separation of the exchange part of the energy results from the assumed one-electron approximation in the wavefunction. For instance, in the case of H_2^+ with a single electron there is still chemical bonding despite the absence of exchange interaction. *The quantum nature of the chemical bond is stipulated by the quantum-mechanical description of the motions of the electrons and nuclei.*

The quantum origin of chemical bonding contributes directly to an understanding of the main property of a chemical compound—its existence and stability. Therefore, *in the study of the correlation composition–structure–properties, electronic structure plays a key role.* Note that in general the term *electronic structure* implies that in addition to the ground-state energy and electron distribution (the wavefunction), the *excited states* are also known. The latter allow one to describe the behavior of the system under the influence of external perturbations, including intermolecular interactions and chemical reactions (Sections 10.1 and 11.1).

But the electronic structure does not describe all the properties of the compound. In particular, the temperature dependence of the properties may be determined by *the dynamics in the nuclear subsystem.* An important feature of the system is *the coupling of the electronic distribution to the nuclear configuration and nuclear dynamics (vibronic coupling).*

The electronic structure, vibronic coupling, and nuclear dynamics describe in principle all the properties of isolated molecules. To describe chemical compounds in their different aggregate states — ensembles of interacting molecule — *quantum-statistical, thermodynamic, and kinetics studies should be employed.*

Definition of the Coordination System

The definition of a coordination system (coordination bond, coordination compound) is not trivial and encounters difficulties. Many previous attempts to give a compact definition based on empirical data have been unsuccessful (see, e.g., [1.15] and Section 6.1). In view of the discussion given below in this section, these attempts failed because they tried to define the coordination compound based on the genealogy (prehistory) of its formation, whereas in fact the properties of any molecular system are determined by its structural features, first by its electronic structure, irrelevant to the method of its preparation [1.25].

The usual definition of a coordination system, which can be traced back to the *coordination theory* created by Alfred Werner more than a century ago [1.27], is that a complex or coordination compound is formed by a central atom (CA) or ion M, which can bond one or several ligands (atoms, atomic groups, ions) $L_1, L_2, \ldots$, resulting in the system $ML_1L_2 \cdots L_n$ (all the ligands L_i or some of them may be identical). This definition is so general that any molecular system can be considered as a coordination compound. For instance, methane can be presented as $C^{4+}(M) + 4H^-(4L)$, that is, as a coordination compound ML_4 [1.15]. To avoid this misunderstanding, it was required that the ions M and ligands L_i should be "real"; they should exist under normal chemical conditions and the reaction of complex formation should take place under the usual conditions.

Even with these limitations, the definition above is invalid, and there are many cases when it is misleading. For example, SiF_6^{2-} has many features of coordination compounds (Section 6.1), whereas, presented as required by the definition as $Si^{4+} + 6F^-$, we come into conflict with the fact that Si^{4+} does not exist under the usual conditions. This example can be treated as a more real composition: $SiF_4 + 2F^-$; then, to include it in the definition, we must assume that M can also be a molecule, but this assumption gives rise to new controversies and misunderstandings. This and many other examples show explicitly that *it is impossible to give a general definition of coordination systems based on the genealogy of their formation.*

On the other hand, the properties of molecular compounds are determined by their electronic structure. This statement leads directly to the idea of classification of chemical bonds and definition of coordination systems *based on electronic structure* [1.9, 1.25]. At present, when the electronic structure of coordination compounds is relatively well studied, the tendency to classify the chemical compounds by their methods of preparation looks somewhat old-fashioned. However, it was not old-fashioned at the time when coordination chemistry was rapidly developing, while knowledge of the electronic structure was rather pure and could not serve as a basis of classification. Note also that the way of thinking in chemistry was (and in a great measure is) more appropriate to *preparative chemistry*, but it is gradually changing to *structural chemistry*. It is quite understandable that the definition of coordination systems based on electronic structure is more convenient to discuss after study of the electronic structure. Therefore, classification of chemical bonds and chemical compounds is discussed in more detail in Section 6.1.

According to Section 6.1, chemical bonds can be classified after their electronic structure into three main classes (Table 6.1). The first is that of *localized valence bonds* formed by two electrons with opposite spins, by one from each of the bonding atoms, and these two electrons occupy one localized bonding orbital. These valence bonds follow the usual rules of valencies of organic compounds, which can be described by one valence scheme without the assumption of resonance structures (superposition of valence schemes). Localized double, triple, ..., bonds are included in this class. Compounds with localized valence bonds can be called *valence compounds.*

The second class contains *linearly delocalized bonds with possible ramification,* in which the valence electrons occupy one-electron molecular orbitals that are delocalized over all or a part of (but more than one) interatomic bonds (e.g., conjugated organic molecules, metallic chain structures and solids). These bonds can be called *conjugated, or orbital bonds.* In fact, this class of bonds includes all organic and main-group-element compounds that cannot be described by one valence scheme.

The third class contains *three-dimensional bonds delocalized around a center: coordination bonds.* Distinct from the conjugated bonds delocalized along the bonding line, a coordination bond is *three-dimensionally center delocalized.* In other words, a coordination bond is formed by a coordination center to which the ligands are bonded via electrons that occupy one-electron orbitals, each of which involves all or several ligands. This means that, in general, there are no localized CA–ligand bonds; they are collectivized by three-dimensionally (i.e., along several CA–ligand bonds) delocalized bonding electrons. It can be shown that delocalization of one-electron orbitals is realized via the d or f orbitals of the central atom, which have many lobes differently oriented in space (Section 2.1), whereas s and p electrons can provide only localized or linearly delocalized orbitals.

This definition allows one to discriminate the coordination bonds from valence and conjugated bonds. For instance, the two tetrahedral systems, CH_4 and $CuCl_4^{2-}$, differ essentially in electronic structure: CH_4 has four localized two-electron bonds C—H (hybridized sp^3 valence bonds), whereas in $CuCl_4^{2-}$ the bonding electrons are delocalized over all the ligands via the copper d electrons (coordination bond). Note that by this definition the bonds in NH_4^+ and BH_4^- are valence bonds analogous to CH_4 [1.28]; similarly, BF_3—NH_3 is a valence compound since its electronic structure is analogous to CF_3—CH_3.

In the SiF_6^{2-} example considered above, SiF_4 is a valence compound because of its localized Si—F bonds (analogies of C—F), while the bond in SiF_6^{2-} can be considered as a coordination bond because the octahedral coordination involves partly the low-lying d orbitals of Si, making the one-electron bonding states delocalized (Section 6.1).

Based on this classification of chemical bonds, the following definition of the coordination compound or, more general, coordination system can be given: *A coordination system $ML_1L_2 \cdots L_n$ consists of a coordination atom (coordination center) M ligated to n atoms or groups of atoms (ligands) $L_1, L_2, \ldots, L_n$ by coordination bonds that are delocalized over all or several ligands.* Following this definition two main structural features characterize the coordination system: the coordination center and the center-delocalized (coordination) bond. These features determine the main properties of coordination bonding discussed in this book; for a brief summary, see Table 6.2.

Transition metal compounds are mostly coordination systems: Even in the solid state of ionic crystals (as well as in the pure metallic state) the local features of the system are controlled by coordination centers with properties which, in essence, are quite similar to those of isolated coordination systems. The main reason for this similarity between molecular and local crystal properties lies in the specific role of the d electrons in both cases. Since these electrons may also be active as low-lying excited states, this book devoted to transition metal systems, mostly coordination systems, includes partly pre- and posttransition and rare earth systems. As emphasized in Section 6.1, in principle any atom may serve as a coordination center: active d states lacking in the free atom may occur due to corresponding chemical interaction that results in d state activation and coordination bonding.

1.3. THE SCHRÖDINGER EQUATION

In this section we present some basic notions of quantum chemistry, the Schrödinger equation and the principal approximations used in its solution for molecular systems, and introduce appropriate denotations. There are quite a number of textbooks on this topic (e.g., [1.29–1.31]).

Form of the Equation

Following the formal scheme of quantum mechanics [1.29], each physical quantity L (energy, momentum, coordinate, etc.) is put in correspondence with an *operator* **L** (a symbol that denotes a certain mathematical operation), such that the experimentally observed values of this quantity $L = L_n, n = 1, 2, \ldots$, are the eigenvalues of the following *operator equation*:

$$\mathbf{L}\Psi_n = L_n \Psi_n \qquad n = 1, 2, \ldots \tag{1.1}$$

The eigenfunction Ψ_n (the wavefunction) contains information about all the properties of the system in the state with $L = L_n$.

In quantum chemistry the most important quantity is the energy of the system, E. The operator of energy is the Hamilton operator **H**, called the *Hamiltonian*. Therefore, the operator equation for the energy is

$$\mathbf{H}\Psi_n = E_n \Psi_n \tag{1.2}$$

This is the *Schrödinger equation for stationary states* (for which the energy has a definite value). For nonstationary states that are time dependent, the Schrödinger equation is

$$i\hbar \frac{\partial \Psi}{\partial t} = \mathbf{H}\Psi \tag{1.3}$$

Equation (1.2) is a particular case of Eq. (1.3).

The Hamiltonian H (hereafter we denote operators in italic type) includes the operators of the kinetic energies of the electrons and nuclei, T, and the potential energy of all the interactions between them, U. In the *nonrelativistic approximation* these interactions are pure electrostatic. Taking account of relativistic effects, the dependence of the masses on velocity associated with magnetic spin–orbital and spin–spin interactions should be included. This can be done on the basis of the *Dirac equation* discussed in Sections 2.1 and 5.4 (see also Sections 5.5 and 6.5).

The total kinetic energy is the sum of the kinetic energy operators of each particle, $T = p^2/2\mu$, where $p = -i\hbar\nabla$ is the operator of the momentum [1.29] (do not confuse this imaginary unit i with the summation index, which is often also i) and μ is its mass,

$$T = \sum_i \frac{-\hbar^2 \Delta_i}{2\mu_i} \tag{1.4}$$

In this equation it is assumed that $p^2 = (-i\hbar\nabla)^2 = -\hbar^2\Delta$ ($i^2 = -1$), and ∇ and Δ are the usual differential operators ($\nabla^2 = \Delta$), $\Delta_i\Psi = \partial^2\Psi/\partial x_i^2 + \partial^2\Psi/\partial y_i^2 + \partial^2\Psi/\partial z_i^2$.

The operator U contains the sum of the Coulomb attractions and repulsions. The attraction of the ith electron to the α nucleus is $U_{i\alpha} = -Z_\alpha e^2/r_{i\alpha}$, where e is the charge of the electron, Z_α the order number of the element in the periodic table equal to the charge of the nucleus, and $r_{i\alpha} = |\mathbf{r}_i - \mathbf{R}_\alpha|$ the electron–nucleus distance, where $\mathbf{r}_i$ and $\mathbf{R}_\alpha$ the radius vectors of the electron and nucleus, respectively. The Coulomb repulsion between the electrons is $U_{ij} = e^2/r_{ij}$, and between the nuclei it is $U_{\alpha\beta} = Z_\alpha Z_\beta e^2/R_{\alpha\beta}$, where $r_{ij} = |\mathbf{r}_i - \mathbf{r}_j|$ and $R_{\alpha\beta} = |\mathbf{R}_\alpha - \mathbf{R}_\beta|$ are the interelectron and internuclei distances, respectively.

Thus the Schrödinger equation for a molecular system of n electrons with the mass m and N nuclei with masses M_α can be written in the following form:

$$[T + U]\Psi_k = E_k \Psi_k$$

or in a more explicit form,

$$\left[\sum_i^n \frac{-\hbar^2 \Delta_i}{2m} + \sum_\alpha^N \frac{-\hbar^2 \Delta_\alpha}{2M_\alpha} - \sum_i^n \sum_\alpha^N \frac{Z_\alpha e^2}{r_{i\alpha}} + \sum_{i<j} \frac{e^2}{r_{ij}} + \sum_{\alpha<\beta} \frac{Z_\alpha Z_\beta e^2}{R_{\alpha\beta}} - E_k \right]$$

$$\Psi_k(\mathbf{r}_1, \mathbf{r}_2, \ldots, \mathbf{r}_n; \mathbf{R}_1, \mathbf{R}_2, \ldots, \mathbf{R}_N) = 0 \qquad (1.5)$$

This equation is in fact a linear differential equation of the second order (of elliptical type) with respect to the $3(n + N)$ variables $\mathbf{r}_i$ and $\mathbf{R}_\alpha$. Its solution yields the stationary energy values of the system E_k and their corresponding wavefunctions Ψ_k. As mentioned above, the latter contains information on all the (nonrelativistic) properties of the system in the state with energy E_k. In particular, Ψ_k also contains full information about the electronic and nuclear charge distribution: $|\Psi(\mathbf{r}_1, \mathbf{r}_2, \ldots, \mathbf{R}_1, \mathbf{R}_2, \ldots)|^2$ equals the probability of finding the first electron at $\mathbf{r}_1$, the second at $\mathbf{r}_2$, and so on (Section 5.2).

Exact solution of the Schrödinger equation allows one, in principle, to determine a priori all the properties of any polyatomic system and its behavior in different conditions. Note that in all cases when exact solutions of Eq. (1.5) have been obtained, they were in good agreement with the experimental data, and in many cases the results of calculation have an accuracy rivaling experiment.

Role of Approximations

At early stages of development of quantum mechanics, the Schrödinger equation raised some hopes that it could be used to describe the entire chemistry, making many experimental approaches unnecessary. In a 1929 publication [1.32] one of the founders of quantum mechanics, P. A. M. Dirac, stated that "the underlying physical laws necessary for the mathematical theory of a large part of physics and the whole of chemistry are thus completely known, and the difficulty is only that the exact application of these laws leads to equations much too complicated to be soluble." After 65 years this statement remains valid; there are still principal difficulties in obtaining exact solutions of Eq. (1.5) for molecular systems with many particles, although the achievements in this field are impressive. With the growth of computers, exact solutions of Eq. (1.5) or even more complicated equations that include relativistic effects become possible for a limited number of electrons and nuclei. This number may increase, and is increasing, but for relatively large numbers of particles the results of numerical computations become difficult to perceive and almost impossible to interpret.

For instance, as mentioned above, the wavefunction of a system with tens of particles emerges from numerical calculations spread over thousands of determinants. With an increase in the number of particles, the numerical information yielded by the computer becomes so vast that it is useless. To rationalize these data and to be able to solve Eq. (1.5) for larger molecular systems, *simplifications* by introducing *approximations and/or analytical models* are absolutely necessary (see also the brief discussion at the end of Section 5.6).

Thus *the exact numerical solution of the Schrödinger or Dirac equation for large molecular systems is at present, in general, an irrational task*. The problem of electronic structure can be solved by introducing approximate methods of solution of Eq. (1.5) that allow one to obtain energies E_k and wavefunctions Ψ_k in a convenient form and to evaluate the physical and chemical quantities with required accuracy. The choice of the approximation that is optimal for the solution of a specific problem for a given molecular system and the analysis of the results in view of the approximations made constitute one of the most important (and sometimes most difficult) problems in quantum chemistry.

Most approximations used in modern quantum chemistry are aimed at the separation of variables in Eq. (1.5). These approximations can be divided into three main groups:

1. Separation of the nuclear dynamics from the electronic motions: the adiabatic approximation (Section 7.1)

2. Substitution of the local interactions between the electrons given by the Coulomb terms e^2/r_{ij} by some averaged interaction which is an additive function of r_i and r_j (neglect of correlation effects): the one-electron approximation (Sections 2.2 and 5.3)

3. Presentation of the one-electron function of many centers, molecular orbitals (MOs) by a sum of one-center functions, atomic orbitals (AO's): the approximation of molecular orbitals as linear combinations of atomic orbitals (MO LCAOs) (Section 5.1) or related approximations in the density functional approaches (Section 5.4)

These approximations are discussed in Chapters 2, 4, 5, and 7.

REFERENCES

1.1. W. Heisenberg, *Physics and Philosophy*, Harper, New York, 1958, p. 168

1.2. L. Pauling, *The Nature of the Chemical Bond*, 3rd ed., Cornell University Press, Ithaca, New York, 1960.

1.3. L. E. Orgel, *Introduction to Transition-Metal Chemistry*, Methuen, London, 1960 (2nd ed., 1966).

1.4. J. S. Griffith, *The Theory of Transition-Metal Ions* , Cambridge University Press, Cambridge, 1961, 455 pp.

1.5. C. J. Ballhausen, *Introduction to Ligand Field Theory*, McGraw-Hill, New York, 1962, 298 pp.

1.6. I. B. Bersuker and A. V. Ablov, *The Chemical Bond in Complex Compounds* (Russ.), Stiinta, Kishinev, 1962, 208 pp.

1.7. C. J. Ballhausen and H. B. Gray, *Molecular Orbital Theory*, W. A.Benjamin, New York, 1964.

1.8. H. L. Schlafer and G. Gliemann, *Basic Principles of Ligand Field Theory*, Wiley, New York, 1969.

1.9. I. B. Bersuker, *Structure and Properties of Coordination Compounds* (Russ.), Khimia, Leningrad, 1971 (2nd ed., 1976; 3rd ed., 1986).

1.10. C. K. Jorgensen, *Modern Aspects of Ligand Field Theory*, North-Holland, Amsterdam, 1971, 538 pp.

1.11. C. J. Ballhausen, *Molecular Electronic Structure of Transition Metal Complexes*, McGraw-Hill, New York, 1979.

1.12. J. K. Burdett, *Molecular Shapes: Theoretical Models of Inorganic Stereochemistry* , Wiley, New York, 1980, 287 pp.

1.13. T. A. Albright, J. K. Burdett, and M.-H. Whangbo, *Orbital Interaction in Chemistry*, Wiley, New York, 1985.

1.14. R. Hoffmann, *Solids and Surfaces: A Chemist's View of Bonding in Extended Structures*, VCH, New York, 1988.

1.15. F. A. Cotton and G. Wilkinson, *Advanced Inorganic Chemistry: A Comprehensive Text*, Wiley, New York, 1992.

1.16. I. B. Bersuker, *The Jahn–Teller Effect and Vibronic Interactions in Modern Chemistry*, Plenum Press, New York, 1984.

1.17. A. Veilard, Ed., *Quantum Chemistry: The Challenge of Transition Metal and Coordination Chemistry*, NATO ASI Series C, Vol. 176, D. Reidel, Dordrecht, The Netherlands, 1986.

1.18. M. Dupuis, Ed., *Supercomputer Simulation in Chemistry*, Lecture Notes in Chemistry, Vol. 44, Springer-Verlag, Heidelberg, 1986.

1.19. P. Coppens and M. B. Hall, Eds., *Electron Distribution and the Chemical Bond*, Plenum Press, New York, 1982.

1.20. I. B. Bersuker and V. Z. Polinger, *Vibronic Interactions in Molecules and Crystals*, Springer-Verlag, New York, 1989.

1.21. D. R. Salahub and M. C. Zerner, Eds., *The Challenge of d and f Electrons: Theory and Computation*, ACS Symposium Series 394, American Chemical Society, Washington, D.C., 1989.

1.22. K. Lipkowitz and D. Boyd, Eds., *Reviews in Computational Chemistry*, Vol. 1–5, VCH, New York, 1989–1994.

1.23. D. Braga, P. J. Dyson, F. Grepioni, and B. F. G. Johnson, *Chem. Rev.*, **94**, 1585–1620 (1994).

1.24. L. V. Gurvitz, G. V. Karachevtsev, and V. N. Kontdrat'ev, *Energies of Chemical Bond Cleavage, Ionization Potentials and Electron Affinities — Reference Book* (Russ.), Nauka, Moscow, 1974, 331 pp.

1.25. I. B. Bersuker, in: *Brief Chemical Encyclopedia* (Russ.), Vol. 5, Sovetskaya Entsiklopedia, Moscow, 1967, pp. 627–636.

1.26. W. Heitler and F. London, *Phys. Z.*, **44**, 455–472 (1927).

1.27. G. B. Kauffman, Ed., *Coordination Chemistry. A Century of Progress (1893–1993)*, ACS Symposium Series 565, American Chemical Society, Washington, D.C., 1994.

1.28. R. F. W. Bader and M. E. Stephen, *J. Am. Chem. Soc.*, **97**, 7391 (1975).

1.29. W. Kauzmann, *Quantum Chemistry: An Introduction*, Academic Press, New York, 1957, 560 pp.

1.30. R. McWeeny, *Methods of Molecular Quantum Mechanics*, 2nd ed., Academic Press, New York, 1989, 573 pp.

1.31. I. N. Levine, *Quantum Chemistry*, 3rd ed., Allyn and Bacon, Boston, 1983.

1.32. P. A. M. Dirac, *Proc. Roy. Soc.*, **A123**, 714 (1929).

2 Atomic States

Atomic states are the prime characteristics of the interacting atoms that form the transition metal compound.

In this chapter a brief general presentation of the subject matter and principal formulas used in subsequent chapters is given (for more details on atomic states, see [2.1–2.4]).

2.1. ONE-ELECTRON STATES

Angular and Radial Functions

The one-electron approximation of a multielectron system is based on the assumption that each electron moves independently in an averaged field created by all other electrons and nuclei. This approximation allows us to perform a complete separation of the variables, that is, to describe the motion of an electron by its own coordinates, independent of the coordinates of other electrons. This assumption, although widely used, is not always valid because it neglects *electron correlation effects* (Section 5.3).

For an atom it is also assumed that the averaged field in which the electron moves has spherical symmetry. Under this assumption the one-electron wavefunction can be presented as a product of the radial $R(r)$ and angular $Y(\theta, \phi)$ parts, where r, θ, and ϕ are polar coordinates with the origin at the nucleus. In the approximation of hydrogenlike functions, one-electron states are described by the same quantum numbers n, l, and m as those in the hydrogen atom: $n = 1, 2, \ldots$; $l = 0, 1, 2, \ldots, (n-1)$; $m = 0, \pm 1, \pm 2, \ldots, \pm l$. Then

$$\Psi_{lmn} = R_{nl}(r) Y_{lm}(\theta, \phi) \tag{2.1}$$

where $Y_{lm}(\theta, \phi)$ is a spherical function,

$$Y_{l,m}(\theta, \phi) = \left[\frac{(2l+1)(l-m)!}{4\pi(l+m)!} \right]^{1/2} P_l^m(\cos \theta) e^{im\phi} \tag{2.2}$$

and $P_l^m(x)$ is an associated Legendre polynomial,

$$P_l^m(x) = (1 - x^2)^{m/2} \frac{d^{l+m}}{dx^{l+m}} \frac{(x^2 - 1)^l}{2^l l!} \tag{2.3}$$

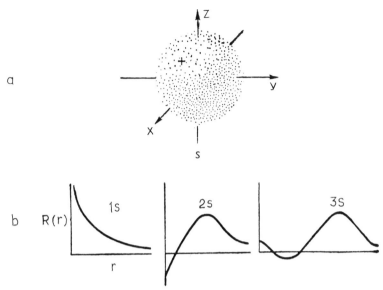

Figure 2.1. Angular (*a*) and radial (*b*) distribution of atomic *s* functions.

For $l = 0$, $m = 0$ (*s* states), the function (2.2) does not depend on the angles θ and ϕ, and the function (2.1) depends only on the distance of the electron to the nucleus r; its diagram in space has the form of a sphere (Fig. 2.1).

For $l = 1$ (*p* states), $m = 0$, ± 1; that is, there are three functions (2.2), two of which (with $m = 1$ and $m = -1$) are complex conjugated. Since for a free atom the one-electron energies ε_{nl} are independent of m and all three states have the same energy (threefold degeneracy), any linear combination of the

Table 2.1. Orthonormalized real angular parts of one-electron *s*, *p*, and *d* functions $Y_{lm}(\theta, \phi)$

	Y_{lm}	
Denotation	Polar Coordinates	Cartesian coordinates
s	$(4\pi)^{-1/2}$	$(4\pi)^{-1/2}$
p_x	$(3/4\pi)^{1/2} \sin \theta \cos \phi$	$(3/4\pi)^{1/2} r^{-1} x$
p_y	$(3/4\pi)^{1/2} \sin \theta \cos \phi$	$(3/4\pi)^{1/2} r^{-1} y$
p_z	$(3/4\pi)^{1/2} \cos \theta$	$(3/4\pi)^{1/2} r^{-1} z$
d_{z^2}	$(5/16\pi)^{1/2} (3 \cos^2 \theta - 1)$	$(5/16\pi)^{1/2} r^{-2} (3z^2 - r^2)$
$d_{x^2-y^2}$	$(15/16\pi)^{1/2} \sin^2 \theta \cos 2\phi$	$(15/16\pi)^{1/2} r^{-2} (x^2 - y^2)$
d_{xy}	$(15/16\pi)^{1/2} \sin^2 \theta \sin 2\phi$	$(15/4\pi)^{1/2} r^{-2} xy$
d_{xz}	$(15/4\pi)^{1/2} \sin \theta \cos \theta \cos \phi$	$(15/4\pi)^{1/2} r^{-2} xz$
d_{yz}	$(15/4\pi)^{1/2} \sin \theta \cos \theta \sin \phi$	$(15/4\pi)^{1/2} r^{-2} yz$

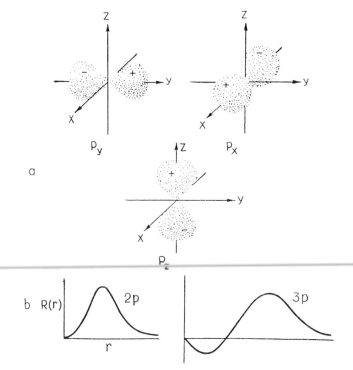

Figure 2.2. Angular (*a*) and radial (*b*) distribution of atomic *p* functions.

three functions is also a wavefunction with the same energy (in any combination only three of these functions are independent). Therefore, it is convenient
to make them real (Table 2.1 and Fig. 2.2).

Similarly, real combinations of functions (2.1) can be chosen for five *d* states
($l = 2$, $m = 0$, ± 1, ± 2), shown in Table 2.1 (Fig. 2.3). Note that diagramming
$p, d, \ldots$ functions becomes possible after the choice of arbitrary linear combinations of the functions (2.2), which are real and allow for visual presentation.
Hence one-electron states with lobes of p, d, $f, \ldots$ states in the absence of
external perturbations are rather conventional; any of the combinations of
$p(d, f)$ states is also a $p(d, f)$ state. Only under external perturbations do the
lobes acquire a specific orientation and a real physical meaning of charge
distribution.

In particular, for seven f functions ($l = 3$, $m = 0$, ± 1, ± 2, ± 3), two sets of
real angular parts are generally used (Table 2.2 and Fig. 2.4). The first set,
called *cubic*, is convenient when atomic states in cubic fields are considered
(Section 4.4). The second set is preferable for lower symmetries. But the
functions of one set can easily be obtained as a linear combination of the
functions of the other set [e.g., $f_{y^3} = -[10^{1/2}f_{y(3x^2-y^2)} + 6^{1/2}f_{yz^2}]/4]$.

The radial functions $R_{nl}(r)$ in (2.1) for one-electron hydrogenlike atoms can
also be presented in an analytic form as follows (Z is the nuclear charge, a_0 is

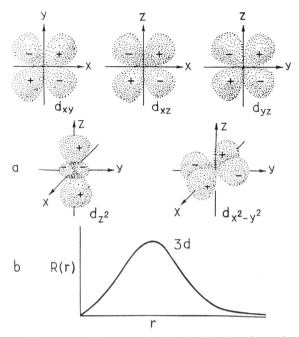

Figure 2.3. Angular (*a*) and radial (*b*) distribution of atomic *d* states.

the Bohr radius):

$$R_{nl}(r) = \left[\frac{(n - l - 1)!}{((n + l)!)^3 2n} \right]^{1/2} \left(\frac{2Z}{na_0} \right)^{l + 3/2} \exp\left(-\frac{Zr}{na_0} \right) r^l L_{n+1}^{2l+1} \frac{2Zr}{na_0} \quad (2.4)$$

where $L_k^i(x)$ is the Laguerre polynomial,

$$L_k^i(x) = \frac{d^i}{dx^i} \left[e^x \frac{d^k}{dx^k} (x^k e^{-x}) \right] \quad (2.5)$$

Expressions for several most usable radial function are given in Table 2.3.

Hydrogenlike functions (2.4) can be used for approximate estimations of some properties of nonhydrogen atoms, provided that the real charge is substituted by the effective charge Z_{eff}. There are various possibilities of choosing Z_{eff} in the wavefunction. One of them is to compare the one-electron energies calculated by this wavefunction with the experimental ionization potentials.

Instead of hydrogenlike function (2.4), Slater-type nodeless atomic functions are more convenient to use in molecular calculations. These functions have the

Table 2.2. Orthonormalized real angular parts of one-electron atomic f functions $Y_{3m}(\theta, \phi)$

Denotation	Y_{3m}	
	Polar Coordinates	Cartesian Coordinates
	Cubic Set	
f_{x^3}	$(7/16\pi)^{1/2}\sin\theta\cos\phi(5\sin^2\theta\cos^2\phi-3)$	$(7/16\pi)^{1/2}r^{-3}x(5x^2-3r^2)$
f_{y^3}	$(7/16\pi)^{1/2}\sin\theta\sin\phi(5\sin^2\theta\sin^2\phi-3)$	$(7/16\pi)^{1/2}r^{-3}y(5y^2-3r^2)$
f_{z^3}	$(7/16\pi)^{1/2}(5\cos^3\theta-3\cos\theta)$	$(7/16\pi)^{1/2}r^{-3}z(5z^2-3r^2)$
f_{xyz}	$(105/16\pi)^{1/2}\sin^2\theta\cos\theta\sin2\phi$	$(105/4\pi)^{1/2}r^{-3}xyz$
$f_{x(z^2-y^2)}$	$(105/16\pi)^{1/2}\sin\theta\cos\phi(\cos^2\theta-\sin^2\theta\sin^2\phi)$	$(105/16\pi)^{1/2}r^{-3}x(z^2-y^2)$
$f_{y(x^2-z^2)}$	$(105/16\pi)^{1/2}\sin\theta\sin\phi(\cos^2\theta-\sin^2\theta\cos^2\phi)$	$(105/16\pi)^{1/2}r^{-3}y(z^2-x^2)$
$f_{z(x^2-y^2)}$	$(105/16\pi)^{1/2}\sin^2\theta\cos\theta\cos2\phi$	$(105/16\pi)^{1/2}r^{-3}z(x^2-y^2)$
	Low-Symmetry Set	
f_{z^3}	$(17/16\pi)^{1/2}(5\cos^3\theta-3\cos\theta)$	$(17/16\pi)^{1/2}r^{-3}z(5z^2-3r^2)$
f_{xz^2}	$(21/32\pi)^{1/2}\sin\theta\cos\phi(5\cos^2\theta-1)$	$(21/32\pi)^{1/2}r^{-3}x(5z^2-r^2)$
f_{yz^2}	$(21/32\pi)^{1/2}\sin\theta\sin\phi(5\cos^2\theta-1)$	$(21/32\pi)^{1/2}r^{-3}y(5z^2-r^2)$
f_{xyz}	$(105/16\pi)^{1/2}\sin^2\theta\cos\theta\sin2\phi$	$(105/4\pi)^{1/2}r^{-3}xyz$
$f_{z(x^2-y^2)}$	$(105/16\pi)^{1/2}\sin^2\theta\cos\theta\cos2\phi$	$(105/16\pi)^{1/2}r^{-3}z(x^2-y^2)$
$f_{x(x^2-3y^2)}$	$(35/32\pi)^{1/2}\sin^3\theta\cos3\phi$	$(35/32\pi)^{1/2}r^{-3}x(x^2-3y^2)$
$f_{y(3x^2-y^2)}$	$(35/32\pi)^{1/2}\sin^3\theta\sin3\phi$	$(35/32\pi)^{1/2}r^{-3}y(3x^2-y^2)$

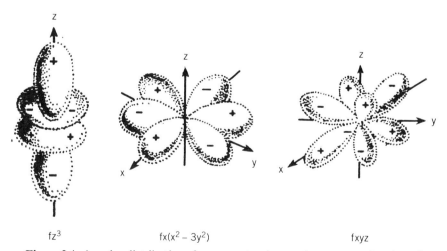

fz^3 $fx(x^2-3y^2)$ $fxyz$

Figure 2.4. Angular distributions for some atomic one-electron wavefunctions f.

Table 2.3. Expressions for some radial parts of one-electron hydrogenlike atomic functions R_{nl} ($\alpha = Z/na_0$)

Function	Expression
R_{10}	$2\alpha^{3/2}e^{-\alpha r}$
R_{10}	$2\alpha^{3/2}e^{-\alpha r}(1 - \alpha r)$
R_{21}	$(2/\sqrt{3})\alpha^{5/2}re^{-\alpha r}$
R_{30}	$\frac{2}{3}\alpha^{3/2}e^{-\alpha r}(3 - 6\alpha r + 2\alpha^2 r^2)$
R_{31}	$(2\sqrt{2}/3)\alpha^{5/2}re^{-\alpha r}(2 - \alpha r)$
R_{32}	$(4/3\sqrt{10})\alpha^{7/2}r^2e^{-\alpha r}$
R_{40}	$\frac{2}{3}\alpha^{3/2}e^{-\alpha r}(3 - 9\alpha r + 6\alpha^2 r^2 - \alpha^2 r^3)$
R_{41}	$(2/\sqrt{15})\alpha^{5/2}re^{-\alpha r}(5 - 5\alpha r + \alpha^2 r^2)$
R_{42}	$(2/3\sqrt{5})\alpha^{7/2}r^2e^{-\alpha r}(3 - \alpha r)$
R_{43}	$(2/3\sqrt{35})\alpha^{9/2}r^3e^{-\alpha r}$

general form

$$R_n(r) = Nr^{n^*-1}\exp\left[-\frac{(Z - \gamma)r}{n^*a_0}\right] \tag{2.6}$$

where N is the normalization constant, n^* the effective principal quantum number, γ the screening constant, and $Z - \gamma$ the effective nuclear charge. The number n^* is assumed to be dependent on n as follows:

n	1	2	3	4	5	6
n^*	1	2	3	3.7	4.0	4.2

To determine γ, all the electrons are divided into several groups: $(1s)$, $(2s, 2p)$, $(3s, 3p)$, $(3d)$, $(4s, 4p)$, $(4f)$, $(5s, 5p)$, $(5d)$, and so on. This partition obeys the following rule: s and p electrons of the same principal quantum number enter the same group, whereas d and f electrons form separate groups. For each (nl) electron of a given group the contribution to γ equals:

- 0 from all the outer electrons that are above the given group (from the electrons of the groups following the given group in the sequence above)
- 0.35 from each of the other electrons of the same group, except for the $1s$ electron, for which the other $1s$ electron contributes 0.30
- 0.85 for each electron of the inner group with the principal quantum number $n - 1$, and 1.00 for each electron of the deeper inner groups $(n - 2, n - 3,$ etc.$)$ - for ns and np electrons
- 1.00 for each inner electron (beginning with the $n - 1$ shell) for nd and nf electrons.

For instance, for the iron atom ($Z = 26$) with the electron configuration $(1s)^2(2s)^2(2p)^6(3s)^2(3p)^6(3d)^6(4s)^2$, we have:

For the $3d$ electron:

$$\gamma = 5 \cdot 0.35 + 18 \cdot 1.00 = 19.75 \qquad Z - \gamma = 6.25$$

For the $4s$ electron:

$$\gamma = 1 \cdot 0.35 + 14 \cdot 0.85 + 10 \cdot 1.00 = 22.25 \qquad Z - \gamma = 3.75$$

Sometimes the parameters of the Slater function (2.6) are considered as variational parameters to be determined from conditions of minimum energy.

In more accurate calculations, numerical Hartree–Fock (HF) wavefunctions (Section 2.2) are used as one-electron states. Computed HF one-electron wavefunctions are available for all atoms and their ionized states [2.5, 2.6]. For practical use, analytical presentations of the numerical functions are more convenient [2.7–2.9].

Atomic orbitals (AOs) are most important in MO LCAO calculations (Chapter 5). There are many aspects concerned with the choice of atomic states to present the atoms in chemical bonding (see *basis sets* in MO calculations, Section 5.3).

Orbital Overlaps and Hybridized Functions

As mentioned above, the well-known and widely used presentation of one-electron states in atoms in the form of s, p, d, and f states is based on the assumption of spherical symmetry of the field in which each electron moves, because only spherical symmetry allows for solutions in the form of spherical functions (2.2). If the field, under external influence, becomes nonspherical, these atomic states are no longer independent. They mix, and the nature of this mixing, or *hybridization*, is determined by the symmetry of the external field. This is one of the obvious consequences of chemical interactions (bonding) in which the atom under consideration takes part.

One of the basic features of two atoms that determines their ability to interact with each other chemically is the "diffusiveness" (extension in space) and mutual orientation of their atomic states in space expressed by their *orbital overlap* (Section 5.2). If the atom A is presented by its AO φ_A, and the atom B by φ_B, the expression

$$S_{AB} = \int \varphi_A \varphi_B \, d\tau \qquad (2.7)$$

is called the *overlap integral*.

It should be noted that it is the integral overlap S_{AB} that characterizes the formation of a chemical bond between neutral atoms. The presence of local regions of overlap (*differential overlap*) is necessary but not sufficient for bond formation, because the local overlaps may have different signs in different regions and thus compensate for each other in the integral overlap. On the other hand, the overlap integral (2.7) itself, although indicating the possibility of bonding, may not be sufficiently informative to fully characterize all the bonding features (Section 5.2).

There are some ways to estimate whether or not the integral (2.7) is zero without its calculation by using symmetry considerations. Group-theoretical rules (Section 3.4) allow us to do it easily. It can also be done using rather simple visual considerations. Indeed, it is well known that any transformation of variables under the sign of the integral does not change its value (because the integral is independent of the coordinate system). This means that if one can find a transformation of the coordinates for which one of the two functions φ_A or φ_B changes it sign whereas the other does not, then $S_{AB} = -S_{AB} = 0$.

For example, take $\varphi_A = s^A$ and $\varphi_B = p_x^B$ (the bonding line is assumed to be along z). Then $S_{AB} = \int s^A p_x^B \, d\tau = 0$ because the coordinate transformation $x' = -x$ (reflection in the plane yz) changes the sign of p_x^B ($p_x' = -p_x$) and does not change the sign of s^A (see Fig. 2.1 and Table 2.1). On the other hand, $\int s^A p_z \, d\tau \neq 0$ since in this case the two functions s^A and p_z are not changed under the symmetry operations of the system.

The symmetry of the region of orbital overlap between two atoms is of great importance for the characterization of their bonding. Depending on this symmetry, one can distinguish the $\sigma, \pi, \delta, \ldots$ bonds. A more rigorous foundation of this classification of chemical bonds in diatomics is based on the quantized values of the projection of the electron momentum on the axis of the

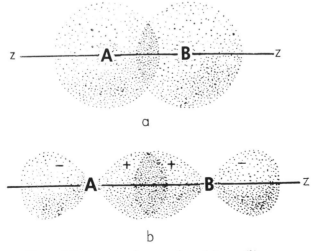

Figure 2.5. Interatomic σ overlaps: (a) s–s; (b) p_σ–p_σ.

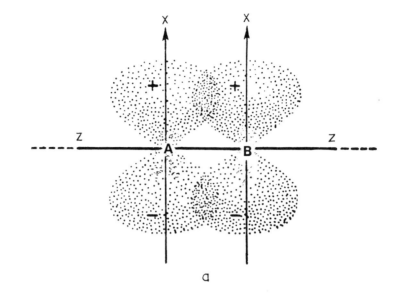

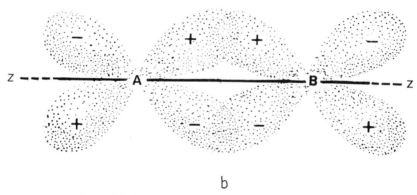

Figure 2.6. Interatomic π overlaps: (a) p_π–p_π; (b) d_π–d_π.

molecule characterized by its quantum numbers $\lambda = 0, 1, 2, \ldots$ which correspond to $\sigma, \pi, \delta, \ldots$ bonds, respectively.

For a σ bond the orbital overlap region is symmetrical with respect to the line of bonding between the atoms, which means that the electron density of the bond (Section 5.2) has axial symmetry (Fig. 2.5). The wavefunction of the system with this bond is independent of the angle of rotation around the molecular axis; the projection of the momentum on this axis (and its quantum number λ) is zero. For a π bond the overlap is symmetrical with respect to the plane comprising the molecular axis (Fig. 2.6), and $\lambda = 1$. In the δ bond there are two mutually perpendicular planes of symmetry that cross at the molecular axis, $\lambda = 2$ (Fig. 2.7; hereafter positive orbital lobes are shadowed).

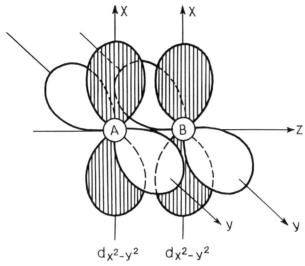

$d_{x^2-y^2}$ $d_{x^2-y^2}$

Figure 2.7. Interatomic $\delta\ d^A_{x^2-y^2} - d^B_{x^2-y^2}$ overlap that has two perpendicular planes of symmetry which cross the axis of bonding.

Obviously, not all atomic functions can be employed in the formation of diatomic bonds of given symmetry. For instance, s functions never form π bonds because they do not have the required symmetry of reflection in the plane crossing the molecular axis. Similarly, p functions cannot form δ bonds since they lack two mutually perpendicular planes of symmetry. On the other hand, under certain conditions d functions can form both δ and π bonds, as well as σ bonds, p functions form π and σ bonds, and s functions form σ bonds only.

The notions of $\sigma, \pi, \delta, \ldots$ bonds, which have a rigorous physical sense in the case of diatomics, may become less definitive and rather conventional in some polyatomic systems, especially in coordination compounds with three-dimensionally delocalized bonds. Indeed, for the latter the same AOs of the CA can take part in the formation of σ bonds with one ligand and π bonds with another (for examples of $\sigma + \pi$ bonds, see Section 6.3). As mentioned above, in the case of transition metal coordination compounds, pure localized (separated) CA–ligand bonds, strictly speaking, make litle sense, but they can be well defined as ligand–complex bonds.

The greater the orbital overlap, *ceteris paribus*, the stronger the bonding (Section 5.2). Therefore, the binding favors spatially oriented orbitals, and this explains the formation of *hybridized orbitals* induced by chemical bonding. As mentioned above, the separation of atomic one-electron orbitals into s, p, d, $f, \ldots$ is a rigorous consequence of the spherical symmetry of the field. Hence in the free atom there are neither hybridized orbitals nor *directed valencies*; they are formed under the influence of the external (bonding) field. Figure. 2.8a

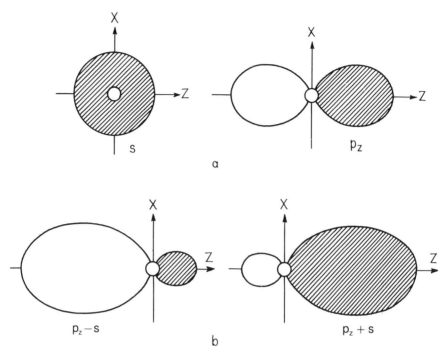

Figure 2.8. sp hybridization: (a) free s and p_z orbitals; (b) hybridized $p_z + s$ and $p_z - s$ orbitals.

shows the two independent s and p_z orbitals, while in Fig. 2.8b the mixed (hybridized) $s + p_z$ and $s - p_z$ functions are illustrated. The hybridized AOs are much more oriented in space along the z and $-z$ axes, respectively, than are the pure s and p_z orbitals, thus favoring stronger chemical bonds with two other atoms in these directions. It is just this enhanced bonding that induces the sp hybridization.

Provided that the geometry of the molecular system formed by the interacting atoms is known, the hybridized orbitals can be obtained from the condition of orthogonality and normalization [2.10]. The following types of hybridized orbitals are widely used: sp, linear coordination; sp^2, trigonal-planar; sp^3 and sd^3, tetrahedral; d^2sp^3, octahedral; and so on. Table 2.4 gives some of the hybridized functions. As stated in Sections 1.2 and 6.1, d orbitals form three-dimensional delocalized bonds rather than localized bonds that require hybridization.

Spin–Orbital Interaction

The relativistic features of atomic orbitals are considered in the next subsection and relativistic effects are discussed in Section 6.5, but one of these effects, spin–orbital interaction, is most important in quantum chemistry and cannot be avoided even in nonrelativistic approaches.

Table 2.4. Some hybridized atomic orbitals

Type of Hybridization	Hybridized Orbitals
sp (along x)	$2^{-1/2}(s + p_x)$, $2^{-1/2}(s - p_x)$
sp^2 (in the plane xy)	$3^{-1/2}(s + \sqrt{2}\,p_x)$
	$3^{-1/2}(s - p_x/\sqrt{2} + \sqrt{3}\,p_y/\sqrt{2})$
	$3^{-1/2}(s - p_x/\sqrt{2} - \sqrt{3}\,p_y/\sqrt{2})$
sp^3 (tetrahedral, with respect to equivalent axes x, y, z)	$\frac{1}{2}(s + p_x + p_y + p_z)$
	$\frac{1}{2}(s + p_x - p_y - p_z)$
	$\frac{1}{2}(s - p_x + p_y - p_z)$
	$\frac{1}{2}(s - p_x - p_y + p_z)$
d^2sp^3	$6^{-1/2}(s + \sqrt{3}\,p_z + \sqrt{2}\,d_{z^2})$
	$6^{-1/2}(s - \sqrt{3}\,p_z + \sqrt{2}\,d_{z^2})$
	$12^{-1/2}(\sqrt{2}\,s + \sqrt{6}\,p_z + d_{z^2} + \sqrt{3}\,d_{x^2-y^2})$
	$12^{-1/2}(\sqrt{2}\,s - \sqrt{6}\,p_z + d_{z^2} + \sqrt{3}\,d_{x^2-y^2})$
	$12^{-1/2}(\sqrt{2}\,s + \sqrt{6}\,p_z + d_{z^2} - \sqrt{3}\,d_{x^2-y^2})$
	$12^{-1/2}(\sqrt{2}\,s - \sqrt{6}\,p_z + d_{z^2} - \sqrt{3}\,d_{x^2-y^2})$

It is known that the electron possesses a spin momentum **s**, its projection having two values: $s_z = \pm\frac{1}{2}\hbar$, corresponding to its two possible positions: spin up and spin down (with respect to the axis of quantization). Taking into account this spin and assuming that it is independent of the orbital motion of the electrons, we multiply the one-electron function (2.1) by the spin function determining the spin state and add the spin quantum numbers $m_s = \pm\frac{1}{2}$ to the three orbital quantum number n, l, and m.

Complications begin when we take into account that the spin momentum is associated with a magnetic moment of the electron (the *Bohr magneton*). Indeed, the orbital motion of the electron creates a magnetic field and the magnetic moment of the electron interacts with the magnetic field of orbital motion, resulting in *spin–orbital interaction*. It is obvious that the spin–orbital interaction is different for the two spin positions, spin-up along the orbital magnetic field and spin-down in the opposite direction, resulting in *spin–orbital splitting* of the energy levels.

The energy of interaction of the electron spin **s** with the magnetic field of its orbital motion, which has the momentum **l** is a relativistic effect [not included in the Schrödinger equation (1.5)] that emerges from Dirac's equation [2.4]. For the operator of this interaction H_{SO}, we have

$$H_{SO} = \xi(r)(\mathbf{l},\ \mathbf{s}) \qquad \xi(r) = -\frac{e}{2m^2c^2r}\frac{dV(r)}{dr} \qquad (2.8)$$

where $V(r)$ is the potential energy of the electron in the atom. In many cases H_{SO} can be considered as a perturbation of the nonrelativistic states. Then the energy correction due to the spin–orbital interaction can be obtained by perturbation theory. In the first order these corrections are equal to the diagonal matrix elements of the perturbation (2.8). Usually, they are characterized by the *one-electron spin–orbital constant* ξ_{nl}:

$$\xi_{nl} = \hbar^2 \int \xi(r) R_{nl}^2(r) r^2 \, dr \tag{2.9}$$

where $R_{nl}(r)$ is the radial part of the atomic wavefunction of the perturbed state (n, l). For a Coulomb field $V = Ze/r$,

$$\xi_{nl} = \frac{Ze^2\hbar^2}{2m^2c^2a_0^3} \int r^{-3} R_{nl}(r) r^2 \, dr$$

$$= \frac{e^2\hbar^2}{2m^2c^2a_0^3 n^3 l(l + 1)(l + \frac{1}{2})} Z^4 \tag{2.10}$$

Here $l \neq 0$; for $l = 0$, $H_{SO} = 0$: s states are not subject to spin–orbital splitting.

From expression (2.10) it is seen that the constant of spin–orbital interaction ξ for an electron in a hydrogenlike atom is strongly dependent on the atomic number Z (proportional to Z^4). For instance, for a $3d$ electron, $\xi_{3d} = 1.4 \cdot 10^{-2} Z^4 \, \text{cm}^{-1}$, and for a $4d$ electron, $\xi_{4d} = 6.1 \cdot 10^{-3} Z^4 \, \text{cm}^{-1}$. It follows that the influence of spin–orbital interactions becomes very strong in heavy atoms, and for large Z values it can no longer be taken as a small perturbation; it should be considered along with other electrostatic interactions in the atom. In this case the wavefunction of the electron cannot be taken as a product of its orbital and spin parts.

Relativistic Atomic Functions

Relativistic atomic states become of great importance to the study of electronic structure and properties of transition metal compounds (Sections 5.4 and 6.5). For large electron speeds v that are realized in heavy atoms, the dependence of the mass of the electron on its speed and the magnetic moment of the electron [not included in the Schrödinger equation (1.5)] should be taken into account. This is done in the Dirac equation, which in the absence of external magnetic fields can be written as follows [2.4]:

$$(\boldsymbol{\alpha}c\mathbf{p} + eU + mc^2)\Psi = E\Psi \tag{2.11}$$

Formally, this equation is somewhat similar to the Schrödinger equation (1.5), but in fact, Eq. (2.11) is much more complicated than (1.5). First, (2.11) is a matrix equation of the fourth rank, which means that there are four coupled equations to be solved. This is due to the matrices α and β. They are

$$\alpha = \begin{pmatrix} 0 & \sigma \\ \sigma & 0 \end{pmatrix} \qquad \beta = \begin{pmatrix} I & 0 \\ 0 & -I \end{pmatrix} \qquad (2.12)$$

where the components of the vector-matrix σ are the well-known Pauli matrices:

$$\sigma_x = \begin{pmatrix} 0 & 1 \\ 1 & 0 \end{pmatrix} \qquad \sigma_y = \begin{pmatrix} 0 & -i \\ i & 0 \end{pmatrix} \qquad \sigma_z = \begin{pmatrix} 1 & 0 \\ 0 & -1 \end{pmatrix} \qquad (2.13)$$

and I is a unit matrix:

$$I = \begin{pmatrix} 1 & 0 \\ 0 & 1 \end{pmatrix} \qquad (2.14)$$

Other notation in (2.11) is as usual: c is the speed of light and $p = -i\hbar\nabla$ [hence $\alpha c p$ means $-i\hbar c(\alpha_x \partial/\partial x + \alpha_y \partial/\partial y + \alpha_z \partial/\partial z)$].

According to the fourth-rank equation (2.11), its solution Ψ, the wavefunction, is also fourth-rank. Indeed, it looks like a column four-vector called a bispinor [2.4]:

$$\psi = \begin{pmatrix} \Psi^{(1)} \\ \Psi^{(2)} \\ \Psi^{(3)} \\ \Psi^{(4)} \end{pmatrix} = \begin{pmatrix} \varphi \\ \chi \end{pmatrix} \qquad (2.15)$$

where φ and χ are spinors composed of the 1, 2 and 3, 4 components, respectively. The presentation of the solution (2.15) as two spinors has a specific physical meaning, since when the ratio v/c is small, the first spinor, φ, is much larger than the second, χ, and the latter can be neglected.

The physical meaning of the wavefunction Ψ of (2.15) that emerges from quantum mechanics is, in general, the same as in the nonrelativistic case: $|\Psi(1)|^2 \, d\tau$ is the probability of finding the electron in the volume $d\tau$ near point 1. Then $|\Psi|^2 = \Psi^*\Psi$, the product of the column-vector Ψ with its complex conjugate, Ψ^* [which is a row vector $(\Psi^{(1)*}, \Phi^{(2)*}, \Psi^{(3)*}, \Psi^{(4)*})$], yields

$$|\Psi|^2 = |\Psi^{(1)}|^2 + |\Psi^{(2)}|^2 + |\Psi^{(3)}|^2 + |\Psi^{(4)}|^2 \qquad (2.16)$$

or

$$|\Psi|^2 = |\varphi|^2 + |\chi|^2 \qquad (2.17)$$

Thus the density of position probability equals the sum of the contributions of all the components of the relativistic wavefunction. In particular, when the electron speed is not very large (e.g., in the region far from the nucleus), the major contribution comes from the large component φ.

To estimate relativistic effects as compared to the nonrelativistic properties, the constant $\alpha = 1/137$, sometimes called *the fine-structure constant*, is introduced (do not confuse this α constant with the α parameter in the nonrelativistic radial functions in Table 2.3). Note that in atomic units the speed of light $c \cong 137$. On the other hand, the electron speed v at the nucleus in the same units is equal to Z, the nuclear charge. Hence the magnitude αZ equals the ratio of the electron speed to the speed of light at the nucleus where this ratio is maximal.

In the spherically symmetric field of the nucleus, the relativistic solution is characterized by four quantum numbers (which are somewhat similar to the nonrelativistic ones):

1. The principal quantum number $n = 1, 2, 3, \ldots,$ which has the same meaning as in the nonrelativistic atom
2. The azimuthal quantum number $l = 0, 1, 2, \ldots, (n - 1)$, which denotes the appropriate states as $s, p, d, \ldots,$ respectively; distinct from the Schrödinger atom, in the relativistic case l is no longer the quantum number of the orbital angular momentum
3. $j = |l \pm \frac{1}{2}|$, the angular momentum quantum number that is always positive (as l is) and half-integer; it takes at most two values
4. m, the magnetic quantum number, which can take all half-odd-integer values from $-j$ to $+j$ [$m = j, j-1, \ldots, -(j-1), -j$].

With these quantum numbers $nljm$, we have for s electrons: $j = \frac{1}{2}, m = \pm\frac{1}{2}$ (two states); for p electrons: $j = \frac{1}{2}$ with $m = \pm\frac{1}{2}$ (two states) and $j = \frac{3}{2}$ with $m = \pm\frac{1}{2}, \pm\frac{3}{2}$ (four states); for d electrons: $j = \frac{3}{2}$ with $m = \pm\frac{1}{2}, \pm\frac{3}{2}$ (four states), and $j = \frac{5}{2}$ with $m = \pm\frac{1}{2}, \pm\frac{3}{2}, \pm\frac{5}{2}$ (six states). Usually, the magnetic quantum number (similar to the nonrelativistic spin quantum number m_s) is not indicated, and hence we have two $ns_{1/2}$ states, two $np_{1/2}$ and four $np_{3/2}$ states, four $nd_{3/2}$ and six $nd_{5/2}$ states, six $nf_{5/2}$ and eight $nf_{7/2}$ states, and so on.

For these states the four-component wavefunctions are [2.4, 2.11]

$$\psi_{nljm} = (N_{nljm})^{1/2} \begin{pmatrix} ga_1 Y_{l,m-1/2} \\ ga_2 Y_{l,m+1/2} \\ -ifa_3 Y_{2j-1,m-1/2} \\ -ifa_4 Y_{2j-1,m+1/2} \end{pmatrix} \tag{2.18}$$

where Y_{lm} are the spherical harmonics (2.2), N_{nljm} is the normalization factor determined from the condition $\int |\Psi_{nljm}|^2 d\tau = 1$ [taking into account Eq.

(2.16)], and a_i are constants as follows:

$$a_1 = \left(\frac{l + \frac{1}{2} \pm m}{2l + 1} \right)^{1/2}$$

$$a_2 = \left(\frac{l + \frac{1}{2} \mp m}{2l + 1)} \right)^{1/2}$$

$$a_3 = \left(\frac{2j - l + \frac{1}{2} \mp m}{4j - 2l + 1)} \right)^{1/2} \qquad (2.19)$$

$$a_4 = \left(\frac{2j - l + \frac{1}{2} \pm m}{4j - 2l + 1} \right)^{1/2}$$

where in the symbol $\pm$ the sign of $j - l$ should be taken, whereas in $\mp$ its opposite is implied, and g and f are the radial parts of the wavefunction: g is the same for the two components $\Psi^{(1)}$ and $\Psi^{(2)}$ of the large spinor, while f is the same for $\Psi^{(3)}$ and $\Psi^{(4)}$ of the small spinor [Eq. (2.18)]. Their general expressions are somewhat awkward [2.4] [they can be presented as a product of an exponent with a polynomial similar to (2.4)], but it is important that *the f component is αZ times smaller than the g component.* Table 2.5 presents some explicit expressions for relativistic s and p functions in a hydrogenlike atom given with an accuracy up to terms αZ included [2.11].

Some interesting features emerge from the relativistic functions as compared with the nonrelativistic functions. First, the relativistic s functions are spherically symmetric, quite similar to the nonrelativistic case. But surprisingly, in contrast to the nonrelativistic p orbitals, the $p_{1/2}$ states are also spherically symmetric. This can easily be confirmed by substituting the four components of any of the two $p_{1/2}$ functions from Table 2.5 into Eq. (2.16). It appears to be a general trend: Relativistic atomic functions with the same j and m quantum numbers have the same angular dependence of the appropriate position density. It means that not only do $s_{1/2}(\frac{1}{2})$ and $p_{1/2}(\frac{1}{2})$ have identical (in this case spherically symmetric) angular dependence of the position density, but $p_{3/2}(\frac{3}{2})$ and $d_{3/2}(\frac{3}{2})$, $d_{5/2}(\frac{5}{2})$, and $f_{5/2}(\frac{5}{2})$ have similar angular electron cloud distribution.

The $p_{3/2}(\frac{1}{2})$ distribution is also different from that of the nonrelativistic p states. Indeed, using Eq. (2.16) with the data in Table 2.5, one can easily find that the angular dependence of the electron distribution in this state is $\cos^2\theta + \frac{1}{3}$ (instead of $\cos^2\theta$ in the p_z function), that is, it has also a $\frac{1}{3}$ weighted s-type spherically symmetric distribution. The $p_{3/2}(\frac{3}{2})$ functions yield an angular distribution identical to the nonrelativistic case.

The s-type distribution in the relativistic p states [$p_{1/2}$ and $p_{3/2}(\frac{1}{2})$] is very important for the bonding properties of corresponding atoms, especially in their ability to form pure σ and π bonds (see below). Another feature of these functions is that they have no angular nodes; it can be shown that all the relativistic atomic functions have no radial nodes, either. Indeed, the radial nodes in the large component g never coincide with those of the smaller component f and hence at the points where g^2 is zero f^2 gives a nonzero

Table 2.5. Relativistic atomic wavefunctions for a hydrogenlike atom to the first order in αZ^a

Atomic State	$\sqrt{N}$	Spinors	Radial Part	Angular Part	
				$m = \frac{1}{2}$	$m = -\frac{1}{2}$
$1s_{1/2}$	1	φ	e^{-r}	1	0
				0	-1
		χ	$\alpha Z e^{-r}$	$\frac{1}{2}i\cos\theta$	$-\frac{1}{2}i\sin\theta\,e^{-i\phi}$
				$\frac{1}{2}i\sin\theta\,e^{i\phi}$	$\frac{1}{2}i\cos\theta$
				$m = \frac{1}{2}$	$m = -\frac{1}{2}$
$2s_{1/2}$	$(\frac{1}{8})^{1/2}$	φ	$e^{-r/2}[1-(r/2)]$	1	0
				0	-1
		χ	$\alpha Z e^{-r/2}[1-(r/4)]$	$\frac{1}{2}i\cos\theta$	$-\frac{1}{2}i\sin\theta\,e^{-i\phi}$
				$\frac{1}{2}i\sin\theta\,e^{i\phi}$	$\frac{1}{2}i\cos\theta$
				$m = \frac{1}{2}$	$m = -\frac{1}{2}$
$2p_{1/2}$	$(\frac{1}{32})^{1/2}$	φ	$re^{-r/2}$	$-(1/\sqrt{3})\cos\theta$	$(1/\sqrt{3})\sin\theta\,e^{-i\phi}$
				$-(1/\sqrt{3})\sin\theta\,e^{i\phi}$	$-(1/\sqrt{3})\cos\theta$
		χ	$\alpha Z e^{-r/2}[1-(r/6)]$	$(\frac{3}{4})^{1/2}i$	0
				0	$(\frac{3}{4})^{1/2}i$
				$m = \frac{1}{2}$	$m = -\frac{1}{2}$
$2p_{3/2}$ $m = \pm\frac{1}{2}$	$(\frac{1}{32})^{1/2}$	φ	$re^{-r/2}$	$(\frac{2}{3})^{1/2}\cos\theta$	$-(\frac{1}{6})^{1/2}\sin\theta\,e^{-i\phi}$
				$-(\frac{1}{6})^{1/2}\sin\theta\,e^{i\phi}$	$-(\frac{2}{3})^{1/2}\cos\theta$
		χ	$\alpha Z r e^{-r/2}$	$(\frac{3}{32})^{1/2}i(\cos^2\theta - \frac{1}{3})$	$-(\frac{3}{32})^{1/2}i\sin\theta\cos\theta\,e^{-i\phi}$
				$(\frac{3}{32})^{1/2}i\sin\theta\cos\theta\,e^{i\phi}$	$(\frac{3}{32})^{1/2}i(\cos^2\theta - \frac{1}{3})$
				$m = \frac{3}{2}$	$m = -\frac{3}{2}$
$2p_{3/2}$ $m = \pm\frac{3}{2}$	$(\frac{1}{32})^{1/2}$	φ	$re^{-r/2}$	$(1/\sqrt{2})\sin\theta\,e^{i\phi}$	0
				0	$(1/\sqrt{2})\sin\theta\,e^{-i\phi}$
		χ	$\alpha Z r e^{-r/2}$	$(\frac{1}{32})^{1/2}i\sin\theta\cos\theta\,e^{i\phi}$	$(\frac{1}{32})^{1/2}i\sin^2\theta\,e^{-2i\phi}$
				$(\frac{1}{32})^{1/2}i\sin^2\theta\,e^{2i\phi}$	$-(\frac{1}{32})^{1/2}i\sin\theta\cos\theta\,e^{-i\phi}$

Source: [2.11].

^{a}The radial coordinate r is given in a_0/Z units, while the function as a whole is in $(Z^3/\pi a_0^3)^{1/2}$ units.

contribution of the order of $(\alpha Z)^2 = (Z/137)^2$. Thus in the relativistic presentation, strictly speaking, there are no points of zero radial distribution of electron position density.

All the foregoing effects in the relativistic atomic states are significant if and only if the spin–orbital interaction is sufficiently strong. For zero (or negligible) spin–orbital splitting all the p states are degenerate (the energies of $p_{1/2}$

and $p_{3/2}$ states coincide), and then any of their linear combinations is also an eigenfunction with the same energy. In particular, the combinations

$$(\tfrac{2}{3})^{1/2}p_{3/2,1/2} - (\tfrac{1}{3})^{1/2}p_{1/2,1/2} \sim p_z$$
$$(\tfrac{1}{2})^{1/2}p_{3/2,3/2} + (\tfrac{1}{3})^{1/2}p_{1/2,-1/2} - (\tfrac{1}{6})^{1/2}p_{3/2,-1/2} \sim p_x \qquad (2.20)$$
$$(\tfrac{1}{2})^{1/2}p_{3/2,3/2} - (\tfrac{1}{3})^{1/2}p_{1/2,-1/2} + (\tfrac{1}{6})^{1/2}p_{3/2,-1/2} \sim p_y$$

have the angular behavior of the appropriate p_z, p_x, and p_y nonrelativistic atomic p functions. A sufficiently large separation between the $p_{1/2}$ and $p_{3/2}$ states caused by the spin–orbital interaction (as compared with the bonding effects) makes this orbital mixing impossible, thus realizing the relativistic behavior of the corresponding atomic states.

More complicated are the orbital overlaps of relativistic functions. When two atoms A and B are characterized by four-component wavefunctions of type (2.15), their overlap, similar to (2.16), is given by the following expression:

$$S_{AB} = \int \Psi_A^* \Psi_B \, d\tau = \int [\Psi_A^{(1)*}\Psi_B^{(1)} + \Psi_A^{(2)*}\Psi_B^{(2)} + \Psi_A^{(3)*}\Psi_B^{(3)} + \Psi_A^{(4)*}\Psi_B^{(4)} \qquad (2.21)$$

With the relativistic functions in Table 2.5, the idea of σ, π, and δ bonds changes significantly. Indeed, only the overlap of $s_{1/2}^A$–$s_{1/2}^B$ functions is of pure σ type, quite similar to the overlap of the corresponding nonrelativistic orbitals. The $p_{1/2}^A$–$p_{1/2}^B$ overlap is essentially different from both p_z^A–p_z^B σ overlap and p_x^A–p_x^B (p_y^A–p_y^B) π overlap. The direct substitution of the four-component functions from Table 2.5 into Eq. (2.21) (in fact, only the large φ component can be tried since the χ spinor is much smaller and, in any case, yields the same angular dependence as the φ spinor) shows that $p_{1/2}^A(\tfrac{1}{2})$–$p_{1/2}^B(\tfrac{1}{2})$ overlap has $\tfrac{1}{3}\sigma$ bonding character and $\tfrac{2}{3}\pi$ antibonding, or vice versa, $\tfrac{1}{3}$ σ antibonding and $\tfrac{2}{3}$ π bonding (see Section 6.5 and Fig. 6.34). Similarly, the $p_{3/2}^A(\tfrac{1}{2})$–$p_{3/2}^B(\tfrac{1}{2})$ overlap comprises $\tfrac{2}{3}$ of σ bonding and $\tfrac{1}{3}$ of π antibonding, or vice versa, $\tfrac{2}{3}$ of σ antibonding and $\tfrac{1}{3}$ of π bonding. The $p_{3/2}^A(\tfrac{3}{2})$–$p_{3/2}^B(\tfrac{3}{2})$ overlap is the same as for nonrelativistic p functions. These conclusions are important to the analysis of electronic structure and bonding in heavy-atom coordination compounds (Section 6.5).

2.2. MULTIELECTRON STATES AND ENERGY TERMS

Electronic Configurations and Terms

For more than one electron in the atom the picture of electronic structure becomes significantly more complicated. In the one-electron approximation the multielectron wavefunction is composed of one-electron functions following specific rules that depend on the magnitudes of the spin–orbital and interelectron interactions. When ignoring these interactions, the electron can be distributed over one-electron states of the type (1.7) with the quantum numbers n, l, m, m_s following *the Pauli principle*. This distribution — the numbers of electrons in s, p, d, f states — forms the *electron configuration* of the atom. With the spin–orbital and interelectron interactions taken into account, there are several (many) states with the same electron configuration that differ by energy and spin state (see also the configuration interaction in Section 5.3).

Methods of determining the states of a given electron configuration are different for different relationships between the spin–orbital and interelectron interactions. For light atoms (approximately to the middle of the periodic table) the spin–orbital interaction is not larger than about 10^3 cm^{-1} (see below in Table 2.7), and hence it is much weaker than the interelectron interaction, which is on the order of 10^4 cm^{-1}. In this case the *Russell–Saunders LS coupling* takes place for which the orbital and spin momenta of electrons $\mathbf{l}_i$ and $\mathbf{s}_i$ are summed up separately into the total orbital momentum $\mathbf{L} = \sum_i \mathbf{l}_i$ and total spin momentum $\mathbf{S} = \sum_i \mathbf{s}_i$; the spin–orbital interaction is considered afterward and results in the total momentum $\mathbf{J} = \mathbf{L} + \mathbf{S}$. In the LS coupling scheme the wavefunction of the atom is a solution of Eq. (1.1) for the operators L^2, L_z, S^2, S_z, and it is characterized by the following quantum numbers:

- L, the total momentum: $L = 0, 1, 2, \ldots$
- $M = \sum_i m_i$, the projection L_z of the total momentum $\mathbf{L}$: $M = L$, $L - 1, \ldots, 0, -1, \ldots, -L$ ($2L + 1$ values)
- S, the total spin
- $M_s = \sum_s m_s$, the projection S_z of the total spin $\mathbf{S}$: $M_S = S, S - 1, \ldots, -S$ ($2S + 1$ values)

The set of states with the same values of L and S, but different M and M_S, all in all $(2L + 1)(2S + 1)$ states, forms the *atomic term*. The terms with $L = 0$, 1, 2, 3, 4, 5, ... are denoted by the capital letters S, P, D, F, G, H, respectively, with indication of the *spin multiplicity* (the number of spin states) equal to $2S + 1$ as a superscript on the left-hand side. For instance, the term with $L = 1$ and $S = 1$ (spin triplet) is denoted by ^3P.

Because of different charge distributions and spin orientations in different one-electron states (Fig. 2.1) and hence different interelectronic interactions, the energies of different terms (even with the same electronic configuration) are significantly different. These differences can be calculated and expressed by the *Slater–Condon*, or *Racah parameters*. In Table 2.6 the relative energies of the terms of d^n configurations expressed by Racah parameters A, B, C are given. They are obtained as explained below.

The analysis shows that in all the cases the state with the lowest energy, the ground state, corresponds to the term of maximum possible (in the configuration under consideration) spin multiplicity and maximum orbital momentum

Table 2.6. Energy terms of electronic d^n configurations expressed by Racah parameters A, B, C

Electronic Configuration	Term[a]	Relative Energy[b]
d^1, d^9	^2D	
d^2, d^8	^3F	$A - 8B$
	^3P	$A + 7B$
	^1G	$A + 4B + 2C$
	^1D	$A - 3B + 2C$
	^1S	$A + 14B + 7C$
d^3, d^7	^4F	$3A - 15B$
	^4P	$3A$
	^2H, ^2P	$3A - 6B + 3C$
	^2G	$3A - 11B + 3C$
	^2F	$3A + 9B + 3C$
	$^2D'$, $^2D''$	$3A + 5B + 5C \pm (193B^2 + 8BC + 4C^2)^{1/2}$
d^4, d^6	^5D	$6A - 21B$
	^3H	$6A - 17B + 4C$
	^3G	$6A - 12B + 4C$
	$^3F'$, $^3F''$	$6A - 5B + \frac{11}{2}C \pm \frac{3}{2}(68B^2 + 4BC + C^2)^{1/2}$
	^3D	$6A - 5B + 4C$
	$^3P'$, $^3P''$	$6A - 5B + \frac{11}{2}C \pm \frac{1}{2}(912B^2 - 24BC + 9C^2)^{1/2}$
	^1I	$6A - 15B + 6C$
	$^1G'$, $^1G''$	$6A - 5B + \frac{15}{2}C \pm \frac{1}{2}(708B^2 - 12BC + 9C^2)^{1/2}$
	^1F	$6A + 6C$
	$^1D'$, $^1D''$	$6A + 9B + \frac{15}{2}C \pm \frac{3}{2}(144B^2 + 8BC + C^2)^{1/2}$
	$^1S'$, $^1S''$	$6A + 10B + 10C \pm 2(193B^2 + 8BC + 4C^2)^{1/2}$
d^5	^6S	$10A - 35B$
	^4G	$10A - 25B + 5C$
	^4F	$10A - 13B + 7C$
	^4D	$10A - 18B + 5C$
	^4P	$10A - 28B + 7C$
	^2I	$10A - 24B + 8C$
	^2H	$10A - 22B + 10C$
	$^2G'$	$10A - 13B + 8C$
	$^2G''$	$10A + 3B + 10C$
	$^2F'$	$10A - 9B + 8C$
	$^2F''$	$10A - 25B + 10C$
	$^2D'$, $^2D''$	$10A - 3B + 11C \pm 3(57B^2 + 2BC + C^2)^{1/2}$
	$^2D'''$	$10A - 4B + 10C$
	^2P	$10A + 20B + 10C$
	^2S	$10A - 3B + 8C$

[a]Terms with the same L and S are distinguished by primes.
[b]The energies of d^{10-n} configurations differ from those of d^n by a constant shift.

for this multiplicity. This is the well-known *Hund rule*. For example, for the electronic configuration $[A](nd)^2$ ($[A]$ denotes the inner closed shell), for instance, V^{3+}, the possible terms, as shown below, are ^3F, ^1D, ^3P, ^1G, and ^1S. Following Hund's rule, the ground term is ^3F, because it has the maximal spin $S = 1$ (spin multiplicity $2S + 1 = 3$), and with the maximal spin it also has the maximal orbital momentum $L = 3$ (the ^3P term also has the maximal spin $S = 1$, but its orbital momentum $L = 1$ is lower).

The origin of Hund's rule can be understood when one takes into account that the maximum spin means that the electrons occupy as many separate one-electron orbitals with parallel spin orientations as possible (in the same orbital two electrons have mutually compensating opposite spins). For such electrons the negative exchange interaction that lowers the energy is maximal (the exchange between electrons with opposite spins is zero). In addition, together with the requirement of maximum orbital momentum, it favors the electron charge distribution over the largest possible volume of the atom, thus lowering the interelectron electrostatic repulsion.

In the approximation of *LS* coupling under consideration the total momentum of the atom is $\mathbf{J} = \mathbf{L} + \mathbf{S}$, its quantum number J acquiring all the values by one from $L + S$ to $|L - S|$: $J = L + S, L + S - 1, \ldots, |L - S|$. When the spin–orbital interaction is included, the energy levels with different J values may be different. The magnitude of this spin–orbital splitting can be obtained by means of perturbation theory. Following Eq. (2.8), the operator of the spin–orbital interaction in a multielectron atom can be written in the following form:

$$H_{SO} = \sum_i \xi(\mathbf{r}_i)(\mathbf{l}_i, \mathbf{s}_i) \tag{2.22}$$

which in the case of *LS* coupling can be transformed to

$$H_{SO} = \lambda(\mathbf{L}, \mathbf{S}) \tag{2.23}$$

where λ is a combination of radial integrals of the type (2.9) called *the spin–orbital constant* of the atom (or ion).

The spin–orbital constant λ plays a significant role in quantum chemistry and the theory of physical methods of investigation of molecules. Distinct from the analogous spin–orbital constant for one electron ξ_{nl}, which is positive [see Eqs. (2.9) and (2.10)], λ can be either positive or negative. Besides theoretical calculations, λ can be obtained from empirical data using *the rule of Landé intervals*. Indeed, $\mathbf{J} = \mathbf{L} + \mathbf{S}$, $J^2 = L^2 + S^2 + 2(\mathbf{L}, \mathbf{S})$ and hence

$$(\mathbf{L}, \mathbf{S}) = \frac{J^2 - L^2 - S^2}{2} \tag{2.24}$$

On the other hand, it is known from quantum mechanics that the mean value of the squares of the momenta in parentheses are $\langle J^2 \rangle = J(J + 1)$,

$\langle L^2 = L(L+1)$, $\langle S^2 \rangle = S(S+1)$, and hence the mean value of the perturbation (2.23), the first correction ΔE_j to the energy level with the appropriate J value of the term LS under consideration is

$$\Delta E_j = \lambda\langle(\mathbf{L},\mathbf{S})\rangle = \frac{\lambda[J(J+1) - L(L+1) - S(S+1)]}{2} \qquad (2.25)$$

Hence for the energy difference between two levels of the same LS term with consequent J values, we have

$$E_{J+1} - E_J = \Delta E_{J+1} - \Delta E_J = \lambda(J+1) \qquad (2.26)$$

This is the rule of Landé intervals that enables us to obtain easily the λ values using the experimental (spectroscopic) data on the energy differences between the components of the multiplet spectrum. Table 2.7 shows some of the values of λ obtained in this way. As one can see, $\lambda > 0$ for electronic configurations d^n with $n < 5$, and $\lambda < 0$ for $n > 5$ (for d^5 $L = 0$ and $\lambda = 0$). In the case of heavy atoms λ increases rapidly with Z, reaching values of several thousand reciprocal centimeters. For rare earth elements there is an approximate empirical formula [2.12]

$$\lambda = 200(Z - 55) \text{ cm}^{-1} \qquad (2.27)$$

For heavy atoms the LS coupling approximation may become invalid because the spin–orbital interaction is not small compared with the interelectron repulsion. In these cases the separate summation of the orbital momenta of the electrons into the total momentum $\mathbf{L}$ and the spin momenta into the total spin

Table 2.7. Constants of spin–orbital coupling λ for some transition metal ions in the ground state

Ion	Electronic Configuration	Ground State	λ (cm^{-1})
Ti^{3+}	d^1	^2D	154
V^{3+}	d^2	^3F	104
V^{2+}	d^3	^4F	55
Cr^{3+}	d^3	^4F	87
Cr^{2+}	d^4	^5D	57
Mn^{3+}	d^4	^5D	85
Mn^{2+}, Fe^{3+}	d^5	^6S	0
Fe^{2+}	d^6	^5D	−100
Co^{2+}	d^7	^4F	−180
Ni^{2+}	d^8	^3F	−335
Cu^{2+}	d^9	^3D	−852

S is ungrounded. For sufficiently strong spin–orbital interactions the opposite limit case is used, in which the orbital $\mathbf{l}_i$ and spin $\mathbf{s}_i$ momenta of the electron are first summed into the total momentum of the electron $\mathbf{j} = \mathbf{l}_i + \mathbf{s}_i$, and then these one-electron total momenta are summarized into the total momentum of the atom $\mathbf{J} = \sum_i \mathbf{j}_i$. This is called the *j-j coupling scheme*. Examples of *j-j* coupling are considered in Section 5.5. In many cases the intermediate picture between the *LS* and *j-j* coupling schemes is more appropriate. With care, these cases can also be handled by the LS approximation.

Multielectron Wavefunctions

In the one-electron approximation to the study of multielectron systems, each electron is described by a wavefunction that does not contain the coordinates of the other electrons, and the total wavefunction is constructed by one-electron functions following certain rules. The main requirement for the total wavefunction is that it is antisymmetrical with respect to the permutation of the (orbital and spin) coordinates of any two electrons. This condition follows from the quantum-mechanical *principle of indistinguishability* of identical particles with half-integer spins and results directly in *the Pauli principle*.

To obey the condition of antisymmetry with respect to electron permutations, the presentation of the full wavefunction Φ in the form of a determinant composed of one-electron functions is most convenient. For *n*-electron systems with closed shells,

$$\Phi(1, 2, \ldots, n) = (n!)^{-1/2} \begin{vmatrix} \psi_1(1) & \psi_1(2) & \cdots & \psi_1(n) \\ \psi_2(1) & \psi_2(2) & \cdots & \psi_2(n) \\ \vdots & \vdots & & \vdots \\ \psi_n(1) & \psi_n(2) & \cdots & \psi_n(n) \end{vmatrix} \qquad (2.28)$$

where each number $1, 2, \ldots, n$ in parentheses denotes the three orbital and one spin coordinates of the corresponding electron, and the factor $(n!)^{-1/2}$ occurs as a result of normalization (the one-electron functions ψ_i are assumed to be orthonormalized).

From (2.28) one can easily verify that the function Φ has the required symmetry properties. Indeed, the interchange of the coordinates of any two electrons, say 1 with 2, is equivalent to the interchange of two columns of the determinant (1 and 2), and this changes the sign of the latter.

If the one-electron functions are characterized by the four quantum numbers n, l, m, m_s, the full wavefunction is determined by the set of quantum numbers of all the occupied one-electron states, which can be denoted as follows:

$$\Phi(n_1 l_1 m_1 m_{s1}; n_2 l_2 m_2 m_{s2}; \ldots; n_n l_n m_n m_{sn}) \qquad (2.29)$$

For electrons of the same shell (nl) the quantum numbers n and l may be omitted, and the two values of the spin quantum number can be denoted by

"+" and "−". For instance, for two equivalent (i.e., with the same n) d electrons with quantum numbers $m_1 = 2$, $m_{s1} = \frac{1}{2}$ and $m_2 = -1$, $m_{s2} = -\frac{1}{2}$, the wavefunction can be written as $\Phi(2^+, 1^-)$. As mentioned above, the one-electron functions in the LS coupling scheme under consideration are presented as a product of the orbital and spin parts, $\psi(nlmm_s) = \varphi_{nlm}(\mathbf{r}_i, \theta, \phi)\eta_i(m_s)$, where φ_{nlm} is determined by Eq. (2.1).

Using these denotations, we demonstrate below how the possible electronic terms of a given configuration can be revealed. Consider an example of two d electrons above the closed shell of an atom (i.e., the configuration $[A](nd)^2$). There are five orbital d states and each of them has two spin states; hence there are 10 states to be occupied by the two electrons under the condition that each state accepts only one electron. The number of possible pairs of states equals the number of combination of 10 by 2: $C_{10}^2 = 10 \cdot 9/2 = 45$. For each of these 45 possible states of the configuration d^2 there is a wavefunction $\Phi(m_1 m_{s1}; m_2 m_{s2})$ determined after Eq. (2.28).

If we neglect the interelectronic interaction, all 45 states have the same energy, and there is a 45-fold degeneracy. In Table 2.8 the wavefunctions for the 45 states are grouped according to the values of the quantum numbers of the projection of the summary momentum of the two electrons $M = m_1 + m_2$ and summary spin $M_s = m_{s1} + m_{s2}$. Such tables can easily be composed for any electron configuration of the atom. Then, to reveal the possible terms, one has to separate the groups of states that have the same L and S but different M and M_s, $(2L + 1)(2S + 1)$ states in each group (each term).

It is seen from Table 2.8 that the largest value $M = 4$ is possible only in combination with $M_s = 0$. Since $M = L, L - 1, \ldots$, and $M_s = S, S - 1, \ldots$, the

Table 2.8. All possible 45 states of the two-electron $(nd)^2$ configuration (in the one-electron approximation) classified by the quantum numbers M and M_s

		M_s	
M	1	0	−1
4		$\Phi(2^+; 2^-)$	
3	$\Phi(2^+; 1^+)$	$\Phi(2^+; 1^-)$, $\Phi(2^-; 1^+)$	$\Phi(2^-; 1^-)$
2	$\Phi(2^+; 0^+)$	$\Phi(2^+; 0^-)$, $\Phi(2^-; 0^+)$, $\Phi(1^+; 1^-)$	$\Phi(2^-; 0^-)$
1	$\Phi(2^+; -1^+)$	$\Phi(2^+; -1^-)$, $\Phi(2^-; -1^+)$	$\Phi(2^-; -1^-)$
	$\Phi(1^+; 0^+)$	$\Phi(1^+; 0^-)$, $\Phi(1^-; 0^+)$	$\Phi(1^-; 0^-)$
0	$\Phi(2^+, -2^+)$	$\Phi(2^+; -2^-)$, $\Phi(2^-, 2^+)$	$\Phi(2^-; -2^-)$
	$\Phi(1^+; -1^+)$	$\Phi(1^+; -1^-)$, $\Phi(1^-; -1^+)$, $\Phi(0^+; 0^-)$	$\Phi(1^-; -1^-)$
−1	$\Phi(1^+; -2^+)$	$\Phi(1^+; -2^-)$, $\Phi(1^-; -2^+)$	$\Phi(1^-; -2^-)$
	$\Phi(0^+; -1^+)$	$\Phi(0^+; -1^-)$, $\Phi(0^-; -1^+)$	$\Phi(0^-; -1^-)$
−2	$\Phi(0^+; -2^+)$	$\Phi(0^+; -2^-)$, $\Phi(0^-; -2^+)$, $\Phi(-1^+; -1^-)$	$\Phi(0^-; -2^-)$
−3	$\Phi(-1^+; -2^+)$	$\Phi(-1^+; -2^-)$, $\Phi(-1^-; -2^+)$	$\Phi(-1^-; -2^-)$
−4		$\Phi(-2^+; -2^-)$	

largest L is also 4 with $S = 0$. Thus the term with the largest L is the spin singlet ^1G. This term has $(2L + 1)(2S + 1) = 9$ states with $M_s = 0$ and $M = 4$, $3, 2, 1, 0, -1, -2, -3, -4$. They can be easily found in Table 2.8 and eliminated.

From the remaining terms the senior one is that with $M = 3$ and $M_s = 1$. It belongs to the term with $L = 3$ and $S = 1$ [i.e., to the ^3F term that has $(2L + 1)(2S + 1) = 21$ states with $M = 3, 2, 1, 0, -1, -2, -3$, and $M_s = 1, 0$, -1 for each of the M values]. Eliminating them from Table 2.8, we again find the senior state with $M = 2$ and $M_s = 0$ that belongs to the ^1D term (five states). In a similar way we also find the ^3P term (nine states) and the ^1S term (one state), thus counting all 45 states. Hence for the configuration $[A](nd)^2$ the terms ^1G, ^3F, ^1D, ^3P, and ^1S are possible. Similarly, the possible terms of all the other configurations d^n listed in Table 2.6 were found.

The wavefunctions of the atomic terms with certain L, M, S, and M_s values $\Psi(LMSM_s)$ can be found as a linear combination of the functions $\Phi(m_1m_{s1}; m_2m_{s2})$ [see (2.28)] that satisfy the condition of being an eigenfunction of the operators L^2, S^2, L_z, S_z. The method of their determination based on the symmetry properties of these operators is given in special manuals [2.1–2.3, 2.13]. In particular, if the functions are known for some of the M and M_S values, they can also be found for other quantum numbers using the following relationships:

$$(L_x \pm iL_y)\Psi(LMSM_s) = \hbar[(L \pm M + 1)(L \mp M)]^{1/2}\Psi(L, M \pm 1, S, M_s)$$

$$(2.30)$$

$$(S_x \pm iS_y)\Psi(LMSM_s) = \hbar[(S \pm M_s + 1)(S \mp M_s)]^{1/2}\Psi(L, M, S, M_s \pm 1)$$

$$(2.31)$$

Table 2.9 gives some of the widely used wavefunctions for the terms ^3F and ^3P of the configurations d^2 and d^8, as well as ^4F and ^4P for the configurations d^3 and d^7. The functions are given for one value of the spin quantum number M_s; for the other values of M_s they can be found from the relation (2.31). The configurations d^1, d^4, d^6, and d^9, using the principle of *complementary electronic configurations* (see below), can be presented as one-electron configurations above a closed shell; their functions $\Psi(L, M, S, M_s)$ coincide with $\psi(nlmm_s)$.

Slater–Condon and Racah Parameters

As mentioned above, without the electron interaction presented by the term $H' = \sum e^2/r_{ij}$ in the Hamiltonian, all the terms of a given electron configuration have the same energy. When the electron interaction is included, the energies of different terms diverge, due to the differences in electron repulsion and exchange interaction. To determine this energy-level splitting, one can consider the electron interaction H' as a perturbation and solve the perturbation theory

Table 2.9. Wavefunctions $\Psi(LMSM_s)$ of some terms of d^2 and d^3 configurations expressed by linear combinations of determinant functions $\Phi(m_1 m_{s1}; m_2 m_{s2}; \ldots)$

Term	$\Psi(LMSM_s)$	$\sum_i C_i \Phi$
$^3F(d^2)$	$\Psi(3\ 3\ 1\ 1)$	$\Phi(2^+; 1^+)$
	$\Psi(3\ 2\ 1\ 1)$	$\Phi(2^+; 0^+)$
	$\Psi(3\ 1\ 1\ 1)$	$(\frac{3}{5})^{1/2}\Phi(2^+; -1^+) + (\frac{2}{5})^{1/2}\Phi(1^+; 0^+)$
	$\Psi(3\ 0\ 1\ 1)$	$(\frac{1}{5})^{1/2}\Phi(2^+; -2^+) + (\frac{4}{5})^{1/2}\Phi(1^+; -1^+)$
	$\Psi(3\ -1\ 1\ 1)$	$(\frac{3}{5})^{1/2}\,{}^1\Phi(1^+; -2^+) + (\frac{2}{5})^{1/2}\Phi(0^+; -1^+)$
	$\Psi(3\ -2\ 1\ 1)$	$\Phi(0^+; -2^+)$
	$\Psi(3\ -3\ 1\ 1)$	$\Phi(-1^+; -2^+)$
$^3P(d^2)$	$\Psi(1\ 1\ 1\ 1)$	$(\frac{2}{5})^{1/2}\Phi(2^+; -1^+) - (\frac{3}{5})^{1/2}\Phi(1^+; 0^+)$
	$\Psi(1\ 0\ 1\ 1)$	$(\frac{4}{5})^{1/2}\Phi(2^+; -2^+) - (\frac{1}{5})^{1/2}\Phi(1^+; -1^+)$
	$\Psi(1\ -1\ 1\ 1)$	$(\frac{2}{5})^{1/2}\Phi(1^+; -2^+) - (\frac{3}{5})^{1/2}\Phi(0^+; -1^+)$
$^4F(d^3)$	$\Psi(3\ 3\ \frac{3}{2}\ \frac{3}{2})$	$\Phi(2^+; 1^+; 0^+)$
	$\Psi(3\ 2\ \frac{3}{2}\ \frac{3}{2})$	$\Phi(2^+; 1^+; -1^+)$
	$\Psi(3\ 1\ \frac{3}{2}\ \frac{3}{2})$	$(\frac{2}{5})^{1/2}\Phi(2^+; 1^+; -2^+) + (\frac{3}{5})^{1/2}\Phi(2^+; 0^+; -1^+)$
	$\Psi(3\ 0\ \frac{3}{2}\ \frac{3}{2})$	$(\frac{4}{5})^{1/2}\Phi(2^+; 0^+; -2^+) + (\frac{1}{5})^{1/2}\Phi(1^+; 0^+; -1^+)$
	$\Psi(3\ -1\ \frac{3}{2}\ \frac{3}{2})$	$(\frac{2}{5})^{1/2}\Phi(2^+; -1; -2^+) + (\frac{3}{5})^{1/2}\Phi(1^+; 0^+; -2^+)$
	$\Psi(3\ -2\ \frac{3}{2}\ \frac{3}{2})$	$\Phi(1^+; -1^+; -2^+)$
	$\Psi(3\ -3\ \frac{3}{2}\ \frac{3}{2})$	$\Phi(0^+; -1^+; -2^+)$
$^4P(d^3)$	$\Psi(1\ 1\ \frac{3}{2}\ \frac{3}{2})$	$(\frac{3}{5})^{1/2}\Phi(2^+; 1^+; -2^+) - (\frac{2}{5})^{1/2}\Phi(2^+; 0^+, -1^+)$
	$\Psi(1\ 0\ \frac{3}{2}\ \frac{3}{2})$	$(\frac{1}{5})^{1/2}\Phi(2^+; 0^+; -2^+) - (\frac{4}{5})^{1/2}\Phi(1^+; 0^+; -1^+)$
	$\Psi(1\ -1\ \frac{3}{2}\ \frac{3}{2})$	$(\frac{3}{5})^{1/2}\Phi(2^+; -1^+; -2^+) - (\frac{2}{5})^{1/2}\Phi(1^+; 0^+; -2^+)$

problem taking the functions $\psi(LMSM_s)$ as zeroth-order functions. The appropriate secular equation of perturbation theory is then essentially simplified, owing to the fact that the operators L^2, S^2, L_z, and S_z commute with H'. It means that with the functions Ψ that are eigenfunctions of the operators above only matrix elements of the perturbation H' that are diagonal in L and S are nonzero, and they are the same for different M and M_s. In other words, the energy corrections ΔE caused by electron interaction are the same for all the states of the same term LS, and they can be found directly as diagonal matrix elements:

$$\Delta E(L, S) = \int \sum \frac{e^2}{r_{ij}} |\Psi(LMSM_s)|^2 \, d\tau \qquad (2.32)$$

The way of calculation of integrals of the type (2.32) is well known. Since Ψ is a linear combination of Φ in Table 2.9, (2.32) can be reduced to integrals of the type

$$\int \Phi^*(m_1, m_{s1}, \dots) \frac{e^2}{r_{ij}} \Phi(m'_1, m'_{s1}, \dots) \, d\tau \tag{2.33}$$

This integral equals zero when the two functions Φ differ by more than two one-electron states. Indeed, different one-electron functions are orthogonal to each other, and since $r_{12} = |\mathbf{r}_1 - \mathbf{r}_2|$ depends on the coordinates of only two electrons, integration over the other coordinates with orthogonal functions yields zero. For instance,

$$\int \Phi^*(2^+; 1^-; 0^+) \sum_{i,j} \frac{1}{r_{ij}} \Phi(1^+; 2^-; 0^-) \, d\tau = 0$$

whereas

$$\int \Phi^*(2^+; 1^-; 0^+) \sum_{i,j} \frac{1}{r_{ij}} \Phi(2^+; 2^-; 1^+) \, d\tau \neq 0$$

Thus only four one-electron functions can be different in the nonzero integrals (2.33); denote them by a, b, c, and d. Then the expression for ΔE is reduced to a sum of two-electron integrals of the following type:

$$[ab \,|\, cd] = \int a^*(1)b(1) \frac{e^2}{r_{12}} c^*(2)d(2) \, d\tau_1 \, d\tau_2 \tag{2.34}$$

In the notation on the left-hand side of Eq. (2.34), widely employed in quantum chemistry, a and b are one-electron functions of the first electron, while c and d contain the coordinates of the second electron. Some of these one-electron functions can coincide. In particular, the integrals

$$[aa \,|\, bb] = I(a, b) = \int a^*(1)a(1) \frac{e^2}{r_{12}} b^*(2)b(2) \, d\tau_1 \, d\tau_2 \tag{2.35}$$

$$[ab \,|\, ba] = K(a, b) = \int a^*(1)b(1) \frac{e^2}{r_{12}} b^*(2)a(2) \, d\tau_1 \, d\tau_2 \tag{2.36}$$

are most usable. The first is called the *Coulomb integral*; it equals the energy of the electrostatic repulsion of two electron clouds, described by the one-electron functions $a(1)$ and $b(2)$, respectively. $K(a, b)$ the *exchange integral*, represents the energy of exchange interaction between the two electrons in the states a and b.

To calculate the integral (2.34), the following expression is useful:

$$\frac{1}{r_{12}} = \sum_{k=0}^{\infty} K_k(r_1, r_2) P_k(\cos \gamma_{12}) \tag{2.37}$$

where

$$K_k(r_1, r_2) = \begin{cases} \dfrac{r_1^k}{r_2^{k+1}} & \text{if } r_1 < r_2 \\ \dfrac{r_2^k}{r_1^{k+1}} & \text{if } r_2 < r_1 \end{cases} \tag{2.38}$$

and $P_k(\cos \gamma_{12})$ is the Legendre polynomial [Eq. (2.3) with $m = 0$] of the argument $\cos \gamma_{12} = \cos \theta_1 \cos \theta_2 + \sin \theta_1 \sin \theta_2 \cos(\phi_1 - \phi_2)$. Substituting the one-electron functions a, b, c, and d by their expressions of the type (2.1) together with (2.37) and (2.38) into (2.34) and using some orthogonality conditions for the spherical functions, we can integrate over the angular coordinates, thus reducing the expression for $[ab|cd]$ to a limited number of two-electron radial integrals of the following type:

$$R^{(k)}(abcd) = e^2 \iint R^*_{n(a)l(a)}(r_1) R_{n(b)l(b)}(r_1) K_k(r_1, r_2)$$

$$\times R^*_{n(c)l(c)}(r_2) R_{n(d)l(d)}(r_2) r_2^2 r_2^2 \, dr_1 \, dr_2 \tag{2.39}$$

For the Coulomb and exchange contributions the following notations are used:

$$R^{(k)}(aabb) = F^{(k)} \tag{2.40}$$

$$R^{(k)}(abba) = G^{(k)} \tag{2.41}$$

The constants $F^{(k)}$ and $G^{(k)}$ are named *Slater–Condon parameters*. For equivalent electrons that have the same n and l values, all the radial integrals $R^{(k)}$ are of the $F^{(k)}$ type. In atomic calculations the parameters F_k are more convenient; they differ from $F^{(k)}$ by a numerical coefficient. For instance, for equivalent d electrons the nonzero F_k parameters are

$$F_0 = F^{(0)}, \quad F_2 = \tfrac{1}{49} F^{(2)} \quad F_4 = \tfrac{1}{144} F^{(4)} \tag{2.42}$$

Finally, in many cases combinations of Slater–Condon parameters called *Racah parameters* are used:

$$A = F_0 - 49F_4 \quad B = F_2 - 5F_4 \quad C = 35F_4 \tag{2.43}$$

The correction to the energy due to electron interactions and hence the relative energies of atomic terms of d^n configuration expressed in terms of Racah parameters A, B, and C are given in Table 2.6.

To calculate the Racah parameters, as seen from expression (2.39), radial wavefunctions of the atomic one-electron states are needed. More often it is preferred to consider the values A, B, and C as empirical parameters and to determine them from a comparison of the calculated energies with those determined experimentally from spectroscopic data. For the electronic configuration $[A](nd)^2$, discussed above, the energy difference between the terms ^3F and ^3P is $E(^3P) - E(^3F) = 15B$ (Table 2.6). On the other hand, the experimental value of this splitting, for instance, for $V^{3+}(3d^2)$ is $\approx 13,000 \text{ cm}^{-1}$, hence $B \approx 870 \text{ cm}^{-1}$. Some values of the Racah parameters for transition metal ions and their complexes are given in Tables 4.7 and 8.3. Sometimes the following estimations for Slater–Condon parameters for f electrons are useful [2.14]: $F_4 \leqslant 0.202F_2$ and $F_6 \leqslant 0.0306F_2$.

In many cases there is no need to calculate in detail the energy terms for all possible electron configurations of the atom because for some of them the electronic terms can be revealed directly from other configurations using the *principle of complementary configurations*. According to this principle, the configuration with n equivalent electrons has the same types of terms as the configuration $N - n$, where N is the number of electrons in the closed shell under consideration. The reason is that formally the problem of n electrons is equivalent to that of $N - n$ "holes," and hence the problem of n particles can be reduced to that of $N - n$ (other) particles. The reduction in computation work achieved by using this principle is significant: For the configuration d^9 the terms are equivalent to those of d^1; for d^8, to d^2; for d^7, to d^3; and so on. The rule is also valid for half-filled shells for which all the orbital states are occupied (orbitally closed shell). This means that for d^4 one can use the configurations of $d^{5-4} = d^1$ (d^5 is a half-filled shell); for d^6, that of $d^{10-6} \to d^4 \to d^1$; and so on. Thus all the terms of d^n configurations can be reduced to those of d^1 and d^2. Tables 2.6 and 2.9 are composed taking into account this principle.

Hartree–Fock Method

The Hartree–Fock method of calculation is of general importance serving as a starting point to many other methods of investigation of multiparticle systems. For multielectron systems (atoms, molecules, crystals) the Hartree–Fock method [2.1–2.3] yields the best solutions (one-electron functions and energies) compatible with the approximation of full separation of the electron variables, that is, ignoring correlation effects (Section 5.3).

To disclose the essence of this method, let us begin with the Hartree approximation. Assume that each electron can be considered as moving independently in an averaged field created by the nucleus and the other electrons. This means that its wavefunction $\varphi_i(\mathbf{r}_i)$ is independent of the variables of other electrons, and the expression $|\varphi_i(\mathbf{r}_i)|^2$ is the probability density to find the ith electron at the point r_i. Since the probability of independent events equals the product of probabilities of each of them, the electronic wavefunction of the system as a whole can be presented in the form of a simple product of the one-electron functions $\varphi_i(\mathbf{r}_i)$:

$$\Phi(\mathbf{r}_i, \mathbf{r}_2, \ldots, \mathbf{r}_n) = \varphi_i(\mathbf{r}_1)\varphi_2(\mathbf{r}_2) \cdots \varphi_n(\mathbf{r}_n) \tag{2.44}$$

This means that $|\Phi|^2$ is the density of the probability to find the electron 1 at $\mathbf{r}_1$, 2 at $\mathbf{r}_2$, and so on. The functions $\varphi_i(\mathbf{r})$ are assumed to be orthonormalized:

$$\int \varphi_i^*(\mathbf{r})\varphi_j(\mathbf{r}) \, d\tau = \delta_{ij} \qquad i,j = 1, 2, \ldots, n \tag{2.45}$$

where hereafter in this book δ_{ij} means the *Kroneker index* (or δ symbol):

$$\delta_{ij} = \begin{cases} 0 & \text{if } i \neq j \\ 1 & \text{if } i = j \end{cases} \tag{2.46}$$

To determine the functions $\varphi(\mathbf{r})$ the *variational principle* can be employed. It states that the functions (2.44) sought for must obey the condition of the minimal total energy E of the system ($\delta E = 0$) calculated with the Hamiltonian H of the Schrödinger equation:

$$E = \int \Phi^* H \Phi \, d\tau \tag{2.47}$$

$$\delta E = \delta \int \Phi^* H \Phi \, d\tau = 0 \tag{2.48}$$

By substituting (2.44) into (2.48) with the Hamiltonian from Section 1.3 and varying the unknown functions $\varphi_i(\mathbf{r}_i)$, one finds the following system of equations for these functions:

$$[H^0 + V_k(\mathbf{r}) - \varepsilon_k]\varphi_k(\mathbf{r}) = 0 \qquad k = 1, 2, \ldots, n, \tag{2.49}$$

where H^0 is the part of the Hamiltonian that contains the kinetic energy T and the energies of interaction with the nuclei U_α:

$$H^0 = T - \sum_\alpha U_\alpha = -\frac{\hbar^2}{2m}\Delta - \sum_\alpha \frac{e^2 Z_\alpha}{|\mathbf{r} - \mathbf{R}_\alpha|} \tag{2.50}$$

and $V_k(\mathbf{r})$ is the sum of the energies of the electrostatic repulsive interaction of the kth electron with the charges of the ith electrons distributed in space with the densities $e|\varphi_i(\mathbf{r}_i)|^2$:

$$V_k(\mathbf{r}) = \sum_{i \neq k} e^2 \int \frac{|\varphi_i(\mathbf{r}')|^2}{|(\mathbf{r} - \mathbf{r}'|} \, d\tau' \tag{2.51}$$

The system of equations (2.49) describes the full separation of the variables of the electrons: Each of its n equations contains the coordinates of only one electron. They are complicated integrodifferential equations (the unknown function is both under differentiation and integration) which can be solved by means of a special iteration procedure named the *method of self-consistent field* (SCF). The grounds of this method are as follows.

Assume that we have a set of functions $\varphi_k(\mathbf{r})$ — the initial set — obtained, for instance, by ignoring the electron interaction, the integral term $V_k(\mathbf{r})$ in (2.49). With these functions, all the terms $V_k(\mathbf{r})$ after (2.51) and their sum (the average potential of interaction of one electron with all the others) can be evaluated. By substituting these terms into the system (2.49), one obtains n independent simple differential equation that can be solved numerically. These solutions yield a new system of functions φ' (and eigenvalues ε') that are more accurate than the initial set φ because of taking the electron interaction into account. With the more accurate functions, one can calculate more accurate potentials of the interelectronic interaction $V_k'(\mathbf{r})$ and again solve (2.49) with these better potentials, and so on, until the new eigenfunctions φ_k and new eigenvalues ε_k coincide with the previous values (within the accuracy required). The solutions obtained in this way are called *self-consistent solutions*. The self-consistent method described above was suggested first by Hartree [2.15].

However, as shown by Fock [2.16], the presentation (2.44) of the wavefunction Φ as a simple product of one-electron functions φ does not satisfy the condition of indistinguishability of the electrons that demands the total wavefunction to be antisymmetrical with respect to the transmutation of the coordinates of any two electrons. To obey this condition in the case of a closed-shell system, the wavefunction Φ should be presented in the form of the determinant (2.28) constructed from the functions $\varphi_i(\mathbf{r})$. The substitution of this determinant function into Eq. (2.48) and the consequent variation of the functions $\varphi_k(\mathbf{r})$ yields the following system of equations instead of (2.49):

$$\left[H^0 + 2 \sum_{i=1}^{n/2} \frac{e^2 \int \varphi_i^*(\mathbf{r}')\varphi_i(\mathbf{r}')}{|\mathbf{r} - \mathbf{r}'|} \, d\tau' - 2\varepsilon_{kk} \right] \varphi_k(\mathbf{r})$$

$$- \sum_{i=1}^{n/2} \left[\frac{e^2 \int \varphi_i^*(\mathbf{r}')\varphi_k(\mathbf{r}')}{|\mathbf{r} - \mathbf{r}'|} \, d\tau' - \varepsilon_{ik} \right] \varphi_i(\mathbf{r}') = 0 \tag{2.52}$$

Here it is taken into account that in accordance with the assumption of a closed shell, the number of electrons n is even, and in each of the orbital states $\varphi_k(\mathbf{r})$ there are two electrons with opposite spins. The second sum in (2.52) is the *exchange correction* that arises additionally [as compared with (2.49)] due to the indistinguishability principle — the exchange interactions. Exchange interactions are nonzero for electrons with the same spin states only, and therefore the number of terms in the second sum in (2.52) is half that in the first sum, the electrostatic interactions.

The term with $i = k$ from the exchange sum that corresponds to the interaction of the electron with itself vanishes automatically since being present in both sums of (2.52) with opposite signs [the remaining term with $i = k$ in the first sum (which has the factor 2) stands for the electrostatic interaction of the ith electron with the other one in the same orbital state $\varphi_i(\mathbf{r})$ but with an opposite spin].

The constants ε_{ik} occur as Lagrange factors at the condition of normalization and orthogonality of the functions φ_k and φ_i (2.45). Most of these constants can be eliminated by substituting for φ_i their linear combinations φ_i'. In other words, we can perform a unitary transformation [see Eq. (3.11)] of the set of functions φ_i to another set φ_i' for which the off-diagonal values ε_{ik}' become zero. Unitary transformations do not change the initial determinant function (2.28) and hence the energy of the system [2.17].

Denoting the remaining diagonal value $\varepsilon_{kk} = E_k$ and introducing the operator P_{ik} that transmutes the indices i and k of the function it acts upon, we have (the prime at φ' is omitted)

$$\left[H^0 + \sum \frac{e^2 \int \varphi_i^*(\mathbf{r}')(2 - P_{ik})\varphi_i(\mathbf{r}')}{|\mathbf{r} - \mathbf{r}'|} \, d\tau' - E_k \right] \varphi_k(\mathbf{r}) = 0 \qquad k = 1, 2, \ldots, n/2$$

(2.53)

This is the *system of Hartree-Fock* (HF) *self-consistent field* (SCF) *equations* for a multielectron system with a closed electronic shell. It can be solved in the same manner as discussed above for the Hartree equations, yielding self-consistent one-electron states: wavefunctions $\varphi_k(\mathbf{r})$ and eigenvalues E_k. The latter has the physical meaning of the one-electron orbital energy in the self-consistent state $\varphi_k(\mathbf{r})$.

Equations (2.53) can be written in the usual Schrödinger equation form:

$$F^k \varphi_k(\mathbf{r}) = E_k \varphi_k(\mathbf{r})$$

(2.54)

where

$$F^k = H^0 + \sum_i \frac{e^2 \int \varphi_i^*(\mathbf{r})(2 - P_{ik})\varphi_i(\mathbf{r}')}{|\mathbf{r} - \mathbf{r}'|} \, d\tau'$$

(2.55)

is the Fock effective one-electron Hamiltonian, sometimes called a *Fockian*. Multiplying Eq. (2.53) or (2.54) by $\varphi_k(r)$ from the left-hand side and integrating over r gives us

$$E_k = H^0_{kk} + \sum_i (2[ii\,|\,kk] - [ik\,|\,ki])$$

$$= H^0_{kk} + \sum_i [2I(i,\,k) - K(i,\,k)] \tag{2.56}$$

where the notation (2.35) for the Coulomb integral $I(i,\,k) = [ii\,|\,kk]$ and (2.36) for the exchange integral $K(i,\,k) = [ik\,|\,ki]$ are used. Taking into account the expression (2.50) for H^0, we come to the conclusion that the self-consistent energy of the kth electron in the one-electron state equals the sum of its kinetic energy, the energy of interaction with the nuclei, and the energy of the Coulomb and exchange interelectron interactions.

In fact, E_k equals the energy that must be applied to withdraw the kth electron from the atom, the *ionization energy*, provided that all the other one-electron states are not changed by the transition to the ionized state. This statement is known as the *Koopmans theorem* [2.18]. For many coordination systems the Koopmans theorem is invalid due to the significant changes in the one-electron states (interelectron interaction) caused by the elimination of one electron; upon ionization the remaining electrons relax to new self-consistent states (Sections 6.2, 6.4, 8.3, etc.).

For the sum of the one-electron energies of the n electrons, we have

$$2 \sum_{k=1}^{n} E_k = 2 \sum_{k-1}^{n/2} H^0_{kk} + 2 \sum_{k=1}^{n/2} [2I(i,\,k) - K(i,\,k)] \tag{2.57}$$

It is important that in this sum the interelectron interaction is included twice: once in the sum of the interaction of a given ith electron with all the others (including the kth electron) and the second time in the sum of the interaction of the kth electron with all the others, including the ith electron. Therefore, to obtain the total energy of the system E, the interelectron interaction should be subtracted once from the expression (2.57). Hence

$$E = 2 \sum_{k-1}^{n/2} H^0_{kk} + \sum_{k=1}^{n/2} [2I(i,\,k) - K(i,\,k)] \tag{2.58}$$

For a system with an open shell of electrons the Hartree–Fock equations (2.53) become much more complicated. For a discussion of this problem, see Section 5.3.

REFERENCES

2.1. H. Eyring, J. Walter, and G. E. Kimball, *Quantum Chemistry*, Wiley, New York, 1947.

2.2. J. C. Slater, *Quantum Theory of Atomic Structure*, McGraw-Hill, New York, 1960.

2.3. P. Gombas, *Theorie und Lösungsmethoden des Mehrteilchenproblems der Wellenmechanik*, Birkhauser, Basel, 1950.

2.4. H. A. Bethe and E. E. Salpeter, *Quantum Mechanics of One- and Two-Electron Atoms*, Springer-Verlag, Berlin, 1957.

2.5. F. Herman and S. Skillman, *Atomic Structure Calculations*, Prentice Hall, Englewood Cliffs, NJ 1963, 438 pp.

2.6. S. Fraga and J. Karwowski, *Tables of Hartree–Fock Atomic Data*, The University of Alberta, Edmonton, Alberta, Canada, 1974, 350 pp.

2.7. E. Clementi, *IBM J. Res. Dev.*, **9**, 2–19 (1965), Supplement, *Tables of Atomic Functions*, IBM, Armork, N.Y., 1965, Tables 1–45

2.8. J. W. Richardson, W. C. Nieuwpoort, R. R. Powell, and M. F. Edgell, *J. Chem. Phys.*, **36**, 1057–1061 (1962).

2.9. J. W. Richardson, P. R. Powell, and W. C. Nieuwpoort, *J. Chem. Phys.*, **38**, 796–801 (1963).

2.10. J. N. Murrell, S. F. A. Kettle, and J. M. Tedder, *Valence Theory*, Wiley, New York, 1965.

2.11. R. E. Powel, *Chem. Educ.*, **45**, 558–563 (1968).

2.12. M. A. El'yashevich, *Rare Earth Spectra* (Russ.), Gostechteorizdat, Moscow, 1953, 456 pp.

2.13. H. L. Schlafer and G. Glieman, *Basic Principles of Ligand Field Theory*, Wiley, New York, 1969.

2.14. C. K. Jorgensen, *Kgl. Danske Vidensk.*, **29**, 1–21 (1955).

2.15. D. Hartree, *The Calculations of Atomic Structures*, Wiley, New York, 1957.

2.16. V. A. Fock, *Z. Phys.*, **61**, 126–148; (1930); **62**, 795–805 (1930).

2.17. J. C. Slater, *Electronic Structure of Molecules*, McGraw-Hill, New York, 1963.

2.18. T. Koopmans, *Physica*, **1**, 104–113 (1933).

3 Symmetry Ideas and Group-Theoretical Description

In the theory of electronic structure and properties of transition metal coordination compounds, the ideas of symmetry are of great, sometimes fundamental importance.

The most complete description of symmetry properties is achieved in mathematical group theory. In this chapter we give minimum information about some rules worked out by group-theoretical methods that are required for a better understanding of the presentation in this book (for more details on group theory, see, e.g., [3.1–3.8]).

3.1. SYMMETRY TRANSFORMATIONS AND MATRICES

Molecular symmetry is the property of a molecule to remain invariant under certain rotations and reflections in space and permutations of identical particles. These rotations, reflections, and permutations are called *symmetry operations* or *symmetry transformations*. For instance, in the equilateral planar–triangular configuration of the molecule V_3, the symmetry transformations are (Fig. 3.1) rotations around the Oz axis perpendicular to the plane of the molecule by angles $2\pi/3$, $4\pi/3$, 2π, and around the axes A_1B_1, A_2B_2, and A_3B_3 by angles π and 2π, as well as reflection in the planes that comprise these axes and are perpendicular to the plane of the molecule.

For the simplest symmetry operations in space, rotation and reflections (for crystals there are also translations), there are conventional notations. The *rotation* around some axis by the angle $2\pi/n$ is denoted by C_n. It is obvious that if after some rotation the molecule is matched with itself, it will behave in the same way when this rotation is repeated an integer number of times.

The result of *consequent performance of two symmetry operations is defined as their product*. If C_n is a symmetry operation of the molecule, then $C_n \cdot C_n = C_n^2$ is also a symmetry operation. In general, C^p with any integer p is also a symmetry operation. In particular, C_n^n is a rotation by an angle 2π that does not change the position of the molecule in space; this is the *identical operation E*. Hence $C_n^n = E$. The axis of rotation in the C_n operation is called the *symmetry axis of order n*, or *n-fold axis*.

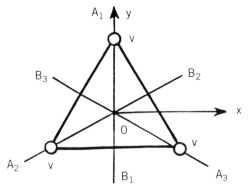

Figure 3.1. Symmetry elements of an equilateral triangle.

The symmetry operation σ denotes *reflection in a plane*; obviously, $\sigma^2 = E$. If there is also an axis of symmetry, the reflection in the plane containing this axis is denoted by σ_v or σ_d, while the reflection in the plane perpendicular to the axis of symmetry is σ_h. The consequent performance of the operations C_n and σ_h, their product $C_n\sigma_h = S_n$, is called the *reflection–rotation operation*. In particular, rotation by an angle $\pi\,(C_2)$, followed by reflection in the plane perpendicular to the rotation axis σ_h, is the *inversion operation* I with respect to the point where the axis crosses the plane: $I = S_2 = C_2\sigma_h$. From these simple operations all the symmetry transformations of molecules can be deduced.

The term *transformations* for these rotations and reflections comes from their mathematical expression by simple linear transformations of the coordinate system. For instance, the rotation of the molecule by an angle ϕ around the axis Oz is equivalent to the transformation of the coordinate system in which the point $M(x, y, z)$ is transfered to $M'(x', y', z')$ and

$$x' = x \cos \phi + y \sin \phi$$

$$y' = -x \sin \phi + y \cos \phi \qquad (3.1)$$

$$z' = z$$

Such transformations are used very often and therefore it becomes necessary to simplify their notation. First, let us write Eq. (3.1) in a more symmetrical form:

$$x' = \cos \phi \cdot x + \sin \phi \cdot y + 0 \cdot z$$

$$y' = -\sin \phi \cdot x + \cos \phi \cdot y + 0 \cdot z \qquad (3.1')$$

$$z' = 0 \cdot x + 0 \cdot y + 1 \cdot z$$

We see that the transformation (3.1) is determined by a set of coefficients that can be arranged in a quadratic table:

$$A = \begin{pmatrix} \cos\phi & \sin\phi & 0 \\ -\sin\phi & \cos\phi & 0 \\ 0 & 0 & 1 \end{pmatrix} \tag{3.2}$$

This table is called the *matrix of the transformation* (3.1). By means of the matrix A the transformation of M into M' can be written in a much simpler way:

$$M' = AM \tag{3.3}$$

where A means that a special operation given by system (3.1) (and denoted by the matrix A) should be performed on the coordinates of M.

The transformation of an arbitrary function of coordinates $f(x, y, z)$ via transformation of the coordinates (3.1) can be written in a similar way:

$$f(x', y', z') = Af(x, y, z) \tag{3.3'}$$

Here A means that the operation (3.1) should apply to all the coordinates, and in this way the entire function f changes. The symbol A indicating a certain operation to be performed over the function that follows A is called an *operator*. Hence each symmetry transformation can be put in correspondence with a certain matrix operator.

If one performs a back rotation (i.e., a rotation by the angle $-\phi$) on the same system, point M' is transferred back to M, and it can be shown using Eqs. (3.1) that the transformation of $M'(x', y', z')$ into $M(x, y, z)$ has the following form:

$$\begin{aligned} x &= x' \cos\phi - y' \sin\phi \\ y &= x' \sin\phi + y' \cos\phi \\ z &= z' \end{aligned} \tag{3.4}$$

with the matrix

$$A' = \begin{pmatrix} \cos\phi & -\sin\phi & 0 \\ \sin\phi & \cos\phi & 0 \\ 0 & 0 & 1 \end{pmatrix} \tag{3.5}$$

This transformation can be written compactly as

$$M = A'M' \tag{3.6}$$

Obviously, the consequent application of operations A and A' to point M transfers this point into itself, hence

$$AA'M = M \tag{3.7}$$

or $AA' = E$, the identical transformation. It can be shown that E is always a diagonal unity matrix with zero off-diagonal elements and units in the diagonal positions. In particular, in the three-dimensional case discussed above,

$$E = \begin{pmatrix} 1 & 0 & 0 \\ 0 & 1 & 0 \\ 0 & 0 & 1 \end{pmatrix} \tag{3.8}$$

and it corresponds to the transformation

$$
\begin{aligned}
x &= x' \\
y &= y' \\
z &= z'
\end{aligned}
\tag{3.9}
$$

The condition $AA' = E$ means that A and A' are *mutually inverse matrices*. If some transformation is described by the matrix A, the inverse transformation A' is described by the inverse matrix A^{-1}. The latter can be obtained from the former by means of a simple transposition of the columns by rows [cf. A' after (3.5) with A after (3.2)]. This operation is called a *transposition* of the matrix. For any real and unitary (see below) transformation, the inverse matrix is just the transposed matrix.

The reflection operation can be presented in a manner quite similar to rotations. For instance, the reflection in the plane xy corresponds to the transformation $x' = x$, $y' = y$, and $z' = -z$ with the matrix

$$A = \begin{pmatrix} 1 & 0 & 0 \\ 0 & 1 & 0 \\ 0 & 0 & -1 \end{pmatrix} \tag{3.9'}$$

while the inversion is

$$
\begin{aligned}
x' &= -x \\
y' &= -y \\
z' &= -z
\end{aligned}
\qquad
A = \begin{pmatrix} -1 & 0 & 0 \\ 0 & -1 & 0 \\ 0 & 0 & -1 \end{pmatrix}
\tag{3.10}
$$

In the general case of n coordinates $x_1, x_2, \ldots, x_n$ (n-dimensional space) their linear transformation, similar to (3.1) or (3.10) for the three-dimensional space, can be written conveniently as follows:

$$x_i' = \sum_j a_{ij} x_j \qquad i = 1, 2, \ldots, n \tag{3.11}$$

where a_{ij} are the coefficients of the transformation. The matrix A corresponding to this transformation is

$$A = \begin{pmatrix} a_{11} & a_{12} & \cdots & a_{1n} \\ a_{21} & a_{22} & \cdots & a_{2n} \\ \vdots & \vdots & & \vdots \\ a_{n1} & a_{n2} & \cdots & a_{nn} \end{pmatrix} \tag{3.12}$$

The matrices (3.2), (3.5), and (3.8) to (3.10) are particular cases of the matrix (3.12) for three-dimensional transformations.

It can be shown that for rotations and reflections the matrix elements a_{ij} obey certain relationships of orthogonality and normalization:

$$\sum_j a_{jk} a_{jl} = \delta_{kl} \tag{3.13}$$

where δ_{kl} is the Kronecker index (2.46): $\delta_{kl} = 1$ if $k = l$, and $\delta_{kl} = 0$ if $k \neq l$.

The linear transformations that obey condition (3.13) are called *unitary transformations*. It can easily be checked that the three-dimensional transformations (rotations and reflections) are unitary transformations. Thus all the molecular space symmetry operations are unitary transformations.

Let us perform two consequent unitary transformations:

$$x_j' = \sum_{k=1}^{n} a_{jk} x_k \qquad j = 1, 2, \ldots, n$$

$$x_i'' = \sum_j b_{ij} x_j' \qquad i = 1, 2, \ldots, n \tag{3.14}$$

or in compact form,

$$M' = AM$$
$$M'' = BM' \tag{3.14'}$$

The result of the two transformations can be also obtained directly by one transformation of x into x'':

$$x_i'' = \sum_{k=1}^{n} c_{ik} x_k \qquad i = 1, 2, \ldots, n \tag{3.15}$$

or

$$M'' = CM \tag{3.15'}$$

where C is the matrix of this transformation. On the other hand, by substitution of the first of (3.14') into the second, we have

$$M'' = BM' = BAM$$

Thus [cf. (3.15')]

$$C = B \cdot A \tag{3.16}$$

Determined in this way, *the matrix C equals the product of the matrices B and A*. Its elements can be found by substituting the first of equations (3.14) into the second:

$$x'' = \sum_j b_{ij} x'_j = \sum_j b_{ij} \sum_k a_{jk} x_k = \sum_{j,k} b_{ij} a_{jk} x_k \tag{3.15''}$$

Hence

$$c_{ik} = \sum_j b_{ij} a_{jk} \tag{3.17}$$

In other words, the matrix elements of the product of two matrices equals the sum of the products of the elements of each row of the multiplier with the elements of the corresponding column of the multiplicand. In this context it is worthwhile to emphasize that the product of two matrices (two symmetry operations) is, in general, not commutative (the multiplier and multiplicand are not interchangeable):

$$B \cdot A \neq A \cdot B \tag{3.18}$$

Thus the consecutive application of two transformations (two symmetry operations) can be described by one matrix equal to the product of the matrices of the two transformations. Obviously, the identical transformation E multiplied by any matrix A does not change it: $E \cdot A = A$.

To conclude this section:

1. Each symmetry transformation of the molecule corresponds to some matrix operator.
2. Inverse transformation corresponds to the inverse matrix.

3. The identical operation corresponds to the unity matrix.
4. The consecutive application of two symmetry operations corresponds to the product of their matrices.

Thus the geometric properties of symmetry are translated into the language of matrices; this is important for direct use of group-theoretical considerations.

3.2. GROUPS OF SYMMETRY TRANSFORMATIONS

One of the most important properties of a molecule is that its *symmetry transformations form a group in the mathematical sense of this word*. A group is a set of elements (of any nature) that satisfies the following conditions:

1. The operation of multiplication of two elements is defined, the product of any of two elements also being an element of the set under consideration.
2. The multiplication obeys the law of associativity: $(AB) \cdot C = A \cdot (BC)$, where A, B, and C are elements of the set.
3. Among the elements there is an identical one, E, that is, an element which being multiplied with another element does not change the latter.
4. For each element there is an inverse element that satisfies the condition $A \cdot A^{-1} = E$.

It can be shown that the set of symmetry operations of a molecule satisfies all the conditions of a group, listed above. Indeed, the product of two symmetry operations defined as their consecutive application is also a symmetry operation: Each of them matches the molecule with itself (by definition), and hence their product also matches the molecule with itself. Then the identical element is the identical transformation (say, the rotation by an angle 2π), and the inverse element, as shown above, is given by the inverse transformation. The associative law can also be easily checked.

Thus the set of symmetry transformations forms a group in the mathematical sense of the word, and this means that all the results of the mathematical theory of groups can be used directly to reveal the properties of symmetry operations and their role in the physical and chemical properties of the molecule. Next, formulate briefly some useful group-theoretical rules.

The number of elements in the group is called the *order of the group*. For a molecule the order of the group can be either a finite or an infinite number. The latter is realized, for instance, in linear molecules for which the number of rotations is infinite, since rotations by any angle around the axis of the molecule is a symmetry operation (in this case the group is continuous). If the order of the group is not a simple number, some subgroups can be separated from the group, which means that within the set of elements of the group there are smaller sets that form smaller groups.

Two elements A and B are called *conjugated* if there is such a third element P for which

$$A = P \cdot B \cdot P^{-1} \tag{3.19}$$

It can be shown that if A is conjugated with B, and B is conjugated with C, then A is conjugated with C. This property allows us to separate the elements of the group into smaller sets in which all the elements are mutually conjugated. These sets are called *classes*.

There is a simple geometric rule, by means of which one can easily separate the symmetry operations into classes. Denote by A the rotation around the Oa axis by an angle ϕ, and by B the rotation by the same angle around another axis Ob. Then if there is an element P of the group that transforms axis Ob into Oa, then $A = PBP^{-1}$, and A and B belong to the same class. Proof of this rule emerges directly from geometrical ideas and can easily be checked. Separation of the group elements into classes is very important for applications.

Note that all the molecular symmetry groups are *point groups* for which any set of consecutive symmetry operations leaves at least one point of the system unchanged. Otherwise, these operations are not symmetry operations because they displace the molecule and it does not match with itself. For crystals there is also translation symmetry, but the local properties can still be described by point groups.

The simplest point groups are as follows (Fig. 3.2). The groups C_n, $n = 1$, $2,\ldots$, contain one axis of symmetry of nth order and n elements: C_n, C_n^2, $C_n^3,\ldots, C_n^n = E$. The groups C_{nh} and C_{nv} can be obtained by adding reflection planes σ_h and σ_v, respectively, to the nth-order axis (in $C_{\infty h}$ and $C_{\infty v}$ the order of the symmetry axis is infinite). The group D_{nh} is obtained by adding a

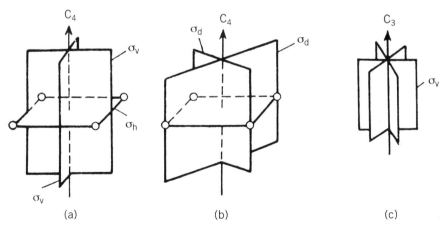

(a)   (b)   (c)

Figure 3.2. Symmetry elements of some point groups: (*a*) fourth-order axis C_4 and symmetry planes σ_v and σ_h of the C_{4h} group; (*b*) σ_d planes of the same group C_{4h}; (*c*) trifold axis C_3 and symmetry planes σ_v of the C_{3v} group.

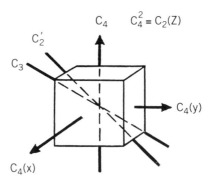

Figure 3.3. Symmetry axes of a cube.

perpendicular second-order axis to the main nth axis (this also creates reflection planes; see below). Most important for coordination compounds are the cubic groups, tetrahedral T, T_d, and T_h, and octahedral O and O_h, as well as icosahedral groups I and I_h.

The description of these groups in detail is given in special manuals mentioned above. Consider as an example the octahedral O_h group, which corresponds to the symmetry operations of an octahedral complex with identical ligands. It includes all the symmetry elements of a cube (Fig. 3.3):

1. Three axes of the fourth-order, C_4, lining the centers of opposite faces of the cube (in octahedral complexes they line two ligands in the *trans* position) and three axes of the second order that coincide with the C_4 axes: $C_2 = C_4^2$

2. Four axes of the third order, C_3, that coincide with the diagonals of the cube

3. Six axes of the second order, C_2', connecting the middle points of the opposite edges

4. A center of symmetry (inversion operation I)

5. Three reflection planes parallel to the faces of the cube σ_h, and six reflection planes σ_d, each containing two opposite edges.

To determine the number of elements (the order of the group) and their distribution over the classes, it is convenient to begin with the subgroup O of the O_h group, which is also known as a separate group of all the rotations of the octahedron (without reflections). According to the geometric rule of distribution of the symmetry operations into classes formulated above, rotations by the same angle around different axes enter the same class if among the elements of the group O under consideration there are rotations that transfer one axis into another. Using this rule, one can establish that in the O group there are 24 elements that form five classes:

1. The class E, which contains one element (the identical element always forms a separate class).

2. The class C_4, which contains six elements, three rotations C_4 and three rotations C_4^3 around the fourth-order axes; C_4^3 indicates three rotations by the angle $2\pi/4$, which is equivalent to C_4^{-1}, the element inverse to C_4, describing the rotation around C_4 by an angle of $-2\pi/4$.

3. The class $C_4^2 = C_2$, which contains three rotations by an angle π around the fourth-order axis C_4.

4. The class C_3, which contains eight rotations by angles $2\pi/3$ and $-2\pi/3$ around the third-order axis (elements C_3 and C_3^2).

5. The class C_2', which contains six rotations by the angle π around the second-order axis.

The elements of the group O_h can be obtained from the elements of the O group by multiplying each of them by the element $S_2 = I$, the inversion in the center of the cube. We have $E \cdot I = I$, $C_3 \cdot I = S_6$, $C_4 \cdot I = S_4$, $C_2' \cdot I = \sigma_d$, and $C_2 \cdot I = -\sigma_h$. This results in a doubling of the number of elements and classes. Hence the O_h group has 48 elements distributed in 10 classes (the number of elements is indicated in parentheses): $E(1)$, $C_4(6)$, $C_2(3)$, $C_3(8)$, $C_2'(6)$, $I(1)$, $S_4(6)$, $\sigma_h(3)$, $S_6(8)$, and $\sigma_d(6)$. Each of these symmetry transformations corresponds to a matrix, and all these matrices form the same group as the transformations. All the relationships obtained above for symmetry transformations are equally valid for the appropriate matrix operators.

3.3. REPRESENTATIONS OF GROUPS AND MATRICES OF REPRESENTATIONS

As is known from quantum mechanics, the state of the system with given energy ε (the eigenstate) is determined by the wavefunction ψ or, more generally, by a set of wavefunctions $\psi_1, \psi_2, \ldots, \psi_f$, that satisfy the Schrödinger equation (1.5):

$$H\psi_n = \varepsilon\psi_n \qquad n = 1, 2, \ldots, f \qquad (3.20)$$

If $f > 1$, there is more than one state with the same energy and the term is f-fold degenerate (the functions ψ_n are assumed to be linearly independent). It is easy to check that if all these functions satisfy Eq. (3.20), any of the linear combinations $a_1\psi_1 + a_2\psi_2 + \cdots + a_f\psi_f$ with arbitrary coefficients a_1, $a_2, \ldots, a_f$, satisfies the same equation.

Let us apply a symmetry operation G to both parts of Eq. (3.20):

$$G(H\psi_n) = G(\varepsilon_n \psi_n) \tag{3.21}$$

By definition, G does not change the molecule and hence its Hamiltonian H. It means that $G(H\psi) = H(G\psi)$. On the other hand, ε_n is a constant; hence $G(\varepsilon_n \psi_n) = \varepsilon_n(G\psi_n)$. Thus Eq. (3.21) yields

$$H(G\psi_n) = \varepsilon_n(G\psi_n) \tag{3.21'}$$

It is seen that the function $\psi'_n = G\psi_n$ [i.e., the ψ_n transformed by the symmetry operation G as required by the rule (3.3')] also satisfies the Schrödinger equation with the same energy ε_n. But there are only f such independent functions. Therefore, ψ'_n must coincide with one of them or (more generally) with their linear combination:

$$\psi'_n = \sum_k^f G_{nk}\psi_k \qquad n = 1, 2, \ldots, f \tag{3.22}$$

where the G_{nk} are constants.

We thus obtain f equations that show how the f linearly independent functions $\psi_1, \psi_2, \ldots, \psi_f$ transform to a new system of functions $\psi'_1, \psi'_2, \ldots, \psi'_f$ by the symmetry operation G. The transformation is determined by the constants G_{nk}, which can be written in the form of a matrix, quite similar to the coordinate transformations (3.12) (Section 3.1):

$$G = \begin{pmatrix} G_{11} & G_{12} & \cdots & G_{1f} \\ G_{21} & G_{22} & \cdots & G_{2f} \\ \vdots & \vdots & & \vdots \\ G_{f1} & G_{f2} & \cdots & G_{ff} \end{pmatrix} \tag{3.23}$$

For another symmetry transformation we can similarly obtain another matrix. Counting all the symmetry transformations, that is, all the elements of the symmetry group of the molecule under consideration, we obtain a set of matrices of dimension f; their number is equal to the number of elements in the group. These matrices form *the representation of the group*, f being its dimensionality, while the set of functions $\psi_1, \psi_2, \ldots, \psi_f$, by means of which the matrices were obtained, is named *the basis of the representation*.

It can be shown that to obtain the group representation there is no need to use the set of basis functions which are wavefunctions of the system with given energy. To obtain the transformations (3.22), the only requirement imposed on the basis functions is that they should be independent and transform to each other by symmetry transformations. The three-dimensional matrices of symmetry rotations and reflections, introduced above, can serve as an example of such a representation of the group (for this representation the coordinates x, y, and z serve as basis functions). But it is important that the wavefunctions of the system can be used as a basis of the group representation.

By using different basis sets, one can obtain different representations for the same group; the number of possible representations is thus infinite. However, not all of these representations are independent and not all of the independent representations are important to the applications.

Let us introduce first the notion of equivalent representations. Assume that by means of a set of basis functions $\psi_1, \psi_2, \ldots, \psi_p$ we obtained a representation G of the group. Apply to these functions some linear transformation of the type (3.22) with the transformation constants given by the matrix S. As a result of this transformation, we obtain another set of functions $\psi'_1, \psi'_2, \ldots, \psi'_p$, that can also serve as a basis of another representation, G'. It can be shown that there is a certain relationship between matrices G of the former and G' of the latter representations:

$$G' = S^{-1}GS \tag{3.24}$$

The representations G and G' related by (3.24) are called *equivalent representations*.

By comparison with (3.19), one can see that the relationship between the matrices of equivalent representations coincides with the condition of conjugation (belonging to the same class) of the symmetry operations corresponding to these matrices, provided that S belongs to the same group.

An important feature of equivalent representations is that their matrices, related by (3.24), have the same characters X. *The character of a matrix $X(G)$ equals the sum of its diagonal elements:*

$$X(G) = G_{11} + G_{22} + \cdots + G_{ff} = \sum_i^f G_{ii} \tag{3.25}$$

In particular, the character of the identical transformation E corresponding to the unity matrix with all the diagonal elements $G_{ii} = 1$, $i = 1, 2, \ldots, f$, equals the dimensionality of the representation f:

$$X(E) = f \tag{3.26}$$

Let us show that the characters of the matrices of equivalent representations coincide:

$$X(G) = X(G') \tag{3.27}$$

Indeed, using the rule of multiplication of matrices (3.17), we can obtain from (3.24) the following relations:

$$X(G') = \sum_i G_{ii} = \sum_i \sum_j \sum_k S_{ik}^{-1} G_{kl} S_{li} = \sum_k \sum_l G_{kl} \delta_{kl} = \sum_k G_{kk} = X(G)$$

where the elements of the inverse matrix S^{-1} are denoted by S_{ik}^{-1} and the relation

$$\sum_i S_{ik}^{-1} S_{li} = \delta_{kl}$$

is employed [cf. Eq. 3.13].

Since all the elements of the same class are related by (3.24), the characters of their matrices obey the condition (3.27); that is, *the matrices of the elements of the same class have the same characters.*

Another important feature of the group representation is the possibility to reduce them to *irreducible representations.* When passing from a given basis of functions to another (equivalent) by means of a linear transformation, it may occur that the entire set of functions can be divided into smaller sets of $f_1, f_2, \ldots, f_r$ ($f_1 + f_2 + \cdots + f_r = f$) functions, such that in each smaller set they transform to one another by symmetry transformations not involving the functions of other sets. In other words, each of these sets can serve as the basis of a representation of smaller dimensionality. The larger representation is thus *reducible.* If such a separation of the basis sets cannot be carried out by linear transformations, the representation is called *irreducible.*

By means of a linear transformation of the basis set, the matrices of a reducible representation can be reduced to a form in which the nonzero matrix elements lie within some smaller square tables (submatrices) that occupy diagonal positions, out of which all the matrix elements are zero as follows:

$$\begin{pmatrix} \begin{matrix} G_{ij} \\ i,j=1,2,\ldots,f_1 \end{matrix} & & & 0 \\ & \begin{matrix} G_{kl} \\ k,l=f_1+1,\ldots,f_1+f_2 \end{matrix} & & \\ & & \ddots & \\ 0 & & & \begin{matrix} G_{md} \\ m,d=f-f_r,\ldots,f \end{matrix} \end{pmatrix} \qquad (3.28)$$

If the dimensionality of these submatrices cannot be lowered by linear transformations of the basis functions, each of them is a matrix of an irreducible representation and the set of such (equal in dimensionality and similar in position) submatrices forms an irreducible representation. Hence the reducible representation of the dimensionality f can be characterized by r sets of matrices of lower dimensionality, $f_1, f_2, \ldots, f_r$ ($f_1 + f_2 + \cdots + f_r = f$), each realizing an irreducible representation of the group. This procedure is called *decomposition of the reducible representation into irreducible parts.*

It follows from the presentation of the matrix of the reducible representation in the form (3.28) that its character equals the sum of the characters of the irreducible representations to which the reducible representation is reduced. Denoting the characters of the reducible and irreducible representations by $X(G)$ and $X^{(\alpha)}(G)$, respectively, we have

$$X(G) = \sum_{\alpha} a^{(\alpha)} X^{(\alpha)}(G) \tag{3.29}$$

where $a^{(\alpha)}$ is the number of times the α irreducible representation enters the reducible representation.

The characters obey some orthogonality relationships. The latter are based on the orthogonality relation for the matrix elements of the matrices of the representations [cf. (3.13)]:

$$\sum_{G} G_{ik}^{(\alpha)} G_{lm}^{(\beta)} = \frac{g}{f_{\alpha}} \delta_{\alpha\beta} \delta_{il} \delta_{km} \tag{3.30}$$

which means that if the two irreducible representations are different, $\alpha \neq \beta$, the sum above is zero. In particular, assuming that β is a unity representation, we have:

$$\sum_{G} G_{ik}^{(\alpha)} = 0 \tag{3.31}$$

for any nonunity representation α. Several interesting consequences follow from Eq. (3.30) [for direct use of Eq. (3.31), see Section 3.4]. In particular, the sum (over all the symmetry operations G) of the products of characters $X^{(\alpha)}(G)$ and $X^{(\beta)}(G)$ of two irreducible representations α and β is equal to zero if the two representations are different, and to the order of the group g if they coincide:

$$\sum_{G} X^{(\alpha)}(G) X^{(\beta)}(G) = g\delta_{\alpha\beta} \tag{3.32}$$

Multiplying the left and the right sides of (3.29) by $X^{(\beta)}(G)$, summing over G, and taking into account the relation (3.30), we get

$$a^{(\beta)} = \frac{1}{g} \sum_{G} X(G) X^{(\beta)}(G) \tag{3.33}$$

This relation is known as the *formula of decomposition of reducible representations into irreducible parts*. It solves directly the problem of finding whether a certain irreducible representation is included in the reducible one, provided that the characters of both the latter and former are known. Many symmetry problems can be solved by this formula.

Let us also list some important rules concerning irreducible group representations and their characters:

1. The number of nonequivalent irreducible representations of a group equals the number of its classes r
2. The sum of the squares of the dimensionalities of the irreducible representations equals the order of the group g: $f_1^2 + f_2^2 + \cdots + f_r^2 = g$.
3. The dimensionalities of the irreducible representations of the group are divisors of its order.
4. Among the irreducible representations there is the unity representation, realized by one basis function which is totally symmetrical with respect to all the symmetry operations of the group; all the characters of the matrices of the unity representation are equal to 1.
5. The characters of the matrices of the irreducible representation Γ, which is a direct product of the representations α and β, equals the products of the characters of the corresponding matrices of these representations.

The term *direct product of representations* needs some clarification. Denote the sets of basis functions of the representations α and β by ψ_α and ψ_β, respectively. Then the direct product of the representations α and β is the representation γ realized by the set of functions from the products $\psi^{(\alpha)}\psi^{(\beta)}$. If $\psi^{(\alpha)}$ and $\psi^{(\beta)}$ coincide, the new representation with the characters $[X(G)]^2$ decomposes into two representations [3.1, 3.7]:

1. Representation of the symmetrical product for which

$$[X]^2(G) = \tfrac{1}{2}\{[X(G)]^2 + X(G^2)\} \qquad (3.34)$$

2. Representation of the antisymmetrical product for which

$$\{X\}^2(G) = \tfrac{1}{2}\{[X(G)]^2 - X(G^2)\} \qquad (3.35)$$

The two products, symmetrical and antisymmetrical, play an important role in the description of physical properties of molecules.

The general properties of irreducible representations and their characters, especially the orthonormalization conditions of the type (3.30), allow one to calculate the values of all the characters. As an example, the irreducible representations and their characters for the octahedral group O_h, widely used in coordination chemistry, are given in Table 3.1. Each column of this table

Table 3.1. Irreducible representations and characters of the octahedral point group O_h^a

O^h	E	$6C_4$	$3C_4^2=C_2$	$8C_3$	$6C_2'$	$S_2=1$	$6S_4$	$3\sigma_h$	$8S_6$	$6\sigma_d$
A_{1g}	+1	+1	+1	+1	+1	+1	+1	+1	+1	+1
A_{1u}	+1	+1	+1	+1	+1	−1	−1	−1	−1	−1
A_{2g}	+1	−1	+1	+1	−1	+1	−1	+1	+1	−1
A_{2u}	+1	−1	+1	+1	−1	−1	+1	−1	−1	+1
E_g	+2	0	+2	−1	0	+2	0	+2	−1	0
E_u	+2	0	+2	−1	0	−2	0	−2	+1	0
T_{1g}	+3	+1	−1	0	−1	+3	+1	−1	0	−1
T_{1u}	+3	+1	−1	0	−1	−3	−1	+1	0	+1
T_{2g}	+3	−1	−1	0	+1	+3	−1	−1	0	+1
T_{2u}	+3	−1	−1	0	+1	−3	+1	+1	0	−1

^{a}See also Table A1.11.

presents one of the 10 classes of this group, discussed above. The number of its elements is indicated by a coefficient to the denotation of the class, while the rows correspond to the 10 irreducible representations. The notation is that of Mulliken [3.9]: one-dimensional representations are denoted by A and B, two- and three-dimensional reparesentations by E and T, respectively. In practice, the notation of Bethe [3.5] is also used, in which only one letter Γ with indices is employed (Section 3.6). If the group includes the inversion operation I, all the representations have an additional index, either g or u, indicating the parity of the basis functions with respect to the operation of inversion: g represents evenness (*gerade*) and u, oddness (*ungerade*).

Knowledge of the characters of the irreducible representations of the symmetry group enables us to describe all the properties of the molecule related to its symmetry. Tables of characters of irreducible representations of the most usable groups, including information about symmetry properties of Cartesian coordinates and rotations, are given in Appendix 1.

3.4. CLASSIFICATION OF MOLECULAR TERMS, SELECTION RULES, AND WIGNER–ECKART THEOREM

In this and the following sections, some important applications of the group-theoretical rules discussed above are considered briefly. The *classification of molecular terms on symmetry* is at present unavoidable in any rational description of electronic structure and properties of molecules. As shown above, the set of wavefunctions of an eigenstate forms a basis of the representation of the group of symmetry transformations of the molecule. It can be shown that this representation is irreducible. Hence we have a direct correspondence of the energy terms of the molecular system and their wavefunctions to the irreducible representations of its group of symmetry transformations. This forms the basis for the classification of the molecular states on symmetry.

Each energy term is put in correspondence with an irreducible representation of the group of symmetry of the molecular system. *The dimensionality of the representation indicates the degeneracy of the term*, that is, the number of functions that transform to each other by the symmetry operations and hence the number of states with the same energy, while the characters of the matrices of the irreducible representations describe the symmetry properties of these wavefunctions. For instance, for a molecular system with the symmetry described by the group O_h (Table 3.1), the following types of energy terms are possible: nondegenerate, A_{1g}, A_{1u}, A_{2g}, and A_{2u}; double degenerate, E_g and E_u; and threefold degenerate, T_{1g}, T_{1u}, T_{2g}, and T_{2u}.

The classification of molecular terms on symmetry is most important in different applications. Besides the description of the symmetry properties of the molecular states, it allows one to solve many problems, including construction of molecular orbitals with given symmetry (Section 3.5), energy-level splitting in external fields (Section 4.2), selection rules for matrix elements describing observable properties [e.g., light absorption (Section 8.1)], and so on.

Consider the *selection rules for matrix elements*. The most general form of a matrix element standing for molecular properties is:

$$\int \Psi_1^* \mathbf{f} \Psi_2 \, d\tau \tag{3.36}$$

where $\mathbf{f}$ is a scalar, vector, or tensor operator describing the physical property under study (Section 1.3), and Ψ_1 and Ψ_2 are the wavefunctions of two states of the system. In particular, $\Psi_1 = \Psi_2$ for the diagonal matrix element.

Direct calculation of (3.36) is often difficult mainly because the wavefunctions are unknown. However, the *selection* rules, the general rules indicating whether the matrix element (3.36) is zero or nonzero, can be solved using symmetry ideas and group-theoretical results without calculation of the integral. For this purpose, knowledge of the symmetry properties of the operator $\mathbf{f}$ and the wavefunctions Ψ_1 and Ψ_2 is needed. Using the symmetry classification above, Ψ_1 and Ψ_2 can be attributed directly to the corresponding irreducible representations, and the same can be done for $\mathbf{f}$ by considering its behavior under the operations of the symmetry group.

Equation (3.36) is a definite integral, a number, and as such it is independent of coordinate transformations G performed under the sign of the integral. This means that

$$\int F_i \, d\tau = \int G F_i \, d\tau \tag{3.37}$$

where we set $F_i = \Psi_1 \mathbf{f} \Psi_2$. On the other hand, considering F_i as one of the functions of a basis realizing some representation γ of the group, we obtain, after Eq. (3.22),

$$GF_i = \sum_k G_{ik}^{(\gamma)} F_k$$

and hence

$$\int F_i \, d\tau = \int \sum_k G_{ik}^{(\gamma)} F_k \, d\tau$$

Now summing up the left- and right-hand sides of this equation over all the symmetry operations G of the group and taking into account that the integral at the left, being independent of G, becomes just multiplied by the number of elements g, we have

$$g \int \Psi_1^* \mathbf{f} \Psi_2 \, d\tau = \sum_k \int F_k \sum_G G_{ik}^{(\gamma)} \, d\tau \qquad (3.38)$$

But the sum $\Sigma_G G_{ik}^{(\gamma)}$, following (3.31), is nonzero only when γ is a unity representation, and it is zero for all other representations. On the other hand, the representation γ of the product $F = \Psi_1^* \mathbf{f} \Psi_2$ equals the product of representations of Ψ_1, $\mathbf{f}$, and Ψ_2. Thus we arrived at the fundamental rule: *The matrix element* (3.36) *is nonzero if and only if the product of representations to which* Ψ_1, $\mathbf{f}$, *and* Ψ_2 *belong contains the unity representation.*

To obtain selection rules for operators of physical magnitudes using this rule, some simple procedures should be carried out: Provided that the representations of the wavefunctions Ψ_1 and Ψ_2 and the operator $\mathbf{f}$ are known, they should be multiplied, and then, using Eq. (3.33), one should determine whether the product of these three representations contains the unity representation. For the latter all the characters are equal to a unity, while the characters of the product of the representations equal the product of their characters. Note that if the two functions Ψ_1 and Ψ_2 are equal [and hence (3.36) is a diagonal matrix element], the product of their representations must be taken as a symmetrical product with the characters after (3.34): $[X^2](G) = \frac{1}{2}\{[X(G)]^2 + X(G^2)\}$.

From many other group-theoretical rules very useful in applications, we formulate here (without proof) the *Wigner–Eckart theorem*. The degenerate state of the system belongs to the irreducible representation with a dimensionality greater than unity (E, T, etc.), and its matrices have several rows and columns. In this case it is convenient to attribute the wavefunctions of the term to the corresponding rows of the representation. For instance, for the threefold-degenerate T_{1u} term of the O_h group, the three wavefunctions transform under symmetry operations as the three coordinates x, y, z, respectively (Table A1.11). Hence each wavefunction represents a row of the representation T_{1u}.

The three functions of the T_{2g} term (also threefold degenerate) transform as the three products of coordinates xy, xz, and yz, respectively. In general, Γ denotes the representation of the term and γ are its rows.

Consider the matrix element

$$\int \Psi^*(\Gamma_1\gamma_1) f_{\Gamma\gamma} \Psi(\Gamma_2\gamma_2) \, d\tau \equiv \langle \Gamma_1\gamma_1 | f_{\Gamma\gamma} | \Gamma_2\gamma_2 \rangle \tag{3.39}$$

where $f_{\Gamma\gamma}$ is an operator of a physical magnitude that transforms as the γ row of the Γ representation, and $\Gamma_1\gamma_1$, and $\Gamma_2\gamma_2$ are the representations and their rows for the two wavefunctions. The notation of the integral (3.39) given on the right-hand side is used widely in practice. The Wigner–Eckart theorem states that the matrix element (3.39) can be reduced to the following form [3.5]:

$$\langle \Gamma_1\gamma_1 | f_{\Gamma\gamma} | \Gamma_2\gamma_2 \rangle = \langle \Gamma_1 \| f_\Gamma \| \Gamma_2 \rangle \cdot \langle \Gamma_1\gamma_1, \Gamma\gamma | \Gamma_2\gamma_2 \rangle \tag{3.40}$$

where $\langle \Gamma_1 \| f_\Gamma \| \Gamma_2 \rangle$ is the *reduced matrix element*, which depends on the representations Γ_1, Γ, and Γ_2, but not on their rows, and $\langle \Gamma_1\gamma_1, \Gamma\gamma | \Gamma_2\gamma_2 \rangle$ are the *Clebsh–Gordan coefficients*, for which there are ready-made tables.

The essence of the Wigner–Eckart theorem (3.40) is that for degenerate states there is no need to calculate all the matrix elements for all the rows of the representation: If we know at least one matrix element, the reduced matrix element can be obtained from (3.40), and then all the others can be calculated using the known Clebsh–Gordan coefficients. The theorem also limits the number of possible independent parameters of the problem with given symmetry.

3.5. CONSTRUCTION OF SYMMETRIZED MOLECULAR ORBITALS

In the methods of molecular orbital–linear combinations of atomic orbitals (MOLCAO, Chapter 5), as well as in many other related problems, it is necessary to construct LCAOs that transform after the irreducible representations of the symmetry group of the system. Consider as an example an octahedral complex with O_h symmetry. The six ligand σ orbitals (with respect to the ligand–metal bond, Section 2.2) $\sigma_1, \sigma_2, \ldots, \sigma_6$ are shown in Fig. 3.4. By the symmetry operations of the octahedral group O_h, these functions transform to each other; this can be written as follows:

$$\sigma_i' = C_{i1}\sigma_1 + C_{i2}\sigma_2 + \cdots + C_{i6}\sigma_6 \qquad i = 1, 2, \ldots, 6 \tag{3.41}$$

The coefficients C_{ij} of these six transformations form the matrix of the representation for the symmetry operation used, and the number of such

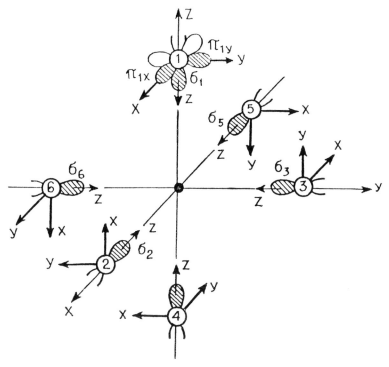

Figure 3.4. Space orientation of the π_{1x}, π_{1y}, and σ_n ($n = 1, 2, \ldots, 6$) ligand orbitals in an octahedral complex (cf. Fig. 5.1).

matrices equals the number of symmetry operations in the group. If all these coefficients are known, it is easy to calculate the characters of the matrices after Eq. (3.25) and then to decompose the reducible representation into irreducible parts using Eq. (3.33). The types of irreducible representations obtained in this way are just those sought for, to which the MOs — linear combinations of the six σ orbitals of the ligand — pertain.

However, to obtain the characters of the symmetry transformation (3.41), there is no need to perform these transformations and to find all the coefficients C_{ij}. Instead, one can use the following rule for σ orbitals: *The character of the representation for a given symmetry transformation equals the number of ligands that remain unmoved by this transformation.*

Let us illustrate this rule by applying the transformation C_4 of the O_h group, the rotation by an angle $\pi/2$ around the axis Oz. As seen from Fig. 3.4, by this rotation two ligands, 1 and 4, remain unmoved from their positions and hence, following the rule above, $X(C_4) = 2$ (for the transformation of the six ligand σ functions). This result can be checked directly. Indeed, by C_4 rotation the new function σ' can easily be expressed by the old ones: $\sigma'_1 = \sigma_1$, $\sigma'_2 = \sigma_6$, $\sigma'_3 = \sigma_2$, and so on. This is convenient to write in the form of a general transformation (3.41):

$$\sigma'_1 = 1 \cdot \sigma_1 + 0 \cdot \sigma_2 + 0 \cdot \sigma_3 + 0 \cdot \sigma_4 + 0 \cdot \sigma_5 + 0 \cdot \sigma_6$$
$$\sigma'_2 = 0 \cdot \sigma_1 + 0 \cdot \sigma_2 + 0 \cdot \sigma_3 + 0 \cdot \sigma_4 + 0 \cdot \sigma_5 + 1 \cdot \sigma_6$$
$$\sigma'_3 = 0 \cdot \sigma_1 + 1 \cdot \sigma_2 + 0 \cdot \sigma_3 + 0 \cdot \sigma_4 + 0 \cdot \sigma_5 + 0 \cdot \sigma_6$$
$$\sigma'_4 = 0 \cdot \sigma_1 + 0 \cdot \sigma_2 + 0 \cdot \sigma_3 + 1 \cdot \sigma_4 + 0 \cdot \sigma_5 + 0 \cdot \sigma_6 \qquad (3.42)$$
$$\sigma'_5 = 0 \cdot \sigma_1 + 0 \cdot \sigma_2 + 1 \cdot \sigma_3 + 0 \cdot \sigma_4 + 0 \cdot \sigma_5 + 0 \cdot \sigma_6$$
$$\sigma'_6 = 0 \cdot \sigma_1 + 0 \cdot \sigma_2 + 0 \cdot \sigma_3 + 0 \cdot \sigma_4 + 1 \cdot \sigma_5 + 0 \cdot \sigma_6$$

The matrix of this transformation is

$$\begin{pmatrix}
1 & 0 & 0 & 0 & 0 & 0 \\
0 & 0 & 0 & 0 & 0 & 1 \\
0 & 1 & 0 & 0 & 0 & 0 \\
0 & 0 & 0 & 1 & 0 & 0 \\
0 & 0 & 1 & 0 & 0 & 0 \\
0 & 0 & 0 & 0 & 1 & 0
\end{pmatrix} \qquad (3.43)$$

and it is seen that the sum of the diagonal elements is indeed $X(C_4) = 2$.

Using this rule, we can obtain directly all the characters of the reducible representation Γ for all the symmetry transformations of the O_h group:

G	E	C'_2	C_4	C_2	C_3	I	S_4	S_6	σ_h	σ_d	
$X(G)$	6	0	2	2	0	0	0	0	4	2	(3.44)

This representation can be decomposed into irreducible parts by means of Eq. (3.33):

$$\Gamma \rightarrow A_{1g} + E_g + T_{1u} \qquad (3.45)$$

Thus the symmetrized (group) orbitals of type σ formed as linear combinations of the σ AOs of the six ligands of an octahedral complex should belong to one of the following irreducible symmetry representations of the O_h group: A_{1g}, E_g, T_{1u}. To construct the appropriate linear combinations of the atomic σ functions that transform after these representations, look first for the simple one, A_{1g}. Since its matrices are one-dimensional and the characters are units, the A_{1g}-type linear combination of σ-type AOs of the ligands for any symmetry operation of the O_h group transforms into itself. The only combination of this type is their sum:

$$\Psi(A_{1g}) = \sigma_1 + \sigma_2 + \sigma_3 + \sigma_4 + \sigma_5 + \sigma_6 \qquad (3.46)$$

Indeed, any symmetry operation transforms σ_i into σ_j and this does not change the sum (3.46).

For degenerate representations E_g ($f = 2$) and T_{1u} ($f = 3$) the procedure is more complicated. To construct the symmetrized orbital $\psi^{(\alpha)}$ that belongs to the α irreducible representation, the *projection formula* can be employed [3.7]:

$$\psi^{(\alpha)} = \frac{f_\alpha}{g} \sum_G X^{(\alpha)}(G)G\psi \tag{3.47}$$

where ψ is an arbitrary function of the basis set (when there are recurring representations of the same symmetry, this relation should be slightly modified [3.4]). To use (3.47), one must choose a function ψ from the basis, apply to this function the symmetry operations G, multiply by its character in the representation α, and sum up the results over all the operations G of the group. The result will be either a linear combination of different functions of the basis sought, or zero. The latter case means that the initial function ψ has not been chosen properly (it does not enter the α representation) and another one should be tried.

Let us apply Eq. (3.47) to find the MO that transforms after the E_g representation of the O_h group. Take σ_1 as the initial ψ. From Table 3.1 one can see that for the E_g representation $X(C_2') = X(C_4) = X(S_4) = X(\sigma_d) = 0$, and hence the corresponding symmetry transformations do not contribute to the sum (3.47). Then:

- For the identical transformation E: $X(E) = 2$, $E\sigma_1 = \sigma_1$
- For eight rotations C_3: $X(C_3) = -1$, and each of the following transformations takes place twice (for rotations by an angle $2\pi/3$ and $-2\pi/3$, respectively): $C_3\sigma_1 = \sigma_2$, $C_3\sigma_1 = \sigma_3$, $C_3\sigma_1 = \sigma_5$, and $C_3\sigma_1 = \sigma_6$ (for four axes C_3, respectively)
- For three rotations: $C_4^2 = C_2$: $X(C_2) = 2$, and σ_1 transforms once in σ_1 and twice in σ_4
- For the inversion operation I: $X(I) = 2$, $I \cdot \sigma_1 = \sigma_4$
- For eight rotation–reflections S_6: $X(S_6) = -1$, and σ_1 transforms consecutively into σ_2, σ_3, σ_5, and σ_6, respectively, twice in each, similar to the transformation C_3
- For three reflections σ_h: $X(\sigma_h) = 2$, and σ_1 transforms once in σ_4 and twice in itself (σ_1)

Taking into account all these transformations and character values and substituting them into Eq. (3.47), we have

$$\psi_1(E_g) = \tfrac{2}{48}[2\sigma_1 - 1 \cdot (2\sigma_2 + 2\sigma_3 + 2\sigma_5 + 2\sigma_6) + 2 \cdot (\sigma_1 + 2\sigma_4)$$
$$+ 2\sigma_4 - 1 \cdot (2\sigma_2 + 2\sigma_3 + 2\sigma_5 + 2\sigma_6) + 2(\sigma_4 + 2\sigma_1)] \tag{3.48}$$
$$= \tfrac{1}{6}(2\sigma_1 + 2\sigma_4 - \sigma_2 - \sigma_3 - \sigma_5 - \sigma_6)$$

This is just one of the E_g functions sought for, with an accuracy up to the normalization factor N; if we neglect ligand–ligand overlap, $N = (\frac{1}{12})^{1/2}$. To get the other E_g function (E_g is twofold degenerate), we should repeat the foregoing procedure, taking σ_2 and σ_3 consecutively as the probe functions ψ. Owing to the equivalency of all the ligands in O_h symmetry, we can conclude directly that for $\psi \sim \sigma_2$ and $\psi \sim \sigma_3$, we obtain, respectively, $\psi'(E_g) = N[2\sigma_2 + 2\sigma_5 - \sigma_1 - \sigma_3 - \sigma_4 - \sigma_6)$, and $\psi_2''(E_g) = N[2\sigma_3 + 2\sigma_6 - \sigma_1 - \sigma_2 - \sigma_3 - \sigma_4]$ (with σ_4, σ_5, σ_6 taken as ψ, we obtain the same three functions ψ_1, ψ_2', ψ_2'').

However, one can easily check that these three E_g functions are not mutually orthogonal and hence are not linearly independent. Therefore, from the two functions ψ_2' and ψ_2'', only one combination should be constructed that transforms after the E_g representation and is orthogonal to ψ_1. Because each of the functions ψ_1' and ψ_2'' transforms after E_g, any of their linear combinations belongs to E_g. It can easily be shown that the combination which is orthogonal to (3.48) is

$$\psi_2(E_g) = \tfrac{1}{2}(\sigma_2 + \sigma_5 - \sigma_3 - \sigma_6) \tag{3.49}$$

The two functions (3.48) and (3.49) represent the E_g state being sought (see Table 5.1).

Moving on to π functions, we should take into account that the procedure above becomes more complicated because the geometric rule of finding the characters of the reducible representation of type (3.44) must be modified. Indeed, as seen from Fig. 3.4, the π orbitals of each ligand have two orientations in space (say, p_{1x} and p_{1y} for the ligand 1), and therefore they do not necessarily transform into themselves under symmetry operations that leave the ligand in its initial position. For instance, under the C_4 rotation of the octahedron, the ligand 1 remains unchanged but its π functions change: $p_{1x}' \rightarrow -p_{1y}$, $p_{1y}' \rightarrow -p_{1x}$ (Fig. 3.4). Therefore, the geometric rule above for finding character values in the case of π functions is modified as follows: *The character of the representation of a given symmetry transformation equals the number of π function that remain unchanged by this transformation minus the number of functions that change their sign.*

Let us apply this rule to the widely met case of 12 π functions of six ligands in octahedral systems of O_h symmetry. Using the notation of Fig 3.4, for the identical transformation all the π functions remain unchanged, and hence $X(E) = 12$. For C_3 rotations no π functions remain unchanged or change sign: $X(C_3) = 0$. The same is true for C_2': $X(C_2') = 0$. For C_4 rotations two ligands (e.g., 1 and 4) remain at their initial positions, but the p_π functions transform to each other, as mentioned above; $X(C_4) = 0$, whereas under the rotation $C_2 = C_4^2$ these functions only change their sign: $p_{1x}' = -p_{1x}$, $p_{1y}' = -p_{1y}$, $p_{4x}' = -p_{4x}$, $p_{4y}' = -p_{4y}$. Hence $X(C_2) = -4$. Similarly, we find all the other characters of the reducible representation:

G	E	C_2'	$8C_3$	$6C_4$	C_2	I	$6S_4$	$8S_6$	$3\sigma_h$	$6\sigma_d$	
$X(G)$	12	0	0	0	-4	0	0	0	0	0	(3.50)

Using Eq.(3.33), one can easily decompose this representation Γ into irreducible parts:

$$\Gamma \to T_{1g} + T_{2u} + T_{2g} + T_{1u} \tag{3.51}$$

Thus the ligand π functions in the O_h group can realize symmetrized orbitals of threefold-degenerate type T_{1g}, T_{1u}, T_{2g}, and T_{2u} only, and hence these functions take part only in such types of MOs.

To find these MOs, use of the projection formula (3.47) is most convenient. Taking as the starting function ψ one of the π functions of a ligand, for instance, π_{2y}, one can easily obtain for the T_{2g} representation,

$$\psi(T_{2g}) = \tfrac{1}{2}(\pi_{2y} + \pi_{3x} + \pi_{5x} + \pi_{6y}) \tag{3.52}$$

Taking the other π orbitals as reference functions in formula (3.47), we obtain the other two T_{2g} functions as well as the T_{1g}, T_{1u}, and T_{2u} combinations: all in all, 12 symmetrized MOs formed by 12 ligand π functions (Table 5.1).

In the same way, any other set of functions can be transformed (projected) to symmetrized combinations that belong to certain irreducible representations of the group of symmetry.

3.6. DOUBLE GROUPS

As mentioned in Section 2.2, with the spin–orbital interaction included, the stationary states of the system are classified by the quantum number J of the total momentum $\mathbf{J} = \mathbf{L} + \mathbf{S}$; $J = L + S, L + S - 1, \ldots, |L - S|$. This classification is of special importance in systems with strong spin–orbital interaction (strong relativistic effects; see Sections 2.1, 5.4, 5.5, and 6.5).

If S is half-integer (for an odd number of electrons), J is also half-integer. The states with half-integer J values are described not by simple functions but by two-component spinors (cf. the relativistic description of atomic states with four-component bispinors in Section 2.1, which in quasi-relativistic approaches are reduced to two-component spinors, Sections 5.4 and 5.5). The two-component spinors, unlike simple functions, under symmetry operations, transform in a special way, realizing *two-valued representations* [3.7]. These are not real representations in the sense described above and do not obey the orthogonality relationships of types (3.30)–(3.33). Therefore, their inclusion into point groups complicates the applications significantly. To avoid these difficulties, the notion of *double groups* is introduced.

Let us consider the usual point group and formally add to its symmetry operations an additional element Q describing a rotation by the angle 2π around an axis. Assume that Q is not identical with the E operation, $Q \neq E$, but that $Q^2 = E$. Then the number of elements of the group is doubled, because each operation G is complemented by a new operation QG. In particular, $C_n^n = Q$, $C_n^{2n} = E$, $\sigma^2 = Q$, $\sigma^4 = E$, and so on. It can be shown that in such a group with formally doubled elements (double group) the two-valued representations describing the symmetry properties of states with half-integer J values (bispinors) decompose into two one-valued representations that have the usual group-theoretical properties. Therefore, if the elements of the double group, its irreducible representations, and their characters are known, states with half-integer spin values can be considered in the same manner as described above, provided that they are attributed to double groups.

The properties of double groups can be obtained in a manner similar to that employed for the simple groups using general theorems of group theory. Doubling the number of elements does not necessarily double the number of classes (and hence representations) of the group. For instance, in the group of rotations of a cube O there are 24 elements and five classes considered above (Section 3.2), while in the corresponding double group O' (the double groups are denoted by the same letters as the simple groups, with a prime) there are 48 elements and eight classes, the new representations being E_1', E_2', and G' (Table A1.13).

Some remarks should to be added about the notations. The representations of the double groups corresponding to the two-valued representations of the simple groups are marked by primes (after Mulliken [3.9]): E' and G' for twofold and fourfold representations, respectively. In the literature, especially for double groups, the Bethe notation [3.5] is also widely used. Therefore, the following *table of correspondence* for these two types of notations may be useful:

Mulliken:	A	B	E	T_1	T_2	E_1'	E_2'	G'
Bethe	Γ_1	Γ_2	Γ_3	Γ_4	Γ_5	Γ_6	Γ_7	Γ_8

$$(3.53)$$

Direct products of irreducible representations of double groups (as well as of simple groups and simple and double groups) are given in Table A1.14.

REFERENCES

3.1. R. Hochstrasser, *Molecular Aspects of Symmetry*, W.A. Benjamin, New York, 1966.

3.2. F. A. Cotton, *Chemical Applications of Group Theory*, 2nd ed., Interscience, New York, 1971.

3.3. J. P. Fackler, Jr., *Symmetry in Coordination Chemistry*, Academic Press, New York, 1979.

3.4. I. G. Kaplan, *Symmetry of Many-Electron Systems*, Academic Press, New York, 1975.

3.5. H. Bethe, *Ann. Phys.*, **3**, 133–208 (1929).

3.6. S. Sugano, Y. Tanabe, and H. Kamimura, *Multiplets of Transition-Metal Ions in Crystals*, Academic Press, New York, 1970, 331 pp.

3.7. L. Landau and E. Liphshitz, *Quantum Mechanics*, 3 ed., Nauka, Moscow, 1974, Chap. 12.

3.8. B. S. Tsukerblat, *Group Theory in Chemistry and Spectroscopy: A Simple Guide to Advanced Usage*, Academic Press, New York, 1994.

3.9. R. S. Mulliken, *Phys. Rev.*, **43**, 279–302 (1933).

4 Crystal Field Theory

*Simple model theories can often be used as a basis for correct understanding of rather complicated phenomena.**

The crystal field theory in its application to transition metal compounds discussed in this chapter may serve as an illustration to this statement.

4.1. INTRODUCTION

Brief History

The basis of crystal field theory (CFT) was created by Bethe in 1929 in his classical work "Term Splitting in Crystals" [4.2]. This publication contains, in essence, all the main elements of modern theory. The first period of development of the CFT in the 1930s is related to the papers by Van Vleck and his co-workers (see [4.1, 4.3] and references therein) in which the origin of the magnetic properties of transition metal ions in crystals was revealed.

The theoretical results of Van Vleck were aimed at "crystalline" effects; their general importance to coordination compounds was not realized at that time. Nevertheless, some important results illustrating the efficiency of the CFT were obtained, including an explanation of magnetic behavior in the cases of weak and strong ligand fields, reduction of the orbital magnetic moment by the crystal field, temperature dependence of the magnetic susceptibility, the Jahn–Teller effect, and so on.

The second period of the development of CFT, the period of its vigorous growth, began in the 1950s. In works by Ilse, Hartmann, Orgel, Moffitt, Ballhausen, Jorgensen, and others (see [4.4–4.9] and references therein), it was shown that CFT successfully explains the origin of absorption spectra in the visible and related regions (the origin of colors), as well as a series of other optical, electric, magnetic, thermodynamic, and electron spin resonance (ESR) properties of coordination systems.

Historically, CFT in its application to inorganic complexes can be considered as a direct extension of the prequantum electrostatic considerations. The electrostatic theory, developed by Kossel [4.10] and Magnus [4.11], is based

---

*There is an expression of Enrico Fermi: "A really good theoretical physicist can obtain right answers even with wrong formulas." Paraphrasing this apothegm, Van Vleck wrote [4.1]: "A really good theoretical chemist can obtain right answers with wrong models."

on the assumption that the central ion and the ligands are kept together by ion–ion or ion–dipole electrostatic interactions. Despite limited possibilities, the pure electrostatic approach was rather stimulating. It was developed further after the creation of quantum mechanics, which allows for a more adequate description of these ideas.

Main Assumptions

In its applications to transition metal coordination compounds, CFT is based on the following main assumptions (basic statements):

1. *The transition metal coordination compound* (as it is described in Section 1.2) *is stable due to the electrostatic interaction between the central atom or ion and ligand–ions or dipoles.*
2. *The central atom (CA) is considered with its detailed electronic structure, while the ligands are assumed to be "structureless" sources of electrostatic fields* (sometimes allowing for their polarization in the field of the CA and the other ligands).
3. *The description of the structure and properties of the system based on this model is carried out by means of quantum mechanics.*

From these three statements, only the first was employed in the old pre-quantum theory. The second and third basic assumptions allow us to consider phenomena that have a quantum nature and take place mainly within the electronic shell of the CA but do not explicitly involve the electronic structure of the ligands. Despite this significant limitation (for more details, see Section 4.6), the CFT, within the limits of its applicability, is a rather efficient means of investigation of many aspects of the electronic structure and properties of transition metal compounds.

In accordance with the basic statements of the CFT, the electronic structure of a coordination system is determined by the Schrödinger equation (1.5) with the Hamiltonian H,

$$H = H_0 + V + W \qquad (4.1)$$

where H_0 includes all the interactions in the CA: the kinetic energy of its n electrons and the interaction between them and with the nucleus; V is the interaction between the CA electrons with the ligands taken as point charges q_i or dipoles μ_i; and W is the electrostatic interaction of the positive charge Ze of the CA nucleus with the ligand charges or dipoles (as above, Z is the order number of the element in the periodic table and e is the absolute value of the electronic charge).

Taking the origin of the polar coordinate system at the nucleus, we can denote the N ligand coordinates by $\mathbf{R}_i(R_i, \vartheta_i, \phi_i)$, $i = 1, 2, \ldots, N$. Then

$$V = -\sum_i^N \sum_j^n \frac{eq_i}{|\mathbf{r}_j - \mathbf{R}_i|} \tag{4.2}$$

$$W = \sum_i^N \frac{Zeq_i}{R_i} \tag{4.3}$$

If $q_i < 0$, the term (4.3) provides the required CA–ligand negative interaction energy due to which the complex is stable. In this case the term V after (4.2) is positive, and hence the electron–ligand interaction destabilizes the complex. This situation is quite usual in real systems: The CA is electropositive and the ligands are electronegative. Formally, there is a possibility that q_i is positive, and then V is negative, providing the necessary stabilization energy, W being destabilizing; this model seems to be unreal except when the electrons are formally substituted by positive holes (Section 4.2).

The term W after (4.3) is usually not considered in CFT explicitly, since the determination of the absolute values of stabilization (bonding) energies is beyond the possibilities of this theory. On the other hand, W is independent of the electron coordinates, and therefore it is constant with respect to the electronic properties, considered in CFT. As usual in quantum chemistry, the valence electrons of the CA are most important in chemical phenomena. Therefore, the number of electrons n in equation (4.2) is often the number of the valence electrons, while Ze is the effective charge of the remaining core.

4.2. SPLITTING OF THE ENERGY LEVELS OF ONE d ELECTRON

Qualitative Aspects and Visual Interpretation

In accordance with the assumptions of CFT, the main effect of complex formation is produced by the ligands' field, which changes the electronic structure of the CA. The problem of calculating atomic states in external fields of different symmetry (different ligand positions) was solved by Bethe [4.2]. *The main effect of ligand influence on the states of the CA is the splitting of its energy levels.* The origin of this splitting is known in quantum mechanics as the *Stark effect.*

Consider the simplest (from the point of view of CFT) case when the central ion of an octahedral complex (Fig. 4.1a) contains only one d electron over the closed shell, as, for instance, in $[Ti(H_2O)_6]^{3+}$. The ground state of the Ti^{3+} ion is ^2D (Table 2.6). This term has an orbital momentum $L = 2$, an orbital degeneracy $2L + 1 = 5$, and a spin $S = 1/2$ (doublet state). The five orbital states are just the five possible angular states of the only d electron given in Table 2.1, with identical radial parts. As seen from Fig. 2.3, the three orbitals

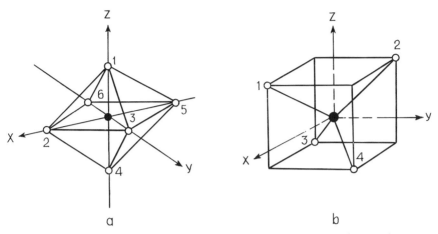

Figure 4.1. Ligand numeration in octahedral (a) and tetrahedral (b) complexes.

d_{xy}, d_{xz}, and d_{yz} (sometimes named d_ε orbitals or, more often, t_{2g} orbitals) are oriented in space such that their distribution maxima (their lobes) fall into the region between the coordinate axes. The remaining two orbitals d_{z^2}, and $d_{x^2-y^2}$ (called d_γ or, more often, e_g orbitals) have their lobs oriented exactly along the axes.

Compare the electron distributions in the two types of d states, e_g and t_{2g}, say $d_{x^2-y^2}$ (e_g) and d_{xy} (t_{2g}), illustrated in Fig. 4.2. Taking into account that the ligands have negative charges (ligand–dipoles are oriented with the negative pole to the CA), we easily come to the conclusion that in octahedral complexes the $d_{x^2-y^2}$ electron is subject to a stronger electrostatic repulsion from the ligands than the d_{xy} electron. Hence the electron energies of these two states (which are equal in the free ion), under the electrostatic influence of the ligands become different: The $d_{x^2-y^2}$ energy level is higher than the d_{xy} level. All three t_{2g} states (d_{xy}, d_{xz}, d_{yz}) are fully equivalent with respect to the six ligands, and therefore they have the same (lower) energy, forming a threefold degenerate term. It can be shown that the two e_g states also have equal energies, forming a twofold-degenerate term.

Thus the five d states that have the same energy in the free atom (or ion) are divided into two groups of states with different energies under the octahedral field of the ligands. In other words, *the fivefold-degenerate D term of the free ion is split in the field of the ligands of an octahedral complex into two terms: threefold-degenerate $^2T_{2g}$ and twofold-degenerate ^2E_g,*

$$^2D \rightarrow {}^2T_{2g} + {}^2E_g \tag{4.4}$$

The denotation here is that of Mulliken (Sections 3.3 and 3.6). Note that one-electron states of the same symmetry are denoted by corresponding lowercase letters (t_{2g}, e_g, etc.) as distinct from many-electron states, represented by capital letters.

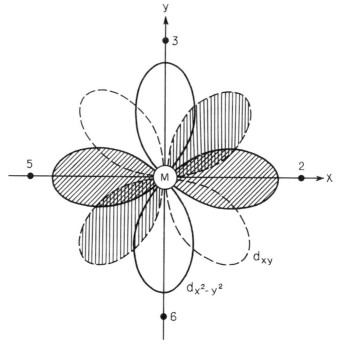

Figure 4.2. Comparison of electron distributions in the two d states, $d_{x^2-y^2}$ (solid lines) and d_{xy} (dashed lines) with respect to the four ligands in the xy plane.

It follows from this consideration that the splitting (4.4) occurs as a result of the smaller repulsion (from the ligands) of the t_{2g} states than e_g states, but all five states are destabilized in the field of the ligands. For applications, it is convenient to present this effect quantitatively as a sum of two effects: *destabilization* and *splitting*. This is illustrated in Fig. 4.3; the corresponding calculations of the destabilization E_0 and splitting Δ values are given below.

Visually, the destabilization equals the energy of repulsion of the CA electrons from the ligand charges when they are assumed to have spherically symmetric distribution around the central atom; this repulsion E_0 is the same for all five d electrons. Obviously, the destabilization must be compensated by the attraction term W between the CA core and ligands after Eq. (4.3); otherwise, the complex is not stable.

For a tetrahedral complex the qualitative picture of the term splitting is inverse to that of the octahedral case. Indeed, in the tetrahedral environment of four ligands the t_{2g} orbitals are oriented with their lobes much closer to the ligands than the e_g orbitals, and hence the former are subject to stronger repulsion than the latter. Therefore, the energy levels of the t_{2g} orbitals are higher than e_g. Symmetry considerations and the calculations, given below, show that in the tetrahedral system, again, the three t_{2g} states, as well as the two e_g states, remain degenerate, forming the T_2 and E terms, respectively (in the case of a

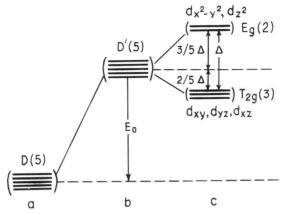

Figure 4.3. Destabilization E_0 and splitting Δ of the atomic D term in the field of six ligands in an octahedral complex (the degeneracies are indicated in parentheses): (a) free atom; (b) spherical averaged ligand field; (c) octahedral field.

tetrahedron there is no inversion symmetry and no classification of the terms by g and u; see Section 3.4 and Appendix 1). Hence in a tetrahedral complex the splitting of the D term is (Fig. 4.4b)

$$D \rightarrow T_2 + E \tag{4.5}$$

which is very similar to that of an octahedral complex (4.4). But as distinct from the octahedral case, the T_2 term is higher in energy than the E term, and the splitting magnitude, as well as the destabilization energy, is smaller.

Both the octahedral and tetrahedral symmetries pertain to the cubic groups of symmetry (Appendix 1). In Fig. 4.4 the splitting of the atomic D term in the field of eight ligands at the corners of a cube is also shown.

If the symmetry of the ligand field is lowered, the terms T_2 and E may be subject to further splitting. Consider the case of a tetragonally distorted octahedron formed by the elongation of the regular octahedron along one of its diagonals, say, z. In this case the energies of the two e_g orbitals, $d_{x^2-y^2}$ and d_{z^2}, are no more equal, since in the latter the repulsion is lower as compared with the former. The three t_{2g} states do not remain equivalent either: Two of them, d_{xz} and d_{yz}, experience (equally) less repulsion from the ligands than does d_{xy}. Therefore, in the tetragonally distorted octahedron the atomic D term splits into four terms, from which only $E_g(d_{xz}, d_{yz})$ remains twofold degenerate (Fig. 4.5). If the symmetry of the ligand field is lowered further (e.g., if the two axes x and y are nonequivalent), the twofold-degenerate term splits, too.

For CA with more than one d electron, the visual interpretation of the splitting becomes more complicated, but the idea remains the same: In the CFT model the energies of the states that are degenerate in the free atom (ion) may differ due to the difference in repulsion from the ligands.

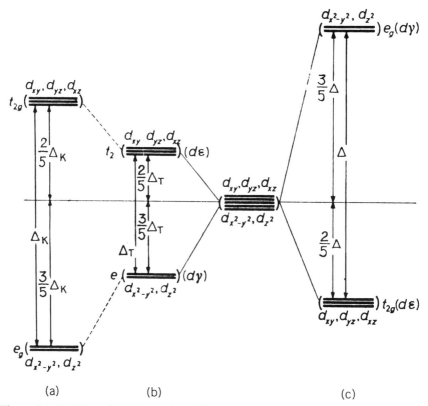

Figure 4.4. Splitting of the D term in cubic (a), tetrahedral (b), and octahedral (c) fields of the ligands.

If the number of d electrons above the closed shell equals nine, the visual interpretation of the splitting becomes possible again; it is quite similar to that of d^1, considered above. Indeed, the d^9 configuration can be formally presented as one positive electronic charge above the closed-shell d^{10} configuration, that is, a d hole in the d^{10} shell (see the principle of complementary configurations [4.15] in Section 2.2). The behavior of the states of the d hole in the field of ligands is the same as for one d electron, with the distinction that the charge of the d hole is opposite that of the d electron, and hence the sign of the interaction with the ligands changes from repulsion to attraction. For these reasons the ground state of the d^9 configuration (the d hole) is also ^2D, as for the d^1 configuration, and its splitting in crystal fields of different symmetries has the same components as d^1, but with their mutual arrangement inverse to that of d^1. *This rule of inverse term splitting for complementary electron configurations is valid for any pair of electronic configuration, d^n and d^{10-n}* ($n = 1, 2, 3, 4$) in which the number of, respectively, d electrons and d holes over the closed shell d^{10} is the same.

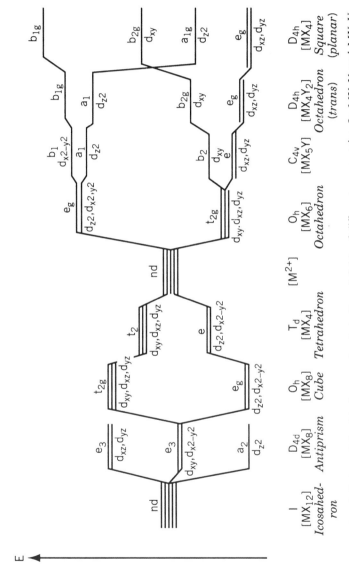

Figure 4.5. Splitting of *d*-orbital energy levels in ligand fields of different symmetries. In MX_5Y and MX_4Y_2 complexes the splitting of the T_{2g} and E_g terms can be inverted depending on the ratio of field strengths X/Y. (After [4.12].)

Calculation of Splitting

The quantitative evaluation of energy-level splitting of the CA in the field of ligands is relatively not difficult. The calculations should be based on the Schrödinger equation (1.5) with the Hamiltonian (4.1). As mentioned above, the term W after (4.3) is not important for splitting since it is independent of electron coordinates. A further assumption of CFT is that the term V, the electron–ligand interaction, is much smaller than the interatomic interactions described by H_0, and hence V can be considered as a perturbation to the solutions of H_0. This assumption is valid if the resulting term splitting obtained in this way is smaller than the energy gap between the terms of the free atom (or ion) solutions of H_0. In Section 4.3 this statement is subject to further discussion.

Consider the case of one d electron above the closed shell, that is, the electron configuration of the CA $[A]d^1$. The solution for the free ion with the Hamiltonian H_0 yields the fivefold orbitally degenerate ^2D term. To reveal the modifications of this term under the perturbation V after (4.2), one has to solve the perturbation theory problem, which in the case of fivefold degeneracy reduces to the secular equation of fifth order with respect to the energy-level corrections ε:

$$\| V_{mm'} - \varepsilon \delta_{mn'} \| = 0 \qquad m, m' = 2, 1, 0, -1, -2 \tag{4.6}$$

or in more detail,

$$\begin{vmatrix} V_{22} - \varepsilon & V_{21} & V_{20} & V_{2-1} & V_{2-2} \\ V_{12} & V_{11} - \varepsilon & V_{10} & V_{1-1} & V_{1-2} \\ V_{02} & V_{01} & V_{00} - \varepsilon & V_{0-1} & V_{0-2} \\ V_{-12} & V_{-11} & V_{-10} & V_{-11} - \varepsilon & V_{-1-2} \\ V_{-22} & V_{-21} & V_{-20} & V_{-2-1} & V_{-22} - \varepsilon \end{vmatrix} = 0 \tag{4.7}$$

where, in accordance with (4.2)

$$V_{mm'} = \sum_i eq_i \int \frac{\psi_m^* \psi_{m'}}{|\mathbf{r} - \mathbf{R}_i|} \, d\tau \tag{4.8}$$

and the functions of the d states are taken in their general form (2.2) (not as real combinations in Table 2.1) with the indices m and m' listed in (4.6).

The general expression for $V_{mm'}$ that is most important in CFT is obtained in Appendix 2 [Eq. (A2.8)]:

$$V_{mm'} = \sum_i eq_i [A_{mm'} F_4(R_i) Y_4^{m-m'}(\vartheta_i, \phi_i) + B_{mm'} F_2(R_i) Y_2^{m-m'}(\vartheta_i, \phi_i)$$

$$+ D_{mm'} F_0(R_i) Y_0^{m-m'}(\vartheta_i, \phi_i) \tag{4.9}$$

Table 4.1. Coefficients $A_{mm'}$ and $B_{mm'}$ in the matrix elements of the crystal field for a d electron (4.9)^a

	$(63/\sqrt{4\pi})A_{mm'}$					$(35/\sqrt{4\pi})B_{mm'}$				
					m'					
m	2	1	0	-1	-2	2	1	0	-1	-2
2	1	$-\sqrt{5}$	$\sqrt{15}$	$-\sqrt{35}$	$\sqrt{70}$	$-\sqrt{20}$	$\sqrt{30}$	$-\sqrt{20}$	0	0
1	$\sqrt{5}$	-4	$\sqrt{30}$	$-\sqrt{40}$	$\sqrt{35}$	$-\sqrt{30}$	$\sqrt{5}$	$\sqrt{5}$	$-\sqrt{30}$	0
0	$\sqrt{15}$	$-\sqrt{30}$	6	$-\sqrt{30}$	$\sqrt{15}$	$-\sqrt{20}$	$-\sqrt{5}$	$\sqrt{20}$	$-\sqrt{5}$	$-\sqrt{20}$
-1	$\sqrt{35}$	$-\sqrt{40}$	$\sqrt{30}$	-4	$\sqrt{5}$	0	$-\sqrt{30}$	$\sqrt{5}$	$\sqrt{5}$	$-\sqrt{30}$
-2	$\sqrt{70}$	$-\sqrt{35}$	$\sqrt{15}$	$-\sqrt{5}$	1	0	0	$-\sqrt{20}$	$\sqrt{30}$	$-\sqrt{20}$

$^a D_{mm'} = (4\pi)^{1/2}\delta_{mm'}$.

where $A_{mm'}$, $B_{mm'}$, and $D_{mm'}$ are constants (determined by corresponding Clebsh–Gordan coefficients) given in Table 4.1, while the functions $F_k(R)$ [not to be confused with the Slater–Condon constants F_k after (2.32)],

$$F_k(R) = R^{-(k+1)} \int_0^R r^k R_{nl}^2(r) r^2 \, dr + R^k \int_R r^{-(k+1)} R_{nl}^2(r) r^2 \, dr \qquad (4.10)$$

with radial functions $R_{n2}(r)$ after Table 2.3 can be calculated and expressed in analytical form (A2.10).

By way of example, let us calculate the splitting of a d electron term in an octahedral field of six identical ligands at the corners of a regular octahedron. The ligand charges and coordinates are (Fig. 4.1a)

$$q_i = q \qquad R_i = R \qquad i = 1, 2, \ldots, 6$$

$$\vartheta_1 = 0 \qquad \vartheta_2 = \vartheta_3 = \vartheta_5 = \vartheta_6 = \frac{\pi}{2} \qquad \vartheta_4 = \pi \qquad (4.11)$$

$$\phi_2 = 0 \qquad \phi_3 = \frac{\pi}{2} \qquad \phi_5 = \pi \qquad \phi_6 = \frac{3\pi}{2}$$

Substituting these values into Eq. (4.9) and taking into account the data of Table 4.1, we find that $V_{21} = V_{20} = V_{2-1} = V_{10} = V_{1-1} = V_{1-2} = V_{0-1} =$

$V_{0-2} = V_{-1-2} = 0$, and only seven matrix elements are nonzero:

$$V_{22} = V_{-2-2} = eq[6F_0(R) + \tfrac{1}{6}F_4(R)]$$
$$V_{11} = V_{-1-1} = eq[6F_0(R) - \tfrac{2}{3}F_4(R)]$$
$$V_{00} = eq[6F_0(R) + F_4(R)] \tag{4.12}$$
$$V_{-22} = V_{2-2} = \tfrac{5}{6}eqF_4(R)$$

With these matrix elements the roots of Eq. (4.7) can be obtained directly:

$$\varepsilon_1 = V_{00}$$
$$\varepsilon_2 = V_{22} + V_{2-2}$$
$$\varepsilon_3 = V_{22} - V_{2-2} \tag{4.13}$$
$$\varepsilon_4 = \varepsilon_5 = V_{11}$$

Then we note that $V_{22} + V_{2-2} = V_{00}$ and $V_{22} - V_{2-2} = V_{11}$. Finally, the roots ε — perturbation corrections to the atomic energy levels — are

$$\varepsilon_1 = \varepsilon_2 = V_{00} = eq[6F_0(R) + F_4(R)]$$
$$\varepsilon_3 = \varepsilon_4 = \varepsilon_5 = V_{11} = eq[6F_0(R) - \tfrac{2}{3}F_4(R)] \tag{4.14}$$

Thus in accordance with the qualitative results described above, the d-electron energy levels (^2D term) are split by the octahedral ligand field into twofold (E_g) and threefold (T_{2g}) degenerate terms. Furthermore, Eqs. (4.14) allow us to obtain the expressions for absolute value and sign of the splitting Δ and destabilization energy E_0. Indeed, from (4.10) it is seen that all the functions $F_k(R)$ are positive, $F_k > 0$. Therefore, $V_{00} > V_{11}$, and the twofold-degenerate E_g term is higher in energy than the T_{2g} term. *The splitting Δ is the main CFT parameter.* We have:

$$\Delta = \varepsilon(E_g) - \varepsilon(T_{2g}) = \tfrac{5}{3}eqF_4(R) \tag{4.15}$$

Then the expressions for the energies (4.14) can be rewritten as follows:

$$\varepsilon(E_g) = E_0 + \tfrac{3}{5}\Delta$$
$$\varepsilon(T_{2g}) = E_0 - \tfrac{2}{5}\Delta \tag{4.16}$$

where

$$E_0 = 6eqF_0(R) \tag{4.17}$$

is just the average energy of repulsion of the d electron from six negative charges q when they are uniformly distributed over a sphere of radius R with the center at the CA, the destabilization energy. Leaving the latter apart, the splitting has some interesting general features. In particular, there is always the *preservation of the center of gravity*: The sum of the energy level displacements (from the E_0 value) multiplied by their degeneracies equals zero. For instance, for the splitting $D \rightarrow T_{2g} + E_g$, we have: $2 \cdot \frac{3}{5}\Delta - 3 \cdot \frac{2}{5}\Delta = 0$. This rule enables us to predict some relations between the energy-level positions with respect to E_0. Similarly, the splitting and destabilization can be calculated for other types of coordination given in Appendix 3.

Compared with the octahedron, the corresponding splitting in the tetrahedron, as mentioned above, is inverted and smaller (Fig. 4.4). Using (A3.14), we have

$$\varepsilon(T_2) = E_0^T + \tfrac{2}{5}\Delta_T \tag{4.18}$$

$$\varepsilon(E) = E_0^T - \tfrac{3}{5}\Delta_T \tag{4.19}$$

where the splitting parameter Δ_T and destabilization energy E_0^T are

$$\Delta_T = \tfrac{20}{27}eqF_4(R) \tag{4.20}$$

$$E_0^T = 4eqF_0(R) \tag{4.21}$$

Note that with the same R and q, $\Delta_T = -\tfrac{4}{9}\Delta$, that is, the splitting magnitude in tetrahedral symmetry, *ceteris paribus*, is $\tfrac{4}{9}$ times the octahedral splitting.

In the cubic coordination (eight ligands at the corners of a cube) the splitting and destabilization energy are qualitatively similar to the tetrahedral case but twice as large in magnitude [(A3.17 and A3.18)]:

$$\Delta_c = \tfrac{40}{27}eqF_4(R) = 2\Delta_T \tag{4.22}$$

$$E_0^c = 8eqF_0(R) = 2E_0^T \tag{4.23}$$

For tetragonally distorted octahedra the description of the splitting requires, in addition to the main CFT parameter Δ, two parameters D_s and D_t [(A3.7)]:

$$D_s = \tfrac{2}{7}eq[F_2(R_2) - F_2(R_1)] \tag{4.24}$$

$$D_t = \tfrac{2}{21}eq[F_4(R_2) - F_4(R_1)] \tag{4.25}$$

where R_1 and R_2 are the distances between the CA and the axial and equatorial ligands, respectively. With these parameters, the energy levels of the d states in the tetragonally distorted octahedron are (Fig. 4.6):

$$
\begin{aligned}
\varepsilon(A_{1g}; d_{z^2}) &= E_0' + \tfrac{3}{5}\Delta - 2D_s - 6D_t \\
\varepsilon(B_{1g}; d_{x^2-y^2}) &= E_0' + \tfrac{3}{5}\Delta + 2D_s - D_t \\
\varepsilon(B_{2g}; d_{xy}) &= E_0' - \tfrac{2}{5}\Delta + 2D_s - D_t \\
\varepsilon(E_g; d_{xz}, d_{yz}) &= E_0' - \tfrac{2}{5}\Delta - D_s + 4D_t
\end{aligned}
\tag{4.26}
$$

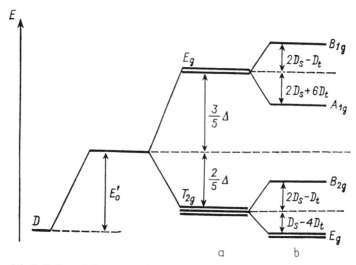

Figure 4.6. Splitting of the energy levels of d states (D term) in regular octahedral field (a), and tetragonally distorted (elongated octahedral) field (b); E_0' is the spherically symmetric destabilization.

where [cf. (4.17)]

$$E_0' = eq[2F_0(R_1) + 4F_0(R_2)] \tag{4.27}$$

and Δ is the same as in Eq. (4.15).

A planar-quadratic complex can be considered as the limit case of a tetragonally distorted octahedron with $R_1 \to \infty$. Then $F_k(R_1) \to 0$, $D_t = \frac{2}{35}\Delta$ (Appendix 3), and only two independent parameters, D_s and Δ, are needed to describe the splitting (Fig. 4.6). After (A3.10):

$$\varepsilon(A_{1g}; d_{z^2}) = E_0'' + \tfrac{9}{35}\Delta - 2D_s$$
$$\varepsilon(B_{1g}; d_{x^2-y^2}) = E_0'' + \tfrac{19}{35}\Delta + 2D_s$$
$$\varepsilon(B_{2g}; d_{xy}) = E_0'' - \tfrac{16}{35}\Delta + 2D_s \tag{4.28}$$
$$\varepsilon(B_g; d_{xz}, d_{yz}) = E_0'' - \tfrac{6}{35}\Delta - D_s$$

with Δ after (4.15) and

$$E_0'' = 4eqF_4(R) \tag{4.29}$$

If the ligand is a point dipole, the term splitting is qualitatively the same, as in the case of point charges. Assuming that the dimensions of the dipole with the dipole moment μ are much smaller than the distance to the CA, one can obtain the following expression for the splitting magnitude in the octahedral case (A3.3):

$$\Delta = -\tfrac{5}{3}e\mu F'_4(R) \tag{4.30}$$

where the prime at F means its derivative [cf. (4.17)]; note that $F'_4 < 0$.

For more complicated coordinations the calculations can be carried out in a similar way, first evaluating the matrix elements $V_{mm'}$ after (4.9) and then solving the secular equation (4.7). For lower symmetries (including those arising due to different ligands), numerical solutions of (4.7) may be necessary. The results for planar trigonal coordination, trigonal biprism and antiprism, and some other cases are given in [4.13]; another example, a semicoordinated octahedron prism of C_{2v} symmetry with $1:4:2$ stereochemistry (of the type NbF_7^{2-}), is considered in [4.14].

Group-Theoretical Analysis

Let us now discuss the problem of term splitting from the point of view of symmetry considerations, given in Chapter 3, without employing the secular equation (4.7). For a free atom the group of symmetry transformations is the group of rotations of a sphere $R(3)$ that has an infinite number of elements and many irreducible representations. The spherical functions $Y_L^M(\vartheta, \phi) = P_L^M(\cos \vartheta)e^{iM\phi}$, where P_L^M is an associated Legendre polynomial [see (2.2)], can be taken as a basis of this group. For each L value there are $2L+1$ spherical functions with different M ($M = 0, \pm 1, \ldots, \pm L$), which transform to each other by the symmetry transformations of the group and realize the irreducible representation of dimensionality $2L + 1$. Hence the atomic terms have a $(2L + 1)$-fold degeneracy.

For the spherical group the characters of the matrices of the representations for a rotation by an angle ϕ can be calculated by the formula [4.2]

$$X(\phi) = \frac{\sin(L + \tfrac{1}{2})\phi}{\sin \tfrac{1}{2}\phi} \tag{4.31}$$

When the atom is introduced in an external field of, for instance, O_h symmetry, the symmetry of the system as a whole (the atom plus the field) becomes O_h, and only those symmetry transformations (rotations) that comply with the O_h restrictions remain. The number of symmetry transformations is thus reduced, which means that the irreducible representation of the spherical group corresponding to the atomic term of $(2L + 1)$-fold degeneracy may become reducible in th O_h group (Section 3.3). The reducible representation can be decomposed into several irreducible representations of smaller dimensionality to which several energy terms of lower degeneracy belong: The term splits.

To determine the term splitting, the representation of the spherical group with the dimensionality $2L + 1$ should be decomposed into irreducible representations of the O_h group. This problem is solved completely by means of the relation (3.33). Consider, for example, the atomic term D in the octahedral complex of O_h symmetry, discussed above. For this term $L = 2$, $2L + 1 = 5$. To employ (3.33), one has to calculate first the characters $X(G)$ of the representation of the spherical group with $L = 2$ for all the operations G of the O_h group. For the first classes, which include rotations only (Table 3.1), the characters can be calculated after (4.31). For the remaining five classes, the symmetry elements are those of the first five classes multiplied by the operation of inversion I (Section 3.2), and hence their characters can be determined as a product of the characters of the two factors. Since the functions of the basis (the spherical functions with $L = 2$) are invariant with respect to the inversion transformation [$X(I) = 1$], the characters for the elements of the five classes with inversion are the same as that without inversion.

For instance, for the element C_2' (rotation by an angle π), using (4.31) with $L = 2$, we obtain

$$X(\pi) = \frac{\sin \frac{5}{2}\pi}{\sin(\pi/2)} = 1$$

and for the element $\sigma_d = C_2'I$,

$$X(\sigma_d) = X(C_2') \cdot X(I) = 1 \cdot 1 = 1$$

In a similar way one can obtain the characters for all the other symmetry operations of the O_h group listed in Table 3.1, as follows:

G	E	$6C_4$	$3C_4^2 = 3C_2$	$8C_3$	$6C_2$	I	$6S_4$	$3\sigma_h$	$8S_6$	$6\sigma_d$
$X(G)$	5	-1	1	-1	1	5	-1	1	-1	1

$$(4.32)$$

Now using Eq. (3.33), one can find out the irreducible representations of the O_h group comprised in the reducible representation (4.32). We have:

$$a^{(A1g)} = \tfrac{1}{48}(5 - 6 + 3 - 8 + 6 + 5 - 6 + 3 - 8 + 6) = 0$$
$$a^{(Eg)} = \tfrac{1}{48}(10 + 0 + 6 + 8 + 0 + 10 + 0 + 6 + 8 + 0) = 1$$

and so on. In this way one finds that $a^{(Eg)} = 1$, $a^{(T2g)} = 1$, and all the other $a^{(\beta)} = 0$. It follows that the fivefold-degenerate D term in the field of O_h symmetry splits into two terms: twofold-degenerate E_g and three-fold-degenerate T_{2g}: $D \rightarrow E_g + T_{2g}$. We can determine the splitting in all other cases of this kind similarly.

Table 4.2 gives the symmetry representations (symmetry types) to which the orbital states of a free atom (spherical functions) belong in fields of different symmetries; a comparison of symmetry types in different groups gives the

Table 4.2. Types of symmetry (irreducible representations of the tables in Appendix 1) to which the atomic states with given quantum numbers $L(l)$ or $J(j)$ belong in the point groups of different symmetry[a]

L or J	O_h	T_d	D_3	D_{4h}	C_{4v}	C_{2v}
0	A_{1g}	A_1	A_1	A_{1g}	A_1	A_1
1	T_{1u}	T_2	$A_2 + E$	$A_{2u} + E_u$	$A_1 + E$	$A_1 + B_1 + B_2$
2	E_g	E	E	$A_{1g} + B_{1g}$	$A_1 + B_1$	$2A_1$
	T_{2g}	T_2	$A_1 + E$	$B_{2g} + E_g$	$B_2 + E$	$A_2 + B_1 + B_2$
3	A_{2u}	A_1	A_2	B_{1u}	B_2	A_2
	T_{1u}	T_2	$A_2 + E$	$A_{2u} + E_u$	$A_1 + E$	$A_1 + B_1 + B_2$
	T_{2u}	T_1	$A_1 + E$	$B_{2u} + E_u$	$B_1 + E$	$A_1 + B_1 + B_2$
4	A_{1g}	A_1	A_1	A_{1g}	A_1	A_1
	E_g	E	E	$A_{1g} + B_{1g}$	$A_1 + B_1$	$2A_1$
	T_{1g}	T_1	$A_2 + E$	$A_{2g} + E_g$	$A_2 + E$	$A_2 + B_1 + B_2$
	T_{2g}	T_2	$A_1 + E$	$B_{2g} + E_g$	$B_2 + E$	$A_2 + B_1 + B_2$
5	E_u	E	E	$A_{1u} + B_{1u}$	$A_2 + B_2$	$2A_2$
	T_{1u}	T_2	$A_2 + E$	$A_{2u} + E_u$	$A_1 + E$	$A_1 + B_1 + B_2$
	T_{2u}	T_1	$A_1 + E$	$B_{2u} + E_u$	$B_1 + E$	$A_1 + B_1 + B_2$

[a]Belonging to several types of symmetry in the cases of $L \neq 0$ can be interpreted as a corresponding splitting.

Table 4.3. Correlations of the irreducible representations of the O_h and D_{4h} groups of symmetry with their subgroups indicating the corresponding symmetry transformations and splitting

Group	Subgroup			Group	Subgroup	
O_h	T_d	D_{4h}	D_3	D_{4h}	C_{4v}	C_{2v}
A_{1g}	A_1	A_{1g}	A_1	A_{1g}	A_1	A_1
A_{1u}	A_2	A_{1u}	A_1	A_{1u}	A_2	A_2
A_{2g}	A_2	B_{1g}	A_2	A_{2g}	A_2	B_1
A_{2u}	A_1	B_{1u}	A_2	A_{2u}	A_1	B_2
E_g	E	$A_{1g} + B_{1g}$	E	B_{1g}	B_1	A_1
E_u	E	$A_{1u} + B_{1u}$	E	B_{1u}	B_2	A_2
T_{1g}	T_1	$A_{2g} + E_g$	$A_2 + E$	B_{2g}	B_2	B_1
T_{1u}	T_2	$A_{2u} + E_u$	$A_2 + E$	B_{2u}	B_1	B_2
T_{2g}	T_2	$B_{2g} + E_g$	$A_1 + E$	E_g	E	$A_2 + B_2$
T_{2u}	T_1	$B_{2u} + E_u$	$A_1 + E$	E_u	E	$A_1 + B_1$

expected splitting of appropriate terms. For instance, the D term ($L = 2$) splits into $E_g + T_{2g}$ in the O_h group, $E + A_1 + E$ in the D_3 group, $A_{1g} + B_{1g} + B_{2g} + E_g$ in the D_{4h} group, and so on. Similar correlations between the symmetry representations which show the possible term splitting when passing from higher symmetries to lower ones in other cases are given in Table 4.3.

4.3. SEVERAL d ELECTRONS

Case of a Weak Field

If the electron configuration of the CA contains more than one d electron above the closed shell, the picture of possible energy terms and their splitting in the ligand fields is significantly complicated by the interaction between the d electrons. If the ligand field is not very strong, the atomic terms can still be classified by the quantum number of the atomic total momentum L, and the influence of the ligands can be taken as a perturbation of the atomic terms; this is the case of the weak ligand field.

The term *weak field* is discussed below in more detail (in particular, see the beginning of Section 4.4 for some different terminology). Here we emphasize that under the influence of the weak field of the ligands the LS coupling between the d electrons (Section 2.2) is not destroyed, and the term with the highest spin is the ground term. Therefore, the complexes with weak ligand field are also called *high-spin complexes*.

For several d electrons the main effect of ligand fields, as for one d electron, is the energy term splitting. However, unlike the d^1 case, visual interpretation of the splitting of the terms of d^n configuration ($n > 1$) is difficult. But the cause of the splitting is the same: in the ligand field, the atomic (multielectron) states that have the same energy in the free atom (ion) are subject to different repulsion from the ligands, owing to their different orientation with respect to these ligands.

Quantitatively, for the electronic configuration of the CA $[A](nd)^2$ in the ligand field, which is weaker than the interaction between the d electrons, we consider first the possible states of the free atom (ion) and find its terms, as is done in Section 2.2, and then determine the influence of the ligand field as a perturbation for each of these terms separately. For two d electrons the possible terms are (Table 2.6) ^3F, ^3P, ^1G, ^1D, and ^1S, the ^3F term belonging to the ground state. Their wavefunctions Ψ (LMSMs) are given in Table 2.8. The perturbation operator, following Eq. (4.2), is

$$V' = \sum_\alpha eq_\alpha \left[\frac{1}{|\mathbf{r}_1 - \mathbf{R}_\alpha|} + \frac{1}{|\mathbf{r}_2 - \mathbf{R}_\alpha|} \right]$$
$$= V(\mathbf{r}_1) + V(\mathbf{r}_2) \tag{4.33}$$

The calculation of the matrix elements of this perturbation is relatively not difficult because the operators $V(\mathbf{r}_i)$ depend on only orbital (not spin) coordinates of only one electron. Presenting the two-electron wavefunction Ψ by the determinant functions $\Phi(m_1 m_{s1}; m_2 m_{s2})$ (Table 2.8), one can see that the matrix element of V' is nonzero if and only if the two spin quantum numbers m_{s1} and m_{s2} and one of the m values (m_1 or m_2)) in the two functions are identical.

This can be expressed as follows:

$$\langle \Psi(m_1 m_{s1}; m_2 m_{s2} | V' | \Psi(m'_1 m'_{s1}; m'_2 m'_{s2}) \rangle$$

$$= \delta_{ms1ms1'} \delta_{ms2ms2'} (V_{m1m1'} \delta_{m2m2'} + V_{m2m2'} \delta_{m1m1'} - V_{m1m2} \delta_{m1'm2'} - V_{m1'm2'} \delta_{m1m2})$$

$$(4.34)$$

Here $V_{mm'}$ are the one-electron matrix elements (4.9), calculated above. Using (4.34), one can obtain the expressions for all the two-electron matrix elements $\langle 1 | V' | 2 \rangle$ by the known one-electron matrix elements $V_{mm'}$; they are given in Appendix 4.

By way of example, let us consider the ground-state term ^3F of the d^2 configuration in the octahedral field of O_h symmetry. The wavefunctions are given in Table 2.8. The term ^3F is orbitally sevenfold degenerate ($L = 3$, $2L + 1 = 7$) and the secular equation of the perturbation theory is of the seventh order [cf. Eq. (4.7)]:

$$\| V'_{ij} - \varepsilon \delta_{ij} \| = 0 \qquad i, j = 1, 2, \ldots, 7 \qquad (4.35)$$

Assuming, as above, that the ligands are point charges and using expressions from Appendix 4 for V'_{ij} and (4.9) for $V_{mm'}$ with the ligand coordinates (4.11), we have: $V'_{12} = V'_{14} = V'_{16} = V'_{17'} = V'_{23} = V'_{24} = V'_{25} = V'_{27} = V'_{34} = V'_{35} = V'_{36} = V'_{45} = V'_{46} = V'_{47} = V'_{56} = V'_{57} = V'_{67} = 0$. Besides

$$V'_{11} = V'_{77} \qquad V'_{22} = V'_{66} \qquad V'_{33} = V'_{55} \qquad (4.36)$$

With these matrix elements the secular equation (4.35) yields the following roots:

$$\varepsilon_1 = V'_{44}$$

$$\varepsilon_{2,3} = V'_{22} + (\tfrac{5}{3})^{1/2} V'_{15} \qquad (4.37)$$

$$\varepsilon_{4,5,6,7} = \tfrac{1}{2} \{ (V'_{11} + V'_{33}) \pm [(V'_{11} - V'_{33})^2 + 4(V'_{15})^2]^{1/2} \}$$

Again substituting V'_{ij} (A4.1) and $V_{mm'}$ according to (4.12), we find the perturbation corrections to the energy levels in the ligand field (the type of symmetry in the O_h group is shown in parentheses):

$$\varepsilon_1(^3A_g) = 2V_{00} = eq[12F_0 + 2F_4]$$

$$\varepsilon_{2,3,4}(^3T_{2g}) = V_{00} + V_{11} = eq[12F_0 + \tfrac{1}{3}F_4] \qquad (4.38)$$

$$\varepsilon_{5,6,7}(^3T_{1g}) = \tfrac{1}{5}V_{00} + \tfrac{9}{5}V_{11} = eq[12F_0 - F_4]$$

Or, introducing the main parameter Δ of CFT after (4.15) and the destabilization energy E_0 after (4.17), we have

$$\varepsilon(^3A_{2g}) = 2E_0 + 6/5\Delta$$

$$\varepsilon(^3T_{2g}) = 2E_0 + \tfrac{1}{5}\Delta \qquad (4.39)$$

$$\varepsilon(^3T_{1g}) = 2E_0 - \tfrac{3}{5}\Delta$$

It follows that the atomic orbitally sevenfold-degenerate ^3F term in the octahedral field of six ligands splits into three terms: one orbitally nondegenerate $^3A_{2g}$ and two threefold-degenerate $^3T_{2g}$ and $^3T_{1g}$:

$$^3F \rightarrow {}^3A_{2g} + {}^3T_{2g} + {}^3T_{1g} \qquad (4.40)$$

Since Δ and E_0 are positive, $^3T_{1g}$ is the ground term, and $^3T_{2g}$ and $^3A_{2g}$ follow consecutively (Fig. 4.7).

The wavefunctions of the terms (4.39) can be obtained as solutions of the secular equation (4.35) in the form of linear combinations of the atomic functions Ψ (LMSM$_s$). They are given in Table 4.4 [for the spin triplet the functions for only one spin value are given; the others can be obtained by the transformation (2.31)].

The splitting of other terms of the $[A](nd)^2$ configuration is obtained similarly:

^1D term:

$$\varepsilon(^1E_g) = eq[12F_0 + \tfrac{4}{7}F_4] = 2E_0 + \tfrac{12}{35}\Delta$$

$$\varepsilon(^1T_{2g}) = eq[12F_0 - \tfrac{8}{21}F_4] = 2E_0 - \tfrac{8}{35}\Delta \qquad (4.41)$$

Table 4.4. Wavefunctions of the component states of the $^3F(d^2)$ term (split in the octahedral field of O_h symmetry) expressed by linear combinations of the atomic functions Ψ(LMSM$_s$) of Table 2.8[a]

Type of Symmetry (Component Terms)	$\sum_i C_i \Psi_i$
A_{2g}	$(\tfrac{1}{2})^{1/2}[\Psi(3\ 2\ 1\ 1) - \Psi(3\ -2\ 1\ 1)]$
T_{1g}	$(\tfrac{5}{8})^{1/2}\Psi(3\ -3\ 1\ 1) + (\tfrac{3}{8})^{1/2}\Psi(3\ 1\ 1\ 1)$
	$\Psi(3\ 0\ 1\ 1)$
	$(\tfrac{5}{8})^{1/2}\Psi(3\ 3\ 1\ 1) + (\tfrac{3}{8})^{1/2}\Psi(3\ -1\ 1\ 1)$
T_{2g}	$(\tfrac{3}{8})^{1/2}\Psi(3\ 3\ 1\ 1) - (\tfrac{5}{8})^{1/2}\Psi(3\ -1\ 1\ 1)$
	$(\tfrac{1}{2})^{1/2}[\Psi(3\ 2\ 1\ 1) + \Psi(3\ -2\ 1\ 1)]$
	$(\tfrac{3}{8})^{1/2}\Psi(3\ -3\ 1\ 1) - (\tfrac{5}{8})^{1/2}\Psi(3\ 1\ 1\ 1)$

[a]The component with $M_s = 1$ only is shown; the states with $M_s = -1$ and $M_s = 0$ can be found using Eq. (2.31).

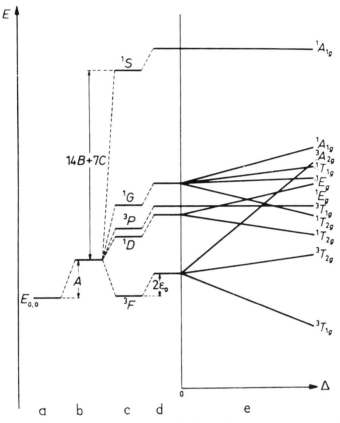

Figure 4.7. Splitting of the terms of the electronic d^2 configuration in octahedral ligand fields (weak field limit): (*a*) *d*-electron energy level; (*b*) interelectron interaction (spherical averaged part); (*c*) interelectron interaction, formation of atomic terms (Section 2.2); (*d*) ligand field destabilization; (*e*) ligand field splitting as a function of Δ.

^3P term:

$$\varepsilon(^3T_{2g}) = 12eqF_0 = 2E_0 \tag{4.42}$$

^1G term:

$$\varepsilon(^1A_{1g}) = eq[12F_0 + \tfrac{2}{3}F_4] = 2E_0 + \tfrac{2}{5}\Delta$$

$$\varepsilon(^1E_g) = eq[12F_0 + \tfrac{2}{21}F_4] = 2E_0 + \tfrac{2}{35}\Delta$$

$$\varepsilon(^1T_{2g}) = eq[12F_0 - \tfrac{13}{21}F_4] = 2E_0 - \tfrac{13}{35}\Delta \tag{4.43}$$

$$\varepsilon(^1T_{1g}) = eq[12F_0 + \tfrac{1}{3}F_2E_0 + \tfrac{1}{5}\Delta$$

^1S term:

$$\varepsilon(^1A_{1g}) = 12eqF_0 = 2E_0 \qquad (4.44)$$

These splittings are illustrated in Fig. 4.7; the destabilization energy — the averaged electron interaction equal to the Racah parameter A (Table 2.6) — is also shown.

The calculations above are carried out in the weak field approximation in which the perturbation theory is applied to each atomic term separately. Therefore, *the criterion of validity of the weak field approximations is that the term splitting is much smaller than the energy gap between the terms.* As seen from Fig. 4.7, for d^2 configurations this criterion is fulfilled if Δ is sufficiently small. For large Δ values the components of the split terms even cross each other, making the approximation of weak field invalid.

For complexes with symmetries lower than O_h, the degenerate terms are subject to further splitting, as shown in Table 4.2. For quantitative estimates one can use the results of Appendix 3. For a tetragonally distorted (elongated) octahedron with the coordinates (A3.4), the nonzero matrix elements V'_{ij} are the same as in the regular octahedron. Hence the roots of the secular equation (4.35) for the splitting of the ^3F term are given by the same general expressions (4.37), in which the matrix elements $V_{mm'}$ are that of Eq. (A3.5), not (4.12). Therefore, as distinct from the regular octahedron, the tetragonally elongated one yields five different roots, two of them being twofold degenerate:

$$\varepsilon_1(^3A_{2g}) = V_{00} + V_{22} + V_{2-2}$$
$$\varepsilon_2(^3B_{2g}) = V_{00} + V_{22} - V_{2-2}$$
$$\varepsilon_3(^3A'_{2g}) = \tfrac{8}{5}V_{11} + \tfrac{2}{5}V_{22} \qquad (4.45)$$
$$\varepsilon_{4,5}(^3E'_g) = V_{11} + \tfrac{4}{5}V_{22} + \tfrac{1}{5}V_{00} - [\tfrac{1}{25}(V_{22} - V_{00})^2 + \tfrac{3}{5}(V_{2-2})^2]^{1/2}$$
$$\varepsilon_{6,7}(^3E''_g) = V_{11} + \tfrac{4}{5}V_{22} + \tfrac{1}{5}V_{00} + [\tfrac{1}{25}(V_{22} - V_{00})^2 + \tfrac{3}{5}(V_{2-2})^2]^{1/2}$$

It is seen that in accordance with the group-theoretical results (Table 4.3), in tetragonal fields T_{2g} and T_{1g} terms of the octahedron undergo further splitting: $T_{2g} \rightarrow B_{2g} + E'_g$, $T_{1g} \rightarrow A'_{2g} + E'_g$. The wavefunctions of these states can be obtained in the usual fashion from the functions in Tables 2.8 and 4.4.

For a square-planar complex with the CA in the center of the square the energy levels are given by Eq. (4.37) with the matrix element $V_{mm'}$ after (A3.9). In the case of tetrahedral symmetry with four point charges at the corners of a tetrahedron and the CA in the center [with the coordinates (A3.11)], the splitting of the ^3F term of the d^2 configuration results in three terms, $^3T_1$, $^3T_2$, and $^3A_2$:

$$\varepsilon(^3T_1) = \tfrac{1}{5}V_{00} + \tfrac{9}{5}V_{11} = eq[8F_0 + \tfrac{4}{9}F_4] = 2E_0^T + \tfrac{3}{5}\Delta_T$$
$$\varepsilon(^3T_2) = V_{00} + V_{11} = eq[8F_0 - \tfrac{4}{27}F_4] = 2E_0^T - \tfrac{1}{5}\Delta_T \qquad (4.46)$$
$$(^3A_2) = 2V_{00} = eq[8F_0 - \tfrac{8}{9}F_4] = 2E_0^T - \tfrac{6}{5}\Delta_T$$

As with one *d* electron, the tetrahedral splitting is similar to the octahedral splitting but with the inverse order of the energy levels: $\varepsilon(^3A_2) < \varepsilon(^3T_2) < \varepsilon(^3T_1)$ [(cf. (4.39)]. Again, $E_0^T = \frac{2}{3}E_0$, and $\Delta_T = -\frac{4}{9}\Delta$, provided that the ligand charges *q* and their distances *R* to the CA are the same, as in the octahedron.

For systems with lower symmetries the calculations are more difficult, but they can be reduced by using the method of equivalent operators [4.15] or irreducible tensor operators [4.16, 4.17]. The tables of spectroscopic coefficients for *p*, *d*, and *f* configurations [4.18] are rather useful for such calculations.

The qualitative picture of term splitting for electronic configurations $[A](nd)^n$ with $n > 2$ in fields of lower symmetry can be found directly by using the complementary rule (Section 2.2): the configurations d^n and d^{10-n} have mutually inverted schemes of term splitting. In the case of weak fields this rule is also valid for the pairs of configurations d^n and d^{5-n}. Therefore, the splitting of the terms of the electronic configuration d^3 can be obtained from that of d^2. In particular, the ground state ^4F has the same three terms, $^4T_{1g}$, $^4T_{2g}$, and $^4A_{2g}$, as the ^3F state of the d^2 configuration, but they are arranged in inverse sequence: $\varepsilon(T_{1g}) > \varepsilon(T_{2g}) > \varepsilon(A_{2g})$, with energy spacing $\frac{3}{5}\Delta$, $-\frac{1}{5}\Delta$, and $-\frac{6}{5}\Delta$, respectively, from the nonsplit level (Fig. 4.8). The average destabilization energy in the case of d^3 is $3E_0$ instead of $2E_0$ for the d^2 configuration (in the approximation under consideration it is proportional to the number of *d* electrons), and again, the preservation rule for the center of gravity of the

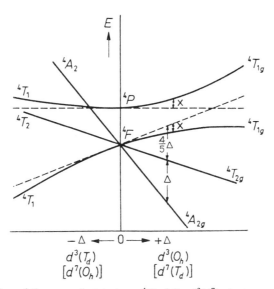

Figure 4.8. Splitting of the ground-state term ^4F of the $d^3(d^7)$ electron configuration in octahedral O_h and tetrahedral T_d ligand fields as a function of the CFT parameter Δ with the $T_{1g}(F)-T_{1g}(P)$ interaction included (x indicates the deviations of the energy terms of the same symmetry due to the nonintersection rule).

multiplet is obeyed (Section 4.2):

$$3\varepsilon(T_{1g}) + 3\varepsilon(T_{2g}) + \varepsilon(A_{2g}) = 0$$

For other configurations the scheme of term splitting can be evaluated similarly: For high-spin configurations the scheme for d^4 corresponds with the inverted one of d^1 (i.e., it is the same as for d^9); for d^5 (term ^6S) there is no splitting; the configuration d^6 corresponds to the inverted d^4, which is analogous to d^1; d^8 corresponds to d^2, and d^7 to d^3 (i.e., to inverted d^2); d^9 is similar to inverted d^1.

Strong Fields and Low- and High-Spin Complexes

In the other limit case, opposite the weak field case, the effect of the ligand field on the states of the CA is strong; it surpasses the electrostatic interaction between electrons. In this case the orbital coupling between electrons is broken and the states with a definitive total momentum quantum numbers L (S, P, D, etc., states), strictly speaking, cease to exist. In other words, each d electron chooses its orientation in space under the influence of the ligand field rather than the other d electrons. A formally similar situation takes place when the orbital coupling between the electrons is broken by the spin–orbital interaction (cf. the jj-coupling scheme in Sections 2.2 and 4.4). This is called the *strong ligand field limit*.

It follows that when the ligand field is strong, it makes no sense to speak about atomic term splitting, since the terms themselves are destroyed. To determine the states in this case, one should first determine the orientations of each of the d states in the ligand field, neglecting the electron interaction, and then evaluate the possible terms of the system taking into account the interaction of the electrons in these crystal-field-oriented electronic states.

As shown in Section 4.2, for one d electron in the octahedral field of the ligands there are two nonequivalent orbital states: the more stable t_{2g} state (d_{xy}, d_{xz}, d_{yz}), in which the electrostatic repulsion from the six ligands is smaller, and the less stable (higher in energy by Δ) state e_g ($d_{x^2-y^2}$, d_{z^2}), in which the repulsion from the ligands is larger. Hence in the strong ligand field, neglecting the electron interaction, the d electrons occupy first the t_{2g} orbitals (maximum six electrons) and then the e_g orbitals (four electrons); the electron configuration is $(t_{2g})^n$ for $n < 6$, and $(t_{2g})^6(e_g)^{n-6}$ for $n > 6$. The energy terms can be obtained from these configurations by including the electron interaction.

Consider the case of the atomic electron configuration $[A](nd)^2$. In strong ligand fields, as stated above, the two d electrons in the ground state of an octahedral complex occupy two t_{2g} orbitals (the state with two electrons in one orbital is higher in energy) forming the $(t_{2g})^2$ configuration. In the excited states one of the two d electrons can occupy the e_g orbital and form the $(t_{2g})^1(e_g)^1$ configuration, which is higher than $(t_{2g})^2$ by Δ, and the two electrons can occupy the e_g orbitals, resulting in the excited $(e_g)^2$ configuration, also higher by Δ than $(t_{2g})^1(e_g)^1$ [and by 2Δ than $(t_{2g})^2$].

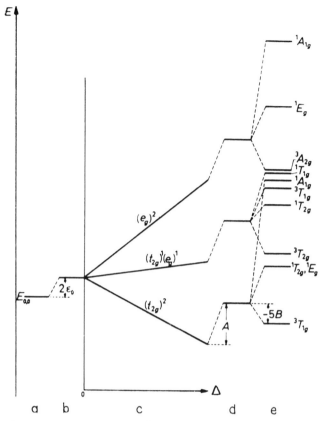

Figure 4.9. Splitting of the terms of the d^2 configuration in strong fields of octahedral symmetry: (a) d-electron energy level; (b) ligand field destabilization; (c) ligand field splitting as a function of Δ; (d) electron interaction destabilization; (e) electron interaction splitting.

Thus the d^2 configuration in the strong octahedral field forms three configurations, $(t_{2g})^2$, $(t_{2g})^1(e_g)^1$, and $(e_g)^2$, situated consecutively with an energy spacing Δ (Fig. 4.9c). In each of these configurations the electron interaction yields several terms, similar to the term formation in the free atom (Section 2.2). The method of evaluation of the energy terms is discussed below; the resulting terms for the $(t_{2g})^2$ configuration expressed by Racah parameters (2.43) are as follows:

$$\varepsilon(^3T_{1g}) = A - 5B$$
$$\varepsilon(^1T_{2g}) = A + B + 2C$$
$$\varepsilon(^1E_g) = A + B + 2C \qquad\qquad (4.47)$$
$$\varepsilon(^1A_{1g}) = A + 10B + 5C$$

Table 4.5. Electronic configuration and ground-state terms of octahedral and tetrahedral complexes in the case of strong ligand fields[a]

Number of d Electrons	Octahedral Complex		Tetrahedral Complex	
	Electronic Configurations	Ground-State Term	Electronic Configuration	Ground-State Term
d^1	t_{2g}	$^2T_{2g}$	e	^2E
d^2	$(t_{2g})^2$	$^3T_{1g}$	$(e)^2$	$^3A_2$
d^3	$(t_{2g})^3$	$^4A_{2g}$	$(e)^3$	^2E
d^4	$(t_{2g})^4$	$^3T_{1g}$	$(e)^4$	$^1A_1$
d^5	$(t_{2g})^5$	$^2T_{2g}$	$(e)^4t_2$	$^2T_2$
d^6	$(t_{2g})^6$	$^1A_{1g}$	$(e)^4(t_2)^2$	$^3T_1$
d^7	$(t_{2g})^6 e_g$	^2E_g	$(e)^4(t_2)^3$	$^4A_2$
d^8	$(t_{2g})(e_g)^2$	$^3A_{2g}$	$(e)^4(t_2)^4$	$^2T_1$
d^9	$(t_{2g})^6(e_g)^3$	^2E_g	$(e)^4(t_2)^5$	$^2T_2$
d^{10}	$(t_{2g})^6(e_g)^4$	$^1A_{1g}$	$(e)^4(t_2)^6$	$^1A_1$

[a]Compare with Tables 6.3 and 6.4.

The splitting of all the configurations above that emerge from d^2 is shown in Fig. 4.9e. In particular, the ground state of the $(t_{2g})^2$ configuration, $^3T_{1g}$, is the same as in the weak field limit. However, the sequence and spacing of the excited states are essentially different. Table 4.5 gives the electronic configurations and the ground-state terms for all the atomic configurations d^n in strong octahedral and tetrahedral ligand fields. By comparison with the corresponding cases of weak fields, one can see that differences occur for $n = 4, 5, 6, 7$ in octahedral symmetry and for $n = 3, 4, 5, 6$ in tetrahedral systems. It is important that in these cases the spin multiplicity of the ground state is always lower in the strong field limit than in the weak field. Therefore, the complexes with strong ligand fields are called *low-spin complexes*, as distinct from complexes with weak ligand fields, which are *high-spin complexes*.

One of the consequences of this result is the statement that there may be complexes of the same metal in the same oxidation state with different (two kinds of) spin multiplicity of the ground state and hence different magnetic (and other) properties which is confirmed experimentally (Section 8.4). Together with elucidation of the origin of the colors (Section 8.2), this explanation of magnetic behavior of transition metal complexes is the most important achievement of CFT. The two cases of ligand fields, weak and strong, remain equally important in MO theory and are discussed again in Section 6.2 (compare Table 4.5 with Tables 6.3 and 6.4).

The criterion of validity of the strong field limit, similar to the case of the weak field, follows from its assumptions. Since the splitting caused by the electron interactions are determined for each of the three configurations $(t_{2g})^2$, $(t_{2g})^1(e_g)^1$, and $(e_g)^2$ separately, the results are valid when the splitting is smaller than the energy gap between them Δ. For splitting of the $(t_{2g})^2$ configuration, the largest distance between its components after (4.47) is $15B + 5C$; hence the condition of validity of the strong field approach is $15B + 5C \ll \Delta$. Otherwise, the terms of the same symmetry from different configurations become strongly mixed [e.g., $^1T_{2g}$ from $(t_{2g})^2$ with $^1T_{2g}$ from $(t_{2g})^1(e_g)^1$], and it is said that there is a *configuration interaction*.

For more than two *d* electrons, the criterion of validity of the strong field approximation can be established similarly. Of special interest are the cases of d^4, d^5, d^6 and d^7 in octahedral complexes and d^3, d^4, d^5, and d^6 in tetrahedral systems, for which the two limit cases differ by the spin of the ground state. Let us introduce the notion of *pairing energy* Π, defined as the difference between the energies of multielectron interactions in the low-spin and high-spin complexes, respectively, divided by the number of pairings destroyed by the low-spin $\rightarrow$ high-spin transition. It is obvious that the low-spin state is preferable if

$$\Pi < \Delta \tag{4.48}$$

On the contrary, if

$$\Pi > \Delta \tag{4.49}$$

the high-spin state is the ground state.

The comparison of the data in Table 2.6 for the relative energies of the terms of d^n configurations with the expressions (4.47) allows one to obtain the following relations [4.15]:

$$\begin{aligned}
\Pi(d^4) &= 6B + 5C \\
\Pi(d^5) &= \tfrac{15}{2}B + 5C \\
\Pi(d^6) &= \tfrac{5}{2}B + 4C \\
\Pi(d^7) &= 4B + 4C
\end{aligned} \tag{4.50}$$

and if the Racah parameters B and C can be assumed the same in different configurations, then

$$\Pi(d^6) < \Pi(d^7) < \Pi(d^4) < \Pi(d^5) \tag{4.51}$$

Some interesting consequences follow from the relations (4.48) through (4.51). First, the pairing energy is the lowest for the d^6 configuration, and hence the low-spin state is preferable in octahedral complexes with this configuration compared with others *ceteris paribus*. Second, taking into account that for tetrahedral complexes Δ is significantly smaller than for octahedral systems, the low-spin configuration for the former is much less probable than for the latter.

Finally, by comparison it was shown [4.15] that for d^6 and d^5 configurations the state with intermediate spins ($S = 1$ for d^6 and $S = \frac{3}{2}$ for d^5) is less probable. All these conclusions have many confirmations in the experimental data (Chapters 8 to 11).

Energy Terms of Strong Field Configurations

Energy terms of strong field configurations can be formed by means of a procedure similar to that used in the formation of atomic terms (Section 2.2). Let us illustrate this by the example of the ground-state configuration $(t_{2g})^2$.

There are three t_{2g} functions of the d electron, d_{xy}, d_{xz}, and d_{yz}, which we denote here by φ_1, φ_2, and φ_3, respectively, and each of them is associated with two spin states, denoted, as above, by "+" and "−". Hence we should distribute the two electrons in six one-electron states, φ_1^+, φ_1^-, φ_2^+, φ_2^-, φ_3^+, φ_3^-, making $C_6^2 = 6 \cdot \frac{5}{2} = 15$ possibilities. Thus there are 15 determinant functions $\Phi(\varphi_i^\pm, \varphi_j^\pm)$ of the type (2.28) that have different total spin projection values: $M_s = 1, 0, -1$. Let us group the 15 functions with respect to M_s (cf. Table 2.8):
$M_s = 1$:

$$\Phi(\varphi_1^+, \varphi_2^+), \ \Phi(\varphi_1^+, \varphi_3^+), \ \Phi(\varphi_2^+, \varphi_3^+) \tag{4.52}$$

$M_s = -1$:

$$\Phi(\varphi_1^-, \ \varphi_2^-), \ \Phi(\varphi_1^-, \ \varphi_3^-), \ \Phi(\varphi_2^-, \ \varphi_3^-) \tag{4.53}$$

$M_s = 0$:

$$\Phi(\varphi_1^+, \varphi_1^-), \ \Phi(\varphi_2^+, \varphi_2^-), \ \Phi(\varphi_3^+, \varphi_3^-)$$
$$\Phi(\varphi_1^-, \varphi_1^+), \ \Phi(\varphi_2^-, \varphi_2^+), \ \Phi(\varphi_3^-, \varphi_3^+) \tag{4.54}$$
$$\Phi(\varphi_1^-, \varphi_2^+), \ \Phi(\varphi_1^-, \varphi_3^+), \ \Phi(\varphi_2^-, \varphi_3^+)$$

Since for spin triplets $S = 1$ and $M_s = 1, 0, -1$, while for singlets $S = 0$, $M_s = 0$, we come to the conclusion that among the 15 functions nine belong to triplets, while the remaining six form singlets. To find them, one can use symmetry considerations. The two one-electron t_{2g} states transform after the T_{2g} representation of the O_h group of symmetry. To determine the possible terms of the $(t_{2g})^2$ configuration, one must find the irreducible representations in the product $T_{2g} \times T_{2g}$ using the relation (3.33) and the characters of the O_h group given in Table 3.1. They are:

$$T_{2g} \times T_{2g} = T_{1g} + T_{2g} + E_g + A_{1g}$$

On the other hand, the symmetry properties of the d functions show that the three functions (4.52) with $M_s = 1$ transform as T_{1g}, the three functions (4.53) with $M_s = -1$ have the same symmetry T_{1g}, and from the functions

(4.53) with $M_s = 0$, one can also form three linear combinations that transform after the same representation T_{1g} (the method to select functions and construct linear combinations that transform after certain types of symmetry is given in Section 3.5). These nine functions form the term $^3T_{1g}$. To satisfy the remaining representations T_{2g}, E_g, and A_{1g} with the six singlet functions from (4.54), we have the only possibility: $^1T_{2g}$, ^1E_g, and $^1A_{1g}$. Thus the $(t_{2g})^2$ configuration yields the terms $^3T_{1g}$, $^1T_{2g}$, ^1E_g, and $^1A_{1g}$. Using the functions (4.52) to (4.54) and the group-theoretical formula (3.47), one can construct the functions for all these terms. They are given in Table 4.6.

The energy difference between the four terms of the d^2 configuration is caused by the corresponding differences in the interelectron interaction described by the operator $\Sigma_{i,j} e^2/r_{ij}$. The latter, in accordance with the strong field approximation, can be considered as a perturbation of the states of the $(t_{2g})^2$ configuration. Taking the 15 states (4.52) to (4.54) as a basis, we solve the secular equation of the perturbation theory to find 15 values of energy corrections ε.

On the other hand, we already know the symmetrized linear combinations of these functions given in Table 4.6, which transform after the irreducible

Table 4.6. Wavefunctions of the states of the configuration $(t_{2g})^2$ in strong ligand fields as linear combinations of the two-electron functions $\Phi_k(\varphi_i^{\pm}, \varphi_j^{\pm})$

Term	M_s	$\sum\limits_{k} C_k \Phi_k$
$^3T_{1g}$	1	$\Phi(\varphi_1^+; \varphi_2^+)$
		$\Phi(\varphi_1^+; \varphi_3^+)$
		$\Phi(\varphi_2^+; \varphi_3^+)$
	0	$(1/\sqrt{2})[\Phi(\varphi_1^+; \varphi_2^-) + \Phi(\varphi_1^-; \varphi_2^+)]$
		$(1/\sqrt{2})[\Phi(\varphi_1^+; \varphi_3^-) + \Phi(\varphi_1^-; \varphi_3^+)]$
		$(1/\sqrt{2})[\Phi(\varphi_2^+; \varphi_3^-) + \Phi(\varphi_2^-; \varphi_3^+)]$
	-1	$\Phi(\varphi_1^-; \varphi_2^-)$
		$\Phi(\varphi_1^-, \varphi_3^-)$
		$\Phi(\varphi_2^-, \varphi_3^-)$
$^1T_{2g}$	0	$(1/\sqrt{2})[\Phi(\varphi_1^+; \varphi_2^-) - \Phi(\varphi_1^-; \varphi_2^+)]$
		$(1/\sqrt{2})[\Phi(\varphi_1^+; \varphi_3^-) - \Phi(\varphi_1^-; \varphi_3^+)]$
		$(1/\sqrt{2})[\Phi(\varphi_2^+; \varphi_3^-) - \Phi(\varphi_2^-; \varphi_3^+)]$
^1E_g	0	$(1/\sqrt{2})[\Phi(\varphi_2^+; \varphi_2^-) - \Phi(\varphi_3^+; \varphi_3^-)]$
		$(1/\sqrt{6})[2\Phi(\varphi_1^+; \varphi_1^-) - \Phi(\varphi_2^+; \varphi_2^-) - \Phi(\varphi_3^+; \varphi_3^-)]$
$^1A_{1g}$	0	$(1/\sqrt{3})[\Phi(\varphi_1^+; \varphi_1^-) + \Phi(\varphi_2^+; \varphi_2^-) + \Phi(\varphi_3^+; \varphi_3^-)]$

representations of the symmetry group of the system, and hence these functions are correct zero-order functions of perturbation theory. With these functions the corrections are equal to the diagonal matrix elements of the electron interaction $\Sigma_{i,j} e^2/r_{ij}$. Methods of calculation of such matrix elements are discussed in Section 2.2. The results can be expressed by Slater–Condon or Racah parameters. In the case under consideration they yield the energies given by Eq. (4.47). For the other configurations, $(t_{2g})^1(e_g)^1$ and $(e_g)^2$ of d^2 the possible energy terms can be found quite similarly, and they result in the scheme of energy terms for the $[A](nd)^2$ states in a strong octahedral field given in Fig. 4.9.

The quantitative criterion of validity of the strong field limit coincides with the criterion of applicability of the perturbation theory: The term splitting must be much smaller than the energy gap Δ between the electronic configurations in the ligand field.

Arbitrary Ligand Fields and Tanabe–Sugano Diagrams

If the ligand field is of intermediate strength for which the criterion for neither a weak field nor a strong field is realized, the problem should be solved with the ligand field and electron interactions considered simultaneously. For each concrete system the calculations can be carried out by numerical computation. However, a general understanding (and sometimes practical results) can be obtained when starting from one of the limit cases, for which the problem can be solved analytically, with subsequent corrections on the above-mentioned term interactions, or configuration interaction.

For instance, for the electronic configuration $[A](nd)^2$, as a result of the splitting in octahedral ligand fields in the weak field limit, some of the terms of the same symmetry (originating from different atomic terms) are quite close in energy (Fig. 4.7): $^1T_{2g}(^1D)$ and $^1T_{2g}(^1G)$, $^1E_g(^1D)$ and $^1E_g(^1G)$. Other terms [e.g., $^3T_{1g}(^3F)$ and $^3T_{1g}(^3P)$, $^1A_{1g}(^1G)$ and $^1A_{1g}(^1S)$] are apparently not so close but may also interact significantly. In the strong field limit examples of close-in-energy terms are $^1A_{1g}(t_{2g})^2$ and $^1A_{1g}(e_g)^2$, $^1T_{2g}(t_{2g})^2$ and $^1T_{2g}(t_{2g})(e_g)$, and so on.

These relatively close-in-energy terms of the same symmetry influence each other; they "interact," and it can be shown that the effect of this interaction is that the interacting energy levels diverge. Therefore, we can say that *there is a repulsion of terms with the same symmetry*. Another formulation of this rule is that *the terms of the same symmetry do not intersect (nonintersection rule)*. Figure 4.8 (page 97) illustrates this effect for the $^4T_{1g}(F) - {}^4T_{1g}(P)$ repulsion in octahedral $[^4T_1(F) - {}^4T_1(P)$ in tetrahedral$] - d^3(d^7)$ complexes.

The magnitude of repulsion (divergence) of two terms of the same symmetry Γ as a result of their interaction $\Delta E(\Gamma)$ can be evaluated by perturbation theory considering the electron interaction $\Sigma e^2/r_{ij}$ and the ligand field potential V after Eq. (4.33) as perturbations. Denote the wavefunctions of the two interacting terms by Ψ_1 and Ψ_2 and the energy gap between them by 2δ. Taking the energy read off in the middle of this gap, we have for the secular equation of the perturbation theory,

$$\begin{vmatrix} -\delta & H_{12} \\ H_{21} & \delta - \varepsilon \end{vmatrix} = 0 \tag{4.55}$$

where

$$H_{12} = \int \Psi_1^* \left[\Sigma \frac{e^2}{r_{ij}} + V \right] \Psi_2 \, d\tau \tag{4.56}$$

is the term interaction energy.

The solutions of (4.55) — corrections to the energies of the interacting terms ε — are

$$\varepsilon_{1,2} = \pm (\delta^2 + H_{12}^2)^{1/2} \tag{4.57}$$

and hence

$$\Delta E = \varepsilon_2 - \varepsilon_1 - 2\delta = 2[(\delta^2 + H_{12}^2)^{1/2} - \delta] \tag{4.58}$$

We emphasize that the matrix element H_{12} after (4.56) equals zero if the two wavefunctions Ψ_1 and Ψ_2 belong to different symmetry types (see the selection rules for matrix elements, Section 3.4). For $H_{12} = 0$, $\Delta E = 0$, and therefore *the terms of different symmetry do not interact.* Certainly, if there are more than two interacting terms, the order of the secular equation (4.55) increases respectively.

With the corrections (4.58) included, the energies of the terms are no longer bound to a certain assumption of the strength of the ligand field and are independent of the reference limit case taken as a starting point. Figure 4.10 shows the correlation of the terms of the $[A](nd)^2$ configuration in octahedral fields of strong, weak, and intermediate strength.

It follows that for arbitrary strength of ligand fields the energy term splitting depends not only on the CFT parameter Δ, but also on the initial energy spacing of the atomic terms. The latter can be defined, as in Table 2.6 for the d^2 configuration, by the three Racah parameters A, B, and C. The parameter A determines the energy of destabilization by the average electron interaction, which is the same for all the terms (see Table 2.6 and Fig. 4.7) and can be excluded by an appropriate choice of the energy read off.

The parameters B and C can be obtained from the empirical spectroscopic data of free atoms and ions (Section 2.2). Table 4.7 lists these parameters for some transition metal ions (most usable in coordination chemistry) together with their ratio $\gamma = C/B$. The value of γ does not differ much for different ions (for rough approximate estimations it may be assumed that $C \approx 4B$). Assuming that γ is known, one can reduce the number of parameters determining the relative energy-level positions to two: Δ and B. Then, by choosing the scale in B units, one gets the energies as a function of only one parameter, Δ.

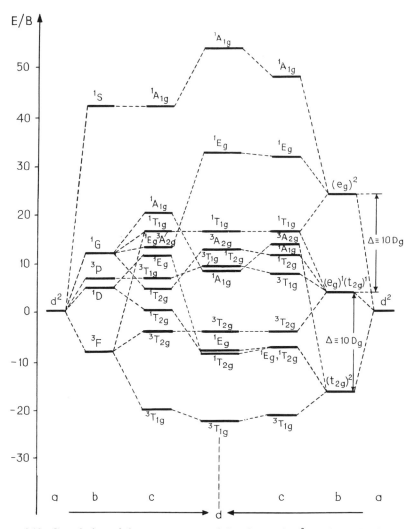

Figure 4.10. Correlation of the energy terms of the electronic d^2 configuration in weak, strong, and intermediate ligand fields of octahedral symmetry. *From left to right* (weak field): *(a)* d levels; *(b)* with interelectron interaction; *(c)* with interelectron interaction in the ligand field. *From right to left* (strong field): *(a)* d levels; *(b)* d-level terms in the ligand field; *(c)* in the ligand field with interelectron interaction. *(d)* Arbitrary field: d-level terms in ligand fields with interelectron and term interactions included. (After [4.7].)

Energy-level diagrams as functions of the CFT parameter Δ for all the d^n configurations ($n = 2, 3, 4, 5, 6, 7, 8$) given in Fig. 4.11 were constructed by Sugano and Tanabe [4.19]. In these diagrams the energy read off is taken at the ground state. Therefore, at a certain value of Δ (more precisely, Δ/B) there is a term crossing, the ground state changes, and all the energy levels on

Table 4.7. Some numerical values for the Racah parameters *B* and *C* (in cm^{-1}) and $\gamma = C/B$ for transition metal ions M^{2+} and M^{3+}

M^{2+}	*B*	*C*	γ	M^{3+}	*B*	*C*	γ
Ti^{2+}	695	2910	4.19	—	—	—	—
V^{2+}	755	3255	4.31	V^{3+}	862	3815	4.43
Cr^{2+}	810	3565	4.40	Cr^{3+}	918	4133	4.50
Mn^{2+}	860	3850	4.78	Mn^{3+}	965	4450	4.61
Fe^{2+}	917	4040	4.41	Fe^{3+}	1015	4800	4.73
Co^{2+}	971	449	4.63	Co^{3+}	1065	5120	4.81
Ni^{2+}	1030	4850	4.71	Ni^{3+}	1115	5450	4.89

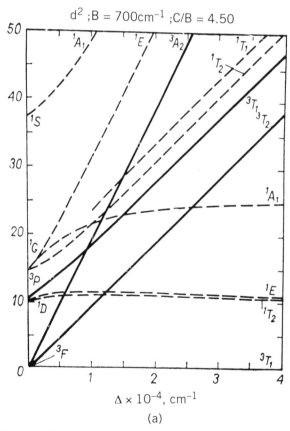

(a)

Figure 4.11. Tanabe–Sugano diagrams: energy levels in 10^3 cm^{-1} as a function of octahedral crystal fields Δ. For convenience, the energy levels with multiplicities different from the ground-state levels are given by dashed lines. The indicies *g* and *u* are omitted. Some levels of minor significance are not shown.

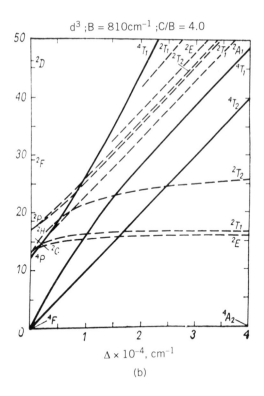

d^3 ; $B = 810\text{cm}^{-1}$; $C/B = 4.0$

(b)

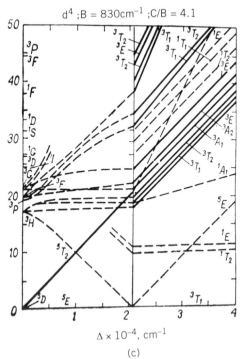

d^4 ; $B = 830\text{cm}^{-1}$; $C/B = 4.1$

(c)

Figure 4.11. (*Continued*).

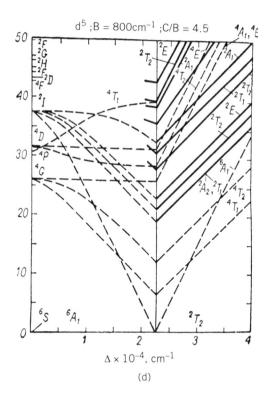

d^5 ;B = 800cm^{-1} ;C/B = 4.5

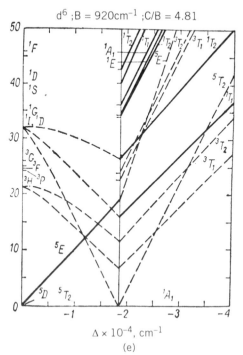

d^6 ;B = 920cm^{-1} ;C/B = 4.81

Figure 4.11. (*Continued*).

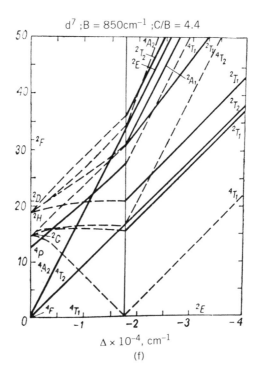

d^7 ;B = 850cm^{-1} ;C/B = 4.4

$\Delta \times 10^{-4}$, cm^{-1}

(f)

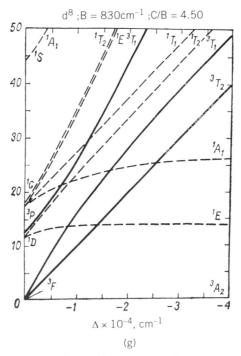

d^8 ;B = 830cm^{-1} ;C/B = 4.50

$\Delta \times 10^{-4}$, cm^{-1}

(g)

Figure 4.11. (*Continued*).

the diagram are subject to a break. Usually at this break the ground-state multiplicity changes, too, and there is a transition from the weak ligand field to the strong field. The Tanabe–Sugano diagrams give the most complete information about the electronic structure of the system in the CFT model. For improvements in these diagrams taking into account the spin–orbital interaction, see, for instance, [4.20].

4.4. *f*-ELECTRON TERM SPLITTING

One of the important features of *f* electrons is that they are usually screened from the ligand field by the outer *s*, *p*, *d* electrons and hence are less affected by the ligands than are the *d* electrons. On the other hand, the *f* electrons are subject to stronger spin–orbital coupling [Eq. (2.27)]. Hence distinct from compounds with *d* electrons, where the spin–orbital coupling is assumed to be smaller than both the ligand field and interelectron interactions, in *f*-electron systems the ligand field is smaller than both the interelectron and spin–orbital interactions.

In fact, following Bethe [4.2], we should compare three important magnitudes: electron interaction, ligand field potential, and spin–orbital interaction. With these three magnitudes, three cases are significant in the CFT:

1. *Weak field.* The ligand potential is weaker than both interelectron and spin–orbital interactions
2. *Intermediate field.* The ligand field is smaller than the interelectron interaction but larger than the spin–orbital coupling
3. *Strong field.* The ligand field potential is stronger than both the electron and spin–orbital interactions.

For *d* electrons case 1 is not important, and therefore it is usually ignored; the remaining two cases, 2 and 3, are called weak and strong field cases, respectively.

For *f* electrons in weak fields the atomic terms should be characterized (in addition to *L* and *S*) by the quantum number $J = L + S - 1, \ldots, |L - S|$, which takes into account the spin–orbital interaction (Section 2.2). Since the total spin *S* can be semi-integer, *J* may also be semi-integer. For instance, the states of one *f* electron with $L = 3$, $S = \frac{1}{2}$, and with the spin–orbital interaction included ($J = \frac{7}{2}, \frac{5}{2}$) are $^2F_{7/2}$ and $^2F_{5/2}$ (the *J* value is indicated as a subscript). The weak field approximation here means that the splitting of each of these terms by the ligand field can be considered separately.

The visual interpretation of the charge distribution in the states with the total momentum quantum number *J* and the splitting of these states in ligand fields of different symmetries is not as straightforward as in the case of *d* electrons, where the spin–orbital interaction can be neglected. Qualitatively, the splitting of f states can be obtained easily by means of the group-theoretical rules (Sections 3.4 and 4.2), while quantitative calculations can be performed by perturbation theory.

However, some grade of understanding of the situation can be reached by using the model of pure orbital states, that is, neglecting the spin–orbital coupling. In particular, considering the angular distributions of atomic f-electron functions from the cubic set, given in Table 2.2 and Fig. 2.4, and the corresponding electrostatic repulsion of the electron in these states from six point charges (or dipoles) of an octahedral complex, one can conclude that in the three states f_{x^3}, f_{y^3}, and f_{z^3}, the repulsion is the greatest (and equal for all of them). In three other states, $f_{x(y^2-z^2)}$, $f_{y(z^2-x^2)}$, $f_{z(x^2-y^2)}$, it is also equal but smaller than in the previous three states, and in the f_{xyz} state it is the smallest (in tetrahedral systems, analogously to d-electron states, this picture is inverted).

Thus the sevenfold orbitally degenerate F term of the free atom (ion) with one f electron is split by the octahedral ligand field into three terms, of which one is nondegenerate and two are threefold degenerate. It can be shown that the symmetries of these states are A_{2u}, T_{2u}, and T_{1u}, respectively (Fig. 4.12). Hence the splitting $F \rightarrow A_{2u} + T_{2u} + T_{1u}$ is similar to that obtained above for the F term of the d^2 configuration in octahedral fields (Figs. 4.7 and 4.8), with an inverted ordering and opposite parity of the terms u instead of g; f states, unlike d states, are odd with respect to reflections.

By way of example we also show the scheme of splitting of the energy levels of an f electron in the field of a hexagonal biprism with the sixfold axis along

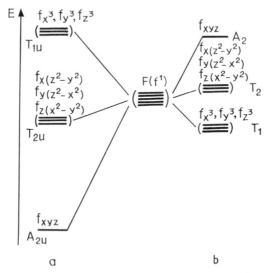

Figure 4.12. Splitting of the atomic energy levels of one f electron in octahedral (a) and tetrahedral (b) ligand fields (the spin–orbital interaction is neglected).

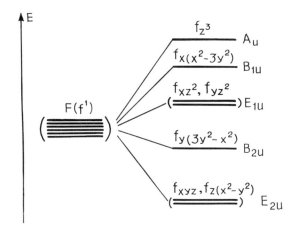

Figure 4.13. Splitting of one ƒ-electron energy levels in the field of a hexagonal biprism.

z (Fig. 4.13) which is realized, for instance, in uranyl complexes [4.21]. Here the low-symmetry set of one-electron atomic angular f functions of Table 2.2 is used. Splitting of the orbital states in other cases in ligand fields of different symmetries can be revealed using Table 4.2.

Quantitative calculations of the energy term splitting of f electrons can be carried out similarly to that of d electrons, considered above. To do this, secular equation (4.6) of the seventh order (for one f electron) must be solved. For the matrix elements $V_{mm'}$ one can obtain expressions of the type (4.9) for arbitrary positions of the ligands [4.22]. They are given in Appendix 5.

However, as mentioned at the beginning of this section, in the case of f electrons the spin–orbital interaction can be stronger than the ligand field, and hence the latter should be considered as a perturbation to the atomic states classified by the quantum number J of the total momentum $\mathbf{J} = \mathbf{L} + \mathbf{S}$. Qualitatively, the splitting of the J term can be obtained by group-theoretical rules given in Section 3.4. If J is an integer, the splitting coincides completely with that expected for the appropriate L value. For semi-integer J values the double groups of symmetry discussed in Section 3.6 should be employed, the terms being classified by their irreducible representations E_1', E_2' (twofold degenerate), G' (fourfold), or, respectively, Γ_6, Γ_7, Γ_8 in Bethe's notations. The splitting is evaluated, as usual, by means of Eq. (3.33), with the characters given in the corresponding tables of Appendix 1. Some results for most usable cases are given in Table 4.8.

For quantitative calculations the ligand field, spin–orbital interaction, and electron interaction should be included in the perturbation simultaneously. Such calculations are carried out numerically (see, e.g., [4.17]). Similar to d electrons, the method of equivalent operators and irreducible tensor operators may be useful for this purpose [4.15–4.18].

Table 4.8. Splitting of atomic terms with semi-integer J values in ligand fields of different symmetries

J	Cubic Symmetry O'	Tetragonal Symmetry D'_4	Hexagonal Symmetry D'_6
$\frac{1}{2}$	E'_1	E'_1	E'_2
$\frac{3}{2}$	G'	$E'_1 + E'_2$	$E'_2 + E'_3$
$\frac{5}{2}$	$E'_2 + G'$	$E'_1 + 2E'_2$	$E'_1 + E'_2 + E'_3$
$\frac{7}{2}$	$E'_1 + E'_2 + G'$	$2E'_1 + 2E'_2$	$E'_1 + 2E'_2 + E'_3$
$\frac{9}{2}$	$E'_1 + 2G'$	$3E'_1 + 2E'_2$	$E'_1 + 2E'_2 + 2E'_3$
$\frac{11}{2}$	$E'_1 + E'_2 + 2G'$	$3E'_1 + 3E'_2$	$2E'_1 + 2E'_2 + 2E'_3$

4.5. CRYSTAL FIELD PARAMETERS AND EXTRASTABILIZATION ENERGY

The most valuable results of CFT are based on its semiempirical versions that yield general qualitative and semiquantitative conclusions. In these versions the electronic structure of the system is described by means of one (or several) parameters, which can be obtained from independent experimental data or from comparison of the theory with the experiments (see Section 4.6). Therefore, the meaning of the CFT parameters is of great importance.

The main parameters of the d^n-term splitting in cubic fields (cube, octahedron, tetrahedron) is the energy gap Δ between the e_g and t_{2g} one-electron states, which is often denoted as $10D_q$. According to Eqs. (4.15) and (4.30),

$$\Delta = \tfrac{5}{3}eqF_4(R) \tag{4.59}$$

for six ligand–point charges q at the corners of a regular octahedron, with R as the interatomic CA–ligand distance, and

$$\Delta = -\tfrac{5}{3}e\mu F'_4(R) \tag{4.60}$$

for the similar case of ligand–dipole moments μ. For a tetrahedron $\Delta_T = -\tfrac{4}{9}\Delta$, while for a cube $\Delta_c = 2\Delta_T = -\tfrac{8}{9}\Delta$ [Eqs. (4.20) and (4.22)].

Direct calculation of Δ after Eqs. (4.59) and (4.60) has not sufficient credibility because of the rough assumptions the CFT is based upon. However, some rules in the relative changes of Δ in a series of similar compounds can be carried out quite satisfactorily from these relations. According to Eqs. (4.59) and (4.60), Δ depends on three parameters. Two of them, q (or μ) and R, characterize ligand charge and position, respectively, while the third, α, is the effective parameter of the CA radial nd function in $F_k(R)$ functions after (4.10), measuring the strength of the coupling of the d electron. In the same group of transition metals (e.g., in the iron group) the α values for different metals in the same oxidation state are close (the ionization potentials determining approximately the α value differ in this series by no more than 10 to 15%). Hence for such a series Δ depends mostly on the ligand field. Therefore, Δ is called the *ligand field parameter*.

When passing from the elements of the first transition group to the second and third groups, the α value decreases significantly and hence the parameter Δ increases. Indeed, following (A2.14), $F_4(R)$ decreases with increasing α (for actual values of interatomic distances R, the derivative $F_4'(x) < 0$, $x = \alpha R$). The data on absorption spectra confirm these conclusions (Section 8.2). For instance, for hydrated transition metal ions of the first transition group, Δ varies in the limits 7500 to 12,500 cm^{-1} and 13,500 to 21,000 cm^{-1} for oxidation states $+2$ and $+3$, respectively, while for the second and third transition groups Δ is 30 to 70% larger. Different ligands can be arranged in a *spectrochemical series* with respect to the splitting Δ they produce in a complex of the same metal (discussed in Section 8.2).

Another parameter of the splitting that may be useful in practice is the *energy of extrastabilization δ*. It equals the energy difference between the initial nonsplit (but destabilized by the average ligand repulsion) term and the new ground term after the splitting. The term *stabilization* for this magnitude may be misleading because the electron–ligand interaction is a repulsion (destabilization). In fact, the new ground state is just less destabilized than the electronic states in average, and it is implied that "less destabilized" can be considered as an additional stabilization [to that produced by the main attractive term (4.3) minus the average destabilization energy E_0 in Eq. (4.17)].

The δ value can easily be obtained from the expression for Δ and the rule of preservation of the center of gravity (Section 4.2). For the electronic configuration $[A](nd)^1$ in an octahedral field $\delta = \frac{2}{5}\Delta$ (Fig. 4.3), while in the tetrahedral fields $\delta_T = \frac{3}{5}\Delta_T = \frac{4}{15}\Delta$, and in the eight-coordinated cube $\delta_c = 2\delta_T = \frac{8}{15}\Delta$ [see Fig. 4.4 and Eqs. (4.20) and (4.22)]. Based on these relations and taking into account the electron occupation of different orbitals in the weak and strong field limits, the extrastabilization energies for different electron configurations d^n are given in Table 4.9.

One of the features of the extrastabilization energy that emerges from this table is its nonmonotonous change with the number of d electrons. In particular, for high-spin complexes δ is zero for d^0, d^5, and d^{10} and has two maxima at d^3 and d^8. This "two-humped" behavior is confirmed qualitatively by many experimental data. By way of an illustrative example, the *heat of formation* of bivalent and trivalent transition metal aqua-complexes as a function of d^n is shown in Fig. 4.14. It is seen that, indeed, ΔH has two maxima and three minima as predicted by the CFT (note that the extrastabilization energy is to be added to the main bonding energy, which is presented in Fig. 4.14 by a solid line).

However, in Table 4.9 the changes in electron interactions by passing from one configuration to another are ignored. Meanwhile, they may be significantly different. For instance, for the d^2 configuration in the limit of a weak field,

Table 4.9. Energy of extrastabilization δ in ligand fields of different symmetries in units of the crystal field parameter Δ^a

Electronic Configuration	Examples of Ions	High Spin			Low Spin		
		Octahedron	Tetrahedron	Cube	Octahedron	Tetrahedron	Cube
d^0	Sc^{2+}	0	0	0	0	0	0
d^1	Ti^{3+}	$\frac{2}{5}$	$\frac{4}{15}$	$\frac{8}{15}$	$\frac{2}{5}$	$\frac{4}{15}$	$\frac{8}{15}$
d^2	V^{3+}	$\frac{4}{5}$	$\frac{8}{15}$	$\frac{16}{15}$	$\frac{4}{5}$	$\frac{8}{15}$	$\frac{16}{15}$
d^3	Cr^{3+}	$\frac{6}{5}$	$\frac{16}{45}$	$\frac{32}{45}$	$\frac{6}{5}$	$\frac{4}{5}$	$\frac{8}{5}$
d^4	Mn^{3+}	$\frac{3}{5}$	$\frac{8}{45}$	$\frac{16}{45}$	$\frac{8}{5}$	$\frac{16}{15}$	$\frac{32}{15}$
d^5	Mn^{2+}, Fe^{3+}	0	0	0	2	$\frac{8}{9}$	$\frac{16}{9}$
d^6	Fe^{2+}, Co^{3+}	$\frac{2}{5}$	$\frac{4}{15}$	$\frac{8}{15}$	$\frac{12}{5}$	$\frac{32}{45}$	$\frac{64}{45}$
d^7	Co^{2+}	$\frac{4}{5}$	$\frac{8}{15}$	$\frac{16}{15}$	$\frac{9}{5}$	$\frac{8}{15}$	$\frac{16}{15}$
d^8	Ni^{2+}	$\frac{6}{5}$	$\frac{16}{45}$	$\frac{32}{45}$	$\frac{6}{5}$	$\frac{16}{45}$	$\frac{32}{45}$
d^9	Cu^{2+}	$\frac{3}{5}$	$\frac{8}{45}$	$\frac{16}{45}$	$\frac{3}{5}$	$\frac{8}{45}$	$\frac{16}{45}$
d^{10}	Zn^{2+}	0	0	0	0	0	0

^{a}For d^n with $n > 1$ the interelectron interaction is neglected.

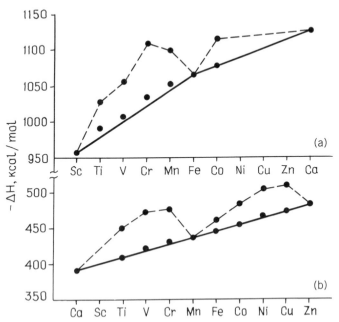

Figure 4.14. "Two-humped" dependence of the heat of formation ΔH of aqua-complexes of divalent (a) and trivalent (b) transition metals. The solid line links the value of ΔH without extrastabilization.

Table 4.10. Relative values of d orbital energies (in Δ units) in crystal fields of different symmetries[a]

Coordination Number	Mode of Coordination (Symmetry)	$d_{x^2-y^2}$	d_{z^2}	d_{xy}	d_{xz}	d_{yz}
1	—	−0.314	0.514	−0.314	0.057	0.057
2	Linear	−0.628	1.028	−0.628	0.114	0.114
3	Trigonal	0.546	−0.321	0.546	−0.386	−0.386
4	Tetrahedral	−0.267	−0.267	0.178	0.178	0.178
	Square-planar	1.228	−0.428	0.228	−0.514	−0.514
5	Square pyramid	0.914	0.086	−0.086	−0.457	−0.457
	Trigonal pyramid	−0.082	0.707	−0.082	−0.272	−0.272
6	Octahedral	0.600	0.600	−0.400	−0.400	−0.400
7	Pentagonal bipyramid	0.282	0.493	0.282	−0.528	−0.528

[a] z is the axis of the highest symmetry.

according to the data in Table 2.6, the energy of interaction between the two electrons in the ground state $^3T_{1g}(^3F)$ (in the $^3T_{1g}$ term originating from the atomic ^3F term) equals $A − 8B$, whereas in the strong field case in the ground state $^3T_{1g}(t_{2g})^2$ [i.e., for the same term $^3T_{1g}$ originating from the strong field configuration $(t_{2g})^2$] it equals $A − 5B$ [see Eqs. (4.47) and Fig. 4.9]. The destabilization energies E_0 are different for different numbers of electrons and ligands, too. These circumstances complicate the quantitative interpretation of the experimental data on ΔH in the CFT.

If the symmetry of the ligand field is lower than cubic, degenerate terms are subject to further splitting; its characterization requires additional parameters. For tetragonal distorted octahedra the additional parameters are D_s and D_t, given, respectively, by Eqs. (4.24) and (4.25) as functions of $F_2(R)$ and $F_4(R)$. A rough simplification, $F_2 \approx 2F_4$, , makes it possible to present approximately all the energy d levels as a function of one parameter Δ. They are given in Table 4.10 for different types of ligand coordination to the CA. The data of this table can be used for qualitative estimations only. For more details on CFT parameters, see [4.23].

4.6. LIMITS OF APPLICABILITY OF CFT

The limits of application of the CFT, as of any other approximate theory, are determined by its main assumptions (postulates) as well as by the additional simplifications introduced for its realization. Usually, the limits of any theory can be established definitively when there is a more general theory for which the one under consideration can be considered as a particular case. In this sense the possibilities and limitations of the CFT are better understood when compared with the conclusions of the (in general) wider MO LCAO theory (Section 5.6).

As indicated above, the assumption of the pure electrostatic nature of the CA–ligand interaction with the ligands as point charges or dipoles a priori excludes the possibility of investigating, by means of the CFT, such important problems as the nature of chemical binding and charge distribution in coordination compounds as well as the phenomena that depend on the details of the electronic structure of the ligands (ligand activation, charge transfer spectra, reactivity, etc.). The properties of coordination compounds that can be analyzed by means of the CFT are limited by those originating from the electronic structure of the central atom influenced by the ligands. This excludes such important problems as complex formation (e.g., π bonding), stereochemistry, reactivity, ligand activation, hyperfine ESR spectra, and others.

Even with these strong limitations the range of CFT remains sufficiently wide. Indeed, it includes the origin of color (electronic absorption spectra in the visible and related regions; Section 8.2), magnetic susceptibility and ESR spectra (without ligand hyperfine structure; Section 8.4), and relative stabilities in solutions, some vibronic interaction effects (Chapter 7), and the number of such problems can be significantly enlarged by including covalence corrections (Section 5.2).

When analyzing the applicability of the CFT, one should distinguish between its qualitative and quantitative aspects. The qualitative aspect of the theory (the analysis of term splitting, symmetry of states, relative energies, spin multiplicities, etc.) covers a much wider spectrum of problems and systems than its quantitative treatment, which requires more accurate energy spectra and wavefunctions. Indeed, the qualitative conclusions of the CFT are based primarily on the symmetry properties of the system, which are independent of the nature of the bonding (and remain the same in all the theories), whereas the quantitative results are bound to the approximation of the electrostatic ligand fields, which is a priori invalid for many systems. Obviously, the qualitative conclusions of the CFT not only have larger limits of application, but are more reliable.

The assumption of the CFT that the ligands are point charges or dipoles seems, at first sight, to be rather rough. Nevertheless, the results obtained in this model, within the limits of its applicability, may be quite reasonable, due primarily to the compensation of errors with opposite signs (Section 5.6). However, the semiempirical versions of the CFT in which the main parameters are obtained from empirical data seem to be more useful.

As mentioned in Section 4.4, f electrons in rare earth complexes are strongly screened from the ligand field by the outer s, p, d electrons. For this reason the covalence with f electron participation is rather weak, and the study of f electron states and related properties by the CFT is quite acceptable also for quantitative calculations (see the case of weak covalence in Section 5.2). This conclusion is confirmed by experimental data [4.17, 4.24, 4.25].

Finally, no less important is the fact that the CFT provided (and provides now) a simple introduction to (a tool for a better understanding of) more sophisticated theories discussed in the next chapters. The relative simplicity and stimulating power of the CFT allowed one of the most significant contributors to this theory, Professor Moffitt, to state [4.26]: "It will be a long time before a method is developed to surpass in simplicity, elegance and power that of crystal field theory. Within its extensive domain it has provided at very least a deep qualitative insight into the behavior of a many-electron system. No other molecular theory, to our knowledge, has provided so many useful numbers which are so nearby correct. And none has a better immediate prospect of extending its chemical applications."

Despite the fact that almost 40 years have passed since this statement was made, only the last two sentences require revision, because of the impressive achievements of MO theories in the last decades.

REFERENCES

4.1. J. H. Van Vleck, in: *22nd International Congress of Pure and Applied Chemistry: Plenary Lectures*, Butterworth, London 1970, p. 235.

4.2. H. Bethe, *Ann. Phys.*, **3**, 133–208 (1928).

4.3. J. H. Van Vleck, *The Theory of Electronic and Magnetic Susceptibilities*, Oxford University Press, London, 1965, 384 pp.

4.4. H. Hartmann, *Theorie der chemischen Binding auf quantum-theoretische Grundlage*, Springer-Verlag, Berlin, 1954, 357 pp.

4.5. L. E. Orgel, *An Introduction to Transition-Metal Chemistry: Ligand Field Theory*, Wiley, New York, 1960.

4.6. C. J. Ballhausen, *Introduction to Ligand Field Theory*, McGraw-Hill, New York, 1962, 298 pp.

4.7. H. L. Schlafer and G. Gliemann, *Basic Principles of Ligand Field Theory*, Wiley, New York, 1969.

4.8. C. K. Jorgensen, *Modern Aspects of Ligand Field Theory*, North-Holland, Amsterdam, 1971, 538 pp.

4.9. I. B. Bersuker, *Electronnoe Stroenie i Svoistva Koordinatsionnykh Soedinenii* (Russ.), Khimia, Leningrad, 2nd ed., 1976; 3rd ed., 1986; 350 pp.

4.10. W. Kossel, *Ann. Phys. (Leipzig)*, **44**, 229 (1916).

4.11. A. Magnus, *Z. Anorg. Chem.*, **124**, 288 (1922).

4.12. H. H. Schmidtke, in: *Physical Methods in Advanced Inorganic Chemistry*, ed. H. A. O. Hill and P. Day, Interscience, London, 1968, pp. 107–166.

4.13. A. L. Companion and M. A. Komarynsky, *J. Chem. Educ.*, **41**, 257–262 (1964).

4.14. M. Randic and Z. Maksic, *Theor. Chim. Acta*, **4**, 145–149 (1966).

4.15. J. S. Griffith, *The Theory of Transition Metal Ions*, University Press, Cambridge, 1962, 455 pp.

4.16. S. E. Harmung and C. E. Schafer, *Struct. and Bonding*, **12**, 201–255, 257–295 (1972).

4.17. B. G. Wyborne, *Spectroscopic Properties of Rare Earth*, Interscience, New York, 1965, 231 pp.

4.18. C. W. Nielson and G. F. Koster, *Spectroscopic Coefficients for the p^n, d^n, and f^n Configurations*, MIT Press, Cambridge, Mass., 1963, 275 pp.

4.19. S. Sugano and Y. Tanabe, *Multiplets of Transition Metal Ions in Crystals*, Academic Press, New York, 1970, 331 pp.

4.20 A. D. Liehr and C. J. Ballhausen, *Ann. Phys.* (USA), **6**, 134–155 (1959); A. D. Liehr, *J. Phys. Chem.*, **67**, 1314–1328 (1963).

4.21. C. A. Coulson and G. R. Lester, *J. Chem. Soc.*, 3650–3659 (1956).

4.22. Z. Maksic and M. Randic, *Theor. Chim. Acta*, **7**, 253–255 (1967).

4.23. M. Gerloch and R. S. Slade, *Ligand Field Parameters*, University Press, Cambridge, 1973, 235 pp.

4.24. M. A. El'ashevich, *Spectry Redkich Zemel'* (Russ.), Gostechteorizdat, Moscow, 1953, 446 pp.

4.25. K. B. Yatsimirski, N. A. Kostromina, and Z. A. Sheka, *Khimia Kompleksnych Soedinenii Redkozemelnykh Elementov*, Naukova Dumka, Kiev, 1966, 493 pp.

4.26. W. Moffitt and C. J. Ballhausen, *Annu. Rev. Phys. Chem.*, **7**, 107–136 (1956).

5 Method of Molecular Orbitals and Related Approaches

Molecular orbitals form the main language of discussion of chemical problems at the electronic level.

Historically, application of the molecular orbital (MO) method to transition metal and rare earth complexes began with improving crystal field theory by including covalency effects, in which case it was called *ligand field theory*. However, in its present form the MO method applied to coordination compounds does not differ basically from that widely used for organic and main group systems, although practically, treatment of coordination systems with this method is much more complicated.

In this chapter the main ideas and special features of the MO approach are presented in a form applicable to transition metal coordination compounds. The general presentation of the MO method is discussed together with methods of numerical calculations (*ab initio*, nonempirical, semiquantitative, and semiempirical), as well as related approaches, including density functional calculations. The application of these methods to the solution of the main problem of coordination chemistry—the origin of chemical bonding—is described in Chapter 6.

5.1. BASIC IDEAS OF THE MO LCAO METHOD

Main Assumptions

The basic idea behind the MO method, as compared with crystal field theory (CFT) discussed in Chapter 4, is to drop the main restricting assumption of CFT that the electronic structure of the ligands can be ignored, and include explicitly all electrons in the quantum-mechanical treatment of the molecular system. The MO approach makes no a priori assumptions about the nature of chemical bonding. Unlike CFT, where atoms or groups of atoms of the complex are assumed to preserve mainly their individual features, in the MO method the coordination system is considered, in principle, as an integral system in which separate atoms lose their individuality.

For example, the complex $Co(NH_3)_6^{3+}$ is considered in MO theory as having a skeleton of six nuclei of nitrogen, 18 nuclei of hydrogen, one nucleus of cobalt, and 84 electrons which move in the field of the nuclei. The motion of each electron is determined by both the nuclear configuration (provided that the latter can be assumed to be fixed; Section 7.1) and the motions of the other 83 electrons. All these motions (including the nuclear motions) are determined by the Schrödinger equation (Section 1.1).

Exact solution of the Schrödinger equation (1.5) for a coordination system is hardly possible at present because of computation difficulties. So far, the only practically acceptable approaches are those based on the adiabatic approximation discussed in Section 7.1, and the one-electron approximation, which assumes that each electron can be considered as moving independently in the mean field created by the nuclei and the remaining electrons (Section 2.2). In the one-electron approximation, the coordination system is described by one-electron states that in general are extended over all the system; they are called *molecular orbitals* (MOs). The MO method was suggested by Hund and Mulliken [5.1]; applications to transition metal coordination compounds were developed by Van Vleck, Orgel, Griffith, Ballhausen, and others (see [5.2–5.6]).

In general, evaluation of one-electron MOs is still a complicated problem. Its solution requires further simplifications, the main one being the LCAO approximation, in which the wavefunction of the MO is presented in the form of a *linear combination of atomic orbitals* (LCAO):

$$\Psi = c_1\psi_1 + c_2\psi_2 + \cdots + c_n\psi_n = \sum_i c_i\psi_i \qquad (5.1)$$

where n is the number of atomic orbitals (note that each atom can be presented by more than one orbital), and ψ_i is the ith one-electron atomic wavefunction. The set of the ψ_i's forms the *LCAO basis*. In fact, the LCAO approximation means that we assume that each MO electron can be found at each of the atoms of the system with a probability determined by the $|c_i|^2$ value; when near the given atom, the MO electron moves as a normal atomic valence electron.

For coordination compounds with heavy atoms further simplifications may be needed. In particular, the inner electrons of atomic closed shells are often considered as localized near the nucleus and hence as not participating directly in the bonding with other atoms, so only valence electrons remain in the MO LCAO treatment. However, separation of inner-core electrons from valence electrons is not straightforward and should be carried out with care, especially in *ab initio* calculations (see Section 5.4).

Secular Equation

Following the variational principle, the coefficients c_i in the probe function (5.1) should satisfy the condition of a minimum total energy value for the system. Actually, the calculation of LCAO coefficients is one of the most important parts of the MO LCAO method and is discussed in more detail in Section 5.3. Here we present some simple analytic relations which, nevertheless, have a basic meaning for all versions of the method under consideration.

In the Hartree–Fock method (Section 2.2), each electron moves independently in an effective self-consistent field of the nuclei and other electrons, with the effective Hamiltonian H given by Eq. (2.44). With the latter the energy of the MO electron (5.1) is

$$E = \frac{\int \Psi^* H \Psi \, d\tau}{\int \Psi^* \Psi \, d\tau} \tag{5.2}$$

Substituting for Ψ its expression (5.1) and introducing the notation

$$S_{ik} = \int \psi_i^* \psi_k \, d\tau \tag{5.3}$$

$$H_{ik} = \int \psi_i^* H \psi_k \, d\tau \tag{5.4}$$

we can easily obtain

$$E \sum_{i,k} c_i^* c_k S_{ik} = \sum_{i,k} c_i^* c_k H_{ik} \tag{5.5}$$

As mentioned earlier (Section 2.1), S_{ik} is the overlap integral between the atomic functions i and k ($S_{ii} = 1$ due to the condition of normalization). The matrix elements (5.4) are discussed in Section 5.3; for $i \neq k$, H_{ik} is called the *resonance integral* and H_{ii} is the *Coulomb integral*. The condition of the energy minimum with respect to the c_i values means that the corresponding first derivative must be zero:

$$\frac{\partial E}{\partial c_i} = 0 \qquad i = 1, 2, \ldots, n \tag{5.6}$$

It yields the following equations with respect to the c_i's:

$$\sum_k c_k (H_{ik} - E S_{ik}) = 0 \qquad i = 1, 2, \ldots, n \tag{5.7}$$

This algebraic system of equations is linear and uniform. It yields nonzero solutions provided that its determinant is zero,

$$\| H_{ik} - E S_{ik} \| = 0$$

or in extended form,

$$
\begin{vmatrix}
H_{11} - E & H_{12} - ES_{12} & \cdots & H_{1n} - ES_{1n} \\
H_{21} - ES_{21} & H_{22} - E & \cdots & H_{2n} - ES_{2n} \\
\vdots & \vdots & \vdots\vdots & \vdots \\
H_{n1} - ES_{n1} & H_{n2} - ES_{n2} & \cdots & H_{nn} - E
\end{vmatrix} = 0
\qquad (5.8)
$$

Condition (5.8) is in fact an algebraic equation of the nth power with respect to the unknown E. Such equations are usually called *secular equations*. In general, (5.8) gives n different solutions E_i; for each of them a set of c_{ik} values, $c_{i1}, c_{i2}, \ldots, c_{in}$, can be obtained from Eqs. (5.7). The uniform system (5.7) yields only $n - 1$ constants c_{ik}, one more being obtained from the condition of normalization, $\int \Psi^* \Psi \, d\tau = 1$.

Thus assuming that the one-electron wavefunctions have the form (5.1), that is, they are linear combinations of n atomic functions, we get n MO energies E_i, $i = 1, 2, \ldots, n$, and n sets of LCAO coefficients for each MO. Equations (5.7) and (5.8) form the basis of all versions of the MO LCAO method.

Classification by Symmetry

One of the special features of coordination compounds concerning use of the MO LCAO method is the importance of symmetry considerations. The latter usually reduce significantly the calculation difficulties. As compared with organic and some main group inorganic compounds, transition metal systems acquire high-symmetry configurations much more often. Provided that one knows the symmetry group of the system (Section 3.2), the possible MOs can be a priori divided on symmetries, each of the MOs being attributed to a certain irreducible representation. Then, using the results of Section 3.5, one can construct the appropriate LCAO of the CA and ligand atomic functions that satisfy the required symmetry properties. With the known symmetry of the MOs, some of the matrix elements of the secular equation (5.8) vanish, thus reducing essentially the order of this equation n.

Let us first specify some notation. Figure 5.1 illustrates the choice of the general and local (ligand) coordinate systems, ligand numeration, and the orientation of their σ and π orbitals in octahedral and tetrahedral complexes. The z axes of the ligand coordinates are directed toward the central atom (for simplicity of overlap integral calculations), while the other axes are arbitrary. The σ orbitals have axial symmetry with respect to the ligand z axis, while the π orbitals lie in the plane that is perpendicular to this axis, and they are oriented along the local x or y axes. The ligand orbitals are also labeled with the number of their ligand. For instance, π_{2x} means the π orbital of the second ligand oriented along the x axis of the local coordinate system.

As shown in Section 3.4, the symmetry of the system determines directly the classification of the MOs on the irreducible representations of the correspond-

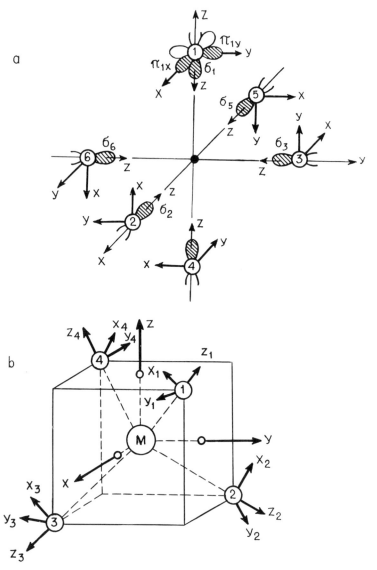

Figure 5.1. General and ligand local coordinate systems for octahedral (*a*) and tetrahedral (*b*) complexes. In the octahedral case the orientation of the π_{1x}, π_{1y}, and σ_n ($n = 1, 2, \ldots, 6$) ligand orbitals is also demonstrated.

ing point group. For example, for an octahedral complex with six identical ligands that has O_h symmetry, the following types of MOs are possible (in parentheses, the degeneracy is indicated): $A_{1g}(1)$, $A_{1u}(1)$, $A_{2g}(1)$, $A_{2u}(1)$, $E_g(2)$, $E_u(2)$, $T_{1g}(3)$, $T_{1u}(3)$, $T_{2g}(3)$, $T_{2u}(3)$ (Table 3.1). These symmetries restrict significantly the number of independent functions that describe the term under consideration (or the MO): It should be equal to the degeneracy (the number in parentheses), and these functions should transform into each other under the symmetry transformations of the O_h group (Section 3.4).

Denote the wavefunction of the MO of appropriate symmetry of a transition metal coordination system by

$$\Psi = a\psi_0 + b\Phi \tag{5.9}$$

where ψ_0 is the atomic orbital of the CA or a linear combination of such orbitals [if there are two or several central atoms, the form (5.9) and the consequent results including Table 5.1, discussed below, should be modified], and Φ is a molecular orbital, a linear combination of the ligand atomic functions ψ_i:

$$\Phi = c_1\psi_1 + c_2\psi_2 + \cdots + c_n\psi_n \tag{5.10}$$

and the LCAO coefficients $a, b, c_1, \ldots, c_n$ should be determined by calculations.

Since the MO (5.9) must belong to one of the irreducible representations of the symmetry group of the system, both ψ_0 and Φ should belong to the same representation, they must have the same symmetry properties. Such symmetrized linear combinations of atomic orbitals that transform after a given type of symmetry (irreducible representation) of the group of symmetry of this system are called *group-symmetrical orbitals*, or simple *group orbitals*, or else *symmetrized (symmetry adopted) orbitals*. The possible group orbitals of the system are thus determined by its symmetry group (Appendix 1).

Symmetrized Orbitals

To determine the coefficients c_i in Eq. (5.10) that satisfy the necessary symmetry conditions for the wavefunction, one has to perform some relatively simple transformations described in Section 3.5. In the way shown there, all the symmetrized MOs can be easily determined. The results for σ and π symmetrized MOs of some important cases of octahedral (O_h symmetry), tetrahedral (T_d), and bipyramidal–tetragonal (D_{4h}) systems are given in Tables 5.1 to 5.3. Table 5.3 also contains the MO functions for square-planar complexes, for which one should put $\sigma_1 = \sigma_4 = \pi_1 = \pi_4 = 0$ in the expressions of the functions for the bipyramidal system.

In all cases in Tables 5.1 to 5.3 it is assumed that the CA takes part in MO formation by its s, p, d, and f atomic orbitals, while the ligands participate by one σ (simple or hybridized) and two π orbitals each. It is seen that, for instance, in an octahedral system the s orbital of the central atom takes part in the MO A_{1g} type only, while the two e_g type orbitals, d_{z^2}, and $d_{x^2-y^2}$, form a two-fold-degenerate σ MO of type E_g, whereas three t_{2g} orbitals, d_{xy}, d_{xz}, and

Table 5.1. Atomic functions of the central atom ψ_0 and symmetrized ligand σ and π orbitals, Φ_σ and Φ_π, for different types of symmetry of the O_h group for octahedral complexes

Type of Symmetry	ψ_0	Φ_σ	Φ_π
A_{1g}	s	$(1/\sqrt{6})(\sigma_1+\sigma_2+\sigma_3+\sigma_4+\sigma_5+\sigma_6)$	—
A_{2u}	f_{xyz}	—	—
T_{1u}	p_x, f_x	$(1/\sqrt{2})(\sigma_2-\sigma_5)$	$\frac{1}{2}(\pi_{1x}-\pi_{4y}-\pi_{3x}+\pi_{6y})$
	p_y, f_y	$(1/\sqrt{2})(\sigma_3-\sigma_6)$	$\frac{1}{2}(\pi_{1y}-\pi_{4x}-\pi_{2y}+\pi_{5x})$
	p_z, f_z	$(1/\sqrt{2})(\sigma_1-\sigma_4)$	$\frac{1}{2}(\pi_{2x}-\pi_{5y}+\pi_{3y}-\pi_{6x})$
E_g	$d_{x^2-y^2}$	$\frac{1}{2}(\sigma_2+\sigma_5-\sigma_3-\sigma_6)$	
	d_{z^2}	$(1/\sqrt{12})(2\sigma_1+2\sigma_4-\sigma_2-\sigma_5-\sigma_3-\sigma_6)$	—
T_{2g}	d_{xy}	—	$\frac{1}{2}(\pi_{2y}+\pi_{5x}+\pi_{3x}+\pi_{6y})$
	d_{xz}	—	$\frac{1}{2}(\pi_{1x}+\pi_{4y}+\pi_{2x}+\pi_{5y})$
	d_{yz}	—	$\frac{1}{2}(\pi_{1y}+\pi_{4x}+\pi_{3y}+\pi_{6x})$
T_{1g}	—	—	$\frac{1}{2}(\pi_{1y}+\pi_{4x}-\pi_{3y}-\pi_{6x})$
	—	—	$\frac{1}{2}(\pi_{2x}+\pi_{5y}-\pi_{1x}-\pi_{4y})$
	—	—	$\frac{1}{2}(\pi_{3x}+\pi_{6y}-\pi_{2y}-\pi_{5x})$
T_{2u}	$f_{x(y^2-z^2)}$	—	$\frac{1}{2}(\pi_{1x}-\pi_{4y}+\pi_{3x}-\pi_{6y})$
	$f_{y(z^2-x^2)}$	—	$\frac{1}{2}(\pi_{2y}-\pi_{5x}+\pi_{1y}-\pi_{4x})$
	$f_{z(x^2-y^2)}$	—	$\frac{1}{2}(\pi_{2x}-\pi_{5y}-\pi_{3y}+\pi_{6x})$

d_{yz}, form only π MOs of T_{2g} type. On the contrary, in the tetrahedral system the E orbitals d_{z^2} and $d_{x^2-y^2}$ form only π MOs, while the T_2 orbitals can, in principle, participate in both σ and π MOs. This is an example of the fact mentioned in Section 2.1—that for coordination compounds separation into σ and π MOs may be conventional: The same atomic orbitals of the CA take part in the formation of MOs with both σ orbitals of (some) ligands and π orbitals of other ligands. Tables 5.1 to 5.3 also show other cases of "$\sigma + \pi$" orbitals (the T_{1u} orbital in the O_h group, A_{2u} and E_u orbitals in D_{4h} systems, etc.). Examples of such systems are discussed in Section 6.3.

Note also that some symmetry-adapted combinations of ligand atomic orbitals do not have appropriate parts on the CA (e.g., T_{1g} states for O_h systems, as well as T_{2u} states for O_h and T_1 for T_d systems in the absence of active f states), and hence these states remain nonbonding. The same is true for the A_{2u} state in O_h systems when the central atom has active f electron orbitals: The latter also remain nonbonding. Tables 5.1 to 5.3 contain all possible group-symmetrical combinations Φ of six σ and 12 π ligand atomic orbitals of octahedral systems, and four σ and eight π orbitals of the ligands of tetrahedral systems.

Table 5.2. Atomic functions of the central atom ψ_0 and the symmetrized ligand σ and π orbitals, Φ_σ and Φ_π, for different types of symmetry of the T_d group for tetrahedral complexes

Type of Symmetry	ψ_0	Φ_σ	Φ_π
A_1	s	$\frac{1}{2}(\sigma_1 + \sigma_2 + \sigma_3 + \sigma_4)$	—
A_2	f_{xyz}	—	—
T_2	$p_x; d_{yz}; f_x$	$\frac{1}{2}(\sigma_1 - \sigma_2 + \sigma_3 - \sigma_4)$	$\frac{1}{4}(\pi_{4x} + \pi_{2x} - \pi_{1x} - \pi_{3x}) +$ $(\sqrt{3}/4)(\pi_{4y} + \pi_{2y} - \pi_{1y} - \pi_{3y})$
	$p_y; d_{xz}; f_y$	$\frac{1}{2}(\sigma_1 + \sigma_2 + \sigma_3 - \sigma_4)$	$\frac{1}{2}(\pi_{1x} + \pi_{2x} - \pi_{3x} - \pi_{4x})$
	$p_z; p_{xy}; f_z$	$\frac{1}{2}(\sigma_1 - \sigma_2 - \sigma_3 - \sigma_4)$	$\frac{1}{4}(\pi_{3x} + \pi_{2x} - \pi_{1x} - \pi_{4x}) +$ $(\sqrt{3}/4)(\pi_{4y} + \pi_{1y} - \pi_{2y} - \pi_{3y})$
E	$d_{x^2-y^2}$	—	$\frac{1}{4}(\pi_{1x} + \pi_{2x} + \pi_{3x} + \pi_{4x}) +$ $(\sqrt{3}/4)(\pi_{1y} + \pi_{2y} + \pi_{3y} + \pi_{4y})$
	d_{z^2}	—	$\frac{1}{4}(\pi_{1x} + \pi_{2x} + \pi_{3x} + \pi_{4x}) -$ $(\sqrt{3}/4)(\pi_{1y} + \pi_{2y} + \pi_{3y} + \pi_{4y})$
T_1	$f_{x(y^2-z^2)}$	—	$\frac{1}{4}(\pi_{2y} + \pi_{4y} - \pi_{3y} - \pi_{1y}) +$ $(\sqrt{3}/4)(\pi_{1x} + \pi_{3x} - 2x - \pi_{4x})$
	$f_{y(z^2-x^2)}$	—	$\frac{1}{2}(\pi_{1y} + \pi_{2y} - \pi_{3y} - \pi_{4y})$
	$f_{z(x^2-y^2)}$	—	$\frac{1}{4}(\pi_{2y} + \pi_{3y} - \pi_{1y} - \pi_{4y}) +$ $(\sqrt{3}/4)(\pi_{2x} + \pi_{3x} - \pi_{1x} - \pi_{4x})$

In Table 5.4 the so-called group overlap integrals G_{0i} are also given:

$$G_{0i} = \int \psi_0^* \Phi_i \, d\tau = \sum_j c_{ij} \int \psi_0^* \psi_j \, d\tau = \sum_j c_{ij} S_{0j} \qquad (5.11)$$

where S_{0j} are usual diatomic overlap integrals that can easily be expressed by the standard tabulated values of the type $S(s, \sigma)$, $S(p, \sigma)$, $S(p, \pi), \ldots$, and the coefficients c_{ij} are those given in Tables 5.1 to 5.3 for symmetry-adapted combinations of the ligand functions.

Simplification of the Secular Equation

With the symmetrized functions ψ_0 and Φ taken as a basis of the MO LCAO method, the secular equation (5.8) can be simplified significantly without reduction of accuracy. Indeed, as shown in Section 3.4, the integrals S_{ik} and H_{ik} after (5.3) and (5.4) are nonzero if and only if the two functions ψ_i and ψ_k have the same symmetry properties (belong to the same symmetry representation).

Table 5.3. Atomic functions of the central atom and the symmetrized ligand σ and π orbitals, Φ_σ and Φ_π, for different types of symmetry of the D_{4h} group for a tetragonally distorted octahedron (for a square-planar D_{4h} system put $\sigma_1 = \sigma_4 = \pi_{1x} = \pi_{1y} = \pi_{4x} = \pi_{4y} = 0$)

Type of Symmetry	ψ_0	Φ_σ	Φ_π
A_{1g}	$s; d_{z^2}$	$(1/\sqrt{2})(\sigma_1 + \sigma_4);$ $\frac{1}{2}(\sigma_2 + \sigma_5 + \sigma_3 + \sigma_6)$	—
B_{1g}	$d_{x^2-y^2}$	$\frac{1}{2}(\sigma_2 + \sigma_5 - \sigma_3 - \sigma_6)$	—
A_{2u}	p_z	$(1/\sqrt{2})(\sigma_1 - \sigma_4)$	$\frac{1}{2}(\pi_{2x} - \pi_{5y} - \pi_{3y} - \pi_{5x})$
E_u	p_x	$(1/\sqrt{2})(\sigma_2 - \sigma_5)$	$(1/\sqrt{2})(\pi_{1x} - \pi_{4y})$
			$(1/\sqrt{2})(\pi_{3x} - \pi_{6y})$
	p_y	$(1/\sqrt{2})(\sigma_3 - \sigma_6)$	$(1/\sqrt{2})(\pi_{1y} - \pi_{4x})$
			$(1/\sqrt{2})(\pi_{2x} - \pi_{5y})$
B_{2g}	d_{xy}	—	$\frac{1}{2}(\pi_{2y} + \pi_{5x} + \pi_{3x} + \pi_{6y})$
E_g	d_{xz}	—	$(1/\sqrt{2})(\pi_{1x} + \pi_{4y})$
			$(1/\sqrt{2})(\pi_{2x} + \pi_{5y})$
	d_{yz}	—	$(1/\sqrt{2})(\pi_{1x} + \pi_{4x})$
			$(1/\sqrt{2})(\pi_{3y} + \pi_{6x})$

This means that in Eq. (5.8), all the off-diagonal elements for any ψ_i and ψ_k that belong to different irreducible representations are zero. After an appropriate grouping of the basic wavefunctions on their symmetries, (5.8) transforms to the following:

$$\begin{vmatrix} H_{ij} - ES_{ij} & & 0 \\ & H_{kl} - ES_{kl} & \\ & & \ddots \\ 0 & & H_{md} - ES_{md} \end{vmatrix} = 0 \qquad (5.12)$$

In this equation it is implied that the first, say, n_1 functions, labeled with i and j, belong to one and the same representation of the symmetry group, the next n_2 functions, labeled with k and l, belong to another representation, and so on, the last group of n_r functions belonging to the rth representation. Hence the r quadratic matrices that occupy the diagonal positions in (5.12) belong to r representations of the symmetry point group of the systems, while the off-diagonal matrix elements between them are zero by symmetry.

Table 5.4. Group-overlap integrals $G_{0\sigma}$ and $G_{0\pi}$ for various types of symmetry of groups O_h, T_d, and D_{4h} for octahedral, tetrahedral, and tetragonally distorted octahedral complexes, respectively, considered in Tables 5.1 to 5.3[a]

Group of Symmetry	Type of Symmetry	ψ_0	$G_{0\sigma}$	$G_{0\pi}$
O_h	A_{1g}	s	$\sqrt{6}\,S(s,\sigma)$	—
	T_{1u}	p_x	$\sqrt{2}\,S(p,\sigma)$	$2S(p,\pi)$
		f_x	$\sqrt{2}\,S(f,\sigma)$	$2S(f,\pi)$
	E_g	$d_{x^2-y^2}$	$\sqrt{3}\,S(d,\sigma)$	—
	T_{2g}	d_{xy}	—	$2S(d,\pi)$
	T_{2u}	$f_{x(y^2-z^2)}$	—	$2S(f,\pi)$
T_d	A_1	s	$2S(s,\sigma)$	—
	T_2	p_x	$(2/\sqrt{3})S(p,\sigma)$	$-(\tfrac{8}{3})^{1/2}S(p,\pi)$
		d_{xy}	$(2/\sqrt{3})S(d,\sigma)$	$(\tfrac{8}{3})^{1/2}S(d,\pi)$
	E	$d_{x^2-y^2}$	—	$(\tfrac{8}{3})^{1/2}S(d,\pi)$
	T_1	$f_{x(y^2-z^2)}$	—	$(2/\sqrt{3})S(f,\pi)$
D_{4h}	A'_{1g}	s	$\sqrt{2}\,S(s,\sigma_1)$	—
		d_{z^2}	$\sqrt{2}\,S(d,\sigma_1)$	—
	A''_{1g}	s	$2S(d,\sigma_2)$	—
		d_{z^2}	$-S(d,\sigma_2)$	—
	B_{1g}	$d_{x^2-y^2}$	$\sqrt{3}\,S(d,\sigma)$	—
	A_{2u}	p_z	$\sqrt{2}\,S(p,\sigma_1)$	$2S(p,\pi_2)$
	E_u	p_x	$\sqrt{2}\,S(p,\sigma_2)$	$2S(p,\pi_1)$; $\sqrt{2}\,S(p,\pi_2)$
	B_{2g}	d_{xy}	—	$2S(d,\pi)$
	E_g	d_{xz}	—	$\sqrt{2}\,S(d,\pi_1)$; $\sqrt{2}\,S(d,\pi_2)$

[a] $S(s,\sigma)$, $S(p,\sigma)$, $S(p,\pi)$, ..., are the usual overlap integrals between the corresponding orbitals of two atoms.

As is well known, the determinant (5.12) equals the product of all the smaller determinants on the diagonal positions. This means that Eq. (5.12) of power n with respect to the MO energy E decomposes into r equations of much lower powers $n_1, n_2, \ldots, n_r (n_1 + n_2 + \cdots + n_r = n)$ that are much easier to solve, thus simplifying significantly the solution of the MO LCAO problem. Hence, if using symmetrized functions, the order of the equation that should be solved is reduced from the whole number of functions in the basis set n to the numbers of them in the partial groups belonging to the same irreducible representation.

For example, for an octahedral complex of O_h symmetry (without f electrons), in accordance with Table 5.1 the secular equation of the MO LCAO method, by using symmetrized functions, decomposes into one equation of second order for the A_{1g} representation [there are only two functions that belong to A_{1g}: ns of the CA and $\Phi_\sigma = (\tfrac{1}{6})^{1/2}(\sigma_1 + \sigma_2 + \cdots + \sigma_6)$ of the ligands], three identical equations of third order for T_{1u}, two identical equations of second order for E_g, and three equations of second order for T_{2g}. One can see that for high-symmetry systems the order of the secular equation in the MO LCAO method is not very high. However, the problem as a whole with many valence electrons is still very complicated because of the interaction between the electrons, due to which the different square blocks in (5.12) are coupled via the integrals H_{ik}, which contain LCAO coefficients of the MOs from all the other blocks. This problem is discussed in more detail in Section 5.3.

5.2. CHARGE DISTRIBUTION AND BONDING IN THE MO LCAO METHOD AND THE CASE OF WEAK COVALENCY

Atomic Charges and Bond Orders

With the MO energies E_i and LCAO coefficients c_{ij} known, one can visualize the electronic structure. In particular, for the charge distribution the Mulliken electron population analysis [5.7] is widely used. This analysis is based on the definition of electronic density. From the quantum-mechanical definition and statistical interpretation of the wavefunction in the coordinate representation, $|\varphi(\mathbf{r})|^2 \, dv$ means the probability to find the electron [that occupies the MO $\varphi(\mathbf{r})$] within the volume dv near the point $\mathbf{r}$ (regardless of its spin value). Therefore, the function $\rho(\mathbf{r})$

$$\rho(\mathbf{r}) = |\varphi(\mathbf{r})|^2 \tag{5.13}$$

represents the probability density (or simply, density) of electronic distribution along the MO $\varphi(\mathbf{r})$.

In a system with n electrons with the wavefunction $\Psi(\mathbf{r}_1, \mathbf{r}_2, \dots, \mathbf{r}_n)$, the probability of finding an electron near $\mathbf{r}_1$ is

$$\rho(\mathbf{r}_1) = n \int |\Psi(\mathbf{r}_1, \mathbf{r}_2, \dots, \mathbf{r}_n)|^2 \, dv_2 \cdots dv_n \tag{5.14}$$

The coefficient n is introduced because any of the n electrons can be at the point $\mathbf{r}_1$. Thus the electronic density and hence the parameters of electronic distribution in the system are completely determined by its wavefunction.

The redistribution of the charge density by the formation of the molecule (chemical bonding) is determined by the function $\Delta\rho$,

$$\Delta\rho = \rho(\mathbf{r}) - \sum_\mu \rho_\mu(\mathbf{r}) \tag{5.15}$$

where $\rho_\mu(\mathbf{r})$ represents the electronic density function of the free atom in the position that it occupies in the molecule. By definition $\Delta\rho$ represents the difference between the electronic density of the molecule and the free atoms, the *deformation density*, which indicates the changes in charge distribution that take place by the binding. The function $\Delta\rho$ is usually presented by equal-density curves which are more informative than the electron density diagrams (Section 8.6).

In the MO LCAO approximation where the MO wavefunction is presented by the LCAO (5.1), the electronic density $\rho(\mathbf{r})$ can be expressed by the LCAO coefficients. Indeed, assume that the ith MO under consideration contains only two basis functions, the AOs of two atoms A and B, ψ_A and ψ_B [e.g., the atomic functions of the central atom and the ligand, denoted in (5.9) as ψ_0 and Φ, respectively]:

$$\varphi_i(\mathbf{r}) = c_{iA}\psi_A + c_{iB}\psi_B \tag{5.16}$$

Then (we assume that the AOs and the LCAO coefficients are real)

$$|\varphi_i(\mathbf{r})|^2 = |c_{iA}|^2|\psi_A(\mathbf{r})|^2 + |c_{iB}|^2|\psi_B(\mathbf{r})|^2 + 2c_{iA}c_{iB}\psi_A\psi_B \tag{5.17}$$

In a more general case with $\varphi_i(\mathbf{r})$ after (5.1),

$$|\varphi_i(\mathbf{r})|^2 = \sum_{\mu,\nu} c_{i\mu}c_{i\nu}\psi_\mu\psi_\nu \tag{5.17'}$$

The first and second terms of expression (5.17) can be interpreted as the probabilities of finding the MO electron at the central atom and ligands, respectively, determined by the module square values of the appropriate LCAO coefficients. The last term has a more complicated nature. Denote the MO occupancy by q_i and define it as the total charge on the MO: $q_i = \int \rho_i(\mathbf{r})\,dv$. Substituting $\rho_i(\mathbf{r}) = q_i|\varphi_i(\mathbf{r})|^2$ and taking into account that the AO functions are normalized, we have:

$$q_i = q_i \sum_{\mu,\nu} c_{i\mu}c_{i\nu}S_{\mu\nu} \tag{5.18}$$

or in the two-center case under consideration,

$$q_i = q_i(|c_{iA}|^2 + |c_{iB}|^2 + 2c_{iA}c_{iB}S_{AB}) \tag{5.18'}$$

where S_{AB} is the overlap integral (or group overlap integral). It is seen from this expression that the whole charge q_i on the MO (the sum in parentheses equals 1) is divided into three parts: the part $q_i|c_{iA}|^2$ originating from the atom A, the $q_i|c_{iB}|^2$ term contributed from the atom B, and the third part, $p_i = q_i \cdot 2c_{iA}c_{iB}S_{AB}$, called the *overlap population*, describing the contribution of the overlap area to the charge distribution.

Mulliken [5.7] proposed defining the electronic charge on the atom by including half the overlap population in each of the two atomic charges:

$$q_{iA} = q_i(|c_{iA}|^2 + c_{iA}c_{iB}S_{AB})$$
$$q_{iB} = q_i(|c_{iB}|^2 + c_{iA}c_{iB}S_{AB})$$

(5.19)

If the atom A participates with its AOs in many MOs of the system, the *effective electronic charge* on the atom after Mulliken is

$$q_A = \sum_i q_{iA}$$

or, in general,

$$q_\mu = \sum_{i,v} c_{i\mu}c_{iv}S_{\mu v}$$

(5.20)

The total (effective) positive charge on the atom equals $Z_A - q_A$, where Z_A is the nuclear charge.

With the charges (5.20), *charge transfers* Δq can also be defined. Denoting the electronic charge on the free atom or group of atoms by q_μ^0, we have:

$$\Delta q_\mu = q_\mu - q_\mu^0$$

(5.20′)

Similarly, for *orbital charge transfer* Δq_i, we get

$$\Delta q_i^A = q_{iA} - q_{iA}^0$$

(5.20″)

where q_{iA}^0 stands for the electronic charge on the ith orbital in the free atom or atomic group A.

In addition to the atomic charges, the *bond orders*, or *density matrix* $P_{\mu v}$,

$$P_{\mu v} = \sum_i q_i c_{i\mu}c_{iv}$$

(5.21)

are used in the characterization of electronic charge distribution.

The Mulliken definition of atomic charges is far from being perfect. First, the overlap population is equally divided into two parts for two interacting atoms A and B, that are, in general, different. Second, the overlap population can be negative, resulting, if large enough, in negative charge densities, or in orbital charges larger than 2. But the most significant fault of the Mulliken atomic charges is that they depend strongly on the basis set of LCAO calculations. It is obvious that the values c_{iA}^2 characterize not so much the probability of the electron to be at the atom A as the probability to be in the

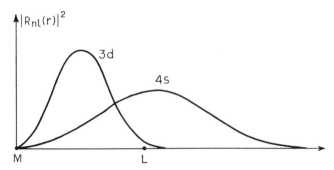

Figure 5.2. Typical radial distribution of atomic $3d$ and $4s$ electron densities of a transition metal M with respect to the position of the ligand L.

·*area where* ψ_A *has its maximum.* Depending on the nature of ψ_A, its maximum can be far outside the atom A.

This argument is especially important for transition metal compounds because their atomic functions are rather diffuse. Figure 5.2 illustrates the radial distribution in the $3d$ and $4s$ Slater type AOs (Section 2.1) of a typical transition metal with respect to the positions of the central atom M and the ligand L. As one can see, while the $3d$ distribution characterizes approximately the electron positions mainly on the CA, the $4s$ AO is far beyond the atomic area, its maximum falling in the region of the ligand.

This means that in the atomic charge q_M calculated after (5.20), the contribution c_{4s}^2 is in fact characterizing not the charge on the central atom but that on the ligand L, thus giving the wrong idea about the real charge on M. Nevertheless, the Mulliken analysis proved to be useful in many cases, especially when not absolute but relative values of charge distributions and their changes along a series of related compounds are considered.

Several other suggestions for more adequate calculations of atomic charges in molecules were made, and among them note the one of Politzer [5.8], in which the atomic charge is determined by direct integration of the electronic density function $\rho(\mathbf{r})$ over the volume of the atom. Again, there is some uncertainty in the choice of the borders of the atom in the molecule. In Section 8.6 a brief discussion of this problem is given in relation to the idea of Bader [8.115] to define the borders of the atom in the molecule as the surface S, where the gradient of the charge density is zero, $\nabla\rho(\mathbf{r}_0)\mathbf{n}(\mathbf{r}_0) = 0$, $\mathbf{r}_0 \in S$, where $\mathbf{n}$ is a unit vector normal to the surface S.

Another rule of calculation of atomic charges was suggested for coordination compounds by Noell [5.9]. For the atomic charge on the atom μ in the general case it is proposed to use the following formula instead of (5.20):

$$q_\mu = \sum_{i,v} (F_\mu F_v)^{1/2} c_{im} c_{iv} S_{\mu v} \tag{5.22}$$

where for the ligands $F_\mu = F_\nu = 1$, while for the CA F_μ is defined by the following procedure. For the central AO

$$\int |\psi_\mu|^2 \, d\tau = 1 = I_\mu + L_\mu + R_\mu \tag{5.23}$$

where I_μ is the portion within the covalent sphere of the metal with the radius R_c $(0 < r < R_c)$, L_μ the ligand portion lying within the ligand cone [the volume with the radius-vector r $(R_c < r < \infty)$ and angles ϑ and ϕ inside the cone of revolution with the apex at the metal and the base determined by the ligand dimensions], and R_μ the remaining part, which enters neither the R_c sphere nor the ligand cone. Then F_μ is defined as follows:

$$F_\mu = I_\mu + R_\mu \tag{5.24}$$

In other words, by introducing the F_μ factor in the definition of the atomic charge of the central atom, one excludes the electronic cloud in the ligand region created by the central atom AO. For instance, for square-planar complexes of Pt and Pd the cones of revolution embracing the four ligands are taken as four 90° cones, while the covalent radius is chosen as $R_c = 1.30 \, \text{Å}$ for both Pt and Pd [5.10]. Examples of calculations are provided in Sections 6.3 and 6.4.

Bonding, Nonbonding, and Antibonding Orbitals

To reveal the mechanism of chemical bonding in the MO LCAO scheme, consider first a simple example. Assume that for a given symmetry representation of the system there are only two basis functions that belong to this representation; one of them ψ_0 is an AO of the CA, while the other one, Φ_1, forms an appropriate linear combination of the ligand functions, as in (5.9). Such a case is realized, for instance, for the A_{1g} representation of an octahedral complex (Table 5.1). The secular equation for the two MOs, $\Psi_1 = c_0'\psi_0 + c_1'\Phi_1$ and $\Psi_2 = c_0''\psi_0 + c_1''\Phi_1$, corresponding to this type of symmetry is

$$\begin{vmatrix} H_{00} - E & H_{01} - EG_{01} \\ H_{10} - EG_{10} & H_{11} - E \end{vmatrix} = 0 \tag{5.25}$$

where $G_{01} = G_{10}$ is the group overlap integral given by Eq. (5.11) and in Table 5.4. This quadratic equation with respect to E can be solved directly. It yields two values of E, E_1 and E_2,

$$E_{1,2} = [H_{00} + H_{11} - 2H_{01}G_{01}]$$
$$\pm \frac{\{(H_{00} - H_{11})^2 + 4[H_{01}^2 + H_{00}H_{11}G_{01}^2 - H_{01}G_{01}(H_{00} + H_{11})]\}^{1/2}}{2(1 - G_{01}^2)}$$
$$\tag{5.26}$$

From this expression it is seen, first, that in the absence of overlap between ψ_0 and Φ_1, that is, when $G_{01} = H_{01} = 0$, we have $E_1 = H_{00}$ and $E_2 = H_{11}$. This means that the two energy levels remain the same as in the atomic states ψ_0 and Φ_1. Substituting these energy values into Eq. (5.7) and then the c_i values obtained into Eqs. (5.1) and (5.9), one can easily find that the wavefunctions for these two states, Ψ_1 and Ψ_2, also remain pure atomic: $\Psi_1 = \psi_0$ and $\Psi_2 = \Phi_1$. The MOs corresponding to these states are called *nonbonding*. Such MOs, as mentioned above, are realized either when some atomic orbitals (or their linear combinations) have no contraparts of the same symmetry on other atoms (as, e.g., in the case of the T_1 combination of ligand orbitals in a tetrahedral system of T_d symmetry without f electrons; Table 5.2), or when the corresponding group overlap integrals G_{ij} and resonance integral H_{ij} are negligible.

In all other cases the values of the two roots (5.26) of Eq.(5.25) are beyond the area between H_{00} and H_{11}: E_1 is smaller than the smallest value of H_{00} and H_{11}, while E_2 is larger than the largest one (Fig. 5.3). It follows that one MO has the energy E_1, which is lower than the atomic energies of the free atoms, and hence this MO is *bonding*. On the contrary, the MO corresponding to the energy E_2 increases the energy compared with the free atoms; this MO is called *antibonding*. Since both G_{ij} and H_{ij} depend on the integral overlap between the ψ_0 and Φ_1 functions, the bonding and antibonding magnitude of the MOs above are also dependent on this overlap.

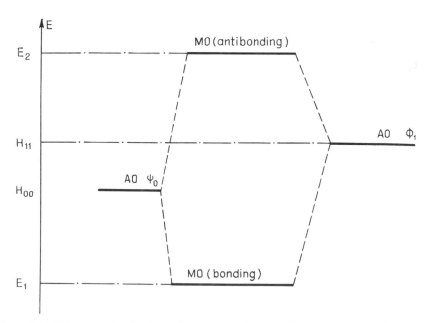

Figure 5.3. MO energy-level scheme for two atomic states from two centers with AOs ψ_0 and Φ_1 that produce bonding and antibonding MOs.

The wavefunctions of these two MOs, their LCAO coefficients c_0 and c_1 at ψ_0 and Φ_1, can easily be evaluated by substituting consecutively the values E_1 and E_2 after (5.26) into Eq. (5.7) and calculating the corresponding coefficients c_0', c_1' and c_0'', c_1'' (taking into account also the normalization condition $c_0^2 + c_1^2 + 2c_1c_0G_{01} = 1$ for each pair of the c_0', c_1' and c_0'', c_1'' coefficients, respectively).

The bonding energy obviously depends on the number of electrons that occupy bonding (antibonding) MOs. Consider the case when each of the bonding parts, the central atom in the state ψ and the ligands in the state Φ_1, are occupied by one electron. Then, after the formation of the MOs that mix (collectivize) these states, the new two states, two MOs, should be populated by the same two electrons. Following the Pauli principle, in the ground state the electrons occupy the lowest MOs. Neglecting the interelectron interaction, the bonding energy ΔE is approximately

$$\Delta E = H_{00} + H_{11} - 2E_1 \tag{5.27}$$

where it is taken into account that in the nonbonding state the energies of the two electrons are H_{00} and H_{11}, respectively. Substituting E_1 from (5.26), we have

$$\Delta E = \frac{2H_{01}G_{01} - G_{01}^2(H_{00}+H_{11}) + [(H_{00}-H_{11})^2 + 4H_{01}^2 + 4H_{00}H_{11}G_{01}^2 - 4H_{01}G_{01}(H_{00}+H_{11})]^{1/2}}{1 - G_{01}^2}$$

$$\tag{5.28}$$

Note that if there are four electrons, two from each bonding part, they must occupy both the bonding and antibonding MOs in the ground state and

$$\Delta E = 2(H_{00} + H_{11}) - 2(E_1 + E_2)$$
$$= (1 - G_{01}^2)^{-1}[4H_{01}G_{01} - 2(H_{00} + H_{11})G_{01}^2] \cong 0 \tag{5.29}$$

Here it is taken into account that within the approximations under consideration, $H_{01} \cong \frac{1}{2}(H_{00} + H_{11})G_{01}$ (Section 5.4).

Equation (5.29) means that bonding and antibonding MOs approximately compensate for each other, so there is almost zero bonding when they are both fully occupied by electrons. The compensation of bonding and antibonding MOs is also confirmed by more rigorous numerical calculations [5.11]. Hence the bonding in the MO scheme occurs only when there is no full compensation of the bonding MOs by the antibonding ones, and the bonding is produced only by those electrons that occupy uncompensated bonding orbitals.

If the number of basis functions that transform according to the given symmetry type (and hence the order of the secular equation and the number of MOs) is larger than two, the possibility of visual interpretation of the origin of the bonding features in the MO LCAO scheme decreases significantly, but the basic ideas of bonding, nonbonding, and antibonding MOs with respect to the corresponding free atomic states remains valid. In Section 7.2, another criterion of MO bonding properties based on their orbital vibronic constants is given.

It is clear that the total bonding energy equals the sum of the individual MO contributions. Taking into account the almost exact compensation of the bonding MOs with the corresponding antibonding ones, we come to the conclusion that *the total bonding energy equals the summary contribution of the uncompensated bonding MOs* (see also Section 6.2). The number of the latter depends on the ordering of the bonding and antibonding MOs and their occupation numbers. *The highest occupied MO is usually abbreviated as HOMO, while the lowest unoccupied MO is LUMO.*

Case of Weak Covalency

In transition metal coordination bonding (including solid-state lattice formation) the notion of *weak covalency* is very important. In fact, this term is applicable to cases when the bonds between the central atom and ligands are rather ionic and the covalency can be considered as a correction. Starting with ionic parts, let us define the condition of weak covalency by assuming that the magnitudes of the overlap integral S (or G) and $H_{01}/H_{11} - H_{00}$ are small compared with unity and their squares can be neglected. This means that $S \ll 1$ and $H_{01} \ll (H_{11} - H_{00})$ (we assume that $H_{11} > H_{00}$). Then one can obtain from Eq. (5.26) the following relations for the energies of the bonding E_1 and antibonding E_2 MO energies:

$$E_1 \cong H_{00} - \frac{(H_{01} - H_{00}G_{01})^2}{H_{11} - H_{00}} \tag{5.30}$$

$$E_2 \cong H_{11} - \frac{(H_{01} - H_{11}G_{01})^2}{H_{11} - H_{00}} \tag{5.31}$$

For the wavefunctions of these two MOs, the following form seems to be convenient:

$$\Psi_1 = N_1(\psi_0 + \gamma\Phi_1) \tag{5.32}$$

$$\Psi_2 = N_2(\psi_0 - \lambda\Phi_1) \tag{5.33}$$

where N_1 and N_2 are the normalization coefficients,

$$N_1 = (1 + \gamma^2 + 2\gamma G_{01})^{-1/2}$$
$$N_2 = (1 + \lambda^2 - 2\gamma G_{01})^{-1/2} \tag{5.34}$$

The relation between λ and γ can be obtained from the condition of orthogonality $\int \Psi_1^* \Psi_2 \, d\tau = 0$:

$$\lambda = \frac{\gamma + G_{01}}{1 + \gamma G_{01}} \tag{5.35}$$

Within the approximation of weak covalency at hand,

$$\gamma \cong -\frac{H_{01} - H_{00}G_{01}}{H_{11} - H_{00}} \tag{5.36}$$

and

$$\lambda \cong -\frac{H_{01} - H_{11}G_{01}}{H_{11} - H_{00}} = \gamma + G_{01} \tag{5.37}$$

From these formulas it is seen, first, that the values γ and λ are also small because both G_{01} and H_{01} are small. Hence the bonding MO after Eq. (5.32) is mostly ψ_0, a CA function with a small contribution of the ligand part Φ_1, while the antibonding MO (5.33) is mainly a ligand function with some contribution from the central atom. If $\gamma = 0$, $\Psi_1 = \psi_0$, and the bonding electron is on the CA. On the other hand, $\gamma = 0$ is possible only when $H_{01} = G_{01} = 0$, and this means, following (5.37), that $\lambda = 0$ and hence $\Psi_2 = \Phi_1$; that is, the antibonding MO is of purely ligand origin. Thus both the MOs in this case are nonbonding and the bond remains purely ionic.

If $\gamma \neq 0$ and $\lambda \neq 0$, the bonding electrons become collectivized, and for $\gamma = \lambda = 1$ the bonding state includes both the CA state Ψ_0 and the ligands state Φ_1, with equal probability. In this case the bonding by this MO can be considered as entirely covalent. Hence the constants γ (for the bonding MO) and λ (for the antibonding MO) may serve as a measure of covalency.

Note that if $0 < \gamma < 1$, then $\lambda \neq \gamma$, and hence the degree of covalency on the bonding and antibonding MOs is in general different. If, in addition, $G_{01} > 0$, then $\lambda > \gamma$ and $N_2 > N_1$. Consequently, in the state of the bonding MO the probability of the electron being at the CA (determined by N_1^2) and at the ligands ($N_1^2 \gamma^2$) is always smaller than, respectively, being at the ligands N_2^2 and at the central atom ($N_2^2 \lambda^2$) in the case of the antibonding MO.

This result is quite understandable if one takes into account that in the case of a bonding MO a part of the electronic cloud determined by the term $2N_1^2 \gamma G_{01}$ (the overlap population) is placed in between the bonding atoms (and in this position the electronic charge attracts the nuclei of both atoms, thus contributing to the bonding), whereas in the antibonding case the corresponding term, $2N_2^2 \lambda G_{01}$, is negative: The antibonding MO produces a negative overlap density, a subtraction of the electronic cloud between the bonding atoms. As shown above, the energy stabilization on the bonding MO and its destabilization on the antibonding MO, within the approximations employed, almost totally compensate for each other.

Angular Overlap Model

The case of weak covalency is rather widespread in coordination systems. In particular, almost all the MOs formed by f electrons in coordination compounds of rare earth elements belong to weak covalency. Indeed, as mentioned in Section 4.4, the f electrons are screened by the outer d electrons, and therefore they form weak covalent bonds with the ligands (but the outer d electrons may still form stronger covalent bonds). Even in d electron complexes in the case of high-oxidation states, covalency is formed mainly by outer s and p orbitals, while the d electron participation in the overlap with ligands may be much weaker (Section 6.1).

For compounds of weak covalency a qualitative and semiquantitative semiempirical method of determination of the main chemically active MO energies was worked out. It is called the *angular overlap model* (sometimes abbreviated as AOM) [5.4, 5.12].

To simplify Eqs. (5.30) and (5.31), let us employ again the proportionality between H_{01} and G_{01}. According to Eq. (5.119) with $k = 2$, $H_{01} = (H_{00} + H_{11})G_{01}$; substituting this expression into Eqs. (5.30) and (5.31), we have

$$E_1 \cong H_{00} - \frac{H_{11}^2 G_{01}^2}{H_{11} - H_{00}} \tag{5.38}$$

$$E_2 \cong H_{11} + \frac{H_{00}^2 G_{01}^2}{H_{11} - H_{00}} \tag{5.39}$$

It is seen from these equations that within the approximation at hand the stabilization of the bonding MO $H_{00} - E_1$ and the destabilization of the antibonding MO $E_2 - H_{11}$ are proportional to the square of the overlap integral G_{01}. This statement is taken as the basic assumption of the AOM.

The destabilization of the antibonding d and f MO energy levels, which is relatively easily observed in electronic spectra (especially in f electron systems), can serve as a measure of covalency of the bonds and of the nephelauxetic effects (Section 8.2). However, the name of the method and its possibilities are concerned more with the presentation of the overlap integral G_{01} as a product of the radial G_{01}^* and angular θ parts [5.12]:

$$G_{01} = G_{01}^* \theta \tag{5.40}$$

Substituting this expression into Eq. (5.39), one finds that the energy of destabilization of the d and f electron states, $E^* = E_2 - H_{11}$, is proportional to the square of the angular overlap θ:

$$E^* = e_\lambda \theta^2 \tag{5.41}$$

where

$$e_\lambda = \frac{H_{00}^2 (G_{01}^*)_\lambda^2}{H_{11} - H_{00}} \tag{5.42}$$

and the index λ is introduced to distinguish among σ, π, δ,..., metal–ligand radial overlaps. From Eq. (5.41) the main conclusion of the AOM method follows: *The destabilization of the atomic states of the CA due to the formation of covalent bonds with the ligands is proportional to the square of the angular part of the overlap integral.* Since the radial part of the overlap integral G_{01}^* depends on the metal–ligand distance, but not on the geometry of the complex, whereas the angular part θ depends strongly on the geometry, the angular overlap model is a good method for studying relative geometries of complexes.

In a more general treatment for molecular systems with low symmetries, both σ and π overlap can be active in the same bond, the π overlap being of two types, $\pi(s)$ and $\pi(a)$, respectively (symmetric and antisymmetric with respect to the reflection in the plane containing the z axis). Also, there may be different types of ligand orbital overlap with the given CA orbitals. In these cases the formula for the destabilization energy of d and f orbitals (5.41) acquires the form

$$E^* = e_\sigma \sum_i k_{0i}^2(\sigma) + e_\pi(s) \sum_i k_{0i}^2[\pi(s)] + e_\pi(a) \sum_i k_{0i}^2[\pi(a)] \tag{5.43}$$

where k_{0i}^2 are the squares of the angular part of the overlap integral with the ith ligand and depend on the coordinates of the latter. Passing to symmetrized group orbitals that include already an appropriate summation over the ligands, one can simplify Eq. (5.43):

$$E^*(\Gamma) = e_\sigma k_{0\Gamma}^2(\sigma) + e_\pi(s) k_{0\Gamma}^2[\pi(s)] + e_\pi(a) k_{0\Gamma}^2[\pi(a)] \tag{5.44}$$

where $k_{0\Gamma}^2$, unlike k_{0i}^2, are the squares of the angular parts of the group overlap integrals for the Γ representation. Table 5.5 gives the formulas for the $k_{0\Gamma}^2$ values for d orbitals [5.13].

As shown in this table, the dependence of $k_{0\Gamma}^2$ and hence E^* on the angular coordinates of the ligands ϑ and ϕ is very strong.

Equation (5.44) shows that with a small number of parameters, three for each different ligand at most [e_σ, $e_\pi(s)$, $e_\pi(a)$], one can describe the destabilization of the metal orbitals that are determining the spectroscopic properties of the complex in the visible and near-ultraviolet regions. Usually, the constants in Eq. (5.44) are considered as empirical parameters; then the relative destabil-

Table 5.5. Squared angular parts of group-overlap integrals with CA d states^a

d States	$k_{0\Gamma}^2(\sigma)$	$k_{0\Gamma}^2[\pi(s)]$	$k_{0\Gamma}^2[\pi(a)]$
d_{z^2}	$N(\cos^2\vartheta - \sin^2\vartheta/2)^2$	$\frac{3}{4}N\sin^2 2\vartheta$	0
$d_{x^2-y^2}$	$\frac{3}{4}\sin^4\vartheta(\Sigma C_n\cos 2\phi_n)^2$	$\frac{1}{4}\sin^2 2\vartheta(\Sigma C_n\cos 2\phi_n)^2$	$\sin^2\vartheta(\Sigma C_n\sin 2\phi_n)^2$
d_{xy}	$\frac{3}{4}\sin^4\vartheta(\Sigma C_n\sin 2\phi_n)^2$	$\frac{1}{4}\sin^2 2\vartheta(\Sigma C_n\sin 2\phi_n)^2$	$\sin^2\vartheta(\Sigma C_n\cos 2\phi_n)^2$
d_{xz}	$\frac{3}{4}\sin^2 2\vartheta(\Sigma C_n\cos \phi_n)^2$	$\cos^2 2\vartheta(\Sigma C_n\cos \phi_n)^2$	$\cos^2\vartheta(\Sigma C_n\sin \phi_n)^2$
d_{yz}	$\frac{3}{4}\sin^2 2\vartheta(\Sigma C_n\sin \phi_n)^2$	$\cos^2 2\vartheta(\Sigma C_n\sin \phi_n)^2$	$\cos^2\vartheta(\Sigma C_n\cos \phi_n)^2$

$^a N$ is the number of ligands, C_n is the LCAO coefficient of the nth ligand in the group orbital Γ (Tables 5.1 to 5.3), and ϑ_n and ϕ_n are the ligands' angular coordinates (all the sums should be taken from $n = 1$ through $n = N$).

ization energies for different geometries of the complex are well defined by this equation. The entire procedure is easily programmed for personal computers, allowing direct estimations of CA orbital destabilization by different geometries of coordination (for more details, see [5.14]).

So far the most useful applications of the model seem to be in spectroscopy of $d - d$ transitions (Section 8.2), especially with f electron participation. Compared with ligand field theory, the AOM is more attractive because it considers partly the electronic structure of the ligands and introduces more realistic parameters. The AOM occupies a position in between crystal field theory and the MO LCAO method.

However, in general, the deficiencies of the AOM are almost the same as those of crystal field theory (Section 4.6). Indeed, although the AOM takes weak covalences into consideration, it does it for the CA states only. Hence in this model, similar to the CFT, ligand electronic states are not involved explicitly, and the MOs of mainly ligand origin are excluded from consideration. This deprives us of the possibility of considering many important chemical phenomena, including the origin of metal–ligand bonding, charge transfer spectra, origin of configuration instability due to the vibronic mixing of CA states with ligand orbitals, and others.

5.3. METHODS OF CALCULATION OF MO ENERGIES AND LCAO COEFFICIENTS

SCF MO LCAO Approximation

The evaluation of the electronic structure of coordination compounds in the MO LCAO approximation, as shown above, can be reduced to the solution of the secular equation (5.8) that determines the MO energies E_i, and Eq. (5.7) for the LCAO coefficients for each E_i value. Equations (5.7) and (5.8) contain the overlap integrals S_{ij} (or the group overlap integrals G_{ij}) and resonance integrals H_{ij}. Provided that the basis set of AOs is chosen, the S_{ij} (or G_{ij}) values can be calculated directly, whereas evaluation of the H_{ij} magnitudes by Eq. (5.4) is rather difficult.

In Eq. (5.4), H is an effective Hamiltonian for the one-electron state; it describes the mean field of all the nuclei and other electrons in which the given electron moves. This field depends essentially on the states of the other electrons described by their MOs. The latter, in turn, depend on the MO of the electron under consideration. This situation, in the framework of the full separation of the variables of the electrons, is best described by the *Hartree–Fock* (HF) *method of self-consistent field* (SCF) described in Section 2.2. In application to molecules in combination with the MO LCAO approximation, the corresponding equations were first obtained by Roothaan [5.15]; the joint method is abbreviated as *SCF MO LCAO*, or *Hartree–Fock–Roothaan* (HFR) *method*.

The deduction of the HFR equations is relatively simple for systems with closed shells when each MO is occupied by two electrons (the total spin is zero) and there is no orbital degeneracy [5.16], but it becomes much more difficult when there are open shells and for systems in excited states [5.16, 5.17].

Following discussion of the Hartree–Fock method in Section 2.2, for a molecule with a closed-shell configuration of n electrons and $n/2$ states occupied by 2 electrons each, the full wavefunction can be presented as a determinant of the type (2.17) formed by one-electron MOs:

$$\Psi_k = \left(\frac{n}{2}\right)! \det \| \varphi_1(\mathbf{r}_1)\varphi_2(\mathbf{r}_2) \cdots \varphi_{n/2}(\mathbf{r}_{n/2}) \| \qquad (5.45)$$

where k denotes the electronic configuration under consideration.

With this presentation of the wavefunction, the self-consistent MOs $\varphi_i(\mathbf{r}_i)$ are determined by Eq. (2.43). If the MOs are taken as LCAO (5.1), we come to the equations for the MO energies and LCAO coefficients, given by Eqs. (5.8) and (5.7), respectively. In the Hartree–Fock method the matrix elements of the effective Hamiltonian H_{ij} of these equations are given by Eq. (2.44). Its more detailed form is as follows.

Denote the MOs and AOs by Greek and Latin labels, respectively. Then for MOs of type (5.1), we have (the asterisk denotes complex-conjugated magnitudes)

$$\varphi_\mu = \sum_i^N c_{\mu l}\psi_l(\mathbf{r})$$

$$\varphi_\mu^* = \sum_k^N c_{\mu k}^*\psi_k^*(\mathbf{r}) \qquad (5.46)$$

where N is the number of functions in the LCAO basis.

The matrix element of the effective Hamiltonian is

$$H_{ij} = F_{ij} = \int \psi_i^* F^j \psi_j \, d\tau \tag{5.47}$$

where the operator F^j (Fockian) after (2.44) contains the functions (5.46). By substitution we get

$$H_{ij} = H_{ij}^0 + \sum_{k,l} P_{kl}([kl|ij] - \tfrac{1}{2}[kj|il]$$

$$= H_{ij}^0 + I_{ij} - K_{ij} \tag{5.48}$$

Here, in accordance with Eq. (2.38),

$$H_{ij}^0 = T_{ij} - \sum_{\alpha} (U_\alpha)_{ij} \tag{5.49}$$

where the notations are the same as in Chapter 2; T is the operator of kinetic energy of the electron and U_α is its potential energy of attraction to the α nucleus,

$$P_{kl} = \sum_{\mu=1}^{n/2} 2c_{\mu k}^* c_{\mu l} \tag{5.50}$$

is an important characteristic of the charge distribution in the system in the MO LCAO approximation — the matrix of bond orders, or density matrix, mentioned in Section 5.2 [Eq. (5.21)] — and

$$I_{ij} = \sum_{k,l} P_{kl}[kl|ij] \qquad K_{ij} = \tfrac{1}{2} \sum_{k,l} P_{kl}[kj|il] \tag{5.51}$$

are the abbreviated denotations for, respectively, the Coulomb and exchange interactions between the electrons. The notation $[kl|ij]$ stands for two-electron integrals (2.21).

Equations (5.48) to (5.50) show explicitly that to evaluate the H_{ij} values in Eqs. (5.7) and (5.8) of the MO LCAO method, one must know the sets of LCAO coefficients $c_{\mu k}$ that are determined by the same equations. In other words, the equations determining the $c_{\mu k}$ values are nonlinear [if one substitutes the expression (5.48) for H_{ij} into (5.7), the latter becomes cubic with respect to $c_{\mu k}$]. In these cases the solution can be obtained by the method of iterations.

Let us assume that based on general knowledge of the system, one chooses some reasonable values of the $c_{\mu k}$ coefficients for all the MOs. With these coefficients the H_{ij} values after (5.48) can be calculated; then the MO energies E_μ and new LCAO coefficients $c'_{\mu k}$ can be evaluated for each MO after (5.8) and (5.7), respectively. With the new coefficients $c'_{\mu k}$, one can calculate new H'_{ij} values and determine with them new $c''_{\mu k}$ values, and so on, until the newly calculated MO energies E'_μ and LCAO coefficients $c''_{\mu k}$ coincide (within the required accuracy) with the previous ones. It is assumed that in this process of iterative calculations the solution converges to the *self-consistent values* of E_μ and $c_{\mu k}$ sought.

For systems with unpaired electrons (open shells), the full wavefunction can be presented in the form of a linear combination of several determinants of the type (5.45), corresponding to the symmetry of the molecular term under consideration. This results in an essential complication of the calculation procedure [5.16, 5.17]. However, in some cases the presentation with one determinant can be preserved. In particular, in the *unrestricted Hartree–Fock* (UHF) method [5.17–5.19] it is assumed that the orbital parts of the wavefunctions of electrons with opposite spins, α (up, ↑) and β (down, ↓), are different (in the *restricted Hartree–Fock* (RHF) *method* they are the same for the two electrons on a given one-electron orbital). Then the expansion (5.46) should be written separately for MOs with spins α and β, φ_i^α and φ_i^β, respectively, and hence the order of the secular equation (5.8) and the number of LCAO coefficients, $c_{\mu k}^\alpha$ and $c_{\mu k}^\beta$, increase, becoming equal to the number of electrons in the system (in the RHF method for closed shells it is half the number of electrons).

Thus the calculations of electronic structure by the SCF MO LCAO (Roothaan) method can be performed by an iterative procedure in which each iteration consists of two important stages:

1. Calculation of the integrals of Coulomb I_{ij} and exchange K_{ij} interactions between the electrons using Eq. (5.51), as well as the one-electron integrals of kinetic and potential energy (5.49) and overlap integrals S_{ij} (or G_{ij})

2. Solution of Eqs. (5.7) and (5.8)

Each of these stages has its own difficulties, which increase with the number of electrons, but most of them are due to the calculation of integrals (first stage). Indeed, in most MO LCAO methods the number of integrals increases with the number of atoms (and hence the number of the basis functions n) as n^4 (see the discussion in Section 5.6). Many of these integrals are multicenter integrals; their evaluation requires the major part of computer time. However, the results of such calculations are far from being excellent. The reason is in the neglect of *correlation effects* and poor *basis sets*, discussed below.

Calculations by the SCF MO LCAO methods discussed above using neither empirical parameters nor restricting computational simplifications are called *ab initio.* When employing some computational simplifications they are also called *nonempirical.* These methods do not use any empirical parameters except the specification of the system by the number of nuclei and the nuclear charges as well as the number of electrons. In the adiabatic approximation under consideration the Hamiltonian (Fockian) F^j in Eq. (5.47) taken after (2.44) also contains the nuclear configuration (geometry) of the molecular system, but this is not necessarily an empirical parameter. Indeed, in sufficiently full calculations the geometry of the molecule can be varied in a special way to reach the minimum point of the AP energy surface. The coordinates of this minimum are assumed to correspond to the real geometry of the molecule. Such calculations exemplify *geometry optimization.* There are computer programs that perform this procedure automatically.

Role of Basis Sets

The main assumption of the MO LCAO method — the presentation of the MO in the form of LCAO (5.1) — may be subjected to criticism. Indeed, in this presentation one tries to approximate the unknown function of the MO with a limited number of known atomic functions. Visually, the latter serve as molds, while the LCAO coefficients are a means for adapting the molds to the complicated MO surface. It is obvious that when the number of molds is small and their form is rigid and is not a good fit for the MO surface, the MO in the form of (5.1) is not close enough to the real charge distribution (despite the best choice of LCAO coefficients by the variational principle).

Mathematically, any function can be presented in the form of an expansion in a series of other functions, provided that the latter form a set which is complete from a mathematical point of view. The requirement of functional completeness is rather difficult; the usual sets of valence atomic orbitals do not satisfy this condition. For example, the set of functions that includes all possible exact solutions of the Schrödinger equation for a given system forms a complete set.

This explains why the results of calculation of the electronic structure of molecular systems obtained by the MO LCAO method depend on the choice of the basis set, larger (more complete) sets being preferable. However, as mentioned above, an increase in the number of basis functions sharply increases the number of integrals to be calculated, making the problem extremely difficult. Nevertheless, using supercomputers, this can be done for systems of moderate size. If the basis set is complete, the results are the same as if the prime Hartree–Fock equations (2.42) were solved exactly. Such results are called *Hartree–Fock limit solutions.*

In the calculations of transition metal compounds performed so far, the basis set is far from being complete. Often, the *minimal basis set* is used which contains the AOs that are occupied by valence electrons in the ground state of the free atoms forming the molecule (the *valence basis*); the *extended basis* also includes unoccupied AOs of excited states. In general, three kinds of atomic functions are usually used to present the AOs in the MO LCAO method:

1. Hydrogen–like orbitals (Lagerr polynomials)
2. Slater orbitals (Section 2.1)
3. Gaussian functions

For the region of interatomic orbital overlap where the chemical bonding is realized, Slater orbitals are most suitable, but they are not good for the region at the nuclei; hydrogenlike functions may be better for this region. In MO LCAO problems, use of the basis of Slater orbitals or, more precisely, *Slater-type orbitals* (STOs) is most widespread.

Both Slater and hydrogenlike functions create many difficulties when used in molecular integral calculations. These integrals can be calculated much more easily when the corresponding STOs are expressed by a Gaussian function [5.20–5.22]. The latter contain the exponential factor $\exp(-\zeta r^2)$ with an r^2 dependence on r [instead of $\exp(-\zeta r)$ in the STO]; this allows for a much simpler evaluation of the two-electron two-center integrals. On the other hand, the presentation of a given STO in the form of a sum of two or several (n) Gaussian functions (G) with different coefficients is relatively simple and can be a priori tabulated. Such functions are termed STO-nG. Usually, n ranges from 2 to 6. The AOs (HF solutions) ψ_k in Eq. (5.46) can be approximated by one STO [5.23] or by two or several STO [5.24, 5.25]. The notation STO-nG refers to the former case. The minimal basis set is usually STO-3G.

In general, there are several Gaussian functions standing for each AO, and the question is whether they should be all represented by the same $c_{\mu k}$ coefficient in the LCAO (one coefficient for the whole AO), or be grouped in two, three,..., with two, three,..., coefficients $c_{\mu k}$. The former is a case of a one-exponential basis; in the case of two groups of orbitals for the same AO (with two variational coefficients $c_{\mu k}$), the basis is termed *two-exponential, or double-zeta* (DZ). Note that while the number of Gaussian functions in the AO determines the number of integrals to be calculated in the MO LCAO procedure, the number of $c_{\mu k}$ coefficients determines the order of the secular equation (5.8).

The angular part of the AOs in all the presentations remains the same; the $s, p, d, f, \ldots$ behavior is an important feature of the LCAO, determining its classification on symmetry discussed in Section 5.1. But the radial part is not so strictly determined. In particular, the exponential parameter ζ of the Slater orbital for free atoms is calculated to fit its charge distribution (atomic radii, interatomic distances, etc.). If the AO is given by several STOs, the latter could be chosen to fit, say, the HF solutions for the free atom.

Thus to obtain the Gaussian basis set, one must perform several numerical calculation procedures: to get HF solutions, to fit them by STO, and then to fit the STO by Gaussian function. Huzinaga [5.26] (see also [5.27–5.29]) suggested avoiding all these lengthy intermediate calculations and determining the Gaussian basis set directly from the condition of the minimum ground-state energy of the atom. This can be done by means of a Roothaan procedure taking the AO as a linear combination of Gaussian orbitals and considering both the c_{ij} coefficients and the exponential parameters ζ as variational parameters. Basis sets for many atoms were tabulated in such a way [5.30, 5.31].

Obviously, the quality of the Huzinaga basis depends on how much individual Gaussian orbitals are taken to form atomic orbitals of the type s, p, d,.... In this basis there is no sense to discriminate $1s$, $2s$, $3s$, and so on, orbitals, because all these orbitals have the same symmetry and enter in the same LCAO. However, different Gaussian s orbitals or groups of them can have different variational LCAO coefficients.

In conclusion, the Gaussian basis set contains in general n_1 s-type functions, n_2 p-type, n_3 d-type, and so on. These are called *primitive functions*, denoted as $(n_1 s \quad n_2 p \quad n_3 d)$. In the LCAO they may be grouped (contracted) to N_1 s-type $(N_1 < n_1)$, N_2 p-type, N_3 d-type, and so on, AOs. This *contracted basis* is usually denoted in brackets, as $[N_1 s \quad N_2 p \quad N_3 d]$. For example, in the sentence "$(9s5p3d)$ contracted to $[4s3p2d]$," $(9s5p3d)$ means that the basis set contains nine primitive s-type Gaussian functions (nine different radial parts with the same s-type angular factor), five p-type and three d-type functions, while $[4s3p2d]$ means that in the LCAO there are only four c_{ij} coefficients presenting the s functions (four groups of Gaussian functions with independent variational coefficients), three groups of p functions, and two groups for d functions; in all there are 17 primitive basis functions with only nine LCAO coefficients.

In the presentation of Pople et al. (see, e.g., [5.32]) the notations for basis sets of the first row atoms have several figures before the G, denoting the numbers of Gaussian functions in the inner ($1s$) state and in the split-valence states ($2s$ and $2p$). For example, 6-31G means that the inner $1s$ function (for nonhydrogen atoms) is represented by six Gaussian functions, while the outer-valence-shell functions are split into two groups ($2s$, $2p$ and $2s'$, $2p'$), the first being represented by a 3G (the same for $2s$ and $2p$) and the second by a single Gaussian function (one function for $2s'$ and $2p'$).

An extension of this basis set is achieved when *polarization functions* are added to give additional flexibility to the description of the MOs. For s functions polarization additions are of p-type, and for p-functions they are of d-type. If polarization functions are added only for nonhydrogen atoms, the basis set is denoted by one star, for instance, 6-31G*, while additional polarization functions for the hydrogen atoms (p functions) are indicated by adding the second star (6-31G**). In [5.32] the parameters are given for an even more sophisticated basis set for the first row atoms, 6-311G**. This means that the valence shell is split into three groups ($2s$, $2p$; $2s'$, $2p'$; $2s''$, $2p''$), represented by 3, 1, and 1 Gaussian functions, respectively, and polarization functions for nonhydrogens and hydrogens are included. In addition to polarization functions, the basis set may be improved by adding, diffuse (usually hybridized sp) functions more extended in space, which are especially important for anions. Diffuse basis functions are marked by " $+$ ": for example, $4\text{-}31+G$.

In the case of transition metal compounds in the description (n_1s n_2p $n_3d/n_1's$ $n_2'p$ $n_3'd/...$), a slash usually divides different basis sets for different atoms of the same system; the first place is for the basis set of the CA, followed by the basis set for the ligands of the first coordination sphere, and so on. The use of different basis sets for the CA and ligands is quite reasonable, taking into account the difference in their electronic configuration, especially the essential role of d electrons, the *d electron heterogeneity* introduced by the transition metal in the otherwise approximately *electronically homogeneous* system (in the sense that it contains electronically similar *nsnp* atoms). In homogeneous systems with more or less electronically similar atoms, the use of different basis sets for different atoms is not recommended because this can result in an artifact of charge redistribution toward the region where a larger basis set is used.

One of the common errors introduced by the use of limited basis sets emerges in MO LCAO calculations of intermolecular interactions, for instance, in chemical reactions. If we calculate two molecular systems A and B separately with poor basis sets, their molecular properties will be evaluated with certain errors which, in principle, can be eliminated by extending the basis. When approaching A to B and calculating them together with the same two basis sets, the number of basis functions in the MO LCAO scheme becomes that of $A + B$, that is, much larger than for A and B taken separately. This superposition of the basis sets, in addition to describing the A–B interaction, *improves the calculated intramolecular interactions in each of the two molecules.* Therefore, the energy of interaction between A and B, E_{int}, is not equal to the difference between the energies of the joint system $E(A$–$B)$ and the sum of the energies of the separate molecules $E(A) + E(B)$ because of the changes of $E(A)$ and $E(B)$ due to the superposition of the two basis sets. We have

$$E(A) + E(B) - E(A-B) = E_{int} + E_{BSSE}$$

where E_{BSSE} is the energy of the *basis set superposition error* (BSSE). The more complete the basis set, the lower the E_{BSSE} value. For example, for the interaction of Li^+ with $C_2H_5^-$, *ab initio* calculations with STO-3G yield $E_{BSSE} = 72.63$ kcal/mol, while with the $6\text{-}31+G$ basis, $E_{BSSE} = 0.20$ kcal/mol [5.33]. For further discussions of basis sets for transition metal systems, see [5.34, 5.35]. Results of calculations with different basis sets are illustrated in Sections 6.3 to 6.5.

Electron Correlation Effects

Even when the best basis sets are used in the MO LCAO calculations (functional complete sets) and the best possible HF (HF limit) solutions are obtained, the results may be far from satisfactory. The discrepancy between full HF calculations and experimental data is caused by the defect in the HF method mentioned above: It does not include *electron correlation effects.*

The origin of electron correlation effects is as follows. In one-electron approximations any given electron is assumed to move independently in the mean field created by all the other electrons and nuclei. Under this assumption the repulsion between the electrons is considered as an average, but not at each moment. In particular, in this scheme there is a probability that two electrons meet at the same point of space, whereas actually this is impossible because of their repulsion, which keeps them as far as possible from each other. Thus the motions of the electrons in time are not independent, but correlated.

For instance, in the simplest case of a helium atom the two correlated $1s$ electrons at any instant occupy diametrically opposite positions on the $1s$ sphere (Fig. 2.1), whereas in the HF description they can instantly be at the same point of the $1s$ distribution. In the case of many electrons, the correlation around each electron produces a "hole" in the charge distribution of all the other electrons, sometimes called the *Fermi hole.*

A more detailed mathematical analysis shows that if the full wavefunction is taken in one-determinant form (5.45) (thus obeying the Pauli principle), the probability of finding two electrons with parallel spins at the same point $\mathbf{r}$ is zero. This means that the motions of spinlike electrons in the HF method are correlated, and there are correlation Fermi holes in the charge distribution of such spinlike electrons. However, electrons with opposite spins remain uncorrelated. In particular, two electrons on the same MO have opposite spins, and hence are uncorrelated.

The correlation energy E_{corr} of the system is determined as the difference between the exact nonrelativistic value E_{ex} and the Hartree–Fock limit E_{HF}:

$$E_{corr} = E_{ex} - E_{HF}$$

The importance of correlation effects for any molecular system is beyond doubt. Even in the case of small molecules it can be significant when important chemical characteristics are calculated. For instance, for the HF molecule (not to be confused with the HF method), the total energy calculated by the SCF MO LCAO method [5.34] equals -2722.65 eV, while the experimental value is -2734.16 eV. At first sight the discrepancy is not very large, less than 1%. But if we calculate the dissociation energy as the difference between the energy of the whole molecule and the free atoms, we have (in eV) $D = -2722.65 + 2718.54 = -4.11$, whereas the experimental value is -6.08 eV; the error is about 30%. Examples for coordination compounds are discussed in Sections 6.3 and 6.4.

The development of methods of calculation of correlation effects is one of the most important and difficult problems in modern quantum chemistry [5.17]. Most widespread are different versions of *configuration interaction* (*CI*). The main idea of CI is to look for the full wavefunction of the system Ψ in the form of a linear combination of the wavefunctions Ψ_k constructed after (5.45) for different electronic configurations that have the same symmetry properties:

$$\Psi = \sum_k A_k \Psi_k \qquad (5.52)$$

where the coefficients A_k are determined by the variational principle (similar to $c_{\mu k}$ in the MO LCAO method). The inclusion of excited configurations in the search for the best fit to the real charge distribution in the system is certainly very helpful, allowing one to improve the HF results significantly. However, in practice, the number of integrals and the size of determinants to be solved increase rapidly with the number of configurations in the expansion (5.52).

In general, the number of possible excited configurations equals the number of determinants of the form (5.45) that can be constructed from the basis set functions of the system. For example, if for the water molecule with 10 electrons one chooses a basis set of 14 AOs, which in the double-zeta approximation yield 28 orbitals in the LCAO, the number of possible configurations equals the number of combinations of 28 in 10 that equals $\sim 1.3 \cdot 10^7$. With such a large number of excited configurations the problem becomes extremely difficult. Fortunately, there are several ways to reduce the number of configurations in the CI procedure, based mostly on symmetry considerations. It is easily seen that only those excited configurations Ψ_k contribute to the CI that have the same symmetry as the ground state. Indeed, since the Hamiltonian is totally symmetric, all its matrix elements are zero if the two functions, ground and excited, belong to different irreducible representations of the point group of the molecule (Section 3.4). For the foregoing example of the water molecule of C_{2v} symmetry with the ground state $^1A_1$, the effective excited configurations must be of orbital A_1 symmetry and with the same number of α and β (spin-up and spin-down) spin orbitals.

But even after the essential reduction of the number of possible excited configurations by symmetry, it remains too large to be manageable. Further reductions should be based on analysis of the possible role of different configurations in order to make reasonable truncations of the expansion (5.52). The convergence of the latter is also dependent on the basis set. A detailed study of the problem resulted in several versions of the CI method, including the *independent electron-pair approximation* (IEPA), *single and double CI* (CISD), *cluster expansion of the wavefunction, coupled-pair many-electron theory* (CPMET), *coupled-electron-pair approximation* (CEPA), and other *coupled-cluster methods* (for details, see [5.17, 5.36–5.39]).

One of the extensions of the CI method is to include the expansion (5.52) in the SCF MO LCAO scheme, that is, to vary the total energy with respect to $c_{\mu k}$ and A_k simultaneously; in this case the method is called *multiconfiguration* SCF (MC SCF). Another method widely used now is CAS SCF (*complete active space SCF*), in which a full CI is included for a set of orbitals chosen as "active." Quite a number of works are devoted to the use of approaches based

Table 5.6. Summary of *ab initio* methods of electronic structure calculations[a]

Method and Denotation	Choice of Variational Ψ	Characterization
Hartree–Fock (HF)	One determinant built from one-electron MO $\varphi(i)$	Labor $\sim n^4$; $\varphi(i)$ adjusted iteratively by the SCF procedure; method is size extensive
Configuration interaction (CI)	Many determinants $\Psi = \Sigma_k A_k \Psi_k$	Labor $\sim n^5$ or greater; method converges slowly and is in general not size extensive
Multiconfiguration SCF (MCSCF)	Ψ same as in CI	Both CI and $\varphi(i)$ optimized simultaneously; not size extensive, convergence often difficult
Complete active space SCF (CAS SCF)	Special case of MCSCF	All determinants (full CI) within the chosen set of "active" orbitals in the MCSCF scheme; size extensive, CI expansion increases rapidly
Generalized valence bond (GVB)	Special case of MCSCF	VB functions (combinations of determinants) are used in the CI; simplest case, perfect pairing (two-electron bonds)
Many-body (Moller–Plesset) perturbation theory (MP2, MP3, etc.)	Ψ same as in CI	A_k determined by perturbation theory; nonvariational, size extensive
Coupled cluster CPMET (CCD, CCDS, IEPA, CEPA, etc.)	Ψ same as in CI	A_k determined by exponential cluster expansion; nonvariational, size extensive
Quantum Monte Carlo (QMC)	Quantum simulation HF or CI "guiding" functions	Small systems; difficult for excited states

Source: After [5.40] with minor changes.
[a]n is the number of AOs in the basis set.

on many-body Rayleigh–Schrödinger perturbation theory (MB-RSPT) often called *Moeller–Plesset (MP) approximation*; MP-2, MP-3,... denote MP approximations of the second-, third-,... order perturbation theory. Table 5.6 presents a summary of different classes of *ab initio* methods of electronic structure calculations [5.40]. Some of these methods are *size extensive*; that is, they scale as the exact energy with the number of particles in the system (in other words, the energies $E_{AB}(R \to \infty) = E_A + E_B$, where A and B are two parts of the bonded system AB and R is the distance between them) [5.39].

5.4. SEMIQUANTITATIVE APPROACHES

Calculations of the electronic structure of coordination compounds by SCF MO LCAO methods is not a routine procedure yet for chemists who are not experts in the field. Each such calculation is still a creative piece of work that requires from the researcher expert and intuitive knowledge of how to choose the method of calculation, the basis set, the iterative and CI procedures, the output, and so on, and to evaluate the meaning of the results. In addition, it requires considerable computer time, thus being expensive, not to say un-manageable for large molecules. Therefore, there is a strong demand for simplifications of the methods of calculation to make them most accessible.

As mentioned above, the SCF version of the MO LCAO method, when carried out directly, is called *ab initio* or *nonempirical* in the sense that it is not based on any assumptions about the parameters of the molecule except the nuclear charges and number of electrons. (This statement is not exactly true because the calculation procedure is never purely mathematical; it requires a great deal of experience, chemical knowledge, and intuition.) For coordination compounds, including heavy metal atoms and polyatomic ligands, to date, simplifications have been almost absolutely necessary. Some can be introduced without using empirical parameters, thus not greatly violating the nonempirical character of the calculations. In these cases the method may be called *semiquantitative*.

An illustrative example of semiquantitative approaches aimed at transition metal systems is provided by the *Fenske–Hall method* [5.41]. It employs an approximate Mulliken (Section 5.5) presentation of the matrix elements of the effective Hamiltonian in the SCF MO LCAO method, which essentially simplifies the calculations without introducing empirical parameters. In the Fenske–Hall approximation these matrix elements are

$$H_{ii} = \varepsilon_i^A - \sum_{\mu \neq A} \frac{e^2(Z_\mu - q_\mu)}{R_{A\mu}}$$

$$H_{ij}^{AB} = (\varepsilon_i^A + \varepsilon_j^B)S_{ij}^{AB} - T_{ij} - \frac{1}{2} e^2 S_{ij}^{AB} \sum_{\mu \neq A,B} \left(\frac{Z_\mu - q_\mu}{R_{A\mu}} + \frac{Z_\mu - q_\mu}{R_{B\mu}} \right) \qquad (5.53)$$

$$H_{ij}^{AA} = 0$$

where ε_i^A is the atomic orbital energy, $R_{A\mu}$ the distance to the atom μ with the effective charge $Z_\mu^* = Z_\mu - q_\mu$, Z_μ the nuclear charge, and q_μ the electronic charge on the atomic center (5.20), $q_\mu = \Sigma_{i,v} q_i c_{i\mu} c_{iv} S_{\mu v}$, and $S_{\mu v}$ is the overlap integral.

With the matrix elements (5.53) the HFR procedure of the MO LCAO method of calculation of the electronic structure is essentially simplified: Instead of the matrix elements (5.48), including one-, two-, three-, and four-center integrals (whose number is $\sim n^4$), mainly one-center (in ε_i), overlap (S_{ij}), and kinetic energy integrals, as well as Mulliken charges (5.20) should be calculated. Unlike other methods that give similar simplicity (see below), the Fenske–Hall method does not involve any adjustable or empirical parameters, except the interatomic distances, that are chosen in a nonarbitrary fashion.

Pseudopotentials or Effective Core Potentials

Among semiquantitative methods, those that use pseudopotentials are most widespread. To avoid all-electron calculations of the electronic structure which for transition metal compounds may be extremely difficult, various approximations were suggested aimed at excluding the inner electrons from the calculations. We present here a recent suggestion [5.42] that has proved useful.

The idea in general is to assume that the inner electrons are not participating directly in the chemical bonding and hence can be taken into account in a simpler way as a source of a special potential for the outer (valence) electrons. Provided that we know this potential exactly, the all-electron problem becomes a valence-electron problem, essentially simplifying the calculations. This potential [often called a *pseudopotential*; in [5.42] it is called an *effective core potential* (ECP) or *relativistic ECP* (RECP)] should be constructed such that it preserves all the features of the interactions of valence electrons with the core, at least in the valence area, where chemical bonding is important. The procedure to do this is as follows.

The electrons of the transition metal atom are divided into valence and core electrons. The core in the first series transition elements is formed by electrons with an Ar-like shell: $[\ldots 3s^2 3p^6] = [\text{Ar}]$. For the second series they are $[\ldots 3d^{10} 4s^2 4p^6] = [\text{Kr}]$, while for the third one the core is $[\ldots 4d^{10} 5s^2 5p^6 4f^{14}] = [\text{Xe} \cdot 4f^{14}]$ with the exception of the La group, where the [Xe] core only is replaced by pseudopotentials. In some cases the outermost orbitals $(ns)^2(np)^6$ cannot be replaced by ECP and should be treated on an equal footing with the valence nd, $(n + 1)s$, and $(n + 1)p$ electrons [5.42].

Each core electron with the angular momentum quantum number l creates an effective potential U_l for the valence electrons; the summary potential of the core electrons is

$$U(r) = U_L(r) + \sum_{l=0}^{L-1} [U_l(r) - U_L(r)] P_l^*$$
(5.54)

where L is a unity greater than the highest l of any core orbital, and P_l^* is the core projection operator:

$$P_l^* = |l\rangle\langle l|$$
(5.55)

where $|l\rangle = \Phi_l$ is an exact (relativistic or non-relativistic) Hartree–Fock atomic function. Actually, P^* means that in the equation for the valence electron wavefunction, for instance, $|l'\rangle$, with the pseudopotential U, only those terms in the sum over l in (5.54) remain effective, for which the wavefunction $|l\rangle$ is not orthogonal to $|l'\rangle$.

To obtain the effective potential U_l, a pseudo Hartree–Fock function Φ_l^p is introduced. It is assumed that Φ_l^p follows Φ_l as close as possible in the outer (valence) region of the atom, and it is nodeless and goes smoothly to zero inside the atom. More precisely, $\Phi_l^p = \Phi_l$ in the region of $r_e > r_c$, where r_c is near the outermost maximum of Φ_l, while $\Phi^p = r^b f(r)$ inside this region; $f(r)$ is a fifth-degree polynomial and $b = l + 3$ (in the relativistic case b is taken somewhat different, see [5.42]). The coefficients of the polynomial are determined to satisfy the conditions of continuity, normalization, nodeless, and so on.

Now the potential U_l is obtained from the condition that if inserted in the one-electron Hartree–Fock equation, it yields the pseudofunction Φ_l^p with the same energy ε_l. This means that $U_l(r)$ obeys the one-electron equation for the atomic radial wavefunction [5.43]:

$$\left[-\frac{\frac{1}{2}d^2}{dr^2} + \frac{l(l+1)}{2r^2} - \frac{Z}{r} + U_l(r) + V_{val}(r) \right] \Phi_l^p = \varepsilon_l \Phi_l^p \tag{5.56}$$

where $V_{val}(r)$ is the Hartree–Fock (Coulomb and exchange) interaction of the l electron with the valence electrons. If we relate this equation to that for the exact Hartree–Fock atomic orbital Φ_l [5.43], we can extract the expression for U_l sought for:

$$U_l(r) = \varepsilon_l - \frac{l(l+1)}{2r^2} + \frac{Z}{r} + \frac{(\Phi_l^p)''}{2\Phi_l} - \frac{V_{val}\Phi_l^p}{\Phi_l} \tag{5.57}$$

The one-electron effective core potential can be relatively easily evaluated, provided that the atomic Hartree–Fock one-electron functions Φ_l and energies ε_l, as well as the pseudoorbitals Φ_l^p and their second derivatives $(\Phi_l^p)''$, are calculated.

Similar to basis wavefunctions (Section 5.3), the ECPs obtained by Eq. (5.57) are presented conveniently in the form of a sum of Gaussian functions with tabulated coefficients for all the important atoms up to Hg for both nonrelativistic and relativistic cases [5.42]. The latter are obtained on the basis of the principles indicated above, including relativistic terms.

The pseudopotentials are thus generated from atomic data by adjustment to orbital energies and orbital densities. Besides, other ways of adjustment to atomic data are possible, especially to excitation and ionization energies taken from experimental data [5.44]. This version of pseudopotentials has some advantages: The total energy differences are physical observables (while orbital energies and densities are not) and the experimental data incorporate all the additional (e.g., relativistic and core-valence correlation) effects.

Density-Functional Approaches, X_α methods

One rapidly developing trend in electronic structure calculations with approaches already widely used is based on *density-functional theory* [5.45–5.49]. It was shown [5.50] that the multielectron problem can be formulated rigorously in terms of electron densities (5.14), instead of wavefunctions, and reduced to Hartree-like equations that can be solved self-consistently. If the corresponding equations are solved exactly, the two presentations of the problem, by wavefunctions and densities, are completely equivalent. Actually, these equations are solved approximately, and the approximations are different in the two approaches. In many circumstances these differences confer considerable advantages to the density functional description in terms of accuracy and ease, compared with the presentation by wavefunctions.

Density functional calculations were performed first for atomic states and then for crystal lattice band structures. Extensive applications to molecular problems began with the X_α methods, which employ the density functional ideas to approximate the interelectron interactions [5.51–5.54]. As compared with modern more sophisticated density functional approaches, the X_α methods still preserve some advantages of simplicity with not much lower accuracy [5.54] if used in proper circumstances. Therefore, we consider first the X_α methods followed by the Kohn–Sham equations of DFT.

There are two main X_α methods:

1. SCF-X_α-SW: self-consistent-field X_α scattered wave method
2. SCF-X_α-DV: SCF X_α differential variation, sometimes called multiple scattering (MS-X_α) method

Similar to the MO LCAO approaches, the X_α methods are based on the one-electron approximation and the Hartree–Fock self-consistent procedure. However, distinct from the MO LCAO approach, the SCF-X_α-SW method declines presentation of the MO by LCAO and uses another approach: The MO is taken as a combination of solutions obtained separately for the regions near the nuclei, in between them, and outside, considering that in each of these regions there is an average electron density distribution (in the DV method the LCAO presentation is employed partially; see below).

Consider a polyatomic system and assume, following the spirit of the one-electron approximation, that each electron can be ascertained by its coordinate wavefunction $\varphi_j(\mathbf{r})$ and spin. Let us define the electron density $\rho(\mathbf{r})$ [Eqs. (5.13) and (5.14)] as the sum of the densities of the electrons with spins up and down (i.e., spins $\frac{1}{2}$ and $-\frac{1}{2}$, respectively) indicated by arrows:

$$\rho(\mathbf{r}) = \rho\uparrow(\mathbf{r}) + \rho\downarrow(\mathbf{r}) \tag{5.58}$$

$$\rho\uparrow(\mathbf{r}) = \sum_{j\uparrow} n_j|\varphi_j(\mathbf{r})|^2,$$

$$\rho\downarrow(\mathbf{r}) = \sum_{j\downarrow} n_j|\varphi_j(\mathbf{r})|^2 \tag{5.59}$$

Here n_j is the number of electrons (occupation number) in the one-electron state $\varphi_j(\mathbf{r})$ which may be different for spins $\uparrow$ and $\downarrow$. In these notations the density functional expression for the total energy of the system (without internuclear interactions) is

$$E_{X\alpha} = \sum_i n_i \int \varphi_i^*(\mathbf{r})H^0\varphi_i(\mathbf{r})\, d\tau + \tfrac{1}{2}e^2 \int \rho(1)\rho(2)\left(\frac{1}{r_{12}}\right) d\tau_1\, d\tau_2$$
$$+ \tfrac{1}{2}e^2 \int \left[\rho\uparrow(1)U_{X\alpha\uparrow}(1) + \rho\downarrow(1)U_{X\alpha\downarrow}(1)\right] d\tau \tag{5.60}$$

where H^0 is the core Hamiltonian (2.38) (the operator of the kinetic energy of the electron and of its interaction with the nuclei), while $U_{X\alpha\uparrow}$ and $U_{X\alpha\downarrow}$ are the corresponding exchange potentials, which are different for different spins $\uparrow$ and $\downarrow$ because the exchange interaction is nonzero only for electrons with the same spin orientations.

The second term in (5.60), the pure Coulomb repulsion of the electrons (calculated as the repulsion of two electronic clouds), also contains the energy of the self-interaction of the electron, which is canceled by the corresponding terms of the third exchange interaction term, which is opposite in sign [quite similar to the Hartree–Fock equation (2.41)].

Equation (5.60) is an exact expression for the total energy of the system provided that $U_{X\alpha}$ stands for the exact potential of exchange interactions. The name of the method under consideration is based on the approximate choice of this potential. It is taken from the statistical (Thomas–Fermi) model of atoms, from which it follows that *the exchange potential can be approximated by a term proportional to the electron density in the $\frac{1}{3}$ power*. Slater [5.51] suggested choosing this potential for molecules as follows [Hartree–Fock–Slater (HFS) method]:

$$U_{X\alpha\uparrow} = -9\alpha\left(\frac{3}{4\pi}\rho\uparrow\right)^{1/3} \tag{5.61}$$

where α is a parameter (for $U_{X\alpha\downarrow}$ the expression is quite similar). The presentation of the exchange potential by Eq. (5.61) is the first significant approximation of the X_α method.

By means of variation of the energy (5.60) [taking $U_{X\alpha}$ from (5.61)] with respect to the functions $\varphi_i(\mathbf{r})$ for fixed n_i, one can find the equations for $\varphi(\mathbf{r})$ that satisfy the minimum energy condition:

$$\left[-\frac{\hbar^2}{2\mu} \Delta + V_C(\mathbf{r}) + V_{X\alpha}(\mathbf{r}) \right] \varphi_i(\mathbf{r}) = \varepsilon_i \varphi_i(\mathbf{r}) \tag{5.62}$$

where V_C is the Coulomb potential of all the electrons and nuclei, and

$$V_{X\alpha\uparrow}(\mathbf{r}) = \tfrac{2}{3} U_{X\alpha\downarrow}(\mathbf{r}) \qquad V_{X\alpha\downarrow}(\mathbf{r}) = \tfrac{2}{3} U_{X\alpha\downarrow}(\mathbf{r}) \tag{5.63}$$

are the exchange potentials, which are different for different groups of electrons having different spins.

In principle, Eq. (5.62) does not differ from the Hartree–Fock equation (2.41) [the choice of exchange potentials in the form (5.61) makes them closer to the Hartree equations (2.37)]. The two X_α methods, mentioned above, differ by the method of solution of these equations. In the X_α-DV method the function $\varphi_i(\mathbf{r})$ is sought for in the form of LCAO (5.46), and the LCAO coefficients are determined by the condition of energy minimum, quite similar to the MO LCAO method considered in Section 5.3. In this case the self-consistency of the calculations is achieved either on the electron density $\rho(\mathbf{r})$ (SCF-X_α-DV method) or on effective atomic charges (SCC-X_α-DV) [5.55]. However, due to the special choice of the potential V_X, the number of integrals to be calculated is proportional to n^2 (n is the number of basis functions) instead of n^4 in the usual MO LCAO methods, thus significantly simplifying the calculations. Pseudopotentials were also introduced in the X_α-DV method [5.56].

In the X_α-SW method the procedure of solution of Eq. (5.71) is quite different from that of the MO LCAO method, mainly by choice of the "*muffin-tin*" *potential*. Surround each nucleus by a sphere of the radius R with the center at the nucleus in such a way that the spheres from near-neighbor atoms were contiguous to each other (in special cases, e.g., for electron lone pairs, additional "intermediate spheres" can be introduced), and the system as a whole with a large sphere that includes all the atomic spheres. Figure 5.4 illustrates the choice of the muffin-tin potential for the MnO_4^- system [5.52]. Divide the entire space of the system into three regions:

   I. Region of intra-atomic spheres
  II. Region between the atomic spheres but inside the big sphere
 III. Region out of the big sphere

The second essential approximation of the X_α-SW method is that Eq. (5.62) is solved for regions I, II, and III separately and then the solutions are "sewed." The potential $V_C + V_X$ is taken as spherically symmetrical for each sphere in

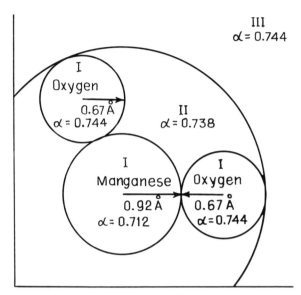

Figure 5.4. Division of the muffin-tin potential of the complex MnO_4^- into three regions and the α values for each of them in the X_α-SW calculations. I, region inside the atomic spheres; II, region in between the atomic spheres but inside the big sphere; III, region outside the big sphere. (From [5.52].)

region I; in region II it is assumed to be constant, averaged over the whole volume of the region; and in region III it is again spherically symmetrical with respect to its center.

With these potentials the solutions of (5.62) in regions I and III are hydrogenlike functions of the type (2.1), with the radial part determined by numerical integration, while in region II the solutions are analytical special functions that can be interpreted as diverging, converging, and standing waves reflected from the atomic spheres I and the outer region III.

To sew these solutions, the function $\varphi_i(\mathbf{r})$ sought for is expanded over the system of solutions in each of the three regions and the coefficients of these expansions are determined from the conditions of continuity of the $\varphi_i(\mathbf{r})$ function and their first derivatives at the borders. These conditions are presented in the form of a set of secular equations that have nontrivial solutions for some definite values of the energies ε—eigenvalues for the one-electron state $\varphi_i(\mathbf{r})$. The procedure of sewing the solutions is called *the method of multiple scattered waves*.

With the wavefunctions $\varphi_i(\mathbf{r})$ known, one can first calculate the electron densities $\rho\uparrow(\mathbf{r})$, $\rho\downarrow(\mathbf{r})$, and $\rho(\mathbf{r})$ after (5.58) and (5.59); then the potential V_c (using the classical Puasson equation) and V_x after Eq. (5.61), as well as their averages (spherically in regions I and III and over the volume in region III) can be evaluated. With these new potentials one can again solve Eq. (5.62) and determine new values of ε' and new (modified) functions $\varphi_i'(\mathbf{r})$, and so on, until complete self-consistency is reached when the new ε' and φ' values do not differ from the previous ones (within the accuracy required).

Note that in the X_α-SW method the relationship between the one-electron energy ε_i and the total energy E is different from that in the Hartree–Fock method. In the X_α method, ε_i equals the derivative of E with respect to the orbital population number n_i [5.51]:

$$\varepsilon_{iX\alpha} = \frac{\partial E_{X\alpha}}{\partial n_i} \tag{5.64}$$

whereas in the Hartree–Fock scheme ε_i equals the energy difference of two states that differ in the occupation numbers:

$$\varepsilon_{iXF} = E_{XF}(n_i = 1) - E_{XF}(n_i = 0) \tag{5.65}$$

The presentation (5.64) has some advantages compared with (5.65). In particular, to determine the energy of a one-electron transition that changes the population numbers of the initial state $n_i \to n_i - 1$ and final state $m_i \to m_i + 1$ in the X_α-SW method, there is no necessity to calculate the total energies of both states, initial and final. Following Eq. (5.64), the energy difference sought for can be evaluated by calculating only one state: the intermediate state with $n_i = n_i - \frac{1}{2}$ and $m_i = m_i + \frac{1}{2}$ [5.51].

Additional explanations for the choice of the sphere radii R_i and the parameter α in the $U_{X\alpha}$ potential after (5.61) are needed. The first suggestion of Slater, to take $\alpha = 1$, was later corrected by Gaspar [5.57], who showed that $\alpha \cong \frac{2}{3}$. More exact values of α can be obtained by comparison of the total energy of the atom calculated with the potential (5.61) with that obtained by the Hartree–Fock method. In this way the α values were tabulated for almost all the atoms [5.58]. There are also other methods of estimation of the α magnitude [5.51]; all of them give rather close values that are in the interval between the Slater value $\alpha = 1$ and the Gaspar value $\alpha = \frac{2}{3}$. In region II α can be taken as an average over the values in the spheres of region I, while in region III it can be taken close to the value for the outer atoms.

The radii of the spheres R_i in region I must obey the condition of continuity of the potential at the border of the contiguous spheres (taking into account the interatomic distances). In the case of a polyatomic molecular system there remains some arbitrariness in the choice of R_i, but fortunately, the results do not depend strongly on small changes of R_i. Instead of contiguity, superposition of spheres can also be allowed. In region III there may be additional border conditions. For instance, for a charged cluster in a crystal, an additional averaged potential produced by the ions of the crystal lattice (similar to the Madelung potential) should be introduced.

One of the advantages of the X_α solution is that it approaches the sum of solutions for noninteracting atoms when increasing the interatomic distances in the system [5.51]. Distinct from this result, the usual MO LCAO method without CI yields solutions that at large interatomic distances include atomic and ionic structures with the same weight (e.g., the structures H + H and $H^+ + H^-$ for the hydrogen molecule). This is obviously due to underestimation of the correlation effects (Section 5.3). Thus the electron correlations are better accounted for in the X_α-SW method than in the MO LCAO method without CI.

In the case of heavy-atom systems, relativistic corrections to the X_α method should be introduced. This can be done by passing from Eq. (5.62) to the relativistic Dirac–Slater equation with four-component functions [5.59] or to the simplified quasi-relativistic version with two-component spinors (see below). Several other improvements were brought to the original X_α methods [5.54].

In general, the main advantage of X_α methods compared with MO LCAO methods is in the significant reduction of labor and computer time required for electronic structure calculations, especially for large molecular systems. In the X_α-SW method there is no problem of choice of the basis set and no n^4 dependence of the number of integrals on the number of functions n in the basis set; in the X_α-DV method it is as n^2.

X_α methods are less suitable for geometry optimization, and this is one of their main shortcomings. Indeed, the X_α results are not very sensitive to geometry changes, especially to the angles between bonds. Another disadvantage of the X_α-SW method is that, in contrast to MO LCAO, there is no direct genealogical linkage of the electronic structure of the system with the atomic structures, which makes visualization of the results and understanding of their chemical origin difficult. Examples of calculations by X_α methods are given in Sections 6.3 and 6.4.

Significant improvement of both the foundation and practical use of DFT methods was achieved in its more recent versions based on the Hohenberg–Kohn theorem and Kohn–Sham equations [5.45–5.50]. In its rigorous formulation [5.50], the main assumption of DFT is that for any polyatomic system, the electron–nuclear and nuclear–nuclear interactions $V(r, Q)$ (Section 7.2) may be regarded as an "external potential" for the electronic subsystem. Under this assumption the Hohenberg–Kohn theorem [5.50] establishes a one-to-one correspondence between the electronic density ρ and the potential V, and hence between ρ and the Hamiltonian H, since the latter is defined completely if V is known.

It follows that the electronic density contains, in principle, the information of all the properties of the system that can be described by the Hamiltonian. But the Hohenberg–Kohn theorem is just a "theorem of existence," it does not show how one can calculate the density of electrons, as well as the Hamiltonian and the other properties of the system, from the densities. Kohn and Sham [5.60] worked out a practical tool to obtain the energies and densities of the system (at least approximately) without solving the many-electron Schrodinger equation for multideterminant wavefunctions.

The Kohn–Sham equations are quite similar to the Hartree-like equation (5.62) for the one-electron functions $\varphi_i(r)$ in the X_α methods, but with a more appropriate potential [instead of $V_{X\alpha}(r)$] which takes into account the correlation effects, in principle, better than in the X_α methods:

$$\left[-\frac{\hbar^2}{2\mu} \nabla^2 + V(r) + \int \frac{\rho(r')}{|r - r'|}\, dr' + \frac{\delta E_{XC}}{\delta \rho} \right] \varphi_i(r) = \varepsilon_i \varphi_i(r) \qquad (5.66)$$

where $\rho(r)$ is given by Eq. (5.58) and E_{XC} is the *exchange correlation potential*, which contains the exchange and correlation terms and the "residual" kinetic energy (the difference between the kinetic energy of interacting and noninteracting electrons). In fact, (5.66) is a Hartree equation for electrons which interact by Coulomb forces "in average," that is, without exchange and correlation effects, but with an additional special potential $\delta E_{XC}/\delta \rho$ which is a functional of ρ and takes into account all the missed effects. (A functional is a function of another function instead of the usual function of variables.)

Equations of the type (5.66) are relatively simple and can easily be solved, provided that the functional $E_{XC}[\rho]$ is known; the expression for ρ in terms of $\varphi_i(r)$ is given by Eq. (5.59). But the functional $E_{XC}[\rho]$ is unknown (in general), and in practice it should be found using some simple reference systems as a model. In particular, such a reference system could be the free gas of electrons or the Thomas–Fermi atom, mentioned above, for which the energy as a function of ρ is known approximately. Concerning practical means to solve Eq. (5.66), the MO LCAO approximation for $\varphi_i(r)$ may be quite suitable. In particular, for transition metal systems the method of linear combinations of Gaussian-type orbitals (LCGTOs), in combination with model core potentials (MCPs) (the LCGTO-MCP-DF method [5.61]) seems to be appropriate.

The X_α method is a particular case of the Kohn–Sham equations (5.66) with the exchange-correlation potential taken in the form (5.61). In this case E_{XC} is a function of density only (and not its derivatives, see below); such presentations are called *local density approximation* (LDA). In more sophisticated calculations with a better account of exchange and correlation effects, the potential E_{XC} also depends on the gradient $\nabla\rho$. For example, in the generalized gradient approximation (GGA) [5.62] we have

$$E_{XC} = -\frac{3}{4}\left(\frac{3}{\pi}\right)^{1/3} \int \rho^{4/3} F(s)\, d\tau \qquad (5.67)$$

where

$$F(s) = (1 + 1.29s^2 + 14s^4 + 0.2s^6)^{1/15}$$

$$s = |\Delta\rho(r)|(24\pi^2)^{1/3}\rho^{4/3}$$

To conclude this brief discussion of DFT, we note that the foundation of the theory is not rigorous because it does not take into account the nuclear motions [5.63]. In particular, the DFT main equations are invalid for electronic degenerate and pseudodegenerate states (Sections 7.3 and 7.4) because in these cases the total wavefunction is not multiplicative with respect to the electronic and nuclear wavefunctions (their motions are not separable), and hence the main assumption of DFT that the electron–nuclear interaction can be regarded as an external potential does not work.

In the wavefunction presentation the special coupling between the electronic and nuclear motions is taken into account in the system of coupled equations (7.6); the latter cannot be evaluated in the DFT because the coupling between different electronic states expressed by off-diagonal elements cannot be calculated in the density presentation. For nondegenerate states that are well separated from the other electronic states, the DFT, although approximate in its foundation (it ignores the dependence of electronic functions on nuclear coordinates [5.63]), yields reasonable good results for energies and charge distributions with less effort than in conventional MO LCAO methods with CI (Section 5.3).

Fragmentary Calculations

The importance of fragmentary calculations is increasing continuously in view of increasing demands for molecular modeling and its applications in various fields, especially in biology and drug design. To avoid bulky calculations of the electronic structure of large molecular systems, it may be useful to expand the molecule into smaller fragments that can be calculated separately and then to construct the solution for the system as a whole from the fragment solutions. Note that in the main versions of MO LCAO calculations, the computation time is approximately proportional to n^4 (n is the number of basis functions), and for large n values and r fragments ($n = n_1 + n_2 + \cdots + n_r$): $n^4 \gg n_1^4 + n_2^4 + \cdots + n_r^4$. Fragmentary calculations are also important in realization of an interface between quantum-mechanical description of local properties in large molecular systems and classical treatment of the system as a whole [5.64].

For nonempirical calculations of large organic molecules with closed shells, molecular fragments can be calculated first and then their wavefunctions can be used as the basis set for a full MO LCAO calculation of the system as a whole [5.65]. For instance, to calculate the ethane molecule in this way, the methane molecule is calculated first and then the solution for the ethane molecule is presented as a linear combination of two solutions for two methane molecules at a certain distance from each other. In so doing, it is assumed that the elimination of two hydrogen atoms is compensated in the basis set by formation of the C—C bond. The method was extended to large organic molecules.

Another method of fragmentary calculations, also aimed at mainly organic molecules but including semiempirical approaches and the possible use of data banks, was suggested in [5.66]. In this method the fragmentation of the large molecule is produced such that the border atoms or atomic groups are included in both neighboring fragments, so each fragment may, in principle, exist separately as a stable molecule. In the large molecule as a whole the fragment solutions are introduced and sewn under the condition that its geometry is maintained (but without additional MO LCAO calculations).

The fragmentary calculations widely used for organic molecules cannot be transferred directly to transition metal coordination compounds because of the d-electron delocalization properties (Sections 1.2 and 6.1) and significant charge transfers between the fragments (see below). A special analysis of the conditions when this separation is valid (at least approximately) yields the following results [5.64, 5.67].

It is obvious that the separation of a molecule into fragments is possible when the interfragment interaction is much smaller than the intrafragment one. To formulate this condition quantitatively, consider an arbitrary molecule cut into two fragments in such a way that there is only one border atom (BA), the border between the fragments crosses this atom (Fig. 5.5), and the BA does not form π bonds with at least one of the two fragments. Denote the σ functions directed to fragments I and II by σ_i and σ_k, respectively, the corresponding σ orbitals of near-neighbor atoms in the fragments being σ_i' and σ_k' (for simplicity, the possible π bond with one of the fragments is omitted). Then the secular equation in the MO LCAO approximation (5.8) with an orthogonalized basis set looks as follows:

$$
\begin{vmatrix}
\text{I} & & & & 0 \\
H_{i'i'} - \varepsilon & H_{i'i} & H_{i'k} & 0 \\
H_{ii'} & H_{ii} - \varepsilon & H_{ik} & H_{ik'} \\
H_{ki'} & H_{ki} & H_{kk} - \varepsilon & H_{kk'} \\
0 & H_{k'i} & H_{k'k} & H_{k'k'} - \varepsilon \\
0 & & & & \text{II}
\end{vmatrix} = 0 \qquad (5.68)
$$

where I and II denote the remaining intrafragment matrix elements (not written explicitly). In this equation the central part, separated by a dashed line, characterizes the condition on the border of the fragments described by the four σ functions above. It is seen that if

$$
H_{ik} = H_{i'k} = H_{ik'} = 0 \qquad (5.69)
$$

Eq. (5.68) decomposes into two independent equations, and the solution of the system as a whole equals the product of the fragment solutions exactly.

From the three off-diagonal matrix elements H_{ik}, $H_{i'k}$, and $H_{ik'}$ the last two are usually small, as caused by nonneighbor overlap (Fig. 5.5), while H_{ik}

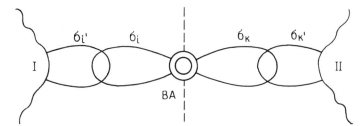

Figure 5.5. Border atom (BA) in the separation of molecular fragments (I and II) and the four σ orbitals, σ_k, σ_l, $\sigma_{k'}$, $\sigma_{l'}$, assumed to be most active in the bonding of the fragments.

depends essentially on the nature of the σ_i and σ_k atomic functions of the BA, namely, on their hybridization with s functions (s content). If there is no s content in the σ_i and σ_k functions (i.e., for pure p, d, f,... functions), $H_{ik} = 0$. In all other cases $H_{ik} \neq 0$, but in certain conditions it can be small. For instance, in the case of sp^3 hybridization $H_{ik} = \frac{1}{4}(H_{ss} - H_{pp})$, while the intra-fragment diagonal element $H_{ii} = \frac{1}{4}(H_{ss} + 3H_{pp})$, and hence H_{ik} can be much smaller than H_{ii}.

On the other hand, if the BA forms active π orbitals with both fragments, there are no grounds to separate the latter by this BA, because the off-diagonal intrafragment and interfragment matrix elements H_{ik} are of the same order of magnitude. Therefore, *the first rule of fragmentation is that the border atom between any two fragments must have only localized σ bonds with at least one of the two fragments* [5.67].

In any fragmentation of the molecule (even when the rule above is fulfilled and H_{ik} is small) the solution obtained differs from the true one for the molecule as a whole, and the problem is to have some criteria of the inaccuracy introduced by such fragmentary calculations. Fortunately, there is a mathematical theorem on localization of the characteristic numbers of matrices proved by Gershgorin [5.68] which contributes significantly to the solution of this problem. Following this theorem, the interval of localization of the roots of the secular equation (5.8) [or (5.66)] of the order n (i.e., the MO energies ε in the MO LCAO method with a basis of n functions) is limited by the following inequalities:

$$|H_{ii} - \varepsilon| \leqslant \sum_{j \neq i} |H_{ij}| \qquad (5.70)$$

(in complex-conjugate numbers these intervals become circles on the complex plane, *Gershgorin circles*). It is seen from this expression that localization of the MO energies ε with respect to the diagonal element (Coulomb integral), H_{ii}, is determined by the sum of absolute values of all the off-diagonal elements (resonance integrals) H_{ij} of the ith row. Therefore, the neglect of one of the matrix elements H_{ik} reduces two intervals of localization of the MO energies in the two fragments [the radii of the ith and kth Gershgorin circles in (5.70)], making them closer to the atomic values H_{ii} and H_{kk}. The effect is thus determined by the magnitude $|H_{ik}|$, compared with the sum of all the other values $|H_{ij}|$ (in the first fragment) or $|H_{kj}|$ (in the second fragment). From these results the second rule of fragmentation emerges: *The greater the intrafragment delocalization (the larger the sum of $|H_{ij}|$) compared with interfragment interaction (H_{ik}), the less the error of fragmentation.*

For small H_{ik} values the fragmentary solutions can be improved by considering H_{ik} as a small perturbation. In the case of n interacting MOs from the two fragments, the value of the new MO energy level ε_α with respect to the unperturbed one ε is given by the following inequality, which is similar to (5.70) [5.67]:

$$|\varepsilon_\alpha - \varepsilon| \leqslant v|c_{\alpha i} H_{ik}| \tag{5.71}$$

where $2v^2 = [(4n + 1)^{1/2} - 1]$ and $c_{\alpha i}$ is the LCAO coefficient in the intrafragment MO formed by the BA orbital. It is seen that the deviation of ε_α from ε under the influence of the near-neighbor fragment is proportional to both H_{ik} and $c_{\alpha i}$. The latter is smaller, the more delocalized the intrafragment MO. Hence, again, *the electron delocalization inside the fragment reduces the influence of the neighbor fragments.* In [5.67] (see also in [5.64]), a procedure of double (intrafragment and interfragment) self-consistent calculations is worked out; it takes into account the influence of interfragment interactions on both the energy levels and charge distribution. The latter is of special importance to transition metal compounds due to significant charge tranfers between the fragments: In these cases the interface between the fragments should be electronically transparent [5.64].

In applications to coordination compounds, the results above allow one to conclude that the central metal atom cannot be used as a BA to perform the fragmentation procedure, due to the rather delocalized nature of the electronic distribution created by the d or f electrons. However, the ligands can be calculated by fragmentary procedures, provided that the conditions above are obeyed. Also, in multinuclear (multicenter) coordination compounds, separations into one-center fragments may be quite necessary to be able to perform the calculations.

Similarly, fragmentation is needed when local states in crystals and localized states on solid surfaces are considered. The difficulties in selecting and separating a local group of atoms around a given center can be overcome, for instance, by choosing the quasi-molecular extended elementary cell [5.69]. In this model the center selected (an absorbed atom on the surface, a defect of the crystal lattice, an impurity atom, etc.), together with its sufficiently large environment, is considered as being periodically (regularly) repeated in the crystal. This allows one to consider the local polyatomic formation as an elementary cell of a model crystal with a large lattice period and to calculate its electronic structure in the same way as for regular crystals.

For multicenter coordination compounds or coordination compounds in ionic crystals, some one-center atomic groups (or a limited number of centers in crystals) can be selected for electronic structure calculations, the cut bonds being substituted by boundary conditions that simulate the influence of the cluster environment in real systems. This can be done, for example, by setting some pseudoatoms on the broken bonds [5.70] (sometimes called dummy atoms) or by using cyclic border conditions for clusters in crystals [5.71], in which the broken bonds on the border are linked to the equivalent bonds on the other border.

Fragmentary calculations of transition metal compounds become most important in modeling organometallic and metallobiochemical systems [5.64]. Indeed, owing to the large organic ligands, the electronic structure of these systems cannot be calculated and their geometry optimized as a whole, while the methods of molecular mechanics widely used for conformation search of large organic systems cannot be applied because of the d electron heterogeneity introduced by the transition metal (Section 9.1). In these conditions the separation of the central fragment (or fragments) containing the metal atom(s) from the remaining organic ligands allows for a treatment of the former by quantum mechanics, while the latter can be optimized by molecular mechanics. The electronically transparent interface [5.64] based on the double (intrafragment and interfragment) self-consistent procedure, mentioned above, allows for charge transfers between the fragments which is of special importance for transition metal systems. Several suggestions of fragmentation and combined quantum mechanical/molecular mechanics calculations for organic systems [5.72–5.75] do not include the option of charge transfer between the fragments and therefore may be inapplicable to transition metal systems.

Relativistic Approaches

One of the most essential differences between the methods of electronic structure calculation for transition metal heavy-atom compounds, and those for organic or simple inorganic molecules, is in the necessity to take into account the relativistic effects in the former, whereas they can often be neglected in the latter. For years relativistic effects have been considered as relativistic correction of nonrelativistic calculations mainly by taking into account the spin–orbital interaction as a perturbation. This is unacceptable in more accurate calculations. For instance, for atoms of the third transition period the spin–orbital interaction is of the same order of magnitude or larger than those of interelectronic and interatomic (chemical) interactions. For this reason, in heavy-atom compounds there is no sense to calculate the electronic terms and chemical bonding without relativistic effects and/or to take into account the spin–orbital interactions as a perturbation, as is done for most light-atom molecules. Extensive developments of relativistic electronic structure calculations in molecules began in the 1970s, although atomic systems were explored much earlier. The latest achievements in this area can be found in several review publications [5.76–5.79].

Relativistic effects in chemical properties of atoms are considered in Section 2.1, and the results obtained there are also used in Section 6.5 in the discussion of chemical bonding. In this section we consider briefly some semiquantitative methods of relativistic electronic structure calculation of coordination compounds. In the next section semiempirical relativistic methods are discussed briefly.

At present the relativistic calculations of transition metal compounds are rather semiquantitative. In a rigorous formulation the relativistic SCF MO LCAO equations are complicated. The main distinction from the nonrelativistic formulation lies in use of the relativistic Dirac Hamiltonian for one-electron operators, in which the velocity of the electrons v, assumed to be comparable with the light velocity c, is taken into account explicitly. For the operator of the interelectronic interactions, the classical expression without relativistic effects is usually used; this does not introduce considerable inaccuracies except for small spin–spin interactions. Thus for the Dirac Hamiltonian [5.43] we have [cf. Eq. (2.11)]

$$H = \sum_i [\boldsymbol{\alpha}_i c \mathbf{p}_i + mc^2 \beta_i + V(i)] + \sum_{i,j'} \frac{e^2}{r_{ij}} \qquad (5.72)$$

where the summation is performed over all the electrons, m is the mass of the electron, $\mathbf{p} = -i\hbar\nabla$ is the operator of the momentum, $V(i) = -\sum_\alpha Z_\alpha e^2/R_{i\alpha}$ and e^2/r_{ij} are the Coulomb electron–nuclei and interelectron interactions, respectively, and α_i and β_i are fourth-rank Dirac matrices given by Eqs. (2.12) to (2.14).

As indicated in Section 2.1, the relativistic Hamiltonian (5.72) has a matrix four-component form. The solutions of the Dirac equation with this Hamiltonian are also four-component one-electron functions—bispinors Ψ composed by two spinors [Eq. (2.15)]—and they obey conditions (2.16) and (2.17). The four-rank matrix equation which determines the one-electron bispinor $\Psi(i)$ is:

$$[\boldsymbol{\alpha}_1 c \mathbf{p}_1 + mc^2 \beta_1 + V(1) + \sum_k \int \Psi_k^+(2) \frac{e^2}{r_{12}} (1 - P_{ik})\Psi_k(2) \, d\tau_2]\Psi_i(1) = E_i \Psi_i(1)$$

$$(5.73)$$

where Ψ^+ indicates the transposed bispinor.

Solving this equation by the MO LCAO self-consistent procedure, one can, in principle, obtain relativistic SCF MO LCAO *ab initio* solutions. However, this is a complicated procedure. A simplification can be reached in the *quasi-relativistic approach* (QRA). In this approach it is assumed that at least for the valence electrons, the relativistic effects can be taken into account approximately up to terms the order $(v/c)^2$, inclusively. Simple estimates show that in the region of the valence electrons, higher powers in v/c are small and can be neglected, confirming the validity of the QRA.

In the quasi-relativistic MO LCAO approximation [5.43, 5.80, 5.81], the wavefunction can be presented by only one two-component spinor $\varphi_i(j)$ because the second one in Eq. (2.15), $\chi_i(j)$, shown in Section 2.1, is on the order of $(v/c)^2$, hence yields higher-order smallness in the energy. In the LCAO approximation φ_i should be presented as an expansion in the atomic spinors Φ_j:

$$\varphi_i = \sum_j c_{ij} \Phi_j \tag{5.74}$$

Performing the calculations of the matrix elements H_{ij} of the secular equation (5.8) with the functions (5.74), one comes to the following formulas [cf. Eq. (5.48)]:

$$H_{ij}^R = H_{ij}^{OR} + I_{ij}^R - K_{ij}^R \tag{5.75}$$

where the quasi-relativistic expressions for H_{ij}^{OR}, I_{ij}^R, and K_{ij}^R are rather different from the nonrelativistic ones [5.81]:

$$H_{ij}^{OR} = H_{ij}^0 + \left\langle i \left| - \frac{p^4}{8m^3c^2} + \frac{\hbar^2 \Delta V}{8m^2c^2} + \frac{\hbar\sigma[\nabla V, \mathbf{p}]}{4m^2c^2} \right| j \right\rangle \tag{5.76}$$

$$I_{ij}^R = I_{ij} + \frac{\hbar}{2m^2c^2} \sum_{k,l} P_{kl} \langle\langle ik | \sigma_1 [\nabla_1 r_{12}^{-1}, \mathbf{p}_1] | lj \rangle\rangle \tag{5.77}$$

$$K_{ij}^R = K_{ij} + \frac{\hbar}{2m^2c^2} \sum_{k,l} P_{kl} \langle\langle ik | \sigma_1 [\nabla_1 r_{12}^{-1}, \mathbf{p}_1] | jl \rangle\rangle \tag{5.78}$$

Here $\langle\langle ik| \, |lj \rangle\rangle$ means integration over r_1 with spinors k and l, and over r_2 with spinors i and j.

As one can see, the QR formula for the matrix elements in the MO LCAO approach contains the usual nonrelativistic terms plus relativistic additions. The physical meaning of all these terms can be revealed by comparison with the relativistic terms that emerge in atomic calculations [5.43]. Similar to the atomic case, three types of relativistic corrections can be distinguished:

1. The term $\sim p^4$, the contribution of the *electron mass dependence on speed*
2. The term $\sim \nabla V$, the *Darvin interaction*, which has no classical interpretation
3. Terms proportional to the Pauli matrices σ, *the spin–orbital interaction*.

As is seen from Eqs. (5.76) to (5.78), the part H_{ij}^{OR} that stands for the kinetic energy and the interaction with the core (nuclei) contains all the three kinds of relativistic corrections, while the Coulomb I_{ij}^R and exchange K_{ij}^R interaction terms contain the correction on spin–orbital interaction only.

Note that the matrix elements (5.76) to (5.78) must be calculated by two-component spinors (all the operators are 2×2 matrices; where not explicitly noted, a 2×2 unit matrix is implied). The latter depend on the basis set. The atomic basis function Φ is an eigenfunction of the operator of the square of the total momentum J^2 and its projection J_Z, and therefore Φ is dependent on additional quantum numbers l, j, M (Sections 2.1 and 2.2):

$$\Phi_{ljM}(\mathbf{r}, \beta) = \begin{pmatrix} \Phi_{ljM}^1 \\ \Phi_{ljM}^2 \end{pmatrix} = R_{lj}(\mathbf{r}) Y_{ljM}(\vartheta, \phi, \beta) \qquad (5.79)$$

Here $R_{lj}(r)$ is a relativistic radial function and $Y_{ljM}(\vartheta, \phi; \beta)$ is a spherical spinor which can be presented as an expansion over spherical functions $Y_{lm}(\vartheta, \phi)$ [5.80]:

$$Y_{ljM}(\vartheta, \phi, \beta) = \sum_{m,\mu} \langle lm \tfrac{1}{2}\mu | jM \rangle Y_{lm}(\vartheta, \phi) q_\mu(\beta) \qquad (5.80)$$

where $q_\mu(\beta)$ is the function of the spin argument β, its two values realizing the two components of the spinor, and $\langle lm \, 1/2\mu | jM \rangle\rangle$ is the corresponding Clebsh–Gordan coefficient. With the functions (5.79) and (5.80), the matrix elements (5.76) to (5.78) are reduced to linear combinations of integrals calculated by usual spherical functions $Y_{lm}(\vartheta, \phi)$ and relativistic radial functions $R_{lj}(\mathbf{r})$.

Since in the relativistic case the spin–orbital interaction is strong and the $j - j$ coupling scheme is valid (Section 2.2), the multielectron states should be classified by the total momentum J, and *the MOs belong to the irreducible representations of the double point-groups* (Section 3.5). The additional terms of relativistic corrections can be calculated in the same approximation as the nonrelativistic terms. The QRA also allows semiempirical simplifications (see Section 5.5).

At present, several methods of relativistic electronic structure calculations are worked out and used in practice [5.76]. These include all-electron fully relativistic *ab initio* calculations [5.82], ECP relativistic and quasi-relativistic calculations and other pseudopotential approaches [5.83–5.86]; relativistic X_α (Dirac–Slater), Dirac–Slater X_α-DV [5.59] and X_α-SW [5.87], quasi-relativistic X_α-SW [5.76, 5.88], the perturbation HFS (Hartree–Fock–Slater), methods [5.79, 5.89]; other density-functional approaches [5.90]; and semiempirical realizations [5.91] (see also Sections 5.5 and 6.5).

5.5. SEMIEMPIRICAL METHODS

Besides the semiquantitative approaches discussed in Section 5.4, semiempirical methods are widely used in practice to simplify electronic structure calculations. In semiempirical MO LCAO methods, the main idea is to substitute as many integrals as possible in matrix elements H_{ii} and H_{ij} of Eqs. (5.7) and (5.8) by empirical parameters. Among these methods one can distinguish two groups which differ as to whether or not they neglect the overlap integrals.

Zero Differential Overlap

The zero differential overlap (ZDO) approach was first suggested independently by Pople and by Pariser and Parr for π electron systems (see [5.92–5.98] and references therein). It assumes that the overlap of different atomic functions ψ_i and ψ_j for any elementary volume—the *differential overlap*—equals zero:

$$\psi_i(\mathbf{r})\psi_j(\mathbf{r})\,d\tau = 0 \qquad i \neq j \tag{5.81}$$

In this approximation all the integrals of the type $[ij\,|\,kl]$ in Eqs. (5.48) and (5.51) are zero except for $i = j$ and $k = l$. This means that all three- and four-center integrals are zero, the interelectronic interaction is described by the two-center integrals of the type $[ii\,|\,kk]$, and the overlap integrals $S_{ij} = \int \psi_i^*(\mathbf{r})\psi_j(\mathbf{r})\,d\tau = 0$. Logically, assumption (5.81) leads to the conclusion that the resonance integral $H_{ij}^0 = \int \psi_i^* H^0 \psi_j\,d\tau$ for $i \neq j$ is also zero, but this gives completely unacceptable results. Therefore, in the ZDO methods it is assumed that $H_{ij}^0 \neq 0$ for i and j from the same or near-neighbor atoms.

At first sight, assumption (5.81) and its inconsistent repudiation when calculating H_{ij}^0 seem to introduce rough approximations. Nevertheless, the ZDO approaches proved to be quite efficient in concrete calculations yielding satisfactory results. Afterward, the method was given additional grounds that also allowed one to clarify the nature of assumption (5.81) and the limits of its validity, and to obtain quantitative relationships for the resonance integral [5.93]. It was shown that the ZDO formulas are valid when neglecting S^2, where S is the overlap integral. More accurate expressions in these methods can be obtained by using the orthogonalized basis [5.94] for the calculation of matrix elements.

In electronic structure calculations of transition metal compounds, the following ZDO methods are in wide use:

1. Complete Neglect of Differential Overlap (CNDO). In this method in the secular equation (5.8):

- All the overlap integrals are zero.

- The diagonal matrix elements H_{ii} are taken from empirical data (see below).

- The off-diagonal core integrals H_{ij}^0 with $i \neq j$ are taken proportional to the overlap integrals S_{ij} when ψ_i and ψ_j are from near-neighbor atoms, and zero otherwise.

• The electron interaction integrals obey the condition

$$[ij\,|\,kl] = [ii\,|\,kk]\delta_{ij}\delta_{kl} \tag{5.82}$$

Under these conditions all three- and four-center integrals vanish, and the matrix element of the SCF MO LCAO method (5.48) is simplified significantly:

$$H_{ij} = H_{ij}^0 + \sum_k P_{kk}[kk\,|\,ii]\delta_{ij} - \tfrac{1}{2}P_{ij}[ii\,|\,jj] \tag{5.83}$$

Denoting, as in all ZDO methods, $\gamma_{ik} = [ii\,|\,kk]$, we have:
For the diagonal element:

$$H_{ii} = H_{ii}^0 + \tfrac{1}{2}P_{ii}\gamma_{ii} + \sum_{k\neq i} P_{kk}\gamma_{ki} \tag{5.84}$$

For the off-diagonal element:

$$H_{ij} = H_{ij}^0 - \tfrac{1}{2}P_{ij}\gamma_{ij} \tag{5.85}$$

However, as has been shown by direct calculations [5.92], in the approximation introduced by (5.84) and (5.85) the results are not invariant with respect to the choice of the local coordinate systems on the atoms. In other words, the results of the calculations depend on the orientations of nonsymmetrical atomic functions ($p, d, f, \ldots$) of a given atom with respect to the others (and this unacceptable feature is inherent to all the methods based on ZDO).

To overcome this principal difficulty, it was suggested [5.92] that all the electron repulsion integrals that depend on orientation be taken equal to each other and to the one calculated with spherical-symmetrical s functions, which do not depend on orientation. In this approximation the repulsion integrals γ_{ij} depend on the type of atom but not on its state. For instance, for the A atom $\gamma_{ij} = \gamma_{AA}$ for all the i labels, and for diatomic integrals of the atoms A and B, $\gamma_{ij} = \gamma_{AB}$ for all i and j. This results in further simplification of the matrix elements H_{ij}^0 (recall that $H^0 = H_A^0 - \Sigma_{B\neq A}U_B$; Section 2.1):

$$\int \psi_{iA}^* H_A^0 \psi_{jA}\,d\tau = \varepsilon_i^A \delta_{ij} \tag{5.86}$$

$$\int \psi_{iA}^* U_B \psi_{jA}\,d\tau = U_B^A \delta_{ij} \tag{5.87}$$

and for near-neighbor atoms A and B

$$\int \psi_{iA}^* H^0 \psi_{jB}\,d\tau = \beta_{AB}^0 S_{ij} \tag{5.88}$$

where ε_i^A characterizes the appropriate atomic orbital energy in the state ψ_i^A [cf. Eq. (5.53)], and β_{AB}^0 is an empirical parameter. Denoting the valence electron density on the atom A by $P_{AA} = \Sigma_k P_{kk}$ [cf. Eq. (5.21)], we obtain the following final expressions for the matrix elements of the CNDO method:

$$H_{ii}^{AA} = \varepsilon_i^A + \left(P_{AA} - \frac{P_{ii}}{2}\right)\gamma_{AA} + \sum_{B \neq A} (P_{BB}\gamma_{AB} - U_B^A) \tag{5.89}$$

$$H_{ij}^{AB} = \beta_{AB}^0 S_{ij} - \frac{P_{ij}\gamma_{AB}}{2} \tag{5.90}$$

$$H_{ij}^{AA} = -\frac{P_{ij}\gamma_{AA}}{2} \tag{5.91}$$

The empirical parameters ε_i^A and β_{AB}^0 are taken as follows. ε_i^A is obtained from the relation

$$-I_{0i}^A = \varepsilon_i^A + (Z_A - 1)\gamma_{AA} \tag{5.92}$$

where I_{0i}^A is the corresponding ionization potential, while $\beta_{AB}^0 = (\frac{1}{2})(\beta_A^0 + \beta_B^0)$ [cf. Eq. (5.100)] and β_A^0 and β_B^0 are appropriate atomic parameters taken, for example, from the data of more accurate MO LCAO calculations for systems containing the atom under consideration.

Another choice of parameters is suggested in the modification of the CNDO method called CNDO/2 (as distinct from the above CNDO/1 method). In CNDO/2 it is assumed that $U_B^A = Z_B \gamma_{AB}$, while ε_i^A is taken from a relation that is somewhat different from (5.92):

$$-(I_{0i}^A + A_i) = \varepsilon_i^A + (Z_A - \tfrac{1}{2})\gamma_{AA} \tag{5.93}$$

where A_i is the corresponding electron affinity.

Thus in the CNDO methods only the overlap integrals S_{ij} and simple repulsion integrals with s functions,

$$\gamma_{AA} = e^2 \int s_A^2(\mathbf{r}_1)s_A^2(\mathbf{r}_2)r_{12}^{-1}\,d\tau_1\,d\tau_2 \tag{5.94}$$

$$\gamma_{AB} = e^2 \int s_A^2(\mathbf{r}_1)s_B^2(\mathbf{r}_2)r_{12}^{-1}\,d\tau_1\,d\tau_2 \tag{5.95}$$

should be calculated; in CNDO/1 the integral

$$U_B^A = Z_B e^2 \int s_A^2(\mathbf{r})|\mathbf{r} - \mathbf{R}|^{-1}\,d\tau \tag{5.96}$$

is also to be computed. The other magnitudes ε_i^A and β_{AB}^0 are taken from empirical data.

2. Neglect of Diatomic Differential Overlap (NDDO). The CNDO methods, described above, contain some essentially restricting approximations that are especially significant in the application to transition metal compounds. In particular, when the ZDO (5.81) is applied to two different orbitals of the same center, the integrals of exchange interactions between the corresponding electrons is neglected. Meanwhile they may reach about 4 to 5 eV.

If the basis of atomic functions contains no more than one function per atom (as, e.g., in π electron approximation for conjugated organic molecules), there are no such integrals. But if there are several functions per atom in the basis set (for a transition metal atom, as mentioned above, they are usually nine), the neglect of one-center repulsion integrals may be a source of considerable errors. To avoid them, the following version of ZDO can be employed, in which approximation (5.81) is applied only to the pairs of orbitals from different atoms, whereas the product of functions $\psi_{iA}(\mathbf{r})\psi_{jA}(\mathbf{r})$ from the same atom is considered nonzero, and the corresponding integrals with these products are calculated explicitly or approximated by parameters. This method of calculation called NDDO results in significant increase of the volume of calculations, as compared with CNDO.

3. Multicenter ZDO (MCZDO). Another difficulty of the CNDO methods is the above-mentioned necessity to introduce the same spherically symmetric (s-type) electronic cloud distribution for all the electrons (p, d, f,...). For molecules containing light (first row) atoms only, this circumstance is not very restricting because the averaged electron distributions in the $2p$ state does not differ very much from that of the $2s$ state. However, in the case of transition metals and rare earth elements, the simplification above becomes unacceptable because of the significant differences in the electron distributions of s, p, d, f orbitals.

In the MCZDO method [5.96] this difficulty is partly overcome while preserving the advantages of the NDDO method compared with the CNDO one. Distinct from the latter, in the MCZDO method:

1. The ZDO approximation (5.81) is not applied to the core integrals H_{ij}^0 and to the one-center integrals of interelectron repulsion.

2. The ZDO approximation is applied to all the other multicenter electron repulsion integrals: $[ij\,|\,kl] = [ii\,|\,kk]\delta_{ij}\delta_{kl}$ for i and j on different atoms.

3. Averaged two-center repulsion integrals are different for different types of electrons; that is, instead of two integrals γ_{AA} and γ_{AB} of the CNDO method, the following are introduced: γ_{ss}, γ_{sp^*} (interaction of s cloud with the averaged p distribution denoted by p^*), γ_{sd^*}, $\gamma_{p^*p^*}$, $\gamma_{p^*d^*}$, $\gamma_{d^*d^*}$, γ_{sf^*}, and so on.

4. Intermediate Neglect of Differential Overlap (INDO) [5.99] **and Modified INDO (MINDO)** [5.100]. These are modified CNDO methods aimed at electronic structure calculations for systems with open shells. In the version for coordination compounds [5.98, 5.101], the INDO method is based on the scheme of the unrestricted Hartree–Fock (UHF) method (Section 5.3), in which each electron is described by its own orbital function (the two electrons with opposite spins on the same orbital of the restricted HF method occupy different orbitals in the UHF method), and on the NDDO approximation, in which the interatomic overlap only is neglected (whereas the intraatomic one is included).

But unlike the NDDO method, the INDO approximation assumes that the two-center integrals of interelectronic interaction $[i^A j^A | k^B l^B] = 0$ if $A \neq B$, $i \neq j$, and $k \neq l$. This simplifies the otherwise rather complicated calculations of the NDDO method, but it still takes into account the main part of electron interactions on each center (including the exchange interaction), which determine the spin states. The one-center interelectron interactions can be approximated by Slater–Condon parameters (Section 2.2).

In final form the matrix elements in this INDO version appear as follows:

$$H_{ii}^{AA} = H_{ii}^0 + \sum_{k,l}^{A} P_{kl}\left([ik|il] - \frac{[ik|li]}{2}\right) \sum_{B \neq A}^{B} \sum_{k} P_{kk}[i^*k^*|i^*k^*] - \sum_{B \neq A} Z_B\langle i^*|\frac{1}{R_B}|i^*\rangle$$

$$(5.97)$$

$$H_{ij}^{AA} = H_{ij}^0 + \sum_{kl}^{A} P_{kl}\left([ik|jl] - \frac{[ik|lj]}{2}\right)$$

$$H_{ij}^{AB} = H_{ij}^0 - P_{ij}[i^*j^*|i^*j^*]$$

$$(5.98)$$

Here the star at the atomic orbital label means that the orbital should be chosen to preserve the corresponding symmetry of the problem, mentioned above (somewhat similar to that in the MCZDO method) [5.98, 5.101].

Different modifications of the INDO method (MINDO/1, MINDO/2, MINDO/3 [5.102, 5.103]) are used primarily for calculations of organic molecules.

Extended Hückel (Hoffmann) Method

The simplest *Hückel method*, where the overlap integrals S_{ij} in the secular equation of the MO LCAO method (5.8) are omitted and the matrix elements are substituted by empirical parameters, having some value in application to organic molecules, is invalid for coordination compounds. By ignoring S_{ij}, one loses the specific role of (the electron heterogeneity introduced by) the d electrons in the electronic structure. In ZDO methods aimed at coordination compounds, for instance, in INDO, this deficiency is compensated by the difference in the corresponding electron repulsion integrals. The *extended Hückel* (EH) and the related *iterative EH* (IEH), or the *self-consistent charge and configuration* (SCCC) *method*, avoid this fault by considering explicitly the overlap integrals in Eq. (5.8). On the other hand, these methods, unlike the ZDO ones, are completely semiempirical in the sense that all the matrix elements in (5.8) are substituted by empirical parameters and overlap integrals S_{ij}, the latter thus being the only magnitudes computed.

In the EH approximation the empirical presentation of the matrix elements is as follows:

$$H_{ij} = -I_i \qquad (5.99)$$

$$H_{ij} = \frac{k}{2}(H_{ii} + H_{jj})S_{ij} \qquad (5.100)$$

where k is a numerical coefficient, which in simple cases is taken equal to 1.67 for σ bonds and 2 for π bonds (see below), and I_i is the energy of ionization of the atomic ith state. The presentation of the diagonal matrix element H_{ii} by the corresponding ionization potential goes back to the simple Hückel method, while formula (5.100) for the off-diagonal element H_{ij} was first used in calculations of transition metal complexes by Wolfsberg and Helmholz [5.105] and widely demonstrated in calculations of organic molecules by Hoffmann [5.104].

It can be shown that Eqs. (5.99) and (5.100) are of the same level of approximation as the semiquantitative expressions (5.53) for the matrix element of the effective Hamiltonian; they differ mainly by the last term in (5.53), often called *crystal field corrections*. The latter improve the interatomic core interactions and were introduced in the semiempirical calculations (see below).

Presentations (5.99) and (5.100) were significantly improved by introducing the dependence of the I_i values on the atomic and orbital charges and the requirement of self-consistency with respect to these charges [5.104]. Consider the diagonal element of the Hamiltonian starting with expression (5.53), $H_{ii}^{AA} = \varepsilon_i^A - \Sigma_{\mu \neq A} e^2(Z_\mu - q_\mu)/R_{A\mu}$. Here ε_i^A has the physical meaning of the orbital energy of the electron in the valence state of atom A in the molecule, that is, when the atom has a certain charge and electron population distribution over the atomic states taking part in the formation of the MO. In other words, ε_i^A is the ith state ionization energy in a specific configuration of other electrons, usually called the *energy of ionization of the valence state* (EIVS): $\varepsilon_i^A = -I_i[A]$, where $[A]$ denotes the appropriate electron configuration of A.

The effective charge of the atoms is $Z_\mu - q_\mu$, where $q_\mu = \Sigma_{i,v}^A q_i c_{i\mu} c_{iv} S_{\mu v}$ is the Mulliken total electronic charge on the μ atom after Eq. (5.20). However, this formula does not take into account that the screening of the nuclei by other electrons is not complete. Therefore, the effective charges of the μ atoms should be taken with a correction constant k_λ that depends on the kind of bonding between A and μ produced by the ith function ($\lambda = \sigma, \pi, \delta, \ldots$). With this correction

$$H_{ii}^A = -I_i[A] - e^2 \sum_{\mu \ne A} \frac{Z_\mu - q_\mu + k_\lambda}{R_{\mu A}} \tag{5.101}$$

In general, EIVS is different from the appropriate ionization energy of the neutral atom because in the valence state the atom may have a considerable charge, and the population of the s, p, d, f orbitals are altered by MO formation. These effects are obviously much more significant for coordination compounds than for organic molecules. To take into account the dependence of ionization energy on the atomic charge, Hoffmann suggested a linear correlation:

$$I_i^A = I_{0i}^A + aq^A \tag{5.102}$$

where I_{0i}^A is the ionization potential of the neutral atom and q^A is the atomic charge (the constant a can be obtained by comparison of I_i^A with one of the next ionization potentials; see below).

The crystal field corrections in the second term in (5.101) are also more important for coordination compounds than for organic molecules because the latter have much smaller atomic charges than the former. These corrections are even more important when the coordination system is in a crystal lattice, where they are summarized into the *Madelung potential*.

Similar corrections should be introduced in the off-diagonal matrix elements (5.53). For semiempirical calculations, several other than (5.100), but more or less equivalent presentations of the off-diagonal element were suggested (Table 5.7). Numerical estimates show that very often all these presentations yield close results [5.106].

Table 5.7. Different presentations of the off-diagonal matrix element of the Hamiltonian H_{ij} by the diagonal elements (U_{ii} and H_{jj}) and the overlap integral S_{ij} in semiempirical methods[a]

Authors	H_{ij}
Wolfsberg and Helmholz	$kS_{ij}(H_{ii} + H_{jj})/2$
Ballhausen and Gray	$-kS_{ij}(H_{ii}H_{jj})^{1/2}$
Cussacs	$(2 - |S_{ij}|)S_{ij}(H_{ii} + H_{jj})/2$
Erraneous	$kS_{ij} \cdot 2H_{ii}H_{jj}/(H_{ii} + H_{jj})$
Morokuma and Fukui	$S_{ij}[(H_{ii} + H_{jj})/2 + k]$
Neuton et al.	$T_{ij} + S_{ij}(U_{ii} + U_{jj})/2$

Source: [5.106].
[a] U_{ii} is the matrix element of the potential energy; if the virial theorem holds, $U_{ii} \approx 2H_{ii}$.

Iterative Extended Hückel Method

The iterative extended Hückel (IEH) method [sometimes called the self-consistent charge and configuration (SCCC) MO LCAO method] is an extension (improvement) of the EH approach important to calculations of electronic structure of coordination compounds. The essence of this improvement is to introduce a self-consistent procedure with respect to the atomic charges and electronic configuration. Indeed, the matrix elements H_{ii} and H_{ij} in Eqs. (5.99) to (5.101) contain these charges and configurations which depend on LCAO coefficients, which, in turn, depend on H_{ii} and H_{ij}, and so on.

Assume that we calculated the matrix elements (5.99) and (5.100) or (5.101), and by solving Eqs. (5.7) and (5.8) we obtained a set of n MO energy levels of the system,

$$\varepsilon_1, \varepsilon_2, \ldots, \varepsilon_n \tag{5.103}$$

and for each of them a set of LCAO coefficients:

$$
\begin{aligned}
&\varepsilon_1: c_{11}, c_{12}, \ldots, c_{1n} \\
&\varepsilon_2: c_{21}, c_{22}, \ldots, c_{2n} \\
&\;\;\vdots \\
&\varepsilon_n: c_{n2}, c_{n2}, \ldots, c_{nn}
\end{aligned}
\tag{5.104}
$$

Each line of (5.104) corresponds to a certain (ith) MO, while each coefficient of the ith MO determines the contribution of the jth AO. By distributing all the electrons over the lowest energies ε_i following the Pauli principle, one finds that the MO population numbers q_i equal 2 (fully occupied MOs), 1 (half occupied), or 0 (unoccupied) (in fragmentary calculations the charges may be fractional [5.64, 5.67]). With these data the atomic charges and, separately, the electronic charges on s, p, d, f, ... orbitals can be evaluated by means of the Mulliken [5.7] population analysis [Section 5.2, Eq. (5.20)]. In units of the electron charge, the electronic charges on the s, p, and d orbitals are

$$q_s^A = \sum_{i,j} q_i c_{is}^A c_{ij} S_{sj}$$

$$q_p^A = \sum_{i,j} \sum_{pA} q_i c_{ip}^A c_{ij} S_{pj} \tag{5.105}$$

$$q_d^A = \sum_{i,j} \sum_{dA} q_i c_{id}^A c_{ij} S_{dj}$$

where the label A in combination with s, p, and d means that the latter denote the corresponding orbitals of atom A. The total (positive) atomic charge Z_A^* is

(m^A is the number of valence electrons on the AO of the neutral atom included in the LCAO)

$$Z_A^* = m^A - (q_s^A + q_p^A + q_d^A) \qquad (5.106)$$

The dependence of the atomic ionization energy on the atomic charge for a given electronic configuration $I_i[A]$ can be well approximated by a three-term square dependence, as follows:

$$I_i[A] = A_i Z^2 + B_i Z + C_i \qquad (5.107)$$

The constants A_i, B_i, and C_i can be found if one knows the ionization potentials I_i for three Z values, for instance, for the neutral ($Z = 0$), ionized ($Z = +1$), and double-ionized ($Z = +2$) states. These data can be obtained from the analysis of the energy terms and ionization potentials of free atoms in the given electronic configuration $[A]$ [5.107]. The latter is taken equal to the weighted-average value for all the terms formed by this configuration (Section 2.2). For instance, the d energy of the configuration d^2 (see Table 2.6) is

$$I_d[d^2] = \tfrac{1}{45}[9E(^1A) + 21E(^3F) + 5E(^1D) + 9E(^3P) + E(^1S)] \qquad (5.108)$$

where the coefficients at the term energies $E(^{2S+1}L)$ equal the term degeneracy $(2S + 1)(2L + 1)$, while the term energies are known from spectroscopic data. The EIVS and the constants A, B, and C for some most usable atoms obtained in this way are given in [5.106–5.109].

However, the atomic orbital charge distribution $d^{q_d} s^{q_s} p^{q_p}$ with q_d, q_p, q_s obtained by Eqs. (5.105) is, in general, fractional (q_d, q_s, and q_p are not integers). For fractional configurations there are no empirical data on ionization energies. They can be approximated by linear combinations of the known values for integer-number configurations with fractional coefficients, which are chosen to make the summary electronic configuration equal to the fractional configuration.

Let us present the configuration $d^{q_d} s^{q_s} p^{q_p}$ with $q_s + q_p + q_d = n(1 - \Delta)$, where n is integer and $n\Delta$ is a small fractional number, by the integer-number configurations d^n, $d^{n-1}s$, $d^{n-1}p$:

$$d^{q_d} s^{q_p} p^{q_s} = ad^n + bd^{n-1}s + cd^{n-1}p \qquad (5.109)$$

Then by equalizing the populations on the same orbitals, we get $c = q_p$, $b = q_s$, and $a = 1 - q_p - q_s - \Delta$. Consequently, the EIVS of a fractional electronic configuration in the case under consideration is

$$I_i[d^{qd}s^{qs}p^{qp}] = (1 - q_p - q_s - \Delta)I_i[d^n] + q_s I_i[d^{n-1}s] + q_p I_i[d^{n-1}p] \qquad (5.110)$$

where

$$\Delta = 1 - \frac{q_s + q_p + q_d}{n} \qquad (5.111)$$

Thus the procedure of self-consistent solutions in the IEH (SCCC) MO LCAO method is as follows. With the LCAO coefficients c_{ij} obtained from Eqs. (5.7) and (5.8), one calculates the atomic and configuration charges after (5.105) and then determines new EIVS by Eq. (5.110), and new matrix elements H_{ii} and H_{ij} by Eqs. (5.99) to (5.101), which allow one to determine new c_{ij} values, new charges, and so on. Provided that this process converges, each new iteration yields more precise results than the previous one, and this process can be continued until the new atomic charges and configuration, within the accuracy required, coincide with the previous ones; that is, they are self-consistent.

The crystal field corrections can be included in the computation program directly. However, in some cases they are omitted. The reason is that usually the atomic charges on the central (metal) atom and ligands have opposite signs, with the positive charge on the CA. Hence for the metal the EIVS value is larger than for the neutral atom, and the crystal field corrections after Eq. (5.101) increase the absolute value of H_{ii}. On the contrary, for the ligand, EIVS is smaller than for the neutral atom, and the crystal field of the positive central atom increases the absolute value of H_{ii}, making it closer to the neutral atom value. Therefore, one can assume that when the ligand charges are not very large, the charge dependence of EIVS of ligand atoms is compensated by crystal field corrections. Hence the self-consistent procedure can be carried out only on the CA charges and configurations, keeping the H_{ii} values for the ligands constant and equal to their neutral atom values.

As to the correction constants k_λ in Eq. (5.101), they remain almost arbitrary, although their choice, within reasonable limits, does not influence strongly the results. In [5.110] two sets of k_λ values were suggested: (1) $k_\sigma = 0.6$, $k_\pi = 0.40$, $k_\delta = 0.2$, and (2) $k_\sigma = 0.5$, $k_\pi = 0.4$, $k_\delta = 0.3$.

One of the main deficiencies of the extended Hückel (Hoffmann) method, described above, is that it does not apply to molecular geometry optimization. Indeed, the Wolfsberg–Helmholz formula (5.100) or its analogs in Table 5.8 do not comprise correctly the (nonoverlap) electrostatic interactions, especially between the atomic cores, and therefore it cannot stand for potential energy curves for stretching modes. Anderson and Hoffmann [5.111] attempt to overcome this difficulty by adding a two-body electrostatic interaction to the energy term in the EH approximation, while Calzaferri et al. [5.112] improved this presentation significantly.

With the two-body electrostatic terms included, there is a possibility to derive the corresponding geometries by energy optimization. The results obtained in this way are strongly dependent on the parametrization of the additional terms [5.112]. For some metalloorganic compounds, appropriate parameters have been suggested recently [5.113].

However, in general, in application to coordination compounds with center-delocalized bonds and nontransferrable metal–ligand bond properties, para-metrization based only on two-body interactions should be handled with much reserve and care. Apparently, such parameters could be applied to groups of compounds, for which the metal–ligand bonds in question are mainly localized and/or they are analogous from the point of view of metal–ligand bond delocalization. The situation here is much similar to that arising in the problem of molecular mechanics, where the intramolecular interactions are approximated by two-body atom–atom potentials with appropriate transferrable parameters (see Section 9.1).

Quasi-relativistic Parametrization

Quasi-relativistic (QR) semiempirical versions are based on the general QR approaches to the MO LCAO scheme, discussed in Section 5.4. Additional difficulties in the QR parametrization of the calculations emerge, in particular, due to the two-component spinor presentation of the wavefunction, and classification of the MO states on the irreducible representations of the double groups of symmetry required by the $j - j$ coupling between the electrons (Sections 2.1 and 3.6).

For the ZDO methods above, the QR expressions for the matrix elements formally remain the same as in the nonrelativistic case [5.114], but the meaning of the parameters is relativistic. In particular, in the CNDO/2 method the overlap integrals and the interelectron repulsion integrals (5.94) and (5.95) must be calculated by QR spinors, while the empirical parameters in (5.93), the ionization potential and the electron affinity, should be taken for atomic states classified in the $j - j$ scheme.

Let us consider in more detail the QR IEH (or SCCC) MO LCAO parametrization. As in the nonrelativistic version, the diagonal matrix elements H_{ii} are given by EIVS expressed in terms of atomic charges and fractional configuration occupation numbers (5.110). But unlike the nonrelativistic case, for which the electron configuration is taken in the form $s^x p^y d^z f^u$, in the QR analog the electronic configuration must be taken after the $j - j$ coupling scheme as $s^a_{1/2} p^b_{1/2} p^c_{3/2} d^l_{3/2} d^k_{5/2} f^m_{7/2} \ldots$, where $x, y, \ldots, a, b, c, \ldots,$ are the fractional occupation numbers.

To illustrate the procedure of the evaluation of the EIVS, consider a simple case of a two-electron system in a sp configuration. For the free atom, two relativistic valence configurations are possible (Section 2.1): $s_{1/2} p_{1/2}$ and $s_{1/2} p_{3/2}$. Their energy-level scheme is shown in Fig. 5.6. Similar to (5.108), the mean energy of these two configurations can be taken as averaged over the

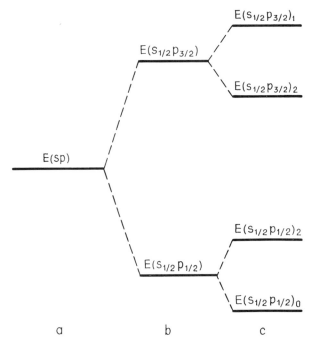

Figure 5.6. Energy-level diagram for a two-electron atom in the *sp* configuration and $j - j$ coupling scheme: (*a*) averaged ("spherical") interelectron interaction; (*b*) spin–orbital interaction added; (*c*) nonspherical ("local") interelectron interaction added (classification on the total momentum $J = 0, 1, 2$).

energy terms with different total momentum quantum numbers J:

$$E(s_{1/2}p_{1/2}) = \tfrac{1}{4}[E(s_{1/2}p_{1/2})_0 + 3E(s_{1/2}p_{1/2})_1] \tag{5.112}$$

$$E(s_{1/2}p_{3/2}) = \tfrac{1}{8}[5E(s_{1/2}p_{3/2})_2 + 3E(s_{1/2}p_{3/2})_1] \tag{5.113}$$

For an ionized atom that has only one valence electron, there are three possible states: $s_{1/2}$, $p_{1/2}$, and $p_{3/2}$ (Fig. 5.7). Then for the EIVS we have

$$-I_{p1/2}[sp] = I_0 + E(s_{1/2}) - E(s_{1/2}p_{1/2})$$

$$-I_{p3/2}[sp] = I_0 + E(s_{1/2}) - E(s_{1/2}p_{3/2}) \tag{5.114}$$

$$-I_{s1/2}[sp] = I_0 + E(p_{1/2}) - E(s_{1/2}p_{1/2})$$

where $I_0 = E_1 - E_0$ is the energy difference between the ground-state energies of the ion and the atom, respectively, from which the configuration energies $E(s_{1/2})$, $E(s_{1/2}p_{1/2})$,..., are read off. In particular, when the ground state of the ion is $s_{1/2}$, $E(s_{1/2}) = 0$.

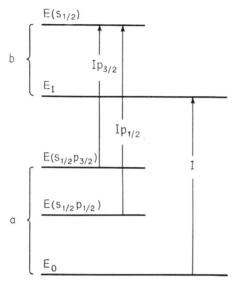

Figure 5.7. Energy-level scheme for a two-electron atom in the *sp* configuration (*a*), and its ion (*b*) (the energies are term averaged). The ionization potentials are shown by arrows.

An example including *d* electrons (for Pt compounds [5.114]) is illustrated in Fig. 5.8. The dependence of QR EIVS on the fractional charges *a, b, c,...*, can be determined in the same way as was done above for the nonrelativistic case (5.110). Semiempirical parametrization for 5*f* elements is given in [5.115]. For reviews of other possibilities, see [5.76–5.79, 5.91, 5.116].

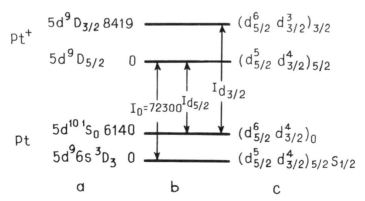

Figure 5.8. Evaluation of EIVS $d_{3/2}$ and $d_{5/2}$ of the Pt atom in the quasi-relativistic approximation to the IEH (SCCC) MO LCAO method (I_0 is the first ionization potential): (*a*) electron configurations and terms of Pt and Pt$^+$; (*b*) energy-level values (in cm^{-1}); (*c*) electron configurations in the strong coupling scheme.

5.6. COMPARISON OF MO METHODS

Discussion of the advantages and disadvantages of MO methods, their mutual comparison, and comparison with other methods is an extensive subject for entire monographs (see, e.g., [5.17, 5.40, 5.117, 5.118]). The goal of this section is to provide a general brief outline of the problem which is aimed at a better understanding and rough estimation of the relative value of semiempirical and nonempirical MO methods in their application to transition metal compounds.

The modern state of the art in MO LCAO methods is rather advanced and, in principle, makes it possible to compute numerically the electronic structure of any molecular system of reasonable size. This statement does not mean that there are no problems in the realization of such computations. Concerning *ab initio* and nonempirical methods, these problems are created by *financial reasons* and difficulties in the *rationalization of the results*.

The financial reasons are due to the increase in the cost of bigger and bigger computers and the computer time required for computation with increase in the number of atoms in the system. For the calculation of a moderate seven-atom octahedral complex of a $3d$ transition metal using a minimal Gaussian basis set, one must compute about 10^9 integrals which can be reduced to 10^7 by preliminary contraction of the basis set (Section 5.3), and the number of integrals usually increases as n^4, where n is the number of functions in the basis set which is larger than the number of atoms (in the X_α-DV method the number of integral is proportional to n^2). For instance, for a 30-atom complex the number of integrals and hence the computer time increase more than 250 times as the basis set increases four times. Obviously, there is a limit to the economic expediency of such calculations.

Another difficulty with nonempirical calculations is that for large systems the computer output information, the wavefunction, spread in thousands of determinants is so vast that without additional rationalization it is useless. In Section 1.2 it is noted that numerical results of electronic structure calculations, taken as such, are rather experimental. Accumulated for series of similar compounds, they can be used for comparisons and theoretical generalizations based on *analytical models*.

From this point of view the semiquantitative and semiempirical methods are more advantageous. First, with these methods the computer time is essentially reduced. For instance, for the MnO_4^- complex with a valence basis set of $3d$, $4s$, and $4p$ AOs of Mn and only p AOs of oxygens, the number of repulsion integrals to be computed in the nonempirical method is 26,796, while in the NDDO method it is reduced to 2415, and in the semiempirical CNDO method they are only 15 [5.96]. In the semiempirical methods the results of numerical computation, as distinct from the *ab initio* data, are quite understandable in terms of visual MOs built up from atomic valence states, the picture of the electronic structure thus being more acceptable for chemists.

However, the semiquantitative, and especially the semiempirical, methods have other faults, the primary ones being *insufficient theoretical foundation, limited quantitative reliability, and low credibility.* As to the theoretical foundation, it is certainly much reduced in semiempirical methods compared with nonempirical methods. For instance, in the extended Hückel (Hoffmann) method (Section 5.5) the coefficient k in the presentation of the off-diagonal matrix element of the Hamiltonian H_{ij} by the diagonal elements, $H_{ij} = \frac{1}{2}k(H_{ii} + H_{jj})$, is usually taken as $k = 2$ for π bonds and $k = 1.67$ for σ bonds, which from the theoretical point of view are almost arbitrary (for further criticism of such methods, see [5.95–5.97]).

Therefore, one cannot expect that the energies and wavefunctions obtained with such coefficients are sufficiently trustworthy. When used with the same k value for a series of compounds, this method yields much *more reliable relative values* in which the systematic error introduced by the arbitrary choice of the basic formula is for the most part excluded. As stated by Hoffmann [5.119], the extended Hückel theory "has the merit of being of the low end of a quality scale of approximate MO calculations. Since all other methods are superior to it, it inculcates in its user a feeling of humility and forces him or her to think about why the calculations come out the way they do. The method is widely applicable and transparent but it has limited quantitative reliability."

Thus semiempirical methods are most reliable when comparative data for series of similar compounds are investigated. For absolute values of semiempirically calculated quantities, the credibility is low, but it increases when there are nonempirical results which can be used as calibration points. As Pyykko put it [5.116]: "It is said that when no such results were yet available, some of the earliest semiempirical conclusions [5.81, 5.114] suggesting that relativistic effects might be important in chemistry, were met with derision." This derision was ungrounded because in the works above, the semiempirical approach was used just for the comparison of relativistic and nonrelativistic results obtained with the same method for the same compounds. Consequent developments confirmed the importance of relativistic effects in chemistry (see Section 6.5).

The low credibility in semiempirical methods is often based on their inappropriate use and the ignorance of *limits of applicability.* All the methods of calculation of electronic structure of molecules have limits of applicability determined by the principles they are based on and the additional assumptions made when the calculations were carried out. The main limitations of the MO and density functional methods are mentioned in Sections 5.3 to 5.5.

In view of the variety of existing methods of calculation, the user may have problems choosing the method that is adequate to his or her problem. It is clear that *the method of calculation of the electronic structure of a given molecular system should satisfy the requirement of minimum labor compatible with the problem to be solved and the accuracy required.*

For instance, to calculate the expected g factors in the ESR spectra of transition metal and rare earth complexes (Section 8.4), the crystal field theory (Chapter 4) may be sufficient, whereas the superhyperfine structure cannot be estimated without taking into account the covalency effects, that is, without using one of the MO approximations (Section 5.2). If the spin density distribution over a larger ligand system is sought, a more complete MO LCAO approach should be employed (semiempirical, semiquantitative, nonempirical, *ab initio*, and their different versions), depending on the required quantitative reliability and accuracy, and technical possibilities.

The problem of *accuracy of the results* of quantum-chemical calculations is not trivial. Although quantum chemistry is, in general, a theoretical discipline based on the background of quantum mechanics, unlike the majority of mathematical disciplines, it has no mathematically rigorous internal criterion of accuracy of calculations. This statement emerges from the basic approximations of quantum chemistry, mentioned in Section 1.3, which are inevitable and have no exact quantitative measure of the errors they introduce.

As a measure of calculation accuracy, the *quantum-mechanical limit*—calculations in which all the interactions are taken into account exactly as it is required by quantum mechanics—could be used. However, strictly speaking, the *exact* quantum-mechanical limit cannot be reached not only because of technical problems, but also in principle, because of the lack of exact Hamiltonians that include the relativistic interaction between the electrons (Section 5.4). Although the latter is important only to some relatively weak magnetic properties, it is a matter of principle that, strictly speaking, the exact quantum-mechanical limit does not exist.

Practically, the quantum-mechanical limit cannot be reached for more prosaic reasons, which are discussed in Sections 1.3, 5.3, and 5.4. For MO and DFT methods, they are: use of limited basis sets and neglect of (a part of) electron correlation effects, adiabatic and nonadiabatic coupling to nuclear motions, and relativistic effects. The extension of the basis set leads to the Hartree–Fock limit. Including the correlation effects (e.g., by CI), one moves significantly to the quantum-mechanical limit but remains far from it. Exact calculations, including nuclear motions and relativistic effects, are beyond the practical possibilities not only for coordination compounds but also for much smaller molecules.

Thus in electronic structure calculations in general, we are far from the quantum-mechanical limit and there are no quantitative estimates of the remaining distance to this limit in each specific case. Therefore, the only practically acceptable criterion of accuracy of the results of calculations is *comparison with experimental data.*

Concerning the comparison with experiments, in many cases the less laborious and less quantitatively reliable semiempirical results may be closer to the experimental data than the nonempirical calculations. This better agreement with experiments is reached not by more accurate calculations but by introducing appropriate empirical and other (sometimes arbitrary, adjustable) parameters. With adjustable parameters, one can obtain results that are more adequate to the experimental data, but they cannot be classified as more accurate results of calculations in the sense of approaching the quantum-mechanical limit.

Table 5.8. Comparison of nonempirical and semiempirical MO LCAO methods in terms of relative advantages and disadvantages denoted by "+" and "−", respectively

Characteristic	Nonempirical Methods	Semiempirical Methods
Theoretical foundation	+	−
Accuracy of results	+	−
Quantitative reliability (credibility)	+	−
Labor consume	−	+
Calculative molecular size	−	+
Adequacy to the experiment	−	+
Flexibility (e.g., the possibility of fragmentary calculations)	−	+
Visualization of the results	−	+
Relation to analytical models	−	+

Source: [5.5].

Table 5.8 compares nonempirical and semiempirical calculations from the point of view of their relative advantages and disadvantages, denoted by "+" and "−", respectively.

The nonempirical methods are ahead in determining the progress in the area as a whole. The recent years' achievements in solving chemical problems (Sections 6.3 to 6.5 and 11.3), based in large part on the use of supercomputers, show the increasing ability of these methods. As mentioned above, nonempirical methods are also used to obtain calibrating points for semiempirical methods, thus influencing their development and serving them as a test.

On the other hand, semiempirical methods have the many advantages of simplicity, visualization, and flexibility; they consume less labor and computer time; and they make possible calculations of much larger systems than do nonempirical methods. These advantages are of particular importance to coordination compounds with heavy atoms and large ligands (e.g., metallobiochemical compounds). Special attention should also be paid to better visualization and more direct relevance to analytical models that allow for better understanding and contribute significantly to the formation of intuitive thinking (Section 1.1). Analytical models also contribute to the rationalization of numerical results and development of the general theory of electronic structure.

As for coordination compounds, some additional remarks should be made on the comparison of MO LCAO approaches with crystal field theory (CFT). CFT provides a satisfactory qualitative description of one of the main effects of coordination, CA term splitting. The part of the MO LCAO scheme concerning the CA states results qualitatively in the same energy-level splitting (Section 6.2), but the absolute value of the main CFT parameter Δ can be correctly calculated only using the MO LCAO approach.

In the late 1950s and early 1960s there was an intensive discussion of whether CFT gives the correct order of magnitude for Δ calculated in the point charge model, and it was concluded that for predominant ionic compounds the covalency contribution to Δ is about 10 to 20% [5.11]. If covalency corrections are introduced, for instance, in the AOM approximation of weak covalency (Section 5.2), the results on energy-level splitting may be quite satisfactory. A more detailed theoretical comparison of CFT, AOM, and MO LCAO methods is given in [5.120].

However, as discussed in Section 4.6, the CFT cannot, in principle, consider phenomena dependent on the details of electronic structure of ligands, including the detailed description of the bonding, π-bond formation, charge transfer spectra, ligand activation by coordination, and so on, and the AOM is also strongly limited to phenomena that take place within the d electron shell (Section 5.2). All these problems should be treated by MO LCAO or density functional methods.

The possibilities of quantum chemistry increase significantly when reasonable combinations of quantum-mechanical, semiclassical, and classical methods are employed. For instance, in a paper by Clementi and Corongiu [5.121], it is shown that very large systems with thousands of atoms can be described by nonempirical quantum-chemical calculations of electronic structure followed by a hierarchy of models, including molecular mechanics, statistical physics and fluid mechanics, and molecular dynamics. The DNA in solutions (more than 4000 atoms) was used as an example. This work shows that the possibilities of theoretical chemistry using electronic structure calculations is practically unlimited and may reach the level of macrosystems, provided that the theory is used with a proper understanding of its possibilities. An essential problem here is to work out appropriate interfaces between different parts of the polyatomic system which are treated by significantly different methods. Attempts have been made to carry out combined quantum-mechanical (QM) calculations with molecular mechanics (MM) treatment (Section 9.1) of organic systems [5.72–5.75]. For instance, for the interaction between reactants and solvent, one can consider the former by QM calculations and the latter by MM. These suggestions do not include the option of charge transfers between its fragments, and hence they are inapplicable to transition metal systems for which such charge transfers are most important. Recently [5.64], a method of QM/MM calculations for transition metal systems has been worked out based on the fragmentation of the system, discussed above (Section 5.4), with QM calculation and geometry optimization of the metal-containing fragment and MM treatment of the organic ligands; the electronic interface based on the double (intrafragment and interfragment) self-consistent procedure allows for interfragment charge transfers.

REFERENCES

5.1. R. S. Mulliken, in: *22nd International Congress of Pure and Applied Chemistry: Plenary Lectures*, Butterworth, London, 1970, pp. 203–215.

5.2. L. E. Orgel, *An Introduction to Transition-Metal Chemistry*, Wiley, New York, 1960.

5.3. K. Ballhausen, *Introduction to Ligand Field Theory*, McGraw-Hill, New York, 1962; *Molecular Electronic Structure of Transition Metal Complexes*, McGraw-Hill, New York, 1979

5.4. C. K. Jorgensen, *Modern Aspects of Ligand Field Theory*, North-Holland, Amsterdam, 1971, 538 pp.

5.5. I. B. Bersuker, *Electronnoe Stroenie i Svoistva Koordinatsionnykh Soedinenii* (Russ.), 3rd ed., Khimia, Leningrad, 1986.

5.6. A. Veillard, Ed., *Quantum Chemistry: The Challenge of Transition Metals and Coordination Chemistry*, NATO ASI Series C, Vol. 176, D. Reidel, Dordrecht, The Netherlands, 1986.

5.7. R. S. Mulliken, *J. Chem. Phys.*, **23**, 1841–1846 (1955).

5.8. P. Politzer and R. R. Harris, *J. Am. Chem. Soc.*, **92**, 6541 (1970); P. Politzer, *Theor. Chim. Acta*, **23**, 203 (1971).

5.9. J. O. Noell, *Inorg. Chem.*, **21**, 11–14 (1982)

5.10. P. J. Hay, *J. Am. Chem. Soc.*, **103**, 1390–1393 (1981).

5.11. R. E. Watson and A. J. Freeman, *Phys. Rev. Ser. A*, **134**, 1526–1546 (1964); E. Simanek and Z. Sroubek, *Phys. Status Solidi*, **4**, 251–259 (1964).

5.12. C. K. Jorgensen, R. Pappalardo, and H. H. Schmidtke, *J. Chem. Phys.*, **39**, 1422 (1963); C. E. Schaffer, *Struct. Bonding*, **14**, 69 (1973).

5.13. D. W. Smith, *Struct. Bonding*, **12**, 49–112 (1972).

5.14. A. B. P. Lever, *Electronic Spectra of Inorganic Compounds*, Elsevier, Amsterdam, 1984.

5.15. C. C. Roothaan, *Rev. Mod. Phys.*, **23**, 69–89 (1951); **32**, 179 (1960).

5.16. J. C. Slater, *Electronic Structure of Molecules*, McGraw-Hill, New York, 1963.

5.17. R. McWeeny, *Methods of Molecular Quantum Mechanics*, 2nd ed., Academic Press, London, 1989.

5.18. J. A. Pople and R. K. Nesbet, *J. Chem. Phys.*, **22**, 571–572 (1954).

5.19. A. T. Amos and G. G. Hall, *Proc. Roy. Soc. London*, **A263**, 483–493 (1961).

5.20. R. McWeeney, *Nature*, **166** (4209), 21–22 (1950).

5.21. S. F. Boys, *Proc. Roy. Soc. London*, **A200**, 542–554 (1950).

5.22. E. Clementi and D. R. Davis, *J. Comput. Phys.*, **1**, 223 (1966).

5.23. E. Clementi and D. C. Raimondi, *J. Chem. Phys.*, **38**, 2686 (1963).

5.24. J. W. Richardson, W. C. Nieuwpoort, R. R. Powell, and W. F. Edgell, *J. Chem. Phys.*, **36**, 1057 (1962).

5.25. E. Clementi, *IBM Res. Dev. Suppl.*, **9**, 2 (1965).

5.26. S. Huzinaga, *J. Chem. Phys.*, **42**, 1293–1302 (1965).

5.27. A. Veillard, *Theor. Chim. Acta*, **12**, 405 (1968).

5.28. B. Roos, A. Veillard, and G. Vinit, *Theor. Chim. Acta*, **20**, 1 (1971).

5.29. B. Roos and K. Siegbahn, *Theor. Chim. Acta*, **17**, 209 (1970).

5.30. S. Huzinaga, *Approximate Atomic Wave Functions*, Vols. I and II, Department of Chemistry Report, University of Alberta, Edmonton, Alberta, Canada, 1971.

5.31. E. Clementi and C. Roetti, *Atomic Data and Nuclear Data Tables*, Vol. 14, Acad. Press, New York, 1974, p. 177–478.

5.32. R. Krishnan, J. S. Binkley, R. Seeger, and J. A. Pople, *J. Chem. Phys.*, **72**, 650–654 (1980).

5.33. S. M. Bachrach and A. Streitwieser, Jr., *J. Am. Chem. Soc.*, **106**, 2283 (1984).

5.34. J. Demuynk, A. Veillard, and U. Wahlgren, *J. Am. Chem. Soc.*, **95**, 5563 (1973); J. Demuynk, Thèse de Doctorat d'Etat, Strasbourg, 1972.

5.35. S. Yoshida, S. Sakaki, and H. Kobayashi, *Electronic Processes in Catalysis: A Quantum Chemical Approach to Catalysis*, VCH, New York, 1994.

5.36. J. Almlof, in: *Lecture Notes in Quantum Chemistry II*, Ed. B. Boss, Lecture Notes in Chemistry, Vol. 64, Springer-Verlag, Berlin, 1994, p. 1.

5.37. P. Carsky and M. Urban, *Ab Initio Calculations: Methods and Applications in Chemistry*, Lecture Notes in Chemistry, Vol. 16, Springer-Verlag, Berlin, 1980. 247 pp.

5.38. W. Kutzelnigg, in: *Modern Theoretical Chemistry*, Vol. 3, ed. H. F. Schaefer III, Plenum Press, New York, 1977.

5.39. P. R. Taylor, in: *Lecture Notes in Quantum Chemistry II*, Ed. B. Boss, Lecture Notes in Chemistry, Vol. 64, Springer-Verlag, Berlin, 1994, p. 125.

5.40. D. R. Salahub and M.C. Zerner, in: *The Challenge of d and f Electrons. Theory and Computation*, ed. D. R. Salahub and M. C. Zerner, ACS Symposium Series 394, American Chemical Society, Washington, D.C., 1989, pp. 1–16.

5.41. M. B. Hall and R. F. Fenske, *Inorg. Chem.*, **11**, 768–775 (1972).

5.42. P. J. Hay and W. R. Wadt, *J. Chem. Phys.*, **82**, 270, 299 (1985); W. R. Wadt and P. J. Hay, *J. Chem. Phys.*, **82**, 284 (1985).

5.43. H. A. Bethe and E. E. Salpeter, *Quantum Mechanics of One- and Two-Electron Atoms*, Springer-Verlag, Berlin, 1957.

5.44. U. Weding, M. Dolg, H. Stoll, and H. Preuss, in: *Quantum Chemistry: A Challenge of Transition Metals and Coordination Chemistry*, ed. A. Veillard, NATO Series C, Vol. 176, D. Reidel, Dordrecht, The Netherlands, 1986, p. 79.

5.45. R. G. Parr and W. Yang, *Density Functional Theory of Atoms and Molecules*, Oxford University Press, New York, 1989.

5.46. J. K. Labanowski and J. Andzelm, Eds., *Density Functional Methods in Chemistry*, Springer-Verlag, New York, 1991; J. Weber, Ed., *Applications of Density Functional Theory in Chemistry and Physics*, *New J. Chem.*, **16**, N 12 (1992).

5.47. N. H. March, *Electron Density Theory of Atoms and Molecules*, Academic Press, New York, 1992.

5.48. S. B. Trickey, Ed., *Density Functional Theory of Many-Fermion Systems*, in: *Adv. Quant. Chem.*, **21** (1990).

5.49. N. C. Handy, in: *Lecture Notes in Quantum Chemistry II*, Ed. B. Ross, Lecture Notes in Chemistry, Vol. 64, Springer-Verlag Berlin, 1994, p. 91.

5.50. P. C. Hohenberg and W. Kohn, *Phys. Rev.*, **136**, B846 (1964); P. C. Hohenberg, W. Kohn, and L. J. Sham, in: Ref. [5.48], p. 7.

5.51. J. C. Slater, *Adv. Quant. Chem.*, **6**, 1–92 (1972); J. C. Slater and K. H. Johnson, *Phys. Rev. B*, **5**, 844–853 (1972).

5.52. K. H. Johnson, *Adv. Quantum Chem.*, **7**, 143–185, (1973).

5.53. K. H. Johnson, and F. C. Smith, Jr., *Phys. Rev. B*, **5**, 813–843 (1972); K. H. Johnson, J. C. Norman, Jr., and J. W. D. Connoly, in: *Computational Method for Large Molecules and Localized States in Solids*, ed. F. Herman, A. D. McLean, and R. K. Nesbet, Plenum Press, New York, 1973, pp. 161–201; K. H. Johnson, *Annu. Rev. Phys. Chem.*, **26**, 39 (1975).

5.54. H. Chermette, *New J. Chem.*, **16**, 1081–1088 (1992).

5.55. A. Rosen, D. E. Ellis, H. Adachi, and F. W. Averill, *J. Chem. Phys.*, **65**, 3629–3634 (1976).

5.56. J. Q. Snijders and E. J. Baerends, *Mol. Phys.*, **33**, 1651–1662 (1977).

5.57. R. Gaspar, *Acta Phys. Hung.*, **3**, 263–286 (1954).

5.58. K. Schwarz, *Phys. Rev. B*, **5**, 2466–2468 (1972).

5.59. A. Rosen and D. E. Ellis, *J. Chem. Phys.*, **62**, 3039–3049 (1973); A. Rosen and D. E. Ellis, *J. Chem. Phys.*, **62**, 3039 (1975); D. E. Ellis and G. L. Goodman, *Int. J. Quantum Chem.*, **25**, 185 (1984).

5.60. W. Kohn and L. J. Sham, *Phys. Rev.*, **140**, A1133 (1965).

5.61. D. R. Salahub, *Adv. Chem. Phys.*, **69**, 447 (1987); D. R. Salahub, R. Fournier, P. Mlynarski, I. Papai, A. St. Amant, and J. Ushio, in: Ref. [5.46a], pp. 77–100.

5.62. J. P. Perdew and W. Yue, *Phys. Rev. B*, **33**, 8800–8882 (1986).

5.63. I. B. Bersuker, *Int. J. Quantum Chem.* (submitted); *Proceedings of the First Electronic Computational Chemistry Conference—CDROM*, Eds, S. M. Bachrach, D. B. Boyd, S. K. Gray, W. Hase, and H. S. Rzepa, ARInternet, Landover, MD, 1995, paper 7.

5.64. I. B. Bersuker and R. S. Pearlman, *Proceedings of the 15th Austin Symposium on Molecular Structure*, The University of Texas, Austin, 1994, p. 17; I. B. Bersuker, M. Leong, J. E. Boggs, and R. S. Pearlman, in: *Proceedings of the First Electronic Computational Chemistry Conference—CDROM*. Eds. S. M. Bachrach, D. B. Boyd, S. K. Gray, W. Hase, and M. S. Rzepa, ARInternet, Landover, MD, 1995, paper 8.

5.65. R. E. Christoffersen, *Adv. Quantum Chem.*, **6**, 333–393 (1972).

5.66. A. S. Dimoglo, *Khim. Farm. Zh.*, **19**, 438 (1985); I. B. Bersuker and A. S. Dimoglo, in: *Reviews in Computational Chemistry*, Vol. 2, ed. K. Lipkowitz and D. Boyd, VCH, New York, 1991, p. 423–460.

5.67. I. B. Bersuker, *Teor. Eksp. Khim.*, **9**, 3–12 (1973) [Engl. Transl.: *Theor. Exp. Chem.*, **9**, 1–7 (1973)].

5.68. S. A. Gershgorin, *Izv. Akad. Nauk SSSR*, Ser. VII, OFMN, No. 6, 749–752 (1931).

5.69. R. A. Evarestov, E. A. Kotomin, and A. N. Ermoshkin, *Molecular Models of Point Defects in Large-Gap Solids* (Russ.), Riga, Zinatne, 1983, 288 pp.

5.70. F. P. Larkins, *J. Phys. Chem.*, **4**, 3065–3082 (1971).

5.71. J. K. Burdett, *Molecular Shapes. Theoretical Models of Inorganic Stereochemistry*, Wiley, New York, 1980.

5.72. A. Warshel and M. Levitt, *J. Mol. Biol.*, **103**, 227–249 (1976).

5.73. U. C. Singh and P. A. Kollman, *J. Comput. Chem.*, **7**, 718–730 (1986).

5.74. J. Field, P. A. Bash, and M. Karplus, *J. Comput. Chem.*, **11**, 700–733 (1990).

5.75. J. Gao and X. Xia, *Science*, **258**, 631–635 (1992); J. Gao, L. W. Chou, and A. Auerbach, *Biophys. J.*, **65**, 43–47 (1993); R. V. Stanton, D. S. Hartsough, and K. M. Merz, Jr., *J. Phys. Chem.*, **97**, 11868-11870 (1993).

5.76. P. Pyykko, *Chem. Rev.*, **88**, 563–594 (1988).

5.77. S. Wilson, Ed., *Methods in Computational Chemistry*, Vol. 2, *Relativistic Effects in Atoms and Molecules*, Plenum Press, New York, 1988.

5.78. D. R. McKelvey, *J. Chem. Educ.*, **60**, 112–116 (1983).

5.79. G. L. Mali, Ed., *Relativistic Effects in Atoms, Molecules and Solids*, Plenum Press, New York, 1983.

5.80. A. I. Akhieser and V. B. Berestetskii, *Quantum Electrodynamics* (Russ.), Nauka, Moscow, 1969, 623 pp.

5.81. I. B. Bersuker, S. S. Budnikov, and B. A. Leizerov, Internat. *J. Quantum. Chem.*, **6**, 849–858 (1972); *Teor. Eksp. Khim.*, **4**, 586–589 (1974); I. B. Bersuker and I. Ya. Ogurtsov, in: *Methods of Quantum Chemistry* (Russ.), Academy of Sciences USSR, Chernogolovka, 1979, pp. 70–81.

5.82. L. Visscher, P. J. C. Aerts, and W.C. Nieuwpoort, *J. Chem. Phys.*, **96**, 2910–2919 (1992).

5.83. L. A. La John, P. A. Christiansen, R. B. Ross, T. Atashroo, and W. C. Ermler, *J. Chem. Phys.*, **87**, 2812 (1987).

5.84. Y. Sakai, E. Miyoshi, M. Klobukowski, and S. Huzinaga, *J. Comput. Chem.*, **8**, 226, 256 (1987).

5.85. M. Dolg, H. Stoll, and H. Preuss, *J. Chem. Phys.*, **90**, 1730–1734 (1989); W. Kuechle, M. Dolg, and H. Preuss, *J. Chem. Phys.*, **100**, 7535 (1994).

5.86. T. R. Cundari and W. Stevens, *J. Chem. Phys.*, **98**, 5555–5565 (1993).

5.87. D. A. Case, *Annu. Rev. Phys. Chem.*, **33**, 151 (1982).

5.88. H. Chermette and A. Goursot, in: *Local Density Approximation in Quantum Chemistry and Solid State Physics*, ed. J. P. Dahl and J. Avery, Plenum Press, New York, 1984, pp. 353–379.

5.89. E. J. Baerends, J. A. Snijders, C. A. de Lange, and G. Jonkers, in: *Local Density Approximation in Quantum Chemistry and Solid State Physics*, ed. J. P. Dahl and J. Avery, Plenum Press, New York, 1984, pp. 415–485.

5.90. E. K. U. Gross and R. M. Dreizler, in: *Local Density Approximation in Quantum Chemistry and Solid State Physics*, ed. J. P. Dahl and J. Avery, Plenum Press, New York, 1984, pp. 353–379.

5.91. P. Pyykko, Report HUKI 1–86, University of Helsinki, 1986.

5.92. J. A. Pople, D. P. Santry, and G. A. Segal, *J. Chem. Phys.*, **43**, 129–135 (1965).

5.93. M. G. Veselov and M. M. Mestechkin, *Litov. Fiz. Sb*, **3**, 269–276 (1963); I. J. Fisher-Hjalmars, *J. Chem. Phys.*, **42**, 1962–1972 (1965); *Adv. Quantum Chem.*, **2**, 25–46 (1965).

5.94. P. O. Lowdin, *Adv. Quantum Chem.*, **5**, 185–199, (1970).

5.95. J. P. Dahl and C. J . Ballhausen, *Adv. Quantum Chem.*, **4**, 170–226 (1969).

5.96. D. A. Brown, W. I. Chambers, and N. I. Fitzpatric, *Inorg. Chim. Acta Rev*, **6**, 7–30 (1972); R. D. Brown and K. R. Roby, *Theor. Chim. Acta*, **16**, 175–193, 194–216 (1970).

5.97. P. G. Burton, *Coord. Chem. Rev.*, **12**, 37–71 (1974).

5.98. M. Zerner, in: *Reviews in Computational Chemistry*, Vol. 2, ed. K. B. Lipkowitz and D. B. Boyd, VCH, New York, 1991, pp. 313–368.

5.99. J. A. Pople, D. L. Beveridge, and P. A. Dobosh, *J. Chem. Phys.*, **47**, 2026–2030 (1967); M. S. Gordon and J. A. Pople, *J. Chem. Phys.*, **49**, 4643–4650 (1968).

5.100. N. C. Baird and M. J. S. Dewar, *J. Am. Chem. Soc.*, **91**, 352 (1969).

5.101. A. D. Bacok and M. C. Zerner, *Teor. Chim. Acta*, **53**, 21–24 (1979).

5.102. R. C. Bingham, M. J. S. Dewar, and D. H. Lo, *J. Am. Chem. Soc.*, **97**, 1285–1293 (1975); **97**, 1307 (1975).

5.103. M. J. S. Dewar, D. H. Lo, and C. A. Ramsdlen, *J. Am. Chem. Soc.*, **97**, 1311–1318 (1975).

5.104. R. Hoffmann, *J. Chem. Phys.*, **39**, 1397–1412 (1963).

5.105. M. Wolfsberg and L. Helmholz, *J. Chem. Phys.*, **20**, 837–843 (1952).

5.106. S. P. McGlynn, L. G. Vanquickenborne, L. G. Kinoshita, and D. G. Carroll, *Introduction to Applied Quantum Chemistry*, Holt, Rinehart and Winston, New York, 1972, 472 pp.

5.107. C. J. Ballhausen and H. B. Gray, *Molecular Orbital Theory*, W.A. Benjamin, New York, 1964, 273 pp.

5.108. H. Bash, A. Viste, and H. B. Gray, *J. Chem. Phys.*, **44**, 10 (1966).

5.109. V. I. Baranovski and A. B. Nikolski, *Theor. Eksp. Khim.*, **3**, 527–533 (1967).

5.110. H. Bash and H.B. Gray, *Inorg. Chem.*, **6**, 630–644 (1967).

5.111. A. B. Anderson and R. Hoffmann, *J. Chem. Phys.*, **60**, 4271 (1974).

5.112. G. Calzaferri, L. Forss, and I. Kamber, *J. Phys. Chem.*, **93**, 5366–5371 (1989); M. Brandle and G. Calzaferri, *Helv. Chim. Acta*, **76**, 924 (1993); **76**, 2350 (1993).

5.113. F. Savary, J. Weber, and G. Calzaferri, *J. Phys. Chem.*, **97**, 3722–3727 (1993).

5.114. I. B. Bersuker, S. S. Budnikov, and B. A. Leizerov, *Int. J. Quantum. Chem.*, **11**, 543–560 (1977).

5.115. P. Pyykko, L. J. Laakkonen, and K. Tatsumi, *Inorg. Chem.*, **28**, 1801–1805 (1989).

5.116. P. Pyykko, in: *Methods in Computational Chemistry*, ed. S. Wilson, Vol. 2, Plenum Press, New York, 1988, pp. 137–226.

5.117. E. R. Davidson, in: *The Challenge of d and f Electrons: Theory and Computation*, ed. D. R. Salahub and M. C. Zerner, ACS Symposium Series 394, American Chemical Society, Washington, D.C., 1989, pp. 153–164.

5.118. M. C. Zerner, in: *Metal–Ligand Interaction: From Atoms, to Clusters, to Surfaces*, ed. D. R. Salahub and N. Russo, Kluwer, Dordrecht, The Netherlands, 1992, pp. 101–123.

5.119. R. Hoffmann, *Pure Appl. Chem.*, **50**, 55–64 (1978).

5.120. M. Gerloch, *Prog. Inorg. Chem.*, **31**, 371–446 (1984).

5.121. E. Clementi and G. Corongiu, in: *Studies in Physical and Theoretical Chemistry*, Vol. 27: *Ions and Molecules in Solutions*, ed. N. Tanaka, H. Ohtaki, and R. Tamamushi, Elsevier, Amsterdam, 1982, pp. 397–431.

6 Electronic Structure and Chemical Bonding

Chemical bonding predetermines the very existence of chemical substances and underlies purposeful synthesis of new compounds with required structure and properties.

This chapter is devoted to the origin of chemical bonding in transition metal coordination compounds as a feature of their electronic structure.

6.1. CLASSIFICATION OF CHEMICAL BONDS BY THEIR ELECTRONIC STRUCTURE AND THE ROLE OF d AND f ELECTRONS IN COORDINATION BONDING

As follows from the electronic nature of the chemical bond defined in Section 1.2 resulting from collectivization of the electrons of interacting atoms, electronic structure plays a key role in the classification of chemical bonds and definition of coordination bonding. In the literature, the analysis of this problem from the viewpoint of the achievements and modern understanding of the origin of chemical bonding has not received due attention [6.1]. In fact, the commonly used attribution of compounds to different classes is based on the historically established classification, carried out when our knowledge of electronic structure was rather poor, and therefore it creates controversies and misunderstandings. Below we discuss the classification of covalent bonds; as mentioned in Section 1.2, pure ionic compounds do not exist (despite the fact that the ionic model may be useful in particular cases).

Criticism of Genealogical Classification

The traditional ("classical") classification of chemical bonds is based on the idea of atomic valency. Following this idea, it is assumed that there is a large class of *valence compounds* in which the chemical bonds have a localized diatomic nature similar to that of the H—H bond in the H_2 molecule. The main assumption is that *the valence bond is formed by the pairing of two electrons supplied by one from each of the binding atoms.* The development of this concept led to the notions of *multiple bonds* and *bond saturation*, as well as to the presentation of a complex molecule by its *valence structure* with single, double, and triple bonds. For many classes of compounds, their presentation in the form of one valence structure proved to be invalid, and to preserve the concept of valency, the idea of *superposition of two or several valence structures* was employed.

In addition to this traditional valence systems, there is a big class of compounds in which the chemical bonding can be presented as formed by two atoms or atomic groups which (one or both) have no unpaired electrons. For these cases it is assumed, following the valence scheme, that the two electrons needed for the formation of the bond are supplied by one of the bonding groups (the donor of electrons), while the other group participates as an acceptor of electrons. Here we have the *donor–acceptor bond*, or *coordination bond*, the two notions thus being assumed identical.

It is seen that in the traditional classification two main types of bonds — valence bonds and donor–acceptor (coordination) bonds — are distinguished based on the possibility of reducing them to local diatomic and two-electron bonding. More precisely, the whole difference between these two types of bonds is reduced to *the genealogy* (origin) of the two bonding electrons in the diatomic bond: In valence compounds (bonds) the bonding electrons are provided by two atoms (or they occur as a superposition of several such possibilities), whereas in donor–acceptor bonds the two electrons are supplied by only one atom, the donor. Therefore, *this classification can be called genealogical*. It is based entirely on the concept of valency (Table 6.1).

Data accumulated from many years of study of chemical compounds show that the genealogical classification does not correlate with their electronic structure and properties. Indeed, two simple systems, CH_4 and NH_4^+, are isoelectronic and quite similar in the distribution of the Ψ cloud of electrons determining the bonding and properties (in both systems there are four tetrahedral two-electron bonds [6.2]). The difference between them is that in NH_4^+ there is an additional proton in the nucleus of the CA, making the hydrogen atoms more electropositive than in CH_4. However, following the genealogical classification, one must assume that CH_4 is a valence compound, while NH_4^+ is a coordination (donor–acceptor) system (NH_3 is the donor and H^+ is the acceptor). On the other hand, between NH_4^+ and, for example, $CuCl_4^{2-}$ there is almost nothing in common either in electronic structure or in properties. The genealogical classification, however, puts them in the same class of coordination compounds.

These deficiencies of the genealogical classification have not been criticized, nor critically studied and analyzed, because the classification of compounds after their origin (their past history) in some cases reflects the real process of the synthesis of the compounds; in the absence of details of electronic structure the genealogical classification was quite reasonable.

Presently, the subjects of investigation in modern chemistry are real compounds with their properties determined by the *actual electronic structure*, irrelevant to the method of preparation and past history (genealogy). As mentioned in Section 1.2, domination by preparative chemistry in the past is rapidly changing to that of structural chemistry. From the structural point of

Table 6.1. Classification of chemical bonds

I. By origin (genealogy) of the bonding electrons

Type of Bond	Origin of Bonding Electrons	Examples	Characteristic Properties
Valence	Each of the two bonding atoms supplies one unpaired electron, or a superposition of several such possibilities (several valence schemes) is considered	CH_4, C_6H_6, diamond, graphite, NO, $CoCl_2(c)$	No characteristic properties in common
Coordination	Both unpaired electrons are supplied by one of the atoms	$CuCl_4^{2-}$, NH_4^+, $BF_3 \cdot NH_3$	No characteristic properties in common

II. By electronic structure and properties

Type of bond	Electronic Structure	Examples	Characteristic Properties
Valence	The one-electron bonding states are localized between the pairs of bonding atoms and are occupied by two paired electrons	CH_4, NH_4^+, diamond	Approximate additive and transferrable bond energies, vibrational frequencies, dipole moments
Orbital or conjugated	The one-electron bonding states are delocalized over many atoms along the line of bonding with possible ramifications	C_6H_6, graphite	Conductivity along the bond; aromaticity
Coordination	The one-electron bonding states are three-dimensionally delocalized in space around a center	$CuCl_4^{2-}$, $CoCl_2(c)$	Nonadditive and nontransferrable bond features; strong mutual influence of CA–ligand bonds; specific color, magnetic, thermodynamic, and reactivity properties

view, classification by the genealogy of the bonding electrons, which attributes compounds with quite different (sometimes opposite) properties to the same class, is unacceptable. The genealogical classification is also not acceptable as a matter of principle because it is based entirely on the concept of valency, which is not a comprehensive characteristic of all the chemical properties of atoms, especially transition metals.

Classification by Electronic Structure and Properties

In most cases the description of the electronic structure of polyatomic systems is given by one-electron MOs which in general are delocalized over the entire system. The total wavefunction is composed of MOs, by means of an appropriate symmetrization procedure (Section 2.2). It was shown by Lennard-Jones [6.3] that in some special cases, discussed below, there is the possibility of transforming the full wavefunction to the equivalent orbitals. The latter are occupied by two electrons, with the electronic charge concentrated mainly in the space between the corresponding pairs of near-neighbor atoms. In this presentation the total energy equals approximately the sum of the bonding energies between the pairs of atoms described by the equivalent orbitals.

This description of the electronic structure can serve as a theoretical foundation of the existence of valence bonds. However, the possibility of describing the system by equivalent orbitals is restricted by specific conditions: in particular, by the requirement that the number of bonding electrons be exactly equal to the number of bonds doubled [6.3]. Chemical bonds in such (and only such) compounds are valence bonds, indeed. By this definition, *valence compounds include all the systems that can be described by one valence scheme* (without conjugation): saturated hydrocarbons, CH_4 and NH_4^+, as well as BH_4^-, BF_3—NH_3 (electronic analog of CF_3—CH_3), and so on.

Similar electronic structure determines similar characteristic features. For valence compounds these are approximate *additivity* with respect to the bond's properties (e.g., the dipole moments, polarizabilities, bond energies, etc.), and *transferrability*, that is, relatively small changes in the properties of a given bond (dipole moments, vibrational frequencies, and energies) by passing from one compound to another (Table 6.1). The transferrability property implies that comparisons are made between bonds of the same type (the same type of hybridization and bond order), in the first approximation. Compared with the case of delocalized bonds — conjugated and coordination bonds (see below) — changes in the properties of localized valence bonds when passing to other compounds are much smaller and are influenced mainly by the near-neighbor atoms only.

All the other compounds cannot be described by localized (between pairs of atoms) two-electron bonds, and hence they cannot be considered as valence compounds. In nonvalence compounds, the one-electron MOs remain essentially delocalized; conjugated systems form a major part of them. The bonds in these systems can be characterized as *"orbital" or "conjugated" bonds* created by delocalized electrons. In fact, *this definition includes all the compounds that cannot be described by one valence scheme, except coordination compounds.*

In nonvalence conjugated bonds the delocalization of electrons takes place along one dimension or one plane of conjugation (as, e.g., in benzene) with possible ramifications (as, e.g., in naphthalene). Beside them is a large class of nonvalence compounds in which *the electronic states are three-dimensionally delocalized around some centers.* These compounds can reasonably be called *coordination compounds.* Thus we define them as *compounds with high coordination and three-dimensional delocalization.* This characteristic is novel: It differentiates the coordination system from other donor–acceptor compounds.

In this description two factors determine the coordination bond: high coordination, which implies the presence of a center of coordination and its environment, and three-dimensional delocalization, that is, collectivized (non-localized) CA–ligand bonds. It can be shown that the latter condition is obeyed when d (or f) *orbitals* of the central atom that have a multilobe form are actively involved in the bonding: Pure s and p orbitals cannot provide the required combination of high coordination and delocalization. Indeed, by means of s and p orbitals one can obtain, at the most, tetrahedral coordination on behalf of sp^3 hybridization. However, hybridization implies localization (Section 2.1), and hence hybridized states are attributed to localized (valence) bonds and not to delocalized coordination bonds. This is the main cause of the differences in electronic structure (and properties) between the valence compound CH_4 (sp^3 hybridization, localized C—H bonds) and the coordination system $CuCl_4{}^{2-}$ (participation of d electrons delocalized over all the Cu—Cl bonds). The participation of d (or f) orbitals in the formation of coordination bonds is thus most important.

The notion of s, p, d, f, ... atomic orbitals in molecules is not rigorous and may be misleading. Indeed, as mentioned in Section 2.1, these atomic states originate from the spherical symmetry of the free atom. Within the molecule the spherical symmetry is necessarily lowered, and the partition of the orbitals into s, p, d, ... is, strictly speaking, no more valid. However, one can find a region near the nuclei of the atoms where the influence of the atomic environment is small compared with the spherically symmetric nuclear field, and in this region the idea of s, p, d, f orbitals in molecules is approximately valid. Therefore, the statement of participation of d (or f) orbitals in the formation of coordination bonds should be understood in the sense that there are occupied one-electron states which in the region near the nucleus of the CA are of d (or f) nature, and at larger distances they become modified by the environment. In fact, the statement that d or f electrons participate in the bonding implies that they participate in occupied MOs of corresponding symmetry that are delocalized over the CA and all or a part of the ligands.

Note also that in coordination compounds of nontransition elements there may be a significant influence of *virtual d states* (see below). The measure of d participation in the bonding in various systems may be rather different [6.4]; it determines a large variety of special properties and the extent to which these systems are coordination compounds.

Although from the point of view of the CA the coordination bond looks delocalized (it does not allow accurate partition of the bonding into separate metal–ligand bonds), from the ligand viewpoint the bond may be quite localized and directed to the CA. However, this will not be a ligand–metal bond but a ligand–remaining complex bond; any change of other ligands, owing to the three-dimensional delocalization, influences the ligand–complex bond under consideration.

The classification of bonds does not necessarily coincide with the classification of compounds. Indeed, some compounds may have different types of bonds in their different parts. Some localized valence bonds become delocalized by excitation. There are atoms that in different conditions can form different types of bonds, especially when there are low-lying unoccupied d orbitals (see below about the coordination bonding by pre- and post-transition elements). It is reasonable to consider the molecular system as a transition metal coordination compound if it possesses at least one coordination center (coordination bond). Similarly, the compound is a conjugated system if it has at least one conjugated region but no coordination centers, and the molecule is a valence compound if it has neither coordination centers nor planar delocalized bonds but only localized bonds.

Coordination compounds include many classes of transition metal and rare earth compounds: complexes, chelates, clusters, metallo-organic compounds, including metallobiochemical systems, solid state and liquid crystals, alloys and solid solutions, chemisorbed surface states, and so on.

Thus based on electronic structure and properties, one can distinguish three main types of chemical bonds (Table 6.1): *valence bonds*, *orbital* (*or conjugated*) *bonds*, and *coordination bonds*. There are no strict demarcation lines between these types of bonds, and in this respect the classification above is conventional; however, outside the border regions, it is quite definitive.

Features of Coordination Bonds

The classification of chemical bonds discussed above and the definition of the coordination bond based on this classification enable us to discriminate some general features of coordination compounds listed in Table 6.2. As mentioned

Table 6.2. Correlation between the features of electronic structure and properties of coordination compounds

Feature of Electronic Structure	Properties
Increasing activity of d and f atomic states when moving down along the periodic table of elements	Increasing tendency to form coordination compounds in the same direction: its almost complete absence in the second period, intermediate position of the third period, and full manifestation of this tendency in the fourth and lower periods
Three-dimensional delocalization of the bonding electron density	Strong interdependence of the CA–ligand bonds and nontransferrability of their properties: energy, bond length, vibrational frequency, dipole moment, polarizability, reactivity, etc.
High symmetry and large coordination numbers, high capacity of d and f orbitals and hence close-in-energy (degenerate and pseudodegenerate) states	Two types of magnetic behavior: high-spin and low-spin, characteristic colors, thermodynamic properties, multiorbital bonds, strong vibronic effects, and so on, in spectroscopy, stereochemistry, crystal chemistry, reactivity, and chemical activation in catalysis

in Section 1.2, the presence of a coordination center allows one to denote the coordination system in a general way as ML_n^p, where M is the d (or f) central atom and L are the n ligands, the latter being either single atoms or groups of them, equal or different, and p is the total charge.

First, the role of d (f) electrons determines the *increasing tendency to form coordination systems when passing from light and main group atoms to transition and rare earth elements*. Second, *the CA–ligand bonds are delocalized, collectivized, and hence strongly interdependent*. Each M—L bond depends strongly on all the other bonds formed by M. Therefore the bond properties should be considered either for group of bonds M—L_1, M—L_2, M—L_3, and so on, or for the bond M—L_1 in the presence of M—L_2, M—L_3,....

Thus *the coordination bonds, in general, are essentially nontransferrable*, and this feature is confirmed by many experimental data. For instance, one of the main bond characteristics, the bond length M—L, where M is a transition metal, and L = N, O, C, Br,..., is strongly dependent on the nature of other bonds formed by M (e.g., the Cu—O bond length varies between 1.8 and 3.0 Å; Table 9.14), whereas the bond length C—L in different valence compounds is almost constant. This feature, formulated in general, does not mean that there are no *particular cases* where some transferrability is possible within a limited group of similar compounds (see "Molecular Modeling" in Section 9.1).

The third main feature includes the *characteristic properties caused by the participation of d (or f) orbitals of the CA*: characteristic color (electronic absorption in the visible and related regions of the spectrum; Section 8.2), magnetic (low-spin and high-spin complexes; Sections 4.3, 6.2, and 8.4), thermodynamic (two-humped dependence of the thermodynamic stability on the number of d electrons; Section 4.5), stereochemical (Chapter 9), and reactivity (Chapter 11) properties. This feature is due to the fact that the usually open shell of d (or f) electrons (formally closed shells but with low-energy excited states have similar properties; Sections 7.3 and 7.4) in combination with high coordination (high symmetry) creates *degenerate or nearly-degenerate (pseudodegenerate) energy terms*, ground or excited, which, in turn, results in a series of special effects and phenomena [6.5].

Coordination Bonding by Pre- and Post-transition Elements

Based on the definition of coordination systems given above, a question emerges as to how to explain the fact that some compounds of post-transition elements are similar in properties to coordination systems. In the definition of coordination bonding, it is required that the CA have active d or f electrons. Nontransition elements have no such electrons in *the free noncoordinated state*. This does not mean that active d or f states cannot occur in the corresponding *oxidation state*, or in the chemical bonding, i.e., in the *valence state*. As emphasized in Sections 6.3 and 11.2, coordination often results in *excitation of the bonding atoms or group of atoms*. Hence, by coordination, the electronic configuration of the bonding groups changes, and in certain conditions these changes may lead to activation of d (or f) electrons of the CA.

Any element has excited d states, but if they are very high in energy, they cannot be activated by the bonding. On the other hand, there may be occupied d states which are deep in energy as compared with the HOMO, and hence they cannot be excited just by bonding. We come to the conclusion that *potential coordination centers of nontransition elements can be found among the immediate pre- or post-transition elements.*

In pre-transition elements of the third period there are active s and p states and higher-in-energy inactive d orbitals. To make the d states active, that is, to lower their energy and to populate them with electrons, strong oxidizing ligands are required which simultaneously are good π donors to the unoccupied d states of the CA (see the discussion of interdependence of σ-acceptor and π-donor properties in Section 6.3). For example, oxygen, sulfur, and chlorine can activate the d states of Al, Ga, Ge, In, Sn, and so on, making them good coordination centers.

For post-transition elements, which are simultaneously pre-transition elements for the next transition group, there are two possibilities of d electron activation. The first is the same as for pre-transition elements: oxidizing ligands with π-donor abilities. The second possibility is to activate the inner occupied d states. For instance, the inactive d electrons of the d^{10} closed shell of Zn^{2+} or Ga^{3+} may become active under the influence of ligands which have significant σ-donor and π-acceptor properties. The former make the d states more diffuse in space due to the additional interelectron repulsion (see the discussion of the nephelauxetic effect in Section 8.2), while the latter allow for π back-donation.

In both cases the d orbitals are involved in the bonding. The measure of d participation is dependent on the nature of the ligands and can vary greatly. In this sense *there are no sharp borders between coordinating and noncoordinating elements.* In principle, *any element can serve as a center of coordination bonding provided that the ligands induce significant d participation.*

Note that even within the transition metal group the extent to which d electrons participate in the coordination bonding may be quite different [6.4]. In high-oxidation states the metal nd electrons are strongly attracted to the nucleus, and hence their overlap with the ligand orbitals is very small; the bonding, in its major part, is realized via $(n + 1)s$ and $(n + 1)p$ orbitals. Donation of electrons from the ligands makes the d states more diffusive, increasing their participation in the bonding. In low-oxidation states (e.g., in carbonyls) the d states are most active in the bonding. This tendency of d-participation in the coordination bonding varies stepwise by moving from early to late transition elements.

Thus *the ability to form coordination compounds with d (or f) participation changes gradually when moving from pre-transition to transition elements, from early to late transition elements, and then to post-transition elements, with strong dependence on the nature of the ligands and without sharp borders between transition and immediate pre- and post-transition elements.*

Following the definition of coordination systems given above based on the electronic structure, pre- and post-transition element compounds are coordination compounds to the extent of d electron participation in the bonding, which makes them similar in properties to the d electron compounds. Experimental data confirm the d orbital participation in the bonding of nontransition elements (see, e.g., [6.6]). In many cases the possible (in principle) d electron participation is in fact minor [6.7] (see also [6.8]).

Direct calculations of the electronic structure to confirm the d electron participation may encounter certain difficulties: the most reliable results of nonempirical calculations are strongly dependent on the choice of the basis set, which in fact predetermines the d-orbital occupation sought for (the more d functions included, the larger d occupation numbers). As stated in Section 5.3, basis-set functions in the expansion of the wavefunction have no direct physical sense and cannot, in principle, be used to relate the electron distribution to the atomic functions. What can be calculated to prove the statements above is the *experimentally observable specific features of coordination compounds* (Table 6.2): for instance, the interdependence of the different metal–ligand bonds' properties in the same compound, their *nontransferrability* to other compounds, and others. So far, to our knowledge, there are no such special-aimed calculations available in literature.

Note again that the division of electrons into s, p, d, f,... types when speaking about electron distribution in molecular systems is conventional since there are no such electrons in polyatomic systems: The interatomic interaction strongly modifies the atomic states. When discussing d orbitals in molecules, one bears in mind the d origin of the corresponding MO that has appropriate symmetry. In this meaning the classification of chemical bonds by their electronic structure is also genealogical, but it is based on the genealogy of the electronic structure of the compound under consideration with respect to its atoms or atomic groups, not on its method of preparation.

6.2. QUALITATIVE ASPECTS AND ELECTRONIC CONFIGURATIONS

Most Probable MO Schemes

Consider a transition metal of the $3d$ group as a CA. Its outer orbitals $3d$, $4s$, and $4p$ in the valence (oxidation) state in complexes form the following sequence of energy levels:

$$E(3d) < E(4s) < E(4p) \tag{6.1}$$

The overlap of the wavefunctions of these orbitals with ligand functions is larger for $4s$ and $4p$ orbitals and smaller for the $3d$ functions. Taking into account the dependence of the bonding (antibonding) properties of the MOs on the magnitude of the overlap integral (5.26), we come to the conclusion that the $3d$ orbitals form less bonding (and antibonding) MOs than do the $4s$ and $4p$ functions.

Consider an octahedral complex of O_h symmetry. The $4s$ orbital belongs to the A_{1g} representation (symmetry) (Table 5.1), $4p$ belongs to T_{1u}, and the five $3d$ orbitals form two groups: two $3d(e_g)$ orbitals (d_{z^2} and $d_{x^2-y^2}$) belonging to the E_g representation and three $3d(t_{2g})$ orbitals (d_{xy}, d_{xz} and d_{yz}) that belong to T_{2g}. The e_g orbitals are directed with their lobes of charge distribution toward the ligands and form σ bonds (Table 5.1), whereas the t_{2g} orbitals can form only π bonds. Hence the overlap of the $3d(t_{2g})$ orbitals with the corresponding ligand functions are smaller than that of $3d(e_g)$. It follows that the largest splitting into bonding and antibonding orbitals is expected in the formation of MOs by the $4s$ and $4p$ orbitals of the CA, smaller splitting comes from the $3d(e_g)$ orbitals, and the smallest one is due to the $3d(t_{2g})$ orbitals. With these ideas we come to the most probable MO energy-level scheme given in Fig. 6.1 for an octahedral coordination compounds of O_h symmetry formed by a $3d$ transition metal CA and ligands that have only s and p active orbitals. By comparison, one can easily make sure that this scheme corresponds to the data in Table 5.1. It is seen that the lowest MOs, a_{1g}, t'_{1u}, e_g, and t_{2g}, are bonding, the t''_{1u}, t_{2u}, and t_{1g} MOs are nonbonding, and the remaining MOs (marked by star) are antibonding.

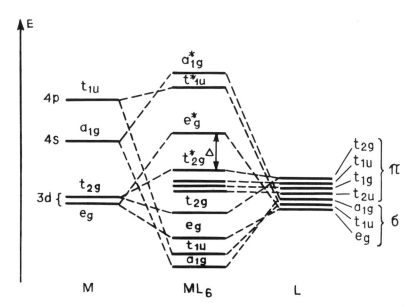

Figure 6.1. Most probable MO energy-level scheme for regular octahedral complexes ML_6 of $3d$ transition metals M with ligands L that have one σ and two π active AOs each.

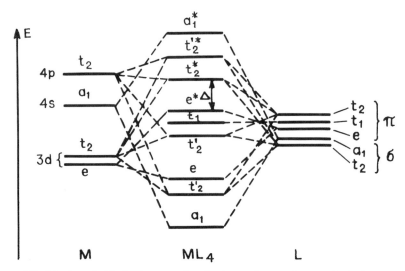

Figure 6.2. Most probable MO energy level scheme for regular tetrahedral complexes ML_4 of $3d$ transition metals M with ligands L that have one σ and two π active AOs each.

For a tetrahedral system of T_d symmetry, on the contrary, the $3d(e)$ orbitals of the central atom are less overlapped with the ligand orbitals than $3d(t_2)$, their role thus changing as compared with the octahedral case: By forming MOs, the energy levels of the e orbitals are less split than the t_2 orbitals (Fig. 6.2).

In systems with lower symmetries the degenerate MO energy levels are split by the ligand field, and the splitting magnitude is dependent on the nature of the ligand. Here the MO energy level scheme becomes complicated and no general ideas can be suggested; the corresponding problems should be solved by more detailed calculations.

Electronic Configurations in Low- and High-Spin Complexes

The next stage in determining the electronic structure of coordination compounds in the MO approximation is to find the electron distribution over the one-electron MOs, *the electronic configuration* of the system. In accordance with the Pauli principle, in the ground state the electrons occupy the lowest-energy MOs, in groups of two with opposite spins on each MO. The number of electrons to be distributed on the MOs equals the number of electrons that occupy the AOs used in the formation of these MOs. For instance, in the $TiCl_6^{3-}$ complex the valence AOs used in the formation of MOs are: one σ hybridized sp and two π-type $3p$ orbitals from each Cl^- ion and $3d$, $4p$, and $4s$ orbitals of Ti^{3+}, a total of $3 \times 6 + 9 = 27$ AOs that form 27 MOs occupied by $6 \times 6 + 1 = 37$ electrons. The number of MOs equals the number of AOs used for their formation, and hence not all MOs will be occupied by two electrons in the ground state.

One can imagine the following picture (Fig. 6.1): The $6 \times 6 = 36$ electrons from the six Cl^- ions occupy the following orbitals (the electron occupation number is indicated in parentheses): $a_{1g}(2)$, $t'_{1u}(6)$, $e_g(4)$, $t_{2g}(6)$, $t''_{1u}(6)$, $t_{2u}(6)$, $t_{1g}(6)$, and hence the $3d$ electron from Ti^{3+} occupies the antibonding t^*_{2g} orbital. This representation has reasonable physical meaning: As shown by calculations, the bonding orbitals a_{1g}, t'_{1u}, e_g, and t_{2g} are basically (and the nonbonding orbitals are completely) ligand orbitals. Of course, the MO occupation scheme remains the same when one starts not from Cl^- and Ti^{3+} ions but from neutral atoms.

Thus the electronic configuration of the complex $TiCl_6^{3-}$ is $(a_{1g})^2(t'_{1u})^6(e_g)^4(t_{2g})^6(t''_{1u})^6(t_{1g})^6(t^*_{2g})^1 - {}^2T_{2g}$, or by denoting the "closed shell" (including all the bonding and nonbonding orbitals fully occupied by electrons) by M in brackets, we have $[M](t^*_{2g})^1 - {}^2T_{2g}$. The full multielectron term ${}^2T_{2g}$ is here determined simply by one unpaired electron in a MO of t_{2g} symmetry, but in more complicated cases it should be determined by multi-electron wavefunction rules (Section 2.2).

The result obtained here is qualitatively the same as it emerges from the more simple treatment of the crystal field theory (CFT) (Section 4.2), where the electronic configuration of the complex at hand is $[A](3d)^1 - {}^2T_{2g}$ (Table 4.5). Indeed, the antibonding t^*_{2g} orbital is localized mostly on the CA, and its difference from the $3d$ orbital of the Ti^{3+} ion is in a small admixture of the π functions of the Cl^- ions. Note that the first excited state in the above MO scheme is $[M](e_g)^1 - {}^2E_g$, which coincides with that obtained in CFT. Hence the long-wavelength electronic transition ${}^2T_{2g} \leftrightarrow E_g$ is analogous to that expected in CFT. However, the energy gap $\Delta = E({}^2E_g) - E({}^2T_{2g})$ that determines the corresponding absorption band position may be quite different in these two theories.

For more than one d electron above the closed shell $[M]$ (e.g., in an octahedral complex of V^{3+} with two d electrons), one encounters complications caused by the interaction between the electrons. Indeed, the approximate MO schemes of Figs. 6.1 and 6.2 are one-electron schemes in which the interelectron interaction is not taken into account explicitly. For two d electrons several possibilities arise for one-electron MO occupation:

1. Two of three different t^*_{2g} orbitals are occupied by the two electrons with parallel spin orientations, resulting in the configuration $(t^*_{2g}\uparrow)(t^*_{2g}\uparrow)$ with a total spin $S = 1$.

2. The same two orbitals are occupied by two electrons with opposite spin orientations, $(t^*_{2g}\uparrow)(t^*_{2g}\downarrow)$, $S = 0$.

3. One t^*_{2g} orbital is occupied by two electrons with opposite spin orientations, $(t^*_{2g})^2$, $S = 0$.

4. One electron is in t^*_{2g}, the other is in the excited orbitals e_g, and so on.

Obviously, the electrostatic and exchange interactions between the electrons in these distributions are different, resulting in different energy terms. The ground-state configuration and spin will be that of lowest energy. Similar situations in free atoms and with ligand crystal fields are considered in detail in Sections 2.2 and 4.3, respectively. Several terms of the same electronic configuration may occur, and all of them can be determined by taking into account the interelectron interactions.

The procedure of distribution of the d electrons of the CA on the antibonding t_{2g}^* and e_g^* MOs with consequent inclusion of the interelectron interactions in each of these distributions is quite similar to that employed in crystal field theory for the strong field limit (Section 4.3). In particular, for two electrons three electronic configurations, $(t_{2g}^*)^2$, $(t_{2g}^*)^1(e_g^*)^1$, and $(e_g^*)^2$, are possible with the same two energy gaps between them equal to the $t_{2g}^* - e_g^*$ separation Δ (disregarding the interelectron interaction). The energy splitting of these configurations by the interelectron interaction results in electronic terms similar in spacing to that obtained in the crystal field approximation (4.41), in which the Racah constants A, B, and C should be substituted for by the parameters of interelectron repulsions on the MOs, not AOs, and hence they are no longer determined by Eq. (2.33).

Corrections to the formulas of atomic energy term formation in ligand fields that take into account the transformation of AOs into MOs can be introduced [6.9]. For antibonding orbitals these corrections are negative, which means that the Racah parameters for antibonding MOs are smaller than for AOs. This result can be understood easily: When passing from AOs to MOs the electronic cloud diffuses over a larger volume in which the interelectronic repulsion is obviously reduced.

Thus in the case under consideration the MO energy scheme of a coordination compound remains the same as in the crystal field theory with a different meaning for the Racah constants A, B, and C that become dependent on the covalency parameters (LCAO coefficients). As in the crystal field approximation, the $(t_{2g}^*)^2$ configuration yields four terms, ^3T_1g, $^1T_{2g}$, ^1E_g, and $^1A_{1g}$, with relative energies given in Eq. (4.41). The largest splitting between them is $\Delta = E(^1A_{1g}) - E(^3T_{1g}) = 15B + 5C$. Hence the limit of strong ligand field in which the configurations above can be considered separately is valid if $15B + 5C \ll \Delta$. If this condition is not fulfilled, and in the opposite limit case of weak ligand field ($\Delta < 15B + 5C$), the terms of the same symmetry from different configurations are strongly mixed and all the configurations should be considered simultaneously (Section 4.3).

Special attention must be paid to cases where, owing to the interelectron interactions, the electronic configuration of the ground state changes when passing from the limit of strong fields to that of weak fields, and vice versa. Let us elucidate this situation by the example of an octahedral complex of a d^4 transition metal atom or ion. The first three (from the four) d electrons, in accordance with Hund's rule, occupy the three t_{2g}^* orbitals with parallel spin orientations, the total spin hence being $S = \frac{3}{2}$. The fourth electron has two main possibilities. One of them is to occupy one of the already half populated orbitals with an opposite orientation of the spin, and then the electronic configuration of the complex becomes $[M](t_{2g}^*\uparrow)^3(t_{2g}^*\downarrow)^1$ with a total spin $S = 1$ and ground state $^3T_{2g}$. Note that the total energy of the system is significantly increased here not only by the interelectron repulsion of the electrons (two of them are now on the same orbital), but also by the reduction in the negative energy of exchange interaction (per electron), which is nonzero for electrons with parallel spins only.

The second possibility is to occupy the e_g orbital, which is higher in energy by Δ (Fig. 6.1), but with the same orientation of spin as in the three t_{2g} orbitals. Then the configuration becomes $[M](t_{2g}^*\uparrow)^3(e_g^*\uparrow)^1$ with a total spin $S = 2$, the ground state being ^5E_g. Here we have less interelectron repulsion (the electrons are in different orbitals and hence they occupy different regions in space), and more negative contribution of the exchange interaction (all the electrons have parallel spin orientations), but one of the electrons is higher in orbital energy by Δ. These two possibilities result in two essentially different electronic configurations with different spin values. The latter is a convenient indicator of the differences. Therefore, the two possibilities are usually called *low-spin and high-spin configurations*, respectively, quite similar to the CFT treatment (Section 4.3).

As in Section 4.3, the energy difference between the two configurations, the pairing energy, can be denoted by Π; it can be estimated by the same formulas (4.38) in which, as above, the Racah parameters A, B, and C should be calculated by MO (not AO) functions; that is, they should be taken as modified by covalency. Provided that the pairing energy Π is known, the question of whether the low-spin or high-spin configuration is realized can be solved directly: If $\Pi < \Delta$, the low-spin case is realized, and if $\Pi > \Delta$, the high-spin configuration is preferable. Since Δ characterizes the strength of the ligand field, these two cases can also be called *strong field and weak field limits*, respectively. As in the crystal field approximation, the two possible configurations become important in d^n complexes for $n = 4$, 5, 6, 7 in octahedral environments, and $n = 3$, 4, 5, 6 for tetrahedral systems, resulting in different (high-spin or low-spin) ground states. Tables 6.3 and 6.4 illustrate these results for octahedral and tetrahedral complexes, respectively.

Covalency Electrons and Ionization Potentials

As shown in Section 5.1, in the MO method the bonding occurs as a result of electronic redistribution in the bonding MOs due to the electronic cloud being concentrated mainly in the space between the nuclei welding them together (another source of bonding is the reduction of the electron kinetic energy). The electronic energy on these orbitals is lower than on the corresponding AOs. On the contrary, on the antibonding MOs the electron pushes away the nuclei, its energy being higher than in the corresponding atomic states. Obviously, the chemical bonding takes place when the occupancy of bonding orbitals

Table 6.3. Electronic configurations and ground states of coordination compounds of d^n metals with octahedral O_h symmetry[a]

CA d^n	Example	High Spin (Weak Field)		Low Spin (Strong Field)	
		Electronic Configuration	Ground State	Electronic Configuration	Ground State
d^1	Ti^{3+}	$(t_{2g}^*\uparrow)^1$	$^2T_{2g}$	$(t_{2g}^*\uparrow)^1$	$^2T_{2g}$
d^2	V^{3+}	$(t_{2g}^*\uparrow)^2$	$^3T_{1g}$	$(t_{2g}^*\uparrow)^2$	$^3T_{1g}$
d^3	Cr^{3+}	$(t_{2g}^*\uparrow)^3$	$^4A_{2g}$	$(t_{2g}^*\uparrow)^3$	$^4A_{2g}$
d^4	Mn^{3+}	$(t_{2g}^*\uparrow)^3(e_g^*\uparrow)^1$	^5E_g	$(t_{2g}^*\uparrow)^3(t_{2g}^*\downarrow)^1$	$^3T_{1g}$
d^5	Mn^{2+} Fe^{3+}	$(t_{2g}^*\uparrow)^3(e_g^*\uparrow)^2$	$^6A_{1g}$	$(t_{2g}^*\uparrow)^3(t_{2g}^*\downarrow)^2$	$^2T_{2g}$
d^6	Fe^{2+} Co^{3+}	$(t_{2g}^*\uparrow)^3(e_g^*\uparrow)^2(t_{2g}^*\downarrow)^1$	$^5T_{2g}$	$(t_{2g}^*\uparrow)^3(t_{2g}^*\downarrow)^3$	$^1A_{1g}$
d^7	Co^{2+}	$(t_{2g}^*\uparrow)^3(e_g^*\uparrow)^2(t_{2g}^*\downarrow)^2$	$^4T_{1g}$	$(t_{2g}^*\uparrow)^3(t_{2g}^*\downarrow)^3(e_g^*\uparrow)^1$	^2E_g
d^8	Ni^{2+}	$(t_{2g}^*\uparrow)^3(e_g^*\uparrow)^2(t_{2g}^*\downarrow)^3$	$^3A_{2g}$	$(t_{2g}^*\uparrow)^3(t_{2g}^*\downarrow)^3(e_g^*\uparrow)^2$	$^3A_{2g}$
d^9	Cu^{2+}	$(t_{2g}^*\uparrow)^3(e_g^*\uparrow)^2(t_{2g}^*\downarrow)^3(e_g^*\downarrow)^1$	^2E_g	$(t_{2g}^*\uparrow)^3(t_{2g}^*\downarrow)^3(e_g^*\uparrow)^2(e_g^*\downarrow)^1$	^2E_g
d^{10}	Zn^{2+}	$(t_{2g}^*\uparrow)^3(e_g^*\uparrow)^2(t_{2g}^*\downarrow)^3(e_g^*\downarrow)^2$	$^1A_{1g}$	$(t_{2g}^*\uparrow)^3(t_{2g}^*\downarrow)^3(e_g^*\uparrow)^2(e_g^*\downarrow)^2$	$^1A_{1g}$

[a]The inner closed shells are omitted.

Table 6.4. Electronic configurations and ground states of coordination compounds of d^n metals with tetrahedral T_d symmetry[a]

CA d^n	Example	High Spin (Weak Field)		Low Spin (Strong Field)	
		Electronic Configuration	Ground State	Electronic Configuration	Ground State
d^1	Ti^{3+}	$(e^*\uparrow)^1$	^2E	$(e^*\uparrow)^1$	^2E
d^2	V^{3+}	$(e^*\uparrow)^2$	$^3A_2$	$(e^*\uparrow)^2$	$^3A_2$
d^3	Cr^{3+}	$(e^*\uparrow)^2(t_2^*\uparrow)^1$	$^4T_1$	$(e^*\uparrow)^2(e^*\downarrow)^1$	^2E
d^4	Mn^{2+}	$(e^*\uparrow)^2(t_2^*\uparrow)^2$	$^5T_2$	$(e^*\uparrow)^2(e^*\downarrow)^2$	$^1A_1$
d^5	Mn^{2+} Fe^{3+}	$(e^*\uparrow)^2(t_2^*\uparrow)^3$	$^6A_1$	$(e^*\uparrow)^2(e^*\downarrow)^2(t_2^*\uparrow)^1$	$^2T_2$
d^6	Fe^{2+} Co^{3+}	$(e^*\uparrow)^2(t_2^*\uparrow)^3(e^*\downarrow)^1$	^5E	$(e^*\uparrow)^2(e^*\downarrow)^2(t_2^*\uparrow)^2$	$^3T_1$
d^7	Co^{2+}	$(e^*\uparrow)^2(t_2^*\uparrow)^3(e^*\downarrow)^2$	$^4A_2$	$(e^*\uparrow)^2(e^*\downarrow)^2(t_2^*\uparrow)^3$	$^4A_2$
d^8	Ni^{2+}	$(e^*\uparrow)^2(t_2^*\uparrow)^3(e^*\downarrow)^2(t_2^*\downarrow)^1$	$^3T_1$	$(e^*\uparrow)^2(e^*\downarrow)^2(t_2^*\uparrow)^3(t_2^*\downarrow)^1$	$^3T_1$
d^9	Cu^{2+}	$(e^*\uparrow)^2(t_2^*\uparrow)^3(e^*\downarrow)^2(t_2^*\downarrow)^2$	$^2T_2$	$(e^*\uparrow)^2(e^*\downarrow)^2(t_2^*\uparrow)^3(t_2^*\downarrow)^2$	$^2T_2$
d^{10}	Zn^{2+}	$(e^*\uparrow)^2(t_2^*\uparrow)^3(e^*\downarrow)^2(t_2^*\downarrow)^3$	$^1A_1$	$(e^*\uparrow)^2(e^*\downarrow)^2(t_2^*\uparrow)^3(t_2^*\downarrow)^3$	$^1A_1$

[a]The inner closed shells are omitted.

predominates that of the antibonding MOs. Therefore, it is worthwhile to consider the question of what electrons (more precisely, what occupied one-electron states) are responsible for the covalent bonding in coordination compounds. It can be shown [6.10] (Section 5.1) that in many cases there is an approximate correspondence between the bonding and antibonding orbitals: For each bonding orbital there is an antibonding MO, so their contributions to the binding are mutually compensated (the total contribution being negligible). Provided that this rule is obeyed, the chemical bond is formed by only those bonding MOs that remain uncompensated by the antibonding orbitals.

In Fig. 6.3 the MO energy scheme for a Cu(II) complex is shown with an indication of the MO occupancies (for simplicity, only σ bonds are shown). It is seen that in the ground state the uncompensated bonding MO are a_{1g} (two electrons), t_{1u} (six electrons), and e_g (one electron). Thus, not the valence one-electron states, but *the inner bonding MOs which are not compensated by the outer antibonding MOs form the chemical bond*. In particular, it is ungrounded to try to describe the covalency in coordination compounds by ESR data or by spin density distribution (Sections 8.4 and 8.6). These data characterize the covalency of the one-electron state of the unpaired electron, whereas the bonding is formed by many inner electrons.

Another important MO problem is to determine the ionization potentials equal to the energy of ionization from a given state of the system. These quantities are especially important for interpretation of the experimental data on photoelectron spectra (Section 8.3). As mentioned in Section 2.2, in the

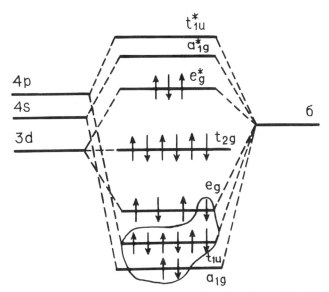

Figure 6.3. Mutual compensation of bonding and antibonding orbitals. The noncompensated bonding orbitals are shown enveloped by a line.

one-electron case the ionization energy, following Koopmans theorem [6.11], equals the energy of the MO state calculated by the self-consistent method with opposite sign. Indeed, in these calculations the MO energy includes the interaction of a given electron with the nuclei and all the other electrons, that is, the energy that should be applied to remove the electron from this MO.

However, this consideration (and the Koopmans theorem) is absolutely true in the case of one electron only. When there are two or many electrons in the system, the ionization of one of them changes the one-electron states of the others. Indeed, reduction of the number of electrons by one reduces their interelectron repulsion, and the one-electron MOs "relax" to new self-consistent states. As a result of these electron relaxations, their energy decreases, and hence the absolute value of the ionization potential decreases by the same amount (the decrease in the interelectron repulsion promotes the ionization).

If the relaxation energy is small, the Koopmans theorem predicts the correct consequence of ionization potentials corresponding to the MO energy-level positions. This is true in many cases, especially for very stable closed-shell molecules, for which the electronic energy-level spacing is sufficiently large. However, as shown by calculations, the Koopmans theorem is far from being valid for other systems, including many coordination compounds (see the discussion in the next sections of this chapter).

The deviations from the prediction of the Koopmans theorem depend on the state to be ionized and change from state to state. For instance, one can expect that the ionization of a (predominantly) d electron from a strong covalent metallo-organic complex results in much stronger reorganization (relaxation) of the other electrons than in the case of the ionization of a ligand nonbonding electron. The exact value of ionization energy should be calculated as the difference between the total energies of the initial (nonionized) and final (ionized) systems evaluated with the interelectron interaction included.

6.3. LIGAND BONDING

General Considerations: Multiorbital Bonds

The origin of ligand bonding is one of the most important problems in coordination chemistry. It provides a basic understanding of a series of properties and processes that involve ligand coordination. To reveal the origin of ligand bonding means to elucidate the electronic structure of the bond, determined by the electronic features of the ligand and the remaining complex, and to establish direct correlations between the bonding properties and electronic structure parameters.

As emphasized in Section 6.1, due to the three-dimensional center-delocalized nature of the coordination bond, all the metal–ligand bonds are, in general, strongly interdependent. Therefore, when considering ligand bonding, one has to investigate not just the metal–ligand bond, but rather, the *complex–ligand bond.* From the ligand side the bond is mostly localized. In other words, ligand bonding means chemical interaction (electron collectivization; Section 1.2) between a ligand and a coordination center that may be strongly dependent on the bonding of the latter to other ligands. This circumstance, together with the significant difference in electronic structure of the metal (*d* electrons) and the ligand (*sp* electrons) — *the d-electron heterogeneity* — makes the metal–ligand bond essentially different from those in organic (and main-group element) compounds.

In the MO LCAO scheme the metal–ligand bond is determined by overlap of the metal and ligand valence orbitals, which, in turn, depends on the valence state of the ligand and the metal and *the mode (geometry) of ligand coordination.* The valence orbitals of transition metals and rare earth elements are given in Section 2.1. Assume that with respect to the ligand under consideration, the complex is characterized by σ and π orbitals formed by a transition metal nd, $(n + 1)s$, and $(n + 1)p$ valence orbitals (which in the presence of other ligands transform into appropriate MOs). Figure 6.4 illustrates one of the σ orbitals formed by the *s*, *p*, and *d* orbitals of the metal, and one π orbital from its d_{xz} (or d_{yz}) orbital (as above, the positive sign of the lobes is shown by shading).

Depending on the oxidation state and the nature of other ligands bound to the coordination center, the valence orbitals of the metal above are differently populated by electrons, and hence they have different bonding abilities. A characteristic case is shown in Fig. 6.4 when the lowest π level of the metal (e.g., d_{xz}) is occupied while its σ AO is unoccupied; the metal (complex) is thus a σ acceptor and a π donor. Of course, this picture is rather simplified: in real cases the complex may have several σ and π levels (see below) and may be both

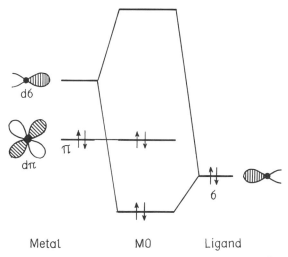

Figure 6.4. σ and π AOs of the metal in a mono-orbital MO bonding with the ligand that has only one active σ orbital.

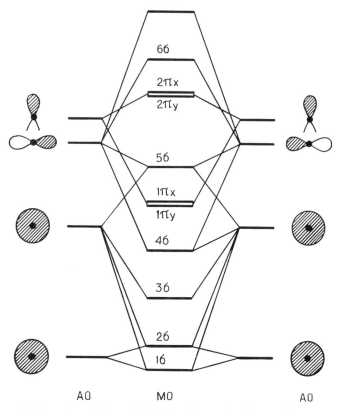

Figure 6.5. Atomic orbitals and MOs for diatomics of second row atoms.

σ and π donor and acceptor on different orbitals, but usually one or two of them are most active in the ligand bonding.

As simple examples of ligands in metal–ligand bonding, free atoms or diatomics formed by the atoms of the second row of the periodic table can be suggested. Figure 6.5 illustrates a typical MO scheme for such a diatomic molecule. Owing to their deep bedding and hence bad overlap, the 1s orbitals form very weak bonding 1σ and antibonding 2σ MOs (in the case of identical atoms they are σ_g and σ_u, respectively), which, being both occupied, do not contribute to the bonding between the atoms.

The 2s orbitals form much stronger bonding and antibonding MOs. The close-in-energy $2p_\sigma$ orbitals of the two atoms have the same σ symmetry and strong overlap with each other and with the 2s orbitals. Therefore, all four orbitals become mixed (hybridized), forming the $3\sigma, 4\sigma, \ldots$ MOs (the 1s orbitals are also mixed in, but this effect is usually negligible).

In addition to the σ orbitals, $2p_\pi$ AOs also overlap (although less than the σ AOs) forming bonding 1π and antibonding 2π MOs. Depending on the nature of the atoms in the diatomic molecule, the mutual position of the close-in-energy MOs 4σ, 1π, and 5σ can vary, and this is important when considering their bonding to the coordination center.

The main features that make the local metal–ligand bond distinct from the corresponding organic bonds are due to the relatively strong asymmetry of the overlaping orbitals produced by the d electron heterogeneity; it causes large orbital charge transfers which are often mutually compensated in the diorbital and multiorbital bonds (see below). Together with the fact that the ligand is bound by the entire complex rather than just by the CA, these distinctions make the usual ideas of multiple bonds insufficient for a full characterization of the bond. For a better description of the ligand bonding, MO-based definition of bond multiplicity is suggested, in which *mono-orbital, diorbital, and multiorbital metal–ligand bonds are distinguished.*

In the MO definition, the multiplicity of the orbital bonding (mono-, di-, and multiorbital) equals the number of complex–ligand bonding MOs uncompensated by the antibonding orbitals (Section 6.2). These types of bonds, as shown below, differ significantly from and are complementary to the well-known single, double, and triple bonds in organic (or main-group) compounds by both definition and effect. The simple multiple bonds are particular cases of the defined multiorbital bonds.

By definition, the number of orbitals that participate in the bonding does not coincide with the bond multiplicity. In the usual definition the multiplicity of the bond equals the number of pairs of electrons that participate in the bonding, whereas the number of bonding orbitals may be smaller, since some of them may be degenerate. In particular, in the triple bond with one σ and two π bonds the latter may be degenerate, and hence the bond is diorbital with one σ MO and one doubly degenerate π MO. Besides, the bonding (antibonding) orbital can be occupied by one electron. The advantage of the MO terminology is seen from the treatment of ligand bonding given later in this section.

An important feature of diorbital and multiorbital metal–ligand bonding is *the mutual compensation of orbital charge transfers.* Consider a diorbital M—L bond in which one of the bonding MOs can be considered a σ MO while the other is of π type. The orbital charge transfers Δq_i defined in Section 5.2 [Eq. (5.20″)], may be both negative and positive. Usually, the orbital charge transfers along the σ MO go from the ligand to the metal: $\Delta q_\sigma < 0$ (negative transfer means reduction of the ligand charge), while for the π MO, $\Delta q_\pi > 0$ (back donation). The total charge transfer is thus

$$\Delta q = \Delta q_\sigma + \Delta q_\pi \qquad (6.2)$$

It is known that Δq cannot be very large because of thermodynamic restrictions (cf. the Pauling electroneutrality principle [6.12]). In mono-orbital binding Δq coincides with the orbital charge transfer and hence they are both small. Distinct from (and contrary to) the mono-orbital bonds, for diorbital bonds *the total charge transfer Δq may be small, while the orbital charge transfers are* *large*, because they may have opposite signs, for which $\Delta q = |\Delta q_\pi| - |\Delta q_\sigma|$. This is a specific feature of the diorbital (multiorbital) bond, as compared with mono-orbital (single) bonds, and it has far-reaching consequences for both bonding energy and changes in the properties of the coordinated ligands. In particular, while the orbital charge transfers compensate each other, their effect on ligand activation may be additive with respect to their absolute values (Section (11.2).

This enhanced coordination by mutual compensated charge transfers is in general not directly related to the usual double bonds in organic compounds. Indeed, in the latter the orbital charge transfers are relatively small and not mutually compensating. Therefore, the usual terminology of single and double bonds which is most useful for organic and main-group systems may not be sufficiently informative for coordination compounds. In the case of diorbital bonding the scheme of mutually compensating charge transfers is qualitatively the same as in back donation, first suggested by Dewar [6.13] and Chatt and Duncanson [6.14].

The charge compensation in the σ-donor and π-dative interactions in the ligand bonding also generates another important effect, the interdependence and mutual enhancing of the two types of charge transfers, Δq_σ and Δq_π: *the π-dative charge transfer to the ligand enhances its donor properties* because of the increase in the interelectron repulsion, and vice versa [6.4, 6.17].

On the other hand, *changes in the occupancies of the ligand MOs may result in its partial excitation*. Indeed, as a result of coordination, the unoccupied excited π MO of the free diatomics becomes populated, while its ground-state σ MO depopulates. *The two effects, orbital charge transfers in opposite directions and partial excitation, strongly influence the properties of the ligands coordinated by diorbital or multiorbital bonds* (Section 11.2).

Note that the total charge transfer to the ligand, $\Delta q = |\Delta q_\pi| - |\Delta q_\sigma|$, reflects neither the orbital charge transfers nor the excitation. Therefore, the attempts to interpret experimental data on coordinated ligands by the Δq value, often reported in publications (e.g., to consider coordinated oxygen in the state of superoxide judging on it stretching vibration frequency; Section 11.2), are ungrounded.

In the qualitative treatment above, it is assumed that the orbital charge transfers Δq_i can be calculated by Eq. (5.20''). In so doing one should take into account the discussion in Section 5.2 concerning the restricted physical meaning of charge distribution given in terms of atomic orbitals and its strong dependence on the basis set employed in the calculations. In many cases the results of sophisticated numerical calculations cannot be visualized in terms of free metal and ligand orbitals, *in terms of the origin of ligand bonding*, as it is done in the qualitative treatment. This is a general feature of more or less exact numerical data on electronic structure discussed in Section 5.6: *the more sophisticated the methods used in accurate numerical calculations of electronic structure, the less visual the possible treatment in terms of atomic orbitals*. Exact wavefunctions cannot be presented by a limited number of atomic orbitals;

usually, they are obtained by extending the basis set and including configuration interactions (Section 5.3).

In some cases the valence AOs (ground and excited) emerge in the resulting MOs with considerable weight, which can be interpreted approximately as a measure of participation of the corresponding orbitals in the bonding. If extended basis sets and/or a superposition of many configurations are used in the calculations, the direct correlation of the results with certain ligand AOs becomes very difficult. However, the MOs can be discriminated by symmetry properties (say, σ, π, δ, etc.), and this allows one to calculate orbital charge transfers for such MOs.

Mono-orbital Bonds: Coordination of NH_3 and H_2O

The case of mono-orbital bonds is simplest. As a rule the mono-orbital bond is realized when the ligand has only one active orbital, which is able to bond to the complex. In this case usually only one σ metal–ligand bond is formed. A simple example of this is the bonding of ammonia, in which the lone pair of electrons of the nitrogen atom occupies its hybridized sp^3 orbital, which is usually shown as the ligand σ orbital. Figure 6.4 (page 211) illustrates the MO scheme for this simple case of ligand bonding.

The charge distribution in this bond is not symmetrical with respect to the metal and ligand; under the assumption that the metal is a σ-receptor and the ligand is a σ-donor, a typical donor–acceptor bond is obtained. This type of ligand bonding generally takes place when either the ligand has no active (sufficiently low in energy) π orbitals that can form bonds with the metal, or the metal π orbitals are weakly active. The former possibility is more probable. In addition to NH_3, which has no active π orbitals, many similar ligands can be suggested (e.g., PH_3, $CH_3{}^-$, H^-, etc.). However, under certain conditions, phosphines (e.g., PF_3), as well as similar compounds of other pre-transition elements, may exhibit π-accepting abilities [6.15].

Mono-orbital bonds can also be realized when both the ligand and metal active π orbitals are occupied. Halogen ions and water can be indicated as examples of ligands that have such π donor orbitals. In these cases, both π-type MOs formed by the ligand and metal orbitals, bonding and antibonding, are occupied by electrons (Fig. 6.6), and hence their total contribution to the bonding is very small, due to mutual compensation (Sections 5.1 and 6.2). In principle, there is a possibility that within this combination of two σ and two π MOs with six electrons the antibonding σ and π MOs are inverted, and the σ MO is occupied instead of the π MO. As a result, the bond remains mono-orbital but of π type. Practically, other MO energy levels may interfere in this region becoming occupied, but then the diorbital or multiorbital bonds must be considered.

In mono-orbital bonds the influence of the complex on the ligand is relatively simple: *polarization*. Indeed, as a result of the bonding, a donor–acceptor shift of the electronic cloud from the ligand to the metal takes place,

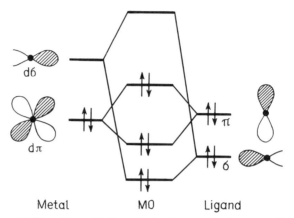

Metal   MO   Ligand

Figure 6.6. Mono-orbital bonding with ligands that have active but occupied π orbitals.

with all the consequences for the properties of the coordinated ligand and the complex as a whole. This effect can also be described in the simpler model of crystal field theory (CFT), provided that the effective charge and polarization of the ligand are taken into account (Section 4.2). Therefore, the analysis of some properties of coordination compounds with simple ligands that form only mono-orbital (single) bonds can often be carried out within the approximation of the CFT.

Mono-orbital bonding, as a rule, does not change the properties of the ligands radically but modifies them, sometimes significantly. For instance, in the case of coordination of the ammonia molecule, mentioned above, the donor–acceptor shift of the electronic charge to the metal weakens the N—H bonds, lowers their stretching vibration frequency, and increases the acidity of the coordinated molecule.

As examples of *ab initio* numerical SCF CI calculations of the electronic structure of mono-orbital bonds, the systems $Ni(H_2O)_n$ and $Ni(PH_3)_n$, $n = 1$, 2 [6.16], can be suggested. In the case of Ni—OH_2, the Ni—O bond, as expected, is mono-orbital of the second kind when the ligand has active π orbitals but they are occupied by electrons (Fig. 6.6). The results of CAS SCF (complete active space SCF; Section 5.3) calculations yield $\Delta q_\sigma = -0.13$, $\Delta q_\pi = -0.03$ (the charges on the atoms are $q_{Ni} = -0.17$, $q_O = -0.69$, $q_H = 0.43$). For H_2O—Ni—OH_2 (linear configuration) the orbital charge transfers are practically the same as in the Ni—OH_2 case: $\Delta q_\sigma = -0.14$, $\Delta q_\pi = -0.04$ ($q_{Ni} = -0.36$, $q_O = -0.70$, $q_H = +0.43$). This confirms that the Ni—O bond is mono-orbital and localized (the small Δq_π value characterizes the slightly asymmetrical charge distribution in the mutual compensating bonding and antibonding π MOs).

To verify the validity of the CFT in this case, the authors [6.16] also calculated the interaction of the $Ni(^1D)$ atom with the dipoles OH_2; the dipole moment was taken equal to 2.4 D formed by the charges -0.860 at a distance of 3.79 a.u. from Ni and $+0.860$ at 4.873 a.u. The resulting bonding energy is

16.0 kcal/mol, which is in good agreement with other more detailed calculations.

Another example of expected mono-orbital bonding is the Ni—P bond in Ni—PH_3. The calculations [6.16] yield $\Delta q_\sigma = -0.32$, $\Delta q_\pi = 0.07$ ($q_{Ni} = -0.24$, $q_P = 0.14$, $q_H = 0.03$). Although the PH_3 molecule at first sight has no active π orbitals, there is a small π back donation from the Ni atom to PH_3. This indicates that unlike the NH_3 case, where there is no back donation, in PH_3 the d orbitals may become weakly active (see also [6.15] and Section 6.1). The dissociation energy (in the $^1A_1$ state) is $D_e = 13.7$ kcal/mol.

For the linear H_3P—Ni—PH_3 system the charge transfers are almost the same, as in Ni—PH_3: $\Delta q_\sigma = -0.26$, $\Delta q_\pi = 0.10$ ($q_{Ni} = 0.32$, $q_P = 0.10$, $q_H = 0.02$), which confirms that the bonding is localized and mono-orbital.

Diorbital Bonds: Coordination of the N_2 Molecule

More widespread and rich in content are *diorbital bonds*. As explained above, this term is used to denote the presence of two types of uncompensated bonding MOs involved in the metal–ligand bonding. The diorbital bond is realized when the ligand possesses, in addition to the σ orbital, a free π orbital that can form an additional π bond with the metal. In the more usual terminology diorbital bonding may result in single, double, triple, and higher-multiplicity bonds, as well as semibonds, depending on the number of electrons on the bonding MOs. The single bond is realized when two bonding electrons occupy one of the two bonding MOs; four electrons on the latter make a double bond. Higher bond multiplicities may occur if one or both bonding MOs are degenerate, for instance, in the case of metal–metal bonds (see below).

As simple examples of ligands in diorbital bonds, the diatomics seem to be most informative. Consider first the most stable diatomic molecule of the series under consideration, *the N_2 molecule*. Figure 6.7 shows its MO energy levels and their wavefunction symmetries with indication of the electron occupancy [6.18]. It is seen that the HOMO is 5σ, while the LUMO is 2π. By comparison with the symmetries of the valence orbital of the metal (Fig. 6.4), one can easily conclude that the possibility of a nonzero orbital overlap and bonding depends on the *geometry of coordination*.

For *linear end-on coordination* when the line of the N_2 molecule coincides with the line of the bond to the metal (Fig. 9.14, page 459), the 5σ MO of N_2 can form further σ MOs with the σ orbital of the metal, while the 2π MO forms π MOs with the metal π orbital. Taking into account the extension of these orbitals in space, one can see that at large distances a better overlap and bonding is achieved with the σ MO, whereas for a good π overlap shorter distances are required.

For *side-on coordination* when the axis of the molecule is perpendicular to the line of the bond to the metal, the overlap with the 5σ MO of N_2 (and hence the possibility of formation of a corresponding σ bond) deteriorates. In this

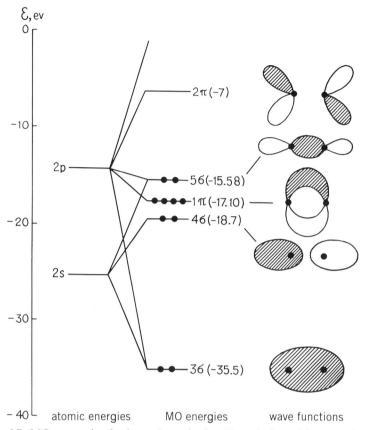

Figure 6.7. MO energy-level scheme (energies in eV are indicated in parentheses) and wavefunction symmetries for the dinitrogen molecule.

geometry the 1π MO has the required σ symmetry with respect to the bond, and despite being deeper in energy, it may become significant. Note that if the metal is a π acceptor, the only possibility of accepting π electrons from the ligand N_2 is provided by the 1π MO in the linear end-on coordination, and by the antibonding 4σ MO in the case of side-on coordination. This example illustrates the general statement of a *strong dependence of ligand bonding on the details of their electronic structure and geometries of coordination.*

If the metal has occupied π and free σ orbitals, dinitrogen coordination produces a diorbital bond with one σ and one π bonding MO in both end-on and side-on geometries, provided that the small contributions of the inner orbitals are neglected (Fig. 6.8).

An example illustrating these general considerations is given by nonempirical calculations of the bonding in FeN_2 [6.19] carried out in the CAS SCF version of MO LCAO (Section 5.3). The basis set is taken as $(14s11p6d)/[8s6p4d]$ for the iron atom and $(9s6p)/[4s4p]$ for N. For FeN_2 with side-on

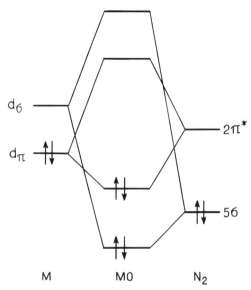

Figure 6.8. Diorbital MO bonding of the N_2 molecule to a σ-acceptor (π-donor) metal M.

coordination, the main orbital charge transfers to the N_2 molecule are $\Delta q_\sigma = -0.13$, $\Delta q_{1\pi} = -0.19$, $\Delta q_{2\pi} = 0.68$. Compared with end-on coordination, the transfer from the 5σ MO is lower, while the transfer from the bonding 1π MO (which in the side-on coordination produces the same σ-type donation; Fig. 6.7) is significant, and the back donation to the antibonding 2π MO is predominant.

Because of opposite signs the Δq_i values compensate each other in the total charge transfer, but they contribute additively toward the weakening of the N—N bond discussed in Section 11.2 (except in special cases where it is important, the multiplicities of the bonds are not indicated). The reduction of the interatomic distance $\Delta R(\text{N—N})$ and the frequency of stretching vibration $\Delta\omega$ by coordination are calculated to be $\Delta R = -0.162$ a.u. and $\Delta\omega = -671$ cm^{-1} (in a bigger CAS SCF version they are $\Delta R = -0.165$ a.u. and $\Delta\omega = -467$ cm^{-1}). This bond weakening is very important in chemical activation of molecules by coordination. In particular, in the case of N_2 this bond weakening is the major problem of nitrogen fixation [6.20]. For other examples of N_2 coordination, see Table 11.2.

Coordination of Carbon Monoxide

The CO molecule is isoelectronic to N_2 and can be characterized qualitatively by the same MO scheme (Fig. 6.9) [6.18]. But unlike N_2, carbon monoxide has other MO energy-level positions and bonding features and nonsymmetrical electronic charge distribution between the C and O atoms: The electronic

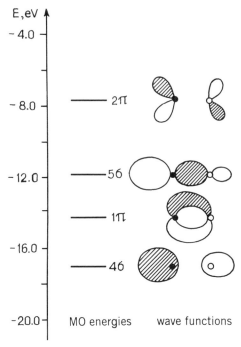

Figure 6.9. MO energy-level scheme and wavefunction symmetries for the CO molecule (cf. with that of N_2 in Fig. 6.7).

charge is attracted more to the oxygen atom and it is more diffusive in the carbon side.

The coordination of CO to the metal results in a diorbital bonding with one σ and one π MOs, quite similar to the N_2 case, but the effect of coordination is different. In particular, the 5σ orbital of CO is slightly antibonding (in N_2 it is bonding), and therefore the charge transfer from this orbital to the metal strengthens the C—O bond. On the other hand, the 2π MO in CO is stronger antibonding than in N_2 (Sections 7.2 and 11.2), and hence the charge transfer to this MO gives a stronger destabilization effect than in N_2. Again, the orbital charge transfers themselves are also different in these two cases.

Many works are devoted to an analysis of CO coordination based on electronic structure calculations. Consider, for instance, the system Pt—C—O, with linear end-on coordination [6.21]. The calculations were carried out in the approximation of the effective core potential with a large basis set, including electronic correlation energy by means of configuration interactions (Section 5.3). In the ground singlet state $^1\Sigma^+$, which corresponds to the dissociation of the system into the Pt atom in the excited singlet state $5d^{10}(^1S)$ and the CO molecule in the ground state, the bonding Pt—CO is rather strong (~ 70.4 kcal/mol), while the equilibrium distance Pt—C is $R_e = 1.707$ Å. With respect to the cleavage into the Pt atom in the ground-state configuration

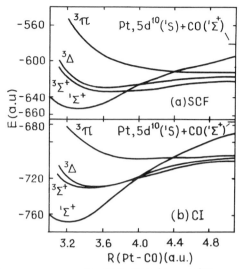

Figure 6.10. Binding energy curves for $^1\Sigma^+$, $^3\Sigma^+$, $^3\Delta$, and $^3\Pi$ states of the linear system Pt—C—O calculated in the SCF MO LCAO approximation without (a) and with (b) CI (energies are given relative to -48 a.u.). (After [6.21].)

$5d^9 6s^1$ and CO in the ground-state, the bonding energy is smaller, 43.8 kcal/mol; in the triplet states $^3\Sigma^+$ and $^3\Delta$ it is 19.4 kcal/mol and 18.2 kcal/mol with equilibrium distances $R_e = 1.820\,\text{Å}$ and $R_e = 1.895\,\text{Å}$, respectively (Fig. 6.10).

Note that in the $^3\Pi$ state obtained from the electronic configuration of the Pt atom with a hole (vacancy) in the d_π orbitals (d_{xz} or d_{yz}), the π-donor properties of Pt are weakened, and there is no Pt—CO bond (Fig. 6.10). This result confirms the considerations, given above, about the role of the diorbital (multiorbital) nature of the bond in the mutual enhancing σ-donor and π-acceptor interactions: The weakening of the π dative interaction Pt—CO reduces the σ donor properties of CO, making bonding impossible.

The orbital charge transfers calculated for the ground state are [6.21] $\Delta q_{5\sigma} = -0.25$ and $\Delta q_{2\pi} = 0.38$ (total for two π MO), and the total transfer $\Delta q = 0.13$, as expected, is not very large. This additional electronic charge on the CO molecule is concentrated mainly on the carbon atom. Calculated in the same approximation, the charges on the atoms in the free CO molecule are $q(C) = 0.12$ and $q(O) = -0.12$, while after coordination they become $q'(C) = 0.02$ and $q'(O) = -0.14$.

The Pt—C bond can be characterized by the overlap population $P(Pt—C)$. It follows from the results of the calculations that, as expected, the major contribution to the Pt—C bonding is due to the σ MO, $P_\sigma(Pt—C) = 0.66$. This includes the contribution of the $5\sigma(CO)$ orbital, $P_{5\sigma} = 0.25$, while that of $5d$ orbitals from the Pt atom is $P_{d\sigma} = 0.41$. For the π MOs, the contribution to the overlap population comes entirely from the Pt atom (the 2π MO of the free

CO molecule is unoccupied), and equals $P_\pi(\text{Pt—C}) = 0.42$, in 0.21 on each of the two π MOs.

Note again (see the discussion in Section 5.2) that the absolute values of charges in different regions of the molecular system (e.g., the atomic charges) in the presence of strong covalent bonds are, in a sense, conventional and depend on the mode of separation of the system into parts and the choice of the wavefunction basis set in the Mulliken population analysis. Some results of the calculation of the systems M—CO, where M = Cr, Fe, Co, Ni, are given in [6.22]. These calculations are quite similar to that of Pt—CO, carried out in the same nonempirical approximation with an effective core potential but with a smaller basis set and for the metal electronic configuration $M(3d^{n-1}4s)$ only. The main bonding features in M—CO are the same as in the case of Pt—CO, namely, the integral charge transfer from the metal to the CO molecule is relatively small, $\Delta q \approx 0.1$ to 0.3, while the orbital charge transfers Δq_σ from CO to the metal and Δq_π from the metal to CO are much larger.

This is also seen from the data in Table 6.5, where the changes of one-electron energy levels by coordination are shown. In particular, the energy of the 5σ level is significantly lowered by coordination, and this confirms once more the antibonding nature of the 5σ MO. Because of the different basis sets used in the calculations, the absolute values of the atomic charges are different from those obtained in [6.21]. This is seen explicitly from a comparison of the data for the free CO molecule: In [6.22] the charges on C and O are $q = \pm 0.33$, whereas from [6.21] it follows that they are $q = \pm 0.12$.

Another example, useful for comparison, is the bonding in Sc—CO calculated [6.23] by the multireference single and double CI method using the pseudopotential approximation. Reasonable results were obtained under the assumption that the starting configuration of the Sc atom is $3d^3$ (term ^4F) and the ground state of linear Sc—CO is $^4\Sigma^-$. The SCF one-electron MO energies are shown in Fig. 6.11. The orbital charge transfers are $\Delta q_\sigma = -0.22$ and $q_{2\pi} = 0.52$, while the total charge transfer from Sc to CO is $\Delta q = 0.30$ ($q_C = 0.07$, $q_O = 0.23$). The orbital charge distribution obtained for the 2σ, 3σ,

Table 6.5. Shifts of the MO energy levels of the CO molecule by coordination to transition metals M (in eV)

| MO | Free CO | M | | | |
		Cr	Fe	Co	Ni
5σ	−15.12	−16.47	−16.67	−17.11	−17.64
1π	−18.15	−17.43	−17.64	−18.07	−18.54
4σ	−21.74	−21.12	−21.25	−21.73	−22.23
3σ	−42.41	−41.26	−41.47	−42.01	−42.57

Source: [6.7].

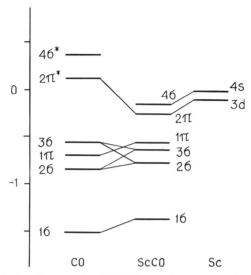

Figure 6.11. Correlation diagram for the SCF one-electron energies in the $^4\Sigma^-$ state of ScCO (atomic units).

4σ, and 2π MOs seem to be very expressive (Fig. 6.12). The dissociation energy for the reaction $ScCO(^4\Sigma^-) \rightarrow Sc(3d^3, {}^4F) + CO(X^1\Sigma^+)$ is 1.14 eV (26.30 kcal/mol). For calculations of Sc—CO see also [5.109].

These data can be compared with similar *ab initio* SCF CI calculations for Ni—CO [6.23]. For the $^1\Sigma^+$ state $\Delta q_\sigma = -0.15$, $\Delta q_\pi = 0.45$, $\Delta q = 0.30$ ($q_{Ni} = 0.30$, $q_C = 0.14$, $q_O = -0.43$), and the dissociation energy into singlet

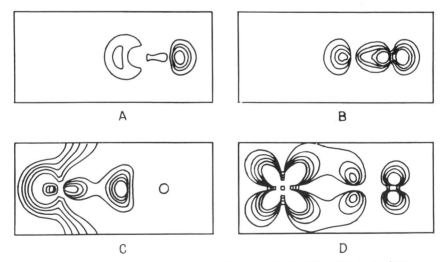

Figure 6.12. Electron density contour maps for the orbitals of Sc—CO in the $^4\Sigma^-$ state: (*A*) 2σ; (*B*) 3σ; (*C*) (4σ; (*D*) 2π. $R(C—O) = 2.2$ Bohr, $R(Sc—C) = 4$ Bohr. (From [6.23].)

nickel, $Ni(^1D)$, is 29.9 kcal/mol. In the case of two ligands $Ni(CO)_2$ in the linear configuration the charge transfers change: $\Delta q_\sigma = -0.24$, $\Delta q_\pi = 0.29$ ($q_{Ni} = 0.11$, $q_C = 0.33$, $q_O = -0.39$), while the energy of dissociation into $Ni(^1D) + 2CO$ is $D_e = 57.1$ kcal/mol. These changes could be expected since the Ni—CO bonds are delocalized.

$\sigma + \pi$ Bonding

So far we have tried to interpret the results of numerical calculations of CO bonding in terms of the diorbital (σ and π) bond. In the cases of M—CO, where M is a metal atom, this interpretation does not encounter difficulties. However, in more complicated cases of ligand bonding, this description may become evidently inadequate. In the general case of complex–ligand bonds, in addition to the σ and π MOs, other MOs of the multielectron system may be involved, making the bonding multiorbital. Let us illustrate this statement by the example of formation of the "$\sigma + \pi$" MO in metal carbonyls in addition to the pure "σ" and "π" MOs [6.24].

Consider the octahedral complex $Cr(CO)_6$ (a general MO scheme and possible electronic configurations for such complexes are given in Sections 6.1 and 6.2). Let us analyze the ligand–complex bond $(CO)_5Cr$—CO. The CO molecule takes part in the end-on coordination by the same 5σ and 2π MOs, as in the simpler system M—CO, and the main bonding σ and π MOs can be found by considering their overlap with the appropriate orbitals of the complex, that is, with the orbitals of the metal modified by the influence of other ligands. As shown below, the presence of the latter is a matter of principle leading to a new type of bonding.

Figure 6.13 shows schematically the cross section of the system in the xz plane comprising the CA and four CO ligands that illustrates the symmetries of the wavefunctions 5σ of the two ligands on the z axis and 1π of the other two ligands on the x axis, as well as the p_z function of the CA. As one can see, the p_z function of the Cr atom has nonzero overlap with both the σ MOs of the two ligands on the z axis and π MOs of the ligands on the x axis, thus forming a common MO which is a σ MO for the former ligands and a π MO for the latter. (There are two more CO ligands on the y axis participating with their 1π MOs in this common MO.) Obviously, an equivalent MO is formed by the σ MOs of the two CO ligands on the x axis and the MOs of the four CO ligands on the z and y axes, overlapping with the p_x orbital of CA, and a third orbital of this kind formed quite similarly by the p_y orbital of the Cr atom. These three MOs belong to the threefold-degenerate T_{1u} representation of the O_h group.

The chemical bonding realized by these MOs mixes the σ bonds of some ligands with the π bonds of others, and vice versa, the bonding thus being $\sigma + \pi$. Figure 6.14 shows the outline of this t_{1u} wavefunction (more precisely of its Γ_8 component after the spin–orbit splitting) obtained by numerical calculations of the $Cr(CO)_6$ system in the approximation of the relativistic method of X_α scattered waves [6.24].

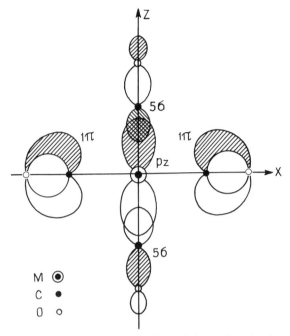

Figure 6.13. $\sigma + \pi$ bonding: the xz cross section of the t_{1u} function in transition metal carbonyls.

Since in the linear combination resulting in the $\sigma + \pi$ MO, three starting functions are used (one p orbital, one combination of the 5σ function of two CO ligands on the same axis, and one combination of the 1π functions of four remaining ligands, all the combinations being t_{1u} symmetrized), three types of t_{1u} MOs (t'_{1u}, t''_{1u}, and t'''_{1u}) emerge from the LCAO calculations [6.24].

The numerical data show that the bonding by the $\sigma + \pi$ MO is rather important; in $Cr(CO)_6$ it has the same order of magnitude as the "pure" σ and π MOs. This is also seen from a comparison of the orbital charge transfers: $\Delta q_{5\sigma} = -0.54$, $\Delta q_{2\pi} = 0.27$ (for each of the 2π MOs), and $\Delta q_{\sigma+\pi} = -0.31$. The last figure contains the transfer of 0.22 electron from the 5σ MO and 0.09 from the 1π MO of each CO ligand to the p orbital of the CA, its integral occupation becoming 1.89. Furthermore, in addition to the antibonding HOMO 5σ and LUMO 2π, the bonding 4σ and 1π MOs take part in the CO binding. In Table 6.6 the orbital charge transfers to and from these orbitals are given [6.24]. The metal–ligand integral charge transfers are different (or even of opposite direction) from those obtained in the simplified diorbital scheme (here again the dependence of the local charges on the method of calculation should be taken into consideration).

The $\sigma + \pi$ MO illustrates visually the statement about the three-dimensional delocalization of the electronic cloud around the CA in coordination compounds. Moreover, as seen from the electron distribution in these orbitals, there is a significant ligand–ligand interaction, which is antibonding in the

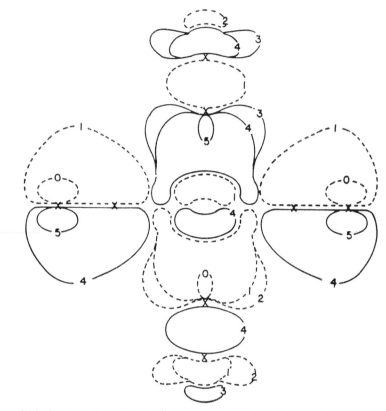

Figure 6.14. Contour lines for the t'_{1u} MO of $Cr(CO)_6$ in the xy plane containing Cr and four CO ligands slightly modified by spin–orbital interaction (Γ_8 component). The $\sigma + \pi$ nature of the MO is shown explicitly (cf. Fig. 6.13). (After [6.24].)

higher MO (Fig. 6.14) and bonding in the lower one [6.24]. This also shows that, strictly speaking, the ligand bonding is not of pure local ligand–metal nature from the point of view of the ligand either: it is combined ligand–metal, ligand–ligand, and so on, although the local ligand–metal contribution is predominant.

Note again that from the viewpoint of the complex as a whole, the t_{1u} orbital is a usual MO, its possible formation resulting from the general MO scheme for octahedral complexes with O_h symmetry (Fig. 6.1 and Table 5.1). The $\sigma + \pi$ description emerges when one tries to consider the metal–ligand bond separately (as a local property). Such a description is useful in revealing the properties of coordinated ligands and ligand–ligand mutual interactions.

The orbital charge transfers are important to the analysis of the properties of coordinated ligands and their activation by coordination (Section 11.2). In the case of CO, the donation of electrons from the slightly antibonding 5σ MO strengthens the C—O bond, while the back donation to the strongly antibond-

Table 6.6. Orbital charge transfers Δq_i in hexacarbonyls of Cr, Mo, and W (in electronic charge units per ligand)

Δq_i	$Cr(CO)_6$	$Mo(CO)_6$	$W(CO)_6$
$\Delta q_{1\pi}$	-0.07	-0.08	-0.06
$\Delta q_{5\sigma}$	-0.54	-0.56	-0.53
$\Delta q_{4\sigma}$	-0.17	-0.11	-0.18
$\Delta q_{2\pi}$	0.27	0.28	0.31

ing 2π MO weakens this bond. Since the 5σ MO is more extended in space than the 2π MO, we come to the conclusion that in the process of approaching CO to M, at large distances where the σ overlap becomes essential (while the π one is still small), the C—O bond is expected to be strengthened (and the interatomic distance shortened) compared with the free molecule. By further approaching the metal, the increase in the occupancy of the antibonding 2π orbital weakens the bond C—O. It is this process that was observed in the detailed calculations of the $(CO)_5Cr$—CO bonding [6.25]. The strengthening of the C—O bond by increasing the M—C distance is also confirmed experimentally in the x-ray analysis. For instance, in the compound $Rh_2(O_2CCH_3)_4(CO)_2$ the Rh—C distance is unusually long (2.092 Å), while the C—O one (1.120 Å) is shorter than in the usual cases [6.26].

It follows from the calculations [6.25] that the energy of the process $Cr(CO)_6 \rightarrow Cr(CO)_5 + CO$ equals 49.8 kcal/mol, which is larger than the mean dissociation energy 29.5 kcal/mol. This result can easily be understood if one takes into account the delocalized nature of the coordination bond and hence the dependence of each of the local M—CO bonds on the presence of other bonds. This result again confirms the nontransferrability of the metal–ligand bond parameters discussed in Section 9.1.

CO Bonding on Surfaces

Ligand bonding on surfaces (chemisorption) forms a part of solid-state chemistry. The electronic structure of solids has some distinctive features, one being the *density of states* that occurs instead of the MO energy levels in the molecular case. However, solid-state systems are beyond the scope of this book (an introduction to this field has recently been given by Hoffmann [6.27]).

In what follows we consider some local features of ligand bonding on surfaces. One of these features is that, depending on the surface structure, several types of bonding may occur, including simultaneous bonding to several coordination centers on the surface: *multicenter bonding*. Similar multicenter bridge bonding takes place in multicenter transition metal complexes (Section 11.2). Depending on the nature of the multicenter bonding, the charge transfers and hence ligand activation may be rather different.

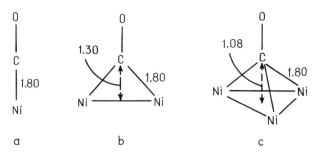

Figure 6.15. Three types of possible CO coordination to the Ni surface: (*a*) one-center; (*b*) two-center; (*c*) tricenter. In the last two cases the C—Ni distance is assumed equal to the one-center case.

The CO bonding on surfaces may be of *one-center, two-center, or tricenter type* (Fig. 6.15). Let us compare first some results of one-center coordination on different faces of different metals, Ti(0001), Cr(110), Fe(110), Co(0001), Ni(100), and Ni(111) [6.28], obtained by the extended Hückel method (Section 5.5). In these calculations it is assumed that the CO molecules are linear end-on coordinated and each center on the surface bounds one CO molecule. The last assumption is not restrictive, since it was shown that the dependence of the results on the CO filling on the surface is very weak (the reduction of the surface filling in half does not change the charge transfers).

In Table 6.7, CO bonding energies and the orbital charge transfers to the 2π MO and from the 5σ MO of CO by coordination to the metals, listed above, as well as the changes in M—C and C—O bond orders ΔP as compared with the noncoordinated case (for which P equals 0 and 1.21, respectively), are given (in the table $\Delta q_{2\pi}$ means the charge transfer to one of the two 2π MO, the total transfer thus being twice that in the table). As one can see, by moving along this series from right to left the C—O bond is weakening [following these data, the bond is hardly possible in the case of Ti(0001)], due primarily to the increase in population of the antibonding 2π MO. On the other hand, the M—C bond is strengthening; this is seen from both the bonding energy and bond orders P(M—C).

Table 6.7. Ligand bonding energies ΔE (in eV per one CO ligand), orbital charge transfers $\Delta q_{5\sigma}$ and $\Delta q_{2\pi}$, and bond order changes ΔP(M—C) and ΔP(C—O) by one-center coordination of the CO molecule to different metals and surfaces[a]

	Ti(0001)	Cr(110)	Fe(110)	Co(0001)	Ni(100)	Ni(111)
$\Delta q_{5\sigma}$	−0.27	−0.33	−0.38	−0.40	−0.40	−0.41
$\Delta q_{2\pi}$	1.61	0.74	0.54	0.43	0.39	0.40
ΔP(M—C	1.11	0.93	0.91	0.83	0.78	0.75
ΔP(C—O)	−0.78	−0.34	−0.25	−0.20	−0.18	−0.19
ΔE, eV	−6.77	−3.44	−2.64	−1.98	−1.97	−1.66

[a]In the free CO molecule P(C—O) = 1.21.

Table 6.8. Orbital charge transfers $\Delta q_{5\sigma}$ and $\Delta q_{2\pi}$, bond order changes $\Delta P(\text{Ni—C})$ and $\Delta P(\text{C—O})$, and bonding energies ΔE (in eV per one CO ligand) in one- (μ_1), two- (μ_2), and three-center (μ_3) bonding to the Ni(111) face for two sets of interatomic distances

| | $R(\text{Ni—C}) = 1.80\,\text{Å}$ | | | $R_N = 1.80\,\text{Å}$ | |
	μ_1	μ_2	μ_3	μ_2	μ_3
$\Delta q_{5\sigma}$	-0.41	-0.43	-0.44	-0.43	-0.46
$\Delta q_{2\pi}$	0.40	0.54	0.64	0.41	0.41
$P(\text{Ni—C})$	0.76	0.57	0.41	0.31	0.16
$\Delta P(\text{C—O})$	-0.19	-0.24	-0.28	-0.19	-0.19
ΔE	-1.66	-2.72	-3.24	-0.89	-0.71

For bridged two-center and tricenter coordination, the problem of adequate choice of interatomic distances occurs. In Table 6.8 the comparison of charge transfers, bond populations, and bonding energies of CO coordinated to the Ni surface are given after [6.28] for two cases:

1. The distance $R(\text{Ni—C}) = 1.80\,\text{Å}$ to all the centers (to two centers in the two-center coordination μ_2, and to three centers in the tricenter coordination μ_3; Fig. 6.15) is the same as in the one-center coordination μ_1. Here the distance from the carbon atom to the surface is smaller, namely, $R_N = 1.30\,\text{Å}$ in two-center bonding, and $R_N = 1.08\,\text{Å}$ in the tricenter case.

2. The distance from the carbon atom to the surface $R_N = 1.80\,\text{Å}$, while the distance $R(\text{Ni—C})$ is $2.19\,\text{Å}$ in two-center bonding and $2.30\,\text{Å}$ in tricenter bonding.

Case 1 seems to be more appropriate, since a considerable increase of the Ni—C distance associated with a corresponding bond weakening (Table 6.8) in the absence of steric hindrance is thermodynamically unconvenient.

As one can see, when passing from one-center coordination to μ_1 to μ_2 and μ_3 in case 1, *ceteris paribus*, the orbital charge transfer to the 2π MO increases, while the C—O bond population decreases [the negative $\Delta P(\text{C—O})$ increases]. In other words, by increasing the number of bonding centers for each CO molecule its activation by coordination increases. This multicenter effect in chemical activation by coordination is also discussed in Section 11.2.

Bonding of NO

In CO bonding, considered above, it has been assumed that the linear end-on geometry of coordination is realized, although, in principle, other geometries can also be important. This problem is discussed from the point of view of its influence on stereochemistry of ligand coordination in Section 9.2. Consider the electronic implications of NO coordination on the surface of Ni(111)

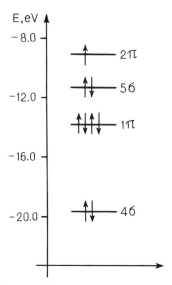

Figure 6.16. MO energy levels of the NO molecule. The wavefunction symmetries (not shown) are quite similar to the CO case in Fig. 6.9 (see also Figs. 6.6 and 6.7).

[6.29]. The electronic structure of the NO molecule differs from that of CO, first, by the positions of the σ and π MO energy levels (Fig. 6.16; cf. Fig. 6.9), which are lower in the NO case (the lower 2π MO makes the NO molecule a good π acceptor), and second, by the population of the (unoccupied in CO) 2π MO by one electron.

Omitting many details of NO coordination to Ni(111) (see [6.29]), we mention here that if the surface is filled in less than (or about) one-third and the coordination is linear end-on normal to the surface, the electronic interaction of the NO molecule with the surface is approximately the same, as in the absence of other NO molecules on the surface. In Table 6.9 the orbital charge transfers Δq_i and changes of bond orders $\Delta P(\text{Ni—N})$ and $\Delta P(\text{N—O})$ for the cases of one-, two-, and tricenter coordination of NO are given. The calculations were carried out in the extended Hückel approximation [6.29] with interatomic distances $R(\text{Ni—N}) = 1.680\,\text{Å}$ for all the cases.

As compared with CO, these calculation are more accurate: The charge transfers from the deeper 4σ and 1π MOs are not neglected, and it is taken into account that for the two-center coordination the two 2π MOs π_x and π_y are not degenerate (the interaction with two Ni atoms in the plane of the π MO and perpendicular to this plane are no longer equal; see Fig. 6.15). We see that in this case, as in the CO case, the increase in the number of coordination centers for each coordinated NO molecule increases its bonding to the surface and decreases the N—O bond order (increases the NO activation). The dependence of the bonding on the angle ϕ between the N—O line and the normal to the surface was also considered [6.29].

Table 6.9. Orbital charge transfers Δq_i, bond order changes, and bonding energies ΔE (in eV per one NO ligand) for one- (μ_1), two- (μ_2), and tricenter (μ_3) bonding of the NO molecule to the Ni(111) face[a]

Δq_i	μ_1	μ_2	μ_3
$\Delta q_{4\sigma}$	−0.156	−0.199	−0.215
$\Delta q_{1\pi x}$	−0.016	−0.097	−0.101
$\Delta q_{1\pi y}$	−0.016	−0.043	−0.101
$\Delta q_{5\sigma}$	−0.276	−0.263	−0.242
$\Delta q_{2\pi x}$	0.408	0.765	0.671
$\Delta q_{2\pi y}$	0.408	0.391	0.671
$\Delta P(\text{Ni—N})$	0.822	0.583	0.483
$\Delta P(\text{NO})$	−0.197	−0.263	−0.308
ΔE	−3.067	−4.130	−4.240

[a] $P(\text{N—O}) = 1.231$, $P(\text{Ni—N}) = 0$.

Coordination of C_2H_4

When polyatomic ligands are coordinated, the bonding picture becomes complicated. However, in some particular cases the problem can be simplified and reduced approximately to coordination of diatomics. Let us consider the coordination of olefines, for instance, ethylene, from this point of view. Simplification is possible in the case of bonding in the scheme of the π *complex* when the C=C bond line is perpendicular to the line of bonding to the metal.

In the planar configuration of C_2H_4 the C=C bond has one σ and one π bonding MO, with the latter lying in the plane perpendicular to that of the molecule (Fig. 6.17). Since in the π complex the C—H bonds do not take part directly in chemical interaction with the metal, one can conclude that the metal–ligand bond in L_nM—C_2H_4 is approximately diorbital. However, this bond is different from those in N_2 and CO coordination: in the last two cases the HOMO is of σ type (bonding in N_2 and weakly antibonding in CO), whereas in C_2H_4 it is a bonding π MO. The LUMO is an antibonding 2π MO in all the cases, but in N_2 and CO there are two MOs of this kind (π_x and π_y), whereas in C_2H_4 there is only one.

In the bonding of C_2H_4 to the metal in the form of a π complex, both of its bonding orbitals (either σ or π) are acting as σ donors, while only one antibonding π^* MO (and less probably, the antibonding σ^* MO) takes part in the metal–ligand diorbital bond. This determines the stronger σ-donor and weaker π-acceptor interaction of ethylene with the metal atom compared with CO and NO. Hence one can expect that the weakening of the C=C bond of an olefine by coordination is promoted by good σ-acceptor properties of the

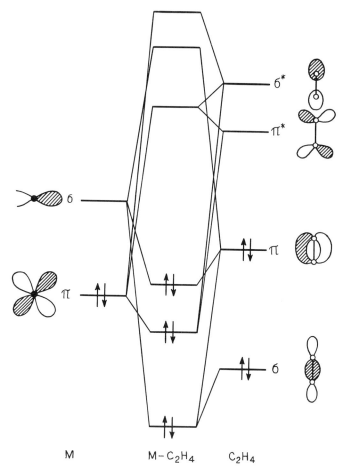

Figure 6.17. MO energy-level scheme for the M—C_2H_4 bonding in the π-type coordination of ethylene to the metal (π complex). Two orbitals of C_2H_4 $\sigma + \pi$ form σ bonds with M, while the other two $\sigma^* + \pi^*$ produce the π bonding to the d_π orbital of M.

metal. In particular, one can anticipate that the C=C bond will be less weakened if the σ-acceptor orbitals of the metal are occupied by electrons.

One of the first well-studied systems of this kind was Ag^+—C_2H_4; its electronic structure was calculated and first reported in [6.30] and then repeated in [6.31] using another AO basis set and comparing the results with the valence approximation (i.e., when the valence orbitals only are taken as a basis set; Section 5.3). Qualitatively, the results of these two works coincide. The C_2H_4 molecule is assumed planar, and the coordination of the Ag^+ ion placed on the twofold C_2 axis perpendicular to the plane of the molecule is energetically more convenient than on a similar axis in the molecular plane [6.30]. The bonding energy (in kcal/mol) is 27.6 [6.30] or 28.9 [6.31], or in the valence approximation, 38.3 [6.31]. The main contribution to bonding is due to the σ-donor charge transfer from the bonding π MO of the ethylene molecule to the σ orbital of the metal (formed mainly from the $5s$ orbital but also from the $5p$ orbital of the Ag atom). A much smaller contribution to the bonding is due to the dative charge transfer to the antibonding π^* MO of ethylene from the $3d_\pi$ orbital of Ag. In the valence approximation the charge on the Ag atom equals $+0.85$ [6.31], and the orbital charge transfers to C_2H_4 is (Fig. 6.17) $\Delta q_{(\sigma+\pi)} = -17$ and $\Delta q_{(\sigma^*+\pi^*)} = 0.02$, respectively.

In the semiempirical CNDO calculations [6.32] the charge transfers are larger and the charge on Ag is $+0.71$, while the $C=C$ bond order is reduced by coordination from 0.96 to 0.88. Approximately half of the Ag^+—C_2H_4 binding (~ 15 kcal/mol [6.30]) is due to the electrostatic polarization of the $C=C$ bond by the Ag^+ ion. The separation of this component of bonding is partly conventional, though.

The calculations of PdC_2H_4, $Pd(C_2H_4)_3$, and $Pd(C_2H_4)_4$ carried out by the same method as Ag^+—C_2H_4 yield similar results [6.31]. In the case of Pd—C_2H_4, $R(Pd$—$C) = 2.2\,\text{Å}$, $R(C=C) = 1.37\,\text{Å}$, and with the planar C_2H_4 molecule coordinated to Pd in the same way as in Ag^+—C_2H_4, the bonding energy 62.8 kcal/mol is much larger than in the latter case despite the absence of pure electrostatic polarization. The bonding is more covalent and multiorbital with orbital charge transfers (Fig. 6.17) $\Delta q_{(\sigma+\pi)} = -0.14$, $\Delta q_{(\sigma^*+\pi^*)} = 0.33$. The total transfer to C_2H_4, $\Delta q = 0.19$, is positive, in contrast to the A^+—C_2H_4 case, where $\Delta q = -0.15$. With the increase in the number of ligands n in the complex $Pd(C_2H_4)_n$, the bonding energy per ligand, ΔE_n, decreases (in kcal/mol): $\Delta E_1 = 62.8$, $\Delta E_3 = 53.6$, and $\Delta E_4 = 45.5$. With electronegative substituents in ethylene (as in C_2F_4), the energy of its antibonding MO lowers and the corresponding metal–ligand charge transfer increases.

More complete calculations of ethylene bonding were carried out for the Zeise salt $PtCl_3(C_2H_4)^-$ (I) and for $PdCl_3(C_2H_4)^-$ (II) [6.33] using nonempirical methods with extended basis sets and relativistic corrections. Figures 6.18 and 6.19 illustrate the results for these two systems in different geometries, including the upright and planar coordination (when the $C=C$ bond is perpendicular to the plane and in the plane of the complex, respectively) and the dissociated state. The experimental data for I obtained by neutron diffraction are in good agreement with these calculations. The Pt—C_2H_4 distance $R = 2.06\,\text{Å}$ (calculated with a basis set that includes the $3d$ function of the carbon atom) is close to the experimental value $R = 2.02\,\text{Å}$.

The planar configuration in the Zeise salt I is 15 kcal/mol higher in energy than the upright one; the bonding energy for the latter is 28.5 kcal/mol. It follows that internal rotation of the ethylene molecule is possible with a rotation barrier of 15 kcal/mol. Such barriers are observed experimentally for related complexes in the region of 10 to 16 kcal/mol [6.34]. The comparison of these results with those obtained by the extended Hückel method [6.34] shows that the latter correctly predicts that the upright coordination is more

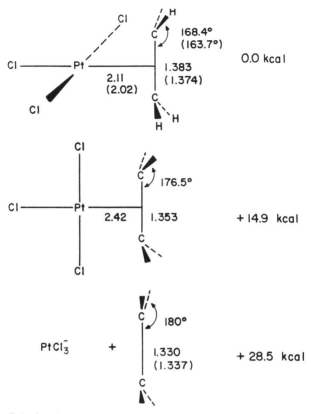

Figure 6.18. Calculated geometries for $PtCl_3(C_2H_4)^-$ for the upright and planar forms of coordinated ethylene along with uncoordinated ethylene (experimental values are indicated in parentheses). (From [6.33].)

stable, but it overestimates the barrier height: 35 kcal/mol in the case when the geometry of the $PtCl_3$ group remains in the planar configuration, and 22 kcal/mol if two Cl atoms deviate from the initial in-plane position by $7°$.

Note that in the calculations with geometry optimization there are no out-of-plane displacements of the Cl atoms: A stronger effect that allows for relaxation of the stressed planar configuration is achieved by increasing the $Pt—C_2H_4$ distance from $R = 2.11 Å$ to $R = 2.42 Å$ (for the latter distance the out-of-plane displacements of the Cl atoms are no more important). In [6.35] other cases of coordination of olefines to metals are considered in detail and the possibility of internal rotations of olefines is discussed. The main contribution to the barrier of rotations is due to the steric repulsion between the CH_2 groups of the olefines and the chlorine atoms in the planar configuration. This result is confirmed by more accurate calculations [6.34].

For the $PdCl_3(C_2H_4)^-$ complex, the bonding energy of the ethylene molecule is 12.3 kcal/mol, and the energy barrier of internal rotations (6.9

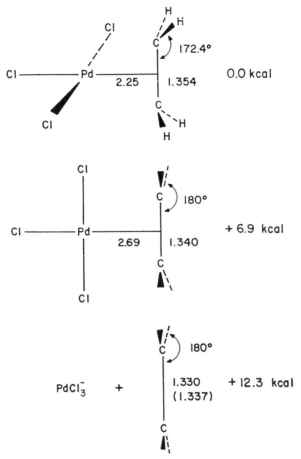

Figure 6.19. Calculated geometries for $PdCl_3(C_2H_4)^-$ for the upright and planar forms of coordinated ethylene along with uncoordinated ethylene. (From [6.33].)

kcal/mol) is much smaller than in the Zeise salt [6.33]. The deformation of the complex due to the ethylene coordination is also smaller. The authors [6.33] believe that this result explains why the Pd complex is more active in olefine oxidation: Although the C=C bond is more activated in the Pt complex, the rather strong $Pt-C_2H_4$ bond decelerates the reaction at the stage of cleavage of the oxidation product.

The orbital charge transfers Δq_σ and Δq_π are given in Table 6.10, in which the values obtained by means of the more accurate Noell method [Eq. (5.23)] are also shown. At first sight the differences in the orbital charge transfers calculated by these two methods are not very large, but the resulting charges on the metal are much different. Indeed, in the Zeise salt the atomic charge on the Pt atom after Mulliken is $q(Pt) = 0.02$, whereas after Noell $q'(Pt) = 1.40$ [similarly, for the Pd complex $q(Pd) = 0.20$ and $q'(Pd) = 1.27$]. Ethylene

Table 6.10. Orbital charge transfers Δq_σ and Δq_π to ethylene in $PtCl_3(C_2H_4)^-$ and $PdCl_3(C_2H_4)^-$ calculated by Mulliken populations analysis and after Noell (in parenthesis)

	$PtCl_3(C_2H_4)^-$	$PdCl_3(C_2H_4)^-$
Δq_σ	−0.37	−0.18
	(−0.24)	(−0.12)
Δq_π	0.23	0.07
	(0.22)	(0.07)

Source: [6.33].

coordination is also discussed in Section 11.3 as the main process in olefin insertion reactions and polymerization with Ziegler–Natta catalysts.

Metal–Metal Bonds and Bridging Ligands

The metal–metal bond may be regarded formally as ligand bonding in which the ligand is a metal. There may be several metal ligands in the case of multicenter (polynuclear) cluster compounds. However, in fact, the metal–metal bond is much different from that of normal ligands [6.36–6.46].

Consider the M—M bond, where M is a transition metal which can participate in the bonding with its s, p, and d orbitals. One of the distinct features of the d states is that they can produce δ bonds (Fig. 2.7) in addition to σ and π bonds. Figure 6.20 illustrates the main types of MOs that are built up from the d AOs of the transition metal M. The σ MO can be formed by s, p_σ, and d_{z^2} AOs, and hence there may be three AOs of σ type (or even more in the case of ligated centers), which produce three bonding and three antibonding MOs instead of the one σ and one σ^* shown in Fig. 6.20. The π bonds are formed by the doubly degenerate d_π (d_{xz}, d_{yz}) AOs with possible admixture of p_π. The δ bond is also doubly degenerate since it is produced by the two degenerate d_δ AOs, d_{xy} and $d_{x^2-y^2}$. The qualitative scheme in Fig. 6.20 is based on the well-known fact that the σ overlap is the strongest possible, producing the largest energy splitting between the bonding and antibonding MOs, the π overlap is smaller, and the δ overlap is the smallest.

As follows from this MO scheme, there are rich possibilities of multiorbital bonding in M—M, depending on the number of electrons n that occupy the MOs. With one or two electrons, we have a mono-orbital single bond. Up to six electrons give one σ and two π bonds (a diorbital triple bond). With more electrons we get triorbital fourfold and fivefold bonds. The last two possibilities are a special feature of M—M bonds which is unknown for other bonds [6.36]. For larger n the number of uncompensated bonding MOs decreases, provided that the additional σ orbitals from the s and p AOs are high in energy and not occupied (see below).

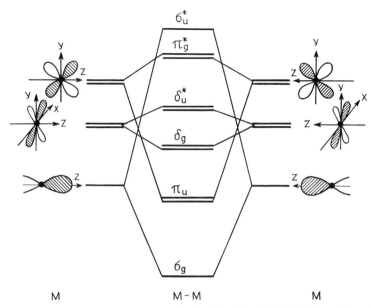

Figure 6.20. σ, π, and δ MOs formed by the d orbitals in the M—M bond: bonding σ and antibonding σ^* (produced by the two d_{z^2} AOs of the two atoms), twofold-degenerate π and π^* (formed by the two d_{xz} and two d_{yz} AOs, respectively), and twofold-degenerate δ and δ^* (emerging from the overlap of, respectively, the two d_{xy} and two $d_{x^2-y^2}$ AOs).

The first identification of a fourfold bond was obtained in 1964 for the Re—Re bond in the $[Re_2Cl_8]^{2-}$ anion in different compounds (e.g., in $K_2Re_2Cl_8 \cdot 2H_2O$) [6.36, 6.40]. This anionic complex is diamagnetic and has the structure shown in Fig. 6.21 [6.41]. Its important feature is that the eight chlorine atoms form a square prism (eclipsed structure), not an antiprism (staggered structure) as expected from the repulsion of the negative Cl^- ions. But the most significant result of these works is that the interatomic Re—Re distance in this complex is very short, about 2.22 Å, shorter than in the metallic state.

The explanation of these data was given qualitatively [6.40] based on the MO scheme of Re—Re bonding given in Fig. 6.20. The eight electrons that remain after formation of the eight Cl^- ions in the $[Re_2Cl_8]^{2-}$ system occupy one σ, two π, and one δ bonding orbitals, resulting in the configuration $\sigma^2\pi^4\delta^2$ with a fourfold bond. The δ bond between the two d_{xy} AOs of the Re atoms also explains the origin of the eclipsed structure of the $[Re_2Cl_8]^{2-}$ anion: In the alternative staggered structure, formation of the δ bond is impossible (the $d_{x^2-y^2}$ orbitals are engaged in the Re—Cl bonds), and it can be assumed that the energy gain in this δ bond is larger than the increase in Cl^-—Cl^- repulsions in the eclipsed structure.

This qualitative explanation of the Re—Re fourfold bond was confirmed afterward by semiempirical and nonempirical calculations. The first SCF

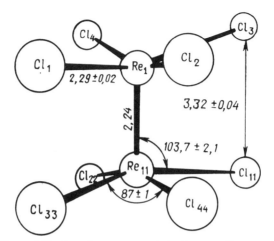

Figure 6.21. Atomic structure of the $[Re_2Cl_8]^{2-}$ ion. (From [6.41].)

X_α-SW computations (Section 5.4) were carried out for $[Mo_2Cl_8]^{4-}$ [6.42] and $[Re_2Cl_8]^{2-}$ [6.43]. The presence of other AOs of the metal and ligands may complicate the pure d-orbital M—M MO energy level scheme of Fig. 6.20. In particular, the additional σ and π MOs formed by the (next to d) s and p orbitals of the metal with the ligands and in the M—M bond may occur between the d–d levels. However, the calculations show that in the oxidation states under consideration, $+2$ and $+3$, the additional states are too high in energy and do not interfere with the levels of the scheme in Fig. 6.20.

The visual presentation of the corresponding wavefunctions by the contour diagrams in Figs. 6.22, 6.23, and 6.24 for σ, π, and δ MOs illustrate their M—M bonding nature; the SCF X_α-SW calculations [6.42, 6.43] confirm the qualitative electronic configuration $\sigma^2\pi^4\delta^2$ with a fourfold bond M—M in $[Mo_2Cl_8]^{4-}$ and $[Re_2Cl_8]^{-2}$. A qualitatively similar result was obtained in the SCF MO LCAO approximation [6.44], but the X_α-SW approach gives better results (lower energies) because it includes a part of the correlation energy. The latter can also be taken into account in the SCF MO LCAO approximation by adding CI (Section 5.3) with higher in energy configurations $\sigma^2\pi^4\sigma^{*2}$, $\sigma^2\pi^2\delta^2\pi^{*2}$, ..., $\sigma^{*2}\pi^{*4}\delta^{*2}$. Without CI the results may yield wrong conclusions. For instance, a single determinant calculation gives no Cr—Cr bonding in $Cr_2(O_2CH)_4(H_2O)_2$ (electronic configuration $\sigma^2\delta^2\delta^{*2}\sigma^{*2}$ with no uncompensated bonding MOs), whereas appropriate CI calculations result in the correct fourfold bonding with an interatomic distance of 2.27 Å [6.45].

Many other M—M bonds in coordination systems were studied and different bond multiplicities revealed, including the fourfold bond W—W in $[W_2(CH_3)_8]^{4-}$, the 3.5-fold bond Tc—Tc in $[Tc_2Cl_8]^{3-}$ with the configuration $\sigma^2\pi^4\delta^2\delta^*$, and Mo—Mo in $[Mo_2(SO_4)_4]^{3-}$ with the configuration $\sigma^2\pi^4\delta$, the triple bond Re—Re in $Re_2Cl_5(dth)_2$ and $Re_2Cl_4(PR_4)_4$ ($\sigma^2\pi^4\delta^2\delta^{*2}$), the shortest fourfold bond Cr—Cr in $Cr_2(C_3H_4)_4$, and others [6.36, 6.46].

Figure 6.22. Contour diagram of the wavefunction of the Mo—Mo σ bonding MO of $[Mo_2Cl_8]^{-4}$; the cross section including the two Mo and four Cl atoms is shown. (After [6.42].)

For the Cr—Cr bond with a very short bond length $R = 1.68\,\text{Å}$ in $Cr_2(C_3H_4)_4$ (cf. $R = 2.55\,\text{Å}$ in the metallic state) and apparently the shortest metal–metal bond Mo—Mo with $R = 1.93\,\text{Å}$ (metallic $R = 2.76\,\text{Å}$), it has been assumed [6.46] that the bond is sixfold, that is, in addition to the triorbital fivefold (one σ, two π, and two δ) bond, a σ bond from the ns orbitals occurs. The X_α calculations [6.47, 6.48] apparently confirm this point of view.

Table 6.11 lists a series of electronic structure calculations of metal–metal bonds in M_2 systems with indication of the ground state, electron configuration, bond order, bond length R_0, and the method of calculation.

Figure 6.23. Contour diagram of the wavefunction of the Mo—Mo π bonding MO in $[Mo_2Cl_8]^{4-}$; the same cross section as in Fig. 6.22 is shown. (From [6.42].)

In some cases the calculations decline the assumptions made on a qualitative base. For example, the Rh—Rh bond in $Rh_2(OCCH_3)_4(H_2O)_2$ was found to be shorter (~ 2.39 Å), than in $Rh_2(dmg)_4(PPh)_2$ (~ 2.94 Å), and this stimulated the assumption [6.49] that in the acetate dimer the Rh—Rh bond is a triple bond with the configuration $\sigma^2\pi^4\delta^2\sigma_n^2\sigma_n'^2\delta^{*2}$, where σ_n and σ_n' are some nonbonding orbitals comprising essentially $5s$ and $5p_\sigma$ AO of Rh. Both semiempirical [6.50] and X_α-SW [6.51] calculations show that in these and some other similar $d^7 - d^7$ Rh dimers the 14 electrons produce the electronic configuration $\sigma^2\pi^4\delta^2\pi^{*4}\delta^{*2}$, leaving only one uncompensated orbital σ. Hence the Rh—Rh bond is mono-orbital (single), and the σ_n, σ_n' orbitals are not active.

For illustration, the energy-level diagram for two such dimers, $Rh_2(O_2CH)_4$ and $Rh_2(O_2CH)_4(PH_3)_2$, obtained by X_α-SW calculations [6.52], is given in Fig. 6.25, which also shows the influence of the axial ligands PH_3. In comparison with the Rh_2 energy-level scheme, it is seen that, indeed, only one bonding σ orbital remains uncompensated in these complexes. The $5s$- and $5p_\sigma$-originating orbitals are much higher in energy; they are not shown.

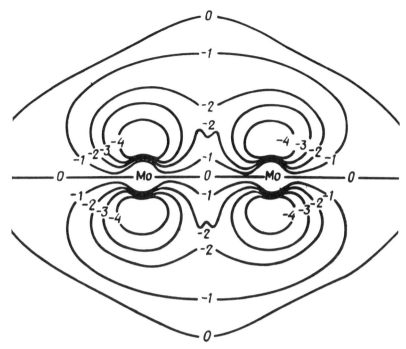

Figure 6.24. Contour diagram of the wavefunction of the Mo—Mo δ bonding MO of $[Mo_2Cl_8]^{4-}$; the cross section containing the two Mo atoms and the positive lobes of their d_{xy} AO (i.e., bisecting two opposite Cl—Mo—Cl right angles in each of the $MoCl_4$ groups) is shown. (After [6.42].)

An important problem related to metal–metal bonds is the role of *bridging and semibridging ligands*. In the presence of ligands that occupy a position close to both metals (or to several metals in polynuclear cluster compounds), a question arises whether there is a direct metal–metal bonding, or it is realized completely through the nearest-neighbor ligands, which thus serve as bonding bridges. An example of this type is discussed briefly in Section 8.4 for the dimeric copper acetate hydrate and other copper carboxylates. The semiempirical IEH (SCCC) MO LCAO calculation (Section 5.5) show [6.53] that in these cases the Cu—Cu bonding is realized via the carboxylate bridges, and there is no direct Cu—Cu bonding.

More accurate calculations of the electronic structure of dimers were carried out recently to reveal the role of bridging ligands. An interesting example is provided by the $Co_2(CO)_8$ complex. The Co—Co interatomic distance in this complex, although not very short, allows one to propose that there is a direct Co—Co single bond, and this assumption is also supported by the *18-electron rule* [6.54] (a closed shell with 18 electrons, similar to [Ar] configuration is assumed to be most stable; see the inert gas rule in Section 9.1). If this is true, the bond can be either linear or bent. The latter possibility is enhanced by the

Table 6.11. Calculated electronic structure of metal–metal bonds in dimer M_2 systems[a]

M_2	Ground State	Electronic Configuration	Bond Order	R, Å	Method of Calculation
Sc_2	$^5\Sigma^-$	$s\sigma_g^2 d\sigma_g^2 \pi_u^2 s\sigma_u^1$	2	2.57	SCF CI
	$^1\Sigma_g^+$	$s\sigma_g^2 d\sigma_g^2 s\sigma_u^2$	1	3.05	SCF
Ti_2	$^1\Sigma_g$	$\sigma_g^2 d\sigma_g^2 \pi_u^2$	4	1.87	SCF
V_2	$^1\Sigma_g$	$s\sigma_g^2 d\sigma_g^2 s\sigma_u^2 \pi_u^4$	3	1.96	SCF
		$\pi_u^4 d\sigma_g^2 s\sigma_g^2 \delta_g^2$	5	1.78	X_α-DV
Mn_2		$d\sigma_g^2 \pi_u^2 \delta_g^4 s\sigma_g^2 \delta_u^2$	5	1.69	X_α-DV
	$^1\Sigma_g$	$d\sigma_g^2 s\sigma_g^2 s\sigma_u^2 \pi_u^4 \delta_u^4$	5	1.52	SCF
Fe_2	$^7\Delta_u$	$d\sigma_g^{1.57} \pi_u^{3.06} \delta_g^{2.53} \delta_u^{2.47} d\sigma_u^{1.49} \pi_g^{2.89} s\sigma_g^2$	1.3	2.40	SCF CI
Co_2	$^5\Sigma_g^+$	$d\sigma_g^2 \pi_u^{3.09} \delta_g^{3.02} \delta_u^{2.98} \pi_g^{2.91} d\sigma_u^2 s\sigma_g^{1.94} s\sigma_u^{0.06}$	1.1	2.40	SCF CI
Ni_2	$^3\Pi_u$	$d\sigma_g^2 \pi_u^2 \delta_g^4 \delta_u^4 s\sigma_g^2 \sigma_u^1 \pi_g^3$	2	2.16	X_α-SW
		$1\sigma_g^2 \sigma_{u2}^2 \sigma_g^2 \pi_u^4 \pi_g \delta_g^3 o_u^3$		2.26	SCF CI
Nb_2	$^1\Sigma_g$	$\sigma_g^2 \pi_u^4 \delta_g^4$	5	1.97	X_α-DV
Tc_2		$\delta\sigma_g^2 \pi_u^4 \delta_g^4 s\sigma_g^2 \delta_u^2$	5	1.92	X_α-DV
Ru_2	$^1\Sigma_g$	$d\sigma_g^2 \pi_u^4 \delta_g^4 \delta_u^4 s\sigma_g^2$	4	1.94	X_α-DV
	$^7\Delta_u$	$d\sigma_g^{1.67} \pi_u^{3.31} \delta_g^{2.73} \delta_u^{2.42} \pi_g^{2.60} d\sigma_u^{1.27} s\sigma_g^{1.94} s\sigma_u^{0.06}$	1.7	2.71	SCF CI

Source: [6.46].

[a]Notations $s\sigma_g$, $d\sigma_g$, π_u, δ_g, ... stand for the corresponding σ, π, δ, MOs formed by $s, d, \ldots$, AOs; u and g orbitals of the same type approximately compensate each other.

symmetry of the system. Figure 6.26 shows these two possibilities together with a third one when there is no direct M—M bond.

Calculations in the IEH approximation [6.55] show that there is no direct Co—Co bond, whereas the CNDO approach yields such a bond [6.56]. This controversy stimulated more accurate *ab initio* SCF computations carried out recently [6.57]. A detailed analysis of the deformation density (Section 8.6) with respect to two $Co(CO)_3$ fragments and two CO molecules and additional topological investigation of the electronic density in the region between the Co atoms shows that there is no direct Co—Co bond in this compound. For related calculations of $Fe_3(CO)_{12}$ and $MoCr(O_2CH)_4$, see [6.58].

In conclusion we make a remark about *semibridging ligands*. This term is usually used to denote a ligand (often a carbonyl, thiocarbonyl) which is asymmetrically coordinated to two metal centers, between which there is direct bonding. Figure 6.27 illustrates this possibility for the CO molecule as a semibridging ligand between two metals (usually, M and M′ are different). The question under discussion in a series of recent papers is about the nature of the bonding to the secondary metal M′. The position of the CO molecule with respect to M′ enables its four bonding electrons on the π orbitals to be σ donors to M′ (see Fig. 6.9). On the other hand, its antibonding π^* MO can serve as a good acceptor of electrons. Hence there are two possibilities of charge transfer, from CO to M′ and from M′ to CO, as shown in Fig. 6.27.

It is clear that the actual charge transfers depend essentially on the nature of the metals M′, M, and other ligands coordinated to these metals. For

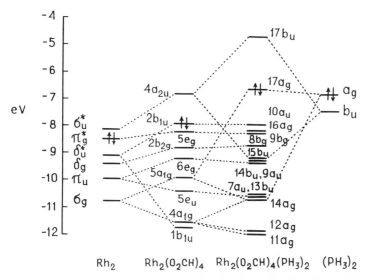

Figure 6.25. SCF X_{α}-SW calculated energy level scheme for the dimers $Rh_2(O_2CH)_4$ and $Rh_2(O_2CH)_4(PH_3)_2$ in correlation with Rh, and PH_3; the highest occupied levels in the ground state are indicated. (After [6.52].)

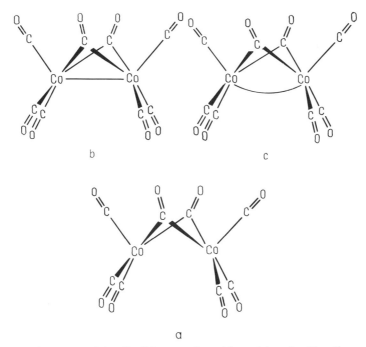

Figure 6.26. Structure of the Co_2CO_8 complex without (*a*) and with a linear (*b*) and bent (*c*) direct Co—Co bond.

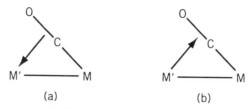

(a)   (b)

Figure 6.27. Semibridging CO ligand with indication of two possible situations of charge transfer (a) from and (b) to the ligand.

instance, if M' is a good σ acceptor and π donor, significant orbital charge transfer to the antibonding π^* MO of CO can be expected, provided that the metal M does not occupy this π^* MO in the bonding with CO; this, in turn, depends on the nature of M and its other ligands. Obviously, the problem should be solved by reliable electronic structure calculations.

Concerning carbonyl complexes, three different types of semibridging ligand bonding can be distinguished: the secondary metal atom M' is an early transition metal (I), M' is an atom from the middle of the series (II), and it is a late transition or posttransition element (III). It was shown in a series of electronic structure calculations carried out for several systems of types II and III that in these cases the CO ligand is a π acceptor (Fig. 6.27b).

For example, for the dimer $(NH_3)_2CuCo(CO)_4$ the electronic structure calculations [6.59] in the Fenske–Hall approximation (Section 5.4) performed for the system as a whole and separately for the two fragments, $(H_3N)_2Cu$ and $Co(CO)_4$, show explicitly that in the system $(H_3N)Cu-Co(CO)_4$ one of the CO ligands is semibridging and accepts about 0.14 electronic charge transfer to its antibonding π^* MO from the copper d_{xz} orbital.

In early transition elements M' (complexes of group I) the d_{xz} orbitals are unoccupied and cannot be electron donors to CO. Detailed calculations of $CpCo(\mu\text{-}CO)_2ZrCp_2$ ($Cp = \eta^5\text{-}C_5H_5$) in the same Fenske–Hall approximation [6.58] show that in the interaction of the two groups, Cp_2Zr and $Co(CO)_2Cp$, the semibridging CO ligand is a four-electron π donor to the electron-deficient Zr atom.

6.4. ENERGIES, GEOMETRIES, AND CHARGE DISTRIBUTIONS

We continue the discussion of electronic structure and bonding in coordination compounds by revealing further features closely related to those of ligand bonding: total and ionization energies, charge distributions, and geometries of coordination.

Ionization Energies

One should be careful when relating the one-electron MO energies to observable spectroscopic properties (ionization potentials, optical, photoelectron, x-ray, etc., spectra; Sections 8.1 to 8.3). In particular, for the ionization potentials I there is the Koopmans theorem, stating that the energy of ionization of an electron from a given electronic state equals the energy of the corresponding MO taken with opposite sign (Sections 2.2 and 6.2). This means that the order and position of the consequent I values or the positions of the peaks of photoelectron spectra, observed experimentally, should correspond to the consequence of the MO energy-level positions from top to bottom. This correspondence takes place indeed in some cases, for instance, in $Ni(CO)_4$ [6.60, 6.61].

However, in many cases, for instance, in metalloorganic compounds and complex anions, the Koopmans theorem does not hold. As emphasized in Sections 2.2 and 6.2, the Koopmans approach does not take into account that when one removes the electron from a given one-electron state, all the other electrons change (relax) to new self-consistent one-electron states (in which the interelectron repulsion is reduced). This relaxation, which may be small or even negligible when the state to be ionized is well localized, becomes very important in the case of coordination compounds where the ionization takes place from MOs containing d states with essentially delocalized charge distribution.

For instance, in the case of ferrocene $Fe(Cp)_2$, the calculations yield the following order of MO energy-level positions [6.62]:

$$a_{1g}(3d) < e_{2g}(\sigma\text{-Cp}) \sim a_{2u}(\pi - Cp) \sim e_{2u}(\sigma - Cp) < e_{2g}(3d)$$
$$< e_{1g}(\pi\text{-Cp}) \sim e_{1u}(\pi\text{-Cp}) \qquad (6.3)$$

The order of ionization potentials I determined experimentally from the photoelectron spectra is different from the sequence (6.3) [6.63]. If one negates the Koopmans theorem, the I values should be calculated as a difference between the energies of the ground state of the $Fe(Cp)_2$ molecule and the final state after ionization $Fe(Cp)_2^+$ (both calculated with full account for interelectron interactions). The calculations carried out in this way [6.62] give another consequence of ionization potentials I:

$$I(e_{2g}) < I(a_{1g}) < I(e_{1u}) < I(e_{1g}) \qquad (6.4)$$

in agreement with the experimental data.

Similarly, for complex anions the HOMO energy levels are often of ligand type (p levels of halogens, π levels of CN, etc.), whereas the first ionization potential calculated as the energy difference between the initial complex and its ionized state corresponds to the elimination of the $3d$ electron of the metal in accordance with the experimental data. In *ab initio* calculations the difference between the self-consistent (SCF) energies of two systems, initial and ionized, is often denoted by ΔSCF.

Another and sometimes not less important reason of the observed deviations from the Koopmans theorem lies in the difference in the energy of electronic correlation (Section 5.3) in these two states. Therefore, a sufficiently full calculation of ionization energies and photoelectron (x-ray) spectra must include the correlation energies (e.g., in the form of CI) in both states, before and after ionization; the corresponding energy difference is then denoted by ΔSCF-CI.

For coordination compounds, as distinct from more simple systems, such a full calculation of ΔSCF-CI values may be very difficult. For instance, for the evaluation of ΔSCF-CI of a relatively small compound, $Ni(C_3H_5)_2$, 10^6 excited states should be included in the CI scheme. This can be done only by means of supercomputers [6.64]. For larger systems, even supercomputers may be insufficient. A modified Green's function method in the form of an extended two-particle-hole approximation of Tamm and Dancoff [6.65] [two-particle-hole Tamm–Dancoff approximation (2phTDA)] has been suggested for these problems and proved to be more convenient and less computer time consuming than CI. The results are illustrated below considering a series of coordination compounds: $Ni(C_3H_5)_2$, $Ni(CN)_4{}^{2-}$, $Co(CN)_6{}^{3-}$, $Fe(CN)_6{}^{4-}$ [6.64, 6.66].

The $Ni(C_3H_5)_2$ molecule was one of the first coordination systems calculated ab initio [6.67, 6.68]. In [6.68], the necessity to take into account the relaxation of electrons by ionization was illustrated explicitly. Afterward it occurred that the calculated SCF values are also not adequate and give an incorrect order of ionization states. In Table 6.12 the ionization energies for the valence states of this system obtained by *ab initio* calculations after Koopmans and by ΔSCF, ΔSCF-CI, and 2eh-DTA methods are listed [6.64, 6.66]. The experimental data were taken from the photoelectron spectra [6.69]. The notations of the MOs are given after the irreducible representations of the symmetry group D_{2h}.

The HOMO $7a_u$ is a pure ligand π MO originating from its a_2 π MO. The next one $6b_g$ is a mixing of an appropriate combination of this a_2 MO with the d_{xz} orbitals of the metal and gives the largest contribution to the bonding. The $11b_u$ MO is a nonbonding antisymmetrical mixing of the ligand π MOs of b_1 type. Their symmetrical combination overlaps with the Ni d_{xy} orbital, yielding the $9a_g$ and $13a_g$ MOs. The $10a_g$, $5b_g$, and $11a_g$ MOs contain, respectively, the metallic $3d_{z^2}$, $3d_{xy}$, and $3d_{x^2-y^2}$ orbitals and are mainly nonbonding. The remaining valence MOs are related to the σ core of the ligands.

As shown by the numerical data of the calculations with CI, the contribution of the basic Hartree–Fock (HF) configuration to the wavefunction of the ground state is 83%, which is significantly less than for compounds without transition metals (where it is greater than 95%). In other words, CI is more important in calculations of the electronic structure of transition metal compounds than for organic compounds. The largest contribution to the CI comes from excited states formed by one- and two-electron excitations $6b_g \rightarrow 7b_g$ and $6b_g^2 \rightarrow 7b_g^2$, where $7b_g$ is the antibonding metal–ligand MO (corresponding to the $6b_g$ bonding MO). In the $7b_g$ MO the contribution of the Ni $3d$ orbital is 43%, while in $6b_g$ it is 38%. Therefore, by including correlation effects, one transfers more electronic charge to the metal.

Table 6.12. Calculated ionization energies of valence MOs of Ni(C₃H₅)₂ (in eV)

		Method of Calculation				
MO	Origin[a]	Koopmans Theorem	ΔSCF	ΔSCF-CI	Extended 2ph-TDA	Experiment[b]
$7a_u$	$\pi(L)$	7.5	6.7	6.7	6.4	7.7(1)
$6b_g$	$3d_{xz}$; $\pi(L)$	9.0	5.6	6.6	7.7	8.1(2)
$11b_u$	$\pi(L)$	11.8	11.0	10.8	10.3	10.3(5)
$13a_g$	$3d_{xy}$; $\pi(L)$	11.7	5.5	6.4	7.6	8.1(2)
$12a_g$	$\sigma(L)$	14.0			13.5	12.7(7)
$5b_g$	$3d_{yz}$	14.0			8.5	
$11a_g$	$3d_{x^2-y^2}$	14.2			8.2	8.5(3)
$6a_u$	$\sigma(L)$	14.6			13.3	
$10b_u$	$\sigma(L)$	14.6			13.4	12.7(7)
$4b_g$	$\sigma(L)$	15.0			13.7	
$10a_g$	$3d_{z^2}$	15.3			8.8	9.4(4)
$9a_g$	$3d_{xy}$; $\pi(L)$	16.4			11.5	11.5(6)
$5a_u$	$\sigma(L)$	16.5			14.9	
$3b_g$	$\sigma(L)$	17.3			15.1	14.2(8)
$9b_u$	$\sigma(L)$	18.0			16.2	
$8a_g$	$\sigma(L)$	19.0			16.5	15.6(9)

Source: [6.64, 6.66].

[a] L denotes the ligand.

As seen from Table 6.12, the Koopmans theorem gives not only incorrect values of ionization energies, but also wrong sequences of the latter, beginning from the third band of the spectrum. The discrepancies are larger for states that include the metal d orbitals (as anticipated) and much less for pure ligand states. The Δ SCF calculation improves essentially the results, but still it gives an incorrect sequence of the ionization bands. The inclusion of correlation effects by the CI method gives a further improvement of the results, while the 2ph-TDA method gives the best fit to the experimental data.

Reasonable results for ionization energies (photoelectron spectra) were obtained by density functional methods. Some earlier works on X_α calculations are reviewed in [6.70, 6.71]. For recent results, see, for instance, [6.72] where a series of transition metal systems, including Ni(CO)₄, Cr(NO)₄, Fe(CO)₂(NO)₂, bis(π-allyl)nickel, and bis(π-allyl)palladium, were calculated by X_α methods, and the results are compared with that of other methods and with experimental data. In [6.73] a more complicated system, ruthenocene, is calculated in the ground and excited states by the so-called linear combinations of Gaussian orbitals — model core potential — density functional (LCGTO-MCP-DF) method [6.74] (Section 5.4). The authors also addressed the issue of term multiplicity (one of the most difficult problems in density functional theories) and obtained absorption and emission frequencies in good agreement with experimental data.

Total and Bonding Energies, Geometries, and Other Properties

Ab initio calculations of relatively simple coordination compounds became possible about 25 years ago. Table 6.13 lists some examples of earlier nonempirical computations of the electronic structure and properties of a series of coordination system (see also [6.75–6.77]). It includes the total energy (for

Table 6.13. Some results on earlier nonempirical calculations of coordination systems

System	Symmetry Group	Electronic State	Total Energy (a.u.)	Bonding Energy (kcal/mol)	Charge on CA	Ref.
$Cr(CO)_6$	O_h	$^1A_{1g}$	−1702.613	240^a	+0.703	[6.60]
$NiF_6{}^{4-}$	O_h	$^3A_{2g}$	−2084.4339			[6.79]
			-2099.1291^b			[6.80]
$NiCl_4{}^{2-}$	D_{4h}	$^1A_{1g}$	−3334.454			[6.78]
	T_d	$^1T_2$	−3334.446			[6.78]
	T_d	$^3T_1$	−3334.5497			[6.78]
$Ni(CN)_4{}^{2-}$	D_{4h}	$^1A_{1g}$	−1872.496	608^a	+0.46	[6.61]
$Ni(CO)_4$	T_d	$^1A_1$	−1953.949	$86^{a,c}$	+0.24	[6.61]
			−1939.436	$144^{a,d}$	+0.466	[6.60]
$CuF_4{}^{2-}$	D_{4h}	$2B_{1g}$				[6.79]
$CuCl_4{}^{2-}$	D_{4h}	$^2B_{1g}$	−3470.577	588^a	+1.28	[6.81]
			-3472.284^b	606^a		[6.81]
	T_d	$^2T_2$	−3470.606			[6.81]
	D_{2d}	$^2B^2$	−3470.608			[6.81]
$VO_4{}^{3-}$	T_d	$^1A_1$	−1229.9361		−0.07	[6.82]
$CrO_4{}^{2-}$	T_d	$^1A_1$	−1329.9148		+0.58	[6.82]
$MnO_4{}^-$	T_d	$^1A_1$	−1435.4853	-186^e	+0.93	[6.82]
			−1448.7571	-167^f		[6.83]
$CrO_8{}^3$	D_{2d}	$^2B_1$	−1598.188	4^f		[6.84]
			−1628.4862	93^f	+2.58	[6.85]
$Fe(C_5H_5)_2$	D_{5h}	^1A_g	−1643.1252		+1.23	[6.62]
$Fe(C_5H_5)_2{}^+$	$^2E_{2g}$	D_{5h}	−1642.821			[6.62]
$Co(C_5H_5)_2$	D_{5h}	$^2E_{1g}$	−1761.8221			[6.68]
		$^2E_{2g}$	−1761.5607			[6.68]
		$^2E_{2u}$	−1761.5084			[6.68]
$Mn(C_5H_5)_2$	D_{5h}	$^2E_{2g}$	−1526.7099			[6.68]
		$^2A_{1g}$	−1526.6521			[6.68]
$Mn(CO)_5H$	C_{4v}	$^1A_1$				[6.86]
$Mn(CO)_5CH_3$	C_{4v}	$^1A_1$				[6.86]
$Ag(C_2H_4)^+$	C_{2v}	$^1A_1$	−6272.1673	28^a	+0.10	[6.30]
$Ni(C_3H_5)_2$	C_{2h}	^1A_g	−1723.8044			[6.68]
$Co(NH_3)_6{}^{3+}$	O_h	$^1A_{1g}$	-189.092^f	654	+1.6	[6.87]

^{a} With respect to the metal and ligands.
^{b} By the extended basis.
^{c} With respect to Ni^0 in the ^1S state.
^{d} With respect to Ni^0 in the ^3F state.
^{e} With respect to the atoms.
^{f} Valence orbitals only.

fixed nuclei), bonding energy (with respect to either the metal in the corresponding oxidation state and the ligands, or the atoms), and the formal charge on the central atom after Mulliken (Section 5.2).

Besides the data presented in Table 6.13, the calculations cited there allow one to determine the electronic distribution in the system, the relative stability of different electronic and geometric configurations (where these configuration were calculated), the one-electron energy-level ordering for the outer electrons, potentials of ionization from different states (expected photoelectron spectra), electron affinities, frequencies and probabilities of electronic transitions with absorption and emission of light, and so on. For instance, Fig. 6.28 illustrates the electronic charge distribution along the Ni—C—N bond in the $Ni(CN)_4^{2-}$ complex [6.61]. Other illustrative examples are as follows.

Several examples are concerned with relative stability of different configurations. For $CuCl_4^{2-}$ the configuration of a compressed (flattened) tetrahedron of D_{2d} symmetry with a Cl—Cu—Cl angle of 120° is most stable (in Table 6.13, the data obtained by the same basis set should be compared) in accordance with the Jahn–Teller effect (Section 7.3).

In the $NiCl_4^{2-}$ complex, the high-spin (triplet) state $^3T_1$ is more stable (by 65 kcal/mol) than the low-spin state $^1T_2$, in agreement with the experimental data. Nonempirical calculations of five-coordinated complexes $CuCl_5^{3-}$, $Cu(H_2O)_5^{2+}$, and $Fe(CO)_5$ [6.88] in the configurations of a square pyramid and of a trigonal bipyramid show that the energies of these two configurations are very close (the difference is on the order of several kcal/mol). This explains their stereochemical softness and the special Jahn–Teller dynamics [6.89]; discussion of these effects is given in Section 9.2. A full SCF calculation with geometry optimization for $Cu(H_2O)_6^{2+}$, $Cr(H_2O)_6^{2+}$, $Mn(H_2O)_6^{3+}$, $Cu(H_2O)_6^+$, and $Mn(H_2O)_6^{2+}$ with correlated pair functional calculations for the former two was carried out recently [6.90]. The results confirm the Jahn–Teller distortions in the first three complexes (and their absence in the last two systems), as predicted by the general theory for orbital degenerate E terms (Sections 7.3 and 9.2).

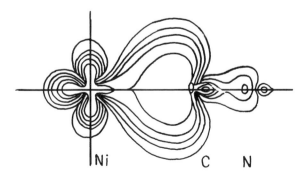

Figure 6.28. Contour map of electronic charge distribution along the Ni—C—N coordinate in $Ni(CN)_4^{2-}$ obtained by nonempirical calculations. (From [6.61].)

Bonding energies are given in Table 6.13 for some cases only. Their comparison with experimental data encounters difficulties, because in many cases it is not clear what should be considered as bonding agents: neutral ligands, their ions, separate atoms, and so on. For instance, the calculated bonding energy in $Ni(CO)_4$ with respect to the ligands and the neutral atom Ni^0 in the singlet state ^1S equals 86 kcal/mol [6.61], whereas in another calculation the bonding energy with respect to the neutral Ni^0 atom in the triplet state ^3F equals 144 kcal/mol [6.60]. The experimental value for the gas phase is 140 kcal/mol.

In $Ni(CN)_4^{2-}$ the bonding energy with respect to the clevage into ions Ni^{2+} and CN^- is 608 kcal/mol. The largest part of it comes from the electrostatic interaction of the ions [6.61], and therefore it cannot be compared with the experimental estimate (43 kcal/mol) obtained under conditions when the ions are solvated. Note that in the case of Ag^+—C_2H_4 discussed in Section 6.3, the largest part of the bonding energy, 15 to 28 kcal/mol, is also due to the electrostatic interaction [6.30].

Further improvement of computers made it possible to calculate the electronic structure of more complicated systems, including metalloporphyrins and other active sites and models of biological systems [6.91–6.96]. For metalloporphyrins [6.91–6.94] the calculations yield the order of the HOMO and LUMO energy levels, ionization potentials, distribution of electronic charge, the metal positions with respect to the porphin plane, the conformation of the ring, and so on. Using a method that combines INDO with RHF SCF (Sections 5.3 and 5.5), parametrized to include transition metals, the electronic structure of a series of heme complexes in model compounds and intact proteins was studied with emphasis on the spin states and electromagnetic properties [6.91]. For such important biological systems as cytochrome P-450, peroxidase, catalase, and metmioglobin, some unresolved questions regarding the resting states have been addressed and interesting features of their functionality were elucidated.

Most difficulties occur in geometry optimization of transition metal systems. A relatively accurate and compact basis set $(11s7p4d/5s3p2d)$ that proved to yield bond lengths with an accuracy of 0.03 to 0.08 Å for $Cr(CO)_6$, $HMn(CO)_5$, $Fe(CO)_5$, $Cr(C_6H_6)_2$, $Fe(C_5H_5)_2$, $Ni(C_4H_4)_2$, $(C_5H_5)Mn(CO)_3$, and $(C_6H_6)Cr(CO)_3$ was suggested [6.97] [for $Ni(CO_4)$ and $Cr(NO)_4$ the basis set used is $(19s9p6d/10s4p4d)$]. The calculations are carried out by the program GAMESS (generalized atomic and molecular electronic structure system). In Table 6.14, the calculated parameters of the geometrical structure of a series of 14 coordination compounds obtained with a relatively not very large supercomputer (FPS-164) using the computer program GAMESS are given together with the experimental data available [6.64]. The calculations were carried out by two basis sets: minimal STO-3G and double-zeta (DZ).

The following conclusions emerge from Table 6.14. First, the results of the calculations, in general, describe correctly the geometry of the compounds under consideration and, with a few exceptions discussed below, predict quantitatively the interatomic distances and angles for these compounds.

Table 6.14. Calculated and experimental parameter values for the equilibrium geometry of a variety of transition metal coordination compounds (distances in Å, angles in degrees)

System	Point Group	Geometry Parameter	Minimal STO-3G	DZ	Experimental Value
			Calculation Basis Set		
ScF_3	D_{3h}	$R(Sc-F)$	1.845	1.879	1.91
$TiCl_4$	T_d	$R(Ti-Cl)$	2.167	2.214	2.17
VF_5	D_{3h}	$R(V-F_{ax})$	1.641	1.744	1.734
		$R(V-F_{eq})$	1.608	1.702	1.708
$VOCl_3$	C_{3v}	$R(V-O)$	1.468	1.518	1.570
		$R(V-Cl)$	2.107	2.177	2.142
		$<(Cl-V-Cl)$	109.9	110.5	111.3
$Cr(NO)_4$	T_d	$R(Cr-N)$	1.576	1.689	1.79
		$R(N-O)$	1.218	1.160	1.16
$Cr(CO)_6$	O_h	$R(Cr-C)$	1.786	1.975	1.92
		$R(C-O)$	1.167	1.142	1.16
$Ni(CO)_4$	T_d	$R(Ni-C)$	1.579	1.900	1.836
$Fe(CO)_2(NO)_2$	C_{2v}	$R(Fe-C)$	1.708	2.198	1.8
		$R(Fe-N)$	1.518	1.822	1.77
		$R(C-O)$	1.154	1.127	1.15
		$R(N-O)$	1.221	1.202	1.12
		$<(C-Fe-C)$	107.7	92.2	—
		$<(N-Fe-N)$	111.8	129.6	—
$Co(CO)_3NO$	C_{3v}	$R(Co-N)$	1.478	1.593	1.76
		$R(Co-C)$	1.664	1.938	1.83
		$R(N-O)$	1.231	1.197	1.10
		$R(C-O)$	1.155	1.132	1.14
		$<(C-Co-N)$	111.9	114.3	—
$Mn(NO)_3CO$	C_{3v}	$R(Mn-C)$	1.751	1.921	1.83
		$R(Mn-N)$	1.513	1.658	1.76
		$R(C-O)$	1.154	1.133	1.14
		$R(N-O)$	1.222	1.164	1.10
		$<(N-Mn-C)$	106.1	104.8	—
$(C_2H_5)NiNO$	C_{5v}	$R(Ni-C)$	2.084	2.211	2.11
		$R(Ni-N)$	1.420	1.571	1.626
		$R(C-C)$	1.420	1.424	1.43
		$R(N-O)$	1.271	1.165	1.165
		$R(C-H)$	1.078	1.067	1.09
$HMn(CO)_5$	C_{4v}	$R(Mn-C_{ax})$	1.725	1.960	1.823
		$R(Mn-C_{eq})$	1.717	1.982	1.823
		$R(Mn-H)$	1.628	1.684	1.50
		$R(C-O_{ax})$	1.162	1.137	1.139
		$R(C-O_{eq})$	1.163	1.137	1.139
		$<(H-Mn-C_{eq})$	72.3	82.3	83.6
		$<(Mn-C-O_{eq})$	171.2	172.8	—
$Mn(CO)_5CN$	C_{4v}	$R(Mn-C-CN)$	2.045		1.98
		$R(Mn-CO_{ax})$	1.825		1.822
		$R(Mn-CO_{eq})$	1.804		1.853
		$R(C-O_{eq})$	1.162		1.134
		$R(C-O_{ax})$	1.152		1.134
		$R(C-N)$	1.156		1.16
$Ni(C_3H_5)_2$	C_{2h}	$R(Ni-C)$	1.744	2.088	—
		$R(Ni-C_t)$	2.183	2.253	2.10
		$R(C_t-C)$	1.405	1.399	1.41
		$R(C_t-H)$	1.074	1.076	1.08
		$R(C-H)$	1.094	1.073	1.08
		$<(C_t-C-C_t)$	128.5	124.0	—

Source: [6.64].

Second, in the overwhelming majority of cases, the better basis set DZ yields better geometry parameter values. Third, in some cases significant discrepancies between the calculated and experimental interatomic distances remain. For instance, in the mixed carbonyls and nitrosyls the calculated M—N distance is 0.1 to 0.2 Å shorter, and the M—C bond 0.1 Å longer, than the experimental values. The calculated Fe—C distance in $Fe(CO)_2(NO)_2$ is completely wrong (by 0.40 Å longer than the experimental value). The Ni—C distance in bis(π-allyl)nickel is also incorrect.

Analysis [6.64] shows that these discrepancies are caused by the neglect of electron correlation effects in the HF calculations of the electronic structure. In particular, the short distance M—N in mixed carbonyls and nitrosyls can be corrected by including CI [6.98, 6.99]. The HF calculated electronic configuration in $Co(CO)_3NO$ is $[core](11e)^4(17a_1)^2(12e)^4$, where $11e$, $17a_1$, and $12e$ are the HOMOs, the former two ($11e$ and $17a_1$) being mainly of d origin (75% and 66%, respectively), while $12e$ represents the π bonding cobalt–nitrosyl.

By taking the CI into account, one has to include additional terms from excited configurations of the same symmetry, whose main term corresponds to the excitation of electrons from the $12e$ to the $13e$ MO. The latter is a π antibonding cobalt–nitrosyl MO, and hence its superposition to the ground state takes into account the electronic correlation along the M—N bond. The calculations show that the configuration $13e$ has a high weight (~ 0.52), and this explains directly the failure of the pure HF calculations: They do not include the antibonding MO $13e$ (which stands for the corresponding correlation effects), thus making the distance Co-N shorter than the experimental value.

Similar difficulties were encountered in calculations of other types of coordination compounds [6.100–6.105, 6.88]: $Ni(CO)_4$ [6.101], sandwich compounds (in ferrocene [6.102] the calculated metal–ligand distance is by 0.23 Å longer than the experimental value, whereas in manganocene [6.104] this distance is correct), $Fe(CO)_5$ [6.105], and so on. In most cases the calculated distances are longer than the observed values. In calculations including electron correlation effects these discrepancies were eliminated. For instance, the HF calculation of the distances to the two axial ligands in $Fe(CO)_5$ [6.88] in its bipyramidal configuration yields a value that is longer by 0.17 Å, than the experimental bond length, whereas for the equatorial distances there is no discrepancy. The calculation [6.105] including large-scale CI with all possible excited configurations (a total of 592,000 configurations compressed in a special way to 7750) and its comparison with the pure HF calculation [6.88] shows that the electron correlation effects increase the orbital charge transfers Δq_i both from the ligand to the metal along the σ bond and from the metal to the ligand on the π MO, thus improving the metal–ligand bond and decreasing its length (cf. Section 6.3). This result is quite understandable since the σ and π orbitals occupy different space regions and the expansion of the charge over a larger space decreases the interelectron repulsion.

This effect takes place for all the bonds, but in $Fe(CO)_5$ it is much stronger for axial ligands than for equatorial. Figure 6.29 illustrates this statement by means of electron deformation densities (Section 8.6) obtained by calculations with and without correlation effects in the planes of the axial (Fig. 6.29a) and equatorial (Fig. 6.29b) ligands [6.105]. It is seen explicitly that the electron correlation effects strongly increase the population of the d_{z^2} orbital of the iron atom at the expense of the 5σ orbitals of the axial CO ligands, as well as the $2\pi^*$ orbitals of the axial and equatorial CO ligands that overlap with the metallic $d_{xz}(d_{yz})$ orbitals (Fig. 6.29a); the population of the in-plane $2\pi^*$ MOs of the equatorial CO ligands is increased about four times less; the donation from the latter is almost unchanged (Fig. 6.29b). That is why, by including the correlation effects, the axial ligands experience a much larger reduction of the distance to the metal than do the equatorial ligands.

The conclusions from these examples have a general meaning. If the ligands possess active π orbitals, the competition between the tendency to form metal–ligand π bonds, which shortens the distance between them, and the opposite tendency to increase this distance, caused by the increasing repulsion on the σ bond for which the equilibrium distance is larger (Section 6.3), is resolved in favor of a shorter distance when the correlation effects are included, because the latter reduce the interelectron repulsion (see also the geometry calculations for $Fe(CO)_5$ and $Ni(CO)_4$ carried out by the X_α method [6.106]).

Density functional calculations compete with the MO LCAO methods in energy calculations — less so in geometries. Figure 6.30 shows the $1e$ one-electron wavefunction in the O—Mn—O plane of the tetrahedral MnO_4^- system obtained in one of the early X_α calculations [6.71]. From other coordination systems calculated by the X_α method, note $TiCl_4$, NiF_6^4, $CuCl_4^{2-}$, $PtCl_4^{2-}$, $Ni(CO)_4$, $Cr(CO)_6$, $Fe(CO)_5$, and others [6.71] (for more recent calculations of carbonils, see, e.g., [6.107]). One of the advantages of the X_α method is that it is applicable to solid-state clusters and polyatomic formations on surfaces. For instance, the cluster NiO_6^{10-} that models the Ni site in the NiO crystal was calculated considering explicitly all its 86 electrons and the influence of the (very strong in this case) crystal field of the environment (the Madelung potential) [6.108]. Besides the good agreement of the results with the photoelectron spectra [6.109], the calculations allow one to address the problem of adsorbed oxygen on metallic nickel surfaces. A comparison of the results of the calculations [6.108] with the photoelectron spectra of adsorbed oxygen and its concentration dependence shows that for sufficiently large concentrations the oxygen atoms penetrate the metal lattice, forming clusters NiO_6^{10-} on the surface [6.110].

The density functional calculations allow one to consider the effect of spin polarization, that is, the difference in the energies of orbitals with spin up and spin down (Section 5.4), which may be significant in complexes with unpaired electrons. For instance, for $MnCl_4^{2-}(d^5)$ and $CoCl_4^{2-}(d^7)$ even the consequence of the t_2, e, and a_1 energy levels is different for the orbitals with unpaired electrons with spins up and down [6.111]. For $CuCl_4^{2-}$ with T_d

(a)

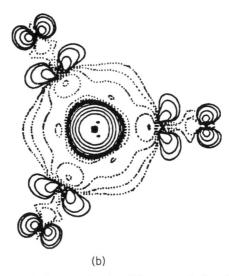

(b)

Figure 6.29. Fe(CO)$_5$ total electron density difference calculated *ab initio* with and without correlation effects: (*a*) in the plane containing the threefold axis; (*b*) in the equatorial plane. Dotted lines indicate negative values (domains of higher density of the SCF calculations without correlation effects). Contours are at 0.0001, 0.0002, 0.0010, 0.0020, 0.0050, 0.0100, 0.0200, and 0.0500 $e/(a.u.)^3$. Crosses indicate the positions of the nuclei; the iron atom is in the center of the plot. (After [6.105].)

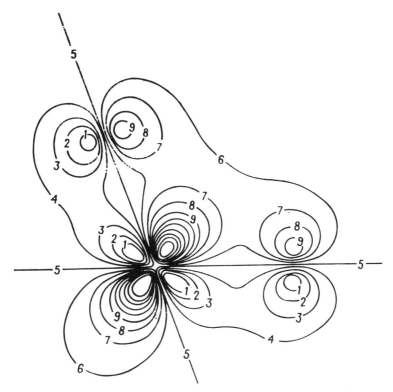

Figure 6.30. Contour map of the $1e$ one-electron wavefunction in the O—Mn—O plane of the tetrahedral MnO_4^- system obtained by SCF X_α-SW calculations. The contours are given with a spacing of 0.05 from -0.2 ($N = 1$) to $+0.2$ ($N = 9$). Along the $N = 5$ line there is the node of the π function. The Mn—O distance is 3.00 a.u. (From [6.71].)

symmetry, the orbitals with opposite spin orientation have the same order of consequence but different energies. More recent work illustrates the advantages of density functional methods' energy calculations of systems with increasing complexity. These include model systems for homogeneous and heterogeneous catalysis [6.112, 6.113], metal–ligand interactions on surfaces [6.114, 6.115], transition metal clusters [6.116], biological systems [6.117], and others.

6.5. RELATIVISTIC EFFECTS

The discussion of relativistic effects (REs) on chemical bonding is given in a separate section for two reasons. First, the topic is of special interest to transition metal coordination compounds. Second, as applied to molecular problems the area as a whole is relatively new. The relativistic Dirac equation (Section 2.1) was used for studies of atoms beginning in the 1930s, whereas its direct application to molecular bonding phenomena (MO theories) started in the late 1960s and early 1970s. This delay is due mainly to the complication of relativistic molecular calculations (Section 5.4), although some reluctance in the study of REs in molecules was also induced by the authoritative statement of Dirac that the REs are not important in chemical bonding, because the average speed, v, of valence electrons is small with respect to the speed of light, c.

In view of recent achievements in this area (see below), this remark is only partially true. Formally, v is small only when the electrons are moving in the valence area. However, s electrons (even-valence s electrons) spend a considerable amount of time near the nuclei, thus being influenced by relativity directly, and all the other electrons are also influenced, indirectly, because of the interelectron interactions. Again, the spin–orbital splitting of energy levels essential for any spectroscopic studies is a purely relativistic effect.

In this section a brief discussion of the main REs relevant to transition metal compounds is given. Relativistic atomic states are considered in Section 2.1, while relativistic approaches to electronic structure calculations are discussed in Sections 5.4 and 5.5. For some reviews, see [6.118–6.121].

Orbital Contraction and Valence Activity

To understand the origin of REs, let us begin with heavy-atom features. One of the most essential REs in atoms, important to chemical properties, is the *relativistic contraction of s and $p_{1/2}$ orbitals*. The Bohr radius of the 1s electron is $R_0 = 4\pi\hbar^2/me^2$, where m is the electron mass. The latter, following the theory of relativity, is dependent on its speed: $m = m_0[1 - (v/c)^2]^{-1/2}$, where m_0 is the rest mass. This means that the relativistic mass is larger than the rest mass of the electron, and hence the radius of its orbital motion becomes smaller, owing to the REs. In atomic units $c \cong 137$, and for 1s electrons $v = Z$, where Z is the atomic number. Thus, for example, for gold, $v/c = 79/137 = 0.58$, and its 1s electron radius R_0 is relativistically contracted by about 19%.

This relativistic orbital contraction of the inner 1s electrons modifies the orbital states of all the other electrons, including the outer valence electrons. Indeed, the higher s orbitals, although having a smaller average speed, must be orthogonal to the inner s states, and therefore their radial distribution undergoes a similar contraction (the p, d, f orbitals are orthogonal to the s orbitals by the angular parts). The calculations show that the relativistic contraction of the valence electrons due to the foregoing effects is, in general, not smaller (it can even be larger) than for the 1s electrons.

Figure 6.31 illustrates the relativistic contraction of the 6s shell of atoms from Cs to Fm obtained from numerical data of relativistic Dirac–Fock (Section 5.4) and Hartree–Fock calculations. Interestingly, it is seen from this figure that the largest relativistic contraction is inherent in the ground state of Au with the electronic configuration $5d^{10}6s^1$. It makes gold a unique element

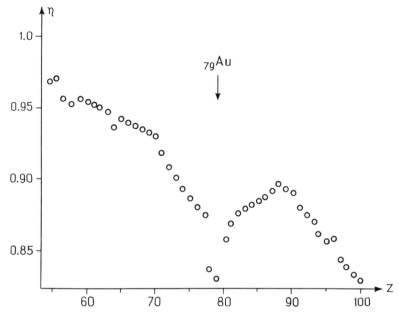

Figure 6.31. Relativistic contraction $\eta = \langle r \rangle_{rel}/\langle r \rangle_{nonrel}$ of the 6s shell in the elements from Cs ($Z = 55$) to Fm ($Z = 100$). (After [6.122].)

and explains its *redox nobility*. A similar relativistic radial contraction is expected for $p_{1/2}$ states (see below).

On the contrary, the d and f electrons are subject to a *radial expansion* and destabilization due to the RE. Indeed, the d and f electron densities are zero at the nuclei (Table 2.3 and Fig. 2.3). Hence these electrons are not expected to have relatively large speeds and direct relativistic reduction of their radii. On the other hand, they are automatically orthogonal to the s orbitals by symmetry (by the angular parts), and hence they are not subject to indirect contraction as are the valence s electrons. However, because of the relativistic contraction of the s and p shells, the d and f electrons become more efficiently screened, and hence they go up in energy and outward radially [6.123].

Thus the main RE in heavy atoms (besides spin–orbital splitting) is *the radial contraction and stabilization of s and $p_{1/2}$ electrons and the radial expansion and destabilization of d and f electrons*. These effects directly influence the possible electronic configurations and *valence states* of heavy atoms [6.124, 6.125]. Indeed, the stabilization of s shells and destabilization of d shells in transition metals obviously favors electronic configurations with more s than d contribution. In particular, this becomes important in comparison of ground-state configurations of the period 5 and period 6 elements in the same group. Table 6.15 illustrates such a comparison for groups 5–10.

As seen from this table, the period 6 elements have more s electrons in the electronic configuration of the free atom than the isoelectronic period 5

Table 6.15. Electronic configurations of the ground-state $4d$ and $5d$ element in groups 5 to 10 of the periodic table

	Group					
Row	5	6	7	8	9	10
5	Nb d^4s^1	Mo d^5s^1	Tc d^5s^2	Ru d^7s^1	Rh d^8s^1	Pd d^{10}
6	Ta d^3s^2	W d^4s^2	Re d^5s^2	Os d^6s^2	Ir d^7s^2	Pt d^9s^1

Source: [6.119].

elements, explicitly demonstrating the increasing relativistic contraction (stabilization) of the s states, compared with the d states, when passing to heavier atoms. In Tc the configuration is d^5s^2 (and hence it does not change when passing to Re) because of the additional stability of half-filled d^5 configurations. Similarly, for Pd, d^{10} is more stable than d^9s^1, whereas in Pt, d^9s^1 is more stable, due to the stronger s contraction.

Similar REs were noted in other atoms and ions [6.119]. For instance, the configuration of Mo$^+$ is $4d^5$, whereas that of W$^+$ is $5d^46s$; the ground state of Lr is $7s^27p^1_{1/2}$ and not $7s^26d^1$, as assumed earlier; the $^3P_0$ ground state of Pb is in 92.5% $(p_{1/2})^2$; and so on.

The change in the ground-state electronic configuration of atoms is definitely very important in their chemical behavior. First, *the formal valency of the atom changes* [cf. Ag(I) and Au(III)]. Although the physical sense of the notion of valency for coordination compounds and d orbital participation (three-dimensional bond delocalization) is far from its prime definition (Sections 1.2 and 6.1), the general understanding of chemical *valence activity* is certainly determined by the electronic configuration, and hence the REs change the valence activity.

A visual example is the difference in chemical behavior of Pd($4d^{10}$) and Pt($5d^96s^1$), especially in their catalytic activity. Consider, for instance, the following oxidative addition and reductive elimination reaction:

$$
\begin{array}{ccc}
\begin{array}{c} PH_3 \\ | \\ Pt \\ | \\ PH_3 \end{array}
& +
\begin{array}{c} H \\ | \\ H \end{array}
\longrightarrow
&
\begin{array}{c} H_3P \diagdown \diagup H \\ Pt \\ H_3P \diagup \diagdown H \end{array}
\end{array}
\qquad (6.5)
$$

I   II

It was shown [6.126] that the system I is favored by the configuration Pt $-3d^{10}$, while II requires $5d^96s^1$. Therefore, the barrier of the transformation (6.5) is low (2.34 kcal/mol). On the contrary, in the case of Pd, this reaction is unknown because Pd prefers the configuration $3d^{10}$ (Table 6.15).

Bond Lengths, Bond Energies, and Vibrational Frequencies

Relativistic effects in bond lengths are not determined entirely by the s orbital contraction and d and f orbital expansion discussed above, although both the former and latter influence the bonding features significantly. Indeed, the bond formation changes the charge distribution in the area between (and around) the bonding atoms, which is very much dependent on their orbitals' overlap and consequent changes in interelectron interaction effects. The attempt to correlate these effects with REs shows that relativistic contraction of orbitals and relativistic contraction of bond lengths are two parallel but largely independent effects [6.119, 6.127].

A more detailed analysis (see [6.120]) shows that the main effect of relativity is the reduction in the kinetic energy of the electrons involved in the bonding. In Section 5.4, general expressions for the contributions of REs to the bonding are given. In particular, as seen from the first of Eqs. (5.81), the first relativistic correction to the electron–core interaction H^0 in the quasi-relativistic approximation $\Delta T^R = -p^4/8m_0^3c^2$ is in fact a correction of the order of $(v/c)^2$ to the nonrelativistic kinetic energy $T = p^2/2m_0$ [note that the rigorous expression for the kinetic energy is $T^R = (p^2c^2 + m_0^2c^4)^{1/2} - m_0c^2$, and $p = m_0v$]. Being negative, this relativistic correction reduces the kinetic energy. The next term in (5.81), the Darvin interaction, is positive and increases the energy, but its value is smaller than the kinetic energy correction, the latter always being predominant. The third correction, the spin–orbital splitting, is discussed below.

The relativistic decrease of the kinetic energy is important to many chemical effects, including the bond length, bond energy, and vibrational frequency. It was shown [6.127] that the negative relativistic contribution increases with the decrease in interatomic distance, and hence *this RE stabilizes and contracts the chemical bond*. Note that this effect becomes important when the RE are strong enough to overcome the increase of the nonrelativistic kinetic energy with the decrease in interatomic distance. The latter is proportional to v^2, whereas the relativistic correction is v^4, and hence for sufficiently large speeds v, contraction and stabilization of the chemical bond take place.

The simplest heavy-atom compounds studied to confirm the foregoing REs are the metal hydrides MH. Table 6.16 illustrates some HFS (Hartree–Fock–Slater; Section 5.3) relativistic calculations of the M—H bond length R, dissociation energy D, and vibrational frequency ω in comparison with the corresponding HFS nonrelativistic (NR) calculations and experimental data from [6.128]. As one can see from this table, all the bonds become relativistically contracted and stabilized by moving toward heavier atoms, and this trend is confirmed by the corresponding experimental data. The REs are strong in Au and Hg; for them, the NR calculations are inadequate.

From general considerations it is known that all the REs increase as some powers of Z. For simple hydrides of metals of the 11, 13, and 14 groups, the bond contraction was shown to be proportional to Z^2 [6.129]:

$$R = R_{NR} - R_R = 17(6) \cdot 10^{-4}Z^2 \text{ pm} \tag{6.6}$$

Table 6.16. Comparison of relativistic (R) and nonrelativistic (NR) calculations of bond lengths R_0, dissociation energies D, and vibrational frequencies ω with experimental data (Exp) for some metal MH hydrides, MH^+ ions, and M_2 molecules

System	R_0 (Å)			D (kcal/mol)			ω (cm^{-1})		
	NR	R	Exp	NR	R	Exp	NR	R	Exp
Cu—H	1.51	1.50	1.46	59	61	66 ± 2	1884	1905	1940
Ag—H	1.71	1.61	1.61	39	47	53 ± 2	1605	1709	1760
Au—H	1.78	1.55	1.52	37	68	74 ± 3	1704	2241	2305
Zn—H$^+$	1.58	1.58	1.52	58	58	65 ± 9	1803	1810	1916
Cd—H$^+$	1.78	1.74	1.68	46	48	48 ± 9	1665	1669	1775
Hg—H$^+$	1.88	1.64	1.59	41	62	53 ± 10	1267	2156	2034
Cu$_2$	2.26	2.24	2.22	51	53	45 ± 2	268	274	266
Ag$_2$	2.67	2.52	—	40	47	37 ± 2	184	203	192
Au$_2$	2.90	2.44	2.47	27	58	52 ± 2	93	201	191

Source: After [6.120].

(1 pm $= 10^{-12}$ m $= 10^{-2}$ Å). For other groups the bond contraction is different and follows roughly the s orbital contraction (Fig. 6.31), going from a minimum for group 1, through a "gold maximum" at group 11, and decreasing further for higher groups. For more complex hydrides the bond contraction is similar. For instance, for CrH_6, MoH_6, and WH_6, ΔR (in pm) equals 0.6, 1.6, and 5.4, respectively [6.130].

Similar bond contraction and stabilization were revealed for metal diatomics Cu_2, Ag_2, and Au_2 (Table 6.16). The NR stability row $Cu_2 > Ag_2 > Au_2$ changes to $Cu_2 > Ag_2 < Au_2$ when the REs are taken into account. Similarly, the REs increase the bonding in W_2 by 25 kcal/mol, changing the NR inequality $Mo_2 > W_2$ to $W_2 > Mo_2$ [6.131]. This trend in the bond strength is observed in various organometallic compounds of groups 6 (Cr > Mo < W), 8 (Fe > Ru < Os), 9 (Co > Rh < Ir), 10 (Ni > Pd < Pt), and 11 (Cu > Ag < Au), and it is attributed to RE [6.132].

On the other hand, multiple metal–metal bonds in binuclear complexes of the type $M_2Cl_4(PR_3)_4$ and M_2X_6, where the bonding involves mostly nd orbitals rather than $(n + 1)s$ orbitals, are much less affected by relativistic corrections (the latter are on the order of 6 to 10 kcal/mol for the $5d$ homologies). In these cases, as well as in similar cases of metal–ligand σ bonds (with mostly d orbitals involved), the bonding energy is determined mainly by the orbital overlap, with minor influence from the REs.

Other examples of bond contraction by passing from $4d$ to $5d$ metals in MX_6 and MX_4 compounds are seen from the experimental data in Table 6.17. Note that the relativistic contractions $R = R_{NR} - R_R$ are much larger than $R(4d) - R(5d)$ since in the absence of REs the atomic radii of the $5d$ metals are larger than those of the $4d$ metals. It is seen from the table that by passing from complexes of the $4d$ triad Ru, Rh, and Pd, to the $5d$ triad Os, Ir, and Pt, with

Table 6.17. Some experimental interatomic distances in 4d and 5d metal MX_6 and MX_4 complexes (in pm)

Compound	M	$R(4d)$	$R(5d)$	$R(4d)-R(5d)$
MF_6	Ru, Os	187.75	183.3	4.7
MF_6	Rh, Ir	187.38	183.0	4.4
MF_6	Pd, Pt	187	182.9	4.1
MCl_4	Zr, Hf	232	231.6	0
MBr_4	Zr, Hf	246.5	245	1.5
MI_4	Zr, Hf	266.0	266.0	0

the same ligands, the bond shortens by about 4 pm (despite the considerable increase in the atomic radii) [6.133]. The experimental contraction for the Zr–Hf pairs is about zero, indicating a considerable relativistic contraction, but smaller than for the triad above.

In $M(CH_3)_2(PR_3)_2$ compounds with M = Pd, Pt, the Pt—C bond is by 3 pm longer, than the Pd—C bond, whereas the Pt—P bond is 4 pm shorter than Pd—P. This is difficult to explain within nonrelativistic ideas, while relativistic pseudopotential calculations reproduce this trend [6.134]. The Pt—P bond is more relativistic contracted than the Pt—C bond because the former is softer than the latter, and hence the same "contraction force" produces a larger effect on the softer Pt—P bond. For linear $M(PH_3)_2$ complexes the Pd—P interatomic distance is 241 pm, while for Pt—P it is 232 pm [6.134], demonstrating a strong relativistic contraction.

Several examples are available to illustrate REs in compounds involving actinides. Table 6.18 shows the results of quasi-relativistic calculations (Sections 5.4 and 5.5) of bond energies in compounds $AMCl_3$ with M = Th, U and A = H, CH_3, compared with corresponding nonrelativistic calculations and experimental data [6.120].

Table 6.18. Calculated relativistic (R) and nonrelativistic (NR) bond energies $D(M—A)$ in $AMCl_3$ complexes and experimental data (Exp) (in kcal/mol)

Compound	$D(M—A)$		
	NR	R	Exp
$HThCl_3$	30.1	76.0	~80
CH_3ThCl_3	35.8	79.8	~80
$HUCl_3$	10.5	70.1	76
CH_3UCl_3	16.8	72.2	72

Source: [6.120].

It is seen that the relativistic corrections here are very large (50 to 60 kcal/mol) and absolutely necessary to reproduce experimental data. On the other hand, the lanthanoid contraction, the contraction of the ionic radii in the lanthanoid series of isostructural crystals with the coordination number 8 by 18.3 pm when passing from La^{III} ($4f^0$) to Lu^{III} ($4f^{14}$), is shown to be only 10%, due to REs [6.119]. For other examples of relativistic bond contraction and stabilization, refer to the review articles [6.119–6.121, 6.132].

Comparative studies of electronic structure with and without REs demonstrating the significance of the latter have been carried out to reveal the origin of specific geometry and color in Bi(C$_6$H$_5$)$_5$ compared with Sb(C$_6$H$_5$)$_5$, P(C$_6$H$_5$)$_5$, and As(C$_6$H$_5$)$_5$ (semiempirical calculations) [6.135], as well as for BiPh$_5$, PbCl$_6{}^{2-}$, and WS$_4{}^{2-}$ (X_α-SW approximation) [6.136]. Nuclear quadrupole coupling and isomer shifts in neptunyl compounds were calculated in the relativistic extended Hückel approximation [6.137].

Interrelation Between Spin–Orbital Splitting and Bonding

Relativistic spin–orbital interaction corrections enter directly the matrix elements of the quasi-relativistic MO LCAO approximation in calculation of the electronic structure [Eqs. (5.81)] and indirectly in the basis set of the LCAO, which should be formed by relativistic atomic functions that follow the jj coupling scheme and double-group representations (Section 2.1; see also Section 5.5 for quasi-relativistic parametrization). In the jj scheme each electron is characterized by its momentum $\mathbf{j}_i = \mathbf{l}_i + \mathbf{s}_i$ ($\mathbf{l}_i$ and $\mathbf{s}_i$ are the orbital and spin momenta, respectively) with quantum numbers $j_i = l_i \pm \frac{1}{2}$; the total momentum of the atom is $\mathbf{J} = \Sigma_i \mathbf{j}_i$ with the quantum number $J = j_1 + j_2$, $j_1 + j_2 - 1, \ldots, |j_1 - j_2|$. Thus in the valence basis set the states $p_{1/2}$, $p_{3/2}$, $d_{3/2}$, $d_{5/2}$, and so on, should be employed in the relativistic calculations instead of the NR functions p, d, f.

However, the energy differences between $p_{1/2}$ and $p_{3/2}$, $d_{1/2}$ and $d_{3/2}$, and so on, are determined by spin–orbital splitting in the corresponding atoms. If this splitting is zero, the appropriate pairs of orbitals are degenerate and there is no sense to distinguish them [see Eq. (2.20)]. The same is true when the spin–orbital splitting is small, compared with the interatomic interaction, because in this case the two component functions may become largely intermixed.

In the case of strong spin–orbital interaction, the two atomic functions with the same orbital quantum number l but different $l + \frac{1}{2}$ and $l - \frac{1}{2}$ values ($l \neq 0$) also become rather independent from the chemical point of view. This statement is especially important for p functions. Consider an atom with one p electron. In the case of strong REs (large spin–orbital splitting), the corresponding relativistic states are $p_{1/2}(\frac{1}{2})$, $p_{3/2}(\frac{1}{2})$, $p_{1/2}(\frac{3}{2})$, and $p_{3/2}(\frac{3}{2})$, where the values $j = \frac{1}{2}, \frac{3}{2}$ are shown in parentheses, while their projections values $\frac{1}{2}, \frac{3}{2}$ are indicated as indices. The total momentum in the $p(\frac{1}{2})$ states ($j = \frac{1}{2}$) is smaller than that of nonrelativistic p electrons ($l = 1$) but larger than for s

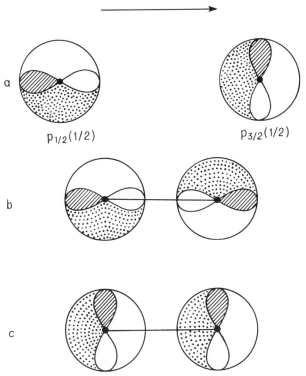

$p_{1/2}(1/2)$ $p_{3/2}(1/2)$

Figure 6.32. Schematic illustrations to the bonding between two atoms with relativistic $p(\frac{1}{2})$ orbitals: (*a*) visual presentation of the $p_{1/2}(\frac{1}{2})$ and $p_{3/2}(\frac{1}{2})$ orbitals; (*b*) $p_{1/2}(\frac{1}{2}) - p_{1/2}(\frac{1}{2})$ overlap representing $\frac{1}{3}$ σ bonding and $\frac{2}{3}$ π antibonding ($p_{1/2}(\frac{1}{2}) + p_{1/2}(\frac{1}{2})$ is $\frac{1}{3}$ σ antibonding and $\frac{2}{3}$ π bonding); (*c*) $p_{3/2}(\frac{1}{2}) + p_{3/2}(\frac{1}{2})$ overlap yielding $\frac{1}{3}$ π bonding and $\frac{2}{3}$ σ antibonding (the arrow indicates the axis of quantization).

electrons ($l = 0$), which have spherical-symmetric distribution. The $p(\frac{1}{2})$ state can be presented visually as having a spherical form (Section 2.1 and Fig. 6.32) which consists of the usual σ-like part (dashed) and a π-like contribution (dotted) that supplements the former to a sphere. The two functions $p_{1/2}(\frac{1}{2})$ and $p_{3/2}(\frac{1}{2})$ differ in their projections on the z axis, that is, by their orientation in space with respect to the axis of quantization.

If two $p_{1/2}(\frac{1}{2})$ functions from two bonding atoms overlap, they form one-third of σ bonding and two-thirds of π antibonding MOs (Fig. 6.32*b*), or vice versa for opposite signs: one-third σ antibonding and two-thirds π bonding. The two $p_{3/2}(\frac{1}{2})$ functions form one-third π bonding and two-thirds σ antibonding, or vice versa [6.138, 6.139]. The $p_{1/2}(\frac{3}{2})$ and $p_{3/2}(\frac{3}{2})$ functions form normal σ and π bonds.

It is evident that in any combination the bonding between two relativistic $p_{1/2}$ states is weaker than in the nonrelativistic case. This weakening is compensated when all the p states are occupied by electrons. Accordingly, the weakening does not occur when all the p states are degenerate or strongly mixed by external influence. Therefore, if not all the p states of the bonding atoms participate with their electrons in the bonding, that is, not all the bonding MOs originating from p states are occupied, *the spin–orbital splitting of p states weakens bonding in which $p(\frac{1}{2})$ states are involved.*

Several experimental facts can be attributed to this RE. For instance, in the series of isostructural (p "isoelectronic") compounds Sb_4, $BiSb_3, \ldots, Bi_4$, the dissociation energy decreases systematically from 9.04 to 6.03 eV [6.140]. Investigation of this RE is just beginning.

On the other hand, there is a back influence of the bonding on the spin–orbital splitting. Indeed, there are two reasons for changes (reduction) in the spin–orbital splitting by chemical bonding. First, the symmetry of the field in which the electron moves decreases and hence the orbital moment of the electron becomes, in general, reduced (Section 8.4); and second, when there is covalent bonding between a heavy atom and a light atom, the relativistic electron of the former becomes delocalized over the region including the light atom, and this delocalization reduces the speed of the electron [6.141].

Consider a heavy-atom (simple) hydride MH and assume that the bonding M—H is realized through overlap of the ns orbital of M (e.g., in the case of Au this orbital is $6s$) with the $1s$ orbital of H. Then the bonding orbital is $N(\psi_{ns}^M + \lambda\psi_{1s}^H)$, with λ as a covalency parameter and N as the normalization constant. The latter is less than unity, $N < 1$, and decreases with increase in λ. Since hydrogen is almost nonrelativistic, the weight of the relativistic wavefunction in the MO is reduced (times N), compared with the AO. In this way all the REs become reduced.

The most (easiest) observable reduction of REs is the spin–orbital splitting, since in principle it can be observed in both the atom and molecules. However, there may be some difficulties concerning the spectroscopic classification of energy levels in the relativistic case.

For some details, let us take as an example the complex $PtCl_6^{2-}$, with the Pt atom having a relatively moderate spin–orbital interaction magnitude (equal to $8.418 \cdot 10^3$ cm^{-1}). The quasi-relativistic (QR) calculations of its electronic structure were performed [6.141] in the semiempirical IEH (SCCC) approximation (Section 5.5), while the appropriate nonrelativistic (NR) calculations were carried out much earlier [6.142].

The comparison allows one to follow, at least qualitatively, how the spin–orbital splitting in the Pt atom is modified under the influence of the bonding. Figure 6.33 illustrates this effect. On the left-hand side (Fig. 6.33a) the $d_{3/2}$–$d_{5/2}$ spin–orbital splitting in the free Pt atom is shown. The crystal field of the ligands and bonding with the appropriate ligand orbitals results in the quasi-relativistic MO LCAO energy levels shown in Fig. 6.33b. Note that the relativistic classification of the MOs is given after the double-group representations (Section 3.6). By comparison, the corresponding nonrelativistic d orbital splitting into T_{2g} and E_g is also shown (Fig. 6.33c).

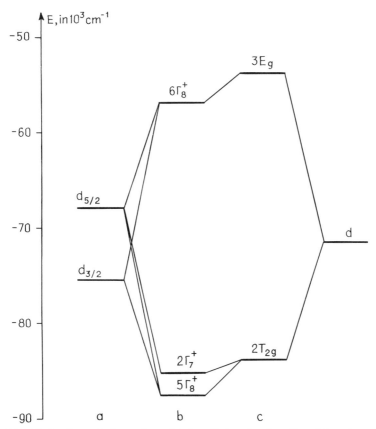

Figure 6.33. Reduction of the spin–orbital splitting by bonding (a) spin–orbital $d_{3/2}$–$d_{5/2}$ splitting in the free Pt atom; (b) MOs formed by these d orbitals with the corresponding ligand counterparts in $PtCl_6^{2-}$; (c) same MOs in the nonrelativistic calculation.

From these data one can see that as far as the MO energy levels under consideration are concerned, when passing from the NR to QR calculations, first, all the energy levels are lowered and second, the T_{2g} level splits into $5\Gamma_8^+$ and $2\Gamma_7^+$. While the former effect can be attributed to contraction and stabilization, discussed above, the latter is obviously due to spin–orbital interaction. However, its magnitude, $1.9 \cdot 10^3$ cm^{-1}, is much smaller than for spin-orbital splitting in the free atom, $8.418 \cdot 10^3$ cm^{-1}. Thus there is an obvious *reduction in spin–orbital splitting by bonding*. The two sources of this effect, mentioned above, are seen from the calculations explicitly: the octahedral crystal field splitting into E_g and T_{2g} mixes the $d_{3/2}$ and $d_{5/2}$ levels of the free atom, and hence the spin–orbital splitting $T_{2g} \rightarrow 5\Gamma_8^+ + 2\Gamma_7^+$ includes both the orbital reduction and covalent delocalization over nonrelativistic atoms (the REs in the Cl atoms are neglected [6.141]).

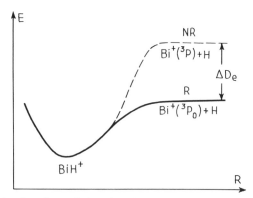

Figure 6.34. Reduction ΔD_e of the dissociation energy limit D_e in BiH$^+$ due to the spin–orbital splitting in the free ion Bi$^+$. In the NR case the dissociation results in Bi$^+$ in the state ^3P which is 2.3 eV higher than the relativistic ground state $^3P_0$.

Spin–orbital splitting in heavy atoms also reduces the dissociation limit D_e in diatomics BiH$^+$, Pb$_2$, PbO, PbH, and Tl$_2$ (for references, see [6.119]). Figure 6.34 explains the origin of this reduction. With zero spin–orbital splitting in the Bi$^+$ ion (i.e., in the NR approximation) the dissociation of BiH$^+$ into Bi$^+$ and H yields the Bi$^+$ ion in the ^3P state, whereas when taking account of the spin–orbital interaction, its ground-state energy $^3P_0$ is 2.3 eV lower. There are also cases when the REs increase the D_e values [6.119].

Other Relativistic Effects

Quite a number of relativistic (mostly quasi-relativistic) calculations of electronic structure and properties of heavy-atom coordination compounds performed to date are listed and reviewed in several publications [6.118–6.121, 6.132, 6.138]. Many works are concerned with photoelectron spectra and bonding. Relativistic and quasi-relativistic X_α-SW calculations for this purpose are reported, for instance, in [6.143, 6.144]. The effect of the central atom "inert lone pair" on the coordination compound stereochemistry, discussed in Section 9.2 (originally suggested by Sidgwick [6.145]), also has some important relativistic aspects. Indeed, since the s states are contracted and stabilized due to relativistic effects, the $(6s)^2$ inert pair effect should increase for heavy metals. This explains the tendency of the $6s^2$ electron pair to remain formally unoxydized in compounds of Tl(I), Pb(II), Bi(III), and so on [6.146]. However, the problem of stereochemistry induced by lone pairs is much more complicated, since it involves excited states via vibronic coupling (Section 9.2).

Basic influences of REs can easily be seen in spectroscopic and optical properties. First is the well-known spin–orbital splitting. One of the special optical properties induced by REs is the yellow color of gold [6.147]. The refractivity responsible for the color is due to the transition from the valence $5d$ band to the mainly $6s$ Fermi level [6.148]. Its sudden onset falls at $hv = 2.4$ eV in accordance with relativistic calculations. Nonrelativistic calculations move this absorption border much higher in energy, out of the visible region. Thus in the nonrelativistic approach, gold is white; only the relativistic effect makes it yellow. Note that for silver the corresponding relativistic rising of $5d$ orbitals and stabilization of the $6s$ states is much smaller than for gold; the absorption border occurs in the ultraviolet at 3.7 eV, making it white. The absorbtion itself is a solid-state effect (the transition $5d^9 6s \rightarrow 5d^8 6s^2$ in atoms is forbidden).

Relativistic calculations of f-element systems are reviewed in [6.149]; for relativistic calculations of solid-state systems, see [6.150–6.153].

REFERENCES

6.1. I. B. Bersuker, in: *Brief Chemical Encyclopedia* (Russ.) Vol. 5, Sovetskaya Entsiklopedia, Moscow, 1967, pp. 627–636

6.2. R. F. W. Bader and M. E. Stephen, *J. Am. Chem. Soc.* , **97**, 7391 (1975).

6.3. J. E. Lennard-Jones, *Proc. Roy. Soc. Ser. A*, **198**, 1–26, 1949

6.4. M. Gerloch, *Coord. Chem. Rev.*, **99**, 117–138 (1990).

6.5. I. B. Bersuker, *The Jahn–Teller Effect and Vibronic Interactions in Modern Chemistry*, Plenum Press, New York, 1964.

6.6. H. Kwart and K. King, *d-Orbitals in the Chemistry of Silicon, Phosphorus and Sulphur*, Springer-Verlag, Berlin, 1977, 220 pp.

6.7. D. V. Korol'kov, *Electronic Structure and Properties of Compounds of Nontransition Elements* (Russ.), Khimia, St. Petersburg, 1992, 312 pp.

6.8. W. Kutzelnigg, *Angew. Chem. Int. Ed. Engl.* , **23**, 272 (1984).

6.9. B. P. Martinenas, K. K. Eringis, and R. S. Dagis, in: *The Theory of Electronic Shells of Atoms and Molecules* (Russ.), Mintis, Vilnius, 1971, pp. 294–296.

6.10. R. E. Watson and A. J. Freeman, *Phys. Rev. Ser. A*, **134**, 1526–1546 (1964); E. Simanek and Z. Sroubek, *Phys. Status Solidi*, **4**, 251–259 (1964).

6.11. T. Koopmans, *Physica*, **1**, 104–113 (1933).

6.12. L. Pauling, *J. Chem. Soc.*, 1461 (1948); in: *Proceedings of the Symposium on Coordination Chemistry*, Copenhagen, 1953, p. 256.

6.13. M. J. C. Dewar, *Bull. Soc. Chim. Fr.*, 79 (1951).

6.14. J. Chatt and L. A. Duncanson, *J. Chem. Soc.*, 2939 (1953).

6.15. D. S. Marynick, *J. Am. Chem. Soc.*, **106**, 4064–4065 (1984).

6.16. M. R. A. Blomberg, U. B. Brandemark, P. E. M. Sieghbahn, K. B. Mathisen, and G. Karlström, *J. Phys. Chem.*, **89**, 2171 (1985).

6.17. I. B. Bersuker, *Zh. Struct. Khim.*, **4**, 461–463, (1963).

6.18. P. A. Christiansen and E. A. McCullogh, Jr., *J. Chem. Phys.*, **67**, 1877–1882, (1977).

6.19. C. W. Bauschlicher, Jr., L. G. M. Petterson, and E. M. Siegbahn, *J. Chem. Phys.*, **87**, 2129–2137 (1987).

6.20. C. Veeger and W. E. Newron, Eds., *Advances in Nitrogen Fixation*, Nijhoff, Boston, 1991.

6.21. H. Bash and D. Cohen, *J. Am. Chem. Soc.*, **105**, 3856–3860 (1983).

6.22. H. Itoh and A. B. Kunz, *Chem. Phys. Lett.*, **64**, 576–578 (1979).

6.23. G. M. Jeung, in: *Quantum Chemistry: The Challenge of Transition Metals and Coordination Chemistry*, NATO ASI Series C, Vol. 176, ed. A. Veilland, D. Reidel, Dordrecht, The Netherlands, 1986, pp. 101–117.

6.24. R. Arratia-Perez and C. Y. Yang, *J. Chem. Phys.*, **83**, 4005–4014 (1985).

6.25. D. E. Sherwood and M. B. Hall, *Inorg. Chem.*, **22**, 93–100 (1983).

6.26. Y. B. Koh and G. G. Christoph, *J. Am. Chem. Soc.*, **101**, 1422 (1979).

6.27. R. Hoffmann, *Solids and Surfaces: A Chemist's View of Bonding in Extended Structures*, VCH, New York, 1988.

6.28. S. Sung and R. Hoffmann, *J. Am. Chem. Soc.*, **107**, 578 (1985).

6.29. S. Sung, R. Hoffmann, and P. A. Thiel, *J. Phys. Chem.*, **90**, 1380–1388 (1986).

6.30. H. Bash, *J. Chem. Phys.*, **56**, 441–450 (1972).

6.31. J. N. Murrell and C. E. Scollary, *J. Chem. Soc. Dalton Trans.*, 1034 (1977)

6.32. S. Sakaki, *Theor. Chim. Acta*, **30**, 159 (1973).

6.33. P. J. Hay, *J. Am. Chem. Soc.*, **103**, 1390–1393 (1981).

6.34. A. A. Bagatur'ants, O. V. Gritsenko, and I. I. Moiseev, *Koord. Khim.*, **4**, 1779–1811 (1978).

6.35. T. A. Albright, R. Hoffmann, J. C. Thibeault and D. L. Thorn, *J. Am. Chem. Soc.*, **101**, 3801–3812 (1979).

6.36. F. A. Cotton and R. A. Walton, *Multiple Bonds Between Metal Atoms*, Clarendon Press, Oxford, 1993.

6.37. W. C. Trogler, *J. Chem. Educ.*, **57**, 424–427 (1980).

6.38. M. Moscowitz, Ed., *Metal Clusters*, Wiley, New York, 1986.

6.39. G. P. Kostikova and D. V. Korol'kov, *Usp. Khim.*, **54**, 591–618 (1985).

6.40. F. A. Cotton, N. F. Curtis, C. B. Harris, B. F. G. Johnson, S. J. Lippard, J. T. Mague, W. R. Robinson and J. S. Wood, *Science*, **145**, 1305 (1964); F. A. Cotton, *Inorg. Chem.*, **3**, 334 (1964); F. A. Cotton, in: *Perspectives in Coordination Chemistry*, ed. A. F. Williams, C. Floriani, and A. E. Merbach, VCH, New York, 1992, pp. 321–332.

6.41. V. G. Kuznetsov, P. A. Koz'min, *Conference on Crystal Chemistry*, Abstracts, Stiintsa, Kishinev, 1961, pp. 74–75; *Zh. Struct. Khim.*, **4**, 55 (1963); F. A. Cotton and C. B. Harris, *Inorg. Chem.*, **4**, 330 (1965).

6.42. J. G. Norman, Jr., and H. J. Kolari, *J. Chem. Soc. Chem. Commun.*, 303 (1974); *J. Am. Chem. Soc.*, **97**, 33 (1975).

6.43. A. P. Mortola, J. W. Moskowitz, and N. Roch, *Int. J. Quantum Chem.*, Symp. 8, 161, (1974); A. P. Mortola, J. W. Moskowitz, N. Roch, C. D. Cowman, and H. B. Gray, *Chem. Phys. Lett.*, **32**, 283 (1964).

6.44. M. Bernard, *J. Am. Chem. Soc.*, **100**, 2354 (1978).

6.45. M. B. Hall, in: *Quantum Chemistry: The Challenge of Transition Metals and Coordination Chemistry*, NATO ASI Series, Vol. 176, ed. A. Veillard, D. Reidel, Dordrecht, The Netherlands, 1986, pp. 391–401.

6.46. A. P. Kl'agina and A. A. Levin, *Koord. Khim.*, **10**, 579 (1984).

6.47. B. Delley, A. J. Freeman, and D. E. Ellis, *Phys. Rev. Lett.* **50**, 488–492 (1983).

6.48. A. P. Kl'agina, G. L. Gutsev, V. D. Lutatskaya, and A. A. Levin, *Zh. Neorg. Khim.*, **29**, 2834 (1984).

6.49. F. A. Cotton, *Acc. Chem. Res.*, **2**, 240 (1969).

6.50. L. Dubicki and R. L. Martin, *Inorg. Chem.*, **9**, 673 (1970).

6.51. J. G. Norman and H. J. Kolari, *J. Am. Chem. Soc.* **100**, 791 (1978).

6.52. B. E. Bursten and F. A. Cotton, *Inorg. Chem.*, **20**, 3042 (1981).

6.53. I. B. Bersuker and Yu. G. Titova, *Teor. Eksp. Khim.*, **6**, 469–478 (1970).

6.54. P. S. Braterman, *Struct. Bonding*, **10**, 57 (1972).

6.55. D. L. Thorn and R. Hoffmann, *Inorg. Chem.*, **17**, 126 (1978); R. H. Summerville and R. Hoffmann, *J. Am. Chem. Soc.*, **101**, 3821 (1979).

6.56. H.-J. Freund and G. Hohlneicher, *Teor. Chim. Acta*, **51**, 145 (1979); H.-J. Freund, B. Dick, and G. Hohlneicher, *Teor. Chim. Acta*, **57**, 181 (1980).

6.57. A. A. Low, K. L. Kunze, P. J. MacDougall and M. B. Hall, *Inorg. Chem.*, **30**, 1079 (1991); **32**, 3880 (1993).

6.58. A. Rosa and E. J. Baerends, *New J. Chem.*, **15**, 815 (1991); R. Wiest, A. Strich, and M. Benard, *New J. Chem.*, **15**, 801 (1991).

6.59. B. J. Morris-Sherwood, C. B. Powell, and M. B. Hall, *J. Am. Chem. Soc.*, **106**, 5079 (1984); A. L. Sargent and M. B. Hall, *J. Am. Chem. Soc.*, **111**, 1563 (1989).

6.60. I. H. Hillier and V. R. Saunders, *Mol. Phys.*, **22**, 1025 (1971).

6.61. J. Demuynck and A. Veillard, *Theor. Chim. Acta*, **28**, 241 (1973).

6.62. M. Coutiere, J. Demuynck, and A. Veillard, *Theor. Chim. Acta*, **27**, 281–287 (1972).

6.63. D. W. Turner, in: *Physical Methods in Advanced Inorganic Chemistry*, ed. H. A. O. Hill and P. Day, Interscience, Oxford, 1968, pp. 79–115

6.64. M. F. Guest, in: *Supercomputer Simulation in Chemistry*, Lecture Notes in Chemistry, Vol. 44, ed. M. Dupuis, Springer-Verlag, Heidelberg, 1986, pp. 98–129.

6.65. J. Schirmer and L. S. Cederbaum, *J. Phys. B*, **11**, 1889 (1978); J. Schirmer, L. S. Cederbaum, and O. Walter, *Phys. Rev.* **28**, 1237 (1983).

6.66. I. M. Hillier, in: *Quantum Chemistry: The Challenge of Transition Metals and Coordination Chemistry*, ed. A. Veillard, D. Reidel, Dordrecht, The Netherlands, 1986, pp. 143–157.

6.67. A. Veillard, *Chem. Commun.*, 1022 (1969); 1427 (1969).

6.68. M.-M. Rohmer and A. Veillard, *J. Chem. Soc. Chem. Commun.*, No. 7, 250–251 (1973).

6.69. C. D. Batich, *J. Am. Chem. Soc.*, **98**, 7585 (1976).

6.70. K. H. Johnson, *Adv. Quantum Chem.*, **7**, 143–185 (1973); *Annu. Rev. Phys. Chem.*, **26**, 39–57 (1975).

6.71. K. H. Johnson, J. C. Norman, Jr., and J. W. D. Connolly, in: *Computational Methods for Large Molecules and Localized States in Solids*, ed. F. Herman, A. D. McLean, and R. K. Nesbet, Plenum Press, New York, 1973, pp. 161–201.

6.72. P. Decleva, G. Fronzoni, and A. Lisini, in: *Density Functional Methods in Chemistry*, ed. J. K. Labanowski and J. W. Andzelm, Springer-Verlag, New York, 1991, pp. 323–336.

6.73. C. Daul, H.-H. Gudel, and J. Weber, *J. Chem. Phys.*, **98**, 4023–4029 (1993).

6.74. D. R. Salahub, *Adv. Chem. Phys.*, **69**, 447 (1987).

6.75. D. A. Brown, W. I. Chambers, and N. I. Fitzpatric, *Inorg. Chim. Acta Rev.*, **6**, 7–30 (1972).

6.76. P. G. Burton, *Coord. Chem. Rev*, **12**, 37–71 (1974).

6.77. E. J. Baerends and P. Ros, *Mol. Phys.*, **30**, 1735–1747 (1975).

6.78. A. Veillard, 15th International Conference on Coordination Chemistry, Moscow, 1973, Section Lecture (preprint), 20 pp.

6.79. J. A. Tossel and W. N. Lipscomb, *J. Am. Chem. Soc.*, **94**, 1505–1517 (1972).

6.80. J. W. Moskowitz, C. Hollister, C. J. Hornback, and H. Bash, *J. Chem. Phys.*, **53**, 2570–2580 (1970).

6.81. J. Demuynk, A. Veillard, and U. Wahlgren, *J. Am. Chem. Soc.*, **95**, 5563–5574 (1973).

6.82. J. A. Connor, I. H. Hillier, V. R. Saunders, M. H. Wood, and M. Barber, *Mol. Phys.*, **24**, 497–509 (1972).

6.83. H. Johansen, *Chem. Phys. Lett.*, **17**, 569–573 (1972).

6.84. P. D. Dacre and H. Elder, *J. Chem. Soc. Dalton Trans.*, No. 13, 1426–1432 (1972).

6.85. J. Fischer, A. Veillard, and R. Weiss, *Theor. Chim. Acta*, **24**, 317–333 (1972).

6.86. M. B. Hall, M. F. Guest, and I. H. Hillier, *Chem. Phys. Lett.*, **15**, 592–593 (1972).

6.87. B. L. Kalman and W. Richardson, *J. Chem. Phys.*, **55**, 4443–4456 (1971).

6.88. J. Denmynck, A. Strich, and A. Veillard, *Nouv. J. Chim.*, **1**, 217 (1977).

6.89. D. Reinen and C. Friebel, *Inorg. Chem.*, **23**, 791 (1984); D. Reinen and M. Atanasov, *Chem. Phys.*, **136**, 27–46 (1989).

6.90. R. Akeson, L. G. M. Petterson, M. Sandstrom, and U. Wahlgren, *J. Phys. Chem.*, **96**, 150–156 (1992).

6.91. A. Veillard, A. Dediew, and M.-M. Rohmer, in: *Horizons of Quantum Chemistry*, ed. K. Fukui and P. Pullman, D. Reidel, Dordrecht, The Netherlands, 1980, pp. 197–225.

6.92. K. Ohno, in: Horizons of Quantum Chemistry, ed. K. Fukui and P. Pullman, D. Reidel, Dordrecht, The Netherlands, 1980, pp. 245–266.

6.93. H. Kashiwagi and S. Obara, *Int. J. Quantum Chem.*, **20**, 843–859 (1981).

6.94. M.-M. Rohmer, A. Dedieu, and A. Veillard, *Chem. Phys.*, **77**, 449–462 (1983).

6.95. F. U. Axe, L. Chantranupong, A. Waleh, J. Collins and G. H. Loew, in: *The Challenge of d and f Electrons: Theory and Computation*, ACS Series 394, American Chemical Society, Washington, D.C., 1989, pp.339–355.

6.96. A. Strich and A. Veillard, *Theor. Chim. Acta*, **60**, 379–383 (1981); *Nouv. J. Chim.*, **7**, 347–352 (1983).

6.97. R. L. Williamson and M. B. Hall, *Int. J. Quantum Chem.*, Symp. 21, 503–512 (1987).

6.98. R. F. Fenske and J. R. Jensen, *J. Chem. Phys.*, **71**, 3374 (1979).

6.99. M. F. Guest, I. H. Hillier, A. A. McDowel, and M. Berry, *Mol. Phys.*, **41**, 519 (1980).

6.100. T. E. Taylor and M. B. Hall, *Chem. Phys. Lett.*, **114**, 338 (1985).

6.101. K. Faegri and J. Almlöf, *Chem. Phys. Lett.*, **107**, 121 (1984).

6.102. H. P. Lüthi, J. H. Ammeter, J. Almöf, and K. Faegri, Jr., *J. Chem. Phys.*, **77**, 2002 (1982).

6.103. J. Almlöf, K. Faegri, Jr., B. E. R. Schilling, and H. P. Lüthi, *Chem. Phys. Lett.*, **106**, 266 (1983).

6.104. K. Faegri, Jr., J. Almlöf, and H. P. Lüthi, *J. Organomet. Chem.*, **249**, 303, (1983).

6.105. H. P. Lüthi, P. E. M. Siegbahn, and J. Almlöf, *J. Phys. Chem.*, **89**, 2156–2161 (1985).

6.106. N. Rösch, H. Jörg, and B. I. Dunlap, in: *Quantum Chemistry: The Challenge of Transition Metals and Coordination Chemistry*, NATO ASI Series C, Vol.176, ed. A. Veillard, D. Reidel, Dordrecht, The Netherlands, 1986, pp. 179–187.

6.107. T. Ziegler and V. Tschinke, in: *Density Functional Methods in Chemistry*, ed. J. K. Labanowski and J. W. Andzelm, Springer-Verlag, New York, 1991, pp. 139–154.

6.108. J. W. D. Connoly, *Solid State Commun.*, **12**, 313–316 (1973).

6.109. G. K. Wertheim and S. Hufner, *Phys. Rev. Lett.*, **28**, 1028–1031 (1972).

6.110. R. P. Messmer, C. W. Tucker, and K. H. Johnson, *Surf. Sci.*, **42**, 341–354 (1974).

6.111. M. Barber, J. Clark, A. Hinchliffe, and D. S. Urch, *J. Chem. Soc. Faraday Trans.*, Part 2, **74**, 681–687 (1978).

6.112. D. R. Salahub, in: *Metal–Ligand Interaction: From Atoms, to Clusters, to Surfaces*, ed. D. R. Salahub and N. Russo, Kluwer, Dordrecht, The Netherlands, 1992, pp. 311–340.

6.113. N. Russo, in: *Metal–Ligand Interation: From Atoms, to Clusters, to Surfaces*, ed. D. R. Salahub and N. Russo, Kluwer, Dordrecht, The Netherlands, 1992, pp. 341–366.

6.114. T. Ziegler, D. R. Salahub and N. Russo, Kluwer, Dordrecht, The Netherlands, 1992, pp. 367–396.

6.115. A. J. Freeman, S. Tang, S. H. Chou, Ye Ling, and B. Delley, in: *Density Functional Methods in Chemistry*, ed. J. K. Labanowski and J. W. Andzelm, Springer-Verlag, New York, 1991, pp. 61–75.

6.116. L. Noodelman, D. A. Case, and E. J. Baerends, in: *Density Functional Methods in Chemistry*, ed. J. K. Labanowski and J. W. Andzelm, Springer-Verlag, New York, 1991, pp. 109–123.

6.117. L. Noodleman and D. A. Case, *Adv. Inorg. Chem.*, **38**, 423–470 (1992); E. I. Solomon, F. Tuczek, D. E. Root, and C. A. Brown, *Chem. Rev.*, **94**, 827–856 (1994).

6.118. D. A. Case, *Annu. Rev. Phys. Chem.*, **33**, 151–171 (1982).

6.119. P. Pyykko, *Chem. Rev.*, **88**, 563–594 (1988).

6.120. T. Ziegler, J. G. Snijders, and E. J. Baerends, in: *The Challenge of d and f Electrons: Theory and Computation*, ed. D. R. Salahub and M. C. Zerner, ACS Symposium Series 394, Aerican Chemical Society, Washington, D.C., 1989, pp. 322–338.

6.121. G. L. Malli, Ed., *Relativistic Effects in Atoms, Molecules and Solids*, G. L. Malli, Ed., Plenum Press, New York, 1983.

6.122. P. Pyykko and J. P. Desclaux, *Acc. Chem. Res.*, **12**, 276 (1979).

6.123. R. G. Boyd, A. C. Larson, and J. T. Waler, *Phys. Rev.*, **129**, 1629 (1963).

6.124. B. Fricke, *Struct. Bonding*, **21**, 89 (1975).

6.125. D. M. Bylander and L. Kleiman, *Phys. Rev. Lett.*, **51**, 889 (1983).

6.126. J. J. Low and W. A. Goddard III, *J. Am. Chem. Soc.*, **106**, 6928 (1984); **108**, 6115 (1986); Organometallics, **5**, 609 (1986).

6.127. P. Pyykko, J. G. Snijders, and E. J. Baerends, *Chem. Phys. Lett.*, **83**, 432 (1981).

6.128. K. S. Krasnov, V. S. Timoshinin, T. G. Danilova, and S. V. Khandozhko, *Handbook of Molecular Constants of Inorganic Compounds*, Israel Program for Scientific Translations, Jerusalem, 1970.

6.129. P. Pyykko, *J. Chem. Soc. Faraday Trans.2*, **75**, 1256 (1979).

6.130. P. Pyykko and J. P. Desclaux, *Chem. Phys.*, **34**, 261 (1978).

6.131. T. Ziegler, V. Tschinke, and A. Becke, *Polyhedron*, **6**, 685 (1987).

6.132. P. Pyykko, in: *The Effects of Relativity in Atoms, Molecules and the Solid State*, ed. S. Willson, I. P. Grant, and B. L. Gyorffy, Plenum Press, New York, 1991.

6.133. A. V. Dzhalavyan, E. G. Rakov, and A. S. Dudin, *Russ. Chem. Rev. (Engl. Transl.)*, **52**, 960 (1983).

6.134. J. M. Wisner, T. J. Bartczak, J. A. Ibers, J. J. Low, and W. A. Goddard, *J. Am. Chem. Soc.*, **108**, 347 (1986).

6.135. A. Schmuck, P. Pyykko, and K. Seppelt, *Angew. Chem. Int. Ed. Engl.*, **29**, 213–215 (1990).

6.136. B. D. El-Issa, P. Pyykko, and H. M. Zanati, *Inorg. Chem.*, **30**, 2781–2787 (1991).

6.137. P. Pyykko and J. Jove, *New J. Chem.*, **15**, 717–720 (1991).

6.138. K. S. Pitzer, *Acc. Chem. Res.*, **12**, 271 (1979).

6.139. K. S. Pitzer, *J. Chem. Phys.*, **63**, 1032 (1975); *Chem. Commun.*, 760 (1975).

6.140. D. Schield, R. Pflaum, K. Sattler, and E. Recknagel, *J. Phys. Chem.*, **91**, 2649 (1987).

6.141. I. B. Bersuker, S. S. Budnikov, and B. A. Leizerov, *Int. J. Quantum Chem.*, **11**, 543–559 (1977).

6.142. F. A. Cotton and C. D. Harris, *Inorg. Chem.*, **6**, 369 (1967).

6.143. R. Arratia-Perez and G. L. Mali, *Chem. Phys. Lett.*, **125**, 143 (1986); *J. Chem. Phys.*, **84**, 5891 (1986); F. Zuloaga, R. Arratia–Perez, and G. L. Mali, *J. Phys. Chem.*, **90**, 4491 (1986).

6.144. H. Chermette and A. Goursot, in: *Local Density Approximation in Quantum Chemistry and Solid State Physics*, ed. J. P. Dahl and J. Avery, Plenum Press, New York, 1984, pp. 635–642.

6.145. N. V. Sidgwick, *The Covalent Link in Chemistry*, Cornell University Press, Ithaca, N.Y., 1933.

6.146. M. L. Cohen and V. Heine, *Solid State Phys.*, **24**, 37 (1970)

6.147. S. Kupratakaln, *J. Phys. C: Solid State Phys.*, **3**, S109 (1970).

6.148. N. E. Christensen and B. O. Seraphin, *Phys. Rev. B*, **4**, 3321 (1971).

6.149. P. Pyykko, *Inorg. Chim. Acta*, **139**, 243–245 (1987).

6.150. G. Trinquier and R. Hoffmann, *J. Phys. Chem.*, **88**, 6696 (1984).

6.151. L. Visscher, P. J. C. Aerts, and W. C. Nieuwpoort, *J. Chem. Phys.*, **96**, 2910–2919 (1992); L. Visscher and W. C. Nieuwpoort, *Theor. Chim. Acta*, **88**, 447–472 (1994).

6.152. M. Dolg, H. Stoll, and H. Preuss, *J. Chem. Phys.*, **90**, 1730–1734 (1989); W. Kuechle, M. Dolg, and H. Preuss, *J. Chem. Phys.*, **100**, 7535 (1994).

6.153. T. R. Cundari and W. Stevens, *J. Chem. Phys.*, **98**, 5555–5565 (1993).

7 Electronic Control of Nuclear Configuration; Vibrations and Vibronic Coupling

Electronic control of molecular shape and nuclear motions via vibronic coupling is one of the current problems in theoretical chemistry which has many applications.

In the fast development of quantum chemistry the motions of the nuclei were paid much less attention than were those of the electrons. Meanwhile, many important chemical phenomena, including chemical transformations, are determined by nuclear displacements induced by electron rearrangements. Any changes in molecular systems under external influences begin with alterations in the less inertial electronic structure, which affects the heavy nuclear framework via vibronic coupling. The latter thus plays an important role in the description of molecular properties.

7.1. MOLECULAR VIBRATIONS

Adiabatic Approximation

The idea of molecular vibrations is based on the assumption that the nuclear and electronic motions can be separated. For stable systems and in the absence of electronic degeneracy or pseudodegeneracy, this assumption can be proved in the *adiabatic approximation*. In view of its general importance, we consider this approximation in some detail, and disclose the criterion of its validity (for more details, see, e.g., [7.1–7.6]).

The adiabatic approximation is based on the strong inequality of the masses and velocities of electrons and nuclei. Since the nuclear mass M is about 2000 times that of the electron m, the velocity of the latter is much greater than that of the former. Therefore, it can be assumed that at every instant position of the nuclei, the electronic distribution is stationary. In other words, because of the relatively slow motions of the nuclei, the electronic state is in time to relax instantly to the changing nuclear positions, and the motions of the nuclei are determined by the averaged field of the electrons. It means that the electron distribution in space is determined by the nuclear coordinates but not by their speed.

This assumption enables us to solve the problem in two stages: (1) to ignore the nuclear motions when solving the electronic part of the problem, and (2) to use the mean electronic energy in a given electronic state as the potential energy for the nuclear motions. The second point in this procedure ignores the nonadiabatic changes in the electronic structure under nuclear displacements, and this is the most restricting part of the adiabatic approximation.

Let us refer to a more rigorous consideration. The Schrödinger equation (1.5) for the system as a whole, including n electrons and N nuclei, can be written as follows:

$$H\Psi(r, Q) = E\Psi(r, Q) \tag{7.1}$$

Divide the total Hamiltonian H into three parts:

$$H = H_r + H_Q + V(r, Q) \tag{7.2}$$

where H_r is the electronic part, including the kinetic energy of the electrons and the interelectron electrostatic interaction, H_Q the kinetic energy of the nuclei, $H_Q = -\Sigma_\alpha(\hbar^2/2M_\alpha)\Delta_\alpha$ (hereafter the index α refers to nuclear coordinates Q_α), and $V(r, Q)$ the energy of the interaction of the electrons with the nuclei plus the internuclei repulsion (r and Q denote the entire set of coordinates of the electrons r_i, $i = 1, 2, \ldots, n$, and nuclei Q_α, $\alpha = 1, 2, \ldots, N$, respectively):

$$V(r, Q) = -\sum_{i,\alpha} \frac{e^2 Z_\alpha}{|\mathbf{r}_i - \mathbf{R}_\alpha|} + \frac{1}{2}\sum_{\alpha,\beta}' \frac{e^2 Z_\alpha Z_\beta}{|\mathbf{R}_\alpha - \mathbf{R}_\beta|} \tag{7.3'}$$

Here $\mathbf{R}_\alpha$ are the vector coordinates of the nuclei. Their relation to the Q_α coordinates is clarified below.

The operator $V(r, Q)$ can be expanded in a series of small displacements of the nuclei about the point $Q_\alpha = Q_{\alpha 0} = 0$ chosen as the origin:

$$V(r, Q) = V(r, 0) + \sum_\alpha \left(\frac{\partial V}{\partial Q_\alpha}\right)_0 Q_\alpha + \frac{1}{2}\sum_{\alpha,\beta} \left(\frac{\partial^2 V}{\partial Q_\alpha \partial Q_\beta}\right)_0 Q_\alpha Q_\beta + \cdots \tag{7.3}$$

Considering the first term of this expansion as the potential energy of the electrons in the field of fixed nuclei, one can solve the electronic part of the Schrödinger equation, the *electronic equation*

$$[H_r + V(r, 0) - \varepsilon_k']\varphi_k(r) = 0 \tag{7.4}$$

and obtain a set of energies ε_k' and wavefunctions $\varphi_k(r)$ for the given nuclear configuration corresponding to point $Q_{\alpha 0}$. To see how these solutions vary under nuclear displacements, the full Schrödinger equation (7.1) must be solved. The total wavefunction $\Psi(r, Q)$ can be sought in the form of an expansion over the set of electronic functions $\varphi_k(r)$:

$$\Psi(r, Q) = \sum_k \chi_k(Q)\varphi_k(r) \tag{7.5}$$

where the expansion coefficients $\chi_k(Q)$ depend on the nuclear coordinates. Substituting (7.5) into Eq. (7.1), we obtain, after some simple transformations, the following infinite system of coupled equations for the functions $\chi_k(Q)$ (the prime at the sum means that the term with $m = k$ is excluded):

$$[H_Q + \varepsilon_k(Q) - E]\chi_k(Q) + {\sum_m}' W_{km}(Q)\chi_m(Q) = 0 \tag{7.6}$$

where $W_{km}(Q)$ denotes the electronic matrix element of the *vibronic interaction* W, that is, the part of the electron–nuclear interaction $V(r, Q)$ that depends on Q:

$$W(r, Q) = V(r, Q) - V(r, 0) = \sum_\alpha \left(\frac{\partial V}{\partial Q_\alpha}\right)_0 Q_\alpha + \frac{1}{2} \sum_{\alpha,\beta} \left(\frac{\partial^2 V}{\partial Q_\alpha \partial Q_\beta}\right)_0 Q_\alpha Q_\beta + \cdots \tag{7.7}$$

and

$$\varepsilon_k(Q) = \varepsilon'_k + W_{kk}(Q) \tag{7.8}$$

is the *adiabatic potential* (AP) equal to the potential energy of the nuclei in the mean field of the electrons in the state $\varphi_k(r)$, provided that the electronic state under consideration is not degenerate or pseudodegenerate (see below).

It is seen from the system of coupled equations (7.6) that if the vibronic mixing of different electronic states is ignored [$W_{km}(Q) = 0$ for $k \neq m$], the coupling between these states vanishes and the system of equations decomposes into a set of simple equations for given k:

$$[H_Q + \varepsilon_k(Q) - E]\chi_k(Q) = 0 \tag{7.9}$$

Each of these equations represents the Schrödinger equation for the nuclei that move in the mean field of the electrons in the state $\varphi_k(r)$, the *equation of nuclear motions*. In other words, the motions of the nuclei and electrons are separated and the problem as a whole can be solved in the two stages mentioned above. In the first stage the electronic states $\varphi_k(r)$ are determined as solutions of Eq. (7.4) and used to calculate the potential energy of the nuclei $\varepsilon_k(Q)$ by Eq. (7.8). In the second stage the wavefunctions $\chi_k(Q)$ and energies E of the nuclei are evaluated after Eq. (7.9), the total wavefunction being $\Psi(r, Q) = \varphi_k(r)\chi_k(Q)$. This is the *simple adiabatic approximation* or *Born–Oppenheimer approximation*.

Thus the simple adiabatic approximation is valid if and only if the terms of the vibronic mixing of different electronic states in Eq. (7.6) can be ignored. It can be shown (see, e.g., [7.1–7.5]) that this is possible if the energy spacing of the vibrations, the vibrational quanta $\hbar\omega$, are much smaller than the electronic energy term spacing:

$$\hbar\omega \ll \varepsilon'_k - \varepsilon'_j \tag{7.10}$$

This inequality can be considered as a *criterion of the adiabatic approximation*; it remains the same in more rigorous treatments [7.1]. If criterion (7.10) is fulfilled, the error introduced by the simple adiabatic approximation is on the order of $(m/M)^{1/2} \sim 2.3 \cdot 10^{-2}$, which is sufficiently small.

In a fuller treatment of the adiabatic approximation, all the terms of $V(r, Q)$ from Eq. (7.3) are introduced into the electronic equation (7.4):

$$[H_r + V(r, Q) - \varepsilon_k(Q)]\varphi_k(r, Q) = 0 \tag{7.11}$$

Here the solutions $\varepsilon_k(Q)$ and $\varphi_k(r, Q)$ depend on Q as parameters, and in the system of coupled equations (7.6) the $W_{km}(Q)$ are replaced by matrix elements of the operator of nonadiabacity, which is also determined by the vibronic interactions W. In this version of the adiabatic approximation, separation of the motions of the electrons and nuclei is ultimately determined by the same type of simplification as in the previous one—by ignoring the vibronic mixing of different electronic states.

However, in detail this approach, called *the adiabatic approximation (or full adiabatic approximation)*, is somewhat different from the Born–Oppenheimer (or simple adiabatic) approximation. The adiabatic approximation, if applicable [i.e., when criterion (7.10) is satisfied], gives more exact results than the simple approach. Some estimates of orders of magnitudes that reveal the measure of the adiabatic approximation may be obtained using the *order of smallness* (or *parameter of smallness*) $a = (m/M)^{1/4} \approx 0.15$. It can be shown [7.5] that the terms neglected in the simple adiabatic approximation, *provided that the criterion (7.10) is satisfied*, are on the order of a, while those of the full adiabatic approximation are on the order of $a^3 \approx 3 \cdot 10^{-3}$, and the ratio of averaged velocities of the nuclei and electrons is on the order of $a^3 \approx 3 \cdot 10^{-3}$. Thus the simple adiabatic approximation is less accurate than the full one.

However, when criterion (7.10) is *not* satisfied, Eq. (7.6), based on the simple adiabatic approximation, is a more suitable starting point to consider the role of electronic states in the origin of special nuclear nonadiabatic (and/or nonvibrational) motions.

The adiabatic approximation is of great importance to chemistry. *Without the adiabatic approximation even the notion of nuclear configuration (molecular shape) loses its usual sense.* Therefore, when a specific molecular configuration is considered, the validity of the adiabatic approximation is implied. However, in many cases this approximation is not valid.

Normal Coordinates and Harmonic Vibrations

For stable molecular systems in nondegenerate states, Eq. (7.9) for the nuclear motions describes molecular vibrations. Indeed, a stable system means that the AP $\varepsilon(Q)$ has a minimum at a certain point $Q = Q_0$; the solution of Eq. (7.9) with such a potential, as shown below, yields nuclear vibrations at this point.

Significant simplification of Eq. (7.9) allowing for its direct solution is reached in the harmonic approximation by means of *normal coordinates*. With N nuclei, (7.9) is an equation of $3N$ coordinates $\mathbf{R}_\alpha(X_\alpha, Y_\alpha, Z_\alpha)$, $\alpha = 1, 2, \ldots, N$; after excluding the coordinates describing the rotation and translation of the system as a whole, it transforms into an equation of $3N - 6$ variables. If the AP as a function of these variables has only one absolute minimum at $\mathbf{R}_\alpha = \mathbf{R}_{\alpha 0}$, the function $\varepsilon(\mathbf{R}_\alpha)$ near this point can be approximated by a paraboloid (quadratic function of R_α). This is the *harmonic approximation*.

In the harmonic approximation the function $\varepsilon(\mathbf{R}_\alpha)$, being a quadratic form of the variables, can be reduced to the canonical form, which means that the variable $\mathbf{R}_\alpha$ can be transformed (by unitary transformations, Section 3.1) such that in the new variables there are only quadrates and no crossing terms in the potential. In other words, the possibility of reducing the quadratic function $\varepsilon(\mathbf{R}_\alpha)$ to its canonical form means that there are new coordinates Q_α, instead of $\mathbf{R}_\alpha(X_\alpha, Y_\alpha, Z_\alpha)$, such that in the new coordinates

$$\varepsilon = \tfrac{1}{2} \sum_\alpha K_\alpha Q_\alpha^2 \tag{7.12}$$

that is, ε does not contain crossing terms of the type $Q_\alpha Q_\beta$, the constant K_α being discussed below. If in the new coordinates realizing the canonical form of the AP (7.12), the operator of kinetic energy of the nuclei $-\Sigma_\alpha (\hbar^2/2M_\alpha)\partial^2/\partial Q_\alpha^2$ is also an additive function of Q_α (i.e., it has no crossing terms), the Q_α coordinates are normal coordinates.

Thus in normal coordinates Q_α, Eq. (7.9) decomposes into $3N - 6$ equations, each being the *equation of the harmonic oscillator*

$$-\frac{\hbar^2}{2M_\alpha}\frac{d^2\chi_\alpha}{dQ_\alpha^2} + \frac{1}{2}\omega_\alpha^2 Q_\alpha^2 \chi_\alpha = E_\alpha \chi_\alpha \tag{7.13}$$

where M_α is the *reduced mass* of the normal vibration α and ω_α is its frequency. Quantum mechanics (e.g., [7.6]) gives a direct solution of this equation with eigenvalues:

$$E_{n\alpha} = \hbar\omega_\alpha(n_\alpha + \tfrac{1}{2}) \qquad n_\alpha = 0, 1, 2, \ldots \tag{7.14}$$

and eigenfunctions $\chi_{n\alpha}(Q_\alpha)$ (the index α is omitted for simplicity):

$$\chi_n(Q) = \left(\frac{a}{2^n n!\, \pi^{1/2}}\right)^{1/2} \exp\left(-\frac{a^2 Q^2}{2}\right) H_n(aQ) \tag{7.15}$$

where $H_n(y)$ is a Hermitian polynomial,

$$H_n(y) = (-1)^n \exp(y^2) \frac{d^n \exp(-y^2)}{dy^n} \tag{7.16}$$

and

$$a = \left(\frac{M\omega}{\hbar}\right)^{1/2} \tag{7.17}$$

In particular, for the ground state $n = 0$,

$$E_0 = \frac{\hbar\omega}{2} \tag{7.18}$$

$$\chi_0(Q) = \left(\frac{M\omega}{\hbar\pi}\right)^{1/4} \exp\left(-\frac{M\omega Q^2}{2\hbar}\right) \tag{7.19}$$

It is important that ω_α is the same frequency as in mechanical vibration with the potential energy $\frac{1}{2}M_\alpha\omega_\alpha^2 Q_\alpha^2$ of the assumed harmonic approximation. This means that ω_α is a solution of the equation of mechanical vibrations:

$$\frac{d^2 Q_\alpha}{dt^2} + \omega_\alpha^2 Q_\alpha^2 = 0 \tag{7.20}$$

In other words, if one approximates the AP by a quadratic term $\varepsilon(Q) = \frac{1}{2}K_\alpha Q_\alpha^2$ (i.e., a parabola that corresponds to the harmonic approximation) then, taking $K_\alpha = M_\alpha\omega_\alpha^2$, one obtains directly the frequency of normal vibrations ω_α. To do this, knowledge of the normal coordinates Q_α, the curvature K_α of the AP at the point of the minimum in the direction Q_α, and the reduced mass M_α is needed.

The methods of determination of normal coordinates are described in detail in manuals on vibrational spectra (e.g., [7.7–7.9]). In coordination compounds with high local symmetry the group-theoretical rules considered in Section 3.4 can be very useful. Indeed, the Hamiltonian in the Schrödinger equation (7.9) for the nuclear motions has the same symmetry as the molecular framework. Therefore, its eigenvalues and eigenfunctions transform as (belong to) the irreducible representations (types of symmetry) of the group of symmetry of the system. In the harmonic approximation, Eq. (7.13) describes normal vibrations (vibrations in normal coordinates), and hence the irreducible representations of the group determine the possible types of symmetry of these vibrations, the shape of the normal coordinates, the degeneracy of the vibrational frequencies, and so on. Using the methods of the theory of symmetry (Section 3.4), one can relatively easily determine these characteristics of the vibrations.

Figures 7.1 to 7.3 show the form of the most important normal vibrations of octahedral, tetrahedral, triangular, and square-planar complexes, while

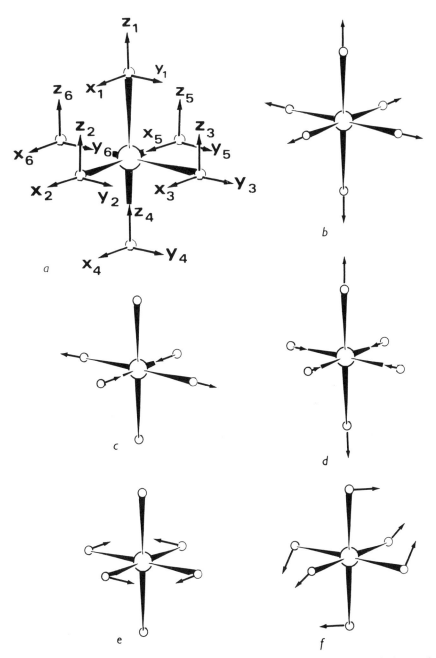

Figure 7.1. Shapes of symmetrized displacements of atoms in an octahedral complex ML_6: (*a*) numbering and orientation of Cartesian displacements; (*b*) totally symmetric A_{1g}, (*c*) E_g-type Q_ε, (*d*) E_g-type Q_θ, and (*e*) T_{2g}-type Q_ξ displacements. For degenerate displacements, any linear combination of them can be realized, e.g., (*f*) $(Q_\zeta + Q_\eta + Q_\xi)/\sqrt{3}$ for T_{2g}. (Note that the ligand local coordinates are chosen different from those of Fig. 5.1.)

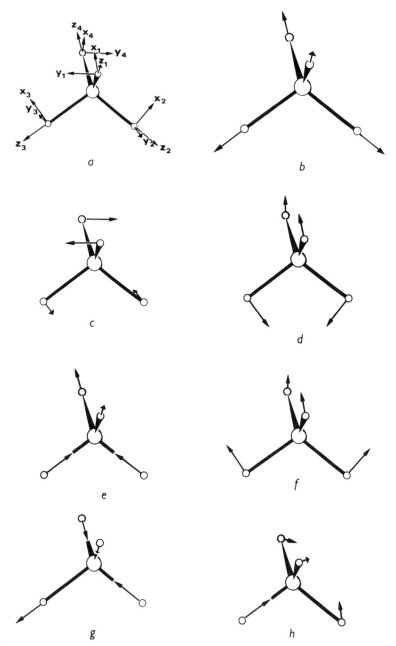

Figure 7.2. Shapes of symmetrized displacements of atoms in a tetrahedral complex: (*a*) numbering and orientation of Cartesian coordinates; (*b*) totally symmetric A_1, (*c*) E-type Q_ε, (*d*) E-type Q_∂; (*e*) T_2-type Q_ξ, and (*f*) T_2'-type Q_ξ^* displacements. In the case of degeneracy any combination of component displacements can be realized [e.g., (*g*)$(Q_\zeta + Q_\eta + Q_\xi)/\sqrt{3}$; (*h*)$(Q_\zeta^* + Q_\eta^* + Q_\xi^*)/\sqrt{3}$].

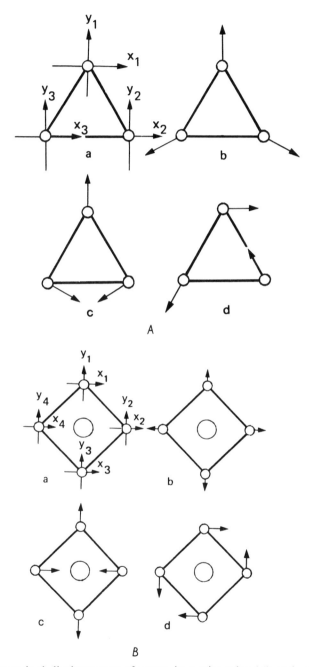

Figure 7.3. Symmetrized displacements of atoms in a triangular (A) and square (B) molecular systems: (a) labeling of Cartesian displacements; (b) totally symmetric displacements of the type A_1; (c) E'-type Q_y in (A) and B_{1g} in (B) displacements; (d) E'-type Q_x in (A) and B_{2g} in (B) displacements.

Table 7.1 gives the expressions of normal coordinates Q_α in Cartesian coordinates X_α, Y_α, Z_α.

Degenerate vibrational modes (vibrations that belong to irreducible representations E, T,...) have the same frequency; similar to the wavefunctions of degenerate states, each normal coordinate of the degenerate set is undefined in the sense that any of their linear combinations is also a normal coordinate. In particular, the shape of twofold $E(Q_2, Q_3)$ (called tetragonal) and threefold $T_2(Q_4, Q_5, Q_6)$ (called trigonal) vibrations given in Table 7.1 and Figs. 7.1 and 7.2, although most usable, are conventional within any linear combination of the two and three degenerate vibrations, respectively.

If there are two or more normal coordinates of the same symmetry (e.g., T_2' and T_2'' in a tetrahedron; Table 7.1), they interact and mix (similar to the case of terms with the same symmetry; Section 4.3). Hence their frequencies are no more independent and the corresponding coordinates are *symmetrized but not normal*. To obtain the normal vibrations, an appropriate perturbation problem must be solved to diagonalize the matrix of interacting coordinates [7.7–7.9] [cf. Eq. (4.55)].

Table 7.1. Symmetrized displacements Q (normal coordinates) expressed by Cartesian coordinates (Figs. 7.1 to 7.3) for some trigonal, tetragonal, tetrahedral, and octahedral systems[a]

Q	Symmetry Type	Transformation Properties	Expressions by Cartesian Coordinates
		Trigonal Systems X_3, Symmetry D_{3h}	
$Q_x^{(t)}$	E'	x	$(X_1 + X_2 + X_3)/\sqrt{3}$
$Q_y^{(t)}$	E'	y	$(Y_1 + Y_2 + Y_3)/\sqrt{3}$
$Q_{a2}^{(r)}$	A_2'	S_z	$(2X_1 - X_2 - \sqrt{3}\,Y_2 - X_3 + \sqrt{3}\,Y_3)/\sqrt{12}$
Q_a	A_1'	$x^2 + y^2$	$(2Y_1 + \sqrt{3}\,X_2 - Y_2 - \sqrt{3}\,X_3 - Y_3)/\sqrt{12}$
Q_x	E'	$2xy$	$(2X_1 - X_2 + \sqrt{3}\,Y_2 - X_3 - \sqrt{3}\,Y_3)/\sqrt{12}$
Q_y	E'	$x^2 - y^2$	$(2Y_1 - \sqrt{3}\,X_2 - Y_2 + \sqrt{3}\,X_3 - Y_3)/\sqrt{12}$
		Square-Planar Systems ML_4, Symmetry D_{4h}	
Q_a	A_{1g}	$x^2 + y^2$	$\frac{1}{2}(Y_1 + X_2 - Y_3 - X_4)$
Q_1	B_{1g}	$x^2 - y^2$	$\frac{1}{2}(Y_1 - X_2 - Y_3 + X_4)$
Q_2	B_{2g}	xy	$\frac{1}{2}(X_1 + Y_2 - X_3 - Y_4)$
Q_a'	A_{2g}	S_z	$\frac{1}{2}(X_1 - Y_2 - X_3 + Y_4)$
Q_x	E_{1u}	x	$\frac{1}{2}(X_1 + X_2 + X_3 + X_4)$
Q_y	E_{1u}	y	$\frac{1}{2}(Y_1 + Y_2 + Y_3 + Y_4)$
Q_x'	E_{1u}'	x	X_0
Q_y'	E_{1u}'	y	Y_0

Table 7.1. (*Continued*)

Q	Symmetry Type	Transformation Properties	Expressions by Cartesian Coordinates
		Tetrahedral Systems ML_4, *Symmetry* T_d	
Q_a	A_1	$x^2 + y^2 + z^2$	$\frac{1}{2}(Z_1 + Z_2 + Z_3 + Z_4)$
Q_ϑ	E	$2z^2 - x^2 - y^2$	$\frac{1}{2}(X_1 - X_2 - X_3 + X_4)$
Q_ε	E	$\sqrt{3}(x^2 - y^2)$	$\frac{1}{2}(Y_1 - Y_2 - Y_3 + Y_4)$
Q'_ξ	T'_2	x, yz	$\frac{1}{2}(Z_1 - Z_2 + Z_3 - Z_4)$
Q'_η	T'_2	y, xz	$\frac{1}{2}(Z_1 + Z_2 - Z_3 - Z_4)$
Q'_ζ	T'_2	z, xy	$\frac{1}{2}(Z_1 - Z_2 - Z_3 + Z_4)$
Q''_ξ	T''_2	x, yz	$\frac{1}{4}(-X_1 + X_2 - X_3 + X_4) +$ $(\sqrt{3}/4)(-Y_1 + Y_2 - Y_3 + Y_4)$
Q''_η	T''_2	y, xz	$\frac{1}{4}(-X_1 - X_2 + X_3 + X_4) +$ $(\sqrt{3}/4)(Y_1 + Y_2 - Y_3 - Y_4)$
Q_ξ	T''_2	z, xy	$\frac{1}{2}(X_1 + X_2 + X_3 + X_4)$
Q_x	T_2	x	X_0
Q_y	T_2	y	Y_0
Q_z	T_2	z	Z_0
		Octahedral Systems ML_6, *Symmetry* O_h	
Q_a	A_{1g}	$x^2 + y^2 + z^2$	$(X_2 - X_5 + Y_3 - Y_6 + Z_1 - Z_4)/\sqrt{6}$
Q_ϑ	E_g	$2z^2 - x^2 - y^2$	$(2Z_1 - 2Z_4 - X_2 + X_5 - Y_3 + Y_6)/2\sqrt{3}$
Q_ε	E_g	$\sqrt{3}(x^2 - y^2)$	$\frac{1}{2}(X_2 - X_5 - Y_3 + Y_6)$
Q_ξ	T_{2g}	yz	$\frac{1}{2}(Z_3 - Z_6 + Y_1 - Y_4)$
Q_η	T_{2g}	xz	$\frac{1}{2}(X_1 - X_4 + Z_2 - Z_5)$
Q_ζ	T_{2g}	xy	$\frac{1}{2}(Y_2 - Y_5 + X_3 - X_6)$
Q'_x	T'_{1u}	x	$\frac{1}{2}(X_1 + X_3 + X_4 + X_6)$
Q'_y	T'_{1u}	y	$\frac{1}{2}(Y_1 + Y_2 + Y_4 + Y_5)$
Q'_z	T'_{1u}	z	$\frac{1}{2}(Z_1 + Z_3 + Z_5 + X_6)$
Q''_x	T''_{1u}	x	$(X_2 + X_5)/\sqrt{2}$
Q''_y	T''_{1u}	y	$(Y_3 + Y_6)/\sqrt{2}$
Q''_z	T''_{1u}	z	$(Z_1 + Z_4)/\sqrt{2}$
Q_x	T_{1u}	x	X_0
Q_y	T_{1u}	y	Y_0
Q_z	T_{1u}	z	Z_0
Q'_ξ	T_{2u}	$x(y^2 - z^2)$	$\frac{1}{2}(X_3 + X_6 - X_1 - X_4)$
Q'_η	T_{2u}	$y(z^2 - x^2)$	$\frac{1}{2}(Y_1 + Y_4 - Y_2 - Y_5)$
Q'_ζ	T_{2u}	$z(x^2 - y^2)$	$\frac{1}{2}(Z_2 + Z_5 - Z_3 - Z_6)$

[a] X_0, Y_0, Z_0 are the Cartesian displacements of the central atom; S_z is an axial vector.

Table 7.2. Classification of symmetrized displacements Γ for several types of molecules with N atoms

N^a	Symmetry	Example, Shape	Γ
4(6)	C_{3v}	NH_3, pyramid	A_1', A_1'', E', E''
5(9)	T_d	MnO_4^-, tetrahedron	A_1, E, T_2', T_2''
7(15)	O_h	CrF_6^{3-}, octahedron	$A_{1g}, E_g, T_{2g}, T_{2u}, T_{1u}', T_{1u}''$
7(15)	D_{4h}	MA_4B_2, tetragonally distorted octahedron	$A_{1g}, A_{1g}'', A_{2u}', A_{2u}'', B_{1g}, B_{2g},$ $B_{2u}, E_g, E_u, E_u', E_u''$
9(21)	O_h	CsF_8, cub	$A_{1g}, A_{2u}, E_g, E_u, T_{2u}, T_{1u}',$ $T_{1u}'', T_{2g}', T_{2g}''$

^{a}The number of normal vibrations $3N - 6$ is indicated in parentheses.

Not all the types of symmetry of the group can be realized as types of vibrations. For instance, in the octahedral O_h system there are no $A_{1u}, A_{2u}, A_{2g}, E_u, T_{1g}$ vibrations; in the tetrahedron there are no A_2, T_1 vibrations; and so on. The types of vibrations symmetries allowed can be established relatively easily by means of group-theoretical rules (Section 3.4). Table 7.2 lists some results on classification of the normal vibrations on symmetry for some most usable types of coordination system.

In conclusion, it is necessary to emphasize that the idea of normal vibrations is based essentially on the harmonic approximation, in which cubic and higher terms in the expansion of the AP in a power series can be neglected. This implies that the AP has one (deep) minimum; at its bottom the amplitude of vibrations is small, provided that only low-vibrational states are populated (low temperatures). Such a potential, in turn, requires that there should be no close-in-energy terms that interact by vibronic coupling. The deviation from the harmonic approximation in the nuclear motions controlled by the electronic structure, *the vibronic anharmonicity*, is discussed in Section 7.4. For anharmonicity corrections to the vibrational frequencies, see [7.7–7.12]).

Special Features of Vibrations in Coordination Systems

The prediction of the number of possible frequencies and shapes of vibrations based on symmetry properties of the system is the more informative the higher the symmetry. This circumstance forms the basis of the qualitative identification of infrared (IR) and Raman spectra of coordination compounds. The higher the symmetry, the easier the identification of the spectra and the analysis of the electronic structure based on these spectra.

For instance, in a regular octahedral system with seven atoms there is a total of $3 \cdot 7 - 6 = 15$ vibrations, which, following Table 7.2, are divided into one A_{1g} vibration, two E_g, and four types of threefold vibrations: $T_{2g}, T_{2u}, T_{1u}', T_{1u}''$. Since degenerate vibrations have the same frequency, only six different frequencies are expected in the vibrational spectra of such systems. For systems with an inversion center, selection rules impose that odd vibrations may be observed in the IR absorption, while even vibrations manifest themselves in the Raman scattering of light. Therefore, three frequencies (T_{2u}, T'_{1u}, and T''_{1u}) are seen in the IR spectra and the other three (A_{1g}, E_g, and T_{2g}) in the Raman spectra [7.9–7.12].

For further analysis and identification it may be useful to consider the splitting of degenerate frequencies in fields of lower symmetry. These splittings are quite similar in nature to the term splitting discussed in Chapter 4. Sections 3.4 and 4.2 illustrate how to determine the splitting of high-symmetry terms (vibrations) under perturbations of lower symmetry, and Tables 4.2 and 4.3 provide some results. If one knows how the degenerate frequencies split in the fields of lower symmetry, vibrational spectroscopy can be used to study the influence of the environment on (or ligand substitution in) the coordination system. Special applications of external uniaxial stress and its influence on the IR line splitting are also employed.

Another important feature of IR spectra is related to the three-dimensional center-delocalized coordination bond (Section 6.1). This delocalization makes the metal–ligand bonds nonspecific and nontransferrable. It means that in the presence of different ligands there is no way to discriminate individual metal–ligand vibrations and to consider that their frequencies remain the same (even approximately) by passing to another complex whth the same metal–ligand bond but with other ligands changed. The three-dimensional delocalization of the bonding electrons makes all the metal–ligand bonds collectivized, and the vibrations, as a rule, are related to the system as a whole. This is one of the main distinctions of vibrational spectroscopy as applied to coordination compounds compared with organic compounds, for which the vibrational frequencies of given atomic bonds are approximately constant.

Finally, the vibrations in coordination compounds are characterized by a great variety of frequencies, ranging from infrared to several tens of cm^{-1}. Detailed analyses of IR and Raman vibrational spectra of transition metal compounds are given in special monographs [7.8–7.12].

7.2. VIBRONIC COUPLING

Vibronic Constants

As stated above, molecular changes under external influence begin with alterations in the less inertial electronic shells, while molecular transformations (and many other properties) are determined by the changes in nuclear motions. *The bridge from electronic to nuclear motions is conveyed by vibronic coupling* [7.1–7.4].

The expression for the operator of vibronic coupling W is given in Eq. (7.7). It describes the interaction of the electrons with the nuclear displacements from the initial configuration, hereafter called the *reference configuration*. The latter is usually taken as the molecular configuration of the stable ground state or any other high-symmetry configuration (see the discussion in [7.1]). With the reference configuration known, one can use symmetrized or normal coordinates in the expansion of $V(r, Q)$ in Q [Eq. (7.3)], which simplifies further treatment.

In Bethe's notations the f-fold-degenerate irreducible representation Γ_i has f rows, denoted by γ. For instance, the twofold-degenerate representation $\Gamma_3 = E$ has two lines $\gamma = \theta, \varepsilon$, while the threefold representation $\Gamma_5 = T_2$ has three lines $\gamma = \xi, \eta, \zeta$. Denoting the symmetrized coordinates by $Q_{\Gamma\gamma}$, we can rewrite the operator of vibronic coupling (7.7) in these coordinates (the reference configuration is taken at $Q_{\Gamma\gamma} = 0$):

$$W(r, Q) = \sum_{\Gamma\gamma} \left(\frac{\partial V}{\partial Q_{\Gamma\gamma}}\right)_0 Q_{\Gamma\gamma} + \frac{1}{2} \sum_{\Gamma\gamma} \sum_{\Gamma'\gamma'} \left(\frac{\partial^2 V}{\partial Q_{\Gamma\gamma} \partial Q_{\Gamma'\gamma'}}\right)_0 Q_{\Gamma\gamma} Q_{\Gamma'\gamma'} + \cdots \quad (7.21)$$

Consider the meaning of the coefficients of this expansion which are derivatives of the electron–nuclear and nuclear–nuclear interaction V. The matrix elements of these coefficients are *the constants of vibronic coupling*, or *vibronic constants*. They are very important in vibronic interaction effects: *Vibronic constants characterize the measure of coupling between the electronic states and nuclear displacements.* In other words, the vibronic constants characterize the influence of the nuclear displacements on the electron distribution or, conversely, the effect of the changes in the electronic structure on the nuclear configuration and dynamics. By means of vibronic constants, the problem of *how electrons control molecular configurations* can be formulated and solved approximately.

Denote the electronic states by the appropriate irreducible representations $\Gamma, \Gamma', \ldots$ of the symmetry group of the molecular system and assume first that states Γ and Γ' are not degenerate. The matrix element

$$F_{\Gamma*}^{(\Gamma\Gamma')} = \langle \Gamma | \left(\frac{\partial V}{\partial Q_{\Gamma*}}\right)_0 | \Gamma' \rangle \quad (7.22)$$

is called the *linear vibronic constant*. Following the rules of group theory, $F_{\Gamma*}^{(\Gamma\Gamma')}$ is nonzero if and only if $\Gamma \times \Gamma' = \Gamma*$. If Γ or Γ', or both, are degenerate (in this case $\Gamma*$ may also be degenerate), a set of linear vibronic constants corresponding to all the lines γ and γ' of the two representations Γ and Γ' and their combinations $F_{\Gamma*\gamma*}^{(\Gamma\gamma\Gamma'\gamma')}$ should be introduced instead of one vibronic constant for nondegenerate states. This can be done easily if we take into account that the matrix elements within a degenerate term differ solely in numerical coefficients (Section 3.4).

Some of the linear vibronic constants have a clear-cut physical meaning. *The diagonal constant of the linear coupling $F_{\Gamma*\gamma*}^{(\Gamma\gamma\Gamma\gamma)} \equiv F_{\Gamma*\gamma*}^{\Gamma\gamma}$ has the sense of the force* *with which the electrons in state $\Gamma\gamma$ affect the nuclei in the direction of symmetrized displacements $Q_{\Gamma*_\gamma*}$.* For instance, $F_{E\theta}^{(E\varepsilon)}$ means the force with which the electrons in the $E\varepsilon$ state distort the nuclear configuration in the direction of the $E\theta$ displacements (see Fig. 7.1).

For degenerate states Γ, according to the group-theoretical rules, the diagonal matrix element $F_{\Gamma*}^{\Gamma}$ is nonzero if the symmetrical product $[\Gamma \times \Gamma]$ contains $\Gamma*$: $\Gamma* \in [\Gamma \times \Gamma]$ (compare with the condition for off-diagonal elements $\Gamma* \in \Gamma \times \Gamma'$). For nondegenerate states, $[\Gamma \times \Gamma] = \Gamma \times \Gamma = A_1$, where A_1, is the totally symmetric representation. It follows that in non-degenerate states $\Gamma* = A_1$, and the electrons can distort the nuclear configuration only in the direction of totally symmetric displacements, for which the symmetry of the system does not change. If the electronic state Γ is degenerate, the symmetrical product $[\Gamma \times \Gamma]$ contains nontotally symmetric (along with symmetrical) representations. Indeed, for cubic symmetry $[E \times E] = A_1 + E$, $[T \times T] = A_1 + E + T_1 + T_2$; for D_{4h} symmetry $[E \times E] = A_1 + B_1 + B_2$; and so on. In these cases Γ may be nontotally symmetric (degenerate E, T_1, T_2, or nondegenerate B_1, B_2, etc.). Thus under the influence of the electrons the nuclear configuration undergoes distortions that are not totally symmetric. It is just these distortions that are predicted by the Jahn–Teller theorem discussed in Section 7.3.

The quadratic (or second-order) vibronic constants can be introduced similarly to the linear constants, although there are some complications. The second derivatives in Eq. (7.21) (which are terms of the type $\partial^2 V / \partial Q_{\Gamma 1} \partial Q_{\Gamma 2}$) can be grouped into a totally symmetric combination and nontotally symmetric parts. The diagonal matrix element of the totally symmetric combination,

$$K_{0\Gamma*}^{\Gamma} = \langle\Gamma|\left(\frac{\partial^2 V}{\partial Q_{\Gamma*}^2}\right)_0|\Gamma\rangle \tag{7.23'}$$

is the nonvibronic contribution to the curvature of the AP, or *primary force constant*. The full curvature $K_{\Gamma*}^{\Gamma}$, which at the minimum of the AP coincides with the *force constant*, is

$$K_{\Gamma*}^{\Gamma} = K_{0\Gamma*}^{\Gamma} - \sum_{\Gamma'}' \frac{|F_{\Gamma*}^{(\Gamma\Gamma')}|^2}{\Delta_{\Gamma'\Gamma}} \tag{7.23}$$

where $\Delta_{\Gamma'\Gamma} = \frac{1}{2}(\varepsilon_{\Gamma'} - \varepsilon_{\Gamma})$ is the corresponding energy semidifference between states Γ' and Γ. The diagonal matrix elements of the nontotally symmetric part of the second derivatives in (7.21) are the *diagonal quadratic (or second-order) vibronic constant* $G_{\Gamma*}^{\Gamma}(\Gamma_1 \times \Gamma_2)$. The off-diagonal matrix elements are the *off-diagonal quadratic vibronic constants* $G_{\Gamma*}^{(\Gamma\Gamma')}$ $(\Gamma_1 \times \Gamma_2)$, where $\Gamma* \in \Gamma_1 \times \Gamma_2$, and Γ_1 and Γ_2 are the representations of the corresponding two displacements [7.1].

Orbital Vibronic Constants

The orbital vibronic constants [7.3, 7.13, 7.14] enable us to consider approximately the influence of each electron independently on the nuclear framework and its dynamics. On the other hand, the introduction of orbital vibronic constants creates a bridge between the idea of vibronic coupling and the MO approach to the investigation of molecular structure and properties. The one-electron MOs supplemented by vibronic coupling constants result in *vibronic molecular orbitals*. The latter, as shown below, present a more refined picture of molecular structure, which includes nuclear dynamic parameters.

Denote the one-electron MO energies by ε_i and their wavefunctions by $\varphi_i(r) = |i\rangle$. Taking into account the additivity of the electron–nuclear interaction operator $V(r, Q)$ with respect to electronic coordinates (7.3'), we have

$$V = \sum_k V'(\mathbf{r}_k) \tag{7.24}$$

$$V'(\mathbf{r}) = -\sum_\alpha \frac{e^2 Z_\alpha}{|\mathbf{r} - \mathbf{R}_\alpha|} + \frac{1}{n}\frac{1}{2} \sum_{\alpha,\beta}{}' \frac{e^2 Z_\alpha Z_\beta}{|\mathbf{R}_\alpha - \mathbf{R}_\beta|} \tag{7.24'}$$

where n is the total number of electrons. Based on these notations, the orbital vibronic constants can be introduced similarly to the vibronic constants for the system as a whole [Eq. (7.22)]. In contradistinction to the orbital vibronic constants, the usual vibronic constants may be called integral vibronic constants.

For the *linear orbital vibronic constants*, we have

$$f_{\Gamma^*}^{(ij)} = \langle i| \left(\frac{\partial V'}{\partial Q_{\Gamma^*}}\right)_0 |j\rangle \tag{7.25}$$

Similarly to the integral case, the totally symmetric part ($\Gamma^* = A_1$) of the orbital diagonal matrix elements of the quadratic terms of vibronic interactions, $k_{0\Gamma^*}^{(ii)} = k_{0\Gamma^*}^i$ ($\Gamma^* = \Gamma_1 = \Gamma_2$) contributes to the orbital force constant $k_{\Gamma^*}^i$ (see below); the nontotally symmetric parts, which are nonzero for degenerate MOs only, form the diagonal second-order orbital vibronic constants $g_{\Gamma^*}^i(\Gamma_1 \times \Gamma_2)$. The off-diagonal matrix elements of these terms are the off-diagonal orbital vibronic constants $g^{(ij)}(\Gamma_1 \times \Gamma_2)$.

The physical meaning of the orbital vibronic constants can be clarified by means of the addition theorem: The linear diagonal integral vibronic constant equals the sum of linear diagonal orbital vibronic constants multiplied by the appropriate MO occupation numbers q_i^Γ:

$$F_{\Gamma^*\gamma^*}^\Gamma = \sum_i q_i^\Gamma f_{\Gamma^*\gamma^*}^i \tag{7.26}$$

The proof of this theorem [7.13] is based on the additivity properties of the vibronic constants mentioned above. Presenting the total wavefunction in the expression of $F_{\Gamma^*}^\Gamma$ after Eq. (7.22) by the determinant (or a linear combination of determinants) of the one-electron MO functions $\varphi_i(r_i)$ and taking into account the expressions (7.24) for V and (7.25) for f_Γ^i, one can deduce formula (7.26).

We thus obtain from Eq. (7.26) that the distorting influence of the electrons on the nuclear framework with a force $F_{\Gamma^*\gamma^*}^\Gamma$ is produced additively by all the appropriate MO single-electron effects $f_{\Gamma^*\gamma^*}^i$. A clear-cut physical meaning of the latter follows immediately: *The linear diagonal orbital vibronic constant* $f_{\Gamma^*\gamma^*}^i$ *equals the force with which the electron of the ith MO distorts the nuclear configuration in the direction of the symmetrized displacements* $Q_{\Gamma^*\gamma^*}$ *minus the proportion of internuclear repulsion in this direction.*

For quadratic vibronic constants, similar to Eq. (7.26), we obtain the following addition formula:

$$G_{\Gamma^*\gamma^*}^\Gamma(\Gamma_1 \times \Gamma_2) = \sum_i q_i^\Gamma g_{\Gamma^*\gamma^*}^i(\Gamma_1 \times \Gamma_2) \tag{7.27}$$

The deduction of similar expressions for the off-diagonal vibronic constants is more difficult because they involve excited states for which calculation of the contribution of the one-electron MOs is complicated. The analysis can be simplified in the "frozen orbital" approximation, equivalent to the approximation of the Koopmans theorem in quantum chemistry (Sections 2.2 and 6.4). In this approximation the one-electron excitation of the system $\Gamma \rightarrow \Gamma'$ is realized by the substitution of only one MO, for instance, i by j, while changes in the other MOs due to alteration of the interelectron repulsion are neglected. In this case there are simple relations between integral and orbital off-diagonal linear vibronic constants:

$$F_{\Gamma^*\gamma^*}^{(\Gamma\Gamma')} = f_{\Gamma^*\gamma^*}^{(ij)} \tag{7.28}$$

For fractional charge transfers, important in applications to chemical problems, the difference in the corresponding electron population numbers q_i and q_j of the two mixing MOs determines the magnitude of the off-diagonal vibronic mixing (Sections 10.3 and 11.2):

$$F_{\Gamma^*\gamma^*}^{(\Gamma\Gamma')} = f_{\Gamma^*\gamma^*}^{(ij)}(q_i - q_j) \qquad \Gamma \neq \Gamma' \tag{7.28'}$$

Using this relation, the AP curvature (or the force constant) $K_{\Gamma^*}^\Gamma$ can be presented as a sum of orbital contributions $k_{\Gamma^*}^i$:

$$\begin{aligned} K_{\Gamma^*}^\Gamma &= \sum_i q_i k_{\Gamma^*}^i \\ k_{\Gamma^*}^i &= k_{0\Gamma^*}^i - \sum_{j'} \frac{|f_{\Gamma^*}^{(ij)}|^2}{\Delta_{ji}} \end{aligned} \tag{7.29}$$

where $\Delta_{ij} = \frac{1}{2}(\varepsilon_i - \varepsilon_j)$. The proof of these relations is analogous to that of Eq. (7.27) [7.3, 7.13].

In the MO approach, the electronic structure of a molecule with fixed nuclei is presented approximately by electron charge distribution in the one-electron

MOs, and by the energies of the latter in the field of the nuclei. This picture is static; it does not characterize sufficiently well the back influence of the electrons on the nuclear framework. If the electronic structure is determined without geometry optimization, the static picture does not indicate whether the nuclear configuration chosen (the reference configuration) will be stable for the electronic structure under consideration. More important, the static picture does not allow us to predict how the nuclear configuration and its dynamics vary under electronic structure alterations.

This deficiency in the MO approach can be overcome by means of the vibronic constants, which characterize the forces (and force constants, anharmonicities, etc.) with which the electrons influence the nuclei. The orbital vibronic constants are of special interest: If these constants and the orbital contributions to the force constants are known, the behavior of the nuclear configuration (changes in its stable configuration, force constants, anharmonicities) under small changes in the electronic structure (changes in electronic MO occupation numbers) can be predicted. These results can be used to predict changes in the reactivity of molecules under electronic rearrangements (Section 11.2). Thus *the orbital vibronic constants complement the static picture of the electronic structure with dynamic parameters.*

In Fig. 7.4 the new parametrization of the molecular structure is illustrated for the HOMO 5σ and the LUMO 2π of two molecules, N_2 and CO, taken as examples (the numerical values are obtained in Section 11.2). In addition to the usual MO energy and wavefunction, it is shown that the electron of the HOMO 5σ in the N_2 molecule tightens the nuclei with the force $f_R^{5\sigma}(N_2) = 3.51 \cdot 10^{-4}$ dyn (R denotes the distance between nuclei), while in the CO molecule the electron of the analogous MO pushes them away with force $f_R^{5\sigma}(CO) = -4.5 \cdot 10^{-4}$ dyn [more precisely, the binding of the nuclei by the electron is less than the corresponding one-electron portion of their repulsion by this amount; see Eq. (7.24′)].

The electron of the LUMO 2π pushes away the nuclei in both the N_2 and CO molecules with force $f_R^{2\pi}(N_2) = -8.18 \cdot 10^{-4}$ dyn and $f_R^{2\pi}(CO) = -12.1 \cdot 10^{-4}$ dyn, respectively. For the appropriate one-electron MO contribution to the force constant $k^{(i)}$-shown in Fig. 7.4 by a spring, we have [Eq. (11.34)]: $k_R^{5\sigma}(N_2) = 0.29 \cdot 10^{-6}$ dyn/cm, $k_R^{2\pi}(N_2) = -0.79 \cdot 10^{-6}$ dyn/cm, $k_R^{5\sigma}(CO) = -0.08 \cdot 10^{-6}$ dyn/cm, and $k_R^{2\pi}(CO) = -0.83 \cdot 10^{-6}$ dyn/cm.

It follows from these data that the orbital 5σ is bonding in N_2 and antibonding in CO, whereas the 2π MO is antibonding in both cases. In general:

For bonding MO:

$$f_{\Gamma^*}^{(i)} > 0, \qquad k_{\Gamma^*}^{(i)} > 0$$

For antibonding MO:

$$f_{\Gamma^*}^{(i)} < 0, \qquad k_{\Gamma^*}^{(i)} < 0$$

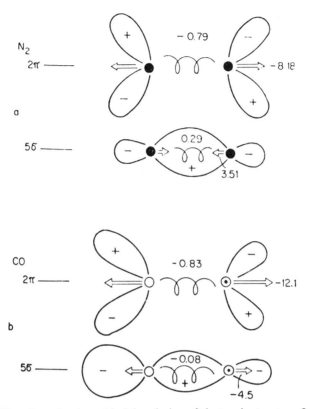

Figure 7.4. Vibronic molecular orbital description of electronic structure. In addition to the MO energies and wavefunctions the orbital vibronic constants characterize the contribution of the orbital electron to the distorting force f (shown by arrows; values are given in 10^{-4} dyn) and force constant coefficient k (shown by springs; values are given in 10^6 dyn/cm): (a) HOMO and LUMO of the nitrogen molecule—the HOMO 5σ is bonding ($f > 0$), whereas the LUMO 2π is antibonding ($f < 0$); (b) HOMO and LUMO of carbon monoxide—the HOMO is weakly antibonding, whereas the LUMO is strongly antibonding.

For nonbonding MO:

$$f_{\Gamma^*}^{(i)} \approx 0, \qquad k_{\Gamma^*}^{(i)} \approx 0$$

and the absolute values of $f^{(i)}$ and $k^{(i)}$ follow approximately the measure of MO contribution to bonding or antibonding.

These relations are not trivial. Indeed, when there are several orbitals of the same type, as in the case under consideration ($5\sigma, 4\sigma, 3\sigma, \ldots$), it is very difficult to reveal the bonding nature of each of them, even qualitatively, from MO calculations only. The orbital vibronic constants provide a semiquantitative measure of the MO bonding or antibonding quality and its contribution to the distorting force, force constants, and anharmonicity constants, thus serving as parameters of electronic control on nuclear configuration and dynamics (Sections 7.4 and 11.2). As mentioned above, the MO complemented with orbital vibronic coupling constants can be termed *vibronic molecular orbital*.

Vibronic MOs are of special importance for analysis of the influence of electronic rearrangements on the nuclear configuration [7.3, 7.13, 7.14]. They are also used in studying intervalence electron transfer [7.15] and can be significant for many other problems in which the details of electronic structure and vibronic coupling are important (see, e.g., [7.16], where they are used in the discussion of the origin of high-T_c superconductivity in fullerens). In polyatomic molecules there is more than one possible symmetrized direction of distortion and softening, and therefore the orbital vibronic constants and the orbital contributions to the force constants and anharmonicity constants contain additional information about molecular distortions (cleavage) by electronic rearrangements.

From the group theory rules (Section 3.4) the linear orbital vibronic constant $f_\Gamma^{(ij)} = \langle i|(\partial V/\partial Q_{\Gamma*})_0|j\rangle$ is nonzero if the direct product of the irreducible representations Γ_i and Γ_j of the i and j MO contains the Γ^* representation of the symmetrized displacements $Q_{\Gamma*}$. Different Γ_i and Γ_j yield Γ^* that may be of any type allowed in the symmetry group of the system under consideration. For the diagonal constant $f_{\Gamma*}^i$, Γ^* must be the component of the symmetrical product $[\Gamma_i \times \Gamma_i]$ [Eq. (3.34)]. Therefore (quite similar to the case of vibronic constants considered above), if Γ_i is nondegenerate, Γ^* is totally symmetric. In other words, the electrons of nondegenerate MO distort the nuclear configuration along Q_{A1} that leaves its symmetry unchanged.

For degenerate MOs the product $[\Gamma_i \times \Gamma_i]$ contains nontotally symmetric representations (in addition to the A_1 representation). Hence the electrons of degenerate MOs distort the nuclear framework, changing its symmetry in accordance with (and in directions determined by) the Jahn–Teller effect (Section 7.4). In addition to this distortion, the orbital electron softens or hardens the nuclear framework according to Eq. (7.29). This equation contains the off-diagonal orbital vibronic constant $f_{\Gamma*}^{(ij)}$, for which Γ may be of any type, and therefore the softening or hardening may be in any direction, depending on the mixing orbitals. For further discussion of orbital vibronic constants, see Section 11.2.

7.3. THE JAHN–TELLER EFFECT

The Jahn–Teller Theorem

In the formulation of the problem given in Section 7.2, the vibronic mixing of different electronic states by nuclear displacements is described by an infinite system of coupled equations (7.6) for the nuclear motions with the AP $\varepsilon_k(Q)$ determined by Eq. (7.8). For an f-fold-degenerate electronic term, which is well separated from other terms, the number of equations can be reduced approximately to f. Qualitatively, many features of the nuclear motions and related experimental observables can be obtained from the AP shapes without solving the vibronic equations. Some general special features of the AP of electronically degenerate states are outlined by the Jahn–Teller theorem [7.17].

The Jahn–Teller theorem is based on group-theoretical analysis of the behavior of the AP of a polyatomic system near the point of electronic degeneracy. Similar to other group-theoretical statements, the Jahn–Teller theorem allows one to deduce qualitative results without performing specific calculations, or essentially reduces the extent of such calculations. However, as opposed to the usual situation in quantum chemistry, in which the group-theoretical treatment is introduced to simplify the calculations, proof of the Jahn–Teller theorem preceded the calculations of APs and stimulated such calculations.

Suppose that by solving the electronic Schrödinger equation (7.4) for the nuclei fixed at the point $Q_{\Gamma\gamma} = Q_{\Gamma\gamma}^0 = 0$, we obtain an f-fold-degenerate electronic term, that is, f states $\varphi_k(r)$, $k = 1, 2, \ldots, f$, with equal energies $\varepsilon_k' = \varepsilon_0$. How do these energy levels change under nuclear displacements $Q_{\Gamma\gamma} \neq 0$? To answer this question, the AP for arbitrary coordinates $Q_{\Gamma\gamma}$ near the point of degeneracy must be determined. This can be done by estimating the effect of vibronic interaction terms $W(r, Q)$ in Eq. (7.24) on the energy levels ε_k'.

For sufficiently small nuclear displacements $Q_{\Gamma\gamma}$, the AP $\varepsilon_k(Q)$ can be obtained as solutions of the secular equation of perturbation theory:

$$
\begin{vmatrix}
W_{11} - \varepsilon & W_{12} & \cdots & W_{1f} \\
W_{21} & W_{22} - \varepsilon & \cdots & W_{2f} \\
\vdots & \vdots & & \vdots \\
W_{f1} & W_{f2} & \cdots & W_{ff} - \varepsilon
\end{vmatrix} = 0 \tag{7.30}
$$

where the W_{ij} are the matrix elements of the vibronic interaction operator calculated with the wavefunctions of the degenerate term. Since the degeneracy is assumed to be caused by the high symmetry of the system, the totally symmetric displacements Q_A (that do not change the symmetry) do not remove the degeneracy and are not considered in this section. Again, second-order terms may also be omitted, due to the assumed small values of $Q_{\Gamma\gamma}$. As a result, the matrix elements of the vibronic interaction contain only linear nontotally symmetric terms $W_{ij} = \Sigma_{\Gamma\gamma} \langle i|(\partial V/\partial Q_{\Gamma\gamma})|j \rangle Q_{\Gamma\gamma}$. If at least one of these terms (e.g., for $Q_{\Gamma*\gamma*}$) is nonzero, at least one of the roots ε of Eq. (7.43) contains linear terms in $Q_{\Gamma*\gamma*}$, and hence the AP $\varepsilon_k(Q)$ has no minimum at the point $Q_{\Gamma*\gamma*}^0 = 0$ with respect to these displacements.

On the other hand, the question of whether the vibronic constant $F_{\Gamma*}^\Gamma = \langle \Gamma|(\partial V/\partial Q_{\Gamma*})_0|\Gamma \rangle$ is zero or not may easily be answered by means of the well-known group-theoretical rule: $F_{\Gamma*}^\Gamma$ is nonzero if and only if the symmetric product $[\Gamma \times \Gamma]$ contains the representation Γ^* of the symmetrized displacement Q_{Γ^*}. For instance, for the usual E term, $[E \times E] = A_1 + E$, and thus if the system under consideration has E vibrations (see the classification of vibrations given in Table 7.2), it has no minimum at the point of degeneracy with respect to the E displacements.

Jahn and Teller [7.17] examined all types of degenerate terms of all symmetry point groups and showed that for any orbital degenerate term of any molecular system, there are nontotally symmetric displacements with respect to which the adiabatic potential of the electronic term (more precisely, at least one of its branches) has no minimum; molecules with a linear arrangement of atoms are exceptions (see below). It was later shown that a similar statement is also valid in the case of spin degeneracy, with the exception of twofold degeneracy for systems with $S = 1/2$ (*Kramers doublets*). The statement about the absence of extremum at the point of degeneracy is just *the Jahn–Teller theorem* [7.17], which may be formulated more rigorously as follows: *If the adiabatic potential of a nonlinear polyatomic system has several ($f > 1$) branches that coincide at one point (f-fold degeneracy), at least one of them has no extremum at this point, Kramers doublets being exceptions.*

The variation of the adiabatic potential in the simplest case of an orbitally doubly degenerate electronic term in the space of only one coordinate Q is shown schematically in Fig. 7.5. It is seen that the two curves intersect at the point of degeneracy. Away from this point the energy term splits and the degeneracy is removed. As a result, the energy is lowered so that the small nuclear displacements Q are of advantage. For larger values of Q the quadratic, cubic, and higher-order terms become important, and further distortion of the system may be inconvenient (Section 7.4).

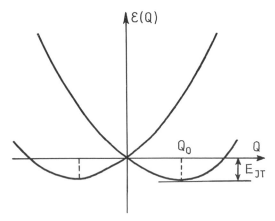

Figure 7.5. Variation of the adiabatic potential of a molecular system in a twofold orbitally degenerate electronic state with respect to one active coordinate Q. At the point of degeneracy $Q = 0$ there is no minimum. E_{JT} is the Jahn–Teller stabilization energy.

The exclusion of linear molecules from the JT statement needs clarification. For linear molecules the nontotally symmetric displacements are of odd type with respect to reflections in the plane comprising the molecular axis, whereas the product of any two wavefunctions of the degenerate states (which have the same parity) is always even with respect to such reflections. This means that the corresponding vibronic constant equals zero, and hence the adiabatic potential at the point of linear configuration, as opposed to nonlinear molecules, has no terms of linear displacements: It is extremal.

However, if the linear terms of the vibronic interactions $W(r, Q)$ are zero, the quadratic terms become of primary importance [see Eq. (7.7)]. They are even, their matrix elements in the case of linear molecules in degenerate states are nonzero, and the solution of the secular equation (7.30) results in a splitting of the energy term and instability. This was shown by Renner [7.18] prior to the publication of the paper of Jahn and Teller [7.17]. The Renner effect is discussed below.

Proof of the JT theorem by means of an examination of all the types of degenerate terms for all the symmetry point groups, one by one [7.17], although rigorous, cannot be considered as sufficiently elegant from the mathematical point of view. More general proofs have been obtained ([7.19]; see also [7.1, 7.6]). However, the search over all the possible cases has its advantages, one of which is to reveal the *Jahn–Teller active modes*, that is, the nuclear displacements Q for which the vibronic constant $F_{\Gamma*}^{\Gamma}$ is nonzero. These active modes (active vibrations) are important basic components of the Jahn–Teller problem, because they form the space of nontotally symmetric nuclear displacements in which the Jahn–Teller theorem is operative.

If the types of possible vibrations of the system under consideration are known (Table 7.2), the Jahn–Teller active nuclear displacements $Q_{\Gamma*}$ may be obtained easily. As indicated above, $F_{\Gamma*}^{\Gamma}$ is nonzero if the symmetric product $[\Gamma \times \Gamma]$ contains $\Gamma*$. Hence the nontotally symmetric components of this product are just the representations $\Gamma*$ of the active displacements $Q_{\Gamma*}$. For instance, for a E_g term in an octahedral system $[E_g \times E_g] = A_{1g} + E_g$, and hence the Jahn–Teller active displacements are of E_g type. Similarly, for a T term in a tetrahedral system, $[T \times T] = A_1 + E + T_2$ and both the E and T_2 displacements are Jahn–Teller active. The Jahn–Teller active displacements for all important point groups can be found in [7.1, 7.17].

The lack of minimum of the AP at the point of electronic degeneracy is usually interpreted as instability of the nuclear configuration at this point. Therefore, the formulation of the JT theorem is often given as follows: A nonlinear polyatomic system in the nuclear configuration with a degenerate electronic term is unstable. This statement of instability is often treated in the sense that the system distorts itself spontaneously so that the electronic term splits and the ground state becomes nondegenerate. Such an interpretation of the Jahn–Teller theorem initiated by its authors [7.17] now appears in monographs and handbooks and has widespread use in general treatments of experimental results. Meanwhile, as shown in a number of publications (see, e.g., [7.1–7.4]), the actual situation in systems with electron degeneracy is much more complicated than is implied by the simple statement of instability. Moreover, taken literally, this statement is not true and may lead to misunderstanding [7.20].

The conclusion about the lack of a minimum of the AP $\varepsilon_k(Q)$ at the point of degeneracy was reached as a consequence of the solution of the electronic part of the Schrödinger equation (7.4), and therefore it cannot be attributed to the nuclear behavior, which is determined by the nuclear motion equations (7.6). The absence of a minimum of the function $\varepsilon_k(Q)$ may generally be interpreted as instability only when there is no degeneracy or pseudodegeneracy. Indeed, in the absence of degeneracy (or in the areas far from the point of degeneracy) the electronic and nuclear motions can be separated in the adiabatic approximation, so that the AP $\varepsilon_k(Q)$ has the meaning of the potential energy of the nuclei in the mean field of the electrons, and hence the derivative $(d\varepsilon_k/dQ)_0$ means the force acting upon the nuclei at the point $Q_{\Gamma\gamma}^0$. Here the condition $(d\varepsilon_k/dQ)_0 \neq 0$ may be interpreted as a nonzero distorting force (in the Q direction) due to which the nuclear configuration becomes unstable.

However, in the presence of electronic degeneracy the AP $\varepsilon_k(Q)$ loses the meaning of the potential energy of the nuclei in the mean field of the electrons, since the motions of the electrons and nuclei near the point of degeneracy cannot he separated. In this area the notion $\varepsilon_k(Q)$ become formal with no definitive physical meaning. Accordingly, the reasoning given above about distorting force and instability is, strictly speaking, invalid. In these cases the term *instability* should be taken formally as an indication of the lack of a minimum for the AP but not as a nuclear feature. As indicated above, the latter must be deduced from the solutions of equations (7.6) of nuclear dynamics.

The distortions of the nuclear configuration of free Jahn–Teller systems are, in general, of dynamic nature in which the quantum mechanically averaged nuclear coordinates remain unchanged [7.20]. *It is not the simple nuclear configuration distortion but special nuclear dynamics which are predicted by the Jahn–Teller theorem in free (unperturbed) molecular systems.* The degeneracy is not removed either; with vibronic interactions included, the electronic degeneracy transforms into vibronic degeneracy.

The specific Jahn–Teller behavior of the AP due to the electronic degeneracy or pseudodegeneracy, similar to many other features of electrostatic origin, can be explained by simple images. Consider, for example, the case of one electron in one of three equivalent p orbitals (p_x, p_y, or p_z) of the CA of a hexacoordinated molecular system MX_6 (term T_{1u}) with the ligands X bearing negative charges (Fig. 7.6). It is clear that if the electron is on the p_x orbital, it interacts more strongly with the nearest ligands 1 and 3 and repels them. As a result, the octahedral complex becomes tetragonally distorted along the x axis.

Similarly, the electron on the p_y orbital repulses ligands 2 and 4, distorting the complex equivalently to the previous case, but along the y axis, and for the electron on the p_z orbital the distortion is along z. It follows that the AP of the

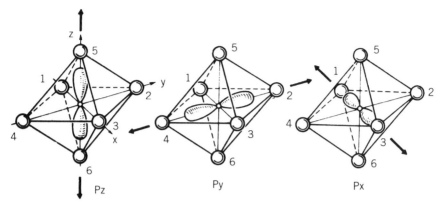

Figure 7.6. Electrostatic origin of Jahn–Teller distortions of the nuclear configuration of an octahedral ML_6 complex in a threefold-degenerate electronic state. If the electron of the CA falls into one of the three equivalent states, it repels (or attracts) the corresponding pair of ligands resulting in a tetragonal distortion. The three equivalent directions of distortion are shown by arrows.

system has three minima, corresponding to the three directions of distortions, and no minimum in the high-symmetry octahedral configuration, due to obvious electrostatic forces (note that the degeneracy of the three p states is lifted by these distortions). The same result emerges from the MO description, with the atomic p orbitals taking part in the corresponding antibonding MOs (for bonding orbitals the distortions have opposite sign).

All the observable effects related to the special features of the adiabatic potential predicted by the Jahn–Teller theorem are called *Jahn–Teller effects* [7.1–7.4]. A large variety of such effects are considered in different sections of this book.

Pseudo Jahn–Teller and Renner Effects

From the discussion above, a question emerges: Are the nonadiabatic and Jahn–Teller effects related to the exact degeneracy of the electronic states, or may the latter be just close in energy? The answer is that electronic states with sufficiently close energy levels are similar in behavior to exact degenerate states [7.21]: Their ground-state adiabatic potential has no minimum at the point of the closest energies, and the nuclear motions are described by coupled equations such as (7.6) instead of (7.9). For reasons given below, the case of sufficiently close energy states is called the *pseudo Jahn–Teller effect* and the corresponding set of energy levels are considered as *pseudodegenerate* (or quasi-degenerate). Pseudo Jahn–Teller nonadiabacity follows directly from condition (7.11), which is not satisfied for sufficiently close energy levels $\varepsilon'_k \approx \varepsilon_j$.

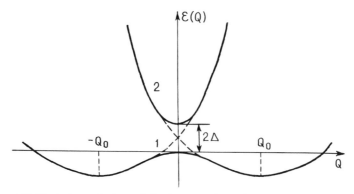

Figure 7.7. Variation of the adiabatic potential of a molecular system with two sufficiently close energy levels that mix under the Q displacements. The picture is similar to that of "avoided crossing" (pseudodegeneracy) shown by dashed lines. The lack of minimum of the ground state at $Q = 0$ (pseudo Jahn–Teller instability) is similar to the Jahn–Teller case in Fig. 7.5, but with quadratic dependence on Q (dynamic instability).

Hence the coupling terms in (7.6) containing the matrix elements of the vibronic interaction W_{km} cannot be neglected, and the adiabatic approximation is invalid.

The similarity with the Jahn–Teller effect is also extended to the behavior of APs near the point of pseudodegeneracy. Detailed calculations and discussion of this point appear in Section 7.4; the simple case of two pseudodegenerate terms and one active coordinate is illustrated in Fig. 7.7. If the pseudocrossing of the two terms shown by the dashed line is taken into account, the analogy with the Jahn–Teller effect becomes quite visual. But distinguished from the Jahn–Teller case, the pseudo Jahn–Teller lack of minimum in the ground state takes place only when a specific inequality is satisfied (Section 7.4).

It follows from the formulation of the Jahn–Teller theorem that it does not refer to linear molecules. The reason is mentioned above: Linear (first-order) low-symmetry distortions of linear molecules are of odd type, and hence the matrix elements of the linear terms of the vibronic coupling W_{ij} are zero due to the selection rules. However, the quadratic terms of W give nonzero matrix elements, since the second-order distortions are even. Hence, when quadratic terms of the vibronic interaction (7.7) are taken into account, linear molecules are subject to effects similar to those described for nonlinear systems. This statement was first proved by Renner [7.17], and it is known as *the Renner effect*.

However, the Renner-type instability is similar but not identical to Jahn–Teller instability. Indeed, quadratic dependence on $Q_{\Gamma*}$ means that there is not intersection (only splitting) of the adiabatic potentials $\varepsilon(Q)$ at the point of degeneracy $Q_{\Gamma*} = 0$. The behavior of $\varepsilon(Q)$ at this point depends strongly on the strength of the vibronic coupling given by the quadratic vibronic constant

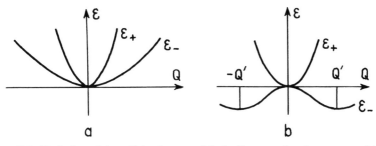

Figure 7.8. Variation of the adiabatic potential of a linear molecular system with respect to odd displacements Q in the case of the Renner effect: (*a*) weak coupling: term splitting without instability; (*b*) strong coupling: dynamic instability of the ground state.

$G_{\Gamma*}^{(\Gamma\Gamma')}$ (Section 7.2). There are two cases (Fig. 7.8):

1. *Weak Renner effect.* The vibronic coupling constant G is sufficiently small, and the vibronic splitting of the adiabatic potential does not result in instability of the ground state (Fig. 7.8a).

2. *Strong Renner effect.* As a result of the vibronic splitting, the ground state becomes unstable (Fig. 7.8b) [the increase in $\varepsilon(Q)$ at larger Q is provided by the higher-order terms in (7.7)].

Thus for linear molecules with electronic degeneracy and strong quadratic vibronic coupling, the adiabatic potential of the ground state has no minimum at the point of degeneracy. However, as opposed to the Jahn–Teller case, the first derivative $(d\varepsilon/dQ_{\Gamma*})_0 = 0$ for all $Q_{\Gamma*}$; that is, the point of degeneracy is extremal, but $(d^2\varepsilon/dQ_{\Gamma*}^2)_0 < 0$, which means that it is a maximum in the $Q_{\Gamma*}$ direction. In terms of instability both the pseudo Jahn–Teller and Renner systems are dynamically (quadratic) unstable, as opposed to the static (linear) instability in the Jahn–Teller case. For further details on the Renner effect, see [7.18, 7.22, 7.23].

The Jahn–Teller Effect in a Twofold-Degenerate State

Jahn–Teller problems — typical cases of electronic degeneracies that strongly influence the nuclear configurations and nuclear motions in molecular systems — are considered in detail in special monographs and books (see [7.1–7.4] and references therein). In this section a short review of some results, important for the study of coordination compounds and partly used in the following sections, are given. Most widespread are doubly degenerate E terms for systems that have at least one axis of symmetry of the third order, C_3, threefold-degenerate T terms (systems with cubic symmetry), and fourfold-degenerate G' (or Γ_8) terms for cubic systems with strong spin–orbital coupling. Two- and many-level pseudo Jahn–Teller effects are considered in the next section.

To solve a Jahn–Teller problem means to find the Jahn–Teller active coordinates for the given f-fold-degenerate electronic term, to determine the AP in the space of these coordinates, and to solve the system of f coupled equations (7.6). The resulting solution includes the energy-level spectrum and wavefunctions that allow one to calculate physical magnitudes. However, each of the foregoing stages of the problem has an independent meaning and gives some qualitative insight into the problem. In particular, the Jahn–Teller active coordinates indicate the space of possible distortions of the system, while the APs describe the possible nuclear motions in the semiclassical approach.

In polyatomic systems with a large number of atoms (N) and hence a large number of possible vibrations $(3N - 6)$, there can be more than one—even a large (in crystals, infinite) number of Jahn–Teller active coordinates. For instance, for a twofold-degenerate E term, the Jahn–Teller active coordinates are also of E type, and the Jahn–Teller problem is $E - e$ (the active vibrations are indicated by lowercase letters). But for large numbers N there may be many E vibrations $(E_1, E_2, \ldots)$ that are Jahn–Teller active. Then we have a *multimode problem* [denoted as $E - (e_1 + e_2 + \cdots)$], distinct from the *ideal problem*, for which there is only one active vibration of given symmetry.

The multimode problem is usually significant for coordination compounds with polyatomic ligands or multicenter systems, although even a simple five-atomic tetrahedral complex MA_4 has two types of vibrations with the same symmetry, T_2' and T_2'' (Table 7.1), and hence its Jahn–Teller problem for a T term interacting with T_2 vibrations is a two-mode problem $[T - (t_2' + t_2'')]$. Under some restrictions, the multimode problem can be reduced to the ideal problem. Later in this section, only ideal problems are considered. Discussion of multimode problems and solutions are given elsewhere (see [7.1] and references therein).

For of a f-fold-degenerate term the f branches of the AP $\varepsilon_k(Q_{\Gamma\gamma})$, $k = 1, 2, \ldots, f$, are determined by Eq. (7.8) with the corrections $W_{kk}(Q_{\Gamma\gamma})$ obtained from the secular equation (7.30), taking W according to (7.21). As mentioned in Section 7.2, there is a totally symmetric combination of the matrix elements of the coefficients at the quadratic terms in W, which is equal to the primary force constant $K_{0\Gamma}$; the corresponding terms can be grouped in expressions of the strain energy $\frac{1}{2}\sum_{\Gamma\gamma} K_{0\Gamma} Q_{\Gamma\gamma}^2$. The matrix elements of the remaining quadratic term coefficients are quadratic vibronic constants. Therefore, we can present the AP (7.8) in the form

$$\varepsilon_k(Q_{\Gamma\gamma}) = \tfrac{1}{2} \sum_{\Gamma\gamma} K_{0\Gamma} Q_{\Gamma\gamma}^2 + \varepsilon_k^v(Q_{\Gamma\gamma}) \tag{7.31}$$

where the vibronic corrections to the electronic term ε^v are the roots of the secular equation

$$\| W_{\gamma\gamma'}^v - \varepsilon^v \| = 0 \qquad \gamma, \gamma' = 1, 2, \ldots, f \tag{7.32}$$

and $W^v_{\gamma\gamma'}$, as distinct from the similar equation (7.30), contains matrix elements of vibronic coupling terms only [without the elasticity term separated in (7.31)].

For a twofold orbitally degenerate E term, the Jahn–Teller active coordinates are either $A_1 + E$ ($[E \times E] = A_1 + E$), or $A_1 + B_1 + B_2$($[E \times E] = A_1 + B_1 + B_2$). The latter case is realized in systems containing symmetry axes of orders multiple by 4 ($C_4, C_8, \ldots$). The totally symmetric vibrations A_1 do not distort the system and can be eliminated by a special choice of the coordinate origin (however, see below). Hence the E term generates either *the $E - e$ problem*, which is a more widespread case, or *the $E - (b_1 + b_2)$ problem* for systems with $4k$-fold symmetry axes, where k is an integer.

Consider *the $E-e$ problem*. The two electronic wavefunctions of the E term may be denoted by $|\vartheta\rangle$ and $|\varepsilon\rangle$, with symmetry properties of the well-known functions $\vartheta \sim 3z^2 - r^2$ and $\varepsilon \sim x^2 - y^2$ (or $\vartheta \sim d_{z^2}$ and $\varepsilon \sim d_{x^2-y^2}$). The two components of the normal E-type (tetragonal) displacements Q_ϑ and Q_ε are illustrated in Fig. 7.1 (see also Fig. 9.27), while their expressions by Cartesian coordinates of the nuclei are given in Table 7.1. Accordingly, the matrix elements $W^v_{\gamma\gamma'}$ and hence $\varepsilon^v_k(Q_{\Gamma\gamma})$ in Eq. (7.32) are dependent on these two coordinates only, and the AP in all the other coordinates after Eq. (7.31) retains a simple parabolic form:

$$\varepsilon_k(Q_{\Gamma\gamma}) = \tfrac{1}{2} \sum{}' K_\Gamma Q^2_{\Gamma\gamma} \qquad (\Gamma \neq E) \tag{7.33}$$

The remaining nonzero matrix elements of the vibronic interaction acquire a simple form if the vibronic constants introduced in Section 7.2 are used. Set

$$F_E = \langle \vartheta | \left(\frac{\partial V}{\partial Q_\vartheta} \right)_0 | \vartheta \rangle \tag{7.34}$$

$$G_E = \frac{1}{2} \langle \vartheta | \left(\frac{\partial^2 V}{\partial Q_\vartheta \partial Q_\varepsilon} \right)_0 | \varepsilon \rangle \tag{7.35}$$

Then, keeping only the linear and second-order vibronic interaction terms, we obtain the explicit form of Eq. (7.32) for the $E - e$ problem:

$$\begin{vmatrix} F_E Q_\theta + G_E(Q^2_\theta - Q^2_\varepsilon) - \varepsilon^v & -F_E Q_\varepsilon + 2G_E Q_\varepsilon Q_\vartheta \\ -F_E Q_\varepsilon + 2G_E Q_\vartheta Q_\varepsilon & -F_E Q_\vartheta - G_E(Q^2_\vartheta - Q^2_\varepsilon) - \varepsilon^v \end{vmatrix} = 0 \tag{7.36}$$

This equation can be solved directly. In polar coordinates

$$Q_\theta = \rho \cos \phi \qquad Q_\varepsilon = \rho \sin \phi \tag{7.37}$$

the solutions are:

$$\varepsilon^v_\pm (\rho, \phi) = \pm \rho [F^2_E + G^2_E \rho^2 + 2F_E G_E \rho \cos 3\phi]^{1/2} \tag{7.38}$$

Inserting these values into Eq. (7.31), we get the following expression for the AP in the space of the Q_ϑ and Q_ε Jahn–Teller active coordinates transformed into (ρ, ϕ):

$$\varepsilon_\pm (\rho, \phi) = \tfrac{1}{2}K_E\rho^2 \pm \rho[F_E^2 + G_E^2\rho^2 + 2F_E G_E\rho \cos 3\phi]^{1/2} \qquad (7.39)$$

In particular, in the linear approximation (i.e., when quadratic terms may be neglected $G_E = 0$), this surface is simplified:

$$\varepsilon_\pm (\rho, \phi) = \tfrac{1}{2}K_E\rho^2 \pm |F_E|\rho \qquad (7.40)$$

Here the adiabatic potential is independent of ϕ; it has the form of a surface of revolution often called the *Mexican hat* (Fig. 7.9). The radius ρ of the circle at the bottom of the trough and its depth reckoned from the degeneracy point at $\rho = 0$ (the *Jahn–Teller stabilization energy* E_{JT}) are given by the relationships

$$\rho_0 = \frac{|F_E|}{K_E} \qquad E_{JT} = \frac{F_E^2}{2K_E} \qquad (7.41)$$

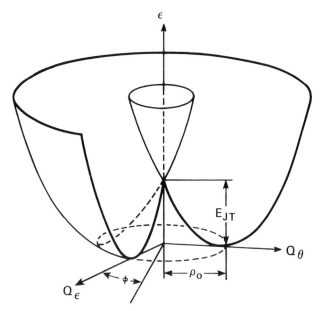

Figure 7.9. AP shape for a twofold-degenerate E term interacting linearly with the twofold-degenerate E-type vibrations described by Q_ϑ and Q_ε coordinates (linear E–e problem, the "Mexican hat"). E_{JT} is the Jahn–Teller stabilization energy.

When taking into account the quadratic terms of vibronic interaction, this surface warps and along the bottom of the trough of the "hat" three wells occur, alternating regularly with three humps (the "tricorn," Figs. 7.10 and 7.11).

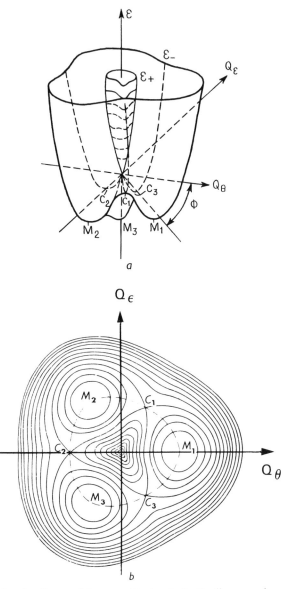

Figure 7.10. AP for the $E-e$ problem considering both the linear and quadratic terms of the vibronic interaction (the "tricorn"): (a) general view; (b) equipotential sections of the lower sheet ε. Three minima (M_1, M_2, M_3) and three saddle points (C_1, C_2, C_3) are linked by the dashed lines of the steepest slope from the latter to the former.

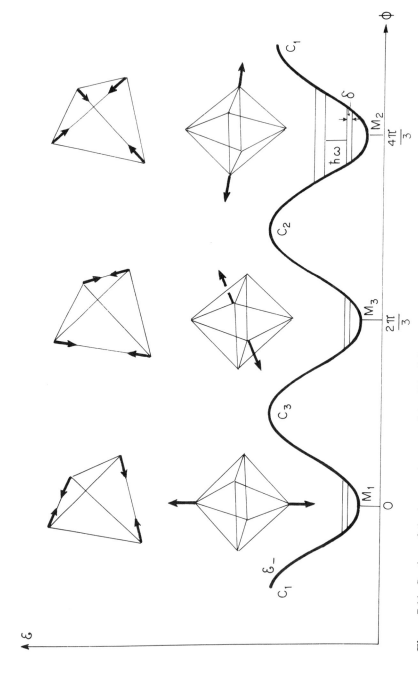

Figure 7.11. Section of the lowest sheet ε_- of the AP of the quadratic $E-e$ problem along the line of steepest slope (dashed lines in Fig. 7.10b) with illustration of the distortions of octahedral and tetrahedral systems at the points of minima and tunneling splitting.

The extremal points of the surface (ρ_0, ϕ_0) are

$$\rho_0 = \pm \frac{F_E}{K_E \mp (-1)^n 2G_E}$$

$$\phi_0 = \frac{n\pi}{3} \qquad n = 0, 1, \ldots, 5$$

(7.42)

with the upper and lower signs corresponding to the cases $F_E > 0$ and $F_E < 0$, respectively. If $F_E G_E > 0$, the points of $n = 0, 2, 4$ are minima and those of $n = 1, 3, 5$ are saddle points, whereas for $F_E G_E < 0$ the two types of extremal points interchange. For the Jahn–Teller stabilization energy E_{JT}, we have

$$E_{JT} = \frac{F_E^2}{K_E - 2|G_E|}$$

(7.43)

and the (minimal) barrier height between the minima is

$$\Delta = \frac{4E_{JT}|G_E|}{K_E + 2|G_E|}$$

(7.44)

In the linear approximation the curvature of the surface along the trough $K_\phi = 0$, and in the perpendicular (radial) direction $K_\rho = K_E$ (note that in the absence of vibronic interactions $K_\phi = K_\rho = K_E$). Taking into account the quadratic terms, one can obtain [7.1]:

$$K_\rho = K_E - 2|G_E| \qquad K_\phi = \frac{9|G_E|(K_E - 2|G_E|)}{K_E - |G_E|}$$

(7.45)

It follows that if $2|G_E| > K_E$, the system has no minima at the point ρ_0 and it decomposes, provided that the higher terms in Q of the vibronic interactions, neglected above, do not stabilize it at larger distances. At the minima points the curvature equals the force constant.

The two wavefunctions $\Psi_\pm$ that correspond to the two sheets of Eq. (7.39), are

$$\Psi_- = \cos\frac{\Omega}{2}|\vartheta\rangle - \sin\frac{\Omega}{2}|\varepsilon\rangle$$

$$\Psi_+ = \sin\frac{\Omega}{2}|\vartheta\rangle + \cos\frac{\Omega}{2}|\varepsilon\rangle$$

(7.46)

where

$$\tan \Omega = \frac{F_E \sin \phi - |G_E|\rho \sin 2\phi}{F_E \cos \phi + |G_E|\rho \cos 2\phi}$$

It is often assumed that $\Omega \equiv \phi$, which is true only in the absence of quadratic terms, $G_E = 0$.

With the shapes of the symmetrical displacements Q_{ϑ} and Q_{ε} and their values at the minima points known, one can evaluate the corresponding Jahn–Teller distortions for different types of molecules.

If the APs are known, some qualitative features of the nuclear behavior can be evaluated in the *semiclassical approach*, that is, considering the nuclei moving along the AP. This approximation is valid when the energy gap between different sheets of the AP is sufficiently large. This is realized for strong vibronic coupling and for nuclear configurations near the minima of the lowest sheet where the energy gap is the largest (in case of the E term this gap equals $4E_{JT}$). If only the linear terms are taken into account and hence the lowest sheet of the AP has the shape of a hat, *the nuclear configuration performs free rotations in the space of the Q_{ϑ} and Q_{ε} coordinates along the circle of minima in the trough*; in this case, each atom, for instance, in a triangle molecule X_3, describes a circle with the radius equal to $\rho_0\sqrt{3}$. The circular motions of these atoms are correlated: The vectors of their displacements are shifted in phase through an angle of $2\pi/3$ (Fig. 7.12). In any instant the equilateral triangle X_3

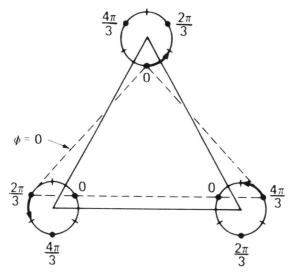

Figure 7.12. Distortions of a triatomic molecule X_3 by moving along the bottom of the trough of the lowest sheet of the adiabatic potential in the linear $E-e$ problem. Each of the three atoms moves along a circle, their phases being concerted. The bold points indicate the minima positions when quadratic terms are taken into account. The dashed triangle corresponds to the point $\phi = 0$ in Fig. 7.10 (the case of compressed triangles is shown; with the opposite sign of the vibronic constant they are elongated).

is distorted to an isosceles triangle, and *the distortion travels as a wave around the geometric center*, performing specific internal rotations.

In an octahedral molecule, which in the trigonal projection looks like two equivalent triangles, the two waves of deformations, traveling around each of the triangles, are opposite in phase. As a result, the octahedron becomes elongated (or compressed) alternatively along each of the three fourfold axes and compressed (elongated) simultaneously along the remaining two axes (Fig. 7.13).

If the quadratic terms of the vibronic interactions are taken into account, the lowest sheet of the adiabatic potential has three minima, in each of which the octahedron is elongated (or compressed) along one of the three axes of order 4 (Fig. 7.11); when allowing for quantum effects, *the nuclear motions along the adiabatic potential surface are likewise hindered rotations and tunneling transitions between the minima*, which may be presented by way of illustrations as *"pulse" motions*.

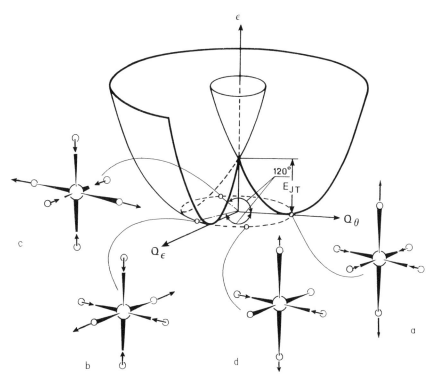

Figure 7.13. Distortions of an octahedral system ML_6 at different points along the bottom of the trough of the "Mexican hat" in the linear $E-e$ problem. At the points $\phi = 0$, $2\pi/3$, $4\pi/3$ the octahedron is tetragonally distorted along the three fourfold axes, respectively (a, b, c). In between these points the configuration has D_{2h} symmetry (d) and varies continuously from one tetragonal configuration to another.

As indicated earlier, since $[E \times E] = A_1 + E$, the totally symmetric displacements A_1 are also Jahn–Teller active in the E state (as in all the other cases): Strictly speaking, the $E–(e + a_1)$ problem should be solved instead of the $E–e$ problem, considered above. However, the totally symmetric displacements do not change the molecular symmetry; they change proportionally only the interatomic distances. Therefore, in many cases it may be assumed that the origin is taken in the new minimum position with regard to A_1 coordinates, so that the interaction with the A_1 displacements, as with all the other Jahn–Teller modes, may be ignored. This cannot be done when one has to compare the vibronic effects in different systems or in a series of systems for which the A_1 displacement contributions may be different (Section 9.4). For this (and other) reasons, the more rigorous expressions for the AP, including the interaction with all the active modes, may be useful.

Denote the linear vibronic constant for totally symmetric displacements ρ_A by F_A and the corresponding force constant by K_A. Then

$$\varepsilon(\rho_a, \rho, \phi) = \tfrac{1}{2}K_A\rho_A^2 - F_A\rho_A + \tfrac{1}{2}K_E\rho^2$$
$$\pm \rho[F_E^2 + G_E^2\rho^2 + 2F_EG_E(\rho\cos 3\phi - 2\sqrt{2}\rho_A)$$
$$+ 4G_E^2(\rho_A(2\rho_A - \sqrt{2}\rho\cos 3\phi)]^{1/2} \tag{7.47}$$

The extremal points for this surface are [cf. Eq. (7.41)]

$$\rho_0 = \pm\frac{F_E + 2\sqrt{2}\eta F_A}{K_E \mp (-1)^n 2G_E - 4\eta G_E}$$

$$\phi_0 = \frac{n\pi}{3} \qquad n = 0, 1, 2, 3, 4, 5 \tag{7.48}$$

$$\rho_{A0} = \frac{F_A}{K_A} \pm 2\sqrt{2}\eta_0 = \rho_{A0}^0 \pm 2\sqrt{2}\rho_0$$

where $\eta = G_E/K_A$ characterizes the role of the quadratic terms of the vibronic interactions with respect to the "homogeneous hardness" K_A of the system.

To obtain the energy spectra and wavefunction in the Jahn–Teller $E–e$ problem, the system of two equations (7.6) should be solved with the potential (7.39) or (7.40). For arbitrary parameter values this can be done numerically. However, for limiting values of strong and weak vibronic coupling, Eq. (7.6) can be solved analytically. Define the quantitative criterion of weak and strong coupling by comparing the Jahn–Teller stabilization energy E_{JT}^Γ [Eq. (7.41) or (7.43)] with the energy of the n_Γ-fold zero-point vibration $n_\Gamma\hbar\omega_\Gamma/2$ (in the case under consideration $\Gamma = E$ and $n_\Gamma = 2$). Get $\lambda_\Gamma = 2E_{JT}^\Gamma/n_\Gamma\hbar\omega_\Gamma$. Then *if $\lambda_\Gamma \ll 1$ ($E_{JT}^\Gamma \ll n_\Gamma\hbar\omega_\Gamma/2$), the vibronic coupling is regarded as weak, and if $\lambda_\Gamma \gg 1$ ($E_{JT}^\Gamma \gg n_\Gamma\hbar\omega/2$), the coupling is strong. λ_Γ is the dimensionless vibronic constant.*

In the limit of weak vibronic coupling the depth of the vibronic minima are much smaller than the zero-vibration energies, and therefore there are no local states in the minima. On the contrary, in the strong coupling limit there are such local states. Nevertheless, in both cases the system is delocalized into all equivalent minima, provided that the stationary states of the free system (not instantaneous or specially prepared states) are considered. Therefore, the terms *dynamic Jahn–Teller effect* for weak coupling and *static Jahn–Teller effect* for strong coupling are unsuitable because, strictly speaking, both cases are dynamic. However, the term *static Jahn–Teller effect* may still be meaningful if used to indicate the situation when in the limit of very deep minima the one-minimum state is manifest in the experiment; this state may be regarded as a quasi-stationary state for the given process of measurement (see the relativity rule concerning the means of observation in Section 9.1).

Consider first the weak coupling limit when $\lambda \ll 1$ and the vibronic interaction W can be considered as a perturbation. The energy levels obtained in the second-order perturbation theory (with respect to linear in W terms) are (Fig. 7.14)

$$E_{nlm} = \hbar\omega_E[n + 1 + 2\lambda_E(l^2 - m^2 - \tfrac{3}{4})] \tag{7.49}$$

$$l = n, n - 2, \ldots, -n + 2, -n; \qquad m = \pm\tfrac{1}{2}, \pm\tfrac{3}{2}, \ldots, \pm[n + \tfrac{1}{2}]$$

where $\lambda_E = (E_{JT}^E/\hbar\omega_E) = F_E^2/2\hbar\omega_E K_E$ [see Eq. (7.41)]. Since l and m are not independent, the energy E_{nlm} depends on only two quantum numbers, say, n and l, $E_{nl} = \hbar\omega_E[n + 1 + 2\lambda_E(\pm l - 1)]$. Hence the $[2(n + 1)$-fold]-degenerate level with a given n splits in $n + 1$ components (l may have $n + 1$ values). Each level remains twofold degenerate, owing to the independency of the energy (7.49) on the sign of m. The ground state is also twofold degenerate, $E_{00, \pm 1/2} = \hbar\omega_E(1 - \lambda_E) = \hbar\omega_E - 2E_{JT}$. However, some of these levels, namely, those with quantum numbers $m = \pm\tfrac{3}{2}, \pm\tfrac{9}{2}, \pm\tfrac{15}{2}, \ldots$ pertain to the symmetry $A_1 + A_2$, which indicates that they are accidentally degenerate (the others being of E symmetry and therefore regularly degenerate).

In the other limit, $E_{JT} > \hbar\omega_E$, the quadratic terms of vibronic coupling may be important. The corresponding AP has the shape given in Fig. 7.10. If the barrier between minima is not very large, $\Delta \ll \hbar\omega_E$, an appropriate separation of the radial (along ρ) and angular (along ϕ) motions is possible: It results in the following equation of motion along ϕ [7.24]:

$$\left(-\alpha \frac{\partial}{\partial\phi^2} + \beta \cos 3\phi - E_m\right)\Phi(\phi) = 0 \tag{7.50}$$

where $\alpha = \hbar^2/3M\rho_0^2$ and $\beta = G_E\rho_0^2$.

This equation has been solved numerically. The energy levels obtained as functions of the ratio β/α are illustrated in Fig. 7.15. It is seen that some of the doublet rotational levels for $\beta = 0$, discussed above, namely those transforming after the irreducible representations $A_1 + A_2$ [i.e., accidental degenerate; cf.

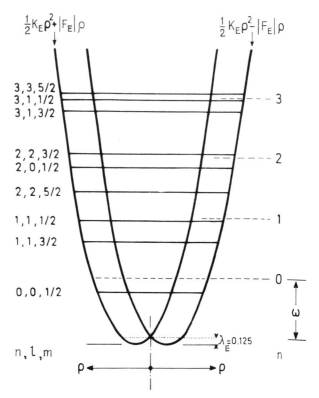

Figure 7.14. Cross section of the AP and vibronic energy levels for the linear $E-e$ problem with weak vibronic coupling, $\lambda_E = 0.125$. The positions of vibrational levels without vibronic coupling are shown by dashed lines.

(7.49)], are split when the quadratic terms of vibronic interactions are taken into account, that is, when $\beta \neq 0$.

If the quadratic terms are nonzero, the motions in the circular trough, as distinct from the linear case, are no longer free rotations. At every point of the trough, as in the linear case, the distorted system performs fast vibrations with the frequency ω_E. When moving along the trough, the distortion of the nuclear configuration changes slowly, assuming a continuous set of geometric figures in the space of E displacements illustrated in Fig. 7.13. However, unlike the linear case, for which the motion of the distorted configuration along the bottom of the trough is uniform, in the presence of quadratic vibronic coupling the changes in nuclear configurations above are hindered (or even reflected) by the quadratic barriers. As a result, the system remains longer at the minima than at the barrier maxima. *The picture as a whole can be characterized as hindered internal rotations of the Jahn–Teller distortions.*

For sufficiently large quadratic terms, when the quadratic barrier Δ according to Eq. (7.44) is on the order of or larger than the vibrational quanta $\hbar\omega_E (\Delta > \hbar\omega_E)$, the $E-e$ problem becomes complicated, with three minima of

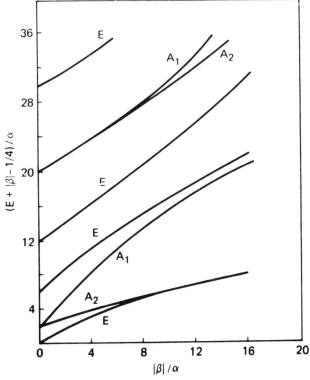

Figure 7.15. Energy levels for the quadratic $E-e$ problem [solutions of Eq. (7.62)] as functions of the ratio of the quadratic to linear vibronic constants, β/α. (After [7.24].)

the adiabatic potential and with high barriers between the minima. In this case the separation of the variables ρ and ϕ is not valid, so the problem should be solved by another technique. Since the minima of the adiabatic potential are deep enough, local quasi-stationary states arise in each of them. Therefore, as far as the lowest vibronic states are concerned, the local states in the three minima can be taken as a starting approximation and then modified due to their interactions by means of perturbation theory. As a result, the local vibrational states in the minima split, and the phenomenon as a whole looks like a quantum-mechanical *tunneling splitting*; it is a special case of tunneling splitting when the electronic and nuclear motions are coupled by the vibronic interaction and the minima are not pure vibrational.

The problem of tunneling splitting in Jahn–Teller systems was solved in different approximations for different cases [7.25, 7.26] (see also [7.1, 7.2, 7.27, 7.28] and references therein). The type of tunneling splitting expected for the $E-e$ problem with strong vibronic coupling is illustrated in Fig. 7.16a. The number of tunneling energy levels equals the number of minima (or more precisely, the number of vibrational states in the minima that interact resulting in the splitting). For the splitting magnitude δ in different states, see [7.1].

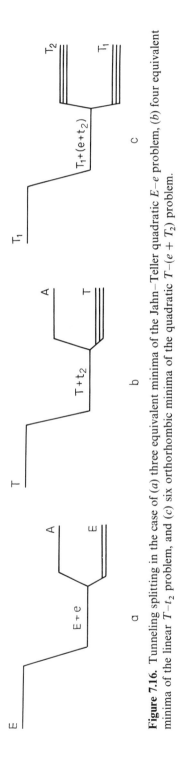

Figure 7.16. Tunneling splitting in the case of (*a*) three equivalent minima of the Jahn–Teller quadratic $E-e$ problem, (*b*) four equivalent minima of the linear $T-t_2$ problem, and (*c*) six orthorhombic minima of the quadratic $T-(e+T_2)$ problem.

Visually, the *transitions between differently oriented distorted configurations may be imagined as pulsating deformations (distortions)*. Assume that as a result of synthesis or external perturbation the system happens to fall into the Jahn–Teller distorted configuration of one of the equivalent minima of the adiabatic potential. Then after a time τ—the lifetime of the system in the minimum (which equals the inverse of the tunneling frequency or the tunneling splitting)—during which the system performs ordinary vibrations, it happens in the equivalent state of another minimum, whose distortion is similar but oriented differently (Fig. 7.11). Then again, after time τ, the system jumps (tunnels) into the third equivalent minimum configuration (or back to the first), and so on.

If the system is octahedral and it is tetragonally distorted in the minimum (as in the $E-e$ problem), for instance, elongated along the fourfold axis, the pulsations of these deformations result in periodic (with a period τ) elongations along each of the three fourfold axes, alternatively. Since $\delta \ll \hbar\omega$, the frequency of pulsating distortions of the molecule is much less than the frequency of the vibrations in the distorted configurations, and the lifetime τ of the latter is much greater than the period of vibrations.

Thus *there are three types of Jahn–Teller dynamics: free internal rotation, hindered rotation, and pulsating motion of Jahn–Teller deformations*.

Triplet States

Threefold orbital degenerate terms are possible for molecular systems that belong to the cubic or icosahedral symmetry point groups (T, T_d, T_h, O, O_h, I, I_h). There are two types of orbital triplets, T_1 and T_2. Since the vibronic effects in these two cases are similar, the consideration of only one of them, say T_2, is sufficient. Denote the three wavefunctions, transforming as the coordinate products yz, xz, xy by $|\xi\rangle$, $|\eta\rangle$, and $|\zeta\rangle$, respectively. In the T-term case, unlike the E term, there are five Jahn–Teller active nontotally symmetric coordinates: two tetragonal Q_θ and Q_ε (E type) and three trigonal Q_ξ, Q_η, and Q_ζ (T_2 type) (see Figs. 7.1 and 7.2 and Table 7.1). For all the other coordinates the adiabatic potential remains parabolic [i.e., they provide no vibronic contributions ε_k^v in Eq. (7.31)]. *The problem is $T - (e + t_2)$*.

The secular equation (7.32) that determines the vibronic parts of the AP $\varepsilon_k^v(Q)$, is of third order. The matrix elements $W_{\gamma\gamma'}^v$ contain two linear vibronic constants, F_E and F_T, and several quadratic constants $G_\Gamma(\Gamma_1 \times \Gamma_2)$ [7.1, 7.2]. Consider first the linear approximation for which $G_\Gamma = 0$. Denote

$$F_E = \langle\zeta|\left(\frac{\partial V}{\partial Q_\theta}\right)_0|\zeta\rangle \qquad F_T = \langle\eta|\left(\frac{\partial V}{\partial Q_\xi}\right)_0|\zeta\rangle \qquad (7.51)$$

According to Eq. (3.44), all the matrix elements of the linear terms of the vibronic interactions can be expressed by means of these two constants, and secular equation (7.32) for $\varepsilon_k^v(Q)$ takes the form

$$
\begin{vmatrix}
F_E(-Q_\theta + \sqrt{3}\,Q_\varepsilon) - 2\varepsilon^v & 2F_T Q_\zeta & 2F_T Q_\eta \\
2F_T Q_\zeta & F_E(-Q_\vartheta - \sqrt{3}\,Q_\varepsilon) - 2\varepsilon^v & 2F_T Q_\xi \\
2F_T Q_\eta & 2F_T Q_\xi & 2F_E Q_\theta - 2\varepsilon^v
\end{vmatrix} = 0 \qquad (7.52)
$$

The three roots of this equation $\varepsilon_k^v(Q)$, $k = 1, 2, 3$, are surfaces in the five-dimensional space of the coordinates $Q_{\Gamma\gamma}$, $\Gamma\gamma = E\vartheta$, $E\varepsilon$, $T\xi$, $T\eta$, $T\zeta$. Together with the parabolic (nonvibronic) parts in Eq. (7.31), they determine the three sheets of the adiabatic potential (in the space of these coordinates), crossing at $Q_{\Gamma\gamma} = 0$:

$$
\varepsilon_k(Q) = \tfrac{1}{2} K_E(Q_\theta^2 + Q_\varepsilon^2) + \tfrac{1}{2} K_T(Q_\xi^2 + Q_\zeta^2 + Q_\eta^2)) + \varepsilon_k^v(Q) \qquad k = 1, 2, 3 \qquad (7.53)
$$

However, the analytical solution of Eq. (7.52) is difficult. Opik and Pryce [7.21] worked out a procedure to determine the extremal points of the surface (7.53) without solving Eq. (7.52). In the particular case when $F_T = 0$, $F_E \neq 0$ (*the T–e problem*), Eq. (7.67) can be solved directly:

$$
\varepsilon_1^v(Q_\theta, Q_\varepsilon) = -F_E Q_\theta
$$

$$
\varepsilon_2^v(Q_\theta, Q_\varepsilon) = \frac{1}{2} F_E Q_\theta + \frac{\sqrt{3}}{2} F_E Q_\varepsilon \qquad (7.54)
$$

$$
\varepsilon_3^v(Q_\theta, Q_\varepsilon) = \frac{1}{2} F_E Q_\theta - \frac{\sqrt{3}}{2} F_E Q_\varepsilon
$$

Substitution of these solutions into Eq. (7.53) yields the AP surface, consisting of a set of paraboloids: Among them, the minima of only those containing the tetragonal Q_θ and Q_ε coordinates are displaced from the origin. In these coordinates the surface has the shape of three equivalent paraboloids intersecting at the point $Q_\theta = Q_\varepsilon = 0$ (Fig. 7.17). The positions of the three minima are given by the coordinates

$$
(Q_0^E, 0) \qquad \left(\frac{1}{2} Q_0^E, \frac{\sqrt{3}}{2} Q_0^E \right) \qquad \left(\frac{1}{2} Q_0^E, \frac{-\sqrt{3}}{2} Q_0^E \right) \qquad (7.55)
$$

where $Q_0^E = F_E/K_E$. For the depth of the minima, the Jahn–Teller stabilization energy, we have:

$$
E_{JT}^E = \frac{F_E^2}{2K} \qquad (7.56)
$$

Note that the relief of the surface sheets near the point of degeneracy (Fig. 7.17) is different from that of the E term (Figs. 7.9 and 7.10): In the case of the

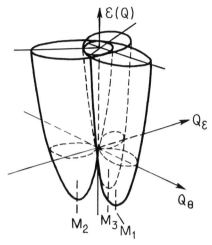

Figure 7.17. AP surface for the Jahn–Teller T–e problem. Three paraboloids intersect at $Q_\theta = Q_\varepsilon = 0$; M_1, M_2, and M_3 are the three minima.

T term there is a real intersection of the surface sheets at the point $Q_\theta = Q_\varepsilon = 0$, whereas for the E term there is a branching at this point instead of an intersection. The wavefunctions for the three paraboloids, $|\zeta\rangle \sim yz$, $|\eta\rangle \sim xz$, and $|\xi\rangle \sim xy$, as distinct from the E-term case, are mutually orthogonal and do not mix by the tetragonal displacements.

In the other particular case, when $F_E = 0$, $F_T \neq 0$ (the $T - t_2$ *problem*) the third-order equation (7.52) cannot be solved directly. Using the method of Opik and Pryce [7.21], one can determine the extremal points of the adiabatic potential without solving Eq. (7.67). For the case in question the surface $\varepsilon(Q_\zeta, Q_\eta, Q_\xi)$ in the space of trigonal coordinates has four minima lying on the C_3 axes of the cubic system at the points $(m_1 Q_0^T, m_2 Q_0^T, m_3 Q_0^T)$, where the four sets of the numbers (m_1, m_2, m_3) are $(1, 1, 1)$, $(-1, 1, -1)$, $(1, -1, -1)$, $(-1, -1, 1)$, and

$$Q_0^T = -\frac{2F_T}{3K_T} \tag{7.57}$$

At these minima the system is distorted along the trigonal axes. The displacements of the atoms corresponding to this distortion for an octahedral system are illustrated in Fig. 7.18. The six ligands, in two sets of three ligands each, move on the circumscribed cube toward two apexes that lie on the appropriate C_3 axes. The depth of the minima, the Jahn–Teller stabilization energy, is

$$E_{JT}^T = \frac{2F_T^2}{3K_T} \tag{7.58}$$

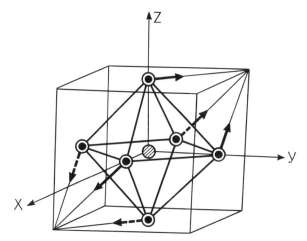

Figure 7.18. Trigonal distortion of an octahedron in an electronic T state (cf. the $Q_\xi + Q_\eta + Q_\zeta$ displacement in Fig. 7.1).

In this case the frequency ω_T of the trigonal T_2 vibrations splits into two [7.1]:

$$\omega_A = \omega_T \qquad\qquad K_A = K_T \qquad\qquad\qquad (7.59)$$

$$\omega_E = (\tfrac{2}{3})^{1/2}\omega_T \qquad K_E = \tfrac{2}{3}K_T \qquad\qquad\qquad (7.60)$$

The electronic wavefunctions in the minima are given by the relation

$$\Psi = \frac{1}{\sqrt{3}}\,(m_1|\xi\rangle + m_2|\eta\rangle + m_3|\zeta\rangle) \qquad\qquad (7.61)$$

with the values of m_1, m_2, and m_3 as given above.

In the T–$(e + t_2)$ *problem*, when the simultaneous interaction with both the tetragonal ($F_E \neq 0$) and trigonal ($F_T \neq 0$) displacements is taken into account, the AP surface in the five-dimensional space of the five coordinates $Q_{\Gamma\gamma}$ is rather complicated, but the extremal points may be obtained using the Opik and Pryce procedure [7.21]. The AP surface for the linear T–$(e + t_2)$ problem has three types of extremal points:

1. Three equivalent tetragonal points, at which only tetragonal coordinates Q_θ and Q_ε are displaced (from the origin $Q_{\Gamma\gamma} = 0$). The coordinates of the minima and their depths are the same as in the linear T–e problem and are given by Eqs. (7.55) and (7.56).

2. Four equivalent trigonal points at which only the trigonal coordinates Q_ξ, Q_η, and Q_ζ are displaced (Fig. 7.18). The minimum coordinates coincide with those obtained in the $T - t_2$ problem [see Eq. (7.57)];

3. Six equivalent orthorhombic points, at each of which one trigonal and one tetragonal coordinate is displaced, and the minima depth is

$$E_{JT}^0 = \tfrac{1}{4}E_{JT}^E + \tfrac{3}{4}E_{JT}^T \tag{7.62}$$

If $E_{JT}^E > E_{JT}^T$, the trigonal points are minima and the tetragonal ones are saddle points; and vice versa, tetragonal points are minima while trigonal ones are saddle points when the opposite inequality holds. The orthorhombic extremal points are always saddle points in the linear approximation (compare with the quadratic case). In particular, when $E_{JT}^E = E_{JT}^T$, all the extremal points (including the orthorhombic ones) have the same depth; in this case a continuum of minima is realized forming a two-dimensional trough on the five-dimensional surface of the AP (in a sense similar to the "hat" in the E-term case) [7.29].

If $K_E = K_T$, the classical motion of an octahedral system along the trough corresponds to the motions of the ligands around identical spheres centered at the apexes of the octahedron [7.1, 7.2]. The displacements of different ligands are correlated: At every instant their radius vectors drawn from the center of the sphere, if shifted to a common origin, form a star whose apexes produce a regular octahedrons rotating around its geometric centre.

In the quadratic approximation the vibronic $T–(e + t_2)$ problem is rather complicated [7.30]. Similarly to the E-term case, the quadratic terms of the vibronic interactions produce significant changes in the shape of the adiabatic potential of the T term. As in the case of linear coupling, there may be either trigonal, or tetragonal minima, but in addition to them, under certain conditions, the orthorhombic extremal points (which are always saddle points in the linear approximation) may become minima. Moreover, there is also a possibility that the orthorhombic minima coexist with either the tetragonal or the trigonal minima [7.1, 7.3].

An important feature of the orbital T terms, as distinct from the E terms, is the large splitting caused by the spin–orbital interaction (in the first order of perturbation theory). Therefore, if the system has unpaired electrons or, more generally, if the total spin S of the state under consideration is nonzero, the problem should be considered for the components that result from spin–orbital splitting of the T term. For instance, the spin doublet ^2T (one unpaired electron) under spin–orbital interaction splits into two components: $^2T = \Gamma_8 + \Gamma_6$, from which the first Γ_8 is a spin quadruplet and the second Γ_6 is a spin doublet. The latter can be treated as a normal doublet, considered above, whereas the quadruplet term Γ_8 requires additional treatment. Since $[\Gamma_8 \times \Gamma_8] = A_1 + E + T_2$, the Jahn–Teller active displacements for the Γ_8 state are E and T_2, and *the problem is* $\Gamma_8–(e + t_2)$, analogous to $T–(e + t_2)$. Many features of the former are similar to the latter but there are also some distinctions. For solutions for $\Gamma_8–(e + t_2)$ as well as further discussion of Jahn–Teller problems, see [7.1–7.4].

Vibronic Reduction of Physical Magnitudes

Reduction in the ground-state physical quantities of electronic nature is one of the important effects of vibronic coupling. It originates from the back influence of the Jahn–Teller nuclear dynamics on the electronic structure and properties.

In the solutions of the vibronic problems obtained above, the ground vibronic state possesses the same type of symmetry, degeneracy, and multiplicity as the initial electronic term in the high-symmetry configuration. This result may be due to the fact that the vibronic coupling terms W [Eq. (7.21)] have the same symmetry as the total Hamiltonian and do not remove the degeneracy of the ground electronic term (in contrast to the rough formulation of the Jahn–Teller theorem, which contains the opposite statement).

The coincidence of the symmetry of the ground-state terms with and without the vibronic coupling allows one to simplify the calculations of many properties. It was shown [7.31] that if taking the vibronic coupling into account, the spin–orbital splitting of the ground state is proportional not only to the spin–orbital coupling constant as in the usual nonvibronic cases, but to this constant multiplied by the overlap integral between the vibrational functions of different minima. Since overlap integrals are always smaller than 1, the vibronic coupling reduces the spin–orbital splitting, sometimes by several orders of magnitude. Ham [7.32] generalized this idea and showed that such a reduction occurs for any physical magnitude provided that its operator depends on electronic coordinates only. This reduction is often called the *Ham effect*.

The following *theorem of vibronic reduction* can be formulated. Suppose that we need to calculate matrix element of the physical magnitude $X_{\Gamma^*\gamma^*}(r)$ (that transforms according to the γ^* line of the Γ^* irreducible representation of the symmetry group of the system) with the functions of the ground vibronic state $\Psi_{\Gamma\gamma}(r, Q)$. Denote the corresponding wavefunctions of the initial electronic term by $\psi_{\Gamma\gamma}(r)$. The theorem of the vibronic reduction states that

$$\int \Psi^*_{\Gamma\gamma 1} X_{\Gamma^*\gamma^*} \Psi_{\Gamma\gamma 2} \, d\tau = K_\Gamma(\Gamma^*) \int \psi_{\Gamma\gamma 1} X_{\Gamma^*\gamma^*} \psi_{\Gamma\gamma 2} \, d\tau \qquad (7.63)$$

where $K_\Gamma(\Gamma^*)$ is a constant that depends on the vibronic properties of the Γ state and only the symmetry Γ^* of the operator X [i.e., $K_\Gamma(\Gamma^*)$ is independent of the nature of X]. The constant $K_\Gamma(\Gamma^*)$ is called the *vibronic reduction factor* (not to be confused with the curvature or force constants, denoted by K_Γ). It follows from this vibronic reduction theory that *if the vibronic reduction factors $K_\Gamma(\Gamma^*)$ are known, there is no necessity to solve the vibronic problem to obtain electronic properties of the ground state; they can be calculated by wavefunctions of the initial electronic term.* In particular, the $K_\Gamma(\Gamma^*)$ constants can be determined from some experimental data and then used for prediction of the vibronic contribution to all the other observables.

Approximate analytic expressions for the vibronic reduction factors in the linear $E-e$ problem, $K_E(A_2)$ and $K_E(E)$, often denoted by p and q, respectively, can be derived directly from the approximate solutions of the problem. It can be shown that for the ideal linear case, the following relation is valid:

$$2q - p = 1 \qquad (7.64)$$

For weak vibronic coupling $E_{\text{JT}} \ll \hbar\omega_E$, we have

$$p \approx \exp\left(-\frac{4E_{\text{JT}}}{\hbar\omega_E}\right) \qquad (7.65)$$

For arbitrary coupling the p value can be derived from numerical solutions; for $0.1 < E_{\text{JT}}/\hbar\omega_E < 3.0$, p obeys the relation [7.32]

$$p = \exp\left[-1.974\left(\frac{E_{\text{JT}}}{\hbar\omega_E}\right)^{0.761}\right] \qquad (7.66)$$

For sufficiently strong vibronic coupling $p = 0$ and $q = \frac{1}{2}$. Note that the relation (7.64) is approximate and has limited applicability. In particular, it fails in the multimode problem as well as when the quadratic vibronic interaction is included. In general, $2q - p \leqslant 1$, and the strength of this inequality may be regarded as an indicator of either multimodal or quadratic (or both) effects. For analytical expressions for the vibronic reduction factors in other cases, see [7.1–7.4].

The expressions above show explicitly that the vibronic reduction of electronic properties depends exponentially on the ratio $E_{\text{JT}}/\hbar\omega$, the dimensionless vibronic constant, which is a measure of the Jahn–Teller effect. The stronger the Jahn–Teller effect, the stronger the reduction. In particular, T_1 operators (to which the orbital angular momentum of the electrons and hence the spin–orbital interaction pertain) are heavily reduced. The essential reduction of the spin–orbital splitting of the ground states (sometimes by one to two orders of magnitude) is one of the clearest manifestations of the Jahn–Teller effect for systems with T terms (the E term of cubic systems is not split by spin–orbital interaction in the first order). Examples of other vibronic reduced quantities are the orbital (anisotropic) part of the Zeeman interaction, interactions between the electronic and nuclear states (dipole–dipole, quadruple, etc.), Coulomb and exchange interactions between Jahn–Teller centers in crystals, and so on.

It must be emphasized that the idea of vibronic reduction loses its simplicity and attractiveness as soon as the physical quantity in question ceases to be defined by the ground state alone. In particular, in the case of strong vibronic coupling the first excited vibronic level Γ_2 is spaced quite close to the ground multiplet Γ_1 (Fig. 7.16), and their mixing may essentially influence the physical properties under consideration. In these cases the vibronic mixing factor $r = K(\Gamma_1|\Gamma|\Gamma_2)$, similar to that in Eq. (7.63), can be introduced [7.32].

7.4. GENERAL TOOL FOR ELECTRONIC CONTROL OF CONFIGURATION INSTABILITY

As mentioned in Section 7.3, the pseudo Jahn–Teller effect may lead to configuration instability which is similar to that of the Jahn–Teller effect. Pseudodegeneracy means a nondegenerate ground state with an excited state relatively low in energy, and the question is how low should the latter be to produce instability. Later in this section we show that there are no limitations on such excited states; moreover, *vibronic (pseudo Jahn–Teller type) mixing of the ground electronic state with the excited states is the only possible source of high-symmetry configuration instability*. This statement can serve as a general tool for considering *how electrons control molecular configurations*.

Nondegenerate (Pseudodegenerate) States

Consider first an easy case of two nondegenerate states Γ and Γ' separated by an energy interval of 2Δ [7.21]. To obtain the adiabatic potential of these states, the vibronic contributions ε_k^v should be evaluated from the solutions of the secular equation (7.32). Assuming that only one coordinate $Q = Q_{\Gamma*}$, $\Gamma* = \Gamma \times \Gamma'$, mixes the two states (in principle, there may be more than one coordinate of the type $\Gamma*$), and taking into account only linear terms in the vibronic interaction W according to Eq. (7.21), we obtain (the energy is read off from the middle of the 2Δ interval between the initial levels)

$$\begin{vmatrix} -\Delta - \varepsilon^v & FQ \\ FQ & \Delta - \varepsilon^v \end{vmatrix} = 0 \tag{7.67}$$

where $F = \langle \Gamma |(\partial V/\partial Q_{\Gamma*})_0| \Gamma' \rangle$ is the off-diagonal linear vibronic constant. Inserting the solutions of Eq. (7.67)

$$\varepsilon^v = \pm (\Delta^2 + F^2 Q^2)^{1/2} \tag{7.68}$$

into Eq. (7.45) and assuming that the force constant is the same in both states $K_0 = K_{0\Gamma} = K_{0\Gamma'}$, we have

$$\varepsilon_{\pm} = \tfrac{1}{2} K_0 Q^2 \pm (\Delta^2 + F^2 Q^2)^{1/2} \tag{7.69}$$

or, after expanding the second term in Q,

$$\varepsilon_{\pm}(Q) = \frac{1}{2}\left(K_0 \pm \frac{F^2}{\Delta}\right)Q^2 \pm \frac{F^4}{\Delta^3}Q^4 \mp \cdots \tag{7.70}$$

It is seen from these expressions that as a result of the vibronic coupling the two adiabatic potential curves change in different ways: In the upper sheet the curvature (the force constant) increases, whereas in the lower one it decreases. If $\Delta > F^2/K_0$, the minima of both states remain at the point $Q = 0$, as in the absence of vibronic mixing. This is the case of *weak pseudo Jahn–Teller effect* (Fig. 7.19a). However, if

$$\Delta < \frac{F^2}{K} \tag{7.71}$$

the curvature of the lower sheet of the adiabatic potential becomes negative and the system is unstable with respect to the Q displacements. This is the *strong pseudo Jahn–Teller effect* (Fig. 7.19b). The minima of the adiabatic potential are given here by $\pm Q_0$,

$$Q_0 = \left(\frac{F^2}{K^2} - \frac{\Delta^2}{F^2}\right)^{1/2} \tag{7.72}$$

with the curvature at the minima points

$$K = K_0\left(1 - \frac{K_0\Delta}{F^2}\right) \tag{7.73}$$

For $\Delta = F^2/K_0$, the curvature of the ground state is zero everywhere.

Consider now a more complicated case of a transition metal complex with the electron configuration d^0, for instance, four-valency titanium in an octahed-

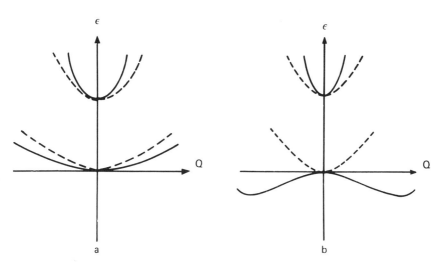

Figure 7.19. AP behavior in the cases of (*a*) weak pseudo Jahn–Teller effect (the ground state is softened but remains stable), and (*b*) strong pseudo Jahn–Teller effect (the ground state becomes unstable at $Q = 0$). The terms without vibronic coupling are shown by dashed lines.

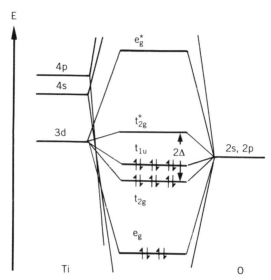

Figure 7.20. HOMO and LUMO for the $TiO_6{}^{8-}$ cluster in the $BaTiO_3$-type crystal.

ral coordination of oxygen atoms, as in the cluster $[TiO_6]^{8-}$ in $BaTiO_3$ [7.1, 7.33]. For an approximate treatment, one may restrict the problem by considering the vibronic mixing of a group of close-in-energy electronic terms that are well separated from the other terms. A typical qualitative scheme of the MO energy levels and electron occupation numbers for this complex is shown in Fig. 7.20. It is seen that for a sufficiently full consideration at least nine MOs, t_{2g}, t_{1u}, and t_{2g}^* (occupied by 12 electrons in the $TiO_6{}^{8-}$ cluster), must be taken into account. With allowance for interelectron repulsion, these states form the ground $^1A_{1g}$ and excited states, from which those of the same multiplicity (singlets) are $^1A_{2u}$, ^1E_u, $^1T_{1u}$, $^1T_{2u}$, $^1A_{2g}$, ^1E_g, $^1T_{1g}$, and $^1T_{2g}$.

In most cases of practical use the energies and wavefunctions, as well as covalency parameters for the MOs, are unknown. For a qualitative analysis of the vibronic effects, the vibronic interaction may be considered at an earlier stage before covalency and multielectron term formation. Then the nine atomic functions, three $3d_\pi$ functions of the Ti^{4+} ion (d_{xy}, d_{xz}, d_{yz}), and six combinations of the $2p_\pi$ functions of the O^{2-} ions (forming the foregoing nine MO t_{2g}, t_{1u}, and t_{2g}^*) may be taken as a basis of the vibronic treatment. These states mix by T_{1u} displacements which have three components, Q_x, Q_y, and Q_z, one of which is shown in Fig. 7.21 [the problem is thus $(A_{1g} + T_{1u}) - t_{1u}$].

According to the Wigner–Eckart theorem (Section 3.4), the matrix elements of the ninth-order secular equation of perturbation theory in the linear approximation with respect to the vibronic coupling contain only one vibronic constant:

$$F = \langle 2p_y | \left(\frac{\partial V}{\partial Q_x} \right)_0 | 3d_{xy} \rangle \tag{7.74}$$

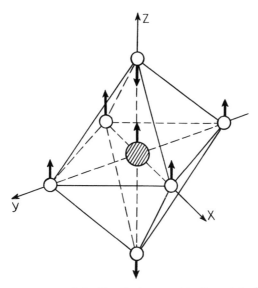

Figure 7.21. One component of the T_{1u} displacement in the octahedral TiO_6 cluster.

Omitting the equation itself, we present its solutions (2Δ is the energy interval between the $3d_{xy}$ and $2p_y$ states):

$$\varepsilon^v_{1,2} = \pm[\Delta^2 + F^2(Q^2_x + Q^2_y)]^{1/2}$$

$$\varepsilon^v_{3,4} = \pm[\Delta^2 + F^2(Q^2_y + Q^2_z)]^{1/2}$$

$$\varepsilon^v_{5,6} = \pm[\Delta^2 + F^2(Q^2_x + Q^2_z)]^{1/2} \qquad (7.75)$$

$$\varepsilon^v_{7,8,9} = -\Delta$$

From these nine levels only the lower six are occupied in the ground state by the 12 electrons above. Substituting ε^v_1, ε^v_3, and ε^v_5 from (7.75) into Eq. (7.31), we obtain the following expression for the ground-state adiabatic potential ($\varepsilon_{7,8,9}$ are independent of Q and can be excluded from further consideration):

$$\varepsilon(Q_x, Q_y, Q_z) = \tfrac{1}{2}K_0(Q^2_x + Q^2_y + Q^2_z) - 2\{[\Delta^2 + F^2(Q^2_x + Q^2_y)]^{1/2}$$

$$+ [\Delta^2 F^2(Q^2_y + Q^2_z)]^{1/2} + [\Delta^2 + F^2(Q^2_x + Q^2_z)]^{1/2}\} \qquad (7.76)$$

The shape of this surface depends on the relation between the constants Δ, F, and K_0. If $\Delta > 4F^2/K_0$, the surface has one minimum at the point $Q_x = Q_y = Q_z = 0$, and the system remains undistorted. This is the weak pseudo Jahn–Teller effect (Fig. 7.19a). However, if

$$\Delta < \frac{4F^2}{K_0} \qquad (7.77)$$

the surface (7.76) acquires a rather complicated shape with four types of extremal points:

1. One maximum at $Q_x = Q_y = Q_z = 0$ (dynamic instability).
2. Eight minima at the points $|Q_x| = |Q_y| = |Q_z| = Q_0^{(1)}$,

$$Q_0^{(1)} = \left(\frac{8F^2}{K^2} - \frac{\Delta^2}{2F^2}\right)^{1/2} \tag{7.78}$$

with the Jahn–Teller stabilization energy

$$E_{JT}^{(1)} = 3\left(\frac{4F^2}{K} + \frac{\Delta^2 K}{4F^2} - 2\Delta\right) \tag{7.79}$$

   At these minima the Ti atom is displaced along the trigonal axes, equally close to three oxygen atoms and removed from the other three.
3. Twelve saddle points at $|Q_p| = |Q_q| \neq 0$, $Q_r = 0$, $p, q, r = x, y, z$ (with a maximum in the section r and minima along p and q). At these points the Ti atom is displaced toward two oxygen atoms lying on the p and q axes, respectively.
4. Six saddle points at $Q_p = Q_q = 0$, $Q_r = Q_0^{(2)}$,

$$Q_0^{(2)} = \left(\frac{16F^2}{K^2} - \frac{\Delta^2}{F^2}\right)^{1/2} \tag{7.80}$$

with a depth

$$E_{JT}^{(2)} = 2\left(\frac{4F^2}{K} - \frac{\Delta^2 K}{4F^2} - 2\Delta\right) \tag{7.81}$$

   With covalency and multielectron term formation included, these results, especially their quantitative expression, are modified, but the main qualitative conclusions do not alter.

   The origin of the instability of the position of the Ti^{4+} ion in the center of the octahedron can be given a visual treatment (similar to that in Fig. 7.22, discussed below). When the Ti atom is in the central position exactly, the overlap of its d_{xy} AO with the appropriate (T_{1u}) combination of the oxygen p_y AO is zero on symmetry (positive overlaps are compensated by negative ones). However, if the Ti atom is shifted toward any of the oxygens, resulting in its off-center position (Fig. 7.21), the overlap becomes nonzero and produces a new covalent bonding that lowers the energy of such distortions.

   Similar treatment is possible for tetrahedral complexes of type MA_4. For them, in the strong vibronic coupling limit and under certain vibronic mixing conditions, four equivalent minima are expected. In each of these minima one bond M-A is longer or shorter than the other three, which remain identical.

In stereochemistry and reactivity problems as well as in spectroscopy and crystal chemistry (Sections 9.2, 9.4, 11.2, etc.), other cases of pseudo Jahn–Teller effects may be significant. In particular, the vibronic mixing of an E term with a A_1 term under E displacements is often encountered [$(E + A_1) - e$ problem]. We consider here this type of mixing for a system with C_{4v} symmetry [7.34]. The two wavefunctions of the E term transform as the x and y coordinates. If we denote the two components of the E mode by Q_x and Q_y and the energy gap between the E and A_1 levels by 2Δ, the secular equation (7.32) in the linear approximation takes the form:

$$\begin{vmatrix} -\Delta - \varepsilon^v & 0 & FQ_x \\ 0 & -\Delta - \varepsilon^v & FQ_y \\ FQ_x & FQ_y & \Delta - \varepsilon^v \end{vmatrix} = 0 \tag{7.82}$$

In polar coordinates (7.37) the roots of this equation are:

$$\varepsilon_{1,3}^v = \pm(\Delta^2 + F^2\rho^2)^{1/2} \qquad \varepsilon_2^v = -\Delta \tag{7.83}$$

and for the adiabatic potential of the ground state, we get

$$\varepsilon(\rho, \phi) = \tfrac{1}{2}K_E\rho^2 - (\Delta^2 + F^2\rho^2)^{1/2} \tag{7.84}$$

It is seen that in the linear approximation the adiabatic potential is independent of the angle ϕ—it is a surface of revolution. Similar to the two-level case considered at the beginning of this subsection, if $\Delta < F^2/K_E$, the surface has a maximum at the point $\rho = 0$ (dynamic instability) and a circular trough at $\rho = \rho_0 = [(F^2/K_E^2) - (\Delta^2/F^2)]^{1/2}$. The depth of the trough [read off the point $\varepsilon(0) = -\Delta$] is

$$E_{\text{PJT}} = \frac{F^2}{2K_E} + \frac{\Delta^2 K_E}{F_E^2 - \Delta}$$

If quadratic terms of the vibronic interaction are taken into account, two quadratic constants G_1 and G_2 must be introduced. The secular equation becomes complicated and the adiabatic potential, unlike the linear case, becomes dependent on the angle ϕ (acquiring the initial symmetry C_{4v}): Four minima regularly alternating with four saddle points occur on the adiabatic potential as a function of ϕ [along the trough (7.84)]. At the extremal points, $\phi = n\pi/4$, where $n = 0, 1, 2, \ldots, 7$. If $G_1 > G_2$, the minima are given by

$$\phi_0(\text{min}) = \frac{n\pi}{4} \qquad n = 1, 3, 5, 7 \tag{7.85}$$

and the saddle points are at $n = 0, 2, 4, 6$. In the opposite case, $G_1 < G_2$, the minima and saddle points interchange, and if $G_1 = G_2$ the surface preserves the trough of the linear approximation.

The examples above show that for relatively close electronic terms (pseudodegeneracy) *the adiabatic potential as a result of the pseudo Jahn–Teller effect acquires features similar to those obtained in the Jahn–Teller effect: instability of the high-symmetry configuration and several equivalent (or a continuum of) minima*. The criterion of the strong effect (or of pseudo Jahn–Teller instability) (7.71) or (7.77) (in every other case this condition looks similar) contains three parameters, Δ, K_0, and F, and therefore it may be "soft" for any of them taken apart. In particular, a strong effect may occur for large values of the energy gap Δ if the force constant K_0 is sufficiently small and the vibronic constant F is large.

Cases when the criterion of strong pseudo Jahn–Teller effect is not satisfied, $\Delta > F^2/K$, are also important. Indeed, although the configuration is stable (the AP has a minimum in this configuration), the curvature of the ground state, as indicated above, is lowered, *the nuclear configuration is softened* in the Q direction, and it is more softened when F is larger and Δ is smaller (for a given K_0 value). This situation is significant, for instance, in the investigation of chemical reaction mechanisms (Section 11.2).

Along with strong similarity, there are essential differences between the Jahn–Teller and pseudo Jahn–Teller effects. An important feature of the pseudoeffect is that the two (or more) mixing electronic states Γ and Γ' may belong to different irreducible representations of the point group of the system, whereas in the Jahn–Teller case they belong to the same representation, $\Gamma = \Gamma'$. Consequently, in the pseudo Jahn–Teller effect the direction of distortion may have any symmetry that is possible in the symmetry group of the system under consideration, whereas in the Jahn–Teller effect these directions are limited by Jahn–Teller active modes (Section 7.3). In particular, for systems with an inversion center, the two mixing states Γ and Γ' in the pseudo Jahn–Teller effect may possess opposite parity. Then the vibronic constant $F_{\Gamma*}^{(\Gamma\Gamma')}$ is nonzero for odd nuclear displacements Q, which remove the inversion center and form a dipole moment (*dipolar instability*). As a result, the system in its minima configurations has a dipole moment (which is zero in the undistorted configuration). This dipolar instability is impossible in Jahn–Teller systems with inversion centers since the mixing electronic terms have the same symmetry $\Gamma = \Gamma'$ and hence the active modes can be even only.

Similarly to Jahn–Teller distortions, the origin of instability of the high-symmetry configuration and its distortion due to the strong pseudo Jahn–Teller effect can be illustrated in simple images. Consider, for instance, a square-planar complex of the type MA_4 with D_{4h} symmetry, and suppose that the d_{z^2} orbital of the metal M (the A_{1g} orbital) is the HOMO, while the LUMO of A_{2u} symmetry is formed by the four p_z orbitals of the ligands (Fig. 7.22) (or vice versa, the d_{z^2} orbital is the LUMO and A_{2u} is the HOMO). In the planar

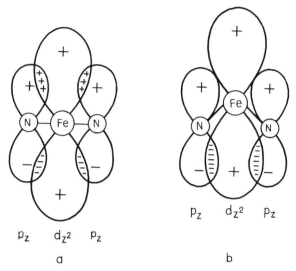

$$p_z \quad d_{z^2} \quad p_z$$

a

$$p_z \quad d_{z^2} \quad p_z$$

b

Figure 7.22. Visual treatment of the origin of the pseudo Jahn–Teller effect using the N—Fe—N fragment of the square-planar FeN_4 group as an example (cf. iron porphyrin, Section 9.2): (*a*) when Fe is in the N_4 plane (on the N—Fe—N line) the d_π–p_π overlap between the HOMO (nitrogen p_π) and LUMO (iron d_{z^2}) orbitals is zero; (*b*) the out-of-plane displacement of the Fe atom results in nonzero d_π–p_π overlap and bond formation which lowers the curvature of the adiabatic potential in the direction of such displacement.

configuration these two MOs do not mix, since the overlap integral S is zero (Fig. 7.22*a*).

However, if the atom M displaces out of and transversely to the plane (in the A_{2u} direction), a nonzero overlap occurs and the two orbitals mix to form a π bond between the metal and the ligands (Fig. 7.22*b*). As a result, the energy of the ground state is lowered (compared with that without mixing) and the curvature of the adiabatic potential in the direction of the A_{2u} displacement decreases (weak effect, Fig. 7.19*a*) or becomes negative (strong effect, Fig. 7.19*b*). The latter case takes place when the covalency of the new bond formed is superior to the changes in strain energy. For more on the origin of vibronic instability, see Eqs. (7.95) to (7.98) and Table 7.4.

Uniqueness of the Vibronic Mechanism of Configuration Instability

Consider in more detail the criterion of the strong pseudo Jahn–Teller effect given by the inequality (7.71). The three parameters Δ, the energy gap between the ground and excited state, F, the vibronic coupling constant, and K_0, the primary force constant, can acquire arbitrary values (or at least a large range of values), which means that the condition (7.71) can be obeyed for any value of one of these parameters, provided that the other two are appropriate. It follows that *the pseudo Jahn–Teller effect can be operative for any polyatomic system without a priori restrictions.*

Note that in the pseudo Jahn–Teller case, all the electrostatic (attraction and repulsion) forces in the reference nuclear configuration are equilibrated; that is, the first derivative of the adiabatic potential in the given direction Q is zero. Thus the pseudo Jahn–Teller effect is rather a case of *unstable equilibrium*, or *dynamic instability*, as distinct from the static instability in the Jahn–Teller effect when the Coulomb forces are not compensated. In both cases the term *configuration instability* means instability of the reference configuration with respect to certain ditortions, which does not necessarily imply absolute instability of the molecular system with respect to its component parts; the system may be stable in the distorted configuration.

The pseudo Jahn–Teller effect provides a mechanism of electronic control of nuclear configuration instability via vibronic coupling. In this mechanism any two electronic states of a given nuclear configuration, ground 1 and excited 2, can cause its instability in a certain (symmetrized) direction Q, provided that inequality (7.71) is satisfied. From this statement an important question emerges: Is this vibronic mechanism of dynamic instability (unstable equilibrium) the only one possible, or are there other mechanisms that control this instability? In other words, *is the condition of instability (7.71) both necessary and sufficient, or it is only sufficient?*

To put this question more rigorously, consider the expression for the curvature of the adiabatic potential of the system at the point $Q_{\Gamma*} = 0$ in the direction of the symmetrized coordinate $Q_{\Gamma*}$ (7.23) (hereafter, for simplicity, the index $\Gamma*$ is omitted):

$$K^{\Gamma} = K_0^{\Gamma} + K_v^{\Gamma} \tag{7.86}$$

where (note that $dH/dQ = dV/dQ$)

$$K_0^{\Gamma} = \langle \Gamma | \left(\frac{\partial^2 H}{\partial Q^2} \right)_0 | \Gamma \rangle \tag{7.87}$$

$$K_v^{\Gamma} = -\sum_{\Gamma}' \frac{|F^{\Gamma\Gamma'}|^2}{\Delta_{\Gamma'\Gamma}} \tag{7.88}$$

and $F^{\Gamma\Gamma'}$ and $\Delta_{\Gamma'\Gamma}$ are denoted above. Equation (7.86) can be obtained by means of second-order perturbation theory, but it is an exact expression for the curvature as the coefficient at $\frac{1}{2}Q^2$ in the dependence of the adiabatic potential on Q; it was discussed first by Bader [7.35, 7.36]. The first term, K_0^{Γ}, according to (7.87), the mean value of the curvature operator in the ground state, has the meaning of the contribution (to K^{Γ}) from the Q displacement of the nuclei in a fixed (frozen) electronic distribution. The second term, K_v^{Γ}, according to (7.88), gives the negative contribution to K^{Γ} that arises because the electrons follow the nuclei (at least partly), and hence the electronic states change under nuclear displacements. This is *the relaxation term*.

On the other hand, this term describes the contribution of the vibronic mixing of the ground state with excited states to the K^Γ value of the former. A similar contribution to K from one excited state as a result of the pseudo Jahn–Teller effect is given by Eq. (7.70). As distinct from this two-level pseudo Jahn–Teller effect, Eq. (7.88) gives the pseudo Jahn–Teller contribution to the ground-state curvature arising from *all* the excited states (*multilevel pseudo Jahn–Teller effect*). K_v^Γ is therefore the *vibronic contribution* to the curvature, as opposed to K_0^Γ, which is the proper or *ground-state contribution*.

Since K_v^Γ is negative, it always lowers the curvature of the ground state: The vibronic coupling destabilizes the system (in the ground state). If $K_0^\Gamma > 0$ and $|K_v| > K_0$, the system possesses a negative total curvature $K^\Gamma < 0$, and hence the vibronic coupling to the excited states causes the instability of the ground state. The condition

$$\sum_{\Gamma'} \frac{|F^{(\Gamma\Gamma')}|^2}{\Delta_{\Gamma'\Gamma}} > K_0^\Gamma \tag{7.89}$$

is thus a multilevel analogy of the criterion of the two-level pseudo Jahn–Teller instability (7.71). In the infinite sum in (7.89), only several terms (in most cases one to three) are significant, the other being negligible, due to either selection rules ($\Gamma^* \in \Gamma \times \Gamma'$) or large denominators $\Delta_{\Gamma'\Gamma}$.

Now, if $K_0^\Gamma < 0$, the system is unstable without vibronic coupling. But if $K_0^\Gamma > 0$, condition (7.89) becomes necessary for the instability. Thus if one proves that

$$K_0^\Gamma > 0 \tag{7.90}$$

the condition of vibronically induced dynamic instability is both necessary and sufficient and the vibronic mechanism of instability is the only possible one in polyatomic systems.

There are several particular proofs of the inequality $K_0 > 0$ (the superscript Γ is omitted) for different polyatomic systems. First the formulation of the problem and a general but not very rigorous proof has been given [7.37]. Then this proof was significantly improved and expanded [7.38]. The first mathematically rigorous proof of the inequality $K_0 > 0$ was obtained for diatomics [7.39], and then it was given for cubic systems (crystals) and some linear chains [7.40, 7.41] (see also [7.1]).

To give an idea of these proofs, consider the expression of K_0 after (7.87) in more detail. The second derivative of the Hamiltonian with respect to the nuclear normal coordinates means the derivative of the electron–nuclear and nuclear–nuclear interactions given by (7.3'). Since the transformation from normal to Cartesian coordinates is unitary, $Q_k = \Sigma_i a_k U_i$ (U_i stands for the coordinates X_i, Y_i, and Z_i),

$$\frac{\partial^2 H}{\partial Q_k^2} = \sum_i a_{ki}^2 \frac{\partial^2 V}{\partial U_i^2} \tag{7.91}$$

For the Coulomb interactions terms V there is a general formula

$$\langle 0| \frac{\partial^2 V}{\partial X_i^2} |0\rangle = \frac{4\pi}{3} e Z_i \rho_i + e Z_i Q_{xx}^{(i)} \tag{7.92}$$

where ρ_i is the numerical value of the electron density at the ith nucleus and $Q_{xx}^{(i)}$ is the gradient of the electric field created by the electrons and other nuclei at the given one i. By definition, $Q_{xx} + Q_{yy} + Q_{zz} = 0$ and $\rho_i > 0$. Hence the first term in (7.93) is always positive, and if the environment of the displacing nuclei under consideration is cubic, $Q_{xx} = Q_{yy} = Q_{zz} = 0$ and

$$K_0 = \langle 0| \left(\frac{\partial^2 H}{\partial Q_k^2}\right)_0 |0\rangle = 4\pi e \sum_\alpha a_{k\alpha}^2 Z_\alpha \rho_\alpha \geqslant 0 \tag{7.93}$$

If the environment is not cubic, the first term in (7.92) is still positive, but the second one, the gradient of the electric field at the given nucleus, is nonzero and it can be either positive or negative; but it was shown that the second term is much smaller than the first [7.1, 7.37, 7.38, 7.41-7.43].

To confirm the analytical deductions and to reveal the excited states that cause the instability of the ground state, *ab initio* calculations were performed for a series of molecular systems in a dynamically unstable configuration where $K < 0$ [7.38]. Some results are presented in Table 7.3. The K_0 values are calculated directly by Eq. (7.87), with the ground-state wavefunctions obtained by numerical calculations in the 3G-STO approximation (Section 5.3). In Table 7.3 the vibronic coupling parameters are also given.

For the systems in this table the distortions (indicated in column 2 by their symmetry) involve displacement of the hydrogen atoms, for which, as follows from the analytical deductions [7.38], the inequality $K_0 > 0$ is less favorable. Nevertheless, the results of the calculations (column 5) show explicitly that indeed $K_0 > 0$ in these cases also. There are many calculations in literature showing that K_0 is always positive. Supported by other types of analytical (including rigorous mathematical) proofs, these results show that the inequality (7.90) is valid in all cases of polyatomic Coulomb equilibrated systems.

Physical Consequences

The proof that $K_0 > 0$ means that configuration instability with $K < 0$ is due to and only to the vibronic coupling to the excited states. For the sake of simplicity, consider the case when only one excited state contributes significantly to inequality (7.89), while the others are negligible. In this case

Table 7.3. *Ab initio* **calculations of vibronic constants** (F^{0i})**, nonvibronic** (K_0)**, and vibronic** (K_v) **contributions to the force constant** (K) **for some molecular hydrides in the unstable configuration with respect to hydrogen displacements**[a]

System, Symmetry	Coordinate of Instability	$F^{(0i)}$ (10^{-4} dyn)	Δ_{i0} (eV)	K_0 (mdyn/Å)	$K_v = \Sigma_i K_v^i$ (mdyn/Å)	$K = K_0 + K_v$ (mdyn/Å)
$H_3, D_{\infty h}$	Π_u	4.84	12.0	0.13	-0.24	-0.11
NH_3, D_{3h}	A_2''	8.76	14.0	0.43	-0.68	-0.25
BH_4, D_{4h}	B_{2u}	6.04	6.8	0.83	-1.34	-0.51
		8.99	22.8			
CH_4, D_{4h}	B_{2u}	6.98	11.1	0.69	-1.27	-0.58
		11.45	25.8			
NH_4^+, D_{4h}	B_{2u}	6.32	14.2	0.44	-0.87	-0.43
		10.55	33.4			
OH_4^{2+}, D_{4h}	B_{2u}	5.14	16.0	0.29	-0.51	-0.22
		9.30	36.4			
AlH_4^-, D_{4h}	B_{2u}	5.34	8.7	1.06	-1.18	-0.12
		9.48	19.8			
SiH_4, D_{4h}	B_{2u}	5.04	7.3	0.71	-1.15	-0.44
		9.15	23.4			
PH_4^+, D_{4h}	B_{2u}	4.53	4.6	0.48	-1.11	-0.63
		8.96	27.2			
SH_4^{2+}, D_{4h}	B_{2u}	3.85	8.4	0.37	-0.66	-0.30
		9.02	25.5			

[a]Two values of $F^{(0i)}$ and Δ_{i0} correspond to two excited states that contribute to the instability.

formulation of the problem is reduced to that of a usual two-level pseudo Jahn–Teller effect considered in Section 7.4. According to Eq. (7.70), the curvatures of the adiabatic potentials of the two states at the point of instability are

$$K_{1,2} = K_0 \pm \frac{F^2}{\Delta} \qquad (7.94)$$

Because $K_0 > 0$, the curvature of the excited state $K_0 + F^2/\Delta$ is positive while that of the ground state $K_0 - F^2/\Delta$, under condition (7.71), is negative. Thus *instability of the ground state is accompanied by a stable excited state,* coupling with which produces the instability. This result—*prediction of the existence of stable excited states in dynamically unstable ground-state configurations*—is one of the general consequences of vibronic instability, which has interesting physical and chemical applications (Chapters 9 to 11).

If more than one excited state contributes to the instability of the ground state in Eq. (7.89), the relation between the K values of all these states becomes more complicated, but the general idea is the same: The negative contribution to the curvature of the ground state, which makes it unstable, equals the sum of positive contributions to the excited states. *As a result of vibronic mixing, the excited states become stabilized.* If the excited state that causes the instability of the ground state is occupied by electrons, the instability disappears: Eq. (7.94) shows that the total change of curvature of the two interacting states equals zero. Therefore, vibronic coupling between fully occupied MOs does not contribute to the instability.

For a further understanding of the origin of vibronic instability, the terms in the sum K_v according to (7.88) may be divided into two groups:

1. The basis wavefunctions $|0\rangle$ and $|i\rangle$ in the matrix element $F^{(0i)}$ are mainly from the same atom. In this case the term $-|F^{(0i)}|^2/\Delta_{0i}$ can be interpreted as the contribution of the polarization of this atom by the displacements of other atoms. For instance, for the instability of the central position of the Ti ion in the octahedron of oxygens in the $\text{TiO}_6{}^{8-}$ cluster of BaTiO_3 with respect to off-center displacements, discussed above, the contribution of the polarization of the oxygen atom by the off-center displacement of the titanium ion is given by the mixing of the oxygen $2p$ and $3s$ atomic function under this displacement [7.40, 7.41]:

$$K_v^{\text{pol}} = -\frac{|\langle 2p_{\sigma z}(\text{O})|(\partial V/\partial Q_z)_0|3s(\text{O})\rangle|^2}{\Delta_{2p3s}} \tag{7.95}$$

Since the integrals $F^{(0i)} = \langle 0|(\partial V/\partial Q)_0|i\rangle$ are calculated with the orthogonal (ground and excited) wavefunctions of the same atom, then, transforming the symmetrized coordinate Q into Cartesian coordinates and taking the corresponding derivative of the Coulomb potential $V = e^2|\mathbf{r} - \mathbf{R}_\beta|^{-1}$, we come to integrals of the type $I_x = \langle 0|(x - X_\beta)/|\mathbf{r} - \mathbf{R}_\beta|^3|i\rangle$, where x are the electronic coordinates of the polarizing atom and X_β are the nuclear coordinates of the displacing atoms. If we assume that approximately R_β is much larger then the atomic size (which is true for the second and next coordination spheres), then $I_x \approx R_\beta^{-3}\langle 0|x|i\rangle$, and the polarization contribution is

$$K_v^{\text{pol}} \sim e^2 \alpha_x R_\beta^{-6} \tag{7.96}$$

where, according to quantum mechanics,

$$\alpha_x = \frac{e^2|\langle 0|x|i\rangle|^2}{\Delta_{0i}} \tag{7.97}$$

is the part of the atomic polarizability in the x direction that is due to the contribution of the ith excited state (the summation over i gives the full atomic polarizability in this direction).

2. The two functions in $F^{(0i)}$ are from two different (near-neighbor) atoms. In this case the vibronic contribution is due to new covalency produced by the distortion. Indeed, in the reference configuration the overlap of these two electronic states is zero (they are orthogonal), hence their vibronic mixing means that a nonzero overlap occurs under the low-symmetry displacements Q.

For the Ti ion's off-center displacements with respect to the oxygen octahedron, the covalent contribution is due to the new overlap of the ground-state t_{1u} combination of the highest occupied $2p_{\pi z}$ functions of the oxygen atoms with the lowest unoccupied d_{xz} function of the titanium ion:

$$K_v^{\text{cov}} = -\frac{|\langle 2p_{\pi z}(\text{O})|(\partial V/\partial Q_x)_0|3d_{xz}(\text{Ti})\rangle|^2}{\Delta_{2p3d}} \tag{7.98}$$

The new overlap (which is forbidden by symmetry in the reference configuration) produces new (additional) covalency. Inequality (7.89), made possible by this term, means that with the new covalency the energy is lower than that of the reference configuration, resulting in instability.

Both kinds of vibronic contribution to instability, new covalency and atomic polarization, may be significant, but the numerical calculations performed so far show that the covalency contribution is an order of magnitude larger than the polarization. Table 7.4 shows three examples of such calculations [7.42]:

Table 7.4. New covalency K_v^{cov} versus polarization K_v^{pol} contributions to the instability of the high-symmetry configuration of several polyatomic systems

	NH_3	$CuCl_5^{3-}$	TiO_6 in $BaTiO_3$
Reference configuration	Planar D_{3h}	Trigonal bipyramidal D_{3h}	Octahedral O_h
Instability coordinate	A_2''	E'	T_{1u}
Ground state[a]	$^1A_1'[2p_z(\text{N})]$	$^2A_1'[3d_{z^2}(\text{Cu})]$	$^1A_{1g}[2p(\text{O})]$
Excited state–cov[b]	$^1A_2''$	$^2E'$	$^1T_{1u}$
	$2p_z(\text{N}) \to 1s(\text{H})$	$3s(\text{Cl}) \to 3d(\text{Cu})$	$2p(\text{O}) \to 3d(\text{Ti})$
K_v^{cov}	$-0.62\,\text{mdyn/Å}$	$-2.85 \cdot 10^{28}\,\text{s}^{-2c}$	
Excited state–pol[b]	$^1A_2''$	$^2E'$	$^1T_{1u}$
	$2p_z(\text{N}) \to 3s(\text{N})$	$3d_{xy}(\text{Cu}) \to 3d_{z^2}$	$2p(\text{O}) \to 3s(\text{O})$
K_v^{pol}	$-0.06\,\text{mdyn/Å}$	$-0.05 \cdot 10^{28}\,\text{s}^{-2c}$	
$K_v^{\text{cov}}/K_v^{\text{pol}}$	$1.03 \cdot 10$	$5.7 \cdot 10$	$1.1 \cdot 10$

[a]The main contributing AOs are indicated in brackets.
[b]The corresponding one-electron excitations are shown.
[c]In mass-weighted units.

the instability of NH_3 in the planar configuration with respect to out-of-plane displacements of the nitrogen atom (toward the stable C_{3v} configuration) [7.38]; $CuCl_5^{3-}$ in the trigonal–bipyramidal configuration with respect to E' displacements (toward a square pyramid) [7.43]; and the TiO_6^{8-} cluster in $BaTiO_3$ with respect to T_{1u} (Ti off-center) displacements initiating the spontaneous polarization of the crystal [7.33, 7.41]. In all these examples the new covalency contribution to the instability is indeed much more significant, by at least an order of magnitude.

The fundamental consequence of the uniqueness of the vibronic instability, the existence of stable excited states that cause the instability of the ground state, has applications in various fields of chemistry. For example, it predicts stable excited states for the unstable ground-state configurations. In stereochemistry (Section 9.2) this means that the low-symmetry configurations of molecular systems have higher symmetry in some excited states (e.g., bent triatomic molecules are expected to be linear in the excited state that is coupled to the ground state via bending displacements). Considering the symmetries of the appropriate MOs, *one can control (manipulate) the geometry (configuration instability) of molecular systems by means of electronic rearrangements: excitation, ionization, MO occupation changes by coordination, redox processes, and so on.* This is also related directly to photochemical reaction (Section 11.3). Some other conclusions from these results are discussed in the following sections: The unstable configuration of the transition state of any chemical reaction may be stable in its excited state (Sections 11.1 and 11.2), internal pseudorotations in coordination compounds are controlled by excited states via vibronic coupling (Section 9.2), and structural phase transformations in condensed media are of vibronic origin (Section 9.4).

Note that in addition to instability of the high-symmetry configuration, the strong vibronic mixing of the ground state with the excited states causes anharmonicity in the nuclear motions of the former [7.1] (*vibronic anharmonicity*). In some cases the vibronic anharmonicity is more important than the proper anharmonicity caused by the higher-order terms in the expansion (7.3).

REFERENCES

7.1. I. B. Bersuker and V. Z. Polinger, *Vibronic Interactions in Molecules and Crystals*, Springer-Verlag, New York, 1989, 422 pp.

7.2. R. Englman, *The Jahn–Teller Effect in Molecules and Crystals*, Wiley, London, 1972, 350 pp.

7.3. I. B. Bersuker, *The Jahn–Teller Effect and Vibronic Interactions in Modern Chemistry*, Plenum Press, New York, 1984, 320 pp.

7.4. G. Fischer, *Vibronic Coupling: The Interaction Between the Electronic and Nuclear Motions*, Academic Press, London, 1984, 222 pp.

7.5. S. Sugano, Y. Tanabe, and H. Kamimura, *Multiplets of Transition Metal Ions in Crystals*, Academic Press, New York, 1970, 331 pp.

7.6. L. D. Landau and E. M. Liphshitz, *Quantum Mechanics: Nonrelativistic Theory*, 3rd ed., Nauka, Moscow, 1974, 752 pp.

7.7. E. B. Wilson, Jr., J. G. Decius, and P. C. Cross, *Molecular Vibrations. The Theory of Infrared and Raman Spectra*, McGraw-Hill, New York, 1958.

7.8. G. Gerzberg, *Infrared and Raman Spectra of Polyatomic Molecules*, Van Nostrand, New York, 1945, 632 pp.

7.9. M. V. Vol'kenshtein, L. A. Gribov, M. A. El'yashevich, and B. I. Stepanov, *Molecular Vibrations* (Russ.), 2nd ed., Nauka, Moscow, 1972, 699 pp.

7.10. K. Nakamoto, *Infrared and Raman Spectra of Inorganic and Coordination Compounds*, 4th ed., Wiley, New York, 1986.

7.11. Yu. Ya. Khariton, Ed., *Vibrational Spectra in Inorganic Chemistry* (Russ.), Nauka, Moscow, 1971, 356 pp.

7.12. D. M. Adams, *Metal–Ligand and Related Vibrations: A critical survey of the infrared and Raman spectra of metallic and organometallic compounds*, Edward Arnold, London, 1967, 379 pp.

7.13. I. B. Bersuker, *Kinet. Katal.*, **18**, 1268–1282 (1977); *Chem. Phys.*, **31**, 85–93 (1978); *Teor. Eksp. Khim.*, **14**, 3–12 (1978); in: (*IUPAC*) *Coordination Chemistry*—20, ed. D. Banerjea, Pergamon Press, New York, 1980, pp. 201—218.

7.14. S. S. Stavrov, I. P. Decusar, and I. B. Bersuker, *New J. Chem.*, **17**, 71 (1993).

7.15. S. B. Piepho, *J. Am. Chem. Soc.*, **110**, 6319–6326 (1988); in: *Mixed-Valency Systems: Applications in Chemistry, Physics and Biology*, ed. K. Prassides, Kluwer, Dordrecht, The Netherlands, 1991, pp. 329–334.

7.16. T.-X. Lu, *Chem. Phys. Lett.*, **194**, 67 (1992); J. C. R. Faulhaber, D. Y. K. Ko, and P. R. Briddon, *Phys. Rev. B*, **48**, 661–664 (1993–I).

7.17. H. A. Jahn and E. Teller, *Proc. Roy. Soc.*, **161**, 220–235 (1937).

7.18. R. Renner, *Z. Phys.*, **92**, 172–193 (1934).

7.19. E. Ruch and A. Schonhofer, *Theor. Chim. Acta*, **3**, 291–304 (1965); E. L. Blount, *J. Math. Phys. (N.Y.)*, **12**, 1890–1896 (1971).

7.20. I. B. Bersuker, *Teor. Eksp. Khim.*, **2**, 518–521 (1966).

7.21. U. Opik and M. H. L. Pryce, *Proc. Roy. Soc.*, **A238**, 425–447 (1957).

7.22. H. Sponer and E. Teller, *Rev. Mod. Phys.*, **13**, 75–170 (1941).

7.23. H. C. Longuet-Higgins, in: *Advances in Spectroscopy*, Vol. 2, Interscience Publishers, New York, 1961, pp. 429–472.

7.24. M. C. M. O'Brien, *Proc. Roy. Soc. (London)*, **A281**, 323 (1964).

7.25. I. B. Bersuker, *Opt. Spectrosc.*, **11**, 319–324 (1961); *Zh. Eksp. Teor. Fiz.*, **43**, 1315–1322 (1962) (Engl. transl.: *Sov. Phys. JETP*).

7.26. I. B. Bersuker, A. G. Vekhter, and I. Ya. Ogurtsov, *Usp. Fiz. Nauk* (*Sov. Phys. Usp*), **116**, 605–641 (1975).

7.27. I. B. Bersuker, Ed., *The Jahn–Teller Effect: A Bibliographic Review*, IFI/Plenum, New York, 1984, 590 pp.

7.28. *Offitsialnii Byuleten' Gosudarstvennogo Komiteta po Delam Izobretenii i Otkrytii* (Russ.), No 40, Publikatsiya o Nauchnom Otkrytii N 202 (I. B. Bersuker), Oct. 1978.

7.29. M. C. M. O'Brien, *Phys. Rev.*, **187**, 407–418 (1969).

7.30. I. B. Bersuker and V. Z. Polinger, *Phys. Lett.*, **44A**, 495–496 (1973); JETP, **66**, 2078–2091 (1974).

7.31. I. B. Bersuker and B. G. Vekhter, *Fiz. Tverd. Tela*, **5**, 2432–2440 (1963) (Engl. Transl.: *Sov. Phys. Solid State*).

7.32. F. S. Ham, Phys.Rev., **A138**, 1727–1740 (1965); in: *Electron Paramagnetic Resonance*, ed. S. Geschwind, Plenum Press, New York, 1972, pp. 1–484.

7.33. I. B. Bersuker, *Phys. Lett.*, **20**, 589 (1966).

7.34. I. B. Bersuker and S. S. Stavrov, *Chem. Phys.*, **54**, 331 (1981).

7.35. R. F. W. Bader, *Mol. Phys.*, **3**, 137–151 (1960); *Can. J. Chem.*, **40**, 1164–1175 (1962).

7.36. R. F. W. Bader and A. D. Bandrauk, *J. Chem. Phys.*, **49**, 1666 (1968).

7.37. I. B. Bersuker, *Nouv. J. Chim.*, **4**, 139–145 (1980); *Teor. Eksp. Khim.*, **16**, 291–299 (1980).

7.38. I. B. Bersuker, N. N. Gorinchoi, and V. Z. Polinger, *Theor. Chim. Acta*, **66**, 161–172 (1984).

7.39. T. K. Rebane, *Teor. Eksp. Khim*, **20**, 532–539 (1984).

7.40. I. B. Bersuker, *Pure Appl. Chem.*, **60**(8), 1167–1174 (1988); *Fiz. Tverd. Tela*, **30**, 1738–1744 (1988).

7.41. I. B. Bersuker, *J. Coord. Chem.*, **34**, 289–338 (1995).

7.42. I. B. Bersuker, *New J. Chem.*, **17**, 3 (1993).

7.43. V. Z. Polinger, N. N. Gorinchoy, and I. B. Bersuker, *Chem. Phys.*, **159**, 75–87 (1992).

8 Electronic Structure Investigated by Physical Methods

Physical methods of investigation provide very powerful sources of information about the electronic structure of transition metal compounds, and the problem is to ascertain direct correlations between the observables and electronic parameters.

The variety of physical methods aimed at experimental study of the electronic structure of coordination compounds can be divided into two groups: *resonance methods*, including all-range spectroscopy from radio through γ-ray frequencies, and *nonresonance methods*, which consist of diffraction methods (x-ray, electron, neutron diffraction) and measurements of magnetic and electric susceptibilities. An important distinction between these two types of methods is that the observables in the resonance methods carry information about at least two states of the system, initial and final, between which a transition takes place (induced by the resonance interaction), whereas the nonresonance method describes, in principle, one electronic state, although the field response in this case may include other states admixed by the external perturbation to the one being considered.

This chapter is not devoted to the systematic presentation of all these methods; for their detailed study the reader is referred to appropriate monographs and handbooks [8.1–8.9]. Instead, we present here an introduction to *the theory of electronic origin of observables in the physical methods of investigation*, with an emphasis on the features related to other properties of transition metal compounds considered in this book. Attention is paid to electronic (including photoelectron) spectra, ESR and EXAFS spectroscopy, as well as magnetic and electric susceptibilities and diffraction methods in electron deformation densities.

8.1. BAND SHAPES OF ELECTRONIC SPECTRA

Qualitative Interpretation of Vibrational Broadening

Electronic spectra result from electronic transitions between two states of the system and carry information about these states. One of their important special features relevant to coordination compounds (as well as to some other molecular systems) is the strong dependence of the electronic energies on the

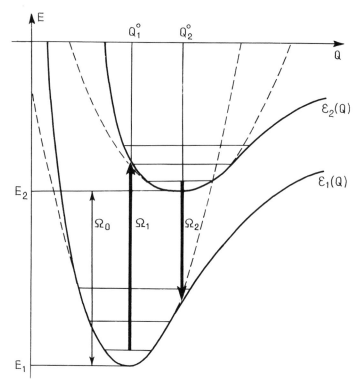

Figure 8.1. APs, vibrational states, and "vertical" transition between two electronic states. Ω_0, Ω_1, and Ω_2 are the pure electronic, absorption, and emission frequencies, respectively.

interatomic distances. Owing to this fact, the stationary states of the system are not purely electronic but electron-vibrational. For free molecular systems in the gas phase, rotational states are also important.

Figure 8.1 illustrates schematically the electronic energies of two non-degenerate states as functions of symmetrized coordinates: adiabatic potentials (Section 7.1) and the corresponding vibrational states in the minima. This presentation by AP alone enables us to obtain many important qualitative features related to the electronic structure. First, take into account the *Franck–Condon principle*, due to which the nuclear configuration does not (is not in time to) change during the electron transition, and hence the latter takes place at the unchanged nuclear configuration of the initial state. Indeed, the electronic transitions are much faster than the vibrational motion; the time of transition τ is approximately inversely proportional to the light frequency Ω, and for visible light $\tau \sim \Omega^{-1} = 10^{-15}$ s, while $\omega^{-1} = 10^{-10}$ to 10^{-12} s.

Due to the Franck–Condon principle, the transitions between the electronic states in Fig. 8.1 are described by vertical arrows starting from the vibrational state in the minimum of the AP of the initial electronic state. It is seen that for most cases, when the minima points of the two electronic states do not coincide, the transition is not pure electronic: It also changes the vibrational states. The transition frequency Ω (the energy of the absorbed or emitted quantum $\hbar\Omega$) also depends on the initial vibrational state. It changes approximately from Ω_1 to Ω_2, depending on the number of vibrational quanta of the ground and excited states involved in the transition. The probability of such transitions that include vibrational components is proportional to the *Franck–Condon factor*, the overlap of the vibrational wavefunctions of the initial and final states.

Due to the thermal population of different vibrational states and the proper width of each transition, the molecular spectra consist of rather wide bell-shaped bands (*vibrational broadening*), distinct from the narrow lines of atomic spectra. The half-width (the width at the half intensity) of the bands is about $\Omega_1 - \Omega_2$. Examining Fig. 8.1, one can see that the width of the band is directly related to the difference in the minima positions Q_1^0 and Q_2^0 of the two AP, that is, to *the shift of equilibrium position by excitation* $\Delta Q = Q_1^0 - Q_2^0$. The greater the value of ΔQ, the wider the transition band.

On the other hand, the difference in the equilibrium interatomic distances in different electronic states that determines ΔQ is strongly dependent on the differences in their electronic configurations. It is known [8.10] that in coordination compounds the ionic radii, which present the interatomic distances directly, depend strongly on the occupation numbers n and m of different d orbitals of the electronic configuration $(t_2)^n(e)^m$ (Sections 4.3 and 6.2). Therefore, change in these occupation numbers, Δn and Δm, may be used for qualitative estimates of changes in the equilibrium interatomic distances by the transition from one electronic state to another. Two groups of electronic transitions can be distinguished:

1. Transition for which $\Delta n \neq 0$ and $\Delta m \neq 0$ with significant changes in the electronic configuration, resulting in *broad bands* of light absorption and luminescence. These bands remain broad down to low temperatures. The calculations (see below) give an estimate of the bandwidth as being from one thousand to several thousand cm^{-1}.

2. Transitions with $\Delta n = 0$, $\Delta m = 0$ for which the electronic configuration remains unchanged (as in spin-forbidden transitions), resulting in *narrow lines* with a width on the order of $100\ cm^{-1}$.

Note that in systems with f electron configurations the dependence of the interatomic equilibrium distances on the number of different f electrons is very weak because their participation in the bonding is weak. Therefore, the transitions that involve changes in only f electron occupation numbers yield mostly narrow lines.

A more quantitative description of the vibrational broadening under consideration can be obtained if one knows the energy diagrams $E = f(\Delta)$ of the energy levels as a function of the crystal field parameter Δ (Section 4.3). As shown in Fig. 8.1, the transition frequency $\hbar\Omega = E_2 - E_1$ is strongly dependent on the minima shift ΔQ; the larger ΔQ, the stronger Ω depends on Q, and the larger the derivative $d(\hbar\Omega)/dQ = (dE_2/dQ) - (dE_1/dQ)$. This magnitude can serve as a rough measure of the bandwidth. On the other hand, the derivative $dE/d\Delta$ characterizes the relative sensitivity of the energy level E to the ligand influence, which in turn depends on the (symmetrized) metal–ligand distance Q. Although the changes in Δ may not be in direct proportionality to changes in Q, for the sake of simplicity one can assume that in the first (linear) approximation $d(\hbar\Omega)/dQ \sim d(\hbar\Omega)/d\Delta$, and hence

$$\frac{d(\hbar\Omega)}{dQ} \sim \frac{d(\hbar\Omega)}{d\Delta} = \frac{dE_2}{d\Delta} - \frac{dE_1}{d\Delta} \tag{8.1}$$

Provided that this relation holds, the division of the electronic transition bands above into two groups can be obtained directly using the results of crystal field theory for the functions $E = f(\Delta)$. Approximately (without taking account of electron interactions explicitly), for the $(t_2)^m(e)^n$ configuration, we have (Sections 4.3 and 4.5)

$$E_1 = \text{const} - m_1(\tfrac{2}{5})\Delta + n_1(\tfrac{3}{5})\Delta$$
$$E_2 = \text{const} - m_2(\tfrac{2}{5})\Delta + n_2(\tfrac{3}{5})\Delta \tag{8.2}$$

and

$$\frac{d(\hbar\Omega)}{d\Delta} = \frac{2}{5}(m_1 - m_2) + \frac{3}{5}(n_2 - n_1) \tag{8.3}$$

If the electronic configuration is not changed by the transition, $m_1 = m_2$, $n_1 = n_2$ and the derivative determining the vibrational broadening is zero —the bands are expected to be narrow. For transitions that change the electronic configuration, the derivative $dE/d\Delta$ is nonzero. For example, for the most studied $t_2 \rightarrow e$ transition $(d \leftrightarrow d)$, $\Delta m = m_1 - m_2 = 1$, $\Delta n = n_1 - n_2 = -1$, $d(\hbar\Omega)/d\Delta = 1$, and the band is broad.

This qualitative reasoning, explaining the origin of two groups of electronic bands, relatively broad and narrow, can be regarded as the first stage in the interpretation of the origin of electronic spectra in coordination complexes. Despite being a rough approximation, it contains significant possibilities for revealing some important features. Consider some examples. Complexes $[\text{TiA}_6]^{3+}$ with one d electron are expected to manifest one d–d transition $^2T_{2g} \rightarrow {}^2E_g$, accompanied by a change in the electronic configuration $(t_{2g})^1 \rightarrow (e_g)^1$ with $\Delta m = 1$, $\Delta n = -1$ that yields a broad band of absorption in accordance with experimental data [8.11]. The transitions $^1A_{1g} \rightarrow {}^1T_{1g}$ and $^1A_{1g} \rightarrow {}^1T_{2g}$, as well as the spin-forbidden transitions $^1A_{1g} \rightarrow {}^3T_{1g}$ and

$^1A_{1g} \to {}^3T_{2g}$ in low-spin octahedral complexes with the electron configuration d^6 (e.g., $[Co(NH_3)_6]^{3+}$) are associated with change in the electronic configuration $(t_{2g})^6 \to (t_{2g})^5(e_g)^1$. Hence the absorption bands should be, and are, broad [8.12]. On the contrary, the spin-forbidden transitions with spin-only changes $(t_{2g}\downarrow)^3 \to (t_{2g}\downarrow)^2(t_{2g}\uparrow)$, for which $\Delta S = 1$ but $\Delta n = \Delta m = 0$ (e.g., the transition $^4A_{2g} \to {}^2E_g$ in $[Cr(H_2O)_6]^{3+}$) yield narrow lines of absorption. A more detailed differentiation of bandwidths is given below.

Theory of Absorption Band Shapes

The main information about the electronic transition between two states described by the wavefunctions ψ_1 and ψ_2 is comprised in the transition dipole moment M_{12}:

$$\mathbf{M}_{12} = \int \psi_1^* \mathbf{M} \psi_2 \, d\tau \tag{8.4}$$

where $\mathbf{M}$ is the operator of either the electric or magnetic dipole moment of the system, depending on the type of transition; similar expressions can be written for quadruple or other multipole transitions. For electric dipole transitions, $\mathbf{M} = \Sigma_i q_i \mathbf{r}_i$, where q_i and $\mathbf{r}_i$ are the charges and radius vectors of the particles, respectively.

The integral (8.4) depends on the wavefunctions ψ_1 and ψ_2, which include vibrational and rotational states involved in the transition. This can be characterized by the dependence $M_{12}(\Omega)$ on the transition frequency Ω. Provided that $M_{12}(\Omega)$ is known, one can easily evaluate the *coefficient of light absorption* $K_{12}(\Omega)$. Usually, it is evaluated from the relation $I = I_0 \exp[-K_{12}(\Omega)l]$, where I_0 and I are the intensities of the incident and transmitted light, and l is the thickness of the absorbing layer of the substance. Then [8.13, 8.14]

$$K_{12}(\Omega) = \frac{4\pi^2 N\Omega}{3hc} |M_{12}(\Omega)|^2 \tag{8.5}$$

where N is the number of absorbing centers in a unit volume and c is the speed of light.

However, if there are close vibrational energy levels with an almost continuous function $M_{12}(\Omega)$, it is convenient to introduce an averaged coefficient of light absorption in a small interval of frequencies $\Delta\Omega$ near Ω. To do this, we introduce the quantity $F_{12}(\Omega)$ proportional to the averaged value $|M_{12}(\Omega)|^2$ over the interval $\Delta\Omega$:

$$F_{12}(\Omega) = (h\Delta\Omega)^{-1} \sum_{vv'}{}' |\langle 1v|\mathbf{M}|2v'\rangle|^2 \rho_{1v} \tag{8.6}$$

where v and v' number the vibrational states of the ground (first) and excited (second) electronic states, respectively, and ρ_{1v} is the probability of population of the initial state $1v$, which depends on temperature.

Thus the function $F_{12}(\Omega)$ determines the dependence of the light absorption coefficient on the frequency Ω and hence the shape of the absorption curve. Therefore, $F_{12}(\Omega)$ is named the *form function of the band*. In the semiclassical approximation (at sufficiently high temperatures), when the nuclei can be assumed to move along the AP, $F_{12}(\Omega)$ can be calculated analytically. Taking the two APs of Fig. 8.1 as shown by dashed lines

$$\varepsilon_1(Q) = \tfrac{1}{2} K Q^2 \tag{8.7}$$

$$\varepsilon_2(Q) = \hbar\Omega_0 + \tfrac{1}{2} K(Q - Q_0)^2 \tag{8.8}$$

where Ω_0 is the frequency of the pure electronic transition, and neglecting the *Dushinski effect* [8.15], that is, taking the force constant K the same in the two states, we obtain the following expression for the form function of the absorption band:

$$F_{12}(\Omega) = \frac{|\mathbf{M}_{12}|^2}{Q_0(2\pi K k T)^{1/2}} \exp\left[-\frac{\hbar^2(\Omega_1 - \Omega)^2}{2Q_0^2 K k T} \right] \tag{8.9}$$

where $\mathbf{M}_{12}$ is the pure electronic transition dipole moment, Ω_1 is the frequency at the band maximum (Fig. 8.1), and k is the Boltzmann constant.

As shown in the expression (8.9), in the semiclassical approximation used above, the absorption band has a Gaussian shape (exponential dependence on the square of the frequency deviations from the maximum value Ω_1), and it is strongly temperature dependent. However, the integral intensity of the band (the area under the absorption curve) is independent of temperature; this can be shown by integrating expression (8.9) over Ω.

A more exact expression for the band shape form function $F_{12}(\Omega)$ can be obtained renouncing the semiclassical approximation and performing a full quantum-mechanical calculation. The latter was carried out under the assumption that only one vibrational frequency ω is active in the vibrational broadening; it yields the following formula [8.14]:

$$F_{12}(\Omega) = |\mathbf{M}_{12}|^2 \exp\left(-a \coth \frac{\beta}{2} + p\beta \right) I_p(Z) \tag{8.10}$$

$$Z = \frac{a}{2} \operatorname{sh} \beta$$

where $p = (\Omega - \Omega_0)/\omega$, $\beta = \hbar\omega/2kT$, $I_p(Z)$ is a Bessel function, and a is the constant of heat release determined by the summary shift of the minima positions $Q^0_{1\alpha} - Q^0_{2\alpha}$ of the ground and excited states. In dimensionless units,

$$a = \sum_\alpha (Q^0_{1\alpha} - Q^0_{2\alpha})^2 \tag{8.11}$$

If only one coordinate is active in the broadening, $a = Q^2_0$.

Band Shapes of Electronic Transitions Between Nondegenerate States; Zero-Phonon Lines

The expressions for the form-function $F_{12}(\Omega)$ obtained above enable us to analyze the expected band shape of electronic absorption of light by transitions between nondegenerate electronic terms. The Bessel function $I_p(Z)$ is nonzero for integer values p only. Hence $F_{12}(\Omega)$ according to (8.10) and the absorption coefficient $K_{12}(\Omega)$ according to (8.5) have the form of a set of equidistant lines with a spacing equal to ω and an envelope that [following the behavior of the exponent in (8.10)] is a slightly asymmetric bell-shaped curve (Fig. 8.2). If one takes into account the natural width of the individual lines and the influence of the crystal or liquid environment, these lines coalesce into a continuous band that has the form of the envelope.

The position of the absorption maximum of the band is determined by the relation

$$\Omega^{abs}_{max} = \Omega_0 + \frac{a\omega}{2} \tag{8.12}$$

An important conclusion emerges from this formula: *The band maximum, in general, does not coincide with the frequency of the pure electronic transition* Ω_0.

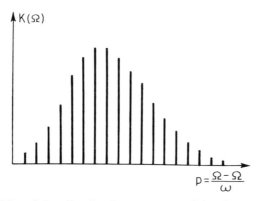

Figure 8.2. Intensities of the vibrational components of the electron transition band. The envelope has a slightly asymmetric Gaussian form.

The temperature dependence of the band shape is also important. At sufficiently high temperatures when the relations $\hbar\omega < 2kT$ and $a \gg \text{sh}(\hbar\omega/2kT)$ hold, the absorption band acquires a symmetrical Gaussian form (8.9). Hence at these temperatures the semiclassical approximation is valid. The half-width $\delta\Omega$ of the band is proportional to $\sqrt{T}$:

$$\delta\Omega = 2\omega \left(\frac{2akT\ln 2}{\hbar\omega}\right)^{1/2} \tag{8.13}$$

In the opposite limit case, at sufficiently low temperatures when $kT \ll \hbar\omega$ and $a \ll 2\,\text{sh}(\hbar\omega/2kT)$, the band is asymmetrical and the half-width is temperature independent:

$$\delta\Omega = 2\omega(a\ln 2)^{1/2} \tag{8.14}$$

Equations (8.10) to (8.14) play a significant role in the interpretation of electronic spectra and allow one to reveal interesting parameters of electronic structure. For example, following (8.13) and (8.14), the band half-width $\delta\Omega$ is proportional to $a^{1/2}$. Hence when there is no large minima shifts and $a = 0$, the absorption bands are narrow, in accordance with the qualitative considerations given above. For most coordination compounds the vibrational frequencies ω are sufficiently large and obey the criterion of low temperatures at room (and lower) temperatures; thus for them, Eq. (8.14) is valid. By combining (8.14) with (8.12), we obtain

$$\Omega^{\text{abs}}_{\text{max}} = \Omega_0 + \frac{(\delta\Omega)^2}{5.5\omega} \tag{8.15}$$

Hence the shift of the band maximum with respect to the frequency of the pure electronic transition is square dependent on the band half-width and can be relatively large. For example, with $\delta\Omega \sim 3000\ \text{cm}^{-1}$ and $\omega \sim 600\ \text{cm}^{-1}$ the shift is $\sim 3000\ \text{cm}^{-1}$, the same as half-width. Note that from the exact formula (8.10) (as also shown in Fig. 8.1)

$$\Omega^{\text{abs}}_{\text{max}} = \Omega_1 \tag{8.16}$$

which means that the frequency at the maximum of absorption coincides with that of the Franck–Condon (vertical) transition from the ground-state vibrational level. It follows that *the frequencies of the band maxima of electronic absorption are determined by the energy gaps between the electronic terms calculated for the equilibrium configuration of the ground state.* This important conclusion means that for the interpretation of the origin of the band maxima positions the simple energy-level scheme obtained in crystal field theory (Chapter 4) or the MO LCAO approach (Chapter 5) yields, in principle, correct results, but *the maximum absorption frequencies are not equal to the energy difference between the ground and excited states, each at their equilibrium positions.*

In the case of luminescence the maximum of the band corresponds to the vertical transition from the minimum of the excited state to the ground state, that is, to Ω_2 (Fig. 8.1). One can easily find that

$$\Omega_2 = \Omega_0 - \frac{a \cdot \omega}{2} \qquad \Omega_1 - \Omega_2 = a \cdot \omega \qquad (8.17)$$

Thus the frequency difference between the absorption and luminescence band maximum (*the Stocks shift*) is due to the difference in the equilibrium configuration of the ground and excited states (Section 8.4).

Note that the transition from the excited electronic state to the ground state (or to another excited state) can also take place as a *radiationless transition* without emission of light. The radiationless transition is a kind of relaxation when the excitation energy is transformed into vibrations, and it is strongly temperature dependent.

An important feature of electronic spectra of polyatomic systems, especially at low temperatures, is the *zero-phonon line* [8.14]. It occurs as an additional acute (resonance) peak at the pure electronic frequency Ω_0 on the background of the broad band, provided that the parameter of heat release a is not very large. As can be concluded from Fig. 8.1, if a (or $Q_1^0 - Q_2^0$) is small, the pure electronic transition with vibrational quantum numbers 0 in both states ($0 \rightarrow 0$, or zero-phonon transition) becomes most probable. Moreover, the transitions $1 \rightarrow 1$, $2 \rightarrow 2$, and so on, which in the absence of anharmonicity have the same frequency Ω_0 as the $0 \rightarrow 0$ line, are also most probable. The sum of all these transitions with the same frequency yields a narrow line of great intensity (the zero-phonon line) on the background of the less intensive one-, two-,... phonon satellites. The latter are much broader and coalesce into a broad phonon wing (mostly of higher frequency). This result emerges from the general formula (8.10) for the form function of the band if we substitute $p = 0$ and $a \rightarrow 0$. The distinct feature of zero-phonon lines is that their frequencies Ω_0 are the same in absorption and luminescence. Figure 8.3 illustrates this statement together with the demonstration of an example spectrum with a zero-phonon line [8.14].

The zero-phonon line carries interesting information about the electronic structure of the system. In particular, it gives the exact value of the energy difference between two electronic terms at their equilibrium nuclear configurations. The role of orbital degeneracies in optical band shapes is discussed in Section 8.2.

Types of Electronic Transitions on Intensity

In expressions for the form function of the band shapes of electronic transition and hence in the absorption coefficient, there is the square of the electronic matrix element $|M_{12}|^2$ that determines the band absolute intensities. They vary

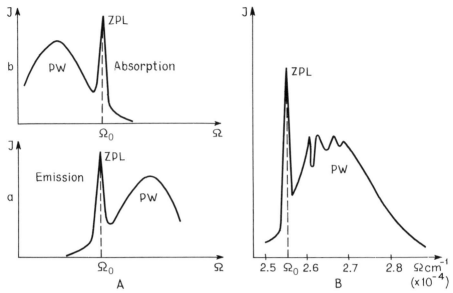

Figure 8.3. (*A*) Absorption (*a*) and emission (*b*) band shapes in the presence of a zero-phonon line (ZPL) (PW is the phonon wing); (*B*) example of experimentally observed absorption spectrum with a ZPL for color centers in NaF.

by several orders of magnitude, dividing the observable spectra into different types well discriminated by intensity.

First, the *selection rules* for the transition moment M_{12} after (8.4) must be taken into account. Since **M** is a vector, they can be obtained directly using group-theoretical procedures, as shown in Section 3.4. The matrix element M_{12} is nonzero if and only if the product of the irreducible representations (Appendix 1) of the wavefunctions ψ_1 and ψ_2 and the operator **M** contains the unity (totally symmetric) representation.

The transitions for which the integral (8.4) is zero by symmetry considerations are called *forbidden transitions*. They can be evaluated in a general way. For electric-dipole transitions in systems with an inversion center the integral (8.4) is zero when the states 1 and 2 have the same parity because **M** changes its sign by inversion. Hence electric-dipole transitions are forbidden as $g \leftarrow / \rightarrow g$ and $u \leftarrow / \rightarrow u$ (*parity-forbidden transitions*) and may be allowed as $g \leftrightarrow u$. Another general case is provided by the *spin-forbidden transitions* sometimes called *intersystem combinations*, when the two states have different spin multiplicities. Provided that the spin–orbital mixing of these states with other states is not taken into account (see below), the integral (8.4) is zero due to the orthogonality of the spin functions (**M** is independent of spin).

For a given transition the three components of M_{12},

$$M_{12}^x = \int \psi_1^* M_x \psi_2 \, d\tau \qquad M_{12}^y = \int \psi_1^* M_y \psi_2 \, d\tau \qquad M_{12}^z = \int \psi_1^* M_z \psi_2 \, d\tau$$

may be subject to different selection rules. Therefore, if the coordination system is in an anisotropic crystal state where all the absorbing centers have the same orientation with respect to the electric (or magnetic) vector of the electromagnetic wave of light, the absorption (emission, reflection) of polarized light may be different in different directions of the crystal (*dichroism or polychroism*). Related phenomena of circular dichroism and magnetic circular dichroism are discussed in Section 8.4.

Selection rules, in general, are discussed in Section 3.4. In application to absorption of polarized light, consider, for example, a complex with D_{4h} symmetry, say, $PtCl_4^{2-}$. Its ground state is $^1A_{1g}$, and the most intensive bands are associated with the charge transfer one-electron transitions from the MOs b_{2u}, e_u, and a_{2u} (formed mainly by the four p_z AOs of the chlorine atoms) to the MO b_{1g} (which is mainly the Pt AO $d_{x^2-y^2}$), resulting in excited $^1A_{2u}$, ^1E_u, and $^1B_{2u}$ states, respectively ($b_{2u} \times b_{1g} = A_{2u}$, $e_u \times b_{1g} = E_u$, $a_{2u} \times b_{1g} = B_{2u}$). Taking into account that in the D_{4h} point group M_z transforms as a z component of a vector and belongs to A_{2u} while M_x and M_y belong to E_u (Appendix 1), we have for the transition $^1A_{1g} \rightarrow {}^1A_{2u}$:

$$A_{1g} \times A_{2u} \times A_{2u} = A_{1g}, \qquad \text{for the } z \text{ component}$$

$$A_{1g} \times E_u \times A_{2u} = E_g, \qquad \text{for the } x \text{ and } y \text{ components}$$

Similarly, for the transition $^1A_{1g} \rightarrow {}^1E_u$: $A_{1g} \times A_{2u} \times E_u = E_g$, $A_{1g} \times E_u \times E_u = A_{1g} + B_{1g} + E_g$, and for $^1A_{1g} \rightarrow {}^1B_{2u}$: $A_{1g} \times A_{2u} \times B_{2u} = B_{1g}$, $A_{1g} \times E_u \times B_{2u} = E_g$. From all these transitions only those are allowed for which the foregoing products of irreducible representations contain the totally symmetric one A_{1g}. It follows that the transition $^1A_{1g} \rightarrow {}^1A_{2u}$ is possible only for light polarized in the z direction, whereas in the x and y directions it is forbidden. On the contrary, the transition $^1A_{1g} \rightarrow {}^1E_u$ is allowed when the light is polarized in the x or y directions (in any direction in the xy plane), whereas it is forbidden in the z direction. The transition $^1A_{1g} \rightarrow {}^1B_{2u}$ is forbidden in any direction (in this approximation).

However, while the selection rules for the matrix elements (8.4) obtained by group-theoretical considerations are exact, the matrix element itself (i.e., the wavefunctions of the initial and final states and the operator of interaction of the system with light) is defined approximately, within a certain model. Therefore, when the matrix element $M_{12} = 0$ by symmetry considerations, it does not mean that the corresponding transition is absolutely forbidden; *it is forbidden within the approximation used*, and hence it can become allowed in the next approximation.

For instance, forbidden electric dipole transitions can become allowed as magnetic-dipole transitions, and if the latter are also forbidden, they can become possible as quadruple transitions. Transitions that are forbidden by parity restrictions become allowed by interaction with odd vibrations, while spin-forbidden transitions are allowed as intersystem combinations when the spin–orbital interaction removes the prohibition, and so on. Depending on the approximation in which the electronic transition becomes allowed, the corresponding intensities of the band have quite different orders of magnitude. Each approximation that makes possible the otherwise forbidden transition can be related to a specific interaction with an order of magnitude determined by the interaction constant. This enables us to estimate the order of magnitude of the intensity of the corresponding transitions.

To characterize the intensities quantitatively, the quantity named *oscillator strength* f_{12} can be employed. The definition of f (not to be confused with f electrons) is as follows:

$$f_{12} = \frac{2m\Omega}{3\hbar e^2} |M_{12}|^2 \tag{8.18}$$

In another presentation:

$$f_{12} = \frac{mc}{2N\pi^2 e^2} \int K_{12}(\Omega)\, d\Omega$$

or

$$f_{12} = \frac{1.15 \cdot 10^3 mc}{\pi^2 e N_A} \int \varepsilon_{12}(\Omega)\, d\Omega \tag{8.19}$$

where $\varepsilon_{12}(\Omega)$ is the *extinction coefficient*, determined from the relation $I = I_0 \cdot 10^{\varepsilon C_0 l}$, where C_0 is the molar concentration (cf. $I = I_0 e^{-kl}$ used above) and N_A is Avogadro's number (Ω is given in s^{-1}).

It can be shown that f cannot be larger than 1: The sum of the oscillator strengths for all the one-electron transitions from a given state equals 1. For different types of transitions, mentioned above, the value of f varies from 1 to 10^{-10} and less. Most intensive are the transitions allowed through the electric-dipole mechanism, that is, between states with opposite parity ($g \leftrightarrow u$ in the case of systems with an inversion center) but with the same spin multiplicity. The oscillator strength of electric dipole bands $f \sim 1$ to 10^{-1}.

In transition metal coordination compounds such transitions are generally known as *charge transfer bands*. In the MO LCAO scheme they correspond to the transitions between bonding and antibonding MO, which in systems with inversion symmetry are of opposite parity. Since these two types of orbitals are usually localized (mainly) on different atomic groups of the system (e.g., the bonding MO is on the ligands, while the antibonding MO is on the CA), the electronic transition between them (transitions of the type $\pi \to d$, $\sigma \to d$, $d \to \pi^*$) are associated with charge transfer from one of these groups to another justifying the name of the transition. An example of charge transfer transition in PtCl$_4{}^{2-}$ was considered above. The transitions $\pi \to d_{z^2}$ (band I) and $\sigma \to d_{z^2}$ (band II) in the spectra of the complexes [Co(NH$_3$)$_5$X]$^{2+}$, X = F, Cl, Br, I may serve as another example of such spectra. Many other examples are given in special monographs [8.2].

Parity-forbidden transitions (transitions forbidden by the parity rule $g \leftarrow/\rightarrow g$, $u \leftarrow/\rightarrow u$) become allowed when taking into account the interaction of the electronic states with odd vibrations which mix odd states with even ones (Section 8.2). For these transitions $f \sim 10^{-4}$ to 10^{-5}. They can also be allowed as *magnetic dipole transitions* with $f \sim 10^{-7}$. Parity-forbidden transitions constitute the majority of observable electronic transitions in coordination compounds in the visible and related regions of light. Among them, the most studied are the d–d transitions considered in Section 8.2.

The next type of "forbidden" transitions is formed by the above-mentioned *spin-forbidden, or intersystem combination transition.* These are transitions between electronic states with different spin multiplicity, for which the integral (8.4) is zero because of the orthogonality of the spin functions of the two states. Spin-forbidden transitions become allowed when the spin–orbital interaction that mixes the states with different spin multiplicity is taken into account. The estimates by perturbation theory show that the mixing terms are on the order of $(\lambda/\Delta)^2$, where λ is the spin–orbital coupling constant (Section 2.1) and Δ is the energy gap between the mixing states. These terms are on the order of 10^{-2} to 10^{-4}, and hence this is the order of the oscillator strength f expected for such spectra.

If the transition is forbidden by both the parity rule and the spin difference, $f \sim 10^{-7}$ when the transition becomes possible due to the interaction with odd vibrations, and $f \sim 10^{-9}$ and $f \sim 10^{-10}$ for magnetic-dipole and electric-dipole transitions, respectively.

Table 8.1 lists the most important types of electronic transitions in coordination compounds with an indication of their oscillator strengths (by order of

Table 8.1. Orders of magnitudes of oscillator strengths f and extinction coefficients in the maximum of the band $\varepsilon(\Omega_{max})$ for different types of electronic transitions

Type of Electronic Transition	f	$\varepsilon(\Omega_{max})$
Electric-dipole	$1-10^{-2}$	10^5-10^3
Parity-forbidden, allowed by odd vibrations	$10^{-4}-10^{-5}$	10^3-10^1
Magnetic-dipole	10^{-6}	1
Electric-quadrupole	10^{-7}	$10-1$
Spin-forbidden (intersystem combination)	$10^{-3}-10^{-5}$	$100-10$
Plus parity-forbidden:		
Allowed by vibrations	$10^{-6}-10^{-7}$	$1-10^{-1}$
Magnetic-dipole	10^{-9}	10^{-3}
Electric-quadrupole	10^{-10}	10^{-4}

magnitude) and coefficients of molar extinction at the maximum intensity of the band $\varepsilon(\Omega_{max})$ [Eq. (8.19)]; the ε values are determined from the relation: $f = 4.32 \cdot 10^{-9} \varepsilon(\Omega_{max}) \delta\Omega$, which is approximately true for symmetrical Gaussian bands.

8.2. $d-d$ TRANSITIONS

Origin and Special Features

In accordance with the role of d and f electrons of the CA in the formation of coordination bonding (Section 6.1), optical transitions that involve these electrons are most important for the study of the electronic structure of transition metal compounds. Of particular interest are the transitions between the states originating from a d^n configuration, often called $d-d$ transitions. These transitions fall into the visible and related regions of the optical spectrum, thus determining the color of the compound. The $d-d$ spectra were subjects of intensive study beginning in the 1920s and, together with the magnetic properties (Section 8.4), served as an experimental basis for the creation of crystal field theory (Chapter 4).

Consider the energy levels of d^n configurations obtained in the CFT (Sections 4.2 and 4.3; in the MO LCAO approach, Section 6.2, the qualitative description is similar). A general picture of d^n energy-level splitting in the cubic field of ligands as a function of the ligand field parameter Δ is given by the Tanabe–Sugano diagrams (Fig. 4.11). These diagrams provide the most important parameter of the spectrum: the positions of the band maxima. To determine the latter, drive a vertical line on the diagram through the point $\Delta = \Delta_0$ relevant to the complex under consideration. Following the conclusions of Section 8.1, the expected band maxima positions are given by the ordinates of the points of intersection of this vertical line with the curves $E_i = f(\Delta_0)$. The value Δ_0 can be found from the experimentally observed position of the maximum of one of the bands.

By way of example, consider the diagram of d^5 energy-level splitting for the Mn^{2+} ion in a cubic weak field, calculated by Orgel [8.16] (Fig. 8.4) and compare it with the experimental absorption spectrum of this hydrated ion (Fig. 8.5) [8.17]. From Fig. 8.4, taking the vertical line at $\Delta_0 \approx 9000$ to $10,000 \text{ cm}^{-1}$, we get the maxima positions of all the bands of the spectrum that can thus be interpreted as corresponding to the electronic transitions from the ground state $^6A_{1g}$ to the excited states $^4T_{1g}$, $^4T_{2g}$, ^4E_g, $^4A_{1g}$, $^4T_{2g}$, ^4E_g, and so on, as indicated in Fig. 8.5; there is relatively good quantitative agreement between the theoretical predictions and experimental values of the maxima frequencies.

The relative band widths of this spectrum can also be explained using the theoretical conclusions obtained in the preceding section. Indeed, the ground state $^6A_{1g}$ originates from the half-filled d shell with the con-

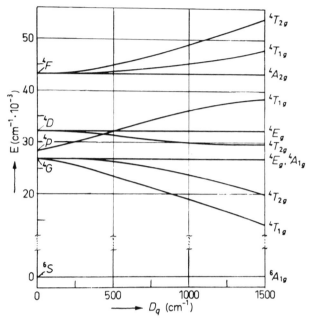

Figure 8.4. Energy-level diagram of the $Mn^{2+}(d^5)$ ion in a cubic crystal field as a function of the parameter $\Delta = 10D_q$. (After [8.16].)

figuration $(t_{2g})^3(e_g)^2$ in the octahedral field. Therefore, the transitions $^6A_{1g} \rightarrow {}^4T_{1g}[(t_{2g})^4(e_g)^1]$ and $^6A_{1g} \rightarrow {}^4T_{2g}[(t_{2g})^4(e_g)^1]$ with $\Delta m = 1$, $\Delta n = -1$, according to (8.3), are expected to yield broad bands in accordance with the experimental data.

On the contrary, the next excited terms, ^4E_g, $^4A_{1g}$, $^4T_{2g}$, and ^4E_g, originate from the same electronic configuration, $(t_{2g})^3(e_g)^2$, as the ground state, and hence the transitions to them are expected to give more narrow lines. This is indeed the case (Fig. 8.5), but the band widths of these transitions differ

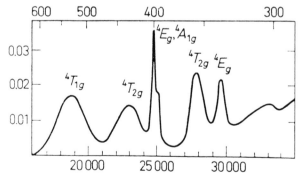

Figure 8.5. Absorption spectrum of the aqua-complex $Mn(H_2O)_6^{2+}$ in cm^{-1} (lower figures) and mμ (upper figures).

significantly from each other, indicating that a more refined analysis is required. It can be performed using the derivatives $d\Omega/d\Delta$ introduced in Section 8.1 [Eq. (8.1)] as an approximate relative measure of the width. Since the ground state is taken as an energy read-off, $dE_1/dQ = 0$ and $d\Omega/d\Delta = dE_2/d\Delta$. The latter can be estimated directly from the slope of the curves $E_i = f(\Delta)$ at $\Delta = \Delta_0$. For the octahedral aqua-complex of Mn^{2+} the relative values of $dE_2/d\Delta$ for the sequence of excited states can be estimated roughly from Fig. 8.4. They are 1, 0.8, 0, 0, 0.2, <0.1, and 0.9 for the terms $^4T_{1g}$, $^4T_{2g}$, ^4E_g, $^4A_{1g}$, $^4T_{2g}$, ^4E_g, and $^4T_{1g}$, respectively. As one can see from Fig. 8.5, the observed band widths follow approximately (although not exactly proportional) the sequence of predicted values.

An important feature of electric-dipole $d-d$ transitions is that they are parity forbidden; that is, they originate from the transitions between two states that have the same parity (Section 8.1). In atoms, $d-d$ transitions are strongly forbidden, but in molecular systems there are several additional interactions that remove this prohibition. Since the parity rule is operative when the system has an inversion center, low-symmetry ligand fields that remove the inversion symmetry make the $d-d$ transition allowed. If there are no low-symmetry fields, the vibronic coupling of the electronic states to odd vibrations that produce the required off-center distortions may induce the $d-d$ transitions (*vibrational-induced transitions*). Provided that the temperature is not very low, this vibronic mechanism is most operative.

In Section 3.4 it is shown how the inclusion of odd vibrations allows for otherwise forbidden $g \leftrightarrow g$ and $u \leftrightarrow u$ transitions. Quantitative estimates of the intensity, band shapes, and temperature dependence of the spectrum can be obtained considering vibronic coupling to odd vibrations explicitly (Section 7.2). Consider a parity-forbidden transition between two even states Ψ_1^g and Ψ_2^g (the same result is valid for two odd states) in a system with inversion symmetry for which the matrix element $\mathbf{M}_{12}$ after (8.4) is zero. Taking into account the vibronic coupling terms (7.3) with odd vibrations Q_u, the wavefunction Ψ^g gets an admixture of an odd function Ψ^u of the excited state:

$$\Psi_1(r, Q_u) = \Psi_1^g + \frac{F^{(gu)}}{\Delta_{ug}} Q_u \Psi^u \qquad (8.20)$$

where for the sake of simplicity only one (odd) linear term of the vibronic interaction (7.3) is taken into account and only one excited state Ψ^u is included; $F^{(g,u)}$ is the constant of vibronic coupling between the two states (7.22) and Δ_{ug} is the energy gap between them.

The additional term in (8.20) that contains the odd wavefunction gives a nonzero contribution on the order $(F/\Delta)Q$ (hereafter the labels g and u are omitted) to the integral (8.4) making the corresponding form function $F_{12}(\Omega)$ on the order of $(F/\Delta)^2$, and modifying the band shape and its temperature dependence. Calculations similar to those resulting in Eq. (8.10) yield an appropriate expression for the form function $F_{12}(\Omega)$ in this case:

$$F_{12}(\Omega) = \left(\frac{F}{\Delta}\right)^2 |M_{12}|^2 \exp\left(\frac{a \coth \beta - a' \coth \beta'}{2}\right) [(n + 1) \exp(-p_1 \beta) I_{p_1}(z)$$

$$+ n \exp(-p_2 \beta) I_{p_2}(z)] \tag{8.21}$$

where in addition to the notations in (8.10), $a' = (F/\hbar\omega')^2$ (in dimensionless units), $p_1 = (\Omega - \Omega_1 + \omega')/\omega$, $p_2 = (\Omega - \Omega_1 - \omega')/\omega$, $\beta' = \hbar\omega'/2kT$, $n = [\exp(\hbar\omega'/kT) - 1]^{-1}$ is the Boltzmann occupation number of the odd vibrations, ω' is their frequency, and Ω_1, the frequency of the band maximum, is:

$$\Omega_1 = \Omega_0 - \left(\frac{F^2}{2}\right) \frac{(2n + 1)\Delta - \hbar\omega'}{\Delta^2 - (\hbar\omega')^2} \tag{8.22}$$

It follows from (8.21) that the odd vibrations which make the bands allowed influence significantly their maximum positions and intensities as well as the temperature dependence of the absorption. In particular, if ω' is not very large, we can assume that for many points of the band $p_1 \approx p_2$ that yields in (8.21) the factor $n + 1/2$ (this factor also remains after integration over Ω). It means that the integral intensity of the band increases with an increase in the occupation of the odd vibrations n, that is, with temperature. This is an important conclusion that enables us to discriminate electronic *d–d* transitions that are allowed by vibrations from other possible mechanisms; *the increase in oscillator strength with temperature is inherent only to these vibrationally induced transitions.*

For non-center-symmetrical (e.g., tetrahedral) ligand fields, mixing of even and odd atomic states of the central atom takes place without vibrations involved. Here the static ligand field gives a much stronger contribution to the *d–d* transition intensity than odd vibrations, but in this case *the intensity does not increase with temperature.* As mentioned above, *d–d* transitions are also allowed through the magnetic-dipole and electric-quadruple mechanisms (with $f \sim 10^{-6}$ and $f \sim 10^{-7}$, respectively), which are practically independent of temperature. These mechanisms may be effective in rare cases when the coupling between the electronic states and odd vibrations is very weak.

Spectrochemical and Nephelauxetic Series

As far as *d–d* transitions are concerned, there are two main parameters that determine the optical band positions in cubic complexes: Δ, the main parameter of ligand field splitting, and B, the Racah parameter of interelectron repulsion in the d states. This can be seen directly from the Tanabe–Sugano diagrams (Fig. 4.11), in which the calculated energy levels of d^n configurations are given as a function of Δ/B. Indeed, if Δ and B are known, all the band frequencies can be estimated from these diagrams, and the example of the

Mn^{2+} complex considered above shows that this way of description of the spectrum is satisfactory for such systems. Let us consider some general rules in the dependence of these two parameters on the nature of the CA and ligands.

The dependence of Δ on the properties of the CA is briefly discussed in Section 4.5. First, Δ strongly increases with the charge (oxidation state) of the CA; that is, it increases in the series $M(II) < M(III) < M(IV)$. For the same oxidation state Δ increases with the principal quantum number of the d electrons: $3d < 4d < 5d$. For $3d$ elements Δ is on the order of $\sim 10,000$ cm^{-1} for M^{2+} and $\sim 20,000$ cm^{-1} for M^{3+}. Passing to $4d$ and $5d$ electrons, Δ increases and reaches about $\sim 40,000$ cm^{-1}. The dependence of Δ on the ligands is weaker but much more diversified.

Table 8.2 shows some examples of Δ values for different ions and ligands (data collected from [8.18, 8.19]). An interesting general feature emerging from these data is that Δ increases when passing from one ligand to another from left to right for all the CA. This allows one to formulate the spectrochemical rule, or the *spectrochemical series*, that characterizes the increase of ligand influence by the increase of Δ:

$$Br^- < Cl^- < \tfrac{1}{2}Ox^{2-} < H_2O < NH_3 < \tfrac{1}{2}En < CN^- \qquad (8.23)$$

A more complete spectrochemical series with indication (in parentheses) of the relative Δ values (in units of that for H_2O) is as follows [8.19] (the

Table 8.2. Values of the crystal field parameter Δ for d^n transition metal complexes with different ligands

		Ligand[a]							
d^n	Ion	$6Br^-$	$6Cl^-$	$3Ox^{2-}$	$6H_2O$	$Enta^{4-}$	$6NH_3$	$3En$	$6CN^-$
$3d^1$	Ti^{3+}	—	—	—	20,300	18,400	—	—	—
$3d^2$	V^{3+}	—	—	16,500	17,700	—	—	—	—
$3d^3$	V^{2+}	—	—	—	12,600	—	—	—	—
	Cr^{3+}	—	13,600	17,400	17,400	18,400	21,600	21,900	26,300
$4d^3$	Mo^{3+}	—	19,200	—	—	—	—	—	—
$3d^4$	Cr^{2+}	—	—	—	13,900	—	—	—	—
	Mn^{3+}	—	—	20,100	21,000	—	—	—	—
$3d^5$	Mn^{2+}	—	—	—	7,800	6,800	—	9,100	—
	Fe^{3+}	—	—	—	13,700	—	—	—	—
$3d^6$	Fe^{2+}	—	—	—	10,400	9,700	—	—	33,000
	Co^{3+}	—	—	18,000	18,600	20,400	23,000	23,300	34,000
$4d^6$	Rh^{3+}	18,900	20,300	26,300	27,000	—	33,900	34,400	—
$5d^6$	Ir^{2+}	23,100	24,900	—	—	—	—	41,200	—
	Pt^{4+}	24,000	29,000	—	—	—	—	—	—
$3d^7$	Co^{2+}	—	—	—	9,300	10,200	10,100	11,000	—
$3d^8$	Ni^{2+}	7,000	7,300	—	8,500	10,100	10,800	11,600	—
$3d^9$	Cu^{2+}	—	—	—	12,600	13,600	15,100	16,400	—

[a] Ox, oxalate; Enta, ethylenetetramine; En, ethylenediamine.

coordinating atom is indicated in bold):

$\mathbf{Br}^-(0.72)$, $(C_2H_5)_2\mathbf{PSe}^-(0.74)$, $\mathbf{SCN}^-(0.75)$, $\mathbf{Cl}^-(0.78)$,

$(CrH_5)_2\mathbf{PS}_2^-(0.78)$, $(CrH_5O)_2\mathbf{PSe}_2^-(0.8)$, $\mathbf{POCl}_3(0.82)$,

$\mathbf{NNN}^-(0.83)$, $(CrH_5O)_2\mathbf{PS}_2^-(0.83)$, $(CrH_5)_2\mathbf{NCSe}_2^-(0.85)$,

$\mathbf{F}^-(0.9)$, $(C_2H_5)_2\mathbf{NCS}_2^-(0.90)$, $(CH_3)_2\mathbf{SO}(0.91)$, $(CH_3)_2\mathbf{CO}(0.92)$,

$CH_3\mathbf{COOH}(0.94)$, $C_2H_5\mathbf{OH}(0.97)$, $(CH_3)_2\mathbf{NCHO}(0.98)$, $CrO_4^{2-}(0.99)$,

$H_2\mathbf{O}(1.00)$, $CH_2(CO_2)^{2-}(1.00)$, $NH_2\mathbf{CS}(1.01)$, $\mathbf{NCS}^-(1.02)$,   (8.24)

$CH_3\mathbf{NH}_2(1.17)$, $NH_2CH_2CO_2^-(1.18)$, $CH_3\mathbf{SCH}_2CH_2\mathbf{SCH}_3(1.22)$,

$CH_3\mathbf{CN}(1.22)$, $C_2H_5\mathbf{N}(1:23)$, $\mathbf{NH}_3(1.25)$,

$\mathbf{NCSH}(1.25)$, $\mathbf{NCSHg}^+(1.25)$, $\mathbf{NH}_2CH_2CH_2\mathbf{NH}_2(1.28)$, $\mathbf{NH}(CH_2NH_2)_2(1.29)$,

$\mathbf{NH}_2OH(1.30)$, $\mathbf{SO}_3^{2-}(1.3)$, $C_6H_4(\mathbf{As}(CH_3)_2(1.33)$, 2,2'-dipyridyl(1.33),

1,10-phenanthroline (1.34), $\mathbf{NO}_2^-(\sim 1.4)$, $\mathbf{CN}^-(\sim 1.7)$

Along with the spectrochemical series, there is a *hypsochromic series* in which the ligands are arranged following the increase of the shift of the first absorption band to higher frequencies (ultraviolet shift):

$$I^- < Br^- < Cl^- \sim SCN^- \sim N_3^- < (C_2H_5O)_2PS_2^- < F^- < C_2H_5)_2NCS_2$$

$$< (NH_2)_2CO < OH^- < COO)_2^{2-} \sim H_2O < NCS^- < NH_2CH_2COO^-$$

$$< NCSHg^+ \sim NH_3 \sim C_5H_5N < NH_2CH_2CH_2NH_2 \sim SO_3^{2-}$$

$$< NH_2OH < NO_2^- < H^- \sim CH_3^- < CN^- \qquad (8.25)$$

Note that these two series, although somewhat similar, are not identical. Indeed, the ultraviolet band shift coincides with the increase in Δ only when the band results from a simple $t_2 \to e$ transition. In more complicated cases with more than one *d* electron, as seen in the Tanabe–Sugano diagrams, the frequency of some transitions may either increase or decrease with an increase in Δ. In particular, in the cubic complex of $Mn^{2+}(d^5)$, discussed above, the frequency of the first band $^6A_g \to {}^4T_{1g}$ [originating from $(t_{2g})^3(e_g)^2 \to (t_{2g})^4(e_g)^1)$] decreases with Δ (Fig. 8.4).

Besides, in the region of the breaks of the curves $E_1 = f(\Delta)$ in Tanabe–Sugano diagrams, a transition from the high-spin to the low-spin spectrum takes place (spin crossover; Section 8.4) which completely changes the type of the first transition. Therefore, the hypsochromic series, in general, does not mean that the parameter Δ increases by the corresponding ligand substitutions from left to right. The rule of hypsochromic increase of Δ is valid only for simple $t_2 \to e$ transitions in complexes with the same spin state.

The other parameter of the d–d transitions, the Racah parameter B (2.43), is also dependent on the nature of the ligands. If one compares the B value of the free ion with that in the complex, one finds that the latter is always smaller than the former. This reduction of B by complex formation, meaning the reduction in the interelectron repulsion in the d states, is obviously caused by the delocalization (expansion) of the electron cloud on larger regions due to the formation of MO, and this interpretation can be confirmed quantitatively by calculations. The effect of ligand influence on interelectron repulsion in the d states characterized by B was first studied by Jorgensen [8.20] (see also [8.18, 8.19]), and it was named the *nephelauxetic effect*, whose Greek roots mean "cloud expanding effect."

The B values for some free ions and their complexes obtained from spectroscopic data are given in Table 8.3. Introducing the *nephelauxetic ratio* $\beta = B_{complex}/B_{ion}$, one can arrange the ligands in a series of increasing β values for the same CA in the same oxidation state, the *nephelauxetic series*:

$$F^- > H_2O > (NH_2)_2CO > NH_3 > (COO)_2^{2-} > NCS^- > Cl^- \sim CN^- > Br^-$$

$$> (C_2H_5O)_2PS_2 \sim S^{2-} \sim I^- > (C_2H_5O)_2PSe_2 \qquad (8.26)$$

An important feature of the nephelauxetic effect is that it is directly related to covalency. Indeed, the delocalization of the d electrons in the complex is caused by the formation of covalent bonds and MOs. Hence the measure of the B reduction in the complex is simultaneously a measure of covalency. This means that following the decrease of β from left to right, the nephelauxetic series (8.26) reflects an increase in covalency in this direction (in the complexes formed by the corresponding ligands). From this point of view the nephelauxetic effect is more informative with respect to chemical bonding than is the spectrochemical effect, although they are both qualitative properties.

Table 8.3. Values of the Racah parameter B for free d^n ions of transition metals and their complexes with different ligands (in cm^{-1})

			Ligand[a]							
d^n	Ion	Free Ion	$6Br^-$	$6Cl^-$	$3Ox^{2-}$	$6H_2O$	Enta^{4-}	$6NH_3$	$3En$	$6CN^-$
$3d^3$	Cr^{3+}	950	—	510	640	750	720	670	620	520
$3d^5$	Mn^{2+}	850	—	—	—	790	760	—	750	—
$3d^5$	Fe^{3+}	~ 1000	—	—	—	770	—	—	—	—
$3d^6$	Co^{3+}	~ 1050	—	—	560	720	660	660	620	440
$4d^6$	Rh^{3+}	~ 800	300	400	—	500	—	460	460	—
$5d^6$	Ir^{3+}	660	250	300	—	—	—	—	—	—
$3d^7$	Co^{2+}	1030	—	—	—	~ 970	~ 940	—	—	—
$3d^8$	Ni^{2+}	1130	760	780	—	940	870	890	840	—

[a]Ox, oxalate; Enta, ethylenetetraamine; En, ethylenediamine.

Only one nephelauxetic ratio β may be insufficient to characterize the nephelauxetic effect in some complexes with many d electrons. Indeed, the value B_{complex} is determined from spectroscopic data (Section 2.2) and can be different for different d–d transitions. Therefore, in a more detailed description of complexes with $(t_2)^m(e)^n$ configurations, three values of B can be introduced: $B(e \rightarrow e)$, $B(e \rightarrow t_2)$, $B(t_2 \rightarrow t_2)$, where $e \rightarrow e$, $e \rightarrow t_2$, and $t_2 \rightarrow t_2$ indicate the type of spectroscopic transition data from which the B value is extracted. In the notation of Bethe, $E = \Gamma_3$ ($e = \gamma_3$) and $T_2 = \Gamma_5$ ($t_2 = \gamma_5$); therefore, the corresponding nephelauxetic ratios are denoted by β_{33}, β_{35}, and β_{55}, respectively. It can be shown that approximately (see [8.19])

$$\frac{\beta_{33}}{\beta_{35}} = \frac{\beta_{35}}{\beta_{55}} \tag{8.27}$$

With this relation only two nephelauxetic parameters remain independent: for example, β_{55} and β_{35}. Table 8.4 presents some of these parameters for a series of hexafluorides.

Table 8.4. Racah's parameters of interelectronic repulsion B_{55} and B_{35} (in K units) and the corresponding nephelauxetic ratios β_{55} and β_{35} for some hexafluoride complexes of d^n transition metals

d^n	Complex	B_{55}	β_{55}	B_{35}	β_{35}
$3d^3$	CrF_6^{3-}	860	0.93	820	0.89
	MnF_6^{2-}	815	0.77	600	0.56
$3d^5$	MnF_6^{4-}	—	—	845	0.94
	FeF_6^{3-}	—	—	845	0.78
$3d^6$	NiF_6^{2-}	—	—	450	0.36
$3d^8$	NiF_6^{4-}	—	—	960	0.92
	CuF_6^{3-}	—	—	650	0.54
$4d^2$	RuF_6	300	0.37	—	—
$4d^3$	$T_cF_6^{2-}$	560	0.79	530	0.76
	RuF_6^{-}	480	0.61	—	—
$4d^6$	RhF_6^{3-}	—	—	460	0.64
	PdF_6^{2-}	—	—	340	0.42
$4d^8$	AgF_6^{3-}	—	—	460	0.60
$5d^2$	OsF_6	380	0.52	—	—
$5d^3$	ReF_6^{2-}	540	0.83	—	—
	OsF_6^{-}	530	0.73	—	—
	IrF_6	380	0.43	—	—
$5d^4$	PtF_6	260	0.30	—	—
$5d^6$	PtF_6^{2-}	—	—	380	0.51

Source: [8.19].

For further details, the number of such parameters can be increased, but this makes little sense; the larger the number of parameters required for the interpretation of experimental data, the less informative the interpretation.

Transitions Involving Orbitally Degenerate States

Most $d-d$ transitions involve orbitally degenerate electronic terms as initial or final states of the transition. Compared with nondegenerate terms, degenerate states may yield quite different types of spectra. Indeed, the band shapes and frequencies are strongly dependent on the APs (Section 8.1), which for degenerate terms are much more complicated [8.21, 8.22]. They are discussed in Section 7.3.

Consider first the case when one of the combining states is an orbitally doubly degenerate E term and the other is nondegenerate ($A \rightarrow E$ and $E \rightarrow A$ transitions). The energy spectrum for the E term can be obtained approximately by solving the linear $E - e$ problem (Section 7.3), while for the nondegenerate electronic A term the vibrational states are usual oscillator states (Section 7.1). Calculating the energy-level differences and transition probabilities for each transition from the oscillator states to the vibronic levels of the E state ($A \rightarrow E$ transition) or vice versa ($E \rightarrow A$ transition), one can obtain all the lines of the expected spectrum. The data given in Fig. 8.6 were obtained numerically in this way. They illustrate the relative intensities of the vibrational components of the bands $A \rightarrow E$ and $E \rightarrow A$ for several values of the dimensionless vibronic coupling constant $\lambda = E_{JT}/\hbar\omega$ (Section 7.3).

It is seen from Fig. 8.6 that the $A \rightarrow E$ band, the envelope of the vibrational components, has a two-humped form. Compared with the band of transitions to a nondegenerate term given in Fig. 8.2, the two maxima can be interpreted as the Jahn–Teller splitting of the nonvibronic band. For transitions $E \rightarrow A$ such splitting does not occur, but as shown below, this is due to neglect of the temperature population of the excited vibrational states of the A term, which is valid for $T = 0$.

A more general picture of the band shape (although less accurate in details) can be obtained in the semiclassical approximation discussed in Section 8.1. Substituting the expressions of the APs (8.7) for the A term and (7.40) for the E term into the expression of the form function $F_{12}(\Omega)$ and performing appropriate integrations, we obtain for $Q_0 = 0$ [8.14, 8.21],

$$F_{12}(\Omega) = \frac{M_{12}^2 K\hbar|\Omega - \Omega_0|}{F^2 2kT} \exp\left[-\frac{k\hbar^2(\Omega - \Omega_0)^2}{F^2 kT} \right] \tag{8.28}$$

The function (8.28) is presented graphically in Fig. 8.7a. It has a symmetrical shape with two humps and a dip at $\Omega = \Omega_0$. Similar to the results above, this band shape can be interpreted as being due to Jahn–Teller splitting of the nonvibronic band. The splitting (the distance between the two maxima) equals $(8E_{JT}kT)^{1/2}$. If one takes into account the contribution of the totally symmetric

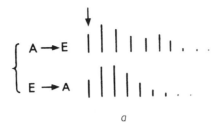

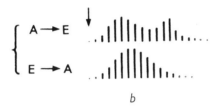

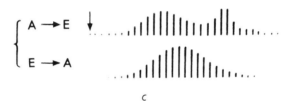

Figure 8.6. Frequencies and relative intensities of vibronic components and the band shape (envelope) for $A \rightarrow E$ and $E \rightarrow A$ transitions calculated at $T = 0$ for the following values of the dimensionless vibronic constant: (*a*) $\lambda = 2.5$; (*b*) $\lambda = 7.5$; (*c*) $\lambda = 15$. The position of the zero-phonon line is shown by an arrow. (After [8.23].)

vibrations to the broadening, the "acute elements" of the curve in Fig. 8.7a are smoothed, and it assumes the form given in either Fig. 8.7*b* or *c*, depending on whether coupling with the totally symmetric vibrations is weaker than that with *e* vibrations, or vice versa. The strength of coupling and the influence of thermal population of the corresponding vibrations is given by the quantities $X_A = F_A^2 \coth(\hbar \omega_A / 2kT)$ and $X_E = F_E^2 \coth(\hbar \omega_E / 2kT)$, where F_A and F_E, and ω_A and ω_E are the corresponding vibronic constants and vibrational frequencies, respectively.

Thus if the totally symmetric vibrations predominate, the dip in the curve is completely filled up and disappears (Fig. 8.7*c*). For comparison, the temperature dependence of the $A \rightarrow E$ band as determined by the numerical solution [8.24] is given in Fig. 8.8. For further details on $A \rightarrow E$, $E \rightarrow A$, and $E \rightarrow E$ transitions, see [8.21, 8.22, 8.29].

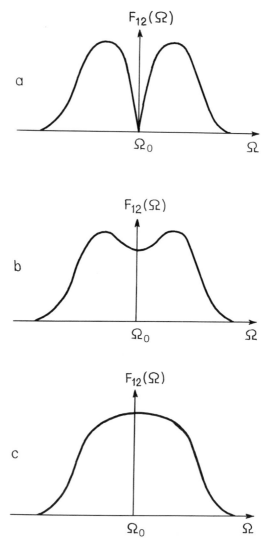

Figure 8.7. Band shape of the $A \to E$ transition calculated in the semiclassical approximation including the linear coupling with E and A vibrations: (a) coupling with totally symmetric vibrations A is neglected; (b) A vibrations are included, but the coupling to E vibrations is predominant; (c) coupling to A vibrations is predominant.

The $A \to T$ transitions have also complicated band shapes except when in the T state the coupling to t_2 vibrations is negligible and the vibronic problem is $T - e$ (Section 7.3). In the latter case no splitting of the $A \to T$ band occurs, although the adiabatic potential of the T term is split (Fig. 7.17). This illustrates how carefully visual pictures should be used in the analysis of complicated phenomena. In general, it can be said that for absorption

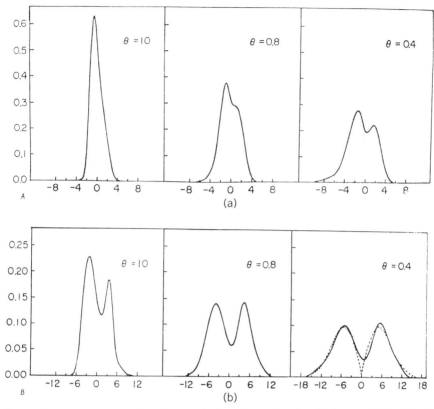

Figure 8.8. Temperature dependence of the absorption band shape of the $A \to E$ transition obtained by numerical solution of the linear $E–e$ problem, including coupling with E and A vibrations with $\lambda_A = 0.5$; $\lambda_E = 0.5$ (a) or $\lambda_E = 5.0$ (b). The frequency with respect to the pure electronic transition $\Omega = 0$ is given in ω_E units, and $\theta = \hbar\omega_E/kT$. For strong vibronic coupling and at larger temperatures the absorption curve approaches the semiclassical limit shown by dashed line (cf. Fig. 8.7).

transitions from nondegenerate to degenerate terms the band does not split if the point of degeneracy on the AP is a point of actual crossing of the surfaces, as in the $T–e$ problem (Fig. 7.17). This is in contrast to the case when the point of degeneracy is a branching point of the surface, as in the $E–e$ problem (Figs. 7.9 and 7.10), for which the band is split.

If the coupling to t_2 vibrations is predominant, the $A \to T$ absorption curve in simple cases has three humps, the band is split into three components, but the temperature dependence and other parameters make them rather nonequivalent. Figure 8.9 presents the numerical results [8.28] obtained for the band shape of this transition for a set of temperatures and vibronic coupling constant. For other examples of transitions and more detailed discussion, see [8.21, 8.22, 8.29].

I

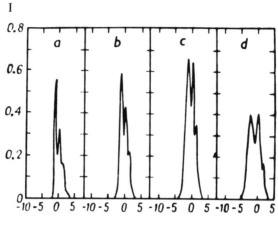

II

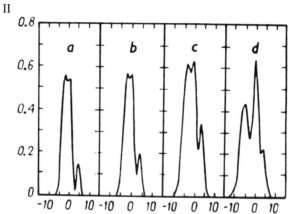

Intensity

III

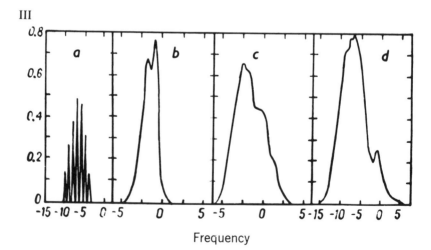

Frequency

The brief discussion in this subsection allows us to conclude that *vibronic effects strongly influence the band shapes of electronic d–d transition* which involve degenerate terms, and that the often used *interpretation of electronic spectra disregarding vibronic interactions may be invalid* if the ground or excited state, or both, are degenerate, or if at least one of them is vibronically coupled to (mixed with) a third one and the vibronic coupling is sufficiently strong.

8.3. X-RAY AND ULTRAVIOLET PHOTOELECTRON SPECTRA; EXAFS

General Ideas

Photoelectron spectroscopy is based on the photoelectric effect, its principle having been disclosed by Einstein in a paper that is one of the cornerstones of quantum mechanics. The energies of light quanta (photons) are limited by their frequency, $E = \hbar\Omega$. Hence there is a threshold frequency Ω_c of the photon corresponding to the minimum energy $e\varphi$ required for removing an electron from a metal to the vacuum, $\hbar\Omega = e\varphi$ (φ is called the work function). For photoemission of electrons from atoms or molecules in the gas phase the minimum energy that creates the threshold equals the binding energy E_B, or the ionization potential I_i from the state i, $\hbar\Omega_c^i \approx I_i$.

If the frequency of the photon Ω is larger than Ω_c, the electron emitted has a nonzero kinetic energy $E_k = \hbar\Omega - I_i$, and

$$I_i = \hbar\Omega - E_k \tag{8.29}$$

Figure 8.9. Band shapes (envelopes of elementary transitions) of the electronic $A \to T$ transitions in absorption and $T \to A$ transitions in emission obtained by numerical solution of the linear $T - t_2$ problem using the following 12 sets of parameters (α is the elementary line width in ω_T units). (After [8.28].)

Figure	I. Absorption $A \to T$ $E_{JT}/\hbar\omega_T = \frac{2}{3}$ $\alpha = 0.5$ $kT/\hbar\omega_T$	II. Absorption $A \to T$ $E_{JT}/\hbar\omega_T = 2$ $\alpha = 1.0$ $kT/\hbar\omega_T$	III. Emission $T \to A$ $E_{JT}/\hbar\omega_T$	α	$kT/\hbar\omega_T$
a	0	0	2	0.1	0
b	0.5	0.5	$\frac{2}{3}$	0.5	0.5
c	1.0	1.0	$\frac{2}{3}$	0.5	4.0
d	4.0	4.0	2	0.5	4.0

Thus if the experiment on photoemission is carried out with sufficiently large (fixed) frequencies of light, the kinetic energy spectrum of the emitted electrons is a replica of the energy distribution of occupied bond states. Photoelectron spectroscopy then becomes a method of direct determination of the electronic states of atoms, molecules, and solids. Obviously, for inner electron (core) states the frequency Ω corresponds to x-rays.

The method was suggested for the optical region by Vilesov, Kurbatov, and Terenin [8.30] and Turner and Al-Joboury [8.31, 8.32] and for the x-ray region by Ziegbahn and co-workers (see [8.3, 8.4, 8.33-8.38] and references therein). It was developed and gained widespread use only when some significant difficulties in exact and high-resolution measurements of electron kinetic energies had been overcome. At present the accuracy of electron energy measurements is about 10^{-2} eV in the optical region and 10^{-1} eV in the x-ray region.

In fact, there are several closely related photoemission and x-ray processes illustrated in Fig. 8.10 which form the basis for four methods of photoelectron spectroscopy:

1. *Ultraviolet photoelectron spectroscopy* (UPS). The light photon ejects the electron from the atomic valence shell or MO to the continuous spectrum (Fig. 8.10a).

2. *X-ray photoelectron spectroscopy* (XPS). The x-ray photon ejects the electron from the inner-shell (core) states of the system (Fig. 8.10b).

3. *Auger electron spectroscopy* (AES). After the formation of a hole in the core shell, radiationless transition of an electron from higher levels to the hole takes place, the excess energy being transferred to another electron which is thus emitted with an appropriate kinetic energy (Fig. 8.10c) (after P. Auger, who first observed such electrons).

4. *X-ray emission spectroscopy* (XES). After the formation of a hole in the inner shell, a transition from the excited one-electron states to the hole state with irradiation of an x-ray photon takes place.

Unlike the UPS and XPS cases, where the kinetic energy of the electrons is a direct consequence of the photoeffect, in the AES the emitted electrons emerge from a post-photoeffect process of radiationless relaxation of the excited state in competition with possible radiation transitions.

Photoelectron spectra are registered in the form of the number (counting rate) of photoelectrons as a function of their kinetic energy (or the ionization potential I_i). Figures 8.11, 8.12, and 8.13 illustrate, by way of examples, a part of the UPS of ferrocene [8.38], two lines of XPS of metallic Pt and Pt in K_2PtCl_6 [8.3], and the AES of TiO_2 [8.3]. In Fig. 8.11 one can see two peaks of the outer $3d(t_{2g})$ orbitals of iron (occupied by six electrons) split by the ligand field of D_{5h} symmetry into a_{1g} and e_g (in fact, these orbitals are MOs, not AOs). The occupation number of the e_g orbitals is twice that of a_{1g} [the

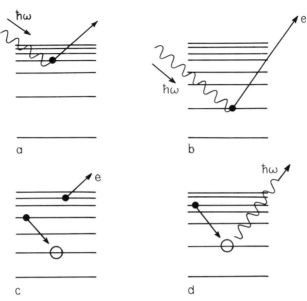

Figure 8.10. Schematic presentation of four types of photoelectron spectroscopy: (*a*) the photon ejects an electron from the valence-shell MO of the system (UPS); (*b*) the photon ejects an electron from the inner (core) states (XPS); (*c*) the hole created by the photon is ra'diationless populated by an electron from higher states, another electron being emitted (AES); (*d*) the inner-shell hole is occupied by a higher-energy electron, an x-ray quantum being emitted (XEP).

outer electron's configuration is $(a_{1g})^2(e_g)^4$]; the intensity of the photoelectron band (the area under the curve) of the former is approximately two times larger than the latter.

Figure 8.12 illustrates the effect of inner-shell line shift in the XPS due to chemical bonding, the *chemical shift*. Indeed, the positions of the two lines of photoelectrons from the N shell ($N_{YI}N_{VII}$ or $4f_{5/2}4f_{7/2}$) of the Pt atom in K_2PtCl_6 is displaced with respect to the same lines in metallic Pt by more than 2 eV toward higher binding energies (lower kinetic energies of the photo-emitted electrons). It means that in the chloroplatinate ion the N shell electrons of Pt are more strongly bonded to the nucleus than in metallic Pt. This is understandable if one takes into account the charge redistribution in K_2PtCl_6: the increasing electron density on the chlorine atoms and its decrease on Pt with respect to the metallic phase [in XPS spectra the source of x-rays (MgK_α) is indicated].

The AES of the KLL transitions in oxygen in TiO_2 (Fig. 8.13) shows the excited configuration terms, from which the radiationless transition produces the emitted electron. In combination with the x-ray photoemission that preceded the Auger effect, the AES determines the energy-level positions of the excited states of inner electrons.

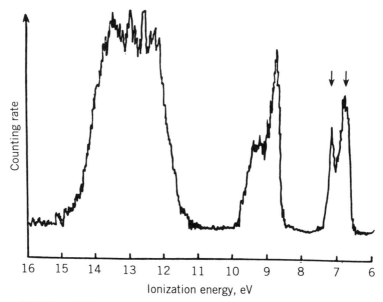

Figure 8.11. Part of the PES of ferrocene. The arrows indicate the 3*d* orbitals of iron. (After [8.38].)

These examples demonstrate the main trends of applications of photo-electron spectroscopy: determination of one-electron energy levels of the system from the positions of photoelectron peaks and their intensities, and evaluation of the parameters of electron density distribution from photo-electron chemical shifts. Both these trends advanced essentially during the last two decades, and new possibilities for the study of electronic structures were elucidated.

Electron Relaxation; Shake-up and CI Satellites

To begin with, attribution of the photoelectron peak positions to electronic energy levels is complicated by several side-effects: electron relaxation, shake-up and shake-off satellites, multiplet splitting, and final-state configuration interaction. Let us consider them sequentially.

The relaxation process is discussed in Section 6.2. In accordance with the Koopmans theorem (Section 2.2), the binding energy of the electron in the given ith MO equals the potential of ionization I_i of this electron taken with the opposite sign:

$$I_i = -\varepsilon_i \tag{8.30}$$

However, the Koopmans theorem does not take into account that when the ith electron is ionized, all the *other electrons relax to new self-consistent states*

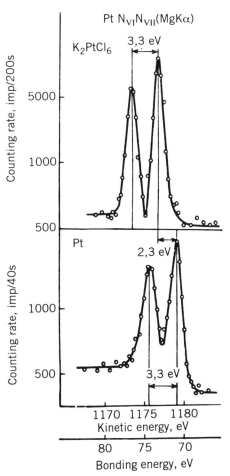

Figure 8.12. Two lines of XPS of the N shell of the Pt atom in K_2PtCl_6 and in metallic platinum. (From [8.3])

in which the interelectron repulsion is reduced by that of one electron (Sections 2.2, 6.2, and 6.4). The relaxation energy may be significantly large, thus reducing the observed ionization potentials compared with those predicted by the Koopmans theorem and, what is even more important, sometimes changing the order of their occurrence in the series of photoelectron peaks. To calculate the ionization potential sufficiently accurately, we must discard the Koopmans theorem and define I_i as the difference between the self-consistent energies of the initial and ionized states calculated independently.

Even when the calculations of I_i are performed with the highest accuracy, there is no a priori evidence that the photoelectron peak positions calculated coincide with the peaks observed because of the *shake-up and shake-off processes*. Their physical meaning is as follows. The electron relaxation during

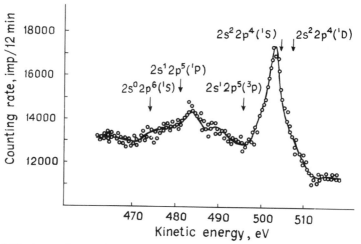

Figure 8.13. Auger electron KLL spectrum of oxygen in TiO$_2$ showing its interpretation as due to transitions from the L shell to the K shell (the roughly estimated position of the appropriate configuration terms are shown by arrows). (From [8.3].)

(or after) the photoionization takes place not instantly but in a certain time scale τ estimated as about $\tau \sim 10^{-17}$ s (cf. the nuclear relaxation time $\sim 10^{-12}$ that permits Franck–Condon transitions; Section 8.1). Therefore, if the photoeffect process has a shorter time scale τ',

$$\tau' < \tau \tag{8.31}$$

the electronic subsystem does not manage to relax during the photoionization, and hence the relaxation energy will not be incorporated (at least completely) into the kinetic energy of the emitted electron. Then what will be seen in the photoelectron experiments under condition (8.31), and how may we relate the spectrum observed to the electronic structure of the system? Quantum mechanics enables us to answer this question.

The nonrelaxed state of the ionized system produced by a very short time-dependent *sudden perturbation* is not a stationary solution of the Schrödinger equation. The theory of sudden perturbations shows that the final nonstationary state Ψ_f can be expanded into a series of the stationary states of the ion Ψ_k:

$$\Psi_f = \sum_k C_k \Psi_k \tag{8.32}$$

Taking into account that Ψ_k are mutually orthonormalized, the probability of the system to fall into one of them is $P_k = |C_k|^2$,

$$C_k = \int \Psi_k^* \Psi_f \, d\tau \tag{8.33}$$

Thus in the case of incomplete relaxation under condition (8.31), there is a probability for the system to occur not in one but in several stationary states Ψ_k producing several peaks of the photoemission electron spectrum (*shake-up satellites*). Visually, the process is as if the system during the photoionization to a nonstationary state were shaken up to one of the stationary states that are excited states of the ion (usually outer-shell excitations). Of course, Eq. (8.32) also includes the ground state of the ion. If Ψ_k is an ionized state, we have a *shake-off process*.

The shake-up transitions described by the transition probability (8.33) are monopole transitions in the sense that the transition operator is scalar (cf. dipole transitions in Section 8.1). This means that all the states Ψ_k that produce the satellite lines in the spectrum must be of the same symmetry and multiplicity [otherwise, the integral (8.33) is zero].

For the shake-up satellites there are some interesting relations estimating qualitatively their possible positions and intensities. Denoting the intensities by I_k^0 (not to be confused with the ionization potential I_k) and their peak energies by E_k (while the Koopmans energies are ε_i) we have

$$\sum_k I_k^0 = I_i^0 \tag{8.34}$$

or

$$\sum_k I_k^0 (E_k - \varepsilon_i)^0 = I_i^0 \tag{8.35}$$

and

$$\sum_k I_k^0 (E_k - \varepsilon_i)^1 = 0 \tag{8.36}$$

These relations, sometimes called *the sum rules*, are based on the invariability of the sum of diagonal matrix elements (the spur of the matrix) with respect to unitary transformations of the basis functions and may be considered as fairly good approximations, provided that the kinetic energy of the electron is large. Expressions of the type (8.35) are called the zero moment (summary value) of the set of quantities I_k^0, while (8.36) is the first moment (center of gravity) of this set. In general,

$$\sum_k I_k^0 (E_k - \varepsilon_i)^n \tag{8.37}$$

is called the nth moment of the distribution.

Thus the zero moment of the set of intensities of the shake-up satellites equals the initially expected (Koopmans) intensity, while their center of gravity (the first moment) coincides with the Koopmans peak. On the other hand, Eq.

(8.34) means that the shake-up satellites borrow intensities from the main line, while (8.36) shows that the Koopmans peak can easily be obtained by finding the center of gravity of the main line plus satellites. Note that the deviation of the satellite line $E_k - \varepsilon_i$ cannot be larger than the relaxation energy. It can be shown that for large deviations the probability $P_k = |C_k|^2$ is small.

Concerning the characteristic time of the photoemission τ', it is strongly dependent on the photoelectron speed and hence on the photon frequency: The larger the latter, the smaller τ' and the more favorable the satellite occurrence. These considerations allow one to formulate some qualitative rules for shake-up satellites given in Table 8.5.

The shake-up satellites are somewhat related to the final-state *configuration interaction satellites*. The problem of configuration interactions (CI) is discussed in Section 5.3 in connection with the problem of correlation effects. Since different electronic configurations yield energy terms with the same symmetry and multiplicity, their linear combination of type (5.62) [or quite a similar one, (8.32)] must be taken as the most general form of the possible final state of the photoelectron transition. This means that there may be more than one final state of the same symmetry that take part in the transition. However, to be observable, the CI states must be comparable in energy and in transition probability determined by the C_k coefficients of the CI expansion. Therefore, only the outer-shell ionized configurations produce observable CI satellites.

Figure 8.14 illustrates the case of a CI satellite to the K^+ $3s$ electron emission line in the UPS of KF, at 14 eV toward greater binding energies. The K^+ ion has a ground state S produced by the electronic configuration $[Ne]3s^2 3p^6$. The photoemission from the $3s$ shell produces a ^2S state with the $[Ne]3s^1 3p^6$ configuration. There are several other relatively close-in-energy configurations that produce the same ^2S term at about 14 eV: $[Ne]3s^2 3p^4 3d^1$, $[Ne]3s^2 3p^4 4s^1, \ldots$, but the calculation shows that only transitions to the M shell have appreciable probabilities, and the satellite observed is identified as the $[Ne]3s^2 4p^4 3d^1$ final state [8.39].

The shake-up satellites are similar to the CI satellites both in nature and appearance. There are still some ways to discriminate the former from the latter. Both types of satellites can be regarded as outer-shell excitations, but the

Table 8.5. Some qualitative rules for shake-up satellite occurrence in PES

Type of Energy Level	Ionization Potential (eV)	Relaxation Energy (eV)	Expected Shake-up Satellites
Outer valence	$< \sim 15$	0.5–5.0	No intensive satellites
Inner valence	15–50	~ 2–5	Intensive and relatively close satellites
Inner core	> 50	> 10	Weak and extended satellite structure

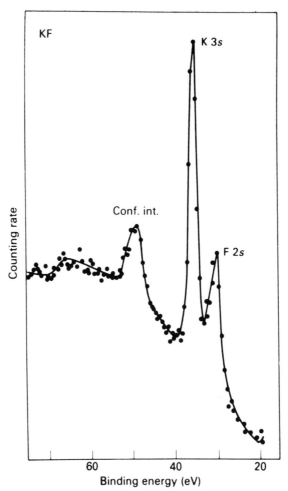

Figure 8.14. Photoelectron spectrum of KF showing a CI satellite of the K $3s$ shell at 14 eV. (From [8.39.)

CI satellites are unique only to the given outer-shell configuration and will not be repeated in other (inner-shell or core) configurations, whereas the same shake-up satellites, being monopole transitions, occur in any (outer, inner, core) photoemission of the system whenever the electron relaxation energy must be compensated. In the example above of photoemission of K^+ in KF, the core lines have no 14-eV satellites ruling out an interpretation based on the monopole mechanism.

Another complication to the interpretation of photoelectron line positions is created by the *multiplet structure* of the core electron energy levels. Consider, for instance, the simple case of the $Mn^{2+}(3d^5)$ ion with the sextet ground state ^6S that has no angular momentum ($L = 0$). The core s-electron photoeffect creates a hole spin $\frac{1}{2}$, which couples to the outer electron spins, producing two terms ^7S and ^5S. They are split by the spin–orbital interaction, and hence two s-core lines are expected instead of one. The two lines have been found, indeed, and they have the intensity ratio $7:5$, as required by the theory (transition probabilities are proportional to degeneracies). Another example is shown in Fig. 8.12, where two lines of the Pt N shell photoelectrons (from $f_{5/2}$ and $f_{7/2}$, respectively) are well resolved. For a nonzero orbital momentum (open shells) the multiplet structure becomes much more complicated.

Sometimes, interpretation of photoelectron spectra is facilitated by studying the *angular dependence of the intensity* (or cross section σ) of photoionization. The differential cross section of photoionization of a given energy level with polarized light is

$$I_\theta = \frac{\sigma_t}{4\pi}\left[1 + \frac{\beta}{2}(3\cos^2\theta - 1)\right] \qquad (8.38)$$

where σ_t is the total cross section of photoionization, θ the angle between the electric vector of the photon and the direction of the outgoing electron, and β— the *parameter of asymmetry*. If the light is not polarized,

$$I_\theta \sim 1 + \frac{\beta}{2}\left(\frac{3}{2}\sin^2\theta - 1\right) \qquad (8.39)$$

where θ is the angle between the radiation beam and outgoing electrons.

The parameter of asymmetry is an unknown function of both the wavefunction of the ionized state and the photoelectron energy. It changes from -1 to $+2$. So far this function has been studied for simple molecules [8.36], not for coordination compounds. Nevertheless, using Eq. (8.38) or (8.39) and experimental I_θ values, one can separate the overlapping bands (components), provided that they have different β values.

As for the band shapes of UPS and XPS, all the results obtained in Sections 8.1 and 8.2 for electronic transitions are valid. Indeed, the ionization process is also an electronic transition in which the final state belongs to the continuous spectrum. The two states can be described by the same APs as in the case of optical transitions (Fig. 8.1), and hence the resulting band shapes and intensities are those discussed in Sections 8.1 and 8.2. In particular, because of the change in the number of electrons, the photoelectron transition is always a spin-forbidden transition (Section 8.1), and in the majority systems it involves orbitally degenerate states, which result in Jahn–Teller splitting (Section 8.2).

Photoelectron band shapes were evaluated by calculating the vibronic states in degenerate and pseudodegenerate excited (ionized) states performed by the Greens function method [8.40]. The results agree well with the experimental data (see, e.g., the theoretical interpretation of the UPS of BF_3 obtained by calculation of the pseudo Jahn–Teller effect in BF_3^+ [8.41]). Other examples are available in [8.42].

Chemical Shift

This notion is widespread in chemistry determining the shift of energy levels due to chemical bonding, and as such it appears in many physical methods of investigation (Mossbauer spectroscopy, NMR, etc.). In the UPS, XPS, and XES under consideration in this section, the chemical shift denotes the shift of the corresponding inner-shell spectral lines caused by the change of the local chemical environment by passing from one compound to another. The shift of the two f electron lines ($f_{5/2}$ and $f_{7/2}$) of Pt in K_2PtCl_6 with respect to their positions in metallic Pt (Fig. 8.12) is an example of this type of chemical shift.

A widespread explanation of the origin of chemical shifts ΔE of inner energy levels is that the formation of chemical bonds redistributes the electronic charge and changes the interelectron repulsion and the screening of the nucleus. The effective atomic charge of the atom A in its valence state in the molecule q^A (Section 5.2) can serve as a rough measure of this charge redistribution. The idea is to use the calculated dependence of the inner energy-level positions in free atoms (ions) as a function of their charges (ionized states) to estimate the atomic charges from chemical shifts in the spectra. Examples of such functions $\Delta E = f(q^A)$ are illustrated in Fig. 8.15 for three atoms, S, P, and Cl.

However, as stated in Section 5.2, the effective atomic charge is not directly observable, and the result of its calculation depends on the definitions and approximations used. For instance, calculated according to Mulliken [Eq. (5.20)], the atomic charges depend on the basis set, whereas Politzer [cf. (5.22)] defined them to depend on the assumed atomic border in the molecule. In addition, the integral charge effect may not be sufficiently informative. Indeed, the contribution of different atomic orbitals to the chemical shift may be quite different (even different in sign, see below), and they may be affected differently by the bonding. Therefore, the dependence of the chemical shift on the integral charge of the atom in the molecule may even be misleading: The chemical shift may be large even when the integral charge is not changed.

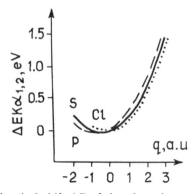

Figure 8.15. Chemical shift ΔE of the photoelectron line $K_{\alpha1,2}$ as a function of the charge of the free atoms (ions) S, P, and Cl. (From [8.37].)

The MO LCAO scheme gives a much more refined possibility to link the observed chemical shifts ΔE with the electronic structure. Consider the orbital charge transfers Δq_i, [Eq. (5.20″)], that characterize the electronic charge transferred to ($\Delta q_i > 0$) or removed from ($\Delta q_i < 0$) each one-electron orbital, and assume that the chemical shift produced by removing one electron from the ith orbital of the atom, the *orbital chemical shift* $\Delta \varepsilon_i$, is known. Then, assuming that Δq_i are not very large, we have approximately

$$\Delta E = - \sum_i \Delta q_i \, \Delta \varepsilon_i \qquad (8.40)$$

The Δq_i values are more significant for the valence electrons that participate in the bonding, and they can be obtained from calculations. The orbital chemical shifts can be obtained relatively easily from empirical data: for instance, from the photoelectron or x-ray emission spectra of free atoms and ions. For the chemical shift of the $K\alpha_1$ line in heavy-atom compounds ($30 < Z < 75$) the authors [8.43], using XES data, obtained the following values of orbital contributions $\Delta \varepsilon_i$ for the valence electrons s, p, d, f (in meV):

$$\Delta \varepsilon_s = \Delta \varepsilon_p = 80 \pm 10$$
$$\Delta \varepsilon_d = -115 \pm 10 \qquad (8.41)$$
$$\Delta \varepsilon_f = -570 \pm 30$$

These data are interesting in several respects. First, they show the most essential differentiation between the contributions of different types of valence electrons to the $K\alpha_1$ line shift: While for s and p electrons it is positive, meaning that they increase the x-ray emission frequency, for d and f electrons it is negative. Second, the absolute values of the effect are also different: They are significantly larger for d and f electrons.

Now consider the chemical shift of the $K_{\alpha 1}$ line in the XES of a nontransition atom in a molecule, which participates in bonding with its valence s and p orbitals ($\Delta \varepsilon_d \approx \Delta \varepsilon_f \approx 0$). Provided that the relation $\Delta \varepsilon_s \approx \Delta \varepsilon_p$ is valid [see (8.41)], we get from (8.40) the following equation for sp elements:

$$\Delta E = -(\Delta q_s + \Delta q_p) \, \Delta \varepsilon_s = -q^A \, \Delta \varepsilon_s \qquad (8.42)$$

where $q^A = \Delta q_s + \Delta q_p$ is the effective charge of the atom under consideration. It is seen that the chemical shift is proportional to the effective charge, in accordance with the above-mentioned widespread ideas, and with Fig. 8.15 (for sufficiently small Δq_i values).

However, for transition metals and rare earth elements in coordination compounds the participation of d electrons in the bonding is most essential, and Eq. (8.42) does not hold. Indeed, with d-electron participation,

$$\Delta E = -(\Delta q_s + \Delta q_p) \, \Delta \varepsilon_s - \Delta q_d \Delta \varepsilon_d \qquad (8.43)$$

and it is seen that the effective charge $q^* = \Delta q_s + \Delta q_p + \Delta q_d$ does not characterize the chemical shift ΔE. On the contrary, if for illustration we take $|\Delta \varepsilon_s| \approx |\Delta \varepsilon_d|$, then

$$\Delta E \approx -(\Delta q_s + \Delta q_p - \Delta q_d)\,\Delta \varepsilon_s \qquad (8.44)$$

As discussed in Section 6.3, in transition metal compounds there are diorbital bonds in which the σ and π bindings give orbital charge transfers Δq_σ and Δq_π of opposite sign ($\Delta q_\sigma < 0$, $\Delta q_\pi > 0$). If, as usual, the σ bonds are realized by s and p orbitals while the π bonds involve d orbitals, the atomic charge $q^A = |\Delta q_s + \Delta q_p| - |\Delta q_d|$ may be very small, even zero, while

$$\Delta E = -(|\Delta q_s + \Delta q_p|\Delta \varepsilon_s| + |\Delta q_d||\Delta \varepsilon_d|) \qquad (8.45)$$

can be very large: *There is no direct correlation between the effective charge on the transition atom and the chemical shift.*

The sensitivity of the chemical shifts to the chemical environment makes the use of the photoelectron method in structural investigations most efficient. Figure 8.16 illustrates an example. It is seen that the position of the line of the nitrogen $1s$ state in the XPS depends essentially on the chemical bonds with this atom and allows us to distinguish between coordinated and outer-sphere positions. In the AES the chemical shifts are significantly larger than those of core levels.

Many examples of applications of UPS, XPS, AES, and XES are considered in the reviews and monographs cited above [8.3, 8.4, 8.33–8.38, 8.44].

EXAFS and Related Methods

Extended x-ray absorption fine structure (*EXAFS*) has become, during the last two decades, a widespread structural tool [8.45–8.49]. The phenomenon of additional modulated absorption beyond the K or L edge of the x-ray absorption spectrum has been known for more than half a century. It is caused by the interference processes that the photoelectron wave undergoes by scattering from the environment of the absorbing atom: The outgoing photoelectron waves propagate to neighboring atoms and scatter back; interference by the initial and scattered waves produces corresponding modifications of the final state of the x-ray transition, resulting in modulated absorption intensity. A modification of EXAFS is known as *x-ray absorption near edge structure* (*XANES*). However, this effect remained unused until it was shown [8.50] that information about the interatomic distances to the neighboring atoms in the modulated absorption intensity can be extracted relatively easily by means of *Fourier transforms*. On the other hand, synchrotron radiation in the x-ray range has become available and shown to be an ideal source for XAFS measurements. Other common sources of x-rays are less effective because the extended absorption is weak.

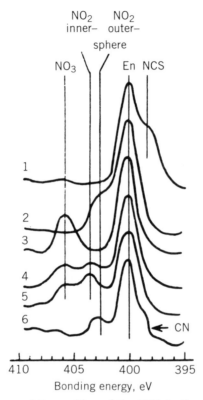

Figure 8.16. Dependence of the position of the XPS 1s line of the nitrogen atom in different groups on the near and distal structure of the molecular environment in (1) [Ir en$_3$] (NCS)$_2$, (2) [Ir en$_3$] (NO$_2$)$_3$, (3) [Ir en$_3$] (NO$_3$)$_3$, (4) [Ir en$_2$ClNO$_2$]NO$_3$, (5) [Ir en$_2$(NO$_2$)$_2$NO$_3$, and (6) [Co(NH$_3$)$_5$CN] (NO$_2$)$_2$. (From [8.44].)

The main advantage of the EXAFS method is that it makes possible determination of the interatomic distances from the absorbing to neighboring atoms of the first and second coordination spheres *in the noncrystalline state*. This makes the method distinct from x-ray diffraction methods and unique when nonsolid and/or nonregular structures are investigated. At present the EXAFS method is applied to study structural features of coordination centers in solutions and on surfaces [surface EXAFS (SEXAFS)], including supported and homogeneous catalysts, biological systems, disordered solids, thin films, and intercalated systems. In addition to general reviews of principles and applications [8.45, 8.48, 8.49, 8.51, 8.52], there are reviews with special emphasis on biology [8.46], chemistry [8.47, 8.54], and catalysis [8.55].

Consider some examples. The dimerization of Mo(IV) in HCl solution was revealed by EXAFS which also made it possible to evaluate the Mo—Mo bond lengths at 2.51 Å [8.56]. In addition, two short Mo—O distances at 1.95 Å and four long ones at 2.15 Å were detected.

For $Cu(H_2O)_6^{2+}$ and $Cr(H_2O)_6^{2+}$ in solution the problem was to determine whether Jahn–Teller tetragonal distortions of the octahedron with an electronic E ground state (Section 7.3) could be detected by EXAFS. It confirmed the prediction obtained through vibronic theory; for axial R_L and equatorial R_S metal–ligand distances, the following values were obtained (in Å): $R_L(Cr—O) = 2.30$, $R_S(Cr—O) = 1.99$, $R_L(Cu—O) = 2.60$, $R_S(Cu—O) = 1.96$ [8.57]. For hydrated ions with nondegenerate terms there is no distortion.

The next example is related to the spin crossover phenomenon (Section 8.4). The transition high spin → low spin is expected to result in contraction of the metal–ligand bond. The EXAFS experiments [8.58] show that in $Fe(phen)_2(NCS)_2$ the transition $S = \frac{5}{2} \to S = \frac{1}{2}$ with a spin change $\Delta S = 2$ yields a mean contraction $\langle \Delta R \rangle = 0.24\,\text{Å}$. In $Co(H_2fsa2en)L_2$, where $H_4fsa2en = N,N'$-ethylenebis(3-carboxysalicylaldimine) and $L = H_2O$ and pyridine, the transition $S = \frac{3}{2} \to S = \frac{1}{2}$ ($\Delta S = 1$) gives $\langle \Delta R \rangle = 0.09\,\text{Å}$.

In biological systems the EXAFS technique allows one to solve some very complicated problems. For nitrogenase [8.46] it was shown that the local environment of Mo is S and Fe (as distinct from another Mo enzyme, sulfite oxidase, where O and S are nearest neighbors). For hemoglobin the EXAFS data [8.52] rule out the large-scale lengthening of the Fe—N bond when passing from oxyhemoglobin to its deoxy form postulated by Perutz and Hoard; in both cases this distance is almost the same, $2.055 \pm 0.01\,\text{Å}$ and $1.98 \pm 0.01\,\text{Å}$, respectively. The iron atom position in hemoglobin and other metal enzymes is discussed in Sections 9.2 and 10.3.

A series of interesting results were obtained in EXAFS measurements for ferroelectric crystals that undergo displacive phase transitions ($BaTiO_3$, $KNbO_3$, $PbTiO_3$) [8.53]. These experiments confirmed that distortions of the crystal lattice are of local origin, as predicted by vibronic theory, and that phase transitions are of order–disorder type, rather than displacive (Section 9.4)

A somewhat related method is *electron energy loss spectroscopy* (EELS) [8.59, 8.60]. In this method a high-energy electron beam (hundreds of keV) is sent through a thin foil (several hundred Å thick) and the electrons scattered in a given direction are energy analyzed. The spectrum of energy loss corresponds to excitation of core and valence electronic states as well as vibrational states. Despite limitations created by demands for sample preparation (thin films), the EELS for not very large Z elements may have the advantage of a better signal-to-noise ratio than for synchrotron absorption, but not many coordination systems have been studied by this technique.

8.4. MAGNETIC PROPERTIES

Magnetic Moment and Quenching of Orbital Contribution

The magnetic properties of any substance can be characterized by its response to the external magnetic field intensity $\mathscr{H}$. This response is best described by the magnetic induction $\mathbf{B}$, defined as the magnetic field intensity inside the matter:

$$\mathbf{B} = \mathscr{H} + 4\pi\mathbf{M} \tag{8.46}$$

where $\mathbf{M}$ is the magnetization, the effective magnetic moment per unit volume. $\mathbf{M}$ also depends on the magnetic field $\mathscr{H}$, and for isotropic media

$$\mathbf{M} = \chi\mathscr{H} \tag{8.47}$$

where χ is the *magnetic susceptibility*. In anisotropic substances χ is not a scalar but a tensor; provided that the axes are chosen appropriately, it has three components χ_x, χ_y, and χ_z that characterize the magnetic susceptibilities in different directions, the *magnetic anisotropy* [for a powder $\chi = (\chi_x + \chi_y + \chi_z)/3$].

The magnetic susceptibility is one of the most important properties of matter, which is related directly to its electronic structure [8.5–8.7, 8.61–8.66]. The magnetization $\mathbf{M}$ of the substance is determined by the magnetic moments μ of its elementary units (its molecules or other magnetic centers) and their cooperative behavior in the magnetic field. If $\mu \neq 0$, the magnetic moments of free molecules (*paramagnetism*) tend to orient along the external magnetic field $\mathscr{H}$, but their chaotic collisions (which increase with temperature T) destroy the orientation and do not allow for full ordering: The magnetization becomes temperature dependent. The Langeven theory predicts that in this case (quite similar to the polarization of electric dipoles) the magnetic susceptibility χ is inversely proportional to the temperature:

$$\chi \sim \frac{N\mu^2}{3kT} \tag{8.48}$$

where N is the number of magnetic units per unit volume. In paramagnetic substances χ varies from 0 to $\sim 10^{-4}$ emu $\cdot$ mol^{-1} (emu, electromagnetic units).

If there is a strong magnetic interaction between the magnetic units μ (usually in crystals), their orientations are no longer free, and at sufficiently low temperatures an ordering of these magnets (a magnetic phase transition) takes place. There are several main types of magnetic ordering: *ferromagnetic* (when all the magnets are parallel and the magnetization is maximal), *antiferromagnetic* (when the near-neighbor elementary magnets occupy alternating directions, the macroscopic total magnetization being zero), *ferrimagnetic* (when in the antiferromagnetic ordering the values of the elementary magnetic moments of opposite orientation are different and the net magnetic moment is nonzero), and others (Table 8.6).

Compared with paramagnets, χ for ferromagnetic materials can be by several orders of magnitude larger, while the antiferromagnets may have the same order of magnitude of χ as paramagnets. Table 8.6 lists the diversity of possible magnetic behavior of substances.

In addition to the magnetization based on the orientation of the elementary magnets, there is also induced magnetization (*diamagnetism*) when under the

Table 8.6. Main types of magnetic response of substances

Type of Magnetism	Origin	$\sim \chi$ at 20°C (cm³/mol)	Temperature Dependence	Examples
DIAMAGNETISM				
	Proper magnetic moment equals zero; induced magnetic moment is opposite to the external field	-10^{-6}	None	KCl, C_6H_6
PARAMAGNETISM				
Atomic, ionic, and molecular	Electronic orbital and/or spin magnetic moment is nonzero	$(1-20)\cdot 10^{-6}$	$1/T$ or $1/(T+\theta)$	H, Ti³⁺, NO, $K_3Fe(CN)_6$
Nuclear	Nuclear magnetic moment is nonzero	10^{-9}	None	H in hydrids
Free electrons	Electronic gas magnetic moment is nonzero	10^{-8}	Weak	Metallic K or Na
Van Vleck, or T-independent	Excited states with other magnetic moment lay higher than kT	10^{-6}	None	$KMnO_4$, Co^{III} amines
COOPERATIVE MAGNETISM (ORDERED SYSTEMS)				
Ferro-magnetism	Crystal lattice with parallel magnetic moment of the centers	$1-10^2$	Complicated; decreases at T_c	Metallic iron
Antiferro-magnetism	Two ferromagnetic sublattices with antiparallel mutual compensating spin	$10^{-4}-10^{-5}$	A maximum at Neel T	$KNiF_3$, MnSe
Ferri-magnetism	Partially compensated anti-parallel spins of different sublattices	10^{-5}	Similar	Fe_3O_4
Meta-magnetism	A kind of ferrimagnetism with complicated ordering of sublattices	10^{-5}	Similar	$NiCl_2$ at liquid H_2

influence of the external magnetic field $\mathscr{H}$ a local circular current occurs with a magnetic field opposite to $\mathscr{H}$. The diamagnetic susceptibility is thus negative; its absolute value is by two to three orders of magnitude smaller than the paramagnetic susceptibility. The diamagnetic behavior (repulsion from magnetic field), although weak, is seemingly the only possible response to the magnetic influence on systems with $\mu = 0$ (however, see the discussion of Van Vleck paramagnetism in the following subsection).

The information about the electronic structure of molecules is contained in its magnetic moment μ. It is known from quantum mechanics that each momentum of the microsystem is associated with a proportional magnetic moment. For the free electron with a spin momentum **S**, its associated magnetic moment is [8.67]

$$\mu = \frac{e\hbar}{mc} \mathbf{S} \tag{8.49}$$

while the magnetic moment of its orbital motion with an orbital momentum **L** (Section 2.2) is

$$\mathbf{\mu}_L = \beta\mathbf{L} \qquad (8.50)$$

where $\beta = eh/2mc$ is the elementary magnetic moment called *Bohr magneton*, $\beta = 9.274 \cdot 10^{-24}$ erg/T. Hence the total magnetic moment of the system is

$$\mathbf{\mu} = \beta(\mathbf{L} + 2\mathbf{S}) \qquad (8.51)$$

This expression is, in fact, the operator of the magnetic moment, while the observable moments are determined as averaged values of the operator $\mathbf{\mu}$ over the states under consideration. In particular, for a free atom with a total momentum $\mathbf{J} = \mathbf{L} + \mathbf{S}$, taking the projection of $\mathbf{\mu}$ on $\mathbf{J}$ and the average $\langle J^2 \rangle = J(J + 1)$, we have [8.67]

$$\mathbf{\mu}_J = g\beta[J(J + 1)]^{1/2} \qquad (8.52)$$

where

$$g = 1 + \frac{J(J + 1) + S(S + 1) - L(L + 1)}{2J(J + 1)} \qquad (8.53)$$

The coefficient g (*Landé factor* or *g-factor*) plays a significant role in molecular magnetism (see below). For a free electron $L = 0$, $J = S$, and $g = 2$. In fact, however, a more correct value for the free electron g-factor is $g = 2.0023$.

Equation (8.52) is valid when there is a single energy level with a given J value, well separated from other levels, that is, when there is sufficiently strong spin–orbital coupling ($\lambda \gg kT$). This case may take place in heavy atoms. In the other limit case when $\lambda \ll kT$ (light atoms and transition metals of the first and second row at not very low temperatures), the energy levels of the same LS term with different J values are almost equally populated, and hence the effective measured magnetic moment, averaged over all the J values, is [8.63]

$$\mu_{\text{eff}} = \beta[4S(S + 1) + L(L + 1)]^{1/2} \qquad (8.54)$$

If the atom or ion is placed in the field of ligands, its electronic structure changes and expressions (8.52) and (8.54), in general, are no longer valid. In particular, the orbital moment is subject to significant changes. The experimental data show that in most cases the magnetic properties of transition metal ions in complexes are as if the orbital contribution vanished, $L = 0$:

$$\mu_{\text{eff}} \approx 2\beta[S(S + 1)]^{1/2} \qquad (8.55)$$

That is, the orbital momentum is reduced to zero and the magnetic moment has a spin-only value.

This *quenching of the orbital contribution to the magnetic moment by the ligand field* can be explained as follows. The effective orbital magnetic moment is due to additional magnetization of the substance (additional magnetic susceptibility), which occurs as a result of free orientation of the magnetic moment of the orbital motion along the external magnetic field. This free orientation is possible in the free atom due to the fact that for $L \neq 0$ the energy term is degenerate, and in the absence of external perturbation there is no fixed direction of the orbital moment—all directions are equivalent.

Now, the main effect of the ligand field on the CA is the splitting of its degenerate energy terms (Section 4.2), as a result of which the orbital motion in the ground state becomes fixed or limited in orientation and cannot freely follow the magnetic field. Therefore, although the orbital momentum of the CA electrons in the field of ligands can be nonzero ($L \neq 0$), it may not be manifest (or it may be only partially manifest) in their magnetic behavior. Hence the magnetic orbital momentum is completely or partially quenched by the ligand field.

The partial quenching of the orbital magnetic moment takes place when the crystal field splitting is not complete, and there are still some possibilities for the orbital magnetic moment to rotate and follow the external magnetic field. These cases are well known from symmetry considerations. Indeed, for the electronic state of a cubic system Ψ_Γ that belongs to the irreducible representation Γ (Section 3.4), the average orbital momentum is determined by the integral

$$\int \Psi_\Gamma^* \mathbf{L} \Psi_\Gamma \, d\tau \tag{8.56}$$

Since the orbital momentum $\mathbf{L}$ (as any other vector) in cubic groups belongs to the T_1 representation, the integral (8.56) is nonzero if and only if the symmetric product $[\Gamma]^2$ contains T_1 [see Eq. (3.34)]. It can easily be shown that this is possible only when $\Gamma = T_1$, or $\Gamma = T_2$. In all other cases the orbital magnetism is completely quenched and the magnetic moment is determined by the spin-only formula (8.55).

The electronic configurations and the ground-state terms of coordination compounds with different coordination geometries are given in Table 6.3. Based on the data in this table one can state that for octahedral complexes with d^3, d^8, d^9, d^{10} as well as high-spin d^4, d^5 and low-spin d^6, d^7 configurations, the spin-only magnetic behavior based on Eq. (8.55) is expected, while for d^1, d^2, high-spin d^6, d^7 and low-spin d^4, d^5 configurations, an orbital contribution to the magnetic moment can be significant. Experimental data [8.7, 8.62, 8.63] confirm these expectations.

The orbital contribution to the magnetic moment is also nonzero when the T term is split by low-symmetry crystal fields or spin–orbital interaction, but the splitting magnitude is not very large compared with kT. If it is on the order of kT, μ_{eff} becomes temperature dependent.

Paramagnetic Susceptibility

The magnetic moments μ of molecular systems can be determined from the magnetic susceptibility $\chi = M/\mathscr{H}$, provided that the relation between μ and χ is known. The simple expression (8.48) for χ obtained from the Langeven theory is valid when all the free rotating elementary molecular magnets have the same magnetic moments. This means that there are no thermally accessible magnetic excited states with $\mu' \neq \mu$.

However, in most cases, and especially in transition metal coordination compounds, there are many close-in-energy states created by spin–orbital and crystal field splitting which have different spins and orbital moments and hence different magnetic moments $\mathbf{\mu}_i$. If these states are thermally populated, the effective magnetic moment in χ equals the averaged moment taken as a sum of Boltzmann-distributed elementary magnets $\mathbf{\mu}_i$. The theory of paramagnetic susceptibility for this case was developed by Van Vleck [8.68]; it yields

$$\chi = N \sum_{n,m} \left(\frac{(\varepsilon_{n,m}^{(1)})^2}{kT} - 2\varepsilon_{n,m}^{(2)} \right) \frac{\exp(-\varepsilon_n^0/kT)}{\sum_{n,m} \exp(-\varepsilon_n^0/kT)} \tag{8.57}$$

where $\varepsilon_{n,m}^{(1)}$ and $\varepsilon_{n,m}^{(2)}$ are the first- and second-order corrections to the energy level ε_n^0 in the magnetic field,

$$\varepsilon_{n,m} = \varepsilon_n^{(0)} + \varepsilon_{n,m}^{(1)}\mathscr{H} + \varepsilon_{n,m}^{(2)}\mathscr{H}^2 \tag{8.58}$$

and it is assumed that the energy splitting in the magnetic field (the differences between $\varepsilon_{n,m}$ with different m) are much smaller than kT (in the absence of the magnetic field $\varepsilon_{n,m} = \varepsilon_n^{(0)}$ is degenerate with respect to m).

If the excited states with $n > 1$ are not thermally populated, $\varepsilon_n^0 - \varepsilon_1^0 \gg kT$, Eq. (8.57) can essentially be simplified:

$$\chi = N \left\{ \sum_m \frac{(\varepsilon_{1,m}^{(1)})^2}{kT} - 2 \sum_m \varepsilon_{1,m}^{(2)} \right\} \tag{8.59}$$

The first term here is the usual paramagnetic susceptibility, which obeys the *Curie law:* $\chi = C/T$. From the Zeeman effect (see below) it is known that the term that is linear in $\mathscr{H}$ in (8.58) is simple related to the effective magnetic moment of the system: $\sum_m (\varepsilon_{1,m}^{(1)})^2 = \mu_{\text{eff}}^2/3$, and hence this part of χ coincides with Eq. (8.48) given by Langeven's theory. However, the second term in (8.58), in contrast to the first, is independent of temperature. Its contribution is *temperature-independent or Van Vleck paramagnetism.* Since the second-order perturbation correction in (8.59) is on the order of $\sum_n |\langle 1|W|n\rangle|^2/(\varepsilon_n^0 - \varepsilon_1^0)$, where W is the energy of interaction of the magnetic moment (8.50) with the external field, $\varepsilon_1^{(2)}$ is larger the more low-lying excited states are admixed to the ground state by the magnetic field. Typical values of the Van Vleck paramagnetism for $3d$ metals range from $60 \cdot 10^{-6}$ emu·mol^{-1} for Cu(II) to $400 \cdot 10^{-6}$ emu·mol^{-1} for Co(II) [8.63] and reach about 10^{-3} emu·mol^{-1} for some rare earth ions.

Calculations of the paramagnetic susceptibilities using the Van Vleck formula (8.57) require knowledge of the thermally populated energy levels in the magnetic field. Thermal population is significant if the energy spacing of these levels is about equal to or smaller than kT. Usually, this means that they are spin sublevels of the same electronic term split by the low-symmetry ligand field and external magnetic field. In this case the spin energy levels can be obtained by means of the method of *spin Hamiltonian* [8.65].

The idea of the spin Hamiltonian is as follows. Provided that the expected spin states emerge from the same electronic term (i.e., no excited electronic states are admixed by the magnetic field), one can average the full Hamiltonian of the system (including all the interactions with the magnetic field) over the electronic and nuclear coordinates and obtain in this way a Hamiltonian that contains explicitly only spin $\mathbf{S}$ and magnetic field $\mathscr{H}$ operators and some averaged parameters. In a general form (without nuclear spins) the spin Hamiltonian appears as follows [8.65]:

$$H = \sum_{i,j} [D_{ij}S_iS_j + \beta g_{ij}\mathscr{H}_iS_j] \tag{8.60}$$

where $i, j = x, y, z$ and D_{ij} and g_{ij} are the tensors of zero-field splitting and the g-factor, respectively. These parameters contain all the information about the electronic structure of the system and can be calculated, provided that the wavefunction of the term under consideration is known. On the other hand, the D_{ij} and g_{ij} values can be obtained from experimental data by comparison of the magnetic properties with those predicted by the spin Hamiltonian (8.60).

The procedure of evaluation of the energy levels ε_i by means of the spin Hamiltonian is to consider $\mathscr{H}$ as a perturbation to the $(2S + 1\text{-fold})$-degenerate spin multiplet. By solving the secular equation

$$\|H_{ij} - \varepsilon\delta_{ij}\| = 0 \tag{8.61}$$

we get the $2S + 1$ energy levels sought, while the matrix elements of the spin operators are determined directly by Eq. (2.3).

The number of independent parameters in the spin Hamiltonian (8.60) depends on the symmetry of the system. *Kramers doublets*, that is, spin doublet states with $S = \frac{1}{2}$, are not split by crystal fields (they split only in magnetic fields). In the crystal field of lower than cubic symmetry, the spin Hamiltonian

Table 8.7. Temperature dependence of the magnetic susceptibility χ/C, $C = Ng^2\beta^2/kT$ ($g = g_l$ for C_l and $g = g_n$ for C_n) for several spin states S of one-center coordination compounds with axial symmetry $(x = D/kT)$

S	
	χ_l/C_l
1	$2e^{-x}/(1 + 2e^{-x})$
$\frac{3}{2}$	$(1 + 9e^{-2x})/4(1 + e^{-2x})$
2	$(2e^{-x} + 8e^{-4x})/(1 + 2e^{-x} + 2e^{-4x})$
$\frac{5}{2}$	$(1 + 9e^{-2x} + 25e^{-6x})/4(1 + e^{-2x} + e^{-6x})$
	χ_n/C_n
1	$(2/x)(1 + e^{-x})(1 + 2e^{-x})$
$\frac{3}{2}$	$[4 + (3/x)(1 - e^{-2x})]/4(1 + 2e^{-2x})$
2	$[(6/x)(1 - e^{-x}) + (4/3x)(e^{-x} - e^{-4x})](1 + 2e^{-x} + 2e^{-4x})$
$\frac{5}{2}$	$[9 + (8/x)(1 - e^{-2x}) + (9/2x)(e^{-2x} - e^{-6x})]/4(1 + e^{-2x} + e^{-6x})$

can be written as follows:

$$H = D[S_z^2 - \tfrac{1}{3}S(S + 1)] + E(S_x^2 - S_y^2) + \beta(g_x \mathscr{H}_x S_x + g_y \mathscr{H}_y S_y + g_z \mathscr{H}_z S_z)$$

$$(8.62)$$

where D and E are the parameters of axial and rhombic distortions, respectively. Depending on the spin value S, this magnetic field Hamiltonian yields a series of energy levels that contribute to χ. In particular, in the case of an axial field $D \neq 0$, $E = 0$, $g_z = g_l$, $g_x = g_y = g_n$, there are two values of χ, χ_l and χ_n, parallel and perpendicular to the axis of symmetry, respectively. Their temperature dependence after (8.57) (without the temperature-independent magnetism) for different total spin S values is given in Table 8.7 [8.64].

Electron Spin Resonance

The electron spin resonance (ESR) method is based on resonance absorption of irradiation in the radio and microwave region associated with transitions between the energy levels of the electronic term split by the external magnetic field. The splitting of energy levels in magnetic fields is known as the *Zeeman effect*. Zeeman splitting can be evaluated using the spin Hamiltonian (8.60) and (8.62) and solving the secular equation (8.61).

In the simplest case, if the total momentum of an atom $\mathbf{J} = \mathbf{L} + \mathbf{S}$ is described by the quantum number J and its projection $m = J, J, J - 1, \ldots, -J,$

the energy levels in the magnetic field $\mathscr{H}$ are given by the simple Zeeman splitting

$$\varepsilon_m = g\beta\mathscr{H} \cdot m \tag{8.63}$$

where g is the Landé factor (g-factor) given by Eq. (8.53).

Electromagnetic transitions between these levels are allowed as magnetic-dipole transition obeying the selection rule $\Delta m = \pm 1$. This is accompanied by absorption of an irradiation quantum

$$\hbar\omega = \varepsilon_{m+1} - \varepsilon_m = g\beta\mathscr{H} \tag{8.64}$$

which for a given external field $\mathscr{H}$, is determined completely by the g-factor. It is seen from Eq. (8.53) that if the spin $S = 0$, but the orbital momentum $L \neq 0$, then $J = L$ and $g = 1$, while for $L = 0$, $S \neq 0$ we have $J = S$ and $g_s = 2$ (as mentioned above, a more precise value is $g_s = 2.0023$).

If the atom is in a transition metal compound, the ligand influence changes the g-factor drastically. To begin with, the orbital contribution to the momentum J, as shown above, may become completely or partially quenched. This reduction of the orbital contribution is determined by the symmetry of the crystal field. There are several other parameters of the electronic structure that contribute significantly to the ESR spectrum: spin–orbital admixing of excited states, symmetry and strength of the ligand fields, yielding *anisotropic g-factors*, admixture of ligand states (*covalence contribution*), splitting of the line due to the splitting of the spin states (*fine structure*), further splitting due to the interaction with the nuclear spin (*hyperfine splitting*), splitting due to the ligand nuclear spin (*super-hyperfine splitting*), reduction of g-factors due to orbital degeneracy (*vibronic reduction*), dependence of the line shape on temperature via interactions with vibrations (*paramagnetic relaxation*), and so on. Thus *the ESR spectra carry very rich information about the electronic structure of transition metal compounds.*

The origin of the covalence contribution can be clarified by a concrete example. Consider the widespread case of tetragonally distorted Cu^{2+} complexes of D_{4h} symmetry. In the approximation of the crystal field theory (Section 4.1) the energy levels of the d^9 configuration in the ligand field are as shown in Fig. 8.17 (cf. Fig. 4.4 and take into account that a hole in the d^{10} configuration has an energy-level diagram inverse to d^1). The B_{1g} ground state corresponds to the atomic orbital $d_{x^2-y^2}$. This state originates from the E term in a cubic field, and hence its orbital contribution is quenched, thus having a pure spin g-factor $g = 2.0023$. However, the spin–orbital interaction mixes this ground state with excited states B_{2g} and E_g; the latter emerge from the T_{2g} term of the cubic system, which has an orbital contribution.

Denote the wavefunctions of the mixing antibonding states B_{1g}, B_{2g}, and E_g by ψ_{0i} with $i = 1, 2, 3$, respectively. In the weak covalence model, following Eq.

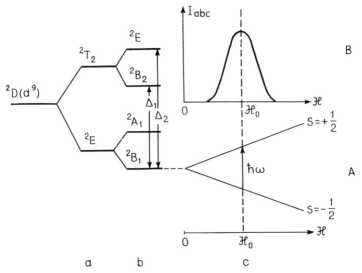

Figure 8.17. The Cu^{2+} $^2D(3d^9)$ term splitting in cubic (*a*) and tetragonal (*b*) crystal fields, and in the external magnetic field $\mathcal{H}$ (*c*, *A*). At $\beta\mathcal{H}_0 = \hbar\omega$ there is a resonance absorption (*B*) that changes the spin state.

(5.33), these wavefunctions are

$$\Psi_i = N_i(\Psi_{0i} - \gamma\Phi_i) \qquad (8.65)$$

where Φ_i is the ligand group orbital, γ the covalence constant, $N_i = (1 + \gamma_i^2 - 2\gamma_i S_i)^{-1}$ the normalization constant, and S_i the overlap integral: $S_i = \langle \psi_{0i} | \Phi_i \rangle$. Then considering the spin–orbital interaction $\lambda(L, S)$ as a perturbation (λ is the spin–orbital coupling constant), we can obtain in second order [8.69]

$$g_l = g_{zz} = 2 - \frac{8\lambda}{\Delta_l} N_1^2 N_2^2$$

$$\qquad (8.66)$$

$$g_n = g_{xx} = g_{yy} = 2 - \frac{2\lambda}{\Delta_2} N_1^2 N_3^2$$

where Δ_1 and Δ_2 are the energy gaps between the ground and the excited states (Fig. 8.17), and terms of the order of $\gamma_i\gamma_j$ and $\gamma_i S_j$ are ignored.

For pure CA states $N_1 = N_2 = N_3 = 1$, and the orbital contributions to g_l and g_n are $-8\lambda/\Delta_1$ and $-2\lambda/\Delta_2$, respectively. Note that for Cu, $\lambda < 0$, and hence these contributions are positive, increasing the g-factors. For nonzero covalence $N_i < 1$, so the orbital contributions to the g-factors are reduced $k_1^2 = N_1^2 N_2^2$ and $k_2^2 = N_1^2 N_3^2$ times for g_l and g_n, respectively. This reduction can be presented as if the spin–orbital constant λ were reduced to $\lambda' = k^2\lambda$.

Provided that Δ_1 and Δ_2 are known (e.g., from spectroscopic data or calculations), we can evaluate λ' from the measured g-factors and estimate k^2, *the covalence reduction.*

The covalent reduction of the spin–orbital constant λ (or the orbital contribution to the g-factor) is similar to the nephelauxetic effect discussed above (Section 8.2), which results from reduction of the interelectron interaction parameter B by formation of covalence bonds. Both these parameter reductions (as well as some others) are due to the quite understandable effects of electron delocalization on larger volumes by coordination. The spin–orbital interaction may be also reduced as a result of special relativistic effects of ligand coordination (Section 6.5).

The two components of the g-factor in Eq. (8.66), g_l and g_n, may differ significantly following the anisotropy of the tetragonal system. For an arbitrary direction of the magnetic field $\mathscr{H}$ with an angle θ to the tetragonal axis,

$$g_\theta = (g_l^2 \cos^2 \theta + g_n^2 \sin^2 \theta)^{1/2} \tag{8.67}$$

For lower symmetries all three components of the g-factor are different. For cubic symmetry, the three components of the g-factor are equivalent and hence no angular dependence of the ESR spectrum is expected. However in many cases, as in the example of Cu^{2+} complexes above, the ground state in cubic fields is orbitally degenerate. This creates quite new circumstances, due to which the ESR spectrum is complicated significantly. Indeed, according to the Jahn–Teller effect and other vibronic interaction effects (Sections 7.3 and 7.4), electronic degeneracy causes a special coupling between the electronic and nuclear motions which makes invalid direct application of the spin Hamiltonian approach.

In particular, for the E term of the example above of Cu^{2+} octahedral complexes with strong vibronic coupling, there are three minima of the AP in which the octahedron is elongated along one of the three fourth-order axes (one for each minimum) (Figs. 7.10 and 7.11). In each of these minima the g-factor corresponds to the tetragonally distorted octahedron.

On the other hand, the system is performing pulsating motions with relatively high-frequency transitions between the equivalent minima. As shown in Section 7.3, these pulsating distortions result in tunneling splitting (Fig. 7.16). When it is sufficiently small (strong vibronic coupling), the tunneling levels are mixed by the external magnetic field, resulting in a special Zeeman splitting that yields a complicated ESR spectrum with a characteristic dependence on temperature and irradiation frequency [8.21]. If the tunneling splitting is larger than the Zeeman splitting ($\sim \beta H$), the ESR spectrum is determined by the ground level only. In this case the g-factor is subject to vibronic reduction (Section 7.3).

Set $g_1 = g_s - (4\lambda/\Delta)$, $g_2 = -(4\lambda/\Delta)$, and let l, m, and n represent the direction cosines of the magnetic field vector $\mathscr{H}$. Then it can be shown that by taking into consideration the vibronic reduction factor $q = K_E(E)$ after (7.64) and (7.66), the angular dependence of the g-factor is

$$g_{1,2} = g_s - g_2 \pm qg_2 f$$
$$f = [1 - 3(l^2 m^2 + l^2 n^2 + m^2 n^2)]^{1/2}$$

(8.68)

For very strong vibronic coupling (deep minima of the adiabatic potential and high barriers between them) $q = \frac{1}{2}$, while in the absence of the reduction, $q = 1$. The angular dependence of g for these two limit cases is shown in Fig. 8.18 together with the experimental data obtained for Cu^2 ions as impurities in MgO at $T = 1.2$ K (see [8.21]). As one can see, the difference between the two spectra (with and without the vibronic coupling) is rather significant and the experimental data confirm unambiguously the importance of the vibronic reduction. For vibronic implications in the ESR spectra of other terms and other conditions, see [8.21, 8.22].

So far we considered a single ESR line, especially its g-factor. As mentioned above, the fine, hyperfine, and super-hyperfine structures of the ESR spectrum may also be important. Consider the $Cr^{3+}(d^3, {}^4F)$ ion in the ligand field of trigonal symmetry D_{3h}, resulting in the orbitally nondegenerate ground-state spin quadruplet ${}^4A_{2g}$ ($S = \frac{3}{2}$, $m_s = \pm\frac{1}{2}, \pm\frac{3}{2}$). The spin–orbital interaction splits this term into two doublets with $m_s = \pm\frac{1}{2}$ and $m_s = \pm\frac{3}{2}$, respectively (e.g., for chromium alum the splitting is $2D = 0.15$ cm^{-1}), which are split further by the external magnetic field $\mathcal{H}$, as shown in Fig. 8.19. As a result, there are three values of $\mathcal{H}$ ($\mathcal{H}_1$, $\mathcal{H}_2$, and $\mathcal{H}_3$) for which the resonance absorption of the same

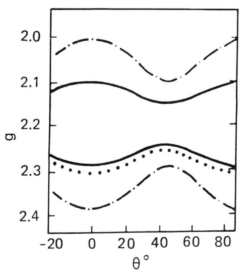

Figure 8.18. Two limiting cases of angular dependence of the g-factor for the Jahn–Teller linear $E-e$ problem with strong ($q = \frac{1}{2}$, solid lines) and without ($q = 1$, dashed lines) vibronic coupling. Experimental data are shown by points.

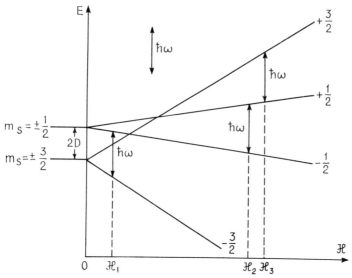

Figure 8.19. Zeeman splitting and ESR transitions for a $^4F(D^3)$ (Cr^{3+} type) term in magnetic fields. The $\mathcal{H}_1$, $\mathcal{H}_2$, and $\mathcal{H}_3$ values indicate the magnetic field intensities for three possible resonance transitions with the same quantum of irradiation $\hbar\omega$.

quantum of irradiation takes place, each of them thus having its own g-factor (fine structure).

The hyperfine structure is described by additional terms H^N in the spin Hamiltonian that take into account the interaction of the electron (**S**) and nuclear (**I**) spins:

$$H^N = \sum_{ij} (A_{ij}S_iI_j + P_{ij}I_iI_j) - g_N\beta_N(\mathbf{I}, \mathcal{H}) \tag{8.69}$$

Here A_{ij} are the constants of the spin–spin (dipole–dipole) magnetic interaction of the electrons and nuclei, P_{ij} are the constants of the quadruple interactions, and the last term describes the interaction of the nuclei with the external magnetic field; g_N and β_N are the nuclear g-factor and Bohr magneton, respectively ($\beta_N \cong \beta/1840$).

While the last term in (8.69) is important for *nuclear magnetic resonance* (NMR) spectra (for a review of NMR spectroscopy in coordination compounds, see [8.71]), the quadruple splitting is most important in Mossbauer spectroscopy. For the spin–spin interaction that yields the hyperfine structure of the ESR spectra, the number of independent constants A_{ij} depends on the symmetry of the system. In cubic systems there is only one constant A, while in axial symmetry there are two, A and B:

$$H^N = AS_zI_z + B(S_xI_x + S_yI_y) \tag{8.70}$$

The constants of hyperfine interaction contain the nuclear magnetic moment proportional to $g_N\beta_N$, which is about 1840 times smaller than the electronic magnetic moment. Therefore, the hyperfine splitting is accordingly smaller than the fine structure. In a sufficiently strong magnetic field each line of the fine structure splits into $2I + 1$ lines, which are equally spaced when $B = 0$ (cubic symmetry). Magnetic-dipole transitions are allowed between any two levels with $\Delta m_s = \pm 1$ and $\Delta M = 0$.

The electron–nuclear dipole–dipole interactions is nonzero only if the electron has an orbital moment $L \neq 0$ [8.65]. However, there is another term of electron–nuclear interaction not included in (8.76), the *contact Fermi interaction*, which is nonzero only for s electrons that have nonzero electron density $|\Psi(0)|^2$ at the nuclei. The contact interaction is much stronger than the dipole–dipole interaction, and it is less dependent on the ligands. Therefore, even when there are no dipole–dipole interactions ($L = 0$) and no direct s orbital participation, the hyperfine structure can still be observed due to configuration interaction (Section 5.3) with s-containing configurations of the same symmetry and multiplicity (*s-configuration interaction*). The hyperfine structure observed in Mn^{2+} salts (d^5, $L = 0$) is an example of this kind.

The hyperfine constants are also linked to the covalency [8.69]. In particular, for N_1 of Eq. (8.65), which is a measure of CA participation in the covalent MO, the following approximate relation holds for systems with axial symmetry:

$$N_1^2 = -\frac{A}{P} + (g_l - 2) + \frac{3}{7}(g_n - 2) + 0.04 \tag{8.71}$$

where $P = 2g_N\beta_N\beta\langle r^{-3}\rangle$ is the constant of the electron–nuclear interaction [8.65]. For Cu^{2+}, $P = 0.36$ cm^{-1}.

The super-hyperfine structure is due to interaction of the unpaired electrons with the nuclear spin of the ligands, which can be described by an additional term in the spin Hamiltonian:

$$H^{N'} = \sum_{N'} A'_{ij}S_iI'_j \tag{8.72}$$

Where N' numbers the ligand nuclei that have nonzero spin I'. Here A'_{ij} are mostly *contact Fermi interaction* constants (the ligands participate in the bonding by mostly s or s-hybridized orbitals). Usually, the stay of the unpaired electron on the ligands is relatively shorter than on the CA. Therefore, A'_{ij} are smaller than A_{ij} and describe further splitting of the hyperfine lines. To compare orders of magnitudes, examples of ESR spectra constants for some Cu^{2+} and V^{4+} complexes in solutions are given in Table 8.8.

Interestingly, the number of super-hyperfine lines and their relative intensities are determined by the number of equivalent ligand nuclei n. In particular,

Table 8.8. ESR spectra parameters for some chelate Cu(II) and V(IV) compounds in solutions

Compound	Solvent[a]	g_l	g_n	$A \cdot 10^4$ (cm^{-1})	$B \cdot 10^4$ (cm^{-1})	$A' \cdot 10^{4\,b}$ (cm^{-1})
Bis(acetylacetonato)Cu(II)	I	2.264	2.036	145.5	29	
Bis(dimethyldithiocarbomato)Cu(II)	I	2.098	2.035	154[c]	40	
				165[c]	43	
	II	2.121	2.040	134[c]	25	
				146.5[c]	27	
Bis(salycilaldoxymato)Cu(II)	III	2.171	2.020	183	41	14
Bis(salycilaldimino)Cu(II)	III	2.14	2.08	168	16	
Bis(8-chinolinato)Cu(II)	III	2.172	2.042	162	25	10
Diclorophenantroline-Cu(II)	III	2.22	2.08	119	29	
Oxo(bisacetylacetonato)V(IV)	III	1.944	1.996	173.5	63.5	

[a] I, 60% toluol, 40% chloroform; II, 40% pyridine, 60% chloroform; III, 40% toluol, 60% chloroform.
[b] Super-hyperfine structure from the ligand nitrogen atoms.
[c] Results for two copper isotopes, ^{63}Cu and ^{65}Cu.
Source: [8.70].

for $I' = \frac{1}{2}$ the line intensities follow exactly the binomial coefficients: 1, n, $n(n - 1)/2, \ldots$, n, 1, the number of lines thus being $n + 1$. Similar rules are established for other values of I' [8.65].

At present the ESR method is one of the most powerful means of investigation of electronic structure of coordination compounds. The main limitations of this method lie in the requirement of unpaired ($S \neq 0$) electrons and not very strong paramagnetic relaxations, which make impossible observation of the ESR spectrum and its resolution. Note also that *the information contained in the ESR spectra refers solely to the states of the unpaired electrons*. In particular, the conclusions about covalency drawn from ESR data [Eq. (8.66)] refer to orbital covalences on the MOs of the unpaired electrons, whereas the covalency of the bond as a whole, as discussed in Section 6.2, is produced by other electrons, namely, by the uncompensated bonding electrons (Fig. 6.3).

Magnetic Exchange Coupling

One of the most important and up-to-date applications of magnetic measurements is for the study of the magnetic exchange interactions in multicenter (polynuclear) coordination compounds [8.6, 8.61, 8.64]. Interaction of two atoms with unpaired spins is a typical problem of covalent bonding. As mentioned in Section 1.2, for the hydrogen molecule it was solved by Heitler and London in 1927, and this solution forms the basis of the *valence bond theory*. In the Heitler–London treatment the H_2 molecule has two states: a bonding spin singlet (the two spins are antiparallel ($\uparrow\downarrow$, $S = 0$) and an antibonding spin triplet ($\uparrow\uparrow$, $S = 1$). The energy gap between these two states is dependent on the two-electron exchange integral K given by Eq. (2.36). In the case of H_2, $K < 0$; if $K > 0$, the triplet state is lower than the singlet. Interaction resulting in a high-spin ground state is called *ferromagnetic interaction* (because it is analogous to the parallel spin ordering in ferromagnets, mentioned above), while that resulting in zero (or lowest) spin is *antiferromagnetic interaction*.

If the two centers with nonzero spins are bonded not directly but via intermediate (bridging) groups, the spin–spin interaction between them becomes more complicated, involving electrons of intermediate centers (*indirect* or *superexchange*). Heisenberg and Dirac suggested a simple spin Hamiltonian model for the magnetic interactions between two centers i and j, afterward modified by Van Vleck (the *HDVV model*), which has the form

$$H = -2JS_i S_j \tag{8.73}$$

where J_{ij} is the *isotropic exchange coupling constant*. The eigenvalues of this Hamiltonian are

$$E(S) = -J[S(S + 1) - S_i(S_i + 1) - S_j(S_j + 1)] \tag{8.74}$$

where the total spin S is determined, as usual, following the rule of addition of moments:

$$S = S_i + S_j, S_i + S_j - 1, \ldots, |S_i - S_j| \tag{8.75}$$

The energy intervals between the levels (8.74) (the Landé intervals, Section 2.2) are

$$\Delta_S = E(S) - E(S - 1) = -2J_{ij}S \tag{8.76}$$

In particular, for two centers with one unpaired electron on each,

$$\Delta_1 = E(1) - E(0) = -2J \tag{8.77}$$

and hence the exchange constant J characterizes the energy gap Δ between different spin states, its sign determining the type of interaction: ferromagnetic for $J > 0$ and antiferromagnetic for $J < 0$.

Calculations (see [8.61]) show that for two directly bonded spin centers a and b (as in the molecule H_2)

$$J = \frac{2(I_{ab} S_{ab}^2 - K_{ab})}{1 - S_{ab}^4} \tag{8.78}$$

where I_{ab} and K_{ab} are, respectively, the Coulomb and exchange integrals given by Eqs. (2.35) and (2.36), and S_{ab} is the overlap integral. For superexchange and for larger spin values on the centers, the expressions for J become much more complicated. Anderson [8.72] first suggested that exchange interaction between metallic ions can be presented as a sum of ferromagnetic J_F and antiferromagnetic J_{AF} contributions:

$$J = J_F + J_{AF} \tag{8.79}$$

where $J_F = 2K_{ab}$, and

$$J_{AF} = 4\beta S_{ab} \tag{8.80}$$

β being the resonance integral (5.4).

Note that while the ferromagnetic contribution (proportional to the Coulomb interelectron repulsion) is always nonzero and positive, the antiferromagnetic term is proportional to the overlap integral S_{ab}, and hence it is nonzero for nonorthogonal orbitals only.

Kahn and co-workers [8.66, 8.73] extended this idea to indirect superexchange interactions in multicenter coordination compounds. In a binuclear complex, as distinct from two metallic atoms, the two one-electron orbitals of the unpaired spins are no longer pure AOs but *magnetic orbitals* defined as single occupied orbitals centered on the two sites a and b, respectively, and partially delocalized toward the terminal and bridging ligands. If the overlap between these two magnetic orbitals is nonzero, they form two MOs with an energy gap $\Delta = -2\beta$. Therefore,

$$J_{AF} = -2\Delta S_{ab} \tag{8.81}$$

in the case of equivalent centers, and

$$J_{AF} = -2(\Delta^2 - \Delta_0^2)^{1/2} S_{ab}, \tag{8.82}$$

in the case of nonequivalent centers; Δ_0 is the energy gap between the reference magnetic orbitals.

A somewhat different suggestion for J_{AF} was given by Hoffmann [8.74]:

$$J_{AF} = -\frac{\Delta^2}{I_{aa} - I_{ab}} \tag{8.83}$$

Since $\Delta \sim S_{ab}$, the two expressions (8.81) and (8.83) are qualitatively not very different.

The main conclusion to be drawn from these formulas is that the antiferromagnetic contribution to the exchange coupling constant J is strongly dependent on the overlap integral between the magnetic orbitals of the interacting centers ($\sim S_{ab}^2$). This explains the dependence of the observed magnetic properties of multicenter compounds on small structural changes [8.75]. In particular, if $S_{ab} = 0$ (orthogonal magnetic orbitals), $J_{AF} = 0$, and the magnetic coupling is expected to be ferromagnetic with a high-spin ground state. On the other hand, the ferromagnetic contribution J_F is also dependent on structural features, mostly on charge density distribution [8.66, 8.75].

An interesting example that contributed significantly to the understanding of exchange coupling in coordination compounds is the binuclear copper(II) acetate hydrate $[Cu(OAc)_2H_2O]_2$. Intensive investigation of this system began in 1952 when Bleaney and Bowers [8.76] showed that its magnetic susceptibility χ and ESR spectrum were unusual for Cu^{2+} complexes (its dimeric structure was not known at that time). Indeed, the temperature dependence of χ has a maximum at $T_{max} = 255$ K [8.77] (for the dehydrated complex $T_{max} = 270$ K) and decreases almost to zero with temperature as shown in Fig. 8.20. Although the g-factors of the ESR spectrum are similar to those of other copper(II) salts, its intensity decreases with temperature and becomes zero at $T = 20$ K, exhibiting a small zero-field absorption at $D = 0.34 \pm 0.03$ cm^{-1}. This magnetic behavior is somewhat like that of Ni(II) compounds with $S = 1$; therefore, Bleaney and Bowers [8.76] assumed the system to be dimeric with two exchange-coupled Cu^{2+} centers in each molecule. The x-ray structural investigation confirmed this assumption (see [8.78]) (Fig. 8.21).

For the dimeric binuclear Cu^{2+} system with two equivalent spin $S = \frac{1}{2}$ magnetic centers, the HDVV Hamiltonian (8.73) yields two energy levels, with $S = 0$ and $S = 1$, respectively. To explain the experimental data, one must assume that the ground state is singlet $S = 0$, which means that $J < 0$ and the magnetic exchange coupling is antiferromagnetic. The magnetic susceptibility is

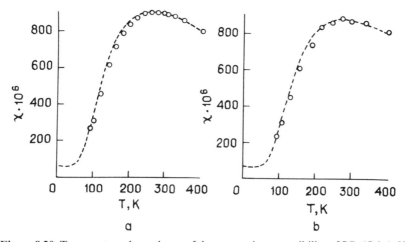

Figure 8.20. Temperature dependence of the magnetic susceptibility of $[Cu(OAc)_2H_2O]_2$ (a) and its dehydrated analog $[Cu(OAc)_2]_2$ (b) calculated by Eq. (8.84) (dashed lines) and obtained from experimental data (points). (From [8.77].)

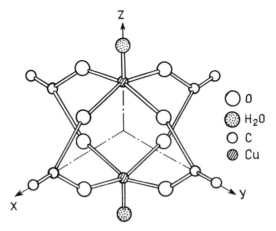

Figure 8.21. Molecular structure of $[Cu(OAc)_2H_2O]_2$.

then determined by the Van Vleck equation (8.57), which in the case of $S = 1$ and $J < 0$ under consideration yields (cf. Table 8.7)

$$\chi = \frac{2Ng^2\beta^2/kT}{3 + \exp(-2J/kT)} \qquad (8.84)$$

This is *the Bleaney–Bowers equation* [8.76]. With $-2J/k = 480$ K $(2J = -334\,\text{cm}^{-1})$ it fits well the experimental data in Fig. 8.20.

The dimeric structure and magnetic exchange coupling between the two copper centers in the acetate and related compounds raised several hypotheses about the electronic structure of these compounds and possible direct Cu—Cu interaction in view of a rather large interatomic distance 2.64 Å: direct δ bonding between the two $d_{x^2-y^2}$ orbitals (Sections 2.1 and 6.3) [8.77], σ bonding between their d_{z^2} orbitals [8.79], superexchange through the bridging carboxyl ions [8.80], and others. The approximate semiempirical IEH calculations (Section 5.5) [8.81] show that the bonding in the copper(II) acetate is much more complicated. The unpaired electron in the triplet state occupies the $d_{x^2-y^2}$ orbital, as indicated by the ESR spectrum (and in this sense the exchange coupling is realized through the two orbitals of a δ bonding), but the mechanism of coupling involves the oxygen atoms, thus being a rather indirect superexchange coupling through the carboxyl bridges. This conclusion agrees well with the weak dependence of the $2J$ constant on the Cu—Cu interatomic distance in different compounds with similar structure [8.82].

The calculations in [8.81] also show that when the electronic charge is reduced from the binuclear binding area, the bonding between the two centers deteriorates; this result can serve as a qualitative explanation of the fact that mono- and dichloroacetates of copper form dimers, whereas trichloroacetates do not. With fluorine, monofluoroacetate only gives dimers. Many other copper dimers were studied from this point of view [8.6]. For a review of magnetic properties of polynuclear carboxylates, see [8.83].

For coordination compounds with more than two magnetic centers the magnetic properties can be revealed in a similar way. Let us consider some examples. For a trinuclear cluster the HDVV Hamiltonian is

$$H = -2J_{12}S_1S_2 - 2J_{13}S_1S_3 - 2J_{23}S_2S_3 \tag{8.85}$$

For equivalent centers in an equilateral triangle, such as in the heteropoly cluster $[SiV_3^{IV}W_9O_{40}]^{-10}$ [8.84], $J_{12} = J_{13} = J_{23} = J$, and for $S(V^{IV}) = \frac{1}{2}$ the temperature dependence of the magnetic susceptibility is

$$\chi = \frac{Ng^2\beta^2}{4kT} \frac{1 + 5\exp(3J/kT)}{1 + \exp(3J/kT)} \tag{8.86}$$

This agrees well with the experimental data. For a similar tetrahedral cluster with equivalent centers with $S = \frac{1}{2}$ and an isotropic exchange J, the energy-level splitting and Van Vleck formulas yield for χ [8.64]:

$$\chi = \frac{2Ng^2\beta^2}{kT} \frac{5 + 3e^{-2J/kT}}{5 + 9e^{-2J/kT} + 2e^{-3J/kT}} \tag{8.87}$$

However, the magnetic behavior of some of the tetrahedral clusters of the type $Cu_4OL_4X_6$, with $X = Cl$, Br and $L = Cl$, Br, pyridine, OPR_3, ONR_3, does not coincide with the predictions of this formula. Indeed, some of these clusters exhibit nonmonotonous behavior of the magnetic moment $\mu(T)$ with a maximum that cannot be explained by internal isotropic exchange with equal J_{ij} parameters ($i, j = 1, 2, 3, 4$); some of the clusters above have not such a maximum of $\mu(T)$. The assumption of static distortions [8.85] explains the presence of the maximum in $\chi(T)$, but it is in contradiction with the ESR measurements.

The solution of the problem has been obtained recently [8.86] based on the Jahn–Teller effect in a tetrahedral four-center system. Using the results of vibronic theory (Sections 7.3 and 7.4), it was shown that the dynamic Jahn–Teller (pseudo Jahn–Teller) distortions on each of the four Cu(II) centers are coupled to each other such that the magnetic moment of the system is that of distorted centers with a maximum on the $\mu(T)$ curve, whereas the ESR spectra correspond with undistorted centers (cf. the relativity rule concerning the means of observation in Section 9.1).

The isotropic magnetic exchange interaction model represented by the HDVV spin Hamiltonian (8.73), although covering most cases of transition metal coordination compounds, is not the only possible one and is not sufficient for the description of all the magnetic properties of multicenter systems [8.61]. If $S > \frac{1}{2}$ and if the paramagnetic centers are not symmetry related, additional terms of the spin Hamiltonian may be required for the description of the magnetic exchange interaction, in particular, the *biquadratic exchange* term:

$$H_{bi} = j_{ij}(S_i S_j)^2 \tag{8.88}$$

Although the constant j_{ij} is much smaller than the isotropic exchange constant J ($j \sim 10^{-2}J$), the biquadratic exchange (8.88) produces new observable effects. In particular, it violates the rule of Landé intervals in spin-level splitting. The biquadratic term is most important in polynuclear compounds with more than two magnetic centers in a high-symmetry arrangement because it splits the otherwise accidentally degenerate spin levels [8.61].

Another term, the *Dzyaloshinsky–Moriya interaction*, or *antisymmetric exchange* [8.61, 8.62, 8.64],

$$H_{DM} = \mathbf{D}_{ij}[\mathbf{S}_i \times \mathbf{S}_j] \tag{8.89}$$

explains the origin of the *spin canting phenomenon* observed in many compounds. The essence of this effect is that the interacting spins in the ordered state (especially in crystals) are not collinear. As seen from Eq. (8.89), because of the vector product $[\mathbf{S}_i \times \mathbf{S}_j]$, which is zero for collinear spins, they try to occupy noncollinear positions. There are symmetry restrictions on this interaction: no inversion center, and at least one magnetic center must have an anisotropic spin component. Again, the constant of antisymmetric exchange $|D| \approx (|g - 2|/g)J$ is much smaller than J.

Finally, the *anisotropic Ising Hamiltonian*

$$H_I = -2J^{ab} S_{az} S_{bz} \tag{8.90}$$

may be significant, especially in multicenter compounds. Its constant J^{ab} is also small:

$$|J^{ab}| \sim \left(\frac{g-2}{g}\right)^2 J \tag{8.91}$$

but it characterizes the dependence of the magnetic properties on the direction of the magnetic field $\mathscr{H}$. For the case of binuclear copper carboxylates above the Hamiltonian (8.90) allows one to obtain for the anisotropic susceptibility,

$$\chi_{\parallel} = \frac{Ng^2\beta^2/2kT}{1 + e^{-J/Kt}} \tag{8.92}$$

which differs from the Bleaney–Bowers equation (8.84).

For comprehensive reviews of different aspects of the theory of magnetochemitry and ESR in a variety of systems, see [8.5–8.7, 8.61–8.66, 8.72, 8.75, 8.78, 8.83]. The problem of *molecular magnets* (see, e.g., [8.66, 8.87]) seems to be one of the most challenging among the modern applications of magnetochmistry.

Spin Crossover

Among many effects and applications of magnetic properties of coordination compounds, we discuss briefly the *spin-crossover phenomenon*. It follows directly from the Tanabe–Sugano diagrams for electronic d^n energy level dependence on the ligand field parameter Δ (Fig. 4.11). It is seen from these diagrams that for the electronic configurations d^4, d^5, d^6, and d^7 there is a certain value of $\Delta = \Delta_0$ for which the ground-state symmetry and multiplicity changes. The energies on the Tanabe–Sugano diagrams are read off from the ground state, and therefore the intersection of the excited term with the ground term looks like an energy-level break.

The two types of ground states, resulting in two types of complexes, high-spin (HS) and low-spin (LS), are discussed in Section 4.3 in the approximation of the crystal field theory, and in Section 6.2 from the viewpoint of the MO scheme. The HS state is the ground state when the inequality (4.49) takes place:

$$\Delta < \Pi \tag{8.93}$$

where Π is the electron pairing energy, that is, the electron interaction energy difference between the LS and HS states per electron, while for $\Delta > \Pi$ the LS state is lower. It means that the crossing of the two spin terms, the *spin crossover, takes place at* $\Delta_0 = \Pi$.

In the approximation of crystal field theory, the Π values for different electronic d^n configurations are given by Eqs. (4.50). In more precise calculations [8.88] [B and C are the Racah parameters given by Eqs. (2.43)],

$$\Delta_0(d^5) = 6.347B + 4.897C$$

$$\Delta_0(d^6) = 2.195B + 3.708C \tag{8.94}$$

$$\Delta_0(d^7) = 5.051B + 3.594\ C$$

Equations (8.94) were derived for cubic-symmetry complexes, but a small reduction of symmetry does not greatly influence the qualitative results of the spin-crossover phenomenon; hereafter the deviation from cubic symmetry is ignored. In tetrahedral complexes the two possible spin states occur in d^3, d^4, d^5, and d^6 configurations. However, because of the smaller Δ values ($\Delta_T = -\frac{4}{9}\Delta_0$), the LS state is usually not realized in these cases.

The point of exact intersection of the two terms HS and LS is a point of accidental degeneracy. At this point the two states coexist with fast transitions

HS ↔ LS. In fact, there is a whole region of coexistence because of vibrations ($\hbar\omega$) and thermal population. It includes the Δ values for which the energy difference between the two terms is $\sim(kT + \hbar\omega)$. Out of this region the excited state may also be populated, and its population is strongly (exponentially) temperature dependent. The increase of temperature always enhances the spin transition (HS → LS if $\Delta < \Delta_0$, and LS → HS if $\Delta > \Delta_0$). In free complexes (gas phase or dilute solution) when the interaction between them is weak, there is a Boltzmann equilibrium between the LS and HS states with continuous dynamic interconversions:

$$\text{LS} \underset{K_{-1}}{\overset{K}{\rightleftharpoons}} \text{HS} \tag{8.95}$$

where the rate constants K and K_{-1}, for Fe(II) complexes, for example, vary between $4 \cdot 10^5$ and $2 \cdot 10^7$ s^{-1} [8.89].

If the interaction between the complexes is sufficiently strong, as it is in crystals, the cooperative effects become most significant. The HS ↔ LS transition is accompanied by a change in the interatomic distances. An example of this is given in Section 8.3, where the contraction of the bond length measured by EXAFS is discussed [8.58]. The spin transition may change the electronic configuration and the degeneracy of the ground state [e.g., $^1A_1 \rightarrow {}^5T_2$ in Fe(II) complexes]. This, in turn, produces distortions in the local environment of the transition center which may be subject to cooperative interactions in the crystal (Section 9.4). The latter favor an ordered state, but the entropy effects (which increase with temperature) destroy this ordering. Hence for sufficiently strong interactions a phase transition at a certain temperature T_c may be expected [8.89, 8.90].

Among the spin-crossover systems in crystals, two kinds of spin transition can be distinguished:

1. *Discontinuous (abrupt) transitions:* occur at a well-defined temperature T_c.
2. *Continuous (gradual) transitions:* take place over an extended range of temperatures; in this case T_c is defined as the temperature for which the fraction of HS systems is half: $n_{HS} = 0.50$.

To distinguish between the two types of spin transitions and to disclose their nature, many experimental methods of investigation were applied, including magnetic, spectroscopic, Mossbauer, x-ray diffraction, EXAFS, and other measurements (see, e.g., [8.89, 8.91–8.94] and references therein). The primary general conclusion is that abrupt transitions are associated with a structural phase transition, while a continuous transition, gradually converting the system from one state to another, does not change the structural phase.

Figure 8.22 illustrates an abrupt HS ↔ LS $S = \frac{3}{2} \leftrightarrow S = \frac{1}{2}$ transition in a Co(II)($3d^7$) compound, Co(H_2fsa$_2$en)(4-t-Bipy)$_2$ shown in the magnetic susceptibility curve $\chi \cdot T$ versus T. The LS state is realized at low temperature, while

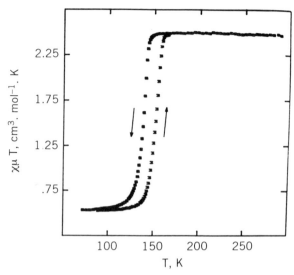

Figure 8.22. Temperature dependence of $\chi_\mu \cdot T$ for $Co(H_2fsa_2en)$ $(4\text{-}t\text{-Bipy})_2$. Downward and upward arrows indicate decreasing and increasing temperatures. (From [8.95].)

the HS state is stable at high temperature, and this is the usual situation in spin-crossover transitions [8.96].

One of the features seen in Fig. 8.22 is *hysteresis*: The transition takes place at slightly different temperatures by heating (arrow up) and cooling (arrow down), respectively. Hysteresis is due to cooperative effects (Section 9.4) and it has interesting applications, mentioned below.

Numerous spin-crossover systems studied so far are mainly iron(II), iron(III), and cobalt(II) complexes, but the phenomenon was also observed in many other metal complexes of Mn(II), Mn(III), Co(III), Ni(II), Mo(II), as well as in systems of the type Nb_6I_{11}, HNb_6I_{11}, and so on. For Fe(II)$(3d^6)$ the two spin states involved in the spin transition are $^1A_1(t_2{}^6)$ and $^5T_2(t_2{}^4e^2)$, while for Fe(III)$(3d^5)$ they are $^2T_2(t_2{}^5)$ and $^6A_1(t_2{}^3e^2)$. For Co(II)$(3d^7)$ the transition occurs between $^2E(t_2{}^6e)$ and $^4T_1(t_2{}^5e^2)$. In NbI_{11} the transition is $S = \frac{1}{2} \leftrightarrow S = \frac{3}{2}$. Besides the temperature dependence (thermally induced spin crossovers), spin transition can be produced by variation of external pressure [8.96, 8.97] and irradiation with light [8.98, 8.99]. The pressure retains the smaller interatomic distances in the LS state, thus increasing T_c; for uniaxial stress an opposite effect may be expected in degenerate HS states that produce low-symmetry distortions. Absorption of light accompanied by an allowed electronic transition from the LS(HS) state to an excited state can be followed by radiationless transitions to the HS(LS) state. This is the effect of *light-induced excited spin-state trapping* (LIESST) [8.98, 8.99], and it was shown to be reversible [8.100].

Interest in spin-crossover systems increased recently due to suggestions to use them as electronic micro-devices in information storage and molecular signal processing based on their property of bistability [8.96]. The two spin states differ essentially in their magnetic properties, and it is possible to control the switch from one state [e.g., the nonmagnetic state $^1A_1$ in Fe(II) complexes] to another (magnetic $^5T_2$), provided that it occurs sufficiently abruptly. On the other hand, the hysteresis of the HS $\leftrightarrow$ LS transition may serve for information storage [8.96].

Magnetic Circular Dichroism

Magnetic circular dichroism is now widely used in studies of coordination compounds, especially in biological systems. In Section 8.1 the effect of dichroism (polychroism) — the dependence of the absorption of polarized light on the direction of polarization with respect to the anisotropic (crystal) system — is mentioned. A related effect, widespread in chemistry, is the *rotation of the plane of polarization* [8.101], often called *optical activity*.

The plane-polarized wave of light can be presented as a sum of two circular-polarized (*cp*) waves: right (*rcp*) and left (*lcp*) [8.1, 8.2, 8.101]. If the absorption coefficients of the right- and left-circular polarized waves K_l and K_r are different (K is determined from the relation $I = I_0 \cdot 10^{-kl}$; Section 8.1), it is said that there is *circular dichroism*. The quantity

$$\varepsilon_l - \varepsilon_r = \frac{K_l - K_r}{c} \tag{8.96}$$

is taken as a measure of the circular dichroism.

Both the optical activity and circular dichroism are due to the asymmetry in charge distribution and polarizability of the compound under consideration. The angle of rotation of the plane of polarized light is

$$\alpha = \frac{n_l - n_r}{\lambda} \tag{8.97}$$

where n_l and n_r are the coefficients of reflection of the *lcp* and *rcp* light, respectively, and λ is the wavelength. While α may be nonzero in the region of transparency, $\varepsilon_l - \varepsilon_r$ is related to the absorption. In the region of absorption all optical active compounds exhibit circular dichroism. An important feature of both optical activity and circular dichroism is thus their *dispersion*, that is, the dependence on light frequency. The curves of dispersion of optical rotation and circular dichroism are used to study molecular structures [8.7].

However, for transition metal compounds the method of *magnetic circular dichroism* (MCD) is more usable [8.2, 8.7]. This method does not demand that the compound to be studied should be optically active: It becomes active under the influence of the external magnetic field. Figure 8.23 illustrates this effect. Consider, for the sake of simplicity, an atomic system with a nondegenerate

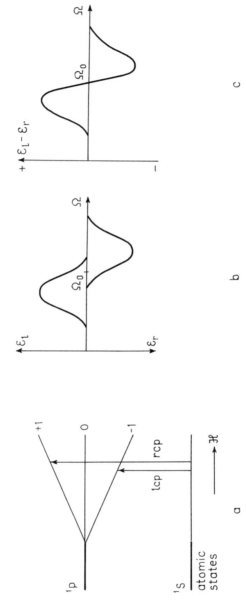

Figure 8.23. Origin of MCD spectrum, term A: (a) atomic ^1S (ground) and ^1P (excited) term splitting in the magnetic field $\mathcal{H}$, and rcp and lcp transitions; (b) lcp and rcp absorption versus light frequency Ω (Ω_0 is the zero-field frequency); (c) summary MCD spectrum.

ground state ^1S and excited threefold-degenerate ^1P state (analogous states in a cubic coordination systems are $^1A_1$ and $^1T_1$). In the magnetic field $\mathscr{H}$ the degenerate P state splits into three components: $\varepsilon_m = mg\beta H$ [Eq. (8.63)] with $m = 0, \pm1$ (Fig. 8.23a).

Now we take into account the selection rules due to which the *lcp* wave is allowed if $\Delta m = -1$, while $\Delta m = +1$ for *rcp*. Hence the maxima of absorption coefficients K_l and K_r are displaced by $2g\beta H$ (Fig. 8.23b), and the $\varepsilon_l - \varepsilon_r$ dispersion curve is as shown in Fig. 8.23c. This is a typical MCD dispersion curve with zero absorption at the zero-field frequency Ω_0; it is called *term A*. Term A is realized when the ground state is nondegenerate while the excited state is degenerate.

Another MCD spectrum, *term C*, is seen when the ground state is degenerate and the excited state is not (Fig. 8.24). The *rcp* component with $\Delta m = +1$ corresponds to the transition from the ground state with $m = -1$, while the *lcp* with $\Delta m = -1$ starts from the excited state $m = +1$ (Fig. 8.24a). Therefore, at extremely low temperatures $T \approx 0$ when the state $m = +1$ is not populated, the *rcp* transition only is observed, and a small *lcp* component occurs when raising the temperature (Fig. 8.24b). Unlike term A, the summary spectrum has no nodal behavior, but the position of its maximum of absorption with respect to Ω_0 is a measure of the magnetic field splitting, which is proportional to $\mathscr{H}$.

The third case of MCD, *term B*, occurs when there is no orbital degeneracy in the two combining states and one of the components of MCD occurs due to the mixing of the ground electronic state with the excited states by the magnetic field. Note that it is precisely this mixing that results in the Van Vleck temperature-independent paramagnetism in Eq. (8.59). Therefore, this MCD spectrum, which is formally similar to the nodeless term C, is distinct from the latter by being independent of temperature. An example of B-term MCD can be found, for instance, in the recent work [8.102] where the MCD spectrum of the $Au_9(PPh_3)_8^{3+}$ and its interpretation in terms of the MO LCAO model is given. Reference [8.103] contains many useful standard definitions and conventions used in the MCD method.

8.5. DETERMINATION OF ELCTRON CHARGE AND SPIN DENSITY DISTRIBUTION BY DIFFRACTION METHODS

In the past two decades experimental determination of electron density distribution in molecular solids reached a high level of accuracy. For molecules containing only light atoms, the charge densities observed are in excellent agreement with sophisticated theoretical calculations. For organic compounds and main-group systems the densities can be predicted, at least qualitatively, from simple valence bond models. This is not the case for transition metal compounds, where the participation of d electrons and their three-dimensional delocalization about the CA (Sections 1.2 and 6.1) makes the expected electron density distribution far less obvious. Therefore, the experimental evaluation of

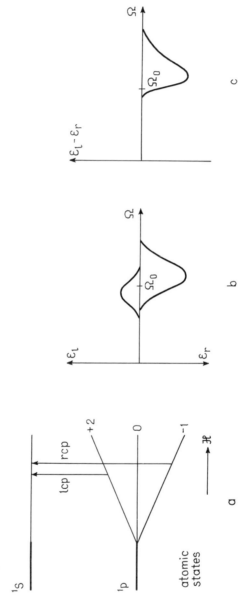

Figure 8.24. Origin of the MCD spectrum, term C: (*a*) atomic ^1P (ground) and ^1S (excited) term splitting in the magnetic field $\mathcal{H}$, and *rcp* and *lcp* transitions; (*b*) *lcp* and *rcp* absorptions versus light frequency Ω (Ω_0 is the zero-field frequency); (*c*) summary MCD spectrum.

electron distribution is an important method of investigation of the electronic structure of transition metal coordination compounds.

Electron densities are usually determined by x-ray or neutron scattering. The technical problems in the experiment itself lie beyond the scope of this book; they can be found elsewhere [8.7, 8.9, 8.104]. Here it is worthwhile to emphasize that in transition metal complexes, as distinct from light-atom compounds, the valence electron distribution contributes a smaller fraction to the total x-ray scattering, thus decreasing the accuracy of the experimental results. Also, experimental problems, such as errors in the corrections for absorption and extinction, become more significant when transition metal or other heavy atoms are present. Nevertheless, the accuracy of such experiments improves continuously.

Method of Deformation Density

The data of high-resolution x-ray intensity measurements are usually presented by a set of experimental *structure factors* F_{obs} with phases obtained from a model structure. The structure factors $F(S)$ are defined as a Fourier transform of the charge density $\rho(r)$ (Section 5.2) [8.9]:

$$F(S) = \int \rho(r) \exp[2\pi i(S, r)] \, dr \tag{8.98}$$

where integration is expanded over the unit cell, S is the scattering vector perpendicular to the diffraction plane, $S = 2 \sin \vartheta / \lambda$, ϑ is the scattering angle, and λ is the x-ray wavelength. If the structure factors F_{obs} are known, the inverse transform allows one to determine the charge density:

$$\rho(r) = \frac{1}{V} \sum_S F_{obs}(S) \exp[-2\pi i(S \cdot r)] \tag{8.99}$$

Here the integral is replaced by a summation over the amplitudes of all diffraction beams allowed by Bragg's law. In simple center-symmetric structures the phases in (8.99) are known, whereas in more complicated (acentric) structures they can be obtained from a model structure; in this case the densities $\rho(r)$ remain model dependent.

The electron density redistribution produced by the chemical bonding can be characterized by the *deformation density* (DD), defined as the difference $\Delta\rho$ between the density for the compounds under consideration $\rho(r)$ and that of the free atoms that occupy the same position, as in the compounds $\rho_{ref}(r)$, $\Delta\rho = \rho(r) - \rho_{ref}(r)$ [Eq. (5.15)]. In terms of the x-ray experiments the DD is

$$\Delta\rho = \frac{1}{V} \sum_S (F_{obs} - F_{ref}) \exp[2\pi i(S \cdot r) \tag{8.100}$$

where F_{ref} are the structure factors of a superposition of neutral, usually spherical atoms (however, see below).

The DD $\Delta\rho(\mathbf{r})$ can also be obtained by calculations. In this case both $\rho(\mathbf{r})$ and $\rho_{ref}(\mathbf{r})$ are computed by means of the methods described in Section 5.3. The densities ρ_{ref} of the free atoms in ground states are known (and tabulated), and the accuracy of $\Delta\rho$ is determined completely by the approximations employed in the calculation of $\rho(\mathbf{r})$ after (5.13). The $\Delta\rho$ values are usually illustrated by means of *deformation density maps*, in which the lines of equal density are given with a definite spacing and with solid lines for positive $\Delta\rho$ values and dashed lines for negative ones.

Knowledge of the DD $\Delta\rho$ allows one to reveal, at least in principle, the electron redistribution due to the formation of the compound, and to relate it to the bonding. However, to do so in practice, many difficulties emerge. Most of them are due to the inaccuracy of the measurements and/or calculations. As mentioned above, in coordination compounds $\Delta\rho$ is a small difference of two large quantities [Eqs. (5.15) and (8.100)] and hence to determine $\Delta\rho$ with high accuracy, the value $\rho(\mathbf{r})$ must be obtained with a much higher accuracy. Meanwhile the experimental resolution in diffraction methods is limited, and so are the calculation possibilities [8.105].

The experimental limitations are due to side effects in intensity measurements (absorption, extinction, multiple, scattering, etc.), scaling of intensities, phasing of structure amplitudes, series truncation, and so on. The theoretical calculations are usually constrained by the adiabatic approximation (Section 7.1), neglect of relativistic effects, finite basis set, and correlation effects (Chapter 5). Additional errors are introduced by neglect of molecular and crystal vibration (temperature smearing).

The results on DD $\Delta\rho$, as mentioned above, are usually given in deformation density maps. Information on electronic structure can be extracted from these maps by three approaches:

1. *Direct inspection*
2. *Modeling with deformation functions*
3. *Comparison with theoretical calculations*

Direct inspection of the experimental DD map enables us to make some qualitative conclusions about the bonding features. In general, positive DD (increase of density as compared with nonbonded atoms) are expected in the region of covalent bonds and strong electronegative atoms. In organic ligands peaks of density are associated with bonding and nonbonding lone pairs. In many cases sharp peaks are also observed around transition metals with partially filled d orbitals because their electron distribution, matching the occupied d states, is nonspherical.

However, in general, a direct correlation of density accumulation with bonding may be wrong, especially when a negative DD is attributed to the absence of bonding. To understand this statement, consider a simple example of the diatomic molecule F_2. In the free atoms all the valence p states have the same occupation, $\frac{5}{3}$ electrons each. By interaction, the two p_σ orbitals from the two atoms form one bonding MO and one antibonding MO, the former only being occupied by two electrons in the ground state. Hence the DD value in the bonding region is approximately $2 - 2(\frac{5}{3}) = -\frac{4}{3}$ electrons.

Thus *despite the two-electron bonding, the deformation density in the bonding region is negative!* This misleading presentation emerges in all cases when there is a bond between atoms with almost completed electronic shells. In particular, it is expected in the case of metal–metal bonds when the metals d shell is occupied in more than half.

These misleading results of the DD method are obviously due to the wrong idea that before entering the bonding the atoms are spherical-symmetric. To understand the situation, consider again the example of the F_2 molecule above. The free fluorine atom has the electronic configuration $1s^2 2s^2 2p^5$ with a p hole in the spherical-symmetric electron distribution that can occupy any of the three p states, p_x, p_y, p_z. Provided that there are no external perturbations, these three one-electron states are equivalent (threefold degeneracy): The hole can occupy any of them or their linear combination with the same probability. As stated in Section 2.1, in the case of degeneracy the real charge distribution remains uncertain until there is an external perturbation that removes the degeneracy (such perturbation is also created by any attempt to observe the charge distribution).

However, *uncertainty (arbitrariness) of the p (or other one-electron) state orientation in space does not mean averaged spherical-symmetric distribution*, although sometimes such a substitution can be justified. For atoms in molecules the uncertain atomic degenerate orbital orientation becomes quite definitive before the bonding. The three-fold degenerate p states of the fluorine atom in the axial field of the other atom split into two levels, A_1 and Π, and it can easily be shown that the ground state is A_1, which corresponds to the p_σ hole (the p_σ orbital occupied by one electron). This is called the *correct zeroth-order function in the case of degeneracy* [8.67].

These zeroth-order states are just a *correct presentation of the interacting atoms before any density redistribution*. They correspond to the lowest-energy configuration of the atom in a given field symmetry. Hence there are only two electrons in the two atomic p_σ states of the two free atoms that form the F_2 molecule, instead of the $2 \cdot 5/3 = 10/3$ electrons in the approximation of averaged spherically symmetric charge distribution. As a result of the MO bonding produced by these two p_σ AOs the charge density increases in the bonding area (Section 5.2), and the deformation density in between the atoms increases, as anticipated. It can be shown that when the atomic shell is less than half-full, the spherical density approximation overestimates the deformation density in the bonding area.

It follows that the approximation of spherical densities of free atoms with open shells is ungrounded and may be misleading. For these systems the correct zeroth-order states should be obtained first (this can be done easily by symmetry considerations, as in the example above), and then the free atom densities can be defined from these states. In other words, the free atom densities p_{ref} should be *prepared* before being subtracted from the measured densities to result in DD.

The failure of the spherical density assumption is a part of a more general failure of the DD approach when used to clarify the origin of chemical bonding without *specification of the reference density* with respect to which the deformation density is determined. As discussed in Section 6.1, coordination bonds are in general not localized, and each metal–ligand bond is strongly dependent on the other bonds formed by the metal. Therefore, atom deformation densities are not sufficiently informative for each metal–ligand bond taken apart. To examine such a bond, the *fragment deformation density*, that is, the difference between the total density and the densities of the ligand and the fragment (including the metal with the other ligands), taken as a whole, should be studied instead of atom deformation densities. An example of fragment DD study is given below.

The fact that charge accumulation in the region between the bonding atoms is not the only cause of chemical bonding (the other one being the reduction of the kinetic energy) is of general importance for the deformation density methods. It means that if there is accumulation of charge (positive $\Delta\rho$) in the bonding region, it can serve as an indication of bonding, but $\Delta\rho \sim 0$ (or even $\Delta\rho < 0$) does not indicate that there is no bonding at all, especially in the cases of delocalized bonds. For coordinated ligands, multiorbital bonds and orbital charge transfers in opposite directions are important (Sections 6.3 and 11.2). Unfortunately, in many (perhaps most) works on DD these important ideas, especially the failure of the atomic spherical densities, are not paid proper attention, the interpretation of the results thus not being sufficiently informative or even misleading.

For quantitative interpretation of the experimental results on DD, the method of *density modeling* can be used. The idea is to represent the electron density $\rho(\mathbf{r})$ approximately by a finite number of analytical probing functions, $g_i(\mathbf{r})$, with corresponding coefficients C_i (populations):

$$\rho(\mathbf{r}) = \sum_i C_i g_i(\mathbf{r}) \qquad (8.101)$$

the C_i being determined by least-square refinement of the x-ray data along with the usual crystallographic parameters (scale factor, positions, thermal parameters, extinctions, etc.). In particular, if the functions $g_i(\mathbf{r})$ have an explicit physical meaning, the model presentation (8.101) allows one to reveal some features of the electron distribution.

For coordination compounds, it is reasonable to model the electron density by d-electron distributions, which means taking the corresponding d-electron spherical harmonics Y_{lm} as the probing functions $g_i(\mathbf{r})$. A simple consideration [8.106] shows that in the crystal field approximation the metal density ρ_M can be presented as

$$\rho_M = \rho_{core} + C_{4s}\rho_{4s} + R(r) \sum_{l,m} C_{lm} Y_{lm} \qquad (8.102)$$

where ρ_{core} includes the density of K, L, and M shells, C_{4s} is the population of the $4s$ orbital, and the last term stands for the d densities ρ_d. On the other hand, these densities can be represented by the sum of the squares of the atomic d functions with appropriate occupancies q_m:

$$\rho_d = [(R_2(r)]^2 \sum_m q_m (Y_{lm})^2 \qquad (8.103)$$

Determining the C_{lm} coefficients from the experimental data and using the known decomposition $(Y_{lm})^2 = \sum_{l'm'} A_{lml'm'} Y_{l'm'}$, we can evaluate the occupancy numbers q_m.

The idea of electronic densities was also used to specify the notion of the *atom in molecules*. It is obvious that by formation of chemical bonds a significant part of the atomic electrons become collectivized (Section 1.2) and the atom loses its individual properties. Therefore, the notion "atom in molecules," strictly speaking, has sense only as indicating the genealogy of the molecule. However, some features of atoms in molecules are important. For instance, atomic charges in molecules are of widespread use in specification of charge distribution. In Section 5.2 the difficulties of defining such atomic charges are discussed: They are due mainly to the lack of atomic borders in molecules.

Bader [8.107] suggested that the atomic borders in molecules should be defined as the surface S at which the gradient of the charge density $\nabla\rho(\mathbf{r})$ equals zero; that is, $\nabla\rho(\mathbf{r})$ changes its sign when moving along the bonding vector:

$$\nabla\rho(\mathbf{r}_0) \cdot \mathbf{n}(\mathbf{r}_0) = 0 \qquad \mathbf{r}_0 \in S \qquad (8.104)$$

where $\mathbf{n}$ is the unit vector normal to S. Based on this presentation, a whole trend of electron density topology has been worked out [8.107] (however, see [8.108]).

Examples

Some examples illustrating the efficiency of the deformation density (DD) studies of electronic structure and bonding in transition metal coordination compounds are given in this section (for other examples, see [8.105, 8.106, 8.109–8.113] and references therein).

Sodium Nitroprusside: Direct Inspection. As a simple example we consider the qualitative conclusions from direct inspection of the DD maps for the slightly distorted octahedral anion $[Fe(CN)_5NO]^{2-}$ of sodium nitroprusside [8.110]. Figure 8.25 shows the main features of the DD in two sections, including the axial (CN—Fe—NO) (Fig. 8.25a) and equatorial (CN—Fe—CN) (Fig. 8.25b) coordinates. First we note the maxima A near the CA situated in between the ligand coordinates. These maxima correspond to the occupied d_{xy} orbital; in the crystal field approximation the high-spin 3d orbital configuration in the

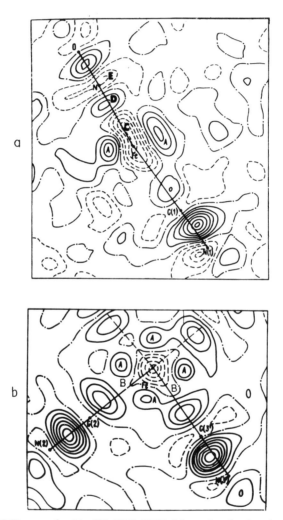

Figure 8.25. DD maps for $Na_2[Fe(CN)_5NO]$: (*a*) section in the plane comprising the CA Fe and the axial ligands NO and CN; (*b*) section in the equatorial plane of Fe and CN groups (only two CN groups are shown). Dot-dashed lines correspond to $\Delta\rho = 0$. Line spacing is equal to 0.1 e/A^3.

tetragonally distorted octahedron is (Fig. 4.5) $d_{xy}^2 d_{xz}^1 d_{yz}^1 d_{z^2}^1 d_{x^2-y^2}^1$. The decreasing density along B near the CA and the maxima in between Fe and C indicate the σ bonds between the iron AO $d_{x^2-y^2}$ and carbon. The minimum of DD (ca -0.30e/A^3) at point C near the Fe atom extended to about $2\,\text{Å}$ in the z direction toward the NO ligand is due to bonding of the d_{z^2} AO with the axial ligands, which is stronger for NO (the atom Fe is $0.185\,\text{Å}$ out of plane toward NO). Along all the ligands and beyond the distal nitrogen there are DD maxima indicating the bonding areas and lone pairs, respectively (beyond oxygen there is no lone pair). This picture of bonding is in qualitative agreement with electronic structure calculations and Mossbauer spectra.

Metal–Metal Bond in $Mn_2(CO)_{10}$: Fragment Deformation Density [8.111].
This is an example of *calculated DD*, which, in addition to the elucidation of the electronic origin of the metal–metal bond, demonstrates the advantages of the fragment deformation density studies.

The structure of $Mn_2(CO)_{10}$ in the eclipsed configuration is

(although the staggered conformer was found in the solid state, the energy difference and the rotation barrier between the two conformers, staggered and eclipsed, is very small [8.111]). For this configuration, the electron density distribution, as well as atomic deformation density (total density minus spherical atoms) and fragment deformation density (total minus two fragments), were calculated by the Fenske–Hall method (Section 5.4). The results are presented in Fig. 8.26.

The fragment DD is equal to the difference between the total density and the density of the two $Mn(CO)_5$ fragments. As mentioned above, owing to the three-dimensional center-delocalized nature of the coordination bond discussed in Section 6.1, the atom deformation density may be less informative with respect to the metal–metal bond than the fragment deformation density. Figure 8.26 confirms this expectation. Indeed, the atomic deformation density (Fig. 8.26b) shows a net loss of density between the two manganese atoms where we expect to find the Mn—Mn bond. In the fragment deformation density (Fig. 8.26c) there is a definitive density accumulation along the Mn—Mn line, demonstrating the formation of the metal–metal bond between the two prepared fragments.

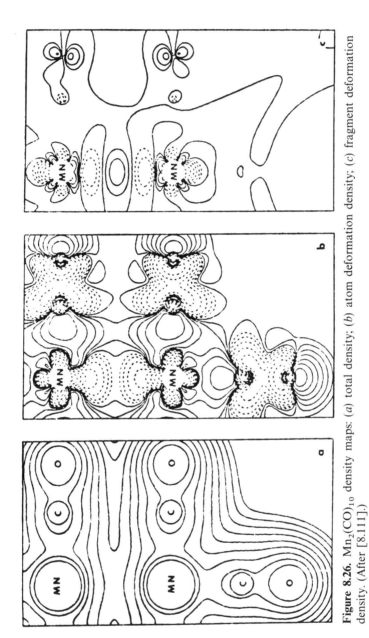

Figure 8.26. $Mn_2(CO)_{10}$ density maps: (*a*) total density; (*b*) atom deformation density; (*c*) fragment deformation density. (After [8.111].)

This example confirms that as discussed above, to get an adequate answer about the bonding origin from DD data the question must be correctly formulated and the bonding species, atoms or fragments, should be properly prepared. As seen from Fig. 8.26b, the deformation density in the region between atoms C and O is also negative. This may be due to the approximation of spherical free atoms (the atomic shell of oxygen is more than half occupied). Another reason is the minimal basis set used in the calculations of the electronic density, which is known to overemphasize the lone pairs beyond the bond [8.111]. The formation of Mn—C bonds is also seen.

Cr(CO)$_6$*, Fe(II)–Phthalocyanine, Co(II)–Tetraphenylporphyrin: Density Modeling.* These examples are given to demonstrate the possibilities of density modeling in the study of electronic structure and bonding by the DD method. The DD for these three compounds have been obtained from diffraction experiments [8.106, 8.113, 8.114], and then presented as fractions of the d orbital densities.

In an earlier work on the octahedral complex $Cr(CO)_6$ [8.113] the difference in the population of the two types of d orbitals e_g and t_{2g} (Section 4.2) has been obtained from nonspherical density distribution around the Cr atom, assuming that its electronic configuration can be either $3d^5 4s$ or $3d^6$. Experimentally, it is difficult to distinguish between these two configurations since the $4s$ state is spherical and the radial distribution difference between $3d$ and $4s$ is very small. In these two models the percentage of e_g character in the d orbital occupancies, $q(e_g)/q(e_g + t_{2g})$, is 24 and 25, respectively, while in the spherical free atom it is 40 and 33 (in the high-spin configurations). Note that the field-oriented configuration of the free atom, specified above (the state with the lowest energy in the octahedral field but without bonding), is $3d^5 4s^1$.

As a result of such modeling, the authors [8.113] obtained numerical estimates of the atomic charges q^A, orbital charge transfers to CO by coordination Δq_σ and Δq_π (Sections 5.2 and 6.3), and the relative e_g occupancy, mentioned above. These data are given in Table 8.9 together with some results of *ab initio* electronic structure calculations [8.115] for comparison.

In Fe(II)–phthalocyanine [8.114] and Co(II)–tetraphenylporphyrin (CoTPP) [8.106], the polyhedron around the CA is square planar, for which the d orbital ordering is $d_{yz} d_{xz} d_{z^2} d_{xy} d_{x^2-y^2}$ (Sections 4.2 and 4.3 and Fig. 4.5). In the spherical-atom high-spin configuration these orbitals are equally occupied by 1.2 electron each in the iron complex and 1.4 in the cobalt complex. In the low-spin configuration the spherical atoms have the configurations $d_{yz}^2 d_{xz}^2 d_{z^2}^2 d_{xy}^0 d_{x^2-y^2}^0$ and $d_{yz}^2 d_{xz}^2 d_{z^2}^2 d_{xy}^{0.5} d_{x^2-y^2}^{0.5}$ for Fe(II) and Co(II), respectively.

However, the prepared oriented (nonspherical) atoms that have the lowest energy in the square-planar field are $d_{yz}^{1.5} d_{xz}^{1.5} d_{z^2}^1 d_{xy}^1 d_{x^2-y^2}^1$ for Fe(II) and $d_{yz}^2 d_{xz}^2 d_{z^2}^1 d_{xy}^1 d_{x^2-y^2}^1$ for Co(II). The DD analysis should be carried out with respect to these prepared atom densities. Table 8.10 illustrates the relevant data, including spherical atom reference densities. It is seen that there is a considerable orbital charge transfer from the field-oriented d_{xz}, d_{yz}, and $d_{x^2-y^2}$

Table 8.9. Charge distributions in $Cr(CO)_6$ obtained from electron deformation density modeling assuming either $3d^5 4s^1$ or $3d^6$ configuration of Cr, in comparison with *ab initio* calculations

Atomic Charges, Orbital Occupancy, Charge Transfers	Experimental		Calculations [8.115]
	$3d^5 4s$	$3d^6$	
q^{Cr}	0.15 ± 0.12		0.70
q^{C}	0.09 ± 0.05		0.23
q^{O}	-0.12 ± 0.05		-0.35
$q(e_g)/q(e_g + t_{2g})$	0.24 ± 0.03	0.25 ± 0.03	0.16
Δq_σ	-0.35 ± 0.04	-0.24 ± 0.04	-0.16
Δq_π	0.38 ± 0.04	0.27 ± 0.04	0.27

orbitals of the iron ion (0.3–0.4 electron from each) to the phthalocyanine and a back donation of about 0.7 electron to the planar d_{xy} orbital; similar charge transfers but smaller in magnitude also occur in the case of CoTPP. Larger charge transfers to and from the CA in iron phthalocyanine as compared with CoTPP can be understood if one takes into account the larger *redox capacitance* (Section 10.1) of phthalocyanine, as compared with the TPP system.

Note that the density modeling (8.102) is based on the crystal field theory and hence is rather qualitative. Therefore, the values of the charge transfers in Tables 8.9 and 8.10 cannot pretend to quantitative interpretation of the DD data. This interpretation is also complicated by the low accuracy of the x-ray density measurement, mentioned above.

Table 8.10. *d*-orbital occupancies q_m in two square-planar complexes of Fe(II) and Co(II) as determined from x-ray electron densities using the crystal field model in comparison with spherical (SPH) ion and field-oriented (OR) values[a]

d AOs	Fe(II)–Phthalocyanine				Co(II)–Tetraphenylporphyrin			
	q_m, SPH Fe(II)		q_m, OR Fe(II)	q_m, exp (Molecule)	q_m, SPH Co(II)		q_m, OR Co(II)	q_m, exp (Molecule)
	HS	LS			HS	LS		
d_{yz}, d_{xz}	2.40	2.00	3.00	2.12	2.80	4.00	4.00	3.7
d_{z^2}	1.20	2.00	1.00	0.93	1.40	2.00	1.00	1.0
d_{xy}	1.20	2.00	1.00	1.68	1.40	0.50	1.00	1.3
$d_{x^2-y^2}$	1.20	0	1.00	0.70	1.40	0.50	1.00	1.0

[a] The differences between oriented ion and molecular values characterize the bonding *deformation density* (HS and LS denote high-spin and low-spin configurations, respectively).

Spin Densities from Neutron Scattering

Unlike x-ray scattering, the *polarized neutron diffraction* experiment reveal *magnetization densities*, or *spin densities*, that is, the density of distribution of magnetic moments created by orbital motion and spin of electrons. The *magnetic structure factor* $\mathbf{M(S)}$ can be presented in the same way as the charge density [Eq. (8.99)] [8.116]:

$$\mathbf{M(S)} = \sum_n f_n(\mathbf{S})\mathbf{M}_n \exp[-2\pi i(\mathbf{S}\cdot\mathbf{r}_n)] \tag{8.105}$$

where $\mathbf{M}_n$ is the magnetic moment of the atom n positioned at $\mathbf{r}_n$, and $f_n(\mathbf{S})$ is the magnetic form factor for this atom,

$$f(\mathbf{S}) = \frac{\int m(\mathbf{r})\exp(i\mathbf{S}\cdot\mathbf{r})\,d\tau}{\int m(\mathbf{r})\,d\tau} \tag{8.106}$$

and $m(\mathbf{r})$ is the magnetization density of the atom.

For a single atomic shell the form factor can be presented as

$$f(\mathbf{S}) = \sum_l A_l(\mathbf{S}) \int R_{nl}^2(r)\, j_l(Sr) r^2\, dr \tag{8.107}$$

where $j_l(Sr)$ is a Bessel function of order l, $R_{nl}^2(r)$ the radial part of the electron wavefunction, and the A_l are expansion coefficients that have been tabulated for many transition metal ions [8.117].

Using these formulas, the observed spin densities, by means of some least-square procedures, can be attributed to certain d state and other MO populations in the system. For instance, from the magnetic structure factors of KNa_2CrF_6, the spin populations were found to be as follows [8.118]: $t_{2g}^{2.66}e_g^{-0.06}4s^{0.4}$ for Cr^{3+} and $2p_{\pi x}^{0.02}2p_{\pi y}^{0.02}2p_\sigma^{-0.02}$ at the fluorine. For the Co^{2+} ion in phthalocyaninato-cobalt [8.119] the spin populations are $3d_{xy}^{0.40}3d_{xz,yz}^{0.17}3d_{z^2}^{0.79}3d_{x^2-y^2}^{-0.21}4s^{-0.44}$, with a total spin -0.17 on the phthalocyanine. In a similar manganese complex the spin populations are different: $3d_{xy}^{0.74}3d_{xz}^{1.17}3d_{ze}^{0.83}3d_{x^2-y^2}^{-0.15}4s^{-0.44}$, with -0.31 on the ligand atoms [8.116].

An example of spin density maps is shown in Fig. 8.27 for phthalocyaninatocobalt(II) [8.119]. Given in two sections, in the molecular plane (Fig. 8.27a) and 0.25 Å above (Fig. 8.27b), the lines of equal spin densities with a spacing of $\pm 2^{n-1}\cdot 10^{-3}e\,\text{Å}^{-3}$ show the variation of the spin distribution in space. From the line shapes it is seen that the unpaired electrons occupy antibonding (or nonbonding) orbitals (the spin density is lower in the bonding region), despite the high covalency of the bonds as a whole. This result

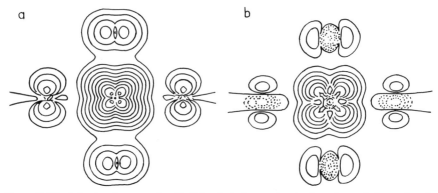

Figure 8.27. Spin density map for Co(II)–phthalocyanine in the plane of the molecule (*a*) and in the section 0.25 Å above (and parallel to) this plane (*b*). The *n*th contour is at the $\pm 2^{n-1} \cdot 10^{-3} e\,\text{Å}^{-3}$ density level from the zero (first continuous) line.

confirms the statements of Section 6.2 (Fig 6.3; see also Section 8.4) that the covalency is realized by the uncompensated bonding MO, which are not necessarily related to the HOMO and LUMO usually occupied by the unpaired electrons.

This conclusion is part of a general feature of spin density maps: The spin density contribution may not be related directly to the chemical bonding. But they are related directly to ESR spectra (Section 8.4), and the spin density method reveals some important features in the nature of these spectra [8.119]. It is also seen from Fig. 8.27*b* that nitrogen π orbitals have negative spin densities.

REFERENCES

8.1. G. Herzberg, *Electronic Spectra and Electronic Structure of Polyatomic Molecules*, Van Nostrand, New York, 1966.

8.2. A. B. P. Lever, *Inorganic Electronic Spectroscopy*, 2nd ed., Elsevier, Amsterdam, 1984.

8.3. K. Ziegbahn, C. Nordling, A. Fahlman, R. Nordberg, K. Hamrin, J. Hedman, G. Johansson, T. Bergmark, S. E. Karlsson, I. Lindgren, and E. Lindberg, *ESCA: Atomic, Molecular and Solid State Structure Studied by Means of Electron Spectroscopy*, Nova Acta Regial Societatis Sceintarum Upsaliensis, Ser. IV, Vol. 20, Upsala, 1967.

8.4. V. I. Nefedov and V. I. Vovna, *Electronic Structure of Chemical Compounds* (Russ.), Nauka, Moscow, 1987, 348 pp.

8.5. A. K. Cheetham and P. Day, Eds., *Solid State Chemistry: Techniques*, Clarendon Press, Oxford, 1988, 398 pp.

8.6. R. L. Carlin, *Magnetochemistry*, Springer-Verlag, Berlin, 1986, 328 pp.

8.7. R. S. Drago, *Physical Methods in Chemistry*, W. B. Saunders, Philadelphia, 1977.

8.8. V. I. Goldanskii and R. H. Herber, Eds. *Chemical Applications of Mossbauer Spectroscopy*, Academic Press, New York, 1968; G. J. Long and F. Grandjean, Eds., *Mossbauer Spectroscopy Applied to Inorganic Chemistry*, Vol. 3, Plenum Press, New York, 1989; D. P. E. Dickson and F. J. Berry, Eds., *Mossbauer Spectroscopy*, Cambridge University Press, Cambridge, 1986, 274 pp.

8.9. P. Coppens, *Concepts of Charge Density Analysis: The Experimental Approach*, in: *Electron Distribution and Chemical Bond*, ed. P. Coppens and H. B. Hall, Plenum Press, New York, 1982, pp. 3–59.

8.10. J. H. Saaten and J. Wieringen, *Recl. Trav. Chim.*, **71**, 420–430 (1952).

8.11. H. Hartman, H. L. Schlafer, and K. H. Hansen, *Z. Anorg. Chem.*, **289**, 40–65 (1957).

8.12. M. Linhard, *Z. Electrochem.*, **50**, 224–238 (1944).

8.13. S. Sugano, Y. Tanabe, and H. Kamimura, *Multiplets of Transition-Metal Ions in Crystals*, Academic Press, New York, 1970, 331 pp.

8.14. Yu. E. Perlin and B. S. Tsukerblat, in: *The Dynamical Jahn–Teller Effect in Localized Systems*, ed. Yu. E. Perlin and M. Wagner, North-Holland, Amsterdam, 1984, pp. 251–346; *Effects of Electron-Vibrational Interactions in Optical Spectra of Impurity Paramagnetic Ions* (Russ.), Shtiintsa, Kishinev, 1974, 368 pp; Yu. E. Perlin, *Usp. Fiz. Nauk* (English Transl.: *Sov. Phys.-Usp.*). **80**, 553–595 (1963).

8.15. F. Dushinsky, *Acta Physicochim. USSR*, **7**, 551–557 (1937).

8.16. L. E. Orgel, *An Introduction to Transition Metal Chemistry*, Methuen, London, 1960.

8.17. C. K. Jorgensen, *Acta Chem. Scand.*, **8**, 1502–1512 (1954).

8.18. C. K. Jorgensen, *Oxidation Numbers and Oxidation States*, Springer-Verlag, Berlin, 1969, 291 pp.

8.19. C. K. Jorgensen, *Modern Aspects of Ligand Field Theory*, North-Holland, Amsterdam, 1971, 538 pp.

8.20. C. K. Jorgensen, *Prog. Inorg. Chem.*, **4**, 73–124 (1962).

8.21. I. B. Bersuker, *The Jahn–Teller Effect and Vibronic Interactions in Modern Chemistry*, Plenum Press, New York, 1984, 319 pp.

8.22. I. B. Bersuker and V. Z. Polinger, *Vibronic Interactions in Molecules and Crystals*, Springer-Verlag, Heidelberg, 1989, 422 pp.

8.23. H. C. Longuet-Higgins, U. Opic, M. H. L. Pryce, and R. A. Sack, *Proc. Roy. Soc. London*, **A244**, 1–16 (1958).

8.24. S. Muramatsu and N. Sakamoto, *J. Phys. Soc. Jpn*, **36**, 839 (1971).

8.25. P. Habitz and W. H. E. Schwarz, *Theor. Chim. Acta*, **28**, 267 (1973).

8.26. V. Loorits, *Izv. Akad. Nauk Est. SSR Fiz. Mat.*, **29**, 208 (1980).

8.27. H. Koppel, E. Haller, L. S. Cederbaum, and W. Domske, *Mol. Phys.*, **41**, 669 (1980).

8.28. R. Englman, R. Caner, and S. Toaff, *J. Phys. Soc. Jpn*, **29**, 306–310 (1970)

8.29. I. B. Bersuker, Ed., *The Jahn–Teller Effect: A Bibliographic Review*, IFI/Plenum, New York, 1984, 589 pp.

8.30. F. I. Vilesov, B. Kurbatov, and A. I. Terenin, *Dokl. Acad. Nauk SSSR*, **138**, 1329–1332 (1961).

8.31. D. W. Turner and M. I. Al-Joboury, *J. Chem. Phys.*, **37**, 3007–3008 (1962).

8.32. M. I. Al–Joboury and D. W. Turner, *J. Chem. Soc.*, 5141–5147 (1963)

8.33. C. S. Fadley, in: *Electron Spectroscopy: Theory, Techniques and Applications*, ed. C. R. Brundle and A. D. Baker, Academic Press, London, Vol. 2, 1978, pp. 1–156.

8.34. D. Briggs, in: *Electron Spectroscopy: Theory, Techniques and Applications*, ed. C. R. Brundle and A. D. Baker, Academic Press, London, Vol. 3, 1979, pp. 305–358.

8.35. T. A. Carlson, *Photoelectron and Auger Spectroscopy*, Plenum Press,, New York, 1975.

8.36. L. N. Mazalov, *X–ray Emission and Electronic Spectroscopy of Molecules* (*Theoretical Bases*) (Russ.), Nauka, Novosibirsk, 1979, 92 pp.

8.37. L. N. Mazalov and V. D. Iumatov, *Electronic Structure of Extragents* (Russ.), Nauka, Novosibirsk, 1984, 200 pp.

8.38. D. W. Turner, in: *Physical Methods in Advanced Inorganic Chemistry*, ed. H. A. O. Hill and P. Day, Interscience, New York, 1968.

8.39. G. K. Wertheim and A. Rosencwaig, *Phys. Rev. Lett.*, **26**, 1179 (1971).

8.40. L. S. Cederbaum and W. Domske, *Adv. Chem. Phys.*, **36**, 205–344 (1977).

8.41. E. Haller, H. Koppel, and L. S. Cederbaum, *J. Chem. Phys.*, **78**, 1359–1370 (1983).

8.42. H. Koppel, W. Domske, and L. S. Cederbaum, *Adv. Chem. Phys.*, **53**, 59–246 (1984).

8.43. O. I. Sumbaev, E. B. Petrovich, Yu. P. Smirnov, I. M. Band, and A. I. Smirnov, *Usp. Fiz. Nauk*, **113**, 360–363 (1974).

8.44. V. I. Nefedov, *Application of X–ray Electron Spectroscopy in Chemistry* (Russ.), in: *Reviews of Science and Technique*, Ser.: *Molecular Structure and Chemical Bond*, Vol. 1, VINITI, Moscow, 1973, 148 pp.

8.45. P. A. Lee, P. H. Citrin, P. H. Eizenberger, and B. M. Kincaid, *Rev. Mod. Phys.*, **53**, 769–806 (1981).

8.46. S. P. Kramer and K. O. Hodgson, *Prog. Inorg. Chem.*, **25**, 1–14 (1979).

8.47. B. K. Teo, *Acc. Chem. Res.*, **13**, 412–420 (1980).

8.48. *EXAFS and Near Edge Structure* (*IV*), Vol. 2, *J. Phys.*, **47**, Coll. C8, No. 12 (1986); B. K. Teo, Ed., *EXAFS: Basic Principles and Data Analysis*, Inorganic Chemistry Concepts, Vol. 9, Springer-Verlag, Berlin, 1986, 349 pp.

8.49. D. C. Koningsberger and R. Prins, Eds., *X–ray Absorption: Principles, Applications, Techniques of EXAFS, SEXAFS, XANES*, Chemical Analysis, Vol. 92, Wiley, New York, 1988, 673 pp.; *X–ray Absorption Fine Structure* (*The 6th International Conference on EXAFS and XANES*), ed. S. S. Husain, Harwood, New York, 1991, 792 pp.

8.50. D. E. Sayers, E. A. Stern, and F. W. Lytle, *Phys. Rev. Lett.*, **27**, 1204 (1971).

8.51. E. A. Stern, *Contemp. Phys.*, **19**, 289 (1978).

8.52. P. M. Eisenberger and B. M. Kinkaid, *Science*, **200**, 1441 (1978).

8.53. O. Hanske–Petitpierre, Y. Yacoby, J. Mustre de Leon, E. A. Stern, and J. J. Rehr, *Phys. Rev. B*, **44**, 6700 (1992); N. Sicron, B. Ravel, Y. Yacoby, E. A. Stern, F. Dogan, and J. J. Rehr, *Phys. Rev. B*, **50**, 13168–13180 (1994-II).

8.54. D. R. Sanstrom and F. W. Lytle, *Annu. Rev. Phys. Chem.*, **30**, 215 (1979).

8.55. F. W. Lytle, H. Via, and J. H. Sinfelt, in: *Synchrotron Radiation Research*, ed. H. Winick and S. Doniach, Plenum Press, New York, 1980.

8.56. S. P. Cramer, H. B. Gray, Z. Dori, and A. Binu, *J. Am. Chem. Soc.*, **101**, 2770 (1979).

8.57. T. K. Sham, J. B. Hastings, and M. L. Perlman, *Chem. Phys. Lett*, **83**, 391–396 (1981).

8.58. C. Cartier, P. Thuery, M. Verdaguer, Z. Zarembowitch, and A. Michailowicz, *J. Phys.*, **47**(C8), 563–568 (1986).

8.59. H. Froitzhein, *Electron Energy Loss Spectroscopy*, in: *Electron Spectroscopy for Surface Analysis*, ed. H. Ibach, Springer-Verlag, Berlin, 1977.

8.60. S. E. Schnatterly, *Solid State Phys.*, **34**, 275–358 (1979).

8.61. B. S. Tsukerblat and M. I. Belinski, *Magnetochemistry and Radiospectroscopy of Exchange Clusters* (Russ.), Shtiintsa, Kishinev, 1983, 280 pp.; B. S. Tsukerblat, M. I. Belinskii, and V. E. Fainzilberg, in: *Soviet Scientific Reviews: Chemistry Reviews*, ed. M. Vol'pin, Harwood, Amsterdam, 1987, pp. 209–277.

8.62. V. T. Kalinikov and Yu. V. Rakitin, *Introduction to Magnetochemistry: The Method of Static Magnetic Susceptibility in Chemistry* (Russ.), Nauka, Moscow, 1980, 302 pp.

8.63. A. Earnshaw, *Introduction to Magnetochemistry*, Academic Press, New York, 1968.

8.64. Ch. J. O'Connor, *Magnetochemistry: Advances in Theory and Experimentation*, in: *Prog. Inorg. Chem.*, **29**, 203–283 (1982).

8.65. A. Abraham and B. Bleaney, *Electron Paramagnetic Resonance of Transition Ions*, Clarendon Press, Oxford, 1970, 912 pp.

8.66. O. Kahn, *Molecular Magnetism*, VCH, New York, 1993, 380pp.; *Magnetism of the Heteropolymetallic Systems*, in: *Struct. Bonding*, **68**, 89–167 (1987).

8.67. L. D. Landau and E. M. Liphshitz, *Quantum Mechanics: Nonrelativistic Theory*, 3rd ed., Nauka, Moscow, 1974, 752 pp.

8.68. J. H. Van Vleck, *The Theory of Electric and Magnetic Susceptibility*, Oxford University Press, Oxford, 1932.

8.69. D. Kivelson and R. Neiman, *J. Chem. Phys.*, **35**, 149 (1961).

8.70. H. R. Gersman and J. D. Swalen, *J. Chem. Phys.*, **36**, 3221 (1961)..

8.71. P. S. Pregosin, Ed., *Transition Metal Nuclear Magnetic Resonance*, Elsevier, Amsterdam, 1991; V. P. Tarasov and V. I. Privalov, *Magnetic Resonance of Heavy Nuclei in the Investigation of Coordination Compounds*, in: *Reviews in Science and Technique, Ser.: Molecular Structure and Chemical Bond*, Vol. 13, VINITI, Moscow, 1989, 136 pp.

8.72. P. W. Anderson, in: *Magnetism*, Vol. 1, ed. G. T. Rado and H. Suhl, Academic Press, New York, 1963, p. 25.

8.73. O. Kahn and B. Briat, *J. Chem.Soc. Faraday Trans. II*, **7**, 268, 1441 (1976).

8.74. P. J. Hay, J. C. Thibeault, and R. Hoffmann, *J. Am. Chem. Soc.*, **97**, 4884 (1975).

8.75. O. Kahn and M.-F. Charlot, in: *Quantum Theory of Chemical Reactions*, Vol. 2, ed. R. Daudel, A. Pullman, L. Salem, and A. Veillard, D. Reidel, Dordrecht, The Netherlands, 1980, pp. 215–240.

8.76. B. Bleaney and K. D. Bowers, *Proc. Roy. Soc. London*, **A214**, 451 (1952).

8.77. B. N. Figgis and R. L. Martin, *J. Chem. Soc.*, 3837 (1956).

8.78. R. L. Martin, in: *New Pathways in Inorganic Chemistry*, ed. E. A. V. Ebsworth, A. G. Maddock, and A. G. Sharpe, Cambridge University Press, London, 1968, Chap. 9.

8.79. L. S. Forster and C. J. Ballhausen, *Acta Chem. Scand.*, **16**, 1385 (1962).

8.80. T. Watanabe, *J. Phys. Soc. Jpn.*, **16**, 1677 (1961).

8.81. I. B. Bersuker and Yu. G. Titova, *Teor. Eksp. Khim.*, **6**, 469–478 (1970).

8.82. M.-F. Charlot, M. Verdaguer, Y. Journaux, P. de Loth, and J. P. Dandey, *Inorg. Chem.*, **23**, 3802 (1984); M. Julve, M. Verdaguer, A. Gleizes, M. Philoche-Levisalles, and O. Kahn, *Inorg. Chem.*, **23**, 3808 (1984).

8.83. J. A. Moreland and R. J. Doedens, *J. Am. Chem. Soc.*, **97**, 508 (1974).

8.84. M. M. Mossova, C. J. O'Connor, M. T. Pope, E. Sinn, G. Herve, and A. Teze, *J. Am. Chem. Soc.*, **102**, 6864 (1980).

8.85. D. H. Jones, J. R. Sams, and R. C. Thompson, *J. Chem. Phys.*, **79**, 3877 (1983).

8.86. I. B. Bersuker, V. Z. Polinger, and L. F. Chibotaru, *Mol. Phys.*, **52**, 1271–1289 (1984).

8.87. T. Mallah, S. Thiebaut, M. Verdaguer, and P. Veillet, *Science*, **262**, 1554 (1993); D. Gatteschi, *Adv. Mat.*, **6**, 635 (1994); J. S. Miller and A. J. Epstein, *Mol. Cryst., Liq. Cryst.*, **271–274** (1995); *Chem. & Eng. News*, 30–41, (October 2, 1995).

8.88. E. Konig and S. Kremer, *Theor. Chim. Acta*, **23**, 12 (1971).

8.89. E. Konig, G. Ritter, and S. K. Kullshreshtha, *Chem. Rev.*, **85**, 219–234 (1985).

8.90. T. Kambara, *J.Chem. Phys.*, **70**, 4199 (1979); N. Sasaki, T. Kambara, *J. Chem. Phys.*, **74**, 3472 (1981).

8.91. E. Konig, *Prog. Inorg. Chem.*, **35**, 527–622 (1987).

8.92. M. Bacci, *Coord. Chem. Rev*, **86**, 245–271 (1988).

8.93. P. Gutlich, *Struct. and Bonding*, **44**, 83–202 (1981).

8.94. C. N. R. Rao, *Int. Rev. Phys. Chem.*, **4**, 19–38 (1985).

8.95. P. Thuery and J. Zarembowitch, *Inorg. Chem.*, **25**, 2001 (1986).

8.96. J. Zarembowitch and O. Kahn, *New J. Chem.*, **15**, 181–190 (1991); J. Zarembowitch, *New J. Chem.*, **16**, 255–267 (1992) .

8.97. C. Roux, J. Zarembowitch, J.-P. Itie, M. Verdaguer, E. Dartyge, A. Fontaine, and H. Tolentino, *Inorg. Chem.*, **30**, 3174 (1991).

8.98. S. Decurtins, P. Gutlich, C.P. Kohler, and H. Spiering, *Chem. Phys. Lett.*, **105**, 1 (1984); *J. Chem. Soc. Chem. Commun.*, 430 (1985).

8.99. S. Decurtins, P. Gutlich, M. Hasselbachk, A. Hauser, and H. Spiering, *Inorg. Chem.*, **24**, 2175 (1985).

8.100. A. Hauser, *Chem. Phys. Lett.*, **124**, 543 (1986).

8.101. S. F. Hason, *Molecular Optical Activity and the Chiral Discrimination*, Cambridge University Press, Cambridge, 1982.

8.102. H.-R. C. Jaw and W. R. Mason, *Inorg. Chem.*, **30**, 275–278 (1991).

8.103. S. B. Piepho and P. N. Schatz, *Group Theory in Spectroscopy with Applications to Magnetic Circular Dichroism*, Wiley-Interscience, New York, 1983.

8.104. G. H. Stout and L. H. Jensen, *X-ray Structure Determination*, Macmillan, New York, 1968.

8.105. M. B. Breitenstein, H. Dannohl, H. Meyer, A. Schweig, and W. Zittlau, in: *Electron Distribution and the Chemical Bond*, ed. P. Coppens and M. B. Hall, Plenum Press, New York, 1982, pp. 255–281.

8.106. E. D. Stevens, in: *Electron Distribution and the Chemical Bond*, ed. P. Coppers and M. B. Hall, Plenum Press, New York, 1987, pp. 331–349

8.107. R. F. W. Bader, *Atoms in Molecules: A Quantum Theory*, Clarendon Press, Oxford, 1990, 439 pp.; R. F. W. Bader, T. T. Nguyen Dang, and Y. Tal, *Rep. Prog. Phys.*, **44**, 893 (1981).

8.108. G. I. Bersuker, C. Peng, and J. E. Boggs, *J. Phys. Chem.*, **97**, 9323 (1993).

8.109. V. G. Tsirelson and M. Yu. Antipin, in: *Crystal Chemistry Problems* (Russ.), ed. M. A. Porai–Koshitz, Nauka, Moscow, 1989, pp. 119–160.

8.110. M. Yu. Antipin and Yu. T. Struchkov, *Metalloorganich. Khim.*, **2**, 128–144 (1989).

8.111. M. B. Hall, in: *Electron Distribution and the Chemical Bond*, ed. P. Coppens and M. B. Hall, Plenum Press, New York, 1982, pp. 205–220.

8.112. P. Becker, Ed., *Electron and Magnetization Densities in Molecules and Crystals*, Plenum Press, New York, 1980, 904 pp.

8.113. B. Rees and A. Mitschler, *J. Am. Chem. Soc.*, **98**, 7918 (1976).

8.114. P. Coppens and L. Li, *J. Chem. Phys.*, **81**, 1983 (1984).

8.115. I. H. Hillier and V. R. Saunders, *Mol. Phys.*, **22**, 1025–1034 (1971).

8.116. R. Mason, in: *Electron Distribution and the Chemical Bond*, ed. P. Coppens and M. B. Hall, Plenum Press, 1982, pp. 351–360.

8.117. R. J. Weiss and A. J. Freeman, *J. Phys. Chem. Solids*, **10**, 147 (1959).

8.118. B. N. Figgis, P. A. Reynolds, and G. A. Williams, *J. Chem. Soc. Dalton Trans.*, 2348 (1980).

8.119. G. A. Williams, B. N. Figgis, and R. Mason, *J. Chem. Soc. Dalton Trans.*, 734 (1981).

9 Stereochemistry and Crystal Chemistry

Stereochemistry underlies chemical intelligence: Without assumptions of molecular shapes there is no way to rationalize molecular structure and chemical transformations.

During a relatively short period, stereochemistry and crystal chemistry of transition metal coordination compounds changed from a (charge) ball-packing treatment to a complicated electronic and vibronic problem.

9.1. SEMICLASSICAL APPROACHES

Definition of Molecular Shape

Stereochemistry deals with spatial arrangement of atoms in molecules: *molecular shapes*. It occupies one of the most important places in the hierarchy of the basic ideas of modern chemistry [9.1–9.6]. Therefore, it is worthwhile to discuss in some detail the physical understanding that underlies the definition of molecular shapes.

The usual assumption that a molecule has a fixed spatial (three-dimensional) arrangement of atoms with small vibrations near some equilibrium positions is not always valid. It excludes some isomers, tautomers, nonrigid molecules, alterdentate ligands, conformers—quite a number of situations when there are large-amplitude nuclear dynamics or intramolecular transformations. In many cases the latter cannot be presented as distinct transitions from one nuclear configuration (conformation) to another.

As stated in Section 1.2, *any rigorous definition of a physical quantity should contain, explicitly or implicitly, an indication of the means of its observation.* This statement follows from the understanding that the quantities which cannot be observed can only pretend to be virtual, but not real. To observe a molecular system in a given configuration, it should have a *lifetime*, τ, larger than the characteristic *time of measurement*, τ', determined by the means of observation:

$$\tau > \tau' \tag{9.1}$$

The lifetime of a given molecular configuration is directly related to the shape of its AP (Section 7.1). If the molecular configuration under consideration corresponds to a sufficiently deep minimum of the AP and there are no other equivalent (or comparable in energy) minima, τ can be assumed to be sufficiently large to define rigorously the molecular shape. If the opposite inequality $\tau \leqslant \tau'$ holds, the molecular (nuclear) configuration becomes uncertain.

By way of example, consider the model of two isomers of rhodium [9.7] shown in Fig. 9.1 (the existence of these isomers is questionable, but this does not influence the discussion below). The two minima corresponding to the two configurations are expected to be different in depths, and we assume that there is an energy barrier between them, ΔE. The essential feature required for the isomers to exist is that *there are localized states in the minima*. The condition for localized states follows from quantum mechanics: The depth of the minimum well should be larger than the kinetic energy of the motions within the well.

If there are localized states in the wells, the system performs vibrations with a frequency ω that can be observed, say, by IR spectra. The characteristic time of measurement τ' is then no less than $T = 2\pi/\omega$. On the other hand, the finite barrier height ΔE allows the system to transfer from one configuration to another with a rate determined either by overcoming the barrier thermally (at sufficiently high temperatures), or tunneling through the barrier. In the latter instance the lifetime of the system at the minimum is determined by the tunneling rate, which is exponentially dependent on the ratio $\Delta E/\hbar\omega$. In general, the observation of a certain configuration is possible when at least

$$\Delta E > \hbar\omega \qquad\qquad (9.2)$$

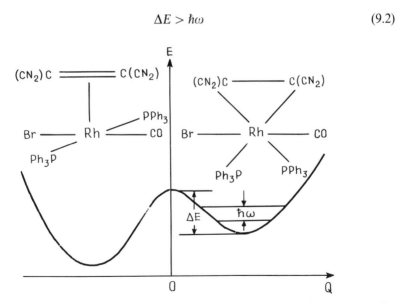

Figure 9.1. Adiabatic potential curve for two feasible isomers of a Rh complex (Ph = phenyl) along the coordinate of their interconversion. The isomers may exist if and only if $\Delta E > \hbar\omega$.

Thus when there are several minima of the AP, the nuclear configurations of each of them may be considered as the corresponding molecular shapes if and only if the energy barriers between them are sufficiently high. Otherwise, when $\Delta E < \hbar\omega$, there are no isomers, and the system performs just large-amplitude vibrations involving the foregoing (and intermediate) configurations.

A similar treatment relates to tautomers. It is generally believed that tautomers differ by electronic distribution only. However, according to the adiabatic approximation, any change in electronic structure is associated with alterations in the nuclear configuration. Tautomer transformations are thus also transitions between the AP minima that are, from this point of view, similar to isomer conversion.

The AP minimum positions that determine the molecular shape can be evaluated by means of direct electronic structure calculations with geometry optimization. Many examples of such calculations are given in Sections 6.3 to 6.5 and 11.3. For small to moderate molecular (including coordination) system, the modern state of the art in numerical quantum chemistry allows one to obtain equilibrium configurations that agree fairly well with experimental data. However, as discussed in Section 5.6, any full numerical quantum-chemical calculation, although contributing significantly to the understanding of the origin of molecular shapes, is in fact a computer experiment, and its results belong almost entirely to the specific system that is subject to experimentation. The numerical results are, in general, not directly transferrable to other systems, and hence they cannot serve as a basis for the formulation of general laws of stereochemistry. Therefore, *general qualitative results remain most important to the understanding and prediction of stereochemistry*, along with quantitative electronic structure calculations with geometry optimization.

Directed Valencies, Localized Electron Pairs, and Valence Shell Electron Pair Repulsion

The existing qualitative models in stereochemistry are based on some assumptions about electronic charge distribution that determines the nuclear configuration. The simplest model employs *hybridized atomic orbitals* (Section 2.1). The idea was developed successfully by Pauling [9.8].

According to the assumption of hybridized orbitals, the atomic pure $s, p, d, \ldots$ orbitals under external influence become mixed. The mixed (hybridized) orbitals, dependent on the mixing coefficients, are spatially oriented and, being occupied by one electron, form *directed valencies*. Table 2.4 shows some examples of sp (Fig. 2.8) and spd hybridization with indication of the directions of the valencies. Kimball [9.9] considered the possible spd hybridizations, resulting in various geometries of directed valencies for coordination numbers from 2 to 8. Formally, these assumed hybridizations include almost all the observed geometries of coordination compounds; for reasons given below, we do not discuss them in more detail (most such hybridizations are listed in [9.1]).

As emphasized in Section 2.1, *hybridization of atomic orbitals that makes them spatially oriented does not mean that free atoms possess such directed orbitals: They are formed as a result of the bonding* (coordination). In the absence of ligands the free atom has spherical symmetry for which $s, p, d, \ldots$ orbitals are well separated in energy and orthogonal in space. Under the influence of ligand fields that destroy the spherical symmetry, a specific type of hybridized orbitals with orientations toward the ligands is produced, provided that the energy gained by the better overlap and bonding by hybridized orbitals is larger than the energy lost in the promotion of electrons from lower to higher orbitals required by the hybridization.

The picture of bonding with hybridized orbitals implies that there are localized metal–ligand bonds realized by appropriate electron pairs; that is, there is *localized electron-pair bonding*. As discussed in Section 6.1, this type of bonding is relevant to the *valence bonds* in *valence compounds*. The valence-pair bonding model is most successful in organic chemistry and main-group compounds, but it fails in coordination chemistry. As shown by rigorous analysis, the more general MO presentation of the electronic states may be reduced to localized orbitals if and only if the valence electrons form a closed shell and their number is twice the number of bonds in the system. Neither of these conditions is fulfilled, in general, in transition metal coordination compounds. On the contrary, the latter differ from simple valence compounds just by their open-shell delocalized bonds, which are due to the participation of d electrons (Section 6.1).

Nevertheless, there are systems, especially among coordination compounds of nontransition elements (not very strong d participation, Section 6.1), for which the description with localized electron pairs is approximately valid. In the stereochemistry of these systems, the approach of *valence shell electron pair repulsion* (VSEPR) may be useful. This approach can be traced back to earlier work of Lewis, Sidgwick, and Powel, and Gillespie and Nyholm (see [9.1–9.3]). The main idea is that as far as all the bonds in the system are formed by localized electron pairs, their repulsion determines the molecular shape.

An important additional circumstance is that some of the localized pairs of electrons may be unshared by the ligands; that is, they are *lone pairs* (analogies of nonbonding orbitals in the MO presentation; see below). The lone pairs (first introduced by Sidgwick [9.3]), although not participating in the bonding, are nevertheless important for stereochemistry because they participate actively in the electron pair repulsion. Usually, lone pairs distort the otherwise more symmetrical arrangement of the localized bonds.

The essence of the model is best understood when one arranges schematically all the electron pairs (including the bonding and lone pairs) on the surface of a sphere around the CA (Fig. 9.2) under the condition that the repulsion energy is minimal [9.2]. Then the ligand and lone-pair positions can be found by calculating the minimum energy arrangement of n point charges constrained to move on the surface of the sphere with the CA in the center and subject to a given interaction potential.

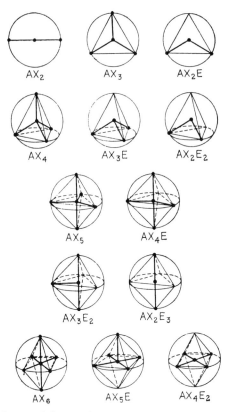

Figure 9.2. Bonding and lone-pair arrangement for different coordination systems AX_nE_n constrained to positions on a spherical surface with the center at the CA (E is the lone pair).

In so doing, not all the electron-pair charges are taken equal. In particular, the following rules are usually employed:

1. A nonbonding pair occupies more space on the surface (its charge is more localized, i.e., less extended in space, and produces stronger repulsion) than bonding pairs.

2. The size of the bonding pairs decreases with the increase in electronegativity of the ligands (due to the same effect of charge extension; localized on the ligands, the electron repulsion decreases due to the increase of the distance between the pairs).

3. Two electron pairs in a double bond, or three electron pairs in a triple bond, occupy more space than the single bond pair.

Some ligand arrangements predicted by this model are shown in Fig. 9.2. As stated above, only limited cases of coordination compounds may be subject to

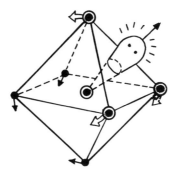

Figure 9.3. Distortion of the octahedral complex XeF_6 produced by the repulsive lone pair in the VSERP model. (From [9.11].)

such treatment. An example of these is the system XeF_6 [9.10, 9.11]. Figure 9.3 shows how the lone pair (which remains after the formation of six Xe—F localized electron-pair bonds) distorts the otherwise octahedral arrangement of six fluorine atoms (it is also assumed that the lone pair is sp hybridized under the ligand field). The lone pair thus occupies a coordination place and the complex becomes seven-coordinated, somewhat similar to IF_7. This situation is subject to more detailed discussion in Section 9.2.

Improved Semiclassical VSEPR Model

We call the approaches of VSEPR type *semiclassical* because they try to avoid microscopic (quantum-chemical) treatment of the electronic structure by means of introducing generalized classical parameters that allow us to reduce the problem of stereochemistry to electrostatic interactions. The simplicity of the VSEPR model, on the one hand, and the complexity of quantum-chemical calculation, on the other hand, explain attempts to improve this model or to work out more sophisticated versions based on the same principles, despite the deficiencies of such approaches, especially for coordination compounds.

One improvement in the VSEPR model has been suggested by Kepert [9.6]. Each metal–ligand bonding pair or lone pair is presented as located at a point in space. The interaction between two electron pairs u_{ij} is taken as a Born repulsion:

$$u_{ij} = \frac{a_{ij}}{r_{ij}^n} \qquad (9.3)$$

where r_{ij} is the distance between the pairs, a_{ij} is a coefficient, and n is an integer, usually between 1 and 12; $n = 6$ is considered as most adequate to the electron density repulsions (in most cases comparative results are given for $n = 1$, 6, and 12). If all the pairs (the bonds) are equivalent, $r_{ij} = r$, the points of the pair location are on a sphere as in the simple VSEPR model, and

$$U = \sum_{ij} u_{ij} = a_n X_n r^{-n} \tag{9.4}$$

where X_n is the coefficient of repulsion that depends on n and the geometry of the coordination polyhedron. For a given n, X_n is obtained from the condition of energy minimization (minimum of U).

However, when there are different ligands and/or there are lone pairs, the interacting electron pairs are no longer equivalent and (9.4) does not hold. To allow for the pair difference, a new parameter $R(i/j)$ equal to the ratio of effective bond lengths is introduced. If two pairs are different, the distances from the CA to the effective point of their location, $r_{\mu i}$ and $r_{\mu j}$, are considered as being different (they are no longer on the same sphere), and

$$R(i/j) = \frac{r_{\mu i}}{r_{\mu j}} \tag{9.5}$$

The coefficients $R(i/j)$ are assumed to be transferrable from one polyhedron to another. Obtained from empirical data, $R(i/j)$ does not exactly follow the electronegativities of the ligands. For instance, for nonmetallic CA's the $R(i/j)$ values, compared with an alkyl ligand with $j = 1$, we have ($n = 6$):

i	Me	Cl	H	F
$R(i/1)$	1.00	1.05	1.09	1.16

For charged ligands X^- bonded to transition metals in moderate oxidation states the effective distance $r_{\mu i}$ is about 20% shorter than for neutral ligands A:

Table 9.1. Coefficients X_n that minimize the repulsion energy [Eq. (9.4)] of N equivalent bonds and/or lone pairs in a coordination system ($n = 6$)

N	X	N	X
2	0.016	8	5.185
3	0.111	9	8.105
4	0.316	10	12.337
5	0.877	11	18.571
6	1.547	12	23.531
7	3.230		

Source: [9.6].

$R(A/X^-) \approx 1.2$ or $R(X^-/A) \approx 0.8$ (for $n = 6$). For higher-charged ligands, say, O^{2-}, $R(O^{2-}/A) \approx 0.9$ for nonmetallic CA's, and $R(O^{2-}/A) \approx 0.7$ for transition metals (all these data for $n = 6$). The lone pairs (l_p) have much smaller distances $r_{\mu 1}$, but the $R(l_p/j)$ values are not exactly transferrable to different complexes; they change with the CA, coordination number, and additional steric effects (mono- or bidentate ligands). For example, for the same l_p and different CAs in the oxidation state $3+$ for $n = 6$:

CA	N	P	As	Sb
$R(l_p /Cl^-)$	0.90	0.59	0.53	0.47

With increase in the coordination number the value $R(l_p/X^-)$ decreases, and it may increase when passing from monodentate to polydentate ligands:

System	$[SbCl_6]^{3-}$	$[Sb(S_2COEt)_3]$
$R(l_p/j)$	0.0	0.15

The coefficients X_n in Eq. (9.4), which minimize the energy U for different coordination numbers N and $n = 6$, are given in Table 9.1. Note that the function $X_6 = f(N)$ is not monotonous: The regular tetrahedron, octahedron, icosahedron are by 10 to 20% more stable than expected for a smooth function. When not all the pairs are equivalent, more than one coefficient X_n should be introduced.

With all the parameters discussed above, some of them being taken from empirical data, the evaluation of molecular shape is reduced to the minimization of a function of several variational parameters (including the r_{ij} values) which determine the geometry of coordination. The problem was considered for many coordination systems [9.6]. By way of example, the scheme for five-coordinated complexes of the type $[MX_5]^{n-}$ is shown in Fig. 9.4.

The most regular polyhedron for this system is either a trigonal bipyramid (TBP) with $\phi_A = 90°$ and $\phi_B = 120°$, or a square pyramid (SP) with $\phi_A = \phi_B$. Calculations [9.6] of the energy [Eq. (9.4)] as a function of ϕ_A and ϕ_B with the parameters discussed above ($n = 6$ and the ratio $X_{ax}/X_{eq} = 1.21$ for the axial and equatorial ligands, respectively), show that there are two minima corresponding to the TBP configurations with $\phi_A = 90°$, $\phi_B = 120°$ and $\phi_A = 120°$, $\phi_B = 90°$, respectively, but the energy barrier between them is negligible. Thus the system can easily convert from one TBP configuration through the SP one to another TBP configuration (*Berry pseudorotation* [9.12]), and all these configurations (including the intermediates along a specific pathway) are equivalent.

In fact, crystal structure data on various systems containing MX_5 polyhedra show a large variety of mainly fixed angles ϕ_A and ϕ_B from 90° to 120° [9.6],

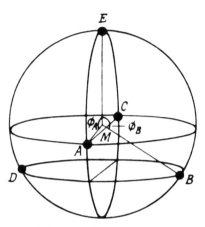

Figure 9.4. Configuration of five-coordinated complex $[MX_5]^{n-}$ of C_{2v} symmetry described by two angles ϕ_A and ϕ_B. (From [9.6].)

and this obviously results from the plasticity effect discussed in Section 9.4. The electronic structure of the MX_5^{n-} systems and electronic control of their nuclear configuration and dynamics, as well as the possible pseudorotation in such systems, is discussed in Section 9.2. In particular, it is shown that the pseudorotation in MX_5^{n-} systems does not follow the Berry rotations scheme of TBP $\rightarrow$ SP $\rightarrow$ TBP interconversion.

Nonbonding Orbitals and Nodal Properties

As seen from the discussion of the VSEPR model above, lone pairs of electrons are important for the geometry of coordination because they occupy a significant region of repulsion (Fig. 9.2), which can be even larger than that for metal–ligand bonds. Each lone pair thus acts stereochemically as a virtual additional ligand, essentially distorting the otherwise more symmetrical polyhedron. However, in practice, the picture may be more complicated since not all lone pairs are stereochemically active; this circumstance is discussed in more detail in Section 9.2. Nevertheless, the problem of how many and what kind of lone pairs are expected in a given coordination compounds is of significant interest.

In a more rigorous MO LCAO formulation, lone pairs, that is, pairs of electrons unshared by the ligands (or unshared by the CA for ligand lone pairs), are formed by two electrons that occupy a nonbonding MO (unshared by ligands means nonbonding). It can be occupied by one electron only, producing stereochemical influence that is qualitatively similar to the lone pair. Hence in the MO description the lone-pair effect can be produced by single electrons.

The possible number of nonbonding orbitals in coordination compounds of the type ML_n and their role in stereochemistry is discussed in [9.13]. It is assumed that the Coulomb integrals $\alpha_i = H_{ii}$ [Eq. (5.4)] for the s, p, d electrons are equal to each other: $\alpha_s = \alpha_p = \alpha_d$, but the results remain valid qualitatively when they are different. An interesting m–n rule was formulated: The number of nonbonding orbitals in a ML_n coordination system equals $|m - n|$, where m is the number of valence orbitals of M, provided that only σ bonds are considered.

The most important feature of nonbonding orbitals in stereochemistry is similar to that of lone pairs in the VSEPR model, but it is formulated more precisely: *The ligands are always located on the nodal lines, planes, or cones of occupied nonbonding orbitals.* Obviously, this orbital effect is due to the same electron repulsion of nonbonding electrons as in the VSEPR model, but it gives more accurate indication of the possible ligand geometry; it is determined by the nodal properties of the nonbonding orbitals, which can be established relatively easily. Table 9.2 gives some information about nonbonding orbitals in different coordination systems of transition elements which participate in coordination with $m = 9$ (one s, three p, and five d) orbitals.

Different types of nonbonding orbitals influence differently the geometry of coordination. First, there are pure CA nonbonding orbitals. With respect to them, the ligands are located either on their nodal planes (as in BH_3 where the hydrogens are in the nodal plane of the p_z orbital; Fig. 2.2), or in the nodal cones formed by, say, the nonbonding d_{xy}, d_{xz}, d_{yz} orbitals in octahedral complexes ML_6 (Fig. 2.3).

Table 9.2. Nonbonding orbitals of the CA in ML_n complexes with nine AOs from the CA and one σ orbital from each ligand

Geometry	p	d	Hybridized
Linear ML_2	p_x, p_y	$d_{x^2-y^2}$, d_{xy}, d_{xz}, d_{yz}	$s - \lambda d_{z^2}$
Trigonal ML_3	p_z	d_{xz}, d_{yz}	$s + \lambda d_{z^2}$, $d_{x^2-y^2} + \lambda_2 p_x$, $d_{xy} + \lambda_3 p_y$
Square-planar ML_4	p_z	d_{xy}, d_{xz}, d_{yz}	$s + \lambda d_{z^2}$
Trigonal bipyramidal ML_5	—	d_{xz}, d_{yz}	$d_{x^2-y^2} + \lambda p_x$, $d_{xy} + \lambda p_y$
Octahedral ML_6	—	d_{xy}, d_{xz}, d_{yz}	—
Trigonal prismatic ML_6	—	—	$s + \lambda_1 d_{z^2}$, $d_{x^2-y^2} + \lambda_2 p_x$, $d_{xy} + \lambda_3 p_y$
Pentagonal bipyramidal ML_7	—	d_{xz}, d_{yz}	—
Square antiprismatic ML_8	—	—	$s + \lambda d_{z^2}$
Dodecahedral ML_8	—	$d_{x^2-y^2}$	—

Source: [9.13].

Nonbonding hybridized orbitals *sp*, *sd*, *dp*,... are of another type. Mainly due to these orbitals, the system avoids higher symmetries in C_{nv} groups. For example, SF_4 has C_{2v} symmetry instead of C_{4v}, D_{4h}, or T_d. In many cases this effect can be presented as resulting from the tendency to avoid occupying metal–ligand strong antibonding orbitals instead of the nonbonding states at lower symmetries: The additional repulsions in the distorted system may be smaller than the energy loss by occupation of antibonding MOs. For instance, in SH_4 two electrons prefer the nonbonding p_x–d_{xy} orbital resulting in C_{2v} symmetry to the antibonding a_1 MO in tetrahedral T_d symmetry; CH_4 is exactly tetrahedral since the critical two antibonding electrons are lacking.

Another example of this type is provided by some sandwich compounds. In the high-symmetry configurations (e.g., D_{5h} or D_{6h}) overlap of the CA *d* orbitals with the ring π MOs of type $d_{xz} - e_1^{\pi}(x)$ produces occupied antibonding orbitals that can be reduced by a simple "slip" distortion (Fig. 9.5). Again, this effect is realized when the antibonding nature of the MOs under consideration is sufficiently strong. Therefore, the first-row metallocenes, for instance, $Co(\eta\text{-}C_5H_5)_2$ and $Ni(\eta\text{-}C_5H_5)_2$, are not distorted [9.14], whereas the second- and third-row sandwich compounds, where the overlap and hence antibonding character of the corresponding MOs are much larger [e.g., $Ru(\eta^6\text{-}C_6Me_6)$ $(\eta^4\text{-}C_6Me_6)$], undergo a slip distortion as shown in Fig. 9.5 [9.15].

There are many other systems where the corresponding MO in the high-symmetry configuration is not sufficiently strong antibonding, and these systems are not distorted, despite the presence of two additional electrons that might be considered as a lone pair. For example, the complexes SbX_6^{3-}, X = Cl, Br, I, MX_6^{2-}, M = Se, Te, have seven electron pairs but remain octahedral, while XeF_8^{2-} has nine electron pairs but is a square antiprism. In these cases it is said that the lone pairs are stereochemically inactive (inert). A

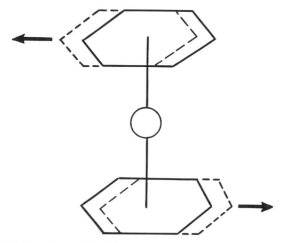

Figure 9.5. Slip distortion of the high-symmetry D_{6h} configuration of sandwich compounds reducing the AO overlap and antibonding character of the $d_{xz} - e_1^{\pi}(x)$ MO.

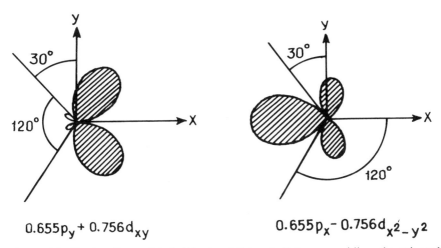

$$0.655p_y + 0.756d_{xy}$$ $$0.655p_x - 0.756d_{x^2-y^2}$$

Figure 9.6. Nonbonding *pd* hybridized orbitals of *E* type providing the trigonal arrangement of the ligands along the nodal lines in trigonal planar and trigonal bipyramidal systems.

more rigorous treatment of these effects, based on the pseudo Jahn–Teller effect, is given in Section 9.2.

Many of the nonbonding orbitals and their nodal properties can be predicted by group-theoretical treatment [9.13]. For example, this predicts that the two *dp* hybridized nonbonding orbitals in the trigonal planar and trigonal bipyramidal configurations may form a coupled *E* representation. Under this condition, calculations yield the coefficients of hybridization: $0.655p_x - 0.756d_{x^2-y^2}$ and $0.655p_y + 0.756d_{xy}$. As shown in Fig. 9.6, their nodal lines provide the necessary trigonal arrangement of the ligands. A related approach to stereochemistry with applications to polyatomic cluster compounds is given in the next subsection.

Complementary Spherical Electron Density Model

A stereochemical model quite similar in spirit to those discussed above in this section has been suggested by Mingos [9.16–9.18]. The essence of the model is as follows. Consider an ML_n complex as a CA surrounded by a sphere of ligand electron densities that are localized into distinct regions of the sphere. In the MO LCAO approximation (Section 5.1) the AOs are taken in appropriate symmetry-adapted linear combinations, separately for the CA and ligands (Table 5.1), and then the linear combinations of the functions of the same symmetry from both parts are combined by the corresponding LCAO coefficients into a MO (5.9). While the AOs ψ_0 of the CA are well-known spherical harmonics $Y_m(\theta, \phi)$ [see Eq.(2.2)] that can be classified as *s*, *p*, *d*,... functions (Section 2.1), Mingos called attention to the fact that the linear combinations of ligand functions Φ can also be presented as spherical harmonics, provided that *the radial distribution is considered the same for all the ligands*, namely, as concentrated on the same sphere with M in the center. This is the most limiting assumption of the model.

Under this assumption the AOs of both the CA and ligands are presented as spherical harmonics Y_{lm} with the same quantum numbers l (azimuthal, $l = 0, 1, 2,\ldots$) and m (magnetic, $m = 0, \pm 1, \pm 2,\ldots$). To deal with real functions, the degenerate states with $m > 0$ are taken in the corresponding real linear combinations, as shown in Section 2.1: $Y_{l,m\pm} = c^{\pm}(Y_{l,m} \pm Y_{l,m-})$.

For example, in MH_4 with spherically distributed hydrogen atoms about M, the corresponding linear combinations of the hydrogen σ function forming spherical harmonics can be presented as follows:

$$\Phi_{lm} = \sum_{i}^{n} c_i \sigma_i = N \sum_{i} Y_{l,m}(\theta_i, \phi_i)\sigma_i \qquad (9.6)$$

where N is the normalization factor and θ_i and ϕ_i are the angular coordinates of the ith ligand on the sphere. For $l = 0$ and $m = 0$, Y_{00} is independent of θ and ϕ, and hence Φ_{00} is the sum of σ_i. If $l = 1$, there are three functions, $\Phi_{1,0}$, $\Phi_{1,1+}$, and $\Phi_{1,1-}$, which can easily be found from (9.6) by substituting the corresponding ligand coordinates. To distinguish between the harmonics Φ_{lm} and that of the CA, the notations S, P, D, and so on, are introduced. They mean that the Φ_{00} function above is S^{σ} (the superscript σ indicates the assumed σ functions of the hydrogen atoms), while Φ_{10}, $\Phi_{1,1+}$, and $\Phi_{1,1-}$ mean P_0^{σ}, P_+^{σ}, and P_-^{σ}, respectively. There are also five D functions, D_0^{σ}, D_{1+}^{σ}, D_{1-}^{σ}, D_{2+}^{σ}, and D_{2-}^{σ}; seven F^{σ} functions; and so on.

The ligand spherical harmonics (9.6) can be found relatively easily, provided that the ligand geometry (the θ_i, and ϕ_i coordinates) is known. In the spherical electron density model under discussion it is assumed that the ligand arrangement should form either the best covering, or the best packing polyhedron on the sphere, or both. The two types of polyhedrons are found as solutions of the following problems [9.18]:

1. *Covering problem:* "If n oil supply depots are available on the surface of the sphere, what is their best arrangement to give the most efficient utilization of oil resources?"

2. *Packing problem:* "If n inimical dictators control the planet, how could they be located on the surface of the sphere so as to maximize the distances between them?"

The results of the solution of covering and packing problems for some most usable coordination numbers n are listed in Table 9.3 (capped polyhedron means a regular polyhedron with an additional capping atom located on a face).

Table 9.3. Best covering and packing polyhedrons ML$_n$ in the spherical ligand electron density model

n	Best Covering Polyhedron	Best Packing Polyhedron
3	Triangle	Triangle
4	Tetrahedron	Tetrahedron
5	Trigonal bipyramid	Square pyramid
6	Octahedron	Octahedron
7	Pentagonal bipyramid	Capped octahedron
8	Dodecahedron	Square antiprism
9	Tricapped trigonal prism	Capped square antiprism

Source: [9.17].

If the geometry of the complex is known, the ligand spherical harmonics can be found directly from Eq. (9.6). For planar complexes with $n = 2, 3, 4$ they are illustrated in Fig. 9.7, and for larger numbers of ligands they are given in [9.17].

One of the most interesting applications of the spherical density model is in the treatment of the *inert gas rule*. This rule states that *coordination systems and main-group molecules with the valence electron configuration of the appropriate inert gas* (i.e., the configurations with 8, 18, 32, etc., valence electrons) *are most*

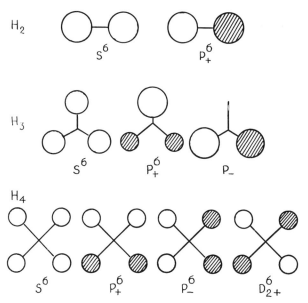

Figure 9.7. Ligand spherical harmonics for planar complexes MH$_n$ with $n = 2, 3, 4$. Open and shaded circles represent the hydrogen 1s distributions taken with positive and negative signs, respectively.

stable. The inert gas rule is widely used in discussions of qualitative molecular structure, but little attention has been paid to its theoretical foundation.

Consider first a hypothetical system in which the ligands form an exact spherical distribution of charge about the CA. It can be shown that in this instance the ligand states, that is, the solutions for a charged ball that has spherical symmetry, are also spherical harmonics $\Phi_{nlm}(\theta, \phi)$ of the type S, P, D, with the same symmetry and nodal properties as s, p, d and with the energy consequence $S < P < D, \ldots$. If there is a considerable overlap of the CA functions with the ligand-sphere functions, appropriate bonding MOs of the type (5.9) or (5.32) are formed:

$$\Psi_{lm} = N_{lm}(\psi_{lm} + \lambda\Phi_{lm}) \tag{9.7}$$

Due to the exact spherical symmetry of the system, the functions ψ_{lm} and $\Phi_{l'm'}$ with different quantum numbers are orthogonal and do not form MOs. Hence the number of bonding MOs equals the number of active valence orbitals of the CA. For example, if the CA has valence s, p_x, p_y, and p_z orbitals, they overlap in pairs with S, P_0, P_+, P_- spherical shell functions to form four bonding MOs that can accommodate eight valence electrons. If the five d electron states are also involved, there are nine bonding MOs with 18 electrons (with f states added there will be 16 bonding MOs with 32 electrons).

If the number of electrons is less than 8, 18, or 32, respectively, not all the bonding MOs are occupied, and the stability of the system is expected to decrease. If the number of electrons is larger, the stability of the system, again, decreases since all the bonding MOs are occupied and the excess electrons occupy antibonding MOs. In sp systems the electrons in excess of eight could, in principle, occupy D states, but they are much higher in energy than S and P. Thus the numbers of electrons of the inert gas are indeed optimal for the stability of the spherical-shell system. This result is hardly surprising in view of the assumed spherical symmetry of the system. The complex of the CA with a spherical-shell charge distribution of ligands is just an extended atom that has qualitatively the same atomic distributions as the free atom. Therefore, in this system the inert gas closed-shell stability rule, well known in free atoms, is obeyed, and for the same reasons as for free atoms: In the dependence of the energy on the hydrogenlike quantum numbers, configurations with $\sum_{l=0}^{l=l^*} 2(2l + 1)$ ($l^* = 0, 1, 2, \ldots$) valence electrons are most stable.

In the spherical density model described above, the ligand charges occupy distinct places on the same sphere, and hence their charge distribution is not spherically symmetric. It means that the orthogonality of the CA and ligand spherical harmonics with different quantum number ψ_{lm} and $\Phi_{l'm'}$ does not hold, and the MO scheme, employed above, should be reconsidered. However, if the nonorthogonality arising due to the nonuniform (not exact spherical) distribution of the ligand charges on the sphere is not very large, one can assume that the qualitative features of the spherical charge distribution above are approximately valid, and this approximation is better the more spherical the ligand arrangement about the central atom. Thus the inert gas rule is expected to hold approximately, and it is more acceptable for a larger number of and more identical ligands.

A classical example of the inert gas rule is provided by the system $[ReH_9]^{2-}$. In the tricapped trigonal prism configuration of the best covering polyhedron (Table 9.3) there are nine ligand spherical harmonics (Table 9.4), which exactly fit the nine CA functions forming nine bonding MOs with an 18-electron closed shell (provided that the above-mentioned nonorthogonality is neglected). A ligand arrangement in a capped square antiprism also generates a complete set of S^σ, P^σ, and D^σ functions, providing an alternative stereochemistry that satisfies the requirement of inert gas formulation (other polyhedrons are nonfit). This means that the configuration of the complex may be flexible within these two polyhedra, and this result is consistent with the stereochemical nonrigidity of $[ReH_9]^{2-}$ observed [9.19]. X_α calculations [9.20] of this system show that the radial distribution of the hydrogens around the CA emulates that of an inert gas.

In the complementary spherical electron density model under consideration this inert gas rule is considerably extended to what can be called *a generalized inert gas rule* [9.17, 9.18]. If the coordination number n is less than 9, the number of ligand spherical harmonics is reduced, and the D^σ functions are successively lost (their number becoming $n - 4$). As a result, the pseudospherical electron distribution assumed for the occupied one S, three P, and five D functions is also lost because of the electron-pair holes in the D^σ shell. However, if the missing D^σ states are compensated for by matching d states occupied by the electrons of the CA, a more spherical electron distribution is attained. For example, a dodecahedral MH_8 system misses one ligand spherical harmonic D_{2+}^σ (Table 9.4). The remaining eight bonding orbitals with 16 electrons do not fit the inert gas rule, which requires 18 bonding electrons. However, if the $d_{x^2-y^2}$ AO of the CA M that approximately matches the D_{2+} orbital by space distribution is occupied by two electrons, spherical symmetry is regained and the generalized inert gas rule of stability is obeyed.

In MH_8 a square-antiprism arrangement of the hydrogens misses the D_0^σ ligand function, which can be compensated by the occupied CA d_{z^2} orbital. Other arrangements, such as a cube or hexagonal bipyramid, require f functions for the compensation which are not available. Hence the generalized inert gas rule serves as a tool for stereochemistry treatments. Table 9.4 presents other examples illustrating the efficiency of the generalized inert gas rule.

To summarize, the complementary spherical electron density model shows that the inert gas rule is valid when there is a complementary interaction between the ligand and central atom electronic states in which the CA nonbonding d electrons compensate the missing ligands, resulting in pseudo-spherical electron distribution. Those polyhedra that give the best covering and/or packing of the ligands on the sphere are most effective in emulating the required spherical distribution by generating the corresponding set of contiguous S^σ, P^σ, D^σ, ... functions. The nonbonding d orbitals of the CA, which

Table 9.4. Illustration of the generalized inert gas rule: The ligand spherical harmonics of the ML_n complex, together with the indicated compensating d orbitals of the CA, form a complete set for an 18-electron closed-shell configuration

n	Structure	S	P_0	P_+	P_-	D_0	D_{1+}	D_{1-}	D_{2+}	D_{2-}	Examples
9	Tricapped trigonal prism	+	+	+	+	+	+	+	+	+	$[ReH_9]^{2-}$
8	Square antiprism	+	+	+	+	d_{z^2}	+	+	+	+	$H_4[W(CN)_8]$
	Dodecahedron	+	+	+	+	+	+	+	$d_{x^2-y^2}$	+	$K_4[W(CN)_8]$, $[MoH_4(PMe_2Ph)_4]$
7	Pentagonal bipyramid	+	+	+	+	+	d_{xz}	d_{yz}	+	d_{xy}	$[OsH_4(PMe_2Ph)_3]$
	Capped octahedron	+	+	+	+	+	+	+	$d_{x^2-y^2}$	d_{xy}	$[W(CO)_4Br_3]$
	Capped trigonal prism	+	+	+	+	+	+	d_{yz}	$d_{x^2-y^2}$	+	$[Mo(CNbut)_7]^{2+}$
6	Octahedron	+	+	+	+	+	d_{xz}	d_{yz}	+	d_{xy}	$[Mo(CO)_6]$
5	Trigonal bipyramid	+	+	+	+	+	d_{xy}	d_{yz}	+	d_{xy}	$[Fe(CO)_5]$
	Square pyramid	+	+	+	+	d_{z^2}	d_{xz}	d_{yz}	+	d_{xy}	$[Ni(CN)_5]^{3-}$
4	Tetrahedron	+	+	+	+	d_{z^2}	d_{xz}	d_{yz}	$d_{x^2-y^2}$	d_{xy}	$[Ni(CO)_4]$

Source: [9.17].

compensate for the missing $9 - n$ ligands (ligand spherical harmonics), can experience additional stabilization through π-bonding effects [9.18].

Other examples and extensions of the complementary spherical electron density model, in particular to cluster compounds, can be found in the literature [9.16–9.18].

Molecular Modeling

An even more classical approach to the problem of stereochemistry of transition metal compounds is possible based on *molecular mechanics* [9.21– 9.23], although it has limited applicability to transition metal compounds. The method of molecular mechanics (MM) is at present widely employed in *molecular modeling*, especially in conformational analysis of organic compounds, including biological systems. In its modern versions (computer programs MM2 and MM3), and used within the limits of validity, MM proved to be very useful in various applications to chemical and biological systems [9.24–9.29].

However, in application to coordination compounds, this method raises serious questions [9.30, 9.31]. *The main assumption of the MM method is that there are well-defined parameters of the bonding between given pairs of atoms that are transferrable from one molecular system to another.* This assumption is approximately valid when organic and related systems are considered. In coordination compounds, as emphasized repeatedly in this book (e.g., Sections 1.2 and 6.1) the metal–ligand bond is, in general, neither localized nor specified: it may depend strongly on the bonds formed by this metal with other ligands. For instance, as seen from Table 9.14, the Cu—O bond length varies from 1.8 Å through 3.0 Å, depending on the other Cu—ligand bonds. Therefore, strictly speaking, there are no fixed transferrable parameters for the metal–ligand bonds to be employed in the molecular force field of MM.

Another implication emerges from the charge redistribution in the ligands by coordination to a transition metal center discussed in Section 6.3, which makes the former electronically excited (the ligand's ground-state HOMO is depopulated, while the excited MOs become populated due to *back donation* from the metal). Therefore, the ligands in transition metal systems *should be modeled in their partially excited state*, the measure of excitation being determined by the coordination center.

Attempts to use the MM method for transition metal systems similar to the way it is used for organic compounds [9.24–9.29], may have some validity in specific cases that allow for sidestepping the foregoing difficulties. For instance, for quite similar coordination centers with the same first coordination sphere and weak influence from the following coordination spheres, we can assume that the force field constants for the same metal—ligand bonds are approximately the same. For example, the Co(III)—N bond parameters in all octahedral complexes with six almost equivalent Co(III)—N bonds which differ slightly in the second and following coordination spheres can be considered approximately transferrable within this set of systems.

The purpose of the MM method is to represent the total energy of the molecular system as a function of its geometry given by interatomic distances and angles between the bonds, and to find the equilibrium geometry from the condition of energy minimum. Unlike quantum-chemical methods in which the total energy is obtained from electronic structure calculations, in the method of MM the total energy is computed as a sum of bond contributions calculated in a classical way by means of empirical parameters. The total energy is presented as follows:

$$E_{total} = E_s + E_b + E_{tors} + E_{vdw} + E_{elec} \tag{9.8}$$

where E_s is the energy of bond stretching, E_b the angle bending energy, E_{tors} the energy of torsional distortion, E_{vdw} van der Waals interaction of non-bonded atoms, and E_{elec} their electrostatic interaction energy. In Fig. 9.8 we give the notation for the coordinates used in the definition of the energies (9.8). The potential of stretching energy (of deformation along the bond) is given by a harmonic term, a parabola with a minimum at the equilibrium (unstrained) interatomic distance R^0:

$$E_s(R_{ij}) = \tfrac{1}{2}K_{ij}(R_{ij} - R_{ij}^0)^2 \tag{9.9}$$

where K_{ij} is the stretching force constant of the i—j bond. Since (9.9) does not include anharmonicity, the strain energy due to bond length deformation is overestimated.

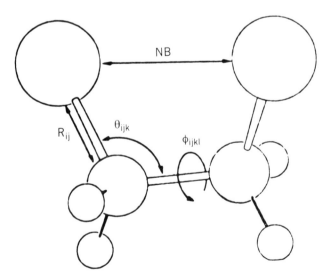

Figure 9.8. Choice of coordinates of intramolecular interactions determining the geometry of molecules in molecular mechanics (NB denotes nonbonded atoms).

Table 9.5. Force field constants K_{ij} and R_{ij}^0 for bond length deformation $E_s(R_{ij}) = \frac{1}{2}K_{ij}(R_{ij} - R_{ij}^0)^2$ in some coordination compounds

Bond Type		K_{ij}	R_{ij}^0
i	j	(mdyn·Å^{-1})	(Å)
Co(III)	N	2.25	1.95
Co(III)	O	2.25	1.93
Co(III)	Cl	1.68	2.30
Ni(II)	N	0.68	2.10
Cu(II)	S	0.50	2.38
Cu(II)	N=	0.89	2.00
Cu(II)	O	0.89	2.00

Source: [9.24].

Equation (9.9) contains two constants, K_{ij} and R_{ij}^0 (sometimes called force field constants), which in the MM method are considered identical for all bonds i—j, taken equal to some values for the free, unstrained bond (unaffected by other bonds). As stated above, in general there are no such constants for specific metal–ligand M—L bonds which could be used for any complex independent of other M—L' bonds formed by the metal M. Nevertheless, in particular cases of quite similar conditions (the same types of bonds in the first coordination sphere), and used with proper reserve and limitations, such constants can be useful. Table 9.5 provides some force field constants for metal–nitrogen interactions as well as some other constants for comparison.

For the bending energy potential, a similar quadratic dependence on the bending angle θ_{ijk} between the two bonds i—j and j—k (Fig. 9.8) is suggested:

$$E_b(\theta_{ijk}) = \frac{1}{2}K_{ijk}(\theta_{ijk} - \theta_{ijk}^0)^2 \qquad (9.10)$$

In a similar way the other terms in (9.8) are presented by classical interaction formulas with constants to be obtained from empirical data [9.24–9.29].

An example illustrating the possibilities of the MM method in the limits of its applicability is demonstrated by the results of the calculations for some macrocycle complexes [9.25]. Figure 9.9 shows the three most probable conformations of transition metal tetraaza macrocycle complexes, and in Fig. 9.10 the total strain energy is given calculated by Eq. (9.8) for two conformers, trans-I and trans-III, of $[M(12\text{-aneN}_4)]^{n+}$ as a function of the M—N bond length. The nature of the metal M is not taken into consideration since it is shown that the results are not sensitive to the assumed M—N force constant, provided that it is sufficiently weak.

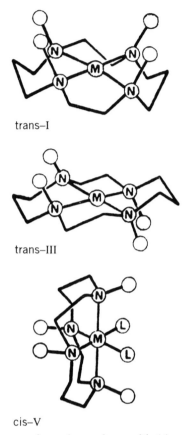

trans–I

trans–III

cis–V

Figure 9.9. Three conformers of metal complexes with 14-aneN$_4$ tetraaza macrocycles. The open circles are hydrogen atoms, while L is another ligand. (After [9.25].)

From the results of MM calculations presented in Fig. 9.10, several conclusions emerge. First, the energy of the *trans*-III conformation with the 12-aneN$_4$ macrocycle is much higher than that of *trans*-I, hence the former is not likely to happen. Second, the best-fit M—N bond length for the more stable conformer is 2.11 Å. Finally, the curvature of the AP at the minimum of the *trans*-I conformer is much smaller than that of *trans*-III, meaning that the former is much more tolerant of the variation in metal ion size.

The relatively good results obtained in some of the MM calculations of transition metal systems show that they are based on a reasonable side-stepping of the difficulties of the MM method, mentioned above. This could be done, in particular, due to the homoligand complexes considered. Attempts to consider heteroligand systems in a similar way, that is, to use the same constants of the force field (9.8) for a M—L$_1$ bond in different systems in which other M—L$_n$ bonds are different, have been shown to be ungrounded [9.30, 9.31]. This statement can be illustrated by a series of specific effects

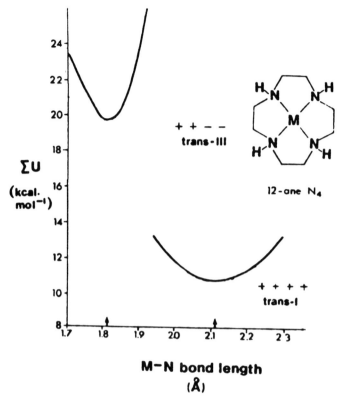

Figure 9.10. Total strain energy calculated by molecular mechanics as a function of the M—N bond length for two conformers of $[M(12\text{-aneN}_4)]^{n+}$, *trans*-I and *trans*-III. The arrows indicate the best-fit M—N sizes for *trans*-III (1.82 Å) and *trans*-I (2.11 Å), their energies at the minima being significantly different. (From [9.25].)

important to transition metal systems but not evolved in the MM method. For example, the well-known effects of mutual influence of ligands (*cis* and *trans* effects; Section 9.3) indicate directly that the local metal–ligand bond in these systems is not specific and cannot be characterized by the same parameters when other bonds are different. Similarly, the Jahn–Teller and pseudo Jahn–Teller effects (Sections 7.3 and 7.4) — for example, the off-center position of the CA (Section 9.2), the plasticity effect (Section 9.4), and so on — are completely beyond the possibilities of the MM methods (note the above-mentioned diversity in Cu—L bond lengths illustrated in Tables 9.14 to 9.17, which obviously cannot be described by fixed parameters).

A combined quantum mechanical/molecular mechanics (QM/MM) method of modeling transition metal compounds, including organometallics and metallobiochemical systems, was suggested recently [9.30, 9.31]. It is devoid of the failures of the MM methods mentioned above, but preserves the simplicity and visualization achieved in modeling organic compounds. The idea is to divide the system into two (or more) fragments, one of which contains the metal center, the others being the remaining organic part. The geometry of the metal fragment can be optimized using semiempirical methods (Section 5.5), while the configuration of the huge organic part can be evaluated by MM methods. Then the use of a special *electronically transparent interface* between QM and MM, discussed in Section 5.4, allows for a self-consistent description of the system as a whole.

The advantage of this fragmentary approach with QM for the metal center and classical description (MM) for the organic ligands is that in the more explicit treatment of the former, the special effect produced by the *d* electrons (including nontransferrability, mutual influence of ligands, Jahn–Teller effects, etc.) are taken into account while the MM treatment of the ligands preserves the simplicity achieved in modeling organic compounds. A full QM optimization of the entire system may be beyond the reasonable possibilities of modern computers and programs: *ab initio* methods are unavailable for large molecules, while semiempirical methods, although affordable for larger (but not very large) systems, are not sufficiently accurate in treating intramolecular nonbonding interactions. In general, when passing from *micro* description to *macro* properties, an interface between quantum description and classical treatment may be inevitable.

Modeling transition metal compounds may serve as a basis for more sophisticated methods of computer screening and design of organometallic compounds with required properties, especially biological activities and *drug design*. This can be done by employing the idea of *structure–activity relationships* (SAR) and *quantitative SAR* (QSAR) [9.32–9.34]. The SAR methods are based on comparison of molecular features of a series of systems with their activities, which reveals the specific features responsible for the activity under consideration. Obviously, the success of these methods depends on the accuracy of the molecular features described as being most appropriate to the activity. Hence appropriate molecular description (modeling) is the main problem of SAR, and there are many attempts to solve this problem, mostly for organic compounds [9.32–9.37].

9.2. VIBRONIC STEREOCHEMISTRY

Nuclear Motion Effects: Relativity to the Means of Observation and Vibronic Amplification of Distortions

From the point of view of quantum mechanics, the methods used in Section 9.1 are semiclassical in the sense that the electron distribution is considered as resulting from approximate (even qualitative) quantum-chemical treatments, while the interaction between localized bonds or lone pairs (nonbonding states) is taken into account as a pure electrostatic repulsion. In so doing, many quantum features of the phenomenon, in particular, vibronic coupling effects, are omitted.

A more rigorous formulation of the problem of expected molecular shapes is given in Chapter 7 based on the vibronic interaction theory. In the adiabatic approximation (Section 7.1) the stable nuclear configuration corresponds to the absolute minimum of the AP. The latter is influenced directly by vibronic coupling between different electronic states (ground and excited). Therefore, the vibronic coupling theory enables us to formulate some general stereochemistry rules, thus serving as an analytical model for the theory of stereochemistry (*vibronic stereochemistry*).

The theory of vibronic coupling is in essence a perturbational approach, and therefore the vibronic effects are of utmost significance in the behavior of the AP surface at the extrema points and points of degeneracy. First, as stated by the Jahn–Teller theorem (Section 7.3), at points of electronic degeneracy, that is, at the nuclear configurations for which the electronic state is degenerate, the adiabatic potential has no minimum. A similar situation emerges in the pseudo Jahn–Teller effect. Following the oversimplified formulation, the lack of minimum of the AP means that the system is unstable in this configuration. In general, this conclusion is not exactly true (Section 7.3).

The lack of minimum of the AP at the point of electronic degeneracy or pseudodegeneracy usually means that there are two or several (or an infinite number of) equivalent minima, at each of which the system is distorted along one of the equivalent symmetry axes. In general, this determines a special type of coupled electronic and nuclear motions. If the energy barriers between the minima of the AP are sufficiently high, the distorted configurations of the system may be observable, provided that the condition (9.1), $\tau > \tau'$, for the means of measurement is satisfied.

The time of measurement τ' is different for different physical methods of measurement. Usually, τ' is inverse proportional to the frequency of external perturbations in the resonance methods of investigation (Sections 8.1 to 8.5). Hence τ' can be quite different in different methods of investigation, whereas τ, the lifetime of the system in one of the equivalent minima, is constant for the given system. It means that for some of the measurements the condition (9.1) is obeyed, $\tau' < \tau$, whereas for others, $\tau' > \tau$. In other words, in a free Jahn–Teller or pseudo Jahn–Teller system with strong vibronic coupling, *the observable nuclear configurations depend on the means of observation and may vary by changing the method of measurement* [9.38].

The relevance of this *relativity rule* concerning the means of observation of molecular shapes is in fact rather widespread, but is sometimes not given due attention. Note that in this rule the distorted configuration is expected to be seen in measurement with higher frequencies ($\tau' < \tau$). This means that we can observe the distorted configuration, say, in optical experiments, and the undistorted configuration (averaged over all the equivalent minima) in the ESR spectra, but *not vice versa*. Sometimes this rule allows us to understand the origin of contradictory empirical data obtained from different experiments. Many examples of its application can be found in literature [9.38].

The situation changes significantly when there are small perturbations slightly lowering the symmetry of the system (differences in the ligands or ligand substituents, or small external fields, including small crystal fields). If these perturbations are sufficiently strong to produce a distortion larger than the Jahn–Teller distortion, they remove the Jahn–Teller effect as such. For smaller perturbations the Jahn–Teller effect is not removed but is modified, with interesting consequences for stereochemistry.

Consider the Jahn–Teller $E-e$ problem (a doubly degenerate electronic E term interacting with e vibrations, Section 7.3) in the linear approximation, that is, a system with an AP in the form of a hat (Fig. 7.9). For the free system the averaged picture displayed in the experiment is an undistorted nuclear configuration. Under the influence of a small distorting perturbation, say, elongating in the Q_θ direction, the circular trough becomes distorted; namely, an additional potential well in the Q_θ direction (and a hump in the opposite $-Q_\theta$ direction) appears. If the depth of the well is greater than the kinetic energy of the circular motion in the trough, the nuclear motions are localized in this well and the corresponding distorted nuclear configuration can be observed in the experiments.

The most exciting result in this picture is that *the magnitude of the distortion is determined mainly by vibronic effects and is almost independent of the perturbation magnitude* [9.38]. Indeed, the additional well is formed at a point of the circular trough with coordinates (ρ_0, ϕ_0), where only ϕ_0 depends significantly on the perturbation. The magnitude of the distortion determined by ρ_0 is almost independent of the perturbation (in fact, the perturbation slightly distorts the trough circle, too; see below). This ρ_0-value distortion is often called the *static limit* of the Jahn–Teller effect.

Thus a small perturbation W acting on a Jahn–Teller system produces a perturbation that is determined by the static limit ρ_0 of the Jahn–Teller effect and stabilizes the static distortions. Since the kinetic energy of the motion along the trough $E_k = \hbar^2/8M\rho_0^2$ is on the order of several cm^{-1} [9.28], the condition $W \geqslant E_k$ is fulfilled even for small perturbations. Meanwhile, the static distortions ρ_0 may be rather large. Hence we obtain the *amplification rule in Jahn–Teller distortions: A small distorting perturbation may be amplified by vibronic effects.*

Let us make some qualitative (or semiquantitative) estimations of the coefficient of vibronic amplification P_a. In the absence of vibronic coupling the distortion magnitude Q_0 can be found from the fact that the perturbation energy W transforms into strain energy: $W = \frac{1}{2}K_E Q_0^2$, where K_E is the force constant for the E distortions under consideration (Section 7.3); hence $Q_0 = (2W/K_E)^{1/2}$. If the vibronic effects are taken into account, $Q_0^{JT} = \rho_0 + Q_0$, and the amplification coefficient, equal to the ratio of the corresponding distortions, is

$$P_a = \frac{Q_0^{JT}}{Q_0} = 1 + \left(\frac{E_{JT}}{W}\right)^{1/2} \tag{9.11}$$

where the relationship $\rho_0 = (2E_{JT}/K_E)^{1/2}$ [see Eq.(7.41)] has been used. The maximum amplification is attained when $W = E_k$:

$$P_a^{max} = 1 + 4\,\frac{E_{JT}}{\hbar\omega_E} = 1 + 4\lambda_E \qquad (9.12)$$

It follows that the vibronic amplification may be very large, since the λ_E value may be substantial. For example, if we assume [as expected for octahedral compounds of Cu(II), high-spin Mn(III), and low-spin Cr(II)] that λ_E is about 5 to 10, then $P_a^{max} \sim 20$ to 40.

In the quadratic E–e problem with a more complicated AP that has three equivalent tetragonal minima (Figs. 7.10 and 7.11), a low-symmetry perturbation makes them nonequivalent and the system becomes "locked" at that minimum, which is deeper. Consequently, the pulsating system under small perturbations becomes strongly distorted statically. Such effects are encountered in many perturbation investigations of Jahn–Teller systems [9.38].

Temperature effects were omitted in the consideration above. It can be shown [9.38] that for a Jahn–Teller system with a threefold-degenerate T term interacting with t_2 nuclear distortions (the T–t_2 problem; Section 7.3) at not very low temperatures, the temperature dependence of the vibronic amplification of external distortions P_a is

$$P_a \approx \frac{E_{JT}^T}{kT} \qquad (9.13)$$

where E_{JT}^T is the Jahn–Teller stabilization energy. For instance, if $E_{JT} \sim 10^3\,cm^{-1}$, then $P_a \sim 10$ even at room temperatures. Other models for the estimation of the magnitude of the effect may be efficient, but the conclusion about an uncommonly large *susceptibility of vibronic systems to distortions* due to vibronic amplification seems to be quite general.

The notion of vibronic amplification contributes to a better understanding of the Jahn–Teller effect on the expected molecular shapes. In particular, it rejects completely the incorrect statements (often encountered in the literature) that the Jahn–Teller effect is not expected in systems where differences in the ligands or other low-symmetry perturbations formally remove the electronic orbital degeneracy. On the contrary, we emphasize in this section that it is only such low-symmetry perturbations (small, but vibronically amplified) that lead to observable distortions (such perturbations can also be created by the process of measurement when $\tau' < \tau$). *In the absence of these perturbations the Jahn–Teller distortions are of dynamic nature and do not manifest themselves in an absolute way* in stereochemistry and crystal chemistry, provided that cooperative effects and structural phase transitions are not essential (Section 9.4).

Qualitative Stereochemical Effects of Jahn–Teller and Pseudo Jahn–Teller Distortions

Provided that the conditions for experimental observation of molecular shapes corresponding to the minima of the adiabatic potential are fulfilled, the distortions of high-symmetry configurations predicted by the vibronic coupling theory (Sections 7.3 and 7.4) have a direct impact on stereochemistry. In the Jahn–Teller effect the distorted configurations are determined by the Jahn–Teller active coordinates. The symmetry of the Jahn–Teller distorted system can be predicted in a general way by means of group-theoretical considerations using the *epikernel principle* [9.39].

Denote the Jahn–Teller active coordinate (the coordinate that distorts the high-symmetry configuration in accordance with the Jahn–Teller theorem; Section 7.3) by Q_α, $\alpha = \vartheta, \varepsilon, \xi, \zeta, \eta, \ldots$ (Table 7.1). If the group of symmetry of the system in the high-symmetry configuration is G, then as a result of the Q_α distortion it reduces to S,

$$G \xrightarrow{Q_\alpha} S \tag{9.14}$$

where S is a subgroup of G in which only those symmetry operations (elements) of G remain that leave Q_α invariant. In other words, in the S subgroup Q_α is totally symmetric, whereas in the G group Q_α is nontotally symmetric, by definition, and it belongs to one of the irreducible representations Γ of G (e.g., in the tetrahedral T_d group Q_θ belongs to the E representation).

The *kernel* of Γ in G, denoted $K(G, \Gamma)$, is the subgroup of G that includes all symmetry elements that are represented in Γ by unit matrices [see Eq. (3.8) and the tables in Appendix 1]. This means that in the kernel subgroup all the basis functions of Γ are totally symmetric. An *epikernel* of Γ in G, denoted $E(G, \Gamma)$ is the subgroup of G that contains all the symmetry elements for which at least one basis function of Γ remains totally symmetric.

The epikernel principle can be formulated as follows [9.39]: *Extrema points on a Jahn–Teller surface prefer epikernels; they prefer maximal epikernels to lower-ranking ones. Stable minima are to be found with the structures of maximal epikernel symmetry.* This statement implies that although forced to distort in order to remove the electronic degeneracy, the system prefers nuclear configurations with higher symmetry compatible with this requirement. In this formulation the epikernel principle is related to the more general statement of Pierre Curie given in 1894 [9.40]: *The symmetry characteristic of a phenomenon is the maximal symmetry compatible with the existence of this phenomenon.*

However, in general it is not excluded that the epikernel principle can be violated. In particular, this can take place when higher-order vibronic interaction terms in Eq. (7.21) are taken into account; however, these terms are usually small.

As stated above, the epikernels can easily be found directly from the character tables of the corresponding point groups: $E(T_d, E) = D_{2d}$, $E(T_d, T_2) = C_{3v}$, C_{2v}, C_s; $E(T_d, E + T_2) = D_{2d}$, D_2, C_{3v}, C_{2v}, C_2, C_s. In the last case it is assumed that the E and T_2 vibrations have the same frequency, forming a fivefold-degenerate Jahn–Teller active space. For the octahedral O_h group $E(O_h, E) = D_{4h}$, C_{4v}, $E(O_h, T_2) = D_3$, C_{3v}, and so on. These distortions are well known from the general formulations of vibronic theory in Section 7.3.

At present, Jahn–Teller distortions in stereochemistry are widely recognized and employed in current investigations (see [9.1, 9.4, 9.38, 9.39, 9.41–9.45] and references therein). Many works are devoted to Cu^{2+} stereochemistry as a reference example for the vibronic stereochemistry as a whole. Since determined mainly from crystal structures, the stereochemistry of Cu^{2+} is discussed in Section 9.4.

As illustrations of the epikernel principle, some specific examples may be mentioned: $Ni^{2+}(d^8)$ and $Cu^{2+}(d^9)$ four-coordinated complexes usually have the D_{2d} structure compatible with an electronic T term and E distortions (similar Zn^{2+} complexes are undistorted tetrahedrons); $Fe(CO)_4$ exhibits C_{2v} distortions, as if resulting from a $T–(e + t_2)$ problem (see the discussion below), while the $Co(CO)_4$ fragment shows trigonal geometry [9.39, 9.42].

Even larger in subject than Jahn–Teller systems are pseudo Jahn–Teller systems. For them the electronic ground state is nondegenerate, but the strong vibronic mixing with excited states makes it unstable with respect to low-symmetry (nontotally symmetric) nuclear displacements. Following the statement (theorem) about the uniqueness of the vibronic mechanism of instability discussed in detail in Section 7.4, the only source of instability (unstable equilibrium) of high-symmetry configurations of molecular systems with a nondegenerate ground state is vibronic mixing with excited states by nuclear displacements of lower symmetry. If one starts with a high-symmetry configuration of the system, its possible instability and directions of distortions are controlled by and only by low-energy and strongly admixing excited states. The number of the latter, which are active in causing the instability of the ground state, is limited by selection rules.

The condition of pseudo Jahn–Teller instability in the direction of the symmetrized coordinate Q is given by Eq. (7.89), which is the basic relation that determines approximately the possible stereochemistry of coordination compounds. This equation reveals the specific excited electronic states that produce a configuration instability in a certain direction. Relations similar to (7.89) were discussed by Bader [9.46] and developed further in application to instability problems by Pearson [9.45]. In the Bader–Pearson treatment the reduction of the curvature of the ground state AP resulting from its vibronic mixing with the excited states is given as one possible explanation of the instability of the ground state. In the vibronic approach, as distinct from the Bader–Pearson treatment, the vibronic mixing of ground and excited electronic states is the only possible source of any configuration instability. This means also that if there is instability, there should be excited states that cause the distortion.

Consider, for example, MX_n systems, where M is a transition element. In a planar MX_4 system the typical electronic configuration is (Section 6.2):

$$\cdots(a_{1g})^2(b_{1g})^2(e_u)^4\{(b_{2g})(e_g)(2a_{1g})(2b_{1g})\}(a_{2u})^0(3a_{1g})^0 \qquad (9.15)$$

where the MOs in braces should be populated by the d electrons. If the ground electronic state is nondegenerate, the distortion of the square toward D_{2d} and tetrahedral T_d symmetry requires strong mixing of the ground state with the excited state by B_{2u} nuclear displacements. This excited state can be obtained by one of the following one-electron transitions: $e_u \to e_g$, $b_{1g} \to a_{2u}$, or $b_{2u} \to 2a_{1g}$ (b_{2u} is an inner MO). Therefore, if the e_g and $2a_{1g}$ MOs are fully occupied by d electrons but the $2b_{1g}$ MO is unoccupied, the square-planar configuration is stable. In other words, low-spin MX_4 d^8 complexes of Ni(II), Pd(II), and Pt(II) are expected to be square planar. On the contrary, high-spin d^5 and d^{10} complexes with an occupied $2b_{1g}$ MO may be unstable in the planar configuration, due to strong mixing with the low-lying $B_{2u}(2b_{1g} \to a_{2u})$ term.

Passing on to octahedral MX_6 systems, let us consider the example of XeF_6 [9.10, 9.11]. Nonempirical calculations of the electronic structure of this molecule [9.47] show that the outer MOs are arranged in the following sequence: $(t_{2g})^6(t_{2u})^6(t_{1u})^6(t_{1g})^6(e_g)^4(a_{1g})^2(2t_{1u})^0$, with an energy gap of about 3.7 eV between the a_{1g} and t_{1u} MOs. This results in instability of the system with respect to T_{1u} displacements. For comparison, the calculations [9.48] performed for a similar system SF_6 yield the following sequence of MOs: $(t_{2g})^6(e_g)^4(t_{1u})^6(t_{2u})^6(t_{1g})^6(a_{1g})^0(2t_{1u})^0$. Since the energy gap between the highest occupied t_{1g} and unoccupied t_{1u} MOs, as distinct from XeF_6, is sufficiently large, the O_h symmetry configuration of SF_6 is stable with respect to odd (dipolar) displacements. A series of investigations of the vibronic effects in the XeF_6 system, including electronographic and spectral measurements and MO LCAO calculations [9.10, 9.11, 9.49], confirm the pseudo Jahn–Teller origin of the instability with respect to the odd T_{1u} displacements — dipolar instability.

Quantitative and Semiquantitative Evaluation: $CuCl_5^{3-}$ Versus $ZnCl_5^{3-}$

By way of example we show here some results of approximate calculations of the K_0, F, Δ values and their contribution to the instability of the D_{3h} trigonal–bipyramidal (TBP) configuration of the pentacoordinated complexes $CuCl_5^{3-}$ (I) and $ZnCl_5^{3-}$ (II) with respect to the two possible types of distortions: E', toward a square-pyramidal (SP) configuration, and A_2'', toward a distorted tetrahedron with an additional ligand on the axis of distortion (Fig. 9.11). These two systems, I and II, are very similar, with the distinction that in II there is an additional d electron of the CA (and an additional proton in the nucleus), making its electron configuration d^{10} instead of the d^9 in I. From the

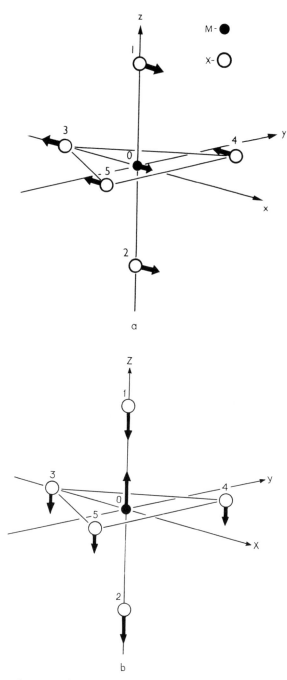

Figure 9.11. Ligand numeration and the main symmetrized distortions in pentacoordinated complexes MX_5^{n-}: (a) E'-type displacements realizing TBP $\rightarrow$ SP conversion; (b) A_2''-type displacement realizing the transformation $MX_5(\text{TBP}) \rightarrow MX_4(\text{tetr.}) + X$.

point of view of electronic structure and vibronic coupling, this difference is most important.

On the other hand, the structure of II in the crystal state also differs radically from I. The $CuCl_5^{3-}$ ion in the $[Co(NH_3)_6][CuCl_5]$ crystal is unstable with respect to E' type distortions describing the conversion TBP → SP (Fig. 9.11a), it has three equivalent almost-SP configurations, and performs fast conversions between them — molecular pseudorotations [9.50–9.53] (see later in this section). The $ZnCl_5^{3-}$ ion in a quite similar crystal $[Co(NH_3)_6][ZnCl_5]$ is also not TBP, but it forms a distorted tetrahedron plus one chlorine ion on (almost) the trigonal axis [9.54], and there are no pseudorotations. This configuration can be regarded as produced by the A_2'' distortion of the TBP configuration, as shown in Fig. 9.11b.

Thus the two pentacoordinated systems I and II in the same crystal environment are differently distorted: along the E' coordinate in I, resulting in three almost-SP configurations and pseudorotations, and along A_2'' in II without pseudorotations. The two different directions of distortions are evidently due to the difference in the electronic configuration of the CA, d^9 (I) versus d^{10} (II), and the question is whether this effect can be described by electronic control via vibronic coupling. Some results of recent attempts to solve this problem [9.55, 9.56] are as follows.

The calculations were carried out by the semiempirical IEH (SCCC) method (Section 5.5), with the excited states taken in the "frozen orbital" approximation; they are formed by the corresponding one-electron excitations of the MOs without recalculation of the corresponding self-consistent states. The numeration of the ligands is shown in Fig. 9.11. The six-atom system has 12 vibrational degrees of freedom, among which in the D_{3h} group three are of E' and two of A_2'' symmetry. By means of special procedures the calculation is reduced to the one-mode (ideal) case.

As a result of the calculations of the electronic structure of $CuCl_5^{3-}$ in the TBP configuration, it was shown that the ground state is $^2A_1'$ of the MO configuration...$(3e'')^4(5e')^4(5a_1')^1$. The highest single occupied MO $5a_1'$ is an antibonding linear combination of the copper AO $3d_{z^2}$ and chlorine AOs $3p_\sigma$:

$$|5a_1'\rangle = 0.60|3d_{z^2}\rangle - 0.58(|3p_{\sigma 1}\rangle + |3p_{\sigma 2}\rangle) - 0.23(|3p_{\sigma 3}\rangle + |3p_{\sigma 4}\rangle + |3p_{\sigma 5}\rangle)$$

$$(9.16)$$

The significant contribution of the metal $3d_{z^2}$ orbital to this MO agrees well with the ESR data.

The MOs $3e''$ and $5e'$ are also antibonding linear combinations of the AOs $3d_{xz}$, $3d_{yz}$, and $3d_{xy}$ of the copper atom with the AOs $3p_\pi$ and $3p_\sigma$ of the chlorine atoms, respectively. Satisfactory MO energy-level ordering was obtained with the crystal field corrections (the Madelung potential) included.

The electronic structure of $ZnCl_5^{3-}$ is qualitatively similar to $CuCl_5^{3-}$, but the MO energy-level positions and occupation numbers, as well as the LCAO coefficients for these two systems, differ significantly. In particular, the HOMO of II is also $5a_1'$, but as distinct from I it has other LCAO coefficients:

$$|5a_1'\rangle = 0.18|3d_{z^2}\rangle - 0.60(|3p_{\sigma1}\rangle + |3p_{\sigma2}\rangle) - 0.33(3p_{\sigma3} + 3p_{\sigma4} + 3p_{\sigma5}) \quad (9.17)$$

and what is even more important, it is occupied by two electrons instead of one in I. This difference eliminates in II the excited states that originate in I from the one-electron transitions to this MO. A similar analysis can be made for the excited-state contributions to the A_2'' displacements [9.56].

By means of these occupied and unoccupied one-electron MOs the excited states of E' and A_2'' symmetry [for which the vibronic constants $F^{A'E'}$ and $F^{A'A''}$ after (7.22) are nonzero] were constructed, and the values K_0, Δ, and F^2/Δ were calculated. Some results are given in Tables 9.6 and 9.7.

Among the excited states of E' type, there are three states in I and two states in II whose vibronic admixture to the ground state gives the major negative contribution to the force constant. The first arises from the excitation of one electron from the fully occupied MO $|1e'\rangle = 0.41(2|3s_3\rangle - |3s_4\rangle - |3s_5\rangle)$ (formed from the $3s$ AOs of the equatorial atoms) to the $5a_1'$ MO. This MO is occupied by one electron in I and by two electrons in II, and therefore in the Zn complex there is no contribution of the $1e' \rightarrow 5a_1'$ excited state.

Although the energy gap to this E' term in I is large, the pseudo Jahn–Teller effect on the E' displacements is rather strong, due to the relatively large vibronic constant $F^{A'E'}$ of the mixing of the A_1' and E' states (which, in turn, is due to the strong dependence of the MOs above on the distortion of the TBP configuration). Indeed, in the TBP configuration the overlap of the $5a_1'$ and $1e'$ orbitals is zero by symmetry, and hence these orbitals do not contribute to metal–ligand bonding. The E' nuclear displacements lower the D_{3h} symmetry of the system to C_{2v}; as a result, the e' MO splits into a_1 and b_2, while the $5a_1'$ MO transforms to a_1. Now the overlap of $a_1(e')$ and $a_1(5a_1')$ is nonzero, resulting in additional bonding of the AO $3d_{z^2}$ of the copper atom with the AOs $3s$ of the three equatorial ligands. This effect is not possible in II.

Similarly, other E' excited states listed in Tables 9.6 and 9.7 can be considered. Some of them give negligible contributions, as seen from the $5e' \rightarrow 5a_1'$ and $5a_1' \rightarrow 6e'$ contributions shown for comparison. These small contributions originate from the *electronic polarization* of the valence shell of the central metal ion by distortion (Section 7.4).

To conclude whether the negative vibronic contribution K_v results in instability, the nonvibronic contribution K_0 must be calculated. But the semiempirical method used in [9.55, 9.56] does not allow one to do that: For calculation of K_0, values of the wavefunction at the nuclei are required [the ρ value in Eq. (7.93)], whereas Slater-type AOs give wrong values of $\Psi(0)$; the calculation of $\Psi(0)$ is in general a rather difficult problem.

However, the computation data obtained above enable one to estimate the relative instabilities induced by vibronic coupling in different symmetrized directions of the same system, for which the K_0 value, in its part depending on

Table 9.6. Excited states of E' and A_2'' symmetry produced by $j \to k$ one-electron excitations, energy gaps $\Delta = E(k) - E(j)$, orbital vibronic constants $F^{(0i)}$, and vibronic contributions $K_v^{(i)}$ and $K_0^{(d)}$ to the relative value of the force constant K_{rel} for the E' and A_2'' displacements in $CuCl_5^{3-}$ (TBP configuration)

$j \to k$	Δ (eV)	$F^{(0i)}$ (10^{-4} dyn)	$K_v^{(i)}$ (mdyn/Å)	$K_v = \Sigma_i K_v^{(i)}$ (mdyn/Å)	$K_0^{(d)}$ (mdyn/Å)	$K_{rel} = K_0^{(d)} + K_v$ (mdyn/Å)
E'						
$1e' \to 5a1'$	11.2	9.03	−0.91			
$1e' \to 6a_1'$	25.0	12.24	−0.75			
$1a_1' \to 6e'$	23.9	10.44	−0.57			
$5e' \to 5a_1'$	2.8	0.58	−0.015			
$5a_1' \to 6e'$	12.1	1.87	−0.036			
				−2.28	−0.35	−2.63
A_2''						
$1a_2'' \to 6a_1'$	24.4	11.68	−0.70			
$3a_2'' \to 6a_1'$	14.4	6.87	−0.41			
$1a_1' \to 4a_2''$	26.5	7.97	−0.30			
$1a_2'' \to 5a_1'$	10.6	8.63	−0.88			
$3a_2'' \to 5a_1'$	0.8	1.72	−0.46			
				−2.75	0.55	−2.20

Table 9.7. Excited states of E' and A_2'' symmetry produced by $j \to k$ one-electron excitations, energy gaps $\Delta = E(k) - E(j)$, orbital vibronic constants $F^{(0i)}$, and vibronic contributions $K_v^{(i)}$ to the relative value of the force constant K_{real} for E' and A_2'' displacements in $ZnCl_5^{3-}$ (TBP configuration)

	$i \to k$	Δ (eV)	$F^{(0i)}$ (10^{-4} dyn)	$K_v^{(i)}$ (mdyn/Å)	$K_v = \Sigma_i K v^{(i)}$ (mdyn/Å)
E'	$1e' \to 6a_1'$	21.6	11.61	-0.78	
	$1a_1' \to 6e'$	24.4	10.07	-0.52	
					-1.30
A_2''	$1a_2'' \to 6a_1'$	20.7	10.84	-0.71	
	$3a_2'' \to 6a_1'$	11.3	6.16	-0.42	
	$1a_1' \to 4a_2''$	26.9	8.92	-0.37	
					-1.51

$\Psi(0)$, can be assumed to be the same. In the cases under consideration, we can compare the vibronic instabilities of the same system, I or II, with respect to the two possible distortions, E' and A_2''. The contribution of the closed-shell core electrons (and all the ns electrons) that make the $\Psi(0)$ quantity nonzero may be assumed to be the same for both distortions, whereas contributions of the d hole to these two displacements are different. The d-state contributions K_0^d to K_0 can be calculated easily [9.56]. The hole in the d^{10} configuration is d_{z^2}, and the contribution is negative for E' distortions and positive for A_2'' (the opposite signs are due to the field gradient created by the d_{z^2} hole, which is positive in the z direction and negative in the xy plane).

The data in Tables 9.6 and 9.7 show explicitly that the $CuCl_5^{3-}$ complex is more unstable with respect to the E' distortion, which transforms the TBP configuration into SP (or near SP) (Fig. 9.11a), whereas in $ZnCl_5^{3-}$ the instability is stronger with respect to the A_2'' distortion, which describes the transformation of the TBP configuration toward a distorted tetrahedron plus one ligand on the trigonal axis at a larger distance (Fig. 9.11b), in full qualitative agreement with the experimental data mentioned above.

These examples show in detail how the electronic structure—the ground and excited states—control the nuclear configuration, under certain conditions, making it unstable with respect to nuclear displacements of specific symmetry.

Off-Center Position of the Central Atom

The displacement of the CA from its geometric center in coordination compounds is a special case of vibronic effects in stereochemistry. An example of this effect is considered in Section (7.4) for the TiO_6^{8-} cluster in perovskites, where it is shown that as a result of the pseudo Jahn–Teller effect (vibronic mixing of the ground A_{1g} state with the excited T_{1u} term), the Ti atom may be displaced from the inversion center, resulting in dipole moment formation. As shown in Section 9.4, this displacement produces the spontaneous polarization of the lattice and ferroelectric phase transitions, provided that cooperative effects in crystals are taken into account. Similar examples are well known for impurity centers in crystals [9.57].

Consider another illustrative example: the out-of-plane displacements of the iron atom, as well as other metal atoms, in metal porphyrins and hemoproteins [9.58]. Besides being of special interest for biology, this example has general significance. It reflects the situation in a great number of corresponding classes of organometallic compounds with close-in-energy states of the metal d electrons and porphyrin (or similar ligand) π electrons. The mixing of these states under nuclear displacements, which shifts the metal atom out of and transverse to the porphyrin ring plane, makes the system soft or even unstable with respect to such displacements. Visually, the out-of-plane displacement of the metal atom with regard to the porphyrin ring is due to the additional π binding of the d orbitals of the metal and π orbitals of the ligands, illustrated in Fig. 7.22.

According to Eq. (7.89), the softening and instability of the high-symmetry configuration due to the pseudo Jahn–Teller effect is determined by the value of F^2/Δ, where $F = \langle 1|(\partial V/\partial Q)_0|2\rangle$ is the vibronic constant of the mixing of states 1 and 2 by the nuclear displacements Q, and 2Δ is the energy gap between them. As a result of the pseudo Jahn–Teller effect the force constant changes from K_0 to $K_0 - (F^2/\Delta)$, and if $(F^2/\Delta) > K_0$, the system becomes

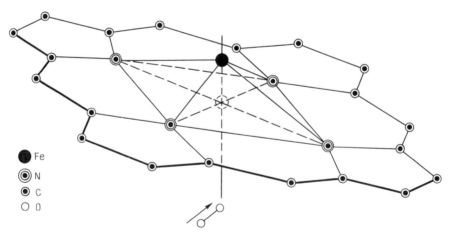

Figure 9.12. Structure of metalloporphyrin. In the D_{4h} symmetry group the out-of-plane displacement of the CA has A_{2u} symmetry. The nondisplaced position of the iron atom is shown by the dashed lines.

unstable with respect to the Q displacements. For metal porphyrins (MP) of D_{4h} symmetry, the out-of-plane displacement is of A_{2u} symmetry (Fig. 9.12), which means that F is nonzero if and only if the product of the representations of the mixing states 1 and 2 contains the A_{2u} representation.

Figure 9.13 shows several HOMO and LUMO energy levels of Mn, Fe, Co, Ni, Cu, and Zn porphyrins, as well as Mn phthalocyanine (MnPc) obtained from calculations for the planar configuration [9.59, 9.60]. It is seen that in FeP, for instance, the calculated ground state is ^3E_g; hence the excited state that couples with the ground state by A_{2u} displacements must be ^3E_u, since in the D_{4h} groups $E_g \times A_{2u} = E_u$. In MnP, MnPc, and CoP the corresponding excited states are $^4A_{2u}$, ^4E_g, and $^4A_{2u}$, respectively. All of them correspond to a one-electron excitation from the $a_{2u}(\pi)$ MO (predominantly from the porphyrin ring) to the empty $a_{1g}(d_{z^2})$ (predominantly from the metal), and the excitation energy $2\Delta = \varepsilon(a_{1g}) - \varepsilon(a_{2u})$ is relatively small. In the remaining metal porphyrins NiP, CuP, and ZnP, the a_{1g} MO is occupied, whereas the next MO of the same symmetry is very high (not shown in Fig. 9.13). It follows that in the last three MeP's there is practically no pseudo Jahn–Teller instability with regard to the out-of-plane displacement of the metal atom, whereas for other cases the effect may be important.

For an estimate of the effect, the calculated values of 2Δ can be used; they are approximately equal to 0.15, 0.20, 0.6, and 1.0 eV in MnP, FeP, CoP, and MnPc, respectively. The vibronic constant has been roughly estimated as

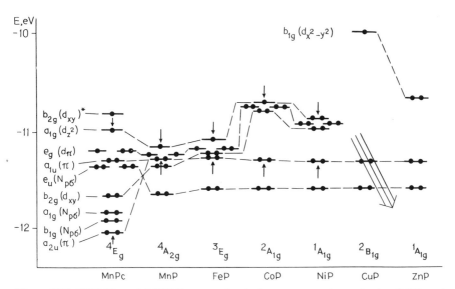

Figure 9.13. HOMO and LUMO energy levels for some metal porphyrins (MP) and manganese phtallocyanine (MnPC) with indication of the ground state for the in-plane position of the metal. The two MO forming the energy gap 2Δ between the ground and excited states that mix under the A_{2u} displacement are shown by arrows.

$F \approx 10^{-4}$ dyn for FeP (for other MeP's with approximately the same wavefunctions, F has the same order of magnitude), while the K_0 value for the A_{2u} displacements can be taken as $K_0 \sim 10^4$ dyn/cm. We have $F^2/K \sim 0.1$ eV, that is, F^2/K and Δ are of the same order of magnitude, and this confirms that the pseudo Jahn–Teller effect may soften the ground state. In MnP and FeP ($\Delta \sim 0.08$ eV and 0.1 eV, respectively) $\Delta < F^2/K$ and the softening transforms into instability, while in CoP and MnPc ($\Delta \sim 0.3$ eV and 0.5 eV, respectively), only a softening but not an instability of the metal position may be expected. These results are in qualitative agreement with the empirical data available.

The metal atom position with respect to the porphyrin ring is significant in determining some biological functions of hemoproteins (Section 10.3). The first explanation of the origin of the out-of-plane position of the iron atom was that in the high-spin configuration the atomic radius of Fe(II) is too large to fit the cavity in the porphine ring [in the low-spin configuration the Fe(II) ion occupies an in-plane position; Section 10.3]. This explanation does not work, in general. Indeed, the cavity in phthalocyanine is considerably smaller than in porphine. Nevertheless, MnPc is planar, whereas MnP is nonplanar. Again, a series of porphyrins of other metals, such as Sn(IV) and Mo(IV), with ionic radii larger than Fe(II), are planar [9.1].

Geometry of Ligand Coordination

Another important problem in the stereochemistry of coordination compounds is the mode of coordination of small ligands to the central atom. This problem, too, can be considered successfully by means of the vibronic approach. Consider, for example, the O_2, CO, and NO molecules coordinated to metal porphyrins and the heme in hemoproteins (for the coordination of NO to other systems, see [9.61] and references therein). Four modes of coordination are observed experimentally (Fig. 9.14): linear end-on, bent end-on, bent side-on, and symmetrical side-on. The study of the mode of coordination with metalloporphyrins (MP) in model compounds shows that depending on the metal, the linear end-on coordination is characteristic for NO and CO, the bent end-on coordination is usually observed for NO or O_2, and the symmetrical side-on geometry can be seen in O_2 coordination. The origin of these geometries has aroused intensive discussion [9.58].

Consider this problem using the vibronic approach. This approach implies that the bent end-on configuration appears as a consequence of the pseudo Jahn–Teller instability of the linear end-on and symmetrical side-on geometries. In metal porphyrins of D_{4h} symmetry with a linear end-on coordination of diatomics at the fifth coordinate, the bending of the ligand is an E-type displacement (for the influence of the imidazole ligand at the sixth coordinate position, e.g., in Hb, see [9.58]). Therefore, the pseudo Jahn–Teller instability with respect to the bending of the diatomic ligand may take place if the product of representations of the ground and (not very high in energy) excited states contains the E representation.

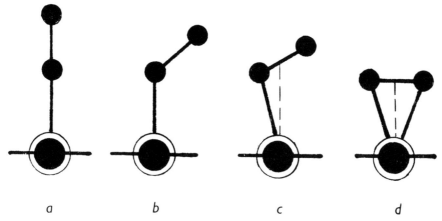

$$a \qquad\qquad b \qquad\qquad c \qquad\qquad d$$

Figure 9.14. Four types of geometries of coordination of diatomics to a coordination center: (*a*) linear end-on; (*b*) bent end-on; (*c*) bent side-on; (*d*) symmetrical side-on.

Several HOMO and LUMO energy levels for the MP–CO, MP–NO, and MP–O_2 systems calculated assuming linear end-on coordination [9.61, 9.62] are given in Fig. 9.15. The qualitative changes in some of these levels due to their vibronic mixing with the excited states formed by one-electron transition from the $e(d_\pi - \pi^*)$ or $e(d_\pi + \pi^*)$ MOs to the $a_1(d_z)$ MO are illustrated in Fig. 9.16. In fact, only the most essential mixing orbitals with sufficiently large vibronic constants should be considered. Large vibronic constants occur when the mixing states contribute substantially to bond formation and are strongly influenced by the E displacements (by the ligand bending). For example, the E displacements formally mix the MO $e(P)$ with $a_1(d_{z^2})$, but since $e(P)$ is the state of the porphine ring not involved in the metal–diatomic bond, the vibronic constant $F = \langle e(P)|(\partial V/\partial Q_E)_0|a_1(d_{z^2})\rangle$ is small to zero.

Denoting the corresponding energy intervals (Fig. 9.15) and vibronic constants by $2\Delta_1$ and $2\Delta_2$ and F_1 and F_2, respectively, and considering the vibronic mixing of each pair of MOs, $e(d_\pi - \pi^*)$ with $a_1(d_{z^2})$, and $e(d_\pi + \pi^*)$ with $a_1(d_{z^2})$, we conclude, using Eq. (9.89), that each mixing lowers the force constant K_E by an amount dependent on the respective MO occupation numbers. Denoting the latter by q_1, q_2, and q_3, respectively, for levels 1, 2, 3 marked with arrows in Fig. 9.15, we obtain the following condition for the instability of the linear end-on configuration:

$$\frac{(q_1 - q_3)F_1^2}{\Delta_1} + \frac{(q_2 - q_3)F_2^2}{\Delta_2} > K_E \qquad (9.18)$$

If this condition is satisfied, the adiabatic potential of the system with respect to the E displacement in question has a maximum for the linear end-on coordination and a continuum of minima, forming a circular trough (7.84)

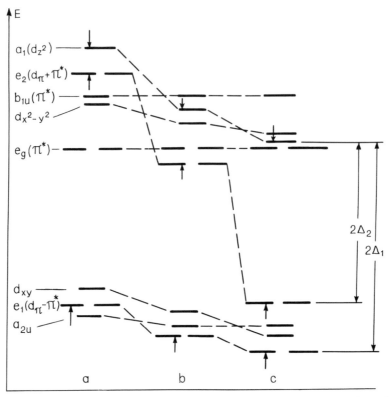

Figure 9.15. MO energy-level diagrams for linear end-on coordination of diatomics to metalloporphyrins: (*a*) CO; (*b*) NO; (*c*) O_2. The arrows indicate the energy gaps $2\Delta_1$ and $2\Delta_2$ between the e and a_1 MO that mix under the bending (*E* type) displacement of the ligand.

[9.58]. Each point of the latter corresponds to a bent end-on coordination at a certain angle to the linear end-on line with arbitrary orientation around this line. If the quadratic terms of the vibronic interaction are included in the calculation, four additional minima (7.85) are formed along the bottom of the trough, regularly alternating with four saddle points. In this case the bent end-on coordination of diatomics has four preferable orientations with respect to the pyrrole ring: either toward the nitrogen atoms, or between them, depending on the sign of the quadratic vibronic constant (see the $E-A$ mixing problem discussed in Section 7.4).

Consider now the MP—CO system, for which $2\Delta_1 = 4.0\,\text{eV}$ and $2\Delta_2 = 0.3\,\text{eV}$ [9.62, 9.63] (Fig. 9.15). Using the known order of magnitudes of F and K_E, we come to the conclusion that if only the lower $e(d_\pi + \pi^*)$ state is occupied by electrons and contributes to softening of the system (i.e., $q_2 - q_3 = 0$), the inequality (9.20) is not satisfied (Δ_1 is large) and the linear coordination is stable. If the higher MO $e(d_\pi - \pi^*)$ is occupied, the inequality

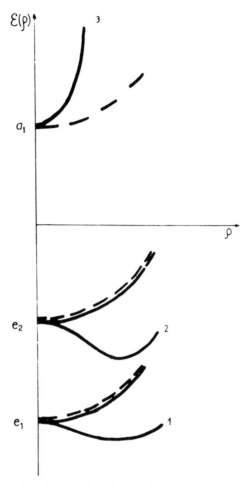

Figure 9.16. MO energies of some of the levels of Fig. 9.16 as a function of the bending angle ρ with (solid line) and without (dashed line) the pseudo Jahn–Teller mixing e–a_1.

(9.24) is satisfied due to the small value Δ_2, and the linear end-on coordination becomes unstable. In MP—CO with linear end-on geometry the state $e(d_\pi - \pi^*)$ remains unoccupied for all $3d^n$ metals, including $3d^{10}$. Hence in $3d^n$ transition metal porphyrins the CO molecule is expected to be linearly end-on coordinated. This prediction of the vibronic theory agrees well with the experimental data mentioned.

For the MP—NO system, the values $2\Delta_1$ and $2\Delta_2$ have the same orders of magnitude as in the previous instance of MP—CO, but the energy-level ordering is different (Fig. 9.15): the $e(d_\pi - \pi^*)$ MO becomes occupied when the number of d electrons plus the antibonding π^* electrons of the NO molecule exceeds six. Thus in the $3d^n$ metal porphyrins the NO molecule in the linear end-on configuration is stable for $n < 5$ and unstable for $n > 5$. This conclusion, too, agrees with the experimental data.

Finally, in MP—O_2 both e MO levels are at about the same distance from the a_1 level (Fig. 9.15) ($2\Delta_1 = 3\,eV$ and $2\Delta_2 = 2.5\,eV$) and may give a comparable contribution to the softening of the linear geometry M—O—O with respect to E bending. Therefore, it can be assumed that the cumulative effect suffices to provide essential softness or even instability of the linear coordination. The softening is expected to be approximately doubled when the number of d electrons plus two antibonding π^* electrons of the O_2 molecule exceed six. This result also agrees qualitatively with the experimental data.

Stability of the symmetric side-on coordination of small ligands can be considered similarly. The displacements toward bent coordination are of B_1 (or B_2) symmetry, for which the nonzero vibronic constant corresponds to the mixing of the $b_1(d_{xz})$ [or $b_2(d_{yz})$] with the a_1 MO. Both these MOs are nondegenerate, and hence the pseudo Jahn–Teller instability can be obtained directly using Eq. (9.22). Estimations for the M—O_2 system show that the criterion $(F^2/\Delta) > K_E$ is not satisfied, and the symmetric side-on coordination of O_2 is expected to be stable.

Another feature directly related to the geometry of ligand coordination is the change in geometry (distortion) of the ligand itself by coordination. The only possible distortion in diatomics is the change in interatomic distance. It takes place, indeed, and this topic is discussed in more detail in Section 11.2, together with chemical activation by coordination. For ligands with three or more atoms the distortion of the nuclear configuration that lowers its symmetry may take place due to the electronic rearrangement induced by coordination.

Consider, for example, the C_2H_4 molecule coordinated as a ligand to a transition metal complex (cf. Sections 6.3 and 11.3). The free molecule in the ground state is planar and has a nondegenerate ground state $^1A_{1g}$ (valence electron configuration $\pi^2\pi^{*0}$), while in the first excited state $^1B_u(\pi^1\pi^{*1})$ it is nonplanar, the two CH_2 groups being in mutual perpendicular planes. By coordination like a π complex to transition metals in low oxidation states, an orbital charge transfer from the bonding π orbital of C_2H_4 to the metal together with a strong π back donation to the free π^* orbital of C_2H_4 take place, and the local charge distribution in C_2H_4 becomes similar to $\pi^{2-\delta}\pi^{*\delta}$. The population of the π^* orbital distorts the ground-state planar configuration of the molecule, making it similar to the excited state. The experimental data show that, indeed, the geometry of coordinated C_2H_4 is intermediate between those of the ground (planar) and excited states, the latter having the configuration with two mutually perpendicular CH_2 groups [9.64]. Approximate calculations confirm this trend [9.65].

For other examples, refer to [9.1, 9.38, 9.42, 9.45, 9.61]. In Section 11.2, other aspects of ligand coordination are considered in connection with their activation.

Stereochemically Active and Inert Lone Pairs

In Section 9.1 the stereochemistry of lone pairs is discussed. In the semiclassical approach the lone pairs are considered as repulsion units alongside bond pairs, and as such they occupy a coordination place distorting the otherwise symmetrical coordination polyhedron. However, in some systems the polyhedron is not distorted, despite the presence of a lone pair, and in these cases the latter is called an inactive or *inert lone pair* [9.66].

Very often the lone pair originates from the CA $(ns)^2$ configuration, with $n = 4, 5, 6$; the post-transition elements In(I), Tl(I), Pb(II), Sb(III), Te(IV), Xe(VI), and so on, are of this type. In $SbBr_6^{3-}$ and $TeCl_6^{2-}$, for example, the $(ns)^2$ lone pair is stereochemically inert (the octahedron is not distorted), whereas in XeF_6, $InCl_6^{5-}$, ... the octahedron is distorted. The $(ns)^2$ pair itself is spherical-symmetrical and does not cause distortions. Hence using the VSEPR model (Section 9.1) to explain the origin of distortion, we must assume that there is a strong hybridization of the ns states, with the np ones resulting in a directed lone pair [9.67, 9.68].

On the other hand, as stated above (Sections 2.1 and 9.1), hybridization is in fact not the cause of the distortions but rather, its consequence. It is mentioned in Section 9.1 that in the more general MO LCAO scheme, distortion may occur if in the high-symmetry configuration the two electrons occupy a strongly antibonding MO which under the distortion transforms to a lone pair. These considerations are qualitatively true, but they give no overall solution of the problem that would contribute to a more general understanding. Such a solution is provided by the vibronic approach [9.69].

Consider the general MO LCAO scheme (Fig. 9.17) for an undistorted octahedral system MX_6, which in the representation of local M—X bonds has a $(ns)^2$ electron pair above the six bonding pairs (in fact, each M—X bond may have more than one bonding electron pair, as in the case of multiple bonds). In this scheme the two ns electrons occupy the antibonding MO a_{1g} (π MOs and ligand nonbonding MOs are not indicated). The ground state of the system as a whole A_{1g} is nondegenerate, but the excited T_{1u} states formed by one-electron excitations $a_{1g} \rightarrow t_{1u}$ are relatively close in energy.

In the vibronic approach the stability or instability of the regular octahedral configuration under consideration is determined by Eq. (7.89), which gives the relationship of the parameters for which the system is stable or unstable and the direction of instability in the latter case. Pseudo Jahn–Teller mixing of the ground $^1A_{1g}$ state with the excited $^1T_{1u}$ by T_{1u} nuclear displacements results in instability of the ground state with respect to T_{1u} distortions (Section 7.4), provided that the condition of instability is satisfied. This distortion is somewhat similar to the dipolar instability produced by the same A_{1g}–T_{1u} mixing in the TiO_6^{8-} octahedron (Section 7.4), but the change from d electrons in Ti to sp electrons in the MX_6 systems under study in this section introduces significant alterations. It can be shown that in the linear approximation with

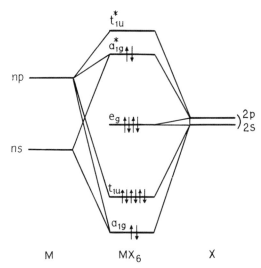

Figure 9.17. MO LCAO energy level scheme for an MX_6 system with a $(ns)^2$ configuration above the closed shells of M and X. Six ligand $\sigma(sp)$ AO form with the CA ns and np AO the bonding a_{1g} and t_{1u} MO and the nonbonding e_g MO occupied by 12 electrons, while the antibonding MO a_{1g}^* is occupied by the two $(ns)^2$ electrons (π MO and ligand nonbonding MO are not indicated).

respect to the vibronic coupling terms in the Hamiltonian, the $s-p$ $(A_{1g}-T_{1u})$ vibronic mixing $[(A_{1g} + T_{1u}) - t_{1u}$ problem] results in a trough of minima in the space of T_{1u} distortions [i.e., all the distortions corresponding to any combination of the three T_{1u} coordinates (Table 7.1) has the same energy], and only the second-order terms make the eight trigonal directions preferable.

However, this problem may be complicated by the fact that the excited T_{1u} term is degenerate and hence is subject to the Jahn–Teller $T_{1u}-(e_g + t_{2g})$ effect [9.69]. If the latter is taken into account, the vibronic problem as a whole is a combined pseudo Jahn–Teller and Jahn–Teller problem $(A_{1g} + T_{1u})-$ $(t_{1u} + e_g + t_{2g})$, meaning that there may be distortions of three types: T_{1u}, E_g, and T_{2g}. The solutions obtained in the linear approximation with respect to the vibronic coupling show that depending on the vibronic coupling constants and the energy gap 4Δ between the ground A_{1g} and excited T_{1u} states, several possibilities arise. Assume that F, F_E, and F_T are the coupling constants to the T_{1u}, E, and T_{2g} displacements and K_0, K_0^E, and K_0^T are the respective primary force constants (i.e., without the vibronic contributions; Section 7.4), and the energies of the two mixing terms are $E(^1A_{1g}) = -3\Delta$ and $E(^1T_{1u}) = \Delta$. Let

$$f = \frac{F^2}{K_0\Delta} \tag{9.19}$$

$$e = \frac{F_E^2}{K_E \Delta} \qquad (9.20)$$

$$t = \frac{F_T^2}{3K_T \Delta} \qquad (9.21)$$

These constants have the physical meaning of the corresponding Jahn–Teller stabilization energies in units of Δ, taken from an appropriate read-off.

In these denotations the expected distortions of the regular octahedron in MX_6 systems under consideration can be evaluated analytically (in [9.69], totally symmetric distortion are also included). The results are illustrated in Fig. 9.18, which in fact comprises two coinciding schemes, one for e versus f and the other for t versus f. The meaning of these schemes is as follows. The area limited by the axes e and f (or t and f) is divided into three domains, E_1, E_2, and E_3, that have different kinds of adiabatic potential minima. In the first domain (dotted area), where $f < 1$, $e < 1$, and $t < 1$, the Jahn–Teller and pseudo Jahn–Teller stabilization energies are smaller than the corresponding

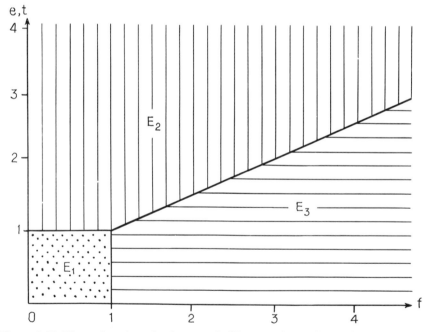

Figure 9.18. Three domains of existence of different Jahn–Teller and pseudo Jahn–Teller distortions in MX_6 systems with $(ns)^2$ lone pairs in each of the two $e-f$ and $t-f$ planes: E_1, no distortions (inert lone pair); E_2, combined dipolar T_{1u} and either tetragonal E_g (in the $e-f$ plane) or trigonal T_{2g} (in the $t-f$ plane) distortions; E_3, pure Jahn–Teller either tetragonal (in the $e-f$ plane) or trigonal (in the $t-f$ plane) distortions.

threshold given by Eq. (7.89), and hence there are neither Jahn–Teller nor pseudo Jahn–Teller distortions. This is the case of inert lone pairs.

In the area where $f > 1$, the pseudo Jahn–Teller T_{1u} (dipolar) distortion becomes operative, but the admixing of the excited T_{1u} term also involves Jahn–Teller T_{2g} and E_g distortions. According to the calculations [9.69], the relative energies of the minima in the three domains in Fig. 9.18 for the t–f scheme are (in Δ units):

$$E_1 = -6$$

$$E_2 = -6 - \frac{2(f-1)^2}{f-t} \tag{9.22}$$

$$E_3 = 2 - 8t$$

with the same relations with e instead of t for the e–f plane. Therefore, the trigonal dipolar (pseudo Jahn–Teller) distortions T_{1u}, with an admixture of either E_g or T_{2g} distortions, are preferable when $E_2 < E_3$, which yields, in addition to $f > 1$,

$$f + 1 > 2t \tag{9.23}$$

and

$$f + 1 > 2e \tag{9.24}$$

respectively. If the opposite inequalities hold,

$$2e > f + 1 \tag{9.25}$$

or

$$2t > f + 1 \tag{9.26}$$

the tetragonal D_{4h} or trigonal D_{3h} Jahn–Teller minima of the excited state T_{1u} are lower in energy (than the assumed ground state), and they are active in stereochemistry. The preference between tetragonal or trigonal minima is the same as in the usual Jahn–Teller T_{1u}–$(e + t_{2g})$ problem (Section 7.3); namely, trigonal distortions are preferable if $t > e$, and tetragonal distortions occur for the opposite inequality, $e > t$.

All these distorted configurations of MX_6 systems with $(ns)^2$ lone-pair configurations are found in different systems (see [9.1–9.6, 9.11–9.13, 9.16, 9.66]). Moreover, the combined distortions, described above, explain the origin of complicated (helicoidal) crystal structures (Section 9.4). In particular, in the InCl crystal ($InCl_6^{5-}$ units) both types of distortions in the E_2 area of Fig. 9.18, trigonal T_{2g} plus dipolar T_{1u}, and tetragonal E_g plus dipolar T_{1u}, are observed in two phases of the crystal, yellow and red, respectively [9.70].

Similar treatment, in principle, is possible for many other types of systems. For instance, in multicenter transition metal clusters the change of geometry from regular tetrahedral in 60-electron tetraclusters to butterfly geometry in similar 62-electron clusters was subject to discussion from the point of view of a vibronic problem on two electronic terms $T_1 + T_2$ mixing via $E_g + T_{2g}$ distortions [9.39, 9.71]. Many observed cluster geometries can be explained in this way.

Pseudorotations in Coordination Systems

As shown in Sections 7.3 and 7.4, the Jahn–Teller and/or pseudo Jahn–Teller effects in the high-symmetry configuration of a molecular system result in the formation of two or several (or an infinite number of) equivalent minima of its adiabatic potential which correspond to equivalently distorted polyhedra. If the energy barriers between these minima are not very high (or even zero as in the case of a trough of minima), the free system performs continuous transitions between the minima. These transitions are classified in Section 7.3 as internal *free rotations, hindered rotations, and pulsating motions.*

As emphasized above, the transitions between the configurations of equivalent minima are never real rotations, although outwardly they look like rotations. For instance, for tetragonal distortion of the octahedron in the quadratic $E-e$ problem, the transition from the configuration elongated along the Z axis to that elongated along the X axis (Figs. 7.11 and 7.13), owing to the identity of the ligands, looks like a rotation on $\pi/2$ along the Y axis. In fact, however, such a rotation does not take place, as the nuclear motions in the transition under consideration are more complicated [9.38] (Fig. 7.13). They can be observed experimentally in the NMR spectra, isotopic substitutions experiments, central atom NQR spectra, tunneling splitting (Section 7.3), and other spectroscopic measurements. These internal motions can be called *pseudorotations.*

In coordination compounds, pseudorotations are sometimes called *flexional behavior* [9.72]. As seen from the examples that follow, in many cases the Jahn–Teller and pseudo Jahn–Teller dynamics can be interpreted visually as a continuously changing "flexional" configuration. One of the first observations of such behavior relates to Cu(II) compounds [9.73]. Another kind of flexionality takes place in coordination compounds with *alterdentate ligands*; they offer to the metal ion more than one equivalent coordination sites, and hence under certain conditions the metal can resonate between them (e.g., in the alloxan radical anion) [9.74].

Pseudorotations in molecular systems, in general, have been known for a long time. Berry [9.12] assumed that the AP of such systems has several equivalent minima (without specifying their origin) with small energy barriers between them, and the observed pseudorotations are due to the transitions

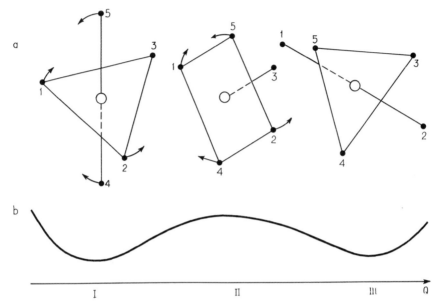

Figure 9.19. Berry pseudorotation of a trigonal bipyramidal molecule: (*a*) distortions shown by arrows (combined *E* type) convert the configuration I into an equivalent III via the intermediate square-pyramidal structure II that has a higher energy; (*b*) AP curve along the distortion coordinate.

between the minima. For instance, the pseudorotation molecule PF_5 is assumed to have an energy minimum in the trigonal–bipyramidal (TBP) configuration I, with F atoms 4 and 5 in axial positions (Fig. 9.19*a*), and as a result of a displacement of type Q (combined E displacements) transforms into the square-pyramidal (SP) configuration II. By further transformation, II converts again into the TBP configuration III but with other atoms F on the threefold axis (1 and 2 instead of 4 and 5). With identical F atoms the transformation I→III looks like a 90° rotation of the threefold axis.

Another example is the SF_4 molecule with minima at C_{2v} symmetry, which can be considered as either a strongly distorted tetrahedron or a less distorted square with two angles F—S—F of 183° and 104°. Here the Berry rotation consists of transitions between two equivalent distorted configurations via the intermediate unstable square-planar geometry (Fig. 9.20); this mechanism of interconversion in SF_4 is confirmed by dynamic NMR experiments [9.75] and by direct electronic structure calculations [9.76].

In both examples the intermediate configuration has a maximum of the AP of the nondegenerate ground state with respect to a specific symmetrized direction Q. Based on the conclusions of vibronic theory, the instability of the high-symmetric intermediate configuration is due to the vibronic mixing of its electronic ground state, with some excited states of required symmetry deter-

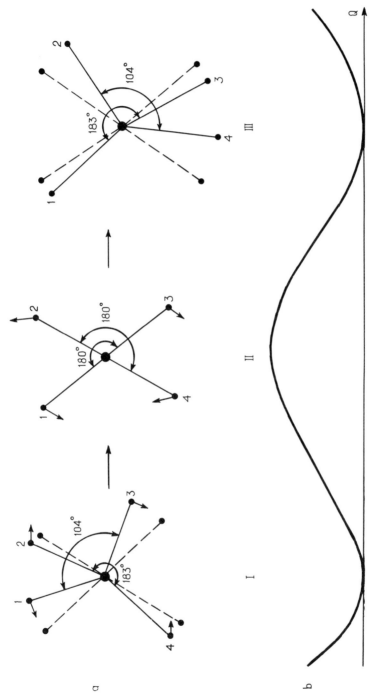

Figure 9.20. Berry interconversion of equivalent distorted square-planar (or distorted tetrahedral) SF$_4$ configurations I and III via square-planar intermediates of higher energy II (shown by dashed lines in I and III). The interconversion coordinate is A_{2u} of the square-planar configuration.

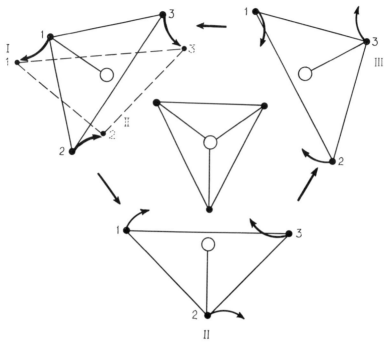

Figure 9.21. Pseudorotation in the $E-e$ Jahn–Teller $CuCl_5^{3+}$ type system with three equivalent equilibrium configurations of SP (or near SP) symmetry (top view); the interconversion between them goes beyond the TBP configuration. The arrows show (schematically) the displacements of the equatorial ligands transforming one configuration into another (the axial ligand displacements are not shown).

mined by the nonzero vibronic constant. The Berry rotations under consideration are thus pseudorotations caused by a strong pseudo Jahn–Teller effect.

It is important that *in Jahn–Teller distortions (orbitally degenerate ground states), pseudorotations do not follow Berry rotations.* Indeed, consider an MX_5 system in the TBP configuration outwardly similar to the Berry PF_5 molecule but with a doubly degenerate electronic E term (an example of such systems, $CuCl_5^{3-}$, was discussed earlier in this section). According to the general solution of the $E-e$ problem (Section 7.3), in a D_{3h} system there are three minima of the adiabatic potential in the space of two E-type displacements (Fig. 7.10). In the case under consideration these minima correspond to three SP (or nearly SP) configurations, shown in Fig. 9.21, and this result is confirmed by many experimental and theoretical investigations [9.50–9.56]).

The pseudorotations here are just interconversions among the three SP configurations. Direct examination of AP surfaces in $E-e$ problems (Figs. 7.10 and 7.12) shows that the lowest pathway to overcoming the barriers between minima never goes through the high-symmetry TBP configuration D_{3h}, and hence the latter is not involved in the pseudorotation; the coordinates of interconversion do not include the point $Q_9 = Q_\varepsilon = 0$. Thus the Jahn–Teller pseudorotation cannot be reduced to the direct $TBP \leftrightarrow SP$ interconversions, as in the Berry mechanism.

Apparently, a similar situation takes place in the $MoCl_5$ complex, which in the TBP configuration has the ground electronic E term. The fact of pseudo-rotation in this complex emerges from its electron diffraction spectra [9.77]; the authors interpreted the experimental data as compatible with a picture in which about 56% of the molecules have the SP configuration C_{4v}, while the remaining have D_{3h} (TBP) symmetry with large amplitudes of the correspond-ing vibrations (if normal vibrations are assumed, 18% of the molecules must be considered as dimers). This explanation ignores the vibronic effects and consequent pseudorotation. The experimental results [9.77] are well under-stood as pseudorotations based on the vibronic coupling scheme discussed above: The percentage of different configurations extracted from the experi-mental data may even allow one to estimate the barrier height between SP configurations.

The situation for tetrahedral systems is similar: The pseudorotation that results from Jahn–Teller distortions is different from that predicted by the Berry mechanism. Indeed, in the $Fe(CO)_4$ complex taken as an example

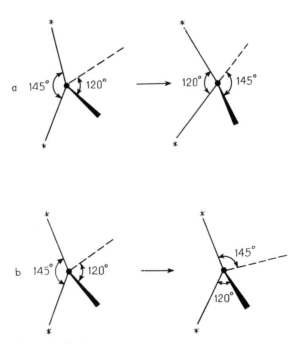

Figure 9.22. Pseudorotation in the $Fe(CO)_4$. The interconversion between two distorted C_{2v} configurations: (a) predicted by the Berry mechanism; (b) observed experimentally. The two isotope ^{13}C and ^{18}O ligands are marked by stars [the complex studied experimentally [9.78] is $Fe(CO)_2(^{13}C^{18}O)_2$].

[9.78, 9.79], the ground-state electronic term T is threefold degenerate and the Jahn–Teller problem is $T–(e + t_2)$. As mentioned above, if both types of displacements, E and T_2, are active (this is determined by the corresponding vibronic constants), the epikernel of the problem is C_{2v}, and this is the symmetry of the six minima for the distorted tetrahedron. The experimental data confirm these distortions: The configuration of the $Fe(CO)_4$ complex is similar to SF_4 with the two angles C—Fe—C at about 145° and 120° (Fig. 9.22).

In the Berry rotation scheme the interconversion between two C_{2v} configurations goes via the intermediate high-symmetry square-planar D_{4h} configuration (Fig. 9.20) or tetrahedral T_d intermediates. The latter seem to be more appropriate for $Fe(CO)_4$ in view of the relatively large angles between the bonds (compared with that of SF_4). However, the experimental data do not confirm such a pathway in the mechanism of interconversion of equivalent distorted configurations in $Fe(CO)_4$.

The pseudorotation in this system has been studied experimentally by means of ligands marked by ^{13}C and ^{18}O isotopes [9.78]. If in the system $Fe(CO)_2(^{13}C^{18}O)_2$ an isomerization is induced by means of an infrared laser which excites the C—O bond, the expected Berry interconversion is that shown in Fig. 9.22a. The observed interconversion is illustrated in Fig. 9.22b; it does not reduce to the Berry rotations. Meanwhile, if one examines the AP surface of the $T–(e + t_2)$ problem (see Section 7.3 and [9.38] for more details), one can see that the observed experimentally isomerization (Fig. 9.22b) corresponds directly to the pathway via the lowest-energy barrier between two nearest-neighbor minima of C_{2v} symmetry. As in the Jahn–Teller $E–e$ problem for MX_5 complexes considered above, *the pathway of the transition between two equivalent minima via the lowest-energy barrier does not cross the configuration of highest symmetry.*

The difference between the Berry (pseudo Jahn–Teller) and Jahn–Teller mechanisms of pseudorotation has an even more important basis than that of energy barriers. To make this statement clear, consider a simple case of an $E–b_1$ Jahn–Teller problem, which is of general interest (see, e.g., Section 9.4). The twofold-degenerate electronic E term, for instance, in square-planar systems, interacts with one Jahn–Teller active coordinate $Q(B_{1g})$, resulting in two minima at $+Q_0$ and $-Q_0$, respectively, as shown in Fig. 9.23. It is important that the electronic functions of these two minima configurations are mutually orthogonal, and therefore in the absence of additional perturbations *no transitions between these two configurations are possible* — they are strictly forbidden. If additional interactions are nonzero [e.g., the interaction with B_{2g} coordinates in the $E–b_{1g}$ problem above that makes it $E–(b_{1g} + b_{2g})$], there is a nonzero probability of transitions between the two configurations via B_{2g} coordinates, but not directly along B_{1g}.

In the Jahn–Teller systems discussed above, the situation is much the same as in the $E–b_1$ problem. Indeed, the Berry rotations in Fig. 9.22b act as a transition between two C_{2v} minima of the adiabatic potential of the tetrahedral

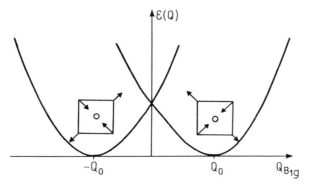

Figure 9.23. AP of the $E-b_{1g}$ problem with two equivalent minima in the space of B_{1g} displacements Q_{B1g} of a square configuration D_{4h}. The intersection at the high-symmetry configuration means that the states in the two minima are exactly orthogonal and no direct interconversion between the two distorted configurations via the square intermediate is possible.

$T-(e + t_2)$ problem. The configurations in these minima can be regarded as tetrahedrons distorted, respectively, to $+Q_{0\zeta}$ and $-Q_{0\zeta}$ from the regular configuration [9.38]. As in the $E-b_1$ problem, the electronic wavefunctions of these two configurations are orthogonal to each other, hence direct transition between them is forbidden. However, *transitions via other coordinates are not forbidden.*

In the PF_5 and SF_4 systems discussed above, the ground state is non-degenerate and hence the electronic states in the minima are not orthogonal. If the excited state producing the instability of the ground state is also nondegenerate, there is only one coordinate of interminimum conversion, the coordinate that mixes the two states. Here the Berry mechanism is the only one possible. If the excited state is degenerate, there may be more than one coordinate of interconversion of the minima configurations, and hence both the Berry and non-Berry mechanisms of pseudorotation are possible.

9.3. MUTUAL INFLUENCE OF LIGANDS

The Model: *trans* and *cis* Influence in Stereochemistry

There is a large amount of experimental and correlation data on the mutual influence of ligands in transition metal coordination compounds. The trend began in the mid-1920s with the work of Chernyaev [9.80], who showed convincingly that in the substitution reactions with square-planar Pt(II) complexes: PtXYZV + U → PtXYZU + V, the ligand V to be substituted first is determined by the properties of the ligand located in the position *trans* to V (*trans influence*), and all the ligands can be arranged in a series according to their *trans*-influence power (Section 11.1). Later it was established that not only the *trans* effect, but also the *cis* effect and other mutual influences of ligands are manifest in many properties of transition metal compounds, including stereochemistry.

The data on mutual influence of ligands are usually divided in two groups:

1. *Static mutual influence*: structural parameters (bond lengths, vibrational frequencies, magnetic resonance parameters, etc.) as functions of mutual influence.

2. *Kinetic mutual influence*: reactivity effects (activation energies and rates of substitution reactions). Sometimes the static effects are called *trans (or cis) influence*, while the kinetic consequences are called *trans (or cis) effect*. Obviously, mutual influence (structural effects) is important for stereochemistry and is considered in this section; the reactivity effects are discussed in Section 11.3.

Stereochemical effects of mutual influence of ligands have been discussed repeatedly (see, e.g., [9.81–9.87]). Illustrative examples of *trans and cis* influence are given in Tables 9.8 to 9.11. The data in Table 9.8 show explicitly that in compounds of Pt(II) with square-planar coordination, the *trans* ligand strongly influences the bond length Pt—Cl. Substitution of the *trans* ligand changes the Pt—Cl bond length from 2.28 to 2.45 Å. The *cis influence* (Table 9.9) is less informative. In comparison with Table 9.8 it is seen that *cis* influence is much weaker than the *trans* influence.

Table 9.8. *Trans* **influence on bond lengths R(Pt—Cl) in some Pt(II) compounds**

Complex	Trans Atom (Ligand)	R(Pt—Cl) (Å)
K[Pt(acac)$_2$Cl]	O	2.28(1)
trans-[(PEt$_3$)$_2$PtCl$_2$]	Cl	2.30(1)
cis-[(p-C$_6$H$_4$S)$_2$PtCl$_2$]	S (of RS⁻)	2.30
trans-[(PEt$_3$)$_2$Pt(CO)Cl]	CO	2.30
cis-[(PEt$_2$PH)Pt(CNEt)Cl$_2$]	RNC	2.314(10)
K$_2$[PtCl$_4$]	Cl	2.316
[Pt(L-methionine H)Cl$_2$)]	S (of R$_2$S)	2.32
K[Pt(NH$_3$)Cl$_3$]H$_2$O	N	2.321(7)
K[Pt(C$_2$H$_4$)Cl$_3$]H$_2$O	C=C	2.327(7)
[Pt(H$_3$NCH$_2$CH=CHCH$_2$NH$_3$)Cl$_3$]	C=C	2.342(2)
cis-[(PMe$_3$)$_2$PtCl$_2$]	P	2.37(1)
cis-[(PEt$_2$Ph)Pt(CNEt)Cl$_2$]	P	2.390(8)
trans-[PMe$_2$Ph)$_2$Pt(CH$_2$SiMe$_3$)Cl]	C (of alkyl)	2.415(5)
trans-[(PPH$_2$Et)$_2$PtHCl]	H	2.42(1)
trans-[PMe$_2$Ph)$_2$Pt(SiMePh$_2$)Cl]	Si	2.45(1)

Source: [9.81].

Table 9.9. *Cis* influence on the bond lengths $R(Pt—Cl)$ in some Pt(II) complexes

Complex	*Cis* Atom (Ligand)	$R(Pt—Cl)$ (Å)
trans-[(PEt$_3$)$_2$PtCl$_2$]	P	2.29
[Pt(*cis*-MeCH=CHCH$_2$NH$_3$)Cl$_3$]	C=C	2.297(6)
K[Pt(C$_2$H$_4$)Cl$_3$]H$_2$O	C=C	2.305
K[Pt(NH$_3$)Cl$_3$]H$_2$O	N	2.315
K$_2$[PtCl$_4$]	Cl	2.316
trans-[Pt(NH$_3$)$_2$Cl$_2$]	N	2.32(1)

Source: [9.81].

Table 9.10 allows us to follow the variation of one of the most informative parameters of the *trans* influence, the *trans* elongation $\Delta = R(M—X_{trans}) - R(M—X_{cis})$, in a series of MOX$_5$ complexes with d^0, d^1, and d^2 configurations [9.88]. It is seen that there is an essential decrease in *trans* influence along the series $d^0 > d^1 > d^2$. Note, however, that in Table 9.10 the change of the d^n configuration is associated with a change of the CA itself. Table 9.11 shows some other structural characteristics for *trans*- and *cis*-PtCl$_2$(PR$_3$)$_2$ (R is mainly Et and sometimes Me): bond lengths $R(Pt—Cl)$ and $R(Pt—P)$, vibrational frequencies $v(Pt—Cl)$ and $v(Pt—P)$, the constants of nuclear spin–spin interaction $J(Pt—P)$ and $J(P—P)$, NMR chemical shift $\delta(^{31}P)$ with respect to H$_3$PO$_4$, NQR frequency $v'(^{35}Cl)$, and the $2p$ electron bonding energy E$_b$ in Cl determined from x-ray photoelectron spectra. A comparison of these data shows that the bonds Pt—Cl in the *trans* complex and Pt—P in the *cis* complex are stronger than the same bonds in the *cis* and *trans* complexes,

Table 9.10. Comparison of *cis* and *trans* bond lengths, $R(M—X_{cis})$ and $R(M—X_{trans})$ and *trans* elongation $\Delta = R(M—X_{trans}) - R(M - X_{cis})$ in some MOX$_5$ complexes with d^0, d^1, and d^2 configurations

Compound	d^n	$R(M—X_{trans})$ (Å)	$R(M—X_{cis})$ (Å)	Δ (Å)
MoOF$_5$	d^0	2.29	1.86	0.43
K$_2$[NbOF$_5$]	d^0	2.06	1.84	0.22
K$_2$[MoOF$_5$]H$_2$O	d^1	2.03	1.87	0.16
OsOF$_5$	d^2	1.72	1.78	−0.06
(NH$_4$)$_2$[MoOBr$_5$]	d^1	2.83	2.55	0.28
K$_2$[MoOCl$_5$]	d^1	2.587	2.39	0.20
K$_2$[ReOCl$_5$]	d^2	2.47	2.39	0.08

Source: [9.88].

**Table 9.11. Comparison of structural quantities in *cis*-
and *trans*-PtCl$_2$(PR$_3$)$_2$ compounds**

Quantity	Cis	Trans
$R(\text{Pt—Cl})$, Å	2.37	2.29
$R(\text{Pt—P})$, Å	2.25	2.30
$v(\text{Pt—Cl})$, cm^{-1}	294	340
$v(\text{Pt—P})$, cm^{-1}	435	419
$J(\text{Pt—P})$, Hz	3520	2400
$J(\text{P—P})$, Hz	-18.7	510
$v'(^{35}\text{Cl})$, MHz	~ 18.0	20.99
$\delta(^{31}\text{P})$, $10^6 d$	24.0	15.8
$E_b(\text{Cl}, 2p)$, eV	198.2	198.6

respectively. This means that PR$_3$ is a stronger *trans*-influencing ligand than Cl, and this *trans* influence is reflected in a number of structural characteristics.

The effects of mutual influence of ligands are obviously of electronic and vibronic origin, and the problem is to formulate some general rules that correlate the observed influences with specific electronic features of the ligands and the CA. There have been many attempts to contribute to the solution of this problem mentioned above; we consider here some general and more recent achievements.

The first attempts to explain the origin of *trans* influence were based on a comparison of σ-donor and π-acceptor properties of the ligands. To begin with, consider a complex MX$_n$ in which the possible π bonding is neglected. If one ligand X is substituted by a better σ donor Y, the positive charge on M decreases and, in general, all the other σ bonds M—X weaken. However, this inductive effect, due to special properties of the wavefunctions of M involved in the bonding and charge transfer, is angular dependent, that is, it has special predominant directions. In a square-planar MX$_4$ complex, for instance, PtX$_4$ with d^8 configuration, four equivalent bonds are formed by the Pt $d_{x^2-y^2}$ orbital, and the directed influence of Y is realized through the MOs that involve the unoccupied p orbitals of Pt. The two ligands Y and X in the *trans* positions to each other are strongly interrelated via these MOs.

A simple expression of this *trans* influence was obtained [9.89] by means of the angular overlap model (AOM) (Section 5.2). In this model, the difference in stabilization energy ΔE of the bond M—X when passing from the *trans* coordinate system X—M—X to the *cis* system Y—M—X is proportional to e_σ and the difference $\Delta e_\sigma = e_\sigma(\text{X}) - e_\sigma(\text{Y})$:

$$\Delta E \sim e_\sigma(\text{X})[e_\sigma(\text{X}) - e_\sigma(\text{Y})] \qquad (9.27)$$

where $e_\sigma(\text{X})$ and $e_\sigma(\text{Y})$ are the AOM parameters for ligands X and Y, respectively [Eqs. (5.42) and (5.44)].

It follows from Eq. (9.27) that if the ligand Y has a larger e_σ value than that of X, $e_\sigma(Y) > e_\sigma(X)$, then $\Delta E < 0$, and the bond in the *trans* position is weakened (elongated). On the contrary, if $e_\sigma(Y) < e_\sigma(X)$, then $\Delta E > 0$ and the *trans* bond M—X is strengthened. Hence the *trans* elongation increases with the $e_\sigma(Y)$ values. If one assumes that the e_σ value increases with the σ-donor properties (decreases with Pauling electronegativities), the *trans* influence (elongation) increases with increasing σ-donor properties or decreasing electronegativities of the ligand Y. Figure 9.24 shows that the experimental data, in general, confirm this trend [9.1, 9.89] (note that the σ-donor effect on the *trans* influence has been suggested much earlier [9.90]).

Moreover, Eq. (9.27) also predicts that the largest *trans* influence is expected not only when the difference $e_\sigma(Y) - e_\sigma(X)$ is large, but also when $e_\sigma(X)$ is large. Thus ligands with larger e_σ values are both more strongly *trans* influencing and more susceptible to the *trans* influence of other ligands. Indeed, the best *trans*-influencing ligands are often also most easily influenced by stronger *trans*-influencing ligands.

The presence of π bonds may be essential for the mutual influence of ligands. The role of π donation in the *trans* effect was first revealed by Chatt, et al. [9.91] and independently by Orgel [9.92], and it is discussed in more detail in Section 11.3. Here we consider some aspects of π bonding that have a direct stereochemical effect. This is, first, the interrelation between σ and π orbital charge transfers.

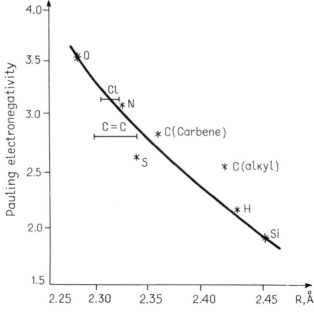

Figure 9.24. Pt—Cl bond length in square-planar Pt(II) complexes (Table 9.8) versus the Pauling electronegativities of the *trans* ligand.

In Section 6.3 the diorbital bonding of the ligand, including one σ and one π bond, is considered (with two π bonds, the effect is similar). It is emphasized there that the total charge transfer $\Delta q = \Delta q_\pi + \Delta q_\sigma$ to the CA ($\Delta q < 0$) or to the ligand ($\Delta q > 0$) may be small, while the orbital charge transfers Δq_π and Δq_σ, can be relatively large, because they may have opposite signs (usually, $\Delta q_\sigma < 0$ and $\Delta q_\pi > 0$: back donation). Note that small Δq values are required by thermodynamic conditions; large total charge transfers are energetically inconvenient due to electron correlation effects (cf. Pauling's principle of electroneutrality [9.8, 9.93]).

For this reason, in the absence of π bonding ($\Delta q_\pi = 0$), Δq_σ cannot be very large. In the presence of a π back donation $\Delta q_\pi > 0$, Δq_σ can be much larger. The π back donation enhances the σ-donor properties of the ligands. But as shown above, the σ-donor properties are directly related to the *trans* influence. Hence the formation of π bonds, enhancing σ-donor properties, increases the *trans* influence [9.94]. This conclusion explains many experimental data on *trans* elongation and other *trans* influences.

At the early stages of discussion the statement about the role of π bonding in *trans* influence has been very stimulating. For example, it was observed that in *cis*-$K_2[Pt(NO_2)_3Cl_3]$ the NO_2^- group, which is able to form strong π bonds with the CA, is relatively weakly *trans* influencing: the Pt—Cl bond length $R \approx 2.34$ Å in the *trans* position to NO_2^- is the same as for the C=C *trans* group in square-planar Pt(II) compounds (Table 9.8), whereas the π-acceptor properties of NO_2^- are much stronger. To explain this discrepancy, it was assumed that the position of the NO_2^- group in the case where it is weakly *trans* influencing is bent with respect to the position in which it forms good π bonds [9.94, 9.95]. X-ray experiments carried out to verify this assumption confirmed the bent position of NO_2^- in the complex above [9.95].

As stated above, directed (*trans*) σ influence is possible when there are such atomic states of the CA (e.g., p states) which, being involved in the bonding, transfer the influence mainly in the *trans* direction. However, this does not mean that *cis* ligands do not participate in the mutual influence. Indeed, since the charge transfer to the CA is limited, all the ligands compete in this process, which means that strong σ-donor ligands in the *cis* position reduce the σ-donor ability of the given ligand and hence its *trans* influence [9.94]. In other words, strong *trans*-influencing (i.e., *trans*-weakening) ligands reduce the similar effect on the *cis* coordinate, thus producing an opposite *cis* influence (*cis* strengthening). Thus in this model *the trans and cis influences of the same ligand have opposite signs.*

This effect is seen explicitly from the data in Table 9.9. Indeed, compare the Pt—Cl bond lengths in two complexes, *trans*-$[Pt(PEt_3)_2Cl_2]$ and *trans*-$[Pt(NH_3)_2Cl_2]$, that differ by the linear fragment in the *cis* position, P—Pt—P and N—Pt—N, respectively. Since the *trans* coordinate Cl—Pt—Cl in these two complexes is the same, the difference in the Pt—Cl bond lengths is due entirely to the *cis* influence. From Table 9.8 it is seen that the *trans* influence of P is stronger than N, and hence the *cis* influence, following the qualitative treatment above, should be opposite. Thus the Pt—Cl bond length is expected to be shorter in the $[Pt(PEt_3)_2Cl_2]$ complex than in $[Pt(NH_3)_2Cl_2]$, in agreement with the experimental data in Table 9.9 (2.29 and 2.32 Å, respectively).

The problem of CA orbitals that promote the directed *trans* and *cis* influences is one of the most important in the prediction of CA that are effective in the mutual influence of ligands. Its full solution requires numerical calculations including all possible active orbitals. However, in general, the results of numerical calculations are not transferrable and cannot be related directly to specific properties of the CA and ligands (Section 6.1); qualitative or semiquantitative treatments remain rather important. More exact but less general results of numerical calculations, and more general but less exact results of analytical (qualitative and semiquantitative) treatments, are complementary to each other.

Vibronic Theory of Mutual Influence of Ligands

The mutual influence of ligands in stereochemistry can be considered as an effect of vibronic influence. Indeed, the substitution of one ligand in the complex by another can be regarded as a change in the electronic structure which produces changes in the nuclear configuration via vibronic coupling. In Section 11.2 detailed consideration of the vibronic effects in ligand coordination is given, which also evaluates the changes of interatomic distances by coordination, and in Section 9.2 the distortion of ligand geometry induced by the coordination is discussed briefly. A somewhat similar idea is used in this section to consider the mutual influence of ligands or, more precisely, the change in mutual influence by changing ligands [9.84–9.86].

Consider a homoligand coordination system of the type MX_n with a nondegenerate electronic ground state. Following Section 7.1, its Hamiltonian can be presented as

$$H = H_{el} + W \tag{9.28}$$

where H_{el} is the electronic part of the Hamiltonian for fixed nuclei [see Eq. (7.9)] and W is vibronic coupling (7.7). In the stable configuration $Q_\alpha = 0$, $\alpha = 1, 2, \ldots, N$, the adiabatic potential $\varepsilon(Q_\alpha)$ has a minimum with respect to all symmetrized coordinates Q_α, and in the harmonic approximation, $\varepsilon(Q_\alpha)$ has the usual quadratic form (7.12), with K_α given by Eqs. (7.86) to (7.88).

Upon substitution of the ligand X by Y, the change in the Hamiltonian (9.28) can be presented just by adding the so-called substitution Hamiltonian H_s equal to the difference between the Hamiltonians of the $MX_{n-1}Y$ and MX_n systems:

$$H = H_{el} + W + H_s \tag{9.29}$$

Now we assume that H_s can be considered as a perturbation. This implies that changes in energy states induced by H_s are small. Then to obtain the adiabatic potential of the system with the Hamiltonian (9.29), following Section 7.1, one must consider two perturbations, H_s and W, instead of W only in the MX_n system. With the two perturbations the adiabatic potential $\varepsilon'(Q_\alpha)$ is [9.84]:

$$\varepsilon'(Q_\alpha) = \varepsilon(Q_\alpha) + h_{00} - \sum_j \frac{h_{0j}^2}{\Delta_{j0}} - 2\sum_{\alpha,j} \frac{h_{0j}F_\alpha^{0j}}{\Delta_{j0}} Q_\alpha \tag{9.30}$$

where we denoted

$$h_{0j} = \langle 0|H_s|j\rangle \tag{9.31}$$

and $F_\alpha^{0j} = \langle 0|\partial H/\partial Q_\alpha)_0|j\rangle$ are the vibronic constants (7.22), Δ_{ij} being the energy gaps. From Eq. (9.30) one can see that in addition to the constant terms h_{00} and $-\sum h_{0j}^2/\Delta_{j0}$, which shift the energy levels, there is a term that is linear in Q_α. Added to the quadratic terms in $\varepsilon(Q_\alpha)$ after (7.12), this linear term displaces the minimum position in the Q_α (or $-Q_\alpha$) directions, the sign of this displacement being determined by the sign of the product $h_{0j}F^{0j}$. The new equivalent positions are

$$Q_\alpha^0 = 2\sum_j \frac{h_{0j}F_\alpha^{0j}}{\Delta_{j0}K_{\alpha 0}} \tag{9.32}$$

where $K_{\alpha 0}$ is given by Eq. (7.87).

Thus the idea of vibronic mutual influence of ligands is that if one substitutes one of the ligands, the changes in the electronic structure are no more consistent with the previous geometry, and other ligands relax to new equilibrium positions (new minima of the AP). To find them, the matrix elements of the vibronic coupling (the vibronic constants F^{0j}) and the substitution operator h_{0j} should be analyzed. This can be done by a model description for more specific types of systems.

Consider, for example, the MO LCAO model for octahedral σ-bonded complexes MX_6 of the following three basic types:

(i) M is a transition element, and there are 12 electrons in the valence σ MOs.

(ii) M is a nontransition element in a low-oxidation state, and there are 14 electrons in the σ MOs (i.e., there is an electron lone pair).

(iii) M is a nontransition element in a higher oxidation state.

The typical MO energy-level schemes for these systems are illustrated in Fig. 9.25. By populating the one-electron MOs with the number of valence electrons

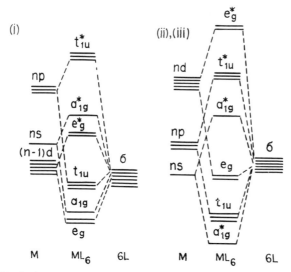

Figure 9.25. Typical σ MO energy-level schemes for octahedral MX_6 complexes of transition (i) and nontransition (ii, iii) elements. The HOMO are: (i) t_{iu}, (ii) a_{1g}^*, (iii) e_g [cf. Fig. 6.1; for the denotations (i), (ii), (iii), see the text].

available, we easily find the HOMO and LUMO. For group (i) the HOMO is t_{1u} and the LUMO is e_g^*; for (ii) the HOMO and LUMO are a_{1g}^* and t_{1u}^*, respectively; and so on (Table 9.12).

The next step in simplification is to restrict the treatment by HOMO and LUMO only. This restriction is often used in quantum chemistry, especially in intermolecular interactions (reactivity problems; Section 11.1). With the HOMO–LUMO constraint, there is only one off-diagonal matrix element (9.31), $h_{01} = h$; only one vibronic constant, $F_{01} = F$; one energy gap, $\Delta_{10} = \Delta$; and Eq. (9.32) simplifies significantly:

$$Q_\alpha = \frac{2hF_\alpha}{\Delta K_\alpha} \tag{9.33}$$

Thus only one type of symmetrized coordinates is nonzero, and this is determined by the selection rules (Section 3.4). Indeed, the vibronic constant is nonzero if the product of the irreducible representations of the ground $|0\rangle$ and excited $|1\rangle$ states contains the irreducible representation of the derivative of the Hamiltonian over Q_α. The representation of this derivative coincides with that of Q_α. Hence for nonzero F_α it is required that $\Gamma_0 \times \Gamma_1$ contain Γ_α. For instance, for complexes of type (i), $|0\rangle$ and $|1\rangle$ are the HOMO and LUMO t_{1u} and e_g^*, respectively, and since $T_{1u} \times E_g = T_{1u} + T_{2u}$, only $Q(T_{1u})$ and/or $Q(T_{2u})$ distortions can, in principle, occur as a result of the substitution $MX_6 \rightarrow MX_5Y$ above.

Table 9.12. Parameters characterizing the distortions in σ-bonded complexes of transition and nontransition elements produced by the substitutions $MX_6 \to MX_5Y$ in octahedral and $MX_4 \to MX_3Y$ in square-planar complexes

Type of Complex	HOMO	LUMO	Distortion Mode for Which $F_\alpha \neq 0$	AOs of CA for Which $h \neq 0$	Distortion Coordinate Q_α
Octahedral					
(i)	t_{1u}	e_g^*	T_{1u}, T_{2u}	p_z, d_{z^2}	$Q_z'' = (Z_1 + Z_4)\sqrt{2}$
(ii)	a_{1g}^*	t_{1u}^*	T_{1u}	$s; p_z$	$Q_z'' = (Z_1 + Z_4)/\sqrt{2}$
(iii)	e_g	a_{1g}^*	E_g	$d_{z^2}; s$	$Q_9 = (2Z_1 - 2Z_4 - X_2 + X_5 - Y_3 + Y_6)/2\sqrt{3}$
Square-Planar					
(i)	e_u	b_{1g}^*	E_u	$p_x; d_{x^2-y^2}$	$Q_x = (X_2 + X_4)/\sqrt{2}$
(ii)	a_{1g}^*	e_u^*	E_u	$s; p_x$	$Q_x = (X_2 + X_4)/\sqrt{2}$

Source: [9.84].

On the other hand, the nonzero distortion after (9.32) also requires that $h = \langle 0|H_s|1\rangle$ should be nonzero, which means that the HOMO and LUMO should contain the atomic orbital of the substituted atom. To obey this condition, if Y is on the z axis, the t_{1u} and e_g^* MOs should contain the p_z and d_{z^2} orbitals of the CA, respectively (see Table 5.1). With these orbitals F is nonzero for $Q(T_{1u})$ only, namely for its Q_z'' component: $Q_z'' = (Z_1 + Z_4)/\sqrt{2}$ (Table 7.1). This distortion displaces the two ligands in the *trans* position 1 and 4 along the z axis (both in the same direction) with respect to the plane of the CA and the other four ligands. In the same manner the distortion coordinates Q_α active in the mutual influence of ligands were obtained for other types of coordination system as shown in Table 9.12.

Thus within the limits of the HOMO–LUMO approximation, the coordinates of mutual influence of ligand can be revealed directly without detailed calculations. As seen from Table 9.12, the distortions induced by ligand substitution are indeed directed, and for the most part these directions are along the *trans* coordinate (*trans* influence), except for case (iii) in octahedral σ-bonded complexes of nontransition elements in high-oxidation states, in which the active Q_9 coordinate yields both *trans* and *cis* displacements with opposite signs, the *cis* influence being twice weaker than the *trans*.

However, the picture as a whole remains incomplete, even qualitatively, without knowledge of the sign of Q_α. Following (9.33), the sign of Q_α is determined by the sign of the product hF_α, so we should analyze the signs of h and F_α according to Eqs. (9.31) and (7.22), respectively. For this purpose, values of the LCAO coefficients for the HOMO and LUMO may be useful. A detailed analysis shows [9.84, 9.85] that this issue can be solved approximately by estimating the value of the matrix element $h \sim \langle 0|H_s|1 \rangle$ for the substitution $X \to Y$, as proportional to the difference between their Coulomb integrals: $h \approx \Delta\alpha = \alpha(Y) - \alpha(X)$, which, in turn, characterize their σ-donor properties (such approximations are quite common in organic chemistry [9.96]). Omitting the calculations, we discuss below some of the main results.

Assume that by substitution $X \to Y$ the σ-donor properties increase, $\Delta\alpha = \alpha(Y) - \alpha(X) > 0$. Then [9.84] for octahedral σ-bonded coordination compounds of group (i) (transition metals with 12 electrons on the six MOs), $h < 0$, $F > 0$, and hence for these systems $Q''_z < 0$. This distortion is equivalent to *trans* elongation because in $Q''_z = (Z_1 + Z_4)/\sqrt{2}$, $Z_1 = Z_4$, and hence $Z_4 < 0$. For $\Delta\alpha < 0$ the *trans* elongation changes to *trans* shortening. Note that in this approximation the *cis* ligands are not involved in the mutual influence.

In complexes of type (ii), $F < 0$ and $h > 0$, and hence $Q''_z < 0$. Thus in octahedral σ-bonded complexes of nontransition elements with 14 electrons on the σ MOs (one lone pair) the distortion by ligand substitution is similar to that expected for group (i): *trans* elongation for increasing σ-donor properties, $\Delta\alpha > 0$, and *trans* shortening when the opposite inequality $\Delta\alpha < 0$ holds. Many examples in the literature confirm these conclusions [9.81–9.83].

In compounds of type (iii), post-transition element complexes in high-oxidation states, the situation is more complicated and further differentiations should be introduced [9.84]. Divide all the MX_6 compounds of this group into two classes according to the position of their σ energy levels of X (Fig. 9.25) with respect to that of M. The first class comprises the compounds of post-transition elements M from the left-hand side of the periodic table for which the energy levels of the ns valence AOs are located about the same level or even above the σ AOs of typical ligands. For the right-hand-side elements, the ns AOs lie below the σ AOs of the ligands, and the compounds of these elements form the second class.

For the first class of group (iii) $F > 0$, and for $\Delta\alpha > 0$, $h < 0$. Hence for these compounds the sign of the Q_9 distortion in Table 9.12 is negative: $Q_9 < 0$. From the form of Q_9 displacements given in Fig. 7.1, one can see that $Q_9 < 0$ means *trans* shortening and *cis* elongation, the shortening being twice the elongation (in contrast to previous cases, for which the distortion by ligand substitution with $\Delta\alpha > 0$ is *trans* elongation). This distortion results in a compressed octahedron with one shorter axis $Y—M—X_{trans}$. Such *trans* shortening may be regarded as a *cis* elongation. Therefore, this kind of ligand mutual influence is often called *cis* influence. Owing to the additivity of the vibronic effects (Section 7.2), the substitution of two ligands in *trans* positions, $MX_6 \to MX_4Y_2$, doubles this effect.

Cis influence of this kind is indeed observed experimentally in octahedral coordination compounds of post-transition elements from the left-hand side of the periodic table. Typical examples are provided by the systems *trans* $SnX_4(CH_3)_2$, with $X = F$, Cl, Br, NCS [9.97], as well as by other Sn(IV) and Ga(III) compounds.

For the second class of type (iii) compounds with post-transition elements from the right-hand side of the periodic table which have sufficiently low ns AOs, F may become negative and the sign of the Q_9 distortion is inverted, $Q_9 > 0$, which means *trans* elongation and *cis* shortening. Although qualitative, the trend as a whole in which *trans* shortening (*cis* elongation) changes to *trans* elongation (*cis* shortening) when passing from left to right in the series of post-transition elements in the periodic table (more precisely, from higher to lower ns orbital energy positions) seems to be true.

In particular, it predicts that for some elements in between the two limits above, the mutual influence is small to zero. Indeed, in SF_5Cl, both the *cis* and *trans* S—F bonds have practically the same lengths as the S—F bond in SF_6 [9.98]. Similarly, there are no differences in the corresponding bond lengths in $SF_5(NCO)$, $SeF_5(NCO)$, and $TeF_5(NCO)$ [9.99]. In contrast to these results, in IOF_5 the *trans* I—F bond is significantly longer than the *cis* bond [9.100] (see also [9.84, 9.85]).

For square-planar σ complexes of type MX_4, the typical HOMO and LUMO, as well as the coordinates of distortion for two types of systems, are given in Table 9.12:

(i) M is a transition metal, and there are eight valence electrons on the σ MOs (typical example $[Pt^{II}X_4]$).

(ii) M is a post-transition element in a low-oxidation state (typical example $[Te^{II}X_4]$).

A procedure quite similar to that used above for octahedral complexes yields [9.84, 9.85] for $\Delta\alpha > 0$ in case (i) $F > 0$, $h < 0$, while in case (ii) $F < 0$, $h > 0$, but in both cases $Fh < 0$, and hence the coordinate of distortion $Q_x < 0$, which indicates *trans* elongation. This is the typical *trans* influence well known for square-planar complexes, in particular, for Pt(II).

The HOMO–LUMO approximation employed above is certainly restricting. If there are several excited states with different symmetries relatively close in energy, this approximation may be invalid. Indeed, as shown above, the symmetry of the excited state (together with that of the ground state) determines the coordinate of distortion, and if only one of the active excited states is taken into account, the real distortion may be lost. For instance, if in the σ-bonded octahedral complexes of post-transition elements in higher oxidation states [type (iii)], one goes beyond the HOMO–LUMO approximation and takes into account vibronic coupling with the next excited electronic state, say t_{1u} (instead of the excited state a_{1g}^* employed in Table 9.12), the coordinate of distortion can be T_{1u}, for instance, $Q_z' = \frac{1}{2}(Z_1 + Z_2 + Z_4 + Z_5)$, which changes the angles between the bonds [9.84–9.86]. For further discussion of vibronic mutual influence of ligands that changes bond angles, see [9.86].

The HOMO–LUMO approximation is also not appropriate when multiple bonds are taken into consideration because they create more close-in-energy

Table 9.13. Experimental data on *cis* elongation and *trans* shortening in some actinide MOX$_5$ compounds

Compound	$R(M—O)$ (Å)	$R(M—X_{trans})$ (Å)	$R(M—X_{cis})$ (Å)
UCl$_6$	—	2.47	2.47
(PPh$_4$)[UOCl$_5$]	1.76	2.43	2.53
(NEt$_4$)$_2$[PaOCl$_5$]	1.74	2.42	2.64
(NEt$_4$)$_2$[UOCl$_4$]	1.75	—	2.67

Source: [9.84].

excited states. However, owing to their additive properties, the vibronic contributions can be considered separately for each active excited state; the final result equals the sum of all these contributions (the distortions should be summed up as vectors). This treatment requires more parameters.

If the ligands produce π bonds, they enhance the *trans* influence, but its sign depends significantly on the coordination center M. Interesting examples of this kind are provided by actinides, where the participation of f electrons in the bonding is important. In complexes of type MX$_6$, where M is an actinide, the HOMO t_{1u} contains both σ and π bonding ($\sigma + \pi$ bonds), while the LUMO is T_{2u}^* (a π MO). The coordinate of distortion by the MX$_6 \rightarrow$ MX$_5$Y substitution is Q_9, and its sign proved to be negative [9.84, 9.86], $Q_9 < 0$. This means that similar to some post-transition [type (iii)] compounds, the distortion in the mutual influence results in *trans* shortening and *cis* elongation. The experimental data (Table 9.13) confirm this prediction. Note that in quite similar complexes of transition metals the mutual influence of ligands results in *trans* elongation. For instance [9.88], in K$_2$[MoOCl$_5$] (I) and K$_2$[ReOCl$_5$] (II), $R(M—X_{trans})$ is significantly larger than $R(M—X_{cis})$ (in Å): 2.587 and 2.39 in (I) and 2.47 and 2.39 in (II), respectively. This illustrates the fact that the electronic structure of the coordinating element M that transfers the mutual influence is most important.

Relative Stability of Isomers

Equations (9.30) to (9.32) enable us to obtain some estimates of relative stabilities of complexes with ligand substitution, in particular, isomers. To do this we have to determine the change in depth of the minimum of the adiabatic potential produced by the corresponding ligand substitutions. By substituting the value of the equilibrium coordinates Q_α^0 after (9.32) into Eq. (9.30), we obtain for the energy difference $\Delta E = \varepsilon'(Q_\alpha^0) - \varepsilon(0)$,

$$\Delta E = h_{00} - \sum_j \frac{h_{0j}^2}{\Delta_{ji}} - 2 \sum_\alpha \frac{(\sum h_{0j} F_\alpha^{0j}/\Delta_{j0})^2}{K_{\alpha 0}} \qquad (9.34)$$

Consider, for example, *cis* and *trans* isomers of square-planar or octahedral complexes. For convenience, denote

$$h_{0j}^{\pm} = h_{0j}^c \pm h_{0j}^t \qquad (9.35)$$

where the superscripts t and c stand for the *trans* and *cis* isomers, respectively. Using Eq. (9.34) for the energy difference $\Delta E(c \to t) = \Delta E_c - \Delta E_t$, we have

$$\Delta E(c \to t) = \sum_j \frac{h_{0j}^+ h_{0j}^-}{\Delta_{0j}} + 2 \sum_\alpha \left(\sum_j \frac{h_{0i}^+ F_\alpha^{0j}}{\Delta_{j0}} \right) \left(\sum_j \frac{h_{0j}^- F_\alpha^{0j}}{\Delta_{j0}} \right) K_{\alpha 0} \qquad (9.36)$$

To make qualitative conclusions, this expression should be analyzed using estimates of the matrix elements similar to those employed above in the study of vibronic mutual influence of ligands in the HOMO–LUMO approximation [9.84]. But occasionally, qualitative results can be obtained by symmetry considerations only. For instance, in square-planar complexes $Pt^{II}X_4$ the HOMO is e_u and LUMO is b_{1g}^*. These two orbitals have opposite parity, which means that the matrix elements h_{0j} according to (9.31) are equal to zero if the substitution operator H_s is even, and nonzero if H_s is odd. In *trans*-substituted complexes $Pt^{II}X_2Y_2$ H_s is even: The system does not change by inversion operation. Hence in Eq. (9.34) in the HOMO–LUMO approximation all the matrix elements $h_{01} = 0$ and the negative terms vanish. On the contrary, in *cis*-PtX_2Y_2, H_s has an odd component and $h_{01} \neq 0$. Therefore, provided that other conditions are identical, *the cis isomer of PtX_2Y_2 is more stable than the trans isomer*.

Since ΔE is quadratic in h_{01}, and h_{01} is proportional to $\Delta \alpha = \alpha(Y) - \alpha(X)$, the relative stability of the *cis* isomer with respect to the *trans* isomer increases rapidly with the difference in the donor properties (*trans* influence) of Y and X. On the contrary, if $\Delta \alpha$ is small, other effects not taken into account in the reasoning above may become essential. In particular, if the ligands X and Y have significantly different charges, the nonvalence electrostatic interaction between the ligands in the two isomers is different, the electrostatic repulsion being lower in the *trans* isomer.

These conclusions are confirmed by experimental data. For example, in MX_2Y_2 with M = Pt(II), Pd(II), Au(III), X = alkyl, PR_3, SR_2, Y = halogen, the relative stability of the *cis* isomer is confirmed by thermochemical and other data, its absolute value decreasing in the same consequence as the *trans* influence: alkyl > PR_3 > SR_2. For $Pt(NH_3)_2Cl_2$, where NH_3 is not very far from Cl in the donor properties, whereas they differ essentially in charge, the *trans* isomer is more stable.

For complexes in degenerate electronic states the vibronic problem of relative stability of isomers is more complicated; in the approximation employed above it can be reduced to a secular equation similar to (7.32). In particular, to compare the two isomers *cis*- and *trans*-$Cu(II)X_4Y_2$, one can start from the regular octahedron $Cu(II)X_6$ with a degenerate electronic E term and introduce, as above, the two interactions, vibronic W and substitution H_s, as two perturbations. The secular equation for the two MO functions $|1\rangle$ and $|2\rangle$ of the E term looks as follows [cf. (7.36)]:

$$\begin{vmatrix} F_E Q_\theta + h_{11} - \varepsilon^v & -F_e Q_\varepsilon + h_{12} \\ -F_E Q_\varepsilon + h_{12} & -F_E Q_\theta + h_{22} - \varepsilon^v \end{vmatrix} = 0 \qquad (9.37)$$

This equation can be solved and analyzed relatively easily [9.85]. The comparison of the minima depths in *cis-* and *trans-*$Cu(II)X_4Y_2$ shows that the *trans* isomer is more stable than the *cis* isomer. Cu(II) isomers are discussed again in Sections 9.4 and 11.2.

9.4. CRYSTAL STEREOCHEMISTRY

In the stereochemistry of transition metal compounds, crystal states are most important because they provide the possibility (often the unique possibility) of direct observation of molecular shapes. Indeed, these compounds rarely exist as free molecules in the gas phase, and the study of molecular shapes by other (e.g., spectroscopic) methods is but indirect observation.

On the other hand, the crystal environment influences the local stereochemistry significantly. One of the principal differences between the crystal chemistry of organic and inorganic compounds is that organic crystals are (mostly) molecular crystals containing conveniently packed organic molecules, whereas inorganic crystals may have nothing to do with molecules, or the entire crystal may be just one giant molecule (compare CH_4 and NaCl crystals). From the point of view of the electronic structure considered in this book, interaction between organic molecules in crystals is mainly nonvalence in nature and can be described by empirical parameters. This cannot be done in many cases of inorganic compounds in which the details of the electronic structure of the crystal components play a key role in crystal lattice formation and stereochemistry (see also the discussion in the last subsection of Section 9.1).

Coordination compounds of transition metals occupy an intermediate place between the two extremes noted above. On the one hand, the coordination system enters the crystal state mostly as a whole entity and preserves many of its molecular features. On the other hand, interaction between the coordination centers in the lattice is relatively strong and is determined by electronic factors that are no less important than nonvalence (steric) interactions.

Classical crystal stereochemistry is based on the idea of ball packing, or its significantly advanced form the VSEPR model, which can be considered as a kind of charge packing (Section 9.1). The crystal structure of coordination compounds is rather complicated, and in most cases it cannot be reduced to charge packing, because of primary importance of quantum effects. As mentioned in Section 9.2, the vibronic approach provides further insight into the problem [9.101]. In particular, it allows one to separate the contribution of local (chemical) forces from that of the long-range interactions in the lattice and to show that the local forces are most essential in determining the polyatomic stereochemistry. In what follows in this section, we emphasize the main nonclassical aspects of stereochemistry in the crystal state.

Plasticity Effect

As mentioned in Section 9.2, the Jahn–Teller and pseudo Jahn–Teller induced distortions in free coordination systems are dynamic in nature; they were classified in Section 7.4 as free rotations, hindered rotations, and pulsating motions. Some of these motions can be observed as pseudorotations, provided that certain conditions considered in Sections 9.1 and 9.2 are satisfied.

In Section 9.1 we discussed the role of small external perturbations that quench the dynamics of the distortions and make them static, so that a very small perturbation may result in strong distortion (vibronic amplification of external perturbations). The symmetry of this static distortion corresponds to that of the external influence, provided that it is uniaxial in the direction of one of the Jahn–Teller active coordinates. It follows that among the many possible equivalent distortions of a coordination system predicted by the vibronic theory (Sections 7.3 and 7.4), the only ones realized in the crystal state are those that correspond to the symmetry of the environment, and in a measure allowed by this environment. Hence *the same coordination polyhedron may have significantly different shapes in different crystals.* This phenomenon looks as if the coordination polyhedron had a soft (plastic) coordination sphere, which in the crystal state takes the form of the crystal environment; it is called *the plasticity effect* [9.38, 9.44] (see also the flexional behavior mentioned in Section 9.2 [9.72]). The first observation of this effect is due to Fackler and Pradilla-Sorzano [9.73, 9.102].

The best examples to illustrate the plasticity effect are octahedral coordination compounds with a twofold-degenerate E term [Cu(II), high-spin Mn(III), and Cr(II), low-spin Co(II), etc.]. For them the adiabatic potential in the case of weak quadratic terms (small G_E constants; Section 7.3) has the form of a hat, which allows for any distortion of the coordination sphere along the symmetrized Q_θ and Q_ε displacements within the limits $Q_\theta^2 + Q_\varepsilon^2 = \text{const.}$ (Fig. 7.13). If the quadratic vibronic terms are significant, only three directions of tetragonal distortions along the fourth-order axes remain equally probable (Fig. 7.10). In other systems with other degenerate terms, distortions with three tetragonal, six orthorhombic, and so on, equivalent directions, as well as continuous sets of distortions (a trough), are possible (see Section 7.3 and [9.38]).

These predictions of the theory are confirmed by a large amount of experimental data. X-ray analysis shows that the six-coordinated polyhedron about the metals with E_g terms is not a regular octahedron even when all the ligands are identical, and in the majority of known cases the octahedron is tetragonally distorted. In Tables 9.14 to 9.17 the crystallographical distances to the two axial (R_L) and four equatorial (R_S) ligands in series of CuO_6 (Table 9.14), CuN_6 (Table 9.15), and other CuX_6 (Table 9.16) polyhedra, as well as in

Table 9.14. Equatorial R_S and axial R_L interatomic distances Cu—O in Cu(II) compounds containing CuO_6 clusters

Compound[a]	R_S (Å)	R_L (Å)
$Cu(C_6H_4OHCOO)_2 \cdot 4H_2O$	1.88	3.00
$Cu(glycollate)_2$	1.92	2.54
$Cu(acac)_2$	1.92	3.08
$Na_2Cu(CO_3)_2$	1.93	2.77
$Na_2Cu(PO_3)_4$	1.94	2.52
$Cu(meso\text{-tartrate}) \cdot H_2O$	1.94	2.54
$Cu(OMPA)_2(ClO_4)_2$	1.94	2.55
$Cu(OH)_2$	1.94	2.63
$Cu_2P_4O_{12}$	1.95	2.38
$Cu(C_8H_5O_4)_2 \cdot 2H_2O$	1.95	2.68
CuO	1.95	2.78
$Ca(Cu, Zn)_4(OH)_6(SO_4)_2 \cdot 3H_2O$	1.96	2.43
$PbCuSO_4(OH)_2$	1.96	2.53
$CuSO_4 \cdot 5H_2O$	1.97	2.41
$Cu(NaSO_4)_2 \cdot 2H_2O$	1.97	2.41
$Cu_6(Si_6O_{19}) \cdot 6H_2O$	1.97	2.68
$Cu(C_2H_5OCH_2COO)_2 \cdot 2H_2O$	1.98	2.38
$CuWO_4$	1.98	2.40
$Tl_2[Cu(SO_3)_2]$	1.99	2.44
$Ba_2Cu(HCOO)_6 \cdot 4H_2O$	2.00	2.18
$Cu(HCOO)_2 9 \cdot 2H_2O$	2.01	2.37
$[C_{14}H_{19}N_2]Cu(hfacac)_3$	2.02	2.18
$CdCu_3(OH)_6(NO_3)_2 \cdot H_2O$	2.03	2.43
$K_2BaCu(NO_2)_6$	2.04	2.29
$Cu_4(NO_3)_2(OH)_6$	2.04	2.34
$Ca(Cu, Zn)_4(OH)_6(SO_4)_2 \cdot 3H_2O$	2.06	2.23
$Cu(OMPA)_3(ClO_4)_2$	2.07	2.07
$Cu(H_2O)_6SiF_6$	2.07	2.07
$Cu(IPCP)_3(ClO_4)_2$	2.07	2.11
$Cu(NO_3)_2HgO \cdot 3H_2O$	2.10	2.10
$Cu(PCP)_3(ClO_4)_2$	2.11	2.04
$Ca(Cu, Zn)_4(OH)_6(SO_4)_2 \cdot 3H_2O$	2.11	2.11
$Cu(ClO_4)_2 \cdot 6H_2O$	2.13	2.28
$(NH_4)Cu(SO_4)_2 \cdot 6H_2O$	2.15	1.97

Source: [9.38, 9.44].

[a]acac, acetoacetate; OMPA, octamethylpyrophosphoramide; hfacac, hexafluoracetylacetonate; IPCP, tetraisopropymethylenediphosphonate; PCP, octamethylmethylenediphosphonic diamide.

Table 9.15. Equatorial R_S and axial R_L interatomic Cu—N distances in Cu(II) compounds containing CuN_6 clusters

Compound[a]	R_S (Å)	R_L (Å)
$Cu(C_4H_7N_5O)_2(ClO_4)_2$	1.97	3.14
$Cu(NH_3)_4(NO_2)_2$	1.99	2.65
$Cu(NH_3)_2(N_3)_2$	2.01	2.62
$Na_4Cu(NH_3)_4Cu(S_2O_3)_2 \cdot NH_3$	2.01	2.88
$Cu(phen)_3(ClO_4)_2$	2.05	2.33
$Cu(dien)_2Br_2 \cdot H_2O$	2.04	2.43
$Cu(N,N'—(CH_3)_2en)_2(NCS)_2$	2.06	2.52
$[Cu(en)_2]Hg(SCN)_4$	2.08	2.58
$Cu(1\text{-}pn)_3Br_2 \cdot 2H_2O$	2.09	2.31
$K_2PbCu(NO_2)_6$	2.11	2.11
$Cu(en)_3SO_4$	2.15	2.15
$Cu(dien)_2(NO_3)_2$	2.22	2.01
$\gamma\text{-}K_2PbCu(NO_2)_6$	2.23	2.05

Source: [9.38, 9.44].

[a]phen, o-phenanthroline; en, ethylenediamine; dien, diethylene-triamine; pn, 1,2-propandiamine.

similar octahedral Mn(III) and Cr(II) systems (Table 9.17) in different compounds, are given [9.44, 9.102, 9.103].

It follows from these tables that the six-atom polyhedra around the Cu(II), Mn(III), and Cr(II) centers in different crystals are mainly elongated octahedra $R_L > R_S$, with two ligands on the long axis and four on the short axes. Although for some of the tabulated compounds the atoms from the second and

Table 9.16. Equatorial R_S and axial R_L distances Cu—X in Cu(III) compounds containing CuX_6 clusters, where X = F, Cl, Br

Compound	R_S (Å)	R_L (Å)
Ba_2CuF_6	1.85; 1.94	2.08
Na_2CuF_4	1.91	2.37
CuF_2	1.93	2.27
K_2CuF_4	1.92	2.22
$KCuF_3$	1.96; 1.89	2.25
$CuCl_2$	2.30	2.95
$(NH_4)_2CuCl_4$	2.30; 2.33	2.79
$CsCuCl_3$	2.28; 2.36	2.78
$CuBr_2$	2.40	3.18

Source: [9.38, 9.44].

Table 9.17. Equatorial R_S and R_L interatomic distances Me—X in some compounds containing MX_6 clusters, where M = Mn(III), Cr(III) and X = O, S, F, Cl, I

Compound	R_S (Å)	R_L (Å)
$K_2MnF_5 \cdot H_2O$	1.83	2.07
$(NH_4)_2MnF_5$	1.85	2.10
K_2NaMnF_6	1.86	2.06
MnF_3	1.79–1.91	2.09
Cs_2KMnF_6	1.92	2.07
$Mn(trop)_3$, I	1.94	2.13
$Mn(trop)_3$, II	1.94–1.99	2.09
CrF_2	1.99	2.43
$Mn(acac)_3$	2.00	1.95
$KCrF_3$	2.14	2.00
$Mn(Et_2dtc)_3$	2.38–2.43	2.55
$CrCl_2$	2.40	2.92
CrI_2	2.74	3.24

Source: [9.38, 9.44].

next coordination spheres are different (and in the crystal state the interatomic distances also depend on the packing of the molecules in the lattice), the large number of these compounds confirms statistically that the deformation of the coordination sphere around Cu(II), Mn(III), and Cr(II) is due to internal forces, that is, to the E-term Jahn–Teller effect. The fact that elongated octahedra are observed confirms the assumption of strong vibronic coupling and strong quadratic vibronic interaction: One of the three adiabatic potential minima is stabilized by the crystal environment.

The octahedron of the first coordination sphere is regular for several compounds, which means that the distortions are not stabilized by the crystal environment or cooperative effects considered below. In other words, the phase transition to lower-symmetry structures due to cooperative effects has not taken place at the temperatures of the x-ray measurements in question; they could be expected at lower temperatures. For some of these compounds [e.g., $CuSiF_6 \cdot 6H_2O$, $KPbCu(NO_2)_6$, etc.] this point of view has been confirmed by ESR and other direct measurements.

Systems reported with tetragonally compressed octahedra in $(NH_4)_2Cu(SO_4)_2 6H_2O$, $Cu(PCP)_3(ClO_4)_2$, $Cu(dien)_2(NO_3)_2$, $Cu(en)_3SO_4$, $Mn(acac)_3$, and Ba_2CuF_6 need thorough investigation. In similar compounds [e.g., in nitrate of bis(terpyridin)Cu(II), $Cu(en)_3Cl_2$, K_2CuF_4, $K_2PbCu(NO_2)_6$] a more careful study shows that in fact the octahedra are tetragonally elongated and antiferrodistortive ordered, which gives an average picture of diffraction similar to that for ferrodistortive-ordered compressed octahedra. A similar conclusion regarding elongated octahedra emerges from ESR investigation [9.104].

In some cases the tetragonally compressed octahedral polyhedron around Cu(II) is presumably controlled by the crystal structure: CuF_6 polyhedra in the Ba_2ZnF_6 crystal may serve as an example [9.105]. Indeed, in Ba_2ZnF_6 the octahedral environment around Zn is compressed due to crystalline effects, and the substitution $Ba_2Zn_{1-x}Cu_xF_6$ with x up to 0.3 does not change the crystal structure and the compressed octahedral environment of Cu(II). One may speculate that in these cases the crystal influence is sufficiently strong; it changes the sign of the quadratic vibronic constant G_E (Section 7.3).

The origin of elongated octahedra (and lack of compressed ones) in Cu(II) compounds has been explained [9.106] as due to the contribution of configuration interaction (Section 5.3) with the excited state, which includes the $4s$ orbital of the copper atom. This negative contribution to the energy in the $^2B_{1g}$ state (elongated octahedron) has been shown to be twice that in the $^2A_{1g}$ state (tetragonally compressed octahedron). Similar understanding has not yet been achieved for Mn(III), Cr(II), and low-spin Co(II) compounds. Recently, additional examples of unusual compressed Cu(II)X_6 polyhedra have been reported [9.107, 9.108].

Another conclusion derived from the data in Tables 9.14 to 9.17 is about the diversity of Cu—O and Cu—N distances, which vary greatly from one system to another, preserving some relationship between R_L and R_S. It has been shown that the numerical values of the vibronic coupling constant can be derived from this relationship [9.38].

Distortion Isomers

In the diversity of crystal environments, there may be cases when not one, but two or several configurations of the coordination sphere are stabilized. If these configurations are sufficiently close in energy but differ in distortion magnitude and direction, they may be observed as different crystal isomers. The distortion isomers of Cu(II) originally synthesized and studied by Gazo et al. [9.44, 9.103] may serve as an example of this kind. These isomers have the same total composition and the same Cu(II) ligand environment but differ in interatomic metal–ligand distances in the distorted coordination sphere. Distortion isomers also differ in their properties, such as color, appearance, crystal form, chemical behavior, solubility, and spectroscopic data. They pass from one to another under the influence of pressure, heating, or long-term storage. In some cases, besides the two principal isomers (usually called α and β), a series of intermediate species have been obtained.

One of the simplest compounds that has distortion isomers is $Cu(NH_3)_2X_2$, where X = Cl, Br. Their possible Jahn–Teller origin was suggested when the isomers have been discovered, but the true understanding of their local pseudo Jahn–Teller and cooperative crystal nature was reached later [9.44, 9.101]. The results of Section 7.4 give quite a natural explanation of the origin of distortion isomers as being due to the vibronic properties of the Cu(II) center accompanied by the stabilizing influence of the crystal lattice.

Consider $Cu(NH_3)_2X_2$ as an example. This crystal comprises mutually parallel chains, each of which is arranged as illustrated in Fig. 9.26a, where all the X atoms occupy equivalent bridge positions. There is a strong interaction between the Cu(II) centers through bridging atoms inside the chain, while the bonding between the chains realized by van der Waals interactions and/or hydrogen bonds is weak. Each copper atom is surrounded by four X atoms in the plane of the square and by two NH_3 groups in the axial *trans* positions. The degeneracy of the ground Cu(II) state in the octahedron is removed as a consequence of the difference between X and NH_3, and the ^2E_g term is split into $^2A_{1g}$ and $^2B_{1g}$. Denote the splitting by 2Δ.

If we assume that the X atoms in the plane form a regular square, the polyhedron $Cu(NH_3)_2X_4$ is a tetragonally distorted octahedron that belongs to D_{4h} symmetry. Consider the pseudo Jahn–Teller effect on the A_{1g} and B_{1g} terms. This case fully corresponds to the two-level system considered in Section 7.4, so the results obtained there may be applied directly. In particular, the normal coordinate Q that mixes states A_{1g} and B_{1g} transforms according to B_{1g} ($A_{1g} \times B_{1g} = B_{1g}$) and the corresponding B_{1g} displacements in the D_{4h} group coincide with the Q_ε displacement of the O_h group given in Fig. 7.1 and Table 7.1 and shown in Fig. 9.27.

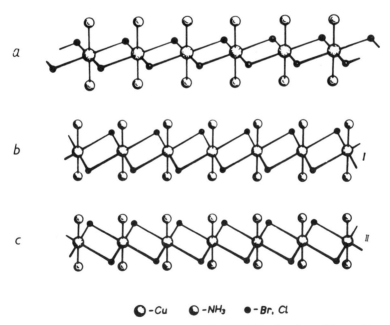

$\bigcirc$-Cu $\bigcirc$-NH_3 $\bullet$-Br, Cl

Figure 9.26. Chain structure of the crystal $Cu(NH_3)_2X_2$ in the cubic undistorted unstable β isomer (a), and in two equivalent configurations of the α isomer, I and II (b and c), resulting from the in-chain cooperative pseudo Jahn–Teller effect. (After [9.38].)

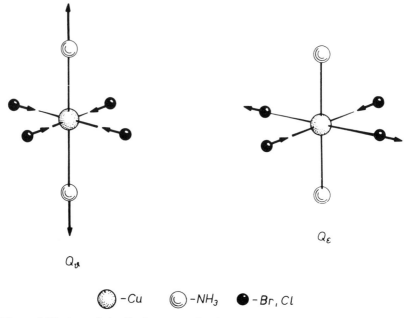

Q_ϑ

Q_ε

$\bigodot - Cu$ $\bigcirc - NH_3$ $\bullet - Br, Cl$

Figure 9.27. Q_θ and Q_ε displacements in the octahedral $Cu(NH_3)X_4$ cluster. The Q_ε distortion has B_{1g} symmetry and mixes the A_{1g} and B_{1g} electronic states.

The vibronic constant of the mixing of the two electronic states is $F = \langle A_{1g}|\partial H/\partial Q)_0|B_{1g}\rangle$, and the AP in the space of the Q coordinate has the form given by Eq. (7.69). If the instability condition (7.71) $\Delta < F^2/K_0$, is satisfied, we obtain two minima on the lower sheet of the adiabatic potential at $\pm Q_0$, determined by Eq. (7.72).

Thus the pseudo Jahn–Teller effect at each center distorts the bipyramidal environment: The equatorial square with four atoms X at the apexes transforms to a rhombus with the major diagonal along Q_x, corresponding to the minimum I in Fig. 9.28 or along Q_y (minimum II).

Due to the strong interaction between the distortions of neighboring centers in the chain through the ligands X in common, a ferrodistortive ordering of these distortions along the chain takes place, and this ordering remains unchanged up to high temperatures. Here the pseudo Jahn–Teller distortions are stabilized by the cooperative effects discussed below in more detail. As a result of this ordering, it can be assumed that at room temperatures each of the chains has two stable configurations, I and II of Fig. 9.26b and c, respectively, that correspond to the two minima in Fig. 9.28 (I and II, respectively). We can conclude that there is a possibility that two equivalent chain structures of $Cu(NH_3)_2X_2$ originate from the pseudo Jahn–Teller effect at each center $Cu(NH_3)_2X_4$, with ferrodistortive ordering of the distortions due to their strong cooperative interactions via the ligands X in common.

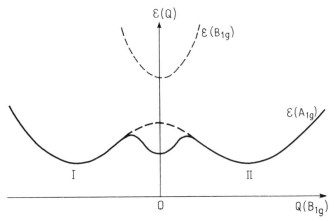

Figure 9.28. Energy curve for the $Cu(NH_3)_2X_2$ crystal as a function of the cooperative ferrodistortive intrachain distortion which is of B_{1g} symmetry on each center. I and II correspond to the pseudo Jahn–Teller minima, while the additional minimum at $Q(B_{1g}) = 0$ corresponds to the best-fit interchain interactions of undistorted chains.

One more important factor, the interaction between the chains in the crystal lattice, remains to be considered. The analysis of the structure of the crystal composed by parallel chains indicates that the intermolecular interaction between the chains is optimal when the chains are not distorted and the entire crystal is cubic. In this cubic state the intermolecular distances (between the strongest interacting atoms of different chains) are minimal, and they increase with intrachain distortions toward configuration I or II. The interchain interaction as a function of the interatomic distances can, in principle, be estimated semiempirically (say, by means of molecular mechanics; Section 9.1). We assume that the interchain interaction in the crystal has a minimum at $Q = 0$ (Fig. 9.28), where Q is the coordinate of the cooperative intrachain distortion corresponding to the B_{1g} distortion at each center. Hence the total energy of the crystal shown schematically in Fig. 9.28 has three minima: Besides two minima I and II for the stable distorted configurations of the chains, there is a third minimum for the undistorted (cubic) crystal.

It is assumed above that all the chains are in the same type of minimum, either I or II, and the intrachain distortions are ordered along the crystal. If different chains are in different types of minima (and the entire crystal is not ordered), the total energy may have additional minima, which, however, are expected to be more shallow than the minima I and II but deeper than the minimum at $Q = 0$ (on the assumption that the intrachain pseudo Jahn–Teller stabilization energy is larger than that of the interchain interactions).

It can be shown that the description above explains qualitatively the origin of all the main features of the distortion isomers in $Cu(NH_3)_2X_2$. The α isomer corresponds to the deepest minimum of the AP (I or II) with the structures illustrated in Fig. 9.26b and c. The unstable β isomer with the cubic structure corresponds to the more shallow minimum at $Q = 0$, and intermediate preparations with noncubic structures correspond to the additional relative minima for the uncorrelated chain distortions. This interpretation agrees well with the experimental features of the isomers, including their behavior under stress and temperature, the dependence on conditions of their preparation, spectral properties, and transitions from one isomer to another [9.44, 9.103].

It is noteworthy that good crystals (single crystals) can be obtained only for the β isomer [9.103], that is, when (as follows from the theory) the intermolecular (interchain) interactions, favoring crystal formation, are at a maximum. On the other hand, the α isomer and intermediate preparations cannot be obtained in the form of good crystals (but only as powders), in agreement with the assumption of poor interchain interaction in these situations. The fact that the β (and only the β) isomer contains the NH_4X compounds as impurities may also be important. The NH_4X system crystallizes in the cubic structure isomorphic to the β isomer of $Cu(NH_3)_2X_2$, and these two systems easily form mixed crystals. The cubic structure of the β isomer is apparently also stabilized by the large amount of NH_4X impurity molecules, which alternately occupy the places of Cu(II) centers in the chains, strongly reducing the intrachain effect. It was shown [9.109] that without the NH_4X admixture the β isomer rapidly converts to the α isomer. The pure β isomer can be prepared and preserved under pressure, however.

The pseudo Jahn–Teller origin of the intrachain distortions has been confirmed by approximate calculations of the pseudo Jahn–Teller distortions of the $Cu(NH_3)_2X_4$ polyhedron. The electronic energy as a function of the $Q(B_{1g})$ distortions has been estimated by means of the angular overlap model (Section 5.2). Using empirical data for the value K_0, the minima positions (I and II) on the AP curve (7.69) have been estimated [9.110]. The results confirm that the distortion is of pseudo Jahn–Teller origin: $Q(B_{1g})$ (calculated) ≈ 0.5 Å, while $Q(B_{1g})$(experimental) $\cong 0.4$ Å.

Temperature-Dependent Solid-State Conformers

The dynamic Jahn–Teller and pseudo Jahn–Teller distortions, under the influence of the crystal environment, may be reduced to static distortions in different ways, resulting in a variety of possible configurations, including temperature-dependent configurations. One of them may be called *temperature-dependent solid-state conformers* [9.111]. They occur when there are two or several rapidly converting distorted configurations which are slightly nonequivalent due to the crystal influence, and the observed averaged configuration is thus temperature dependent.

Consider the system $[Cu(bpy)_2(ONO)]NO_3$ (bpy = bipyridine) with the copper polyhedron cis-CuN_4O_2. In the high-symmetry configuration of $[Cu(bpy)_3]^{2+}$ the system has D_3 symmetry with a twofold-degenerate ^2E ground state. The substitution of one of the bpy groups by ONO reduces the symmetry to C_2 and splits the ^2E term in ^2A and ^2B. Similar to the distortion isomers discussed above, there is the possibility of a pseudo Jahn–Teller effect and consequent instability of the ground state ^2A with respect to B displacements $(A \times B = B)$, provided that the vibronic coupling constant $F = \langle A|\partial H/\partial Q_B|B\rangle$ is sufficiently large and the inequality (7.89) or (7.71) holds. The B displacements emerge from Q_ε (Fig. 9.27 and Table 7.1) and have the same geometry.

Assuming that (7.71) is satisfied, we get for the AP curve as a function of the coordinate $Q(B)$ the picture shown in Fig. 7.19b and reproduced here in Fig. 9.29, together with the indication of the modes of distortion of the CuN_4O_2 polyhedron in the minima. The value of the distortion coordinate Q [the coordinate $Q_\varepsilon = \frac{1}{2}(X_2 - X_5 - Y_3 + Y_6)$; Table 7.1] in our denotations for the cis-CuN_4O_2 polyhedron is

$$Q = \tfrac{1}{2}[\Delta R(Cu-N_1) + \Delta R(Cu-O_1) - \Delta R(Cu-N_2) - \Delta R(Cu-O_2)]$$

$$(9.38)$$

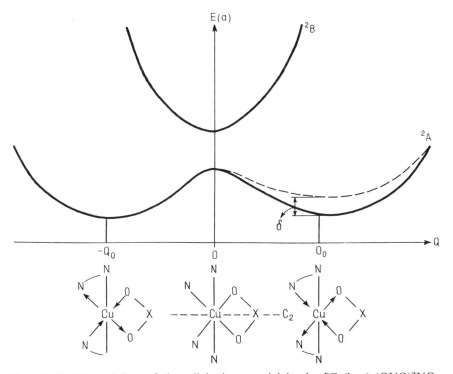

Figure 9.29. Two minima of the adiabatic potential in the [Cu(bpy)$_2$(ONO)]NO$_3$ crystal with a strong pseudo Jahn–Teller effect on each center and slightly different minima depths ($\sim \delta$) due to the crystal environment (influence of the next coordination sphere).

where N_1, N_2, O_1, and O_2 are the two nitrogen and two oxygen atoms in the plane containing the C_2 axis (Fig. 9.29) and ΔR denotes elongation of the bond with respect to that in the high-symmetry configuration. It is seen that for linear distortions, $\Delta R(Cu\text{—}N_1) - \Delta R(Cu\text{—}N_2) = R(N_1) - R(N_2)$ and $\Delta R(Cu\text{—}O_1) - \Delta R(Cu\text{—}O_2) = R(O_1) - R(O_2)$, where $R(X)$ indicates the bond length Cu—X. Hence

$$Q = \tfrac{1}{2}[R(N_1) - R(N_2) + R(O_1) - R(O_2)] \tag{9.39}$$

In the unstable high-symmetry configuration, $Q = 0$, while in the two minima, $Q = \pm Q_0 \neq 0$. If the energy barrier between the minima (i.e., the pseudo Jahn–Teller stabilization energy) is not very large, the system converts rapidly between the minima configurations, and in some experimental measurements (say, by x-ray diffraction methods) the average undistorted configuration will be observed (if the minima are shallow and have no local states, the average configuration will be observed in all the methods; Section 9.1).

The situation changes when, due to the crystal environment, the two minima become slightly nonequivalent, as shown in Fig. 9.29 by the dashed line, but the energy difference δ between the minima is smaller than the barrier height (otherwise, the second minimum disappears). In this case there is no complete averaging over the two configurations because the two minima are not equally populated. Denote the relative populations of the two configurations by n_1 and n_2, respectively. According to Boltzmann's thermal populations,

$$n_2 = n_1 \exp\left(-\frac{\delta}{kT}\right) \tag{9.40}$$

with the normalization

$$n_1 + n_2 = 1 \tag{9.41}$$

Then the observed thermal averaged distortion is

$$Q_{av} = (n_1 - n_2)Q_0 \tag{9.42}$$

or

$$Q_{av} = \frac{Q_0[1 - \exp(-\delta/kT)]}{1 + \exp(-\delta/kT)} \tag{9.43}$$

Thus the observable averaged distortion O_{av} is temperature dependent, its absolute value being determined by both Q_0 and δ.

These distorted configurations of the same compound which change gradually with temperature can be called *temperature-dependent solid-state conformers*. At high temperature when $kT \gg \delta$, in the first order with respect to δ/kT, $\exp(-\delta/kT) \approx 1 - \delta/kT$ and

$$Q_{av} = \frac{Q_0 \delta}{2kT} \qquad (9.44)$$

In the opposite limit case when $kT \ll \delta$, $\exp(-\delta/kT) \approx 0$ and $Q_{av} \approx Q_0$.

The observed distortions in $[Cu(bpy)_2(ONO)]NO_3$ and similar systems [9.111] follow these rules rather well. The atomic structure of this compound has been determined in four temperature regions: 20, 100, 165, and 296 K. Table 9.18 shows the corresponding interatomic distances and the value Q_{av} calculated after Eq. (9.39) for different temperatures as well as the same data for $[Zn(bpy)_2(ONO)]NO_3$ at 295 K, for comparison. In the Zn compound there is no pseudo Jahn–Teller effect of the type present in $[Cu(bpy)_2(ONO)]NO_3$, and hence no temperature dependent conformers are expected.

The data in Table 9.18 are very illustrative for temperature-dependent solid-state conformers; the temperature dependence of Q_{av} closely follows Eq. (9.43). In particular, if we assume that at $T = 20$ K, $Q_0 \approx Q_{av} = 0.3$, while at $T = 296$ K, Eq. (9.44) holds, we obtain directly $\delta = (Q_{av}/Q_0)2kT \approx 69$ cm^{-1}. The author [9.111] performed a more exact estimation: $\delta = 74$ cm^{-1}. Note that in the Zn(II) compound, Q_{av}, as expected, is small to zero, and the absence of conformers in the Zn(II) compound is also seen from the temperature factor in x-ray experiments [9.111].

The solid-state conformers under consideration differ from each other in interatomic distances, and in this sense they are similar to distortion isomers, discussed above. However, the latter coexist at the same temperature, whereas

Table 9.18. Bond lengths $R(X) = R(M—X)$ (in Å) and distortion coordinate values Q_{av} in the MN_4O_2 polyhedra of $[M(bpy)_2(ONO)]NO_3$ with M = Cu at different temperatures, and for M = Zn

	M = Cu				M = Zn
	20 K	100 K	165 K	296 K	295 K
$R(N_1)$	2.142(2)	2.110(2)	2.098(2)	2.085(2)	2.085(2)
$R(N_2)$	2.028(2)	2.060(2)	2.071(2)	2.074(4)	2.082(3)
$R(O_1)$	2.536(2)	2.414(2)	2.351(3)	2.320(5)	2.204(3)
$R(O_2)$	2.051(2)	2.155(2)	2.204(3)	2.230(5)	2.223(3)
Q_{av}	0.299	0.155	0.087	0.050	0.008

different conformers are observed at different temperatures, and the larger the temperature difference, the larger the structural differences of the conformers. Also important is the fact that the structural changes with temperature in conformers take place gradually, in contrast to structural phase transitions, which take place abruptly, at a certain temperature (see below).

Cooperative Effects: Order–Disorder and Displacive Phase Transitions: Helicoidal Structures

Cooperative effects are very important in the crystal stereochemistry of transition metal compounds. Later in this section we consider first briefly a general picture of ordering of local distortions in crystals (cooperative vibronic effects), followed by a discussion of the relative role of long-range (lattice) and short-range (chemical) forces in local stereochemistry and lattice formation using the problem of origin of ferroelectricity as an example.

If the crystal is formed from ready-made entities (molecules) that interact by stericlike forces, the energy of their interaction in the lattice U depends on the mutual orientation of these entities. In such cases the minimum of free energy $F = U - TS$ (S is the entropy) requires at $T = 0$ an ordering of the interacting molecules in which their mutual orientations make U minimal. At higher temperatures the entropy term TS becomes large, and at a certain temperature $T = T_c$ (Curie point) a phase transition to the disordered state for which the entropy is a maximum takes place (*order–disorder transitions*).

In inorganic and coordination compounds, in addition to such order–disorder transitions, *displacive phase transitions* may take place. In them, as distinct from order–disorder transitions, the atomic arrangement in the coordination center itself changes. Displacive transitions are related directly to the electronic structure, demonstrating the fact that in crystal stereochemistry of transition metal coordination compounds, the electronic structure of the coordination center plays a key role and cannot be ignored.

Local distortions in crystals and distortion-related structural phase transitions can be considered by means of the vibronic coupling theory (Chapter 7). Let us begin with a simple case when the Jahn–Teller centers in the crystal possess a twofold-orbital degenerate E term vibronically coupled to B_1 vibrations, which results in the $E-b_1$ problem mentioned in Section 9.2 (Fig. 9.23). Such a situation may occur, for example, when the crystal E term centers in question possess local D_{4h} symmetry. For simplicity, assume that the Jahn–Teller centers are square planar. The adiabatic potential curve with two equivalent minima and the rhombic distortions (B_{1g} displacements) of the square at each of the minima are illustrated in Fig. 9.23. In the absence of interaction between the centers, the two configurations [which we denote here by $(+)$ and $(-)$] are equally probable and the Jahn–Teller effect has the above-discussed dynamic nature: The averaged distortion equals zero and the initial symmetry is preserved. Note that in the case in question, averaging becomes possible due to other perturbations involved; without them the transition $(+) \rightarrow (-)$ is forbidden (see the remark at the end of Section 9.2).

The picture changes when the interactions between the centers are taken into account. Let us consider two interacting centers. In each of them the $E-b_1$ problem is realized. Under the assumption of parallel orientations of their squares, there may be four configurations in which the two one-center distortions are correlated: $(++])$, $(--)$, $(+-)$, and $(-+)$ (Fig. 9.30). It is clear that if the interaction lowers the energy for parallel orientations of the distortions (*ferrodistortive interactions*), the energy of the configurations $(++)$ and $(--)$ is lower than that of $(+-)$ and $(-+)$. On the other hand, if the antiparallel distortions are preferable (*antiferrodistortive interactions*), the configurations $(+-)$ and $(-+)$ are lower in energy. In both cases the two configurations "$+$" and "$-$" of each center are no longer equivalent. This conclusion can be generalized to many interaction centers and in the limit, to the entire crystal.

It is evident that at suitable low temperature, the minimum free-energy configuration is realized in which the crystal has statically distorted and distortion-correlated centers. For ferrodistortive interactions, such an ordering of the local distortions (*ferrodistortive ordering*) leads to a macro-deformation of the crystal as a whole. New properties of the crystal arising from the correlation (ordering) of the Jahn–Teller (pseudo Jahn–Teller) center distortions, including the formation of new crystal structures and structural phase

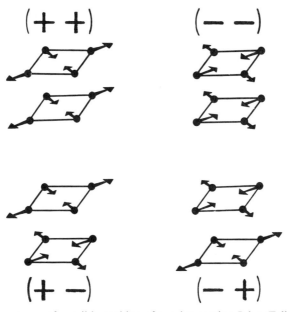

Figure 9.30. Four types of possible packing of two interacting Jahn–Teller centers. For each of them the $E-b_1$ problem of a square-planar system with B_{1g} distortions (Fig. 9.23) is realized.

transitions, are called the *cooperative Jahn–Teller effect* or *cooperative pseudo Jahn–Teller effect*.

In the example above, the two correlated configurations $(+ +)$ and $(- -)$ are equivalent and, in principle, the pair of correlated centers may resonate between the $(+ +)$ and $(- -)$ states, dynamically preserving the initial symmetry. However, in a macroscopic crystal with a large number of centers, the barriers between the equivalently distorted configurations of the entire crystal $(+ + + \cdots)$ and $(- - - - \cdots)$ become so high that transitions between them are practically impossible, and the crystal remains in one of them infinitely long. The situation here is quite similar to that found in ferromagnetics, in which there are also different equivalent directions of magnetization but no spontaneous inversion of the magnetic moment.

In these cases, as distinct from the one-center problem, the symmetry of the ground-state configuration of the crystal is lower than that of the Hamiltonian; this is sometimes called the *effect of broken symmetry*. Strictly speaking, such a crystal state is not stationary, but due to the very large height of the barriers, it may remain there for an infinitely long time.

It is clear that the lattice vibrations and temperature fluctuations tend to destroy the correlation between the Jahn–Teller distortions. With the increase of temperature T the second term in the free energy $F = U - TS$ becomes significant, while the entropy S increases with disorder. Therefore, in principle, for any given energy of interactions of the ordered distortions there is a certain temperature above which the ordering is destroyed and the lattice acquires a more symmetrical structure with independent (uncorrelated) dynamic distortions at each center (provided that the crystal does not melt at lower temperature). This temperature-dependent breakdown of the correlations of the distortions (disordering) is nothing but a structural phase transition. The stronger the distortion at each center and the energy of interactions of distortions, the higher the temperature of the phase transition to the disordered state. *The structural phase transition to the crystal state with disordered Jahn–Teller or pseudo Jahn–Teller local distortions is one of the most important features of the cooperative vibronic effects* [9.38, 9.112, 9.113].

Structural phase transitions in many classes of crystals have been attributed to cooperative vibronic effects. In particular, in a series of tetragonal rare earth zircons of general formula RXO_4, where R is a rare earth (R = Tm, Dy, Tb) and X = V, As, P, a direct correspondence between electronic structure parameters of the Jahn–Teller rare earth ion R^{3+} and the temperatures of structural phase transitions in the cooperative Jahn–Teller approach was established. Other crystals, such as spinel (e.g., $NiCr_2O_4$, $FeCr_2O_4$, $CuCr_2O_4$, FeV_2O_4, $FeCr_2S_4$, Mn_3O_4, $CuFe_2O_4$, $FeTiO_4$), perovskites ($KCuF_3$, $KMnF_3$, $PbFeF_3$), and other structures [$CsCuCl_3$, K_2CuF_4, $(NH_4)_2CoCl_2$, $Copy_2Cl_2$, UO_2, $A_2BCu(NO_2)_6$, $Cu(ONC_5H_6)_6X_2$], were also studied using the vibronic approach (see [9.38, 9.42, 9.112, 9.113] and references therein).

Consider now the displacive mechanism of lattice formation (*displacive phase transitions*), taking *ferroelectricity and ferroelectric phase transitions* in perovskite crystals as an example. In ferroelectric crystals, at certain temperatures, structural phase transitions to a spontaneous polarized (ferroelectric) state take place [9.114]. If there are molecular groups with proper dipole moments in the initial crystal structure, the transition to the ferroelectric phase may be regarded as an ordering of rigid dipoles (order–disorder transitions). However, in perovskite ferroelectrics (e.g., barium titanate) there are no such dipoles: The lattice is assumed to have cubic symmetry in the paraphase (the Ti-site local symmetry is O_h), while the dipole moment is supposed to emerge as a result of the displacive phase transition.

It is evident that if there are no rigid dipole groups in the elementary cells of the crystal, but such groups might occur as a result of Jahn–Teller or pseudo Jahn–Teller effects, *ferroelectric phase transition may take place as a result of the vibronic effects, due to the ordering of dipolar distortions of the interacting centers.* As noted in Section 7.4, the Jahn–Teller effect cannot result in dipolar distortions in systems that have an inversion center, but these distortions can occur as a result of the pseudo Jahn–Teller effect. Again, ferroelectrics are dielectrics that have no degenerate ground states, and hence they are not subject to the Jahn–Teller effect.

A pseudo Jahn–Teller mechanism of spontaneous polarization and ferroelectric phase transitions was suggested three decades ago [9.115]; it led to the vibronic theory of ferroelectricity [9.38, 9.112, 9.116–9.119]. Rigorous versions of the problem were solved by modern methods of theoretical physics. Below we present a qualitative aspect of the problem, mainly the electronic features that led to a physically visual picture of the microscopic origin of ferroelectricity.

Consider first the possible local off-center displacements of the Ti ion with respect to the near-neighbor oxygen octahedron and the remaining lattice in the $BaTiO_3$ crystal. The pseudo Jahn–Teller effect in the octahedral cluster TiO_6^{8-} is considered in Section 7.4; its MO energy-level scheme is given in Fig. 7.20. It is shown there that vibronic mixing of the ground A_{1g} term with the nearest excited term T_{1u} by odd displacements of T_{1u} type (dipolar displacements) under certain conditions makes the ground state unstable with respect to these displacements.

The AP surface in the space of three coordinates Q_x, Q_y, and Q_z (Fig. 7.21), which are three components of the T_{1u} displacement, has a complicated shape given by Eq. (7.76). Its extrema points depend on the relation between three constants: Δ (2Δ is the energy gap between the ground A_{1g} and excited T_{1u} states); F, the vibronic coupling constant given by Eq. (7.74); and K_0, the primary force constant for the T_{1u} displacement. If $\Delta > 4F^2/K_0$, the surface (7.76) has one minimum at $Q_x = Q_y = Q_z = 0$, in which the cluster TiO_6^{8-} is undistorted. But if $\Delta < 4F^2/K_0$, the surface has a maximum at this point with respect to the Q_x, Q_y, and Q_z displacements (dynamic instability), eight equivalent trigonal minima, in which the Ti atom is displaced along one of the eight directions of four trigonal axes with $|Q_x| = |Q_y| = |Q_z| = Q_0^{(1)}$ [Eqs. (7.78) and (7.79)], 12 saddle points of the type $|Q_x| = |Q_y| \neq 0$, $Q_z = 0$, at which the

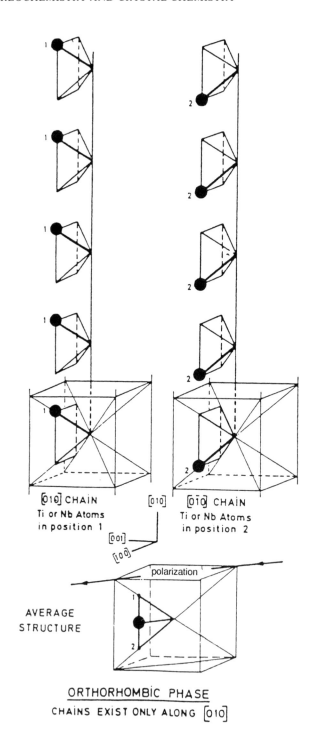

[010] CHAIN
Ti or Nb Atoms
in position 1

[0Ī0] CHAIN
Ti or Nb Atoms
in position 2

[010]

[001]
[100]

polarization

AVERAGE
STRUCTURE

ORTHORHOMBIC PHASE
CHAINS EXIST ONLY ALONG [010]

Ti atom is displaced toward two oxygens, and six saddle points at which, say, $Q_x = Q_y = 0$, $Q_z = Q_0^{(2)}$ [Eqs. (7.80) and (7.81)].

Note that at the minima and saddle points above, the cluster has nonzero dipole moments. Allowing for strong correlation between these clusters in $BaTiO_3$ through common ligands, the cooperative pseudo Jahn–Teller effect results in a structural phase transition to the spontaneously polarized ferroelectric state of the lattice in which all the local dipole moments have the same orientation. Depending on the sign of the interacting distortions, an antiferroelectric ordering with alternating directions of the dipoles is also possible.

A similar analysis of the AP of the crystal as a whole with regard to the dipole displacements of the corresponding sublattices was carried out, and a semiclassical approach was used to consider the motion of the crystal along this adiabatic potential (assuming that all the remaining vibrations may be regarded as a thermal reservoir). By solving the appropriate kinetic equations, it was shown that in this crystal three structural phase transitions should be observed. At low temperatures the crystal is at one of its AP trigonal minima, which corresponds to the ordered displacement of the Ti atoms along one of the trigonal axes. In this case the lattice has rhombohedral structure and is polarized spontaneously along the corresponding trigonal axis.

When the temperature increases, the kinetic energy (or amplitude) of motions in the minimum increases, and the system begins to overcome the lowest barrier between the nearest-neighbor minima through the orthorhombic saddle points. At a specific temperature there is a phase transition to the (macroscopically) orthorhombic structure, resulting from an averaging over the two neighboring minima (Fig. 9.31; note the remark about clustering, given below). In this orthorhombic phase, the pseudo Jahn–Teller dipolar distortions that correspond to the off-center displacements of the Ti atom in the trigonal axis directions C_3 or C_3' are not completely ordered: the lowest barrier between the nearest-neighbor minima is overcome (thermally or by tunneling), so the crystal becomes (dynamically) disordered in the plane of these two minima in the direction C_2'' perpendicular to the C_2 axis of spontaneous polarization. Thus in this phase the distortions are ordered along two of three independent directions, C_2 and C_2', and it is disordered along the third, C_2''.

At higher temperatures, transitions between the minima overcome the higher-barrier saddle points of tetragonal type resulting in a phase transition to the (macroscopically) tetragonal phase. Its dipolar distortions are ordered

---

Figure 9.31. Formation of the orthorombic phase in $BaTiO_3$ and $KNbO_3$. A fragment of the crystal with the C_2 axis of polarization and the off-center displacements of the titanium ion along the near-neighboring trigonal directions in the case of clustering [9.118] is shown. The local fast transitions between the two distorted configurations (or the random orientation of the cluster distortions) makes the crystal structure disordered along the direction perpendicular to C_2 [after [9.131]].

only along one tetragonal axis, and the crystal is polarized in this direction. In the plane perpendicular to this direction, the distortions are dynamically disordered and may be macroscopically described as averaged over four equivalent trigonal minima. Finally, at higher temperatures the maximum point of the AP is overcome by thermal transitions and the crystal becomes completely disordered along all three directions. This corresponds to a phase transition to the paraphase, which is thus a macroscopically averaged state of the dynamically disordered pseudo Jahn–Teller distortions.

This picture obtained by calculations [9.112, 9.115–9.119] qualitatively reproduces all the observed ferroelectric phases in barium titanate, including the consequence of their occurrence with temperature and the symmetry of the crystal with the direction and relative value of polarization in each of its phases. Moreover, it follows that only one of the three ferroelectric phases in $BaTiO_3$ — the low-temperature rhombohedral phase — is ordered completely. The other two, orthorhombic and tetragonal, are partly disordered, the former along one direction and the latter along two directions, while the paraphase is completely disordered in all directions. This conclusion is fundamentally novel and does not follow from any other theories of ferroelectricity. It has been confirmed by quite a number of experiments on diffuse scattering of x-rays [9.120], light scattering [9.121], refined ESR measurements with probe ions [9.122], XAFS experiments [9.123], and so on. Table 9.19 lists some of such publications [9.118]. Similar order–disorder effects are observed in $K_2PbCu(NO_2)_6$ [9.130].

For further consideration, recall that according to the results of Section 7.4, the curvature K of the AP contains two terms, $K = K_0 + K_v$, defined by Eqs. (7.87) and (7.88), respectively. The nonvibronic contribution to the force constant K_0 is always positive, and hence a negative curvature and instability in the Q direction can be realized only when the negative vibronic contribution is sufficiently large, $-K_v > K_0$. It was shown that K_0 *contains all the long-range forces, while* K_v *is essentially of local origin* [9.118]. This follows directly from the more detailed analysis of the origin of the vibronic contribution to the instability, given in Section 7.4. The terms of new covalency of type (7.98) that originate from the vibronic mixing of atomic states from different atoms are essentially determined by the overlap of their wavefunctions $S(2p, 3d)$. Indeed, approximately, $\langle 2p|\partial H/\partial Q|3d\rangle \sim \partial H_{2p3d}/\partial R \sim \partial S(2p, 3d)/\partial R$, where H_{2p3d} is the corresponding resonance integral, S is the overlap integral, and R is the interatomic distance.

Since S decreases exponentially with R, the covalency terms in K_v are negligible for all next-to-near-neighbor interactions because of the near-zero overlap with their atomic wavefunctions. The polarization term is already very small for the near-neighbor interactions (Table 7.4) and, according to Eq. (7.96), it decreases with the interatomic distance as R^{-6}.

Thus the nonvibronic contribution to the curvature of the adiabatic potential K_0 contains all the long-range interactions, while the vibronic contribution K_v is essentially of local origin. Since it has been proved that $K_0 > 0$, only the

Table 9.19. Experimental evidence of local origin of distortions and order–disorder nature of phase transitions in displacive ferroelectrics

Authors, Year	Method, System	Main Result
Comes et al. [9.120], 1968	X-ray, diffuse scattering, $BaTiO_3$	Qualitative confirmation of all the main predictions of the vibronic theory for $BaTiO_3$
Quittet et al. [9.121], 1973	Raman spectra, $BaTiO_3$, $KNbO_3$	Polar distortions in the cubic paraphase
Burns and Dacol [9.125], 1981	Optical refractive index, $BaTiO_3$	Nonvanishing component $\langle p^2 \rangle$ in the cubic phase
Gervais [9.126], 1984	Infrared reflectivity, $BaTiO_3$	
Ehses et al. [9.127], 1981	X-ray, $BaTiO_3$	Strong order–disorder component in the cubic phase
Ito et al. [9.128], 1985	X-ray, $BaTiO_3$	[111] displacement of Ti in the paraphase up to 180 K above T_c
Muller et al. [9.122], 1986	ESR with probing ion $BaTiO_3$, $KNbO_3$	[111] displacements in the rhombohedral phase and reorientations in the orthorhombic phase, $10^{-10} < \tau < 10^{-9}$ (s)
Hanske-Petitpierre et al. [9.123], 1991	XAFS, $KNb_xTa_{1-x}O_3$	[111] displacements in all the three phases for any $x > 0.08$; mean-square displacements much smaller due to dynamics
Dougherty et al. [9.129], 1992	Femtosecond resolution of light scattering, $BaTiO_3$, $KNbO_3$	No relaxational modes which might exclude the local distortion model
Sicron et al. [9.124], 1994	XAFS, $PbTiO_3$	Ti and Pb ions are displaced in the paraphase up to 200 K above T_c

negative contribution of K_v can make the curvature negative and the configuration of the system unstable. Hence *the long-range forces themselves cannot result in instability of the high-symmetry configuration of the lattice.* Together with the very convincing experimental confirmation (Table 9.19), *this result casts doubts on the very existence of displacive phase transitions.*

On the other hand, long-range forces are most important in determining the direction Q for which the inequality $-K_v > K_0$ holds and is strongest. The direction for which K_0 is smaller (the lattice is softer) is most favorable for distortions, and there may be cases when for the actual distortion K_v is smaller than for other distortions because of the lower K_0 value in this direction. For instance, for the off-center displacements of the Ti ion with respect to the oxygen environment and Ba atoms in $BaTiO_3$, discussed above, K_0 is minimal when all the other Ti ions displace in the same way; that is, when the distortions are ordered (in the case in question, along one of the trigonal axes of the crystal). Therefore, actual local distortions at low temperatures often coincide with the direction of the ordering. This creates an illusion that the local distortions are produced by the ordering; that is, they cannot exist without the long-range forces of the lattice. It is the main assumption of the widespread theories of displacive phase transitions. As shown above, the long-range forces themselves may soften the lattice in certain directions, but they cannot produce instabilities without the local vibronic contribution. The interplay between striving for ordering the local distortions in the lattice and the entropy factor that tends to destroy the ordering may result in some clustering of the distortions in the disordered phases; that is, formation of some chain clusters in which the trigonal distortions are ordered as shown in Fig. 9.31 [9.118, 9.122, 9.131].

Qualitatively, an estimation of the Curie temperatures T_c is also possible. They obviously depend on the parameters that determine the instability: the minima depth δ (equal to the Jahn–Teller stabilization energy), given in Eq. (7.79), and the distance between the corresponding saddle points. A simple modeling procedure in which the main features of the adiabatic potential are presented by two generalized parameters [9.132] allows one to make some estimations of T_c. Using the experimental T_c values for two (out of the three) phase transitions in $BaTiO_3$, $T_1 = 393$ K and $T_2 = 278$ K, one can estimate the two parameters of the model and then, using their values, one gets the third T_c value, $T_c = 201$ K, while the experimental value is $T_3 = 183$ K. This result shows that the parameters involved and approximations made are at least not unreasonable.

A qualitative investigation of the variation of the vibronic parameters in the inequality (9.69) along a series of perovskites allows for an examination of the origin of the observed changes in their ferroelectric properties [9.118]. The stronger inequality (9.69), that is, the greater the difference $(4F^2/K_0\Delta) - 1$, the deeper the AP minima and the higher the temperature of the phase transition to the paraphase.

For instance, in the series of titanates of Ca, Sr, and Ba the Ti—O distance increases (due to the increase of the atomic radii from Ca to Sr, to Ba), and hence the d_π–p_π overlap and the covalency of the bond decrease from left to right. As a consequence, the force constant K_0 for the corresponding displacements of the Ti atom, as well as the effective energy gap between the bands, decrease, while the vibronic constant F is less influenced. Thus when passing from calcium titanate to strontium titanate and then to barium titanate, inequality (9.69) is strengthened and the conditions of ferroelectricity occurrence are improved; that is, the temperature of phase transition to the disordered phase increases.

This conclusion is confirmed by empirical data: $CaTiO_3$ is nonferroelectric down to $T = 0$, $SrTiO_3$ is a virtual ferroelectric, and with appropriate impurities or under external perturbation becomes a true ferroelectric at low temperatures, whereas $BaTiO_3$ is ferroelectric up to room temperatures.

In another series of perovskites of Ti, Zr, and Hf, the diffusiveness of the outer d orbitals ($3d$, $4d$, and $5d$, respectively) increases. Consequently, the $d_\pi - p_\pi$ overlap, the covalency of the M—O bond, and hence the K_0 and Δ parameters also increase. Therefore, the ferroelectric properties are expected to deteriorate (the temperature of phase transition is expected to decrease) in this series from left to right in accordance with the experimental data. Similarly, when passing from GeTe to SnTe the $p_\sigma - p_\sigma$ overlap ($4p$–$5p$ and $5p$–$5p$, respectively) increases, and therefore the temperature of phase transition in GeTe is higher than in SnTe.

If the local distortions in crystals do not create dipole moments, the picture of cooperative vibronic interactions and structural phase transitions is the same as considered above, with the distinction that nonpolar distortions produce not ferroelectric, but *ferroelastic phase transition*. Since any local distortion from the high-symmetry configuration is of vibronic nature (Section 7.4), all the structural phase transitions in crystals can be treated in this way, at least in principle [9.133].

One of the interesting features of structural phase transitions is the formation of *helicoidal structures*. Consider the crystal $CsCuCl_3$ [9.134, 9.135]. Its parallel chains in the lattice contain $CuCl_6$ octahedral polyhedra which are interlinked by triple bridges produced by three chlorine atoms, the latter thus forming a common triangular face for two near-neighbor polyhedra (Fig. 9.32). Each Cu(II) center, due to the Jahn–Teller effect, requires a tetragonally distorted octahedron of six chlorine atoms, and there are three equivalent directions of distortions corresponding to three fourfold axes. However, because of the common ligands the distortions of the near-neighbor octahedra are correlated. In particular, if the direction of distortion for a given polyhedron is, say, along the z axis, its neighbor should distort along either x or y axes. From the point of view of the trigonal axis along which the Cu atoms are located in the chain, the directions of the distortions of the two neighboring octahedra are rotated by an angle $\phi = 60°$ (from one apex of the interfacing triangle to the next). This is shown schematically in Fig. 9.33. As one can see, the period of the lattice in the distorted helicoidal screwlike configuration is six times larger than in the undistorted one.

Another example of helicoidal structure is noted in Section 9.2, the InCl crystal [9.136, 9.137]. It has a rock-salt structure in the high-symmetry configuration, but the vibronic interaction of the lone pair $(5s)^2$ with the excited states produces a combined pseudo Jahn–Teller off-center distortion associated with the excited-state Jahn–Teller E_g or T_{2g} distortion [9.69] discussed

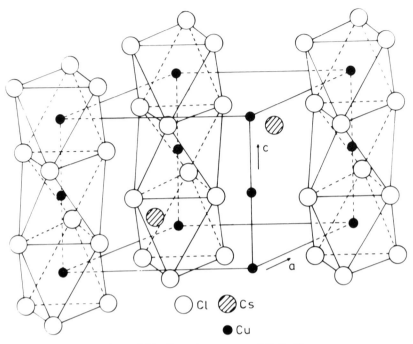

Figure 9.32. Crystal structure of $CsCuCl_3$.

in Section 9.2 in more detail. Owing to the common ligands of the near-neighbor polyhedra, this distortion, for reasons similar to those considered above, produces the helicoidal structure [9.136, 9.137].

In the example of $CsCuCl_3$ above, the new period of the helicoidal distorted lattice is exactly six times that of the undistorted lattice, as shown in Fig. 9.32.

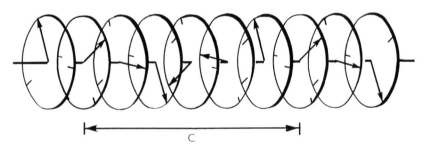

Figure 9.33. Helicoidal structure of the $CsCuCl_3$ crystal. The $CuCl_3$ chain is shown conventionally; the Cu atoms are on the trigonal axis of the local octahedral polyhedron. Due to the common shared Cl ligands the Jahn–Teller tetragonal distortions of the near-neighbor polyhedrons in the chain are shifted in phase by $\phi = 60°$, as shown by arrows, and the lattice period along the chain is six times larger than in the undistorted crystal.

However, there are cases when the ratio of the lattice period in the new phase to that of the higher-symmetry phase is not an integer. In these cases it is said that the new structure is *incommensurate* with the previous one, and the phase transition takes place to an *incommensurate phase*. An interesting example of incommensurate phases is provided by $K_2PbCu(NO_2)_6$ [9.38, 9.130]. A related phenomenon, important for transition metal compounds, is that of *structural–magnetic transitions* [9.130, 9.38].

REFERENCES

9.1. J. K. Burdett, *Molecular Shapes: Theoretical Models of Inorganic Stereochemistry*, Wiley-Interscience, New York, 1980, 287 pp.

9.2. R. G. Gillespie, *Molecular Geometry*, Van Nostrand, Reinhold, London, 1972, 228 pp; R. G. Gillespie and I. Hargittai, *The VSEPR Model of Molecular Geometry*, (International Student Edition), Allyn & Bacon, London, 1991, 240 pp.

9.3. V. V. Sidgwick, *The Covalent Link in Chemistry*, Cornell Univ. Press, Ithaca, New York, 1933.

9.4. *Stereochemistry and Bonding*, in: *Struct. Bonding*, **71** (1989), 198 pp.

9.5. J. E. Fergusson, *Stereochemistry and Bonding in Inorganic Chemistry*, Prentice Hall, Englewood Cliffs, N.J., 1974, 309 pp.

9.6. D. L. Kepert, *Inorganic Stereochemistry*, Springer-Verlag, , Berlin, 1982.

9.7. Yu. S. Varshawskii, T. G. Cherkasova, and M. M. Singh, *Zh. Neorg. Khim.*, **15**, 2746–2751 (1970).

9.8. L. Pauling, *The Nature of the Chemical Bond and the Structure of Molecules and Crystals*, 3rd ed., Cornell University Press, Ithaca N.Y., 1960.

9.9. D. J. Kimball, *J. Chem. Phys.*, **8**, 188–198 (1940).

9.10. R. M. Gavin and L. S. Bartell, *J. Chem. Phys.*, **48**, 2460 (1968).

9.11. L. S. Bartell and R. M. Gavin, *J. Chem. Phys.*, **48**, 2466 (1968).

9.12. R. S. Berry, *J. Chem. Phys.*, **32**, 933 (1960).

9.13. D. M. P. Mingos and L. Zhenyang, *Struct. Bonding*, **71**, 1–56 (1989).

9.14. P. Seiver and J. D. Dunitz, *Acta Crystallogr.*, **B36**, 2255 (1980).

9.15. S. Lange and G. Huttner, *Acta Crystallogr.*, **B28**, 2049 (1972).

9.16. D. M. P. Mingos and D. J. Wales, *Introduction to Cluster Chemistry*, Prentice Hall, Englewood Cliffs, N.J., 1990, 320 pp.

9.17. D. M. P. Mingos, *Pure Appl. Chem.*, **59**, 145–154 (1987).

9.18. D. M. P. Mingos and J. C. Hawes, *Struct. Bonding*, **63**, 1–63 (1985).

9.19. S. C. Abrahams, A. P. Ginsberg, and K. Knox, *Inorg. Chem.*, **3**, 538 (1964).

9.20. A. P. Ginsberg, *Transition Metal Hydrides*, ACS Advances in Chemistry Series 167, American Chemical Society, Washington, D.C., 1978, p. 201.

9.21. U. Burkert and A. L. Allinger, *Molecular Mechanics*, ACS Monograph 177, American Chemical Society, Washington, D.C., 1982.

9.22. J. P. Bowen and N. L. Allinger, in: *Reviews in Computational Chemistry*, Vol. 2, ed. K. B. Lipkowitz and D. B. Boyd, VCH, New York, 1991, pp. 81–97.

9.23. U. Dinur and A. T. Hagler, in: *Reviews in Computational Chemistry*, Vol. 2, ed. K. B. Lipkowitz and D. B. Boyd, VCH, New York, 1991, pp. 99–164.

9.24. G. R. Brubaker and D. W. Johnson, *Coord. Chem. Rev*, **53**, 1–36 (1984).

9.25. R. D. Hancock, *Progr. Inorg. Chem.*, **37**, 187–291 (1989).

9.26. A. K. Rappe, K. S. Colwell, and C. J. Casewit, *Inorg. Chem.*, **32**, 3438–3450 (1993).

9.27. B. P. Hay, *Coord. Chem. Rev.*, **126**, 177–236 (1993).

9.28. G. Wipff, *J. Coord. Chem.*, **27**, 7 (1992).

9.29. H. Marques, *J. Am. Chem. Soc.*, **114**, 7218 (1992).

9.30. I. B. Bersuker and R. S. Pearlman, *15th Austin Symposium on Molecular Structure,* University of Texas, Austin, 1994, *Abstracts P5*, p. 17.

9.31. I. B. Bersuker, M. K. Leong, J. E. Boggs, and R. S. Pearlman, *Proceedings of the First Electronic Computational Chemistry Conference — CDROM*, eds., S. M. Bachrach, D. B. Boyd, S. K. Gray, W. Mase, and H. S. Rzepa, ARInternet, Landover, MD, 1995, paper 8.

9.32. C. Hansch and A. Leo, *Exploting QSAR. Fundamentals and Applications in Chemistry and Biology*, Vol. 1, ACS Professional Reference Book, Washington, 1994; C. Hansch, A. Leo, and D. Hoekman, *Exploring QSAR. Hydrophobic, Electonic, and Steric Constants*, Vol. 2, ACS Professional Reference Book, Washington, 1994.

9.33. A. J. Stuper, W. E. Bruggers, and P. C. Jurs, *Computer-Assisted Studies of Chemical Structure and Biological Function*, Wiley-Interscience, New York, 1979; I. C. Martin, *Quantitative Drug Design. A Critical Introduction*, Marcel Dekker, New York, 1978.

9.34. R. Franke, *Theoretical Drug Design Methods*, Academie-Verlag, Berlin, 1987.

9.35. I. B. Bersuker and A. S. Dimoglo, in: *Reviews in Computational Chemistry*, Eds. K. B. Lipkowitz and D. B. Boyd, Vol. 2, VCH, New York, 1991, pp. 423–460; I. B. Bersuker, A. S. Dimoglo, M. Yu. Gorbachov, M. Pesaro, and P. F. Vlad, *New J. Chem.*, **15**, 371 (1991).

9.36. H. Kubinyi, *QSAR: Hansch Analysis and Related Approaches*, VCH, Weinheim, 1993; *3D QSAR in Drug Design*, Ed. H. Kubinyi, ESCOM, 1994.

9.37. I. D. Kumtz, E. C. Meng, and B. K. Shoichet, *Acc. Chem. Res.*, **27**, 117 (1994).

9.38. I. B. Bersuker, *The Jahn–Teller Effect and Vibronic Interactions in Modern Chemistry*, Plenum Press, New York, 1984, 320 pp.

9.39. A. Ceulemans and L. G. Vanquickenborne, *Struct. Bonding*, **71**, pp. 125–159 (1989).

9.40. P. Curie, *Oeuvres de Pierre Curie*, Gauthiers-Villars, Paris, 1908, pp. 118–141.

9.41. A. Ceulemans, D. Beyens, and L. G. Vanquickenborne, *J. Am. Chem. Soc.*, **106**, 5824–5837 (1984).

9.42. I. B. Bersuker, ed., *The Jahn–Teller Effect: A Bibliographic Review*, IFI/Plenum, New York, 1984, 590 pp.

9.43. B. J. Hathaway, *Struct. Bonding*, **57**, 55–118 (1984).

9.44. J. Gazo, I. B. Bersuker, J. Garaj, M. Kabesova, J. Kohout, H. Langfelderova, M. Melnik, M. Serator, and F. Valach, *Coord. Chem. Rev.*, **19**, 253–297 (1976).

9.45. R. G. Pearson, *Symmetry Rules for Chemical Reactions, Orbital Topology and Elementary Processes*, Wiley, New York, 1976, 548 pp.

9.46. R. F. W. Bader and A. D. Bandrauk, *J. Chem. Phys.*, **49**, 1666–1671 (1968).

9.47. H. Bash, J. W. Moskowitz, C. Holister, and D. Hankin, *J. Chem. Phys.*, **55**, 1922 (1971).

9.48. F. A. Gianturco, C. Quidotti, and U. Lamanna, *J. Chem. Phys.*, **57**, 840 (1972).

9.49. L. S. Bartell, *J. Chem. Phys.*, **73**, 375 (1981).

9.50. D. Reinen and C. Friebel, *Inorg. Chem.*, **23**, 792 (1984).

9.51. D. Reinen and M. Atanasov, *Chem. Phys.*, **136**, 27 (1989).

9.52. D. Reinen and M. Atanasov, *Chem. Phys.*, **155**, 157 (1991).

9.53. M. Atanasov, W. Koenig, M. Craupner, and D. Reinen, *New J. Chem.*, **17**, 115 (1993).

9.54. D. W. Meek and J. A. Ibers, *Inorg. Chem.*, **9**, 405 (1970).

9.55. V. Z. Polinger, N. N. Gorinchoy, and I. B. Bersuker, *Chem. Phys.*, **159**, 75–87 (1992).

9.56. N. N. Gorinchoy, I. B. Bersuker, and V. Z. Polinger, *New J. Chem.*, **17**, 125 (1993).

9.57. G. I. Bersuker and V. Z. Polinger, to be published.

9.58. I. B. Bersuker and S. S. Stavrov, *Coord. Chem. Rev.*, **88**, 1–68 (1988).

9.59. M. Zerner and M. Gouterman, *Theor. Chim. Acta*, **4**, 44 (1966).

9.60. M. Gouterman, in: *The Porphyrins*, ed. D. Dolphin, Vol. III, Academic Press, New York, 1978, p. 1.

9.61. J. H. Enemark, R. D. Feltham, *Coord. Chem. Rev.*, **13**, 339 (1974).

9.62. M. Zerner, M. Gouterman, and H. Kobayashi, *Theor. Chim. Acta*, **6**, 363 (1966).

9.63. W. A. Eaton, L. K. Hanson, P. J. Stephens, J. C. Sutherland, and B. R. Dunn, *J. Am. Chem. Soc.*, **100**, 4991 (1978).

9.64. R. Rericha, *J. Mol. Struct. (Theochem.)*, **227**, 305–310 (1991).

9.65. I. B. Bersuker, S. S. Budnikov, and M. Yu. Gorbachov, unpublished.

9.66. A. F. Wells, *Structural Inorganic Theory*, Clarendon Press, Oxford, 1984.

9.67. L. E. Orgel, *J. Chem. Educ.*, 3815–3819 (1959).

9.68. S. Y. Wang and L. L. Lohr, Jr., *J. Chem. Phys.*, **60**, 3901–3919 (1974); **61**, 4110–4118 (1974).

9.69. W. J. A. Maaskant and I. B. Bersuker, *J. Phys. Condensed Matter*, **3**, 37–47 (1991).

9.70. W. J. A. Maaskant, *New J. Chem.*, **17**, 97 (1993).

9.71. A. Ceulemans, *J. Chem. Phys.*, **84**, 6442–6451 (1986).

9.72. B. J. Hathaway, *Struct. Bonding*, **57**, 55–118 (1984).

9.73. J. Pradilla-Sorzano and J. P. Fackler, *Inorg. Chem.*, **12**, 1182 (1973).

9.74. A. von Zelewsky, *Inorg. Chem.*, **20**, 4448–4449 (1981).

9.75. W. G. Klemperer, D. D. Traficante, and G. M. Whitesides, *J. Am. Chem. Soc.*, **97**, 7023 (1975).

9.76. M. M. L. Chen and R. Hoffmann, *J. Am. Chem. Soc.*, **98**, 1651 (1976).

9.77. J. Brunvoll, A. A. Ischenko, V. P. Spiridonov, and T. G. Strand, *Acta Chem. Scand.*, **A38**, 115–120 (1984).

9.78. B. Daines, A. McNeish, M. Poliakoff, and J. J. Turner, *J. Am. Chem. Soc.*, **99**, 7573–7579 (1977).

9.79. M. Poliakoff and A. Ceulemans, *J. Am. Chem. Soc.*, **106**, 50–54 (1984).

9.80. I. I. Chernyaev, in: *Selected Works: Complex Compounds of Platinum* (Russ.), Nauka, Moscow, 1973.

9.81. F. R. Hartley, *Chem. Soc. Rev.*, **2**, 163 (1973).

9.82. E. M. Shustorovich, M. A. Porai-Koshitz, and Iu. A. Buslaev, *Coord. Chem. Rev.*, **17**, 1 (1975).

9.83. V. I. Nefedov and M. M. Hofman, *Mutual Influence of Ligands in Inorganic Compounds* (Russ.), in: *Itoghi Nauki i Techniki*, Ser.; *Inorganic Chemistry*, Vol. 6, VINITI, Moscow, 1978, 192 pp.

9.84. A. A. Levin, *Sov. Sci. Rev. B. Chem. Rev.*, **9**, 279–335 (1987).

9.85. V. I. Levin and P. N. Dyachkov, *Electronic Structure, Geometry, Isomerism and Transformations of Heteroligand Molecules* (Russ.), Nauka, Moscow, 1990, 256 pp.

9.86. A. A. Levin, *New J. Chem.*, **17**, 31 (1993).

9.87. J. Gazo, R. Boca, E. Jona, M. Kabesova, L. Macaskova, J. Sima, P. Pelikan, and F. Valach, *Coord. Chem. Rev.*, **43**, 87–131 (1982).

9.88. M. A. Porai-Koshitz and L. A. Atovmean, *Crystal Chemistry and Stereochemistry of Molybdenum Coordination Compounds* (Russ.), Nauka, Moscow, 1974.

9.89. J. K. Burdett and T. A. Albright, *Inorg. Chem.*, **18**, 2112–2120 (1979).

9.90. A. A. Grinberg, *Zh. Neorg. Khim.*, **4**, 1517 (1959).

9.91. J. Chatt, L. A. Duncanson, and L. M. Venanzi, *J. Chem. Soc.*, 4456–4469 (1955).

9.92. L. E. Orgel, *J. Inorg. Nucl. Chem.*, **2**, 137–140 (1956).

9.93. L. Pauling, *J. Chem. Soc.*, 1461 (1948); in: *Proceedings of the Symposium on Coordination Chemistry*, Copenhagen, 1953, p. 25.

9.94. I. B. Bersuker, *Zh. Neorg. Khim.*, **9**, 36–41 (1964); *Zh. Struct. Khim.*, **4**, 461 (1963).

9.95. G. B. Bokii and I. B. Bersuker, *Zh. Struct. Khim.*, **4**, 934 (1963).

9.96. M. J. S. Dewar, *The PMO Theory of Organic Chemistry*, Plenum Press, New York, 1975.

9.97. C. W. Hobbs and R. S. Tobias, *Inorg. Chem.*, **9**, 1037 (1970).

9.98. L. S. Bartell, S. Doun, and C. J. Marsden, *J. Mol. Struct.*, **75**, 271–282 (1981).

9.99. H. Oberhammer, K. Seppelt, and R. Mews, *J. Mol. Struct,.* **101**, 325–331 (1983).

9.100. L. S. Bartell, F. B. Clippard, and E. J. Jacob, *Inorg. Chem.*, **15**, 3009 (1976).

9.101. I. B. Bersuker, *J. Coord. Chem.*, **34**, 289–338 (1995).

9.102. J. P. Fackler and A. Avdeev, *Inorg. Chem.*, **13**, 1864–1875 (1974).

9.103. J. Gazo, *Pure Appl. Chem.*, **38**, 279–301 (1974).

9.104. C. Friebel and D. Reinen, *Z. Anorg. Allg. Chem.*, **407**, 193–200 (1974).

9.105. J. K. Burdett, *Inorg. Chem.*, **20**, 1959 (1981).

9.106. H. Yamatera, *Acta Chem. Scand. A*, **33**, 107–111 (1979).

9.107. K. Finnie, L. Dubicki, E. R. Krausz, and M. J. Riley, *J. Inorg. Chem.*, **29**, 3908 (1990).

9.108. D. A. Tucker, P. S. White, K. L. Trojan, M. L. Kirk, and W. E. Hatfield, *Inorg. Chem.*, **30**, 823–826 (1991).

9.109. B. Papankova, M. Serator, J. Strachelsky, and J. Gazo, *Proceedings of the 8th Conference on Coordination Chemistry*, Smolenice, Bratislava, 1980, pp. 321–326.

9.110. T. Obert and I. B. Bersuker, *Proceedings of the 19th International Conference on Coordination Chemistry*, Prague, 1978, Vol. 2, p. 94; *Czech. J. Phys. B*, **33**, 568–573 (1983).

9.111. Ch. J. Simmons, *New. J. Chem.*, **17**, 77–95 (1993).

9.112. I. B. Bersuker and V. Z. Polinger, *Vibronic Interactions in Molecules and Crystals*, Springer-Verlag, Berlin, 1989, 422 pp.

9.113. G. A. Gehring and K. A. Gehring, *Rep. Prog. Phys.*, **38**, 1 (1975).

9.114. M. E. Lines and A. M. Glass, *Principles and Applications of Ferroelectrics and Related Materials*, Clarendon Press, Oxford, 1977.

9.115. I. B. Bersuker, *Phys. Lett.*, **20**, 586 (1966).

9.116. I. B. Bersuker and B. G. Vekhter, *Ferroelectrics*, **19**, 137–150 (1978).

9.117. I. B. Bersuker, *Fiz. Tverd. Tela* (*Sov. Phys. Solid State*), **30**, 1738–1744 (1988).

9.118. I. B. Bersuker, *Ferroelectrics* **164**, 75–100 (1995).

9.119. P. I. Konsin and N. N. Kristofel, in: *Interband Model of Ferroelectricity* (Russ.), ed. E. Bursian, Leningrad Pedagogical Institute, Leningrad, 1987, pp. 32–68.

9.120. R. Comes, M. Lambert, and A. Guinner, *Solid State Commun.*, **6**, 715 (1968)

9.121. A. M. Quitet, M. Lambert, and A. Guinier, *Solid State Commun*, **12**, 1053 (1973); R. Comes, R. Currat, F. Denoyer, M. Lambert, and M. Quittet, *Ferroelectrics*, **12**, 3 (1976).

9.122. K. A. Muller, *Helv. Phys. Acta*, **59**, 874–884 (1986) in: *Nonlinearity in Condensed Matter*, ed. A. R. Bishop, Springer-Verlag, Heidelberg, 1986, p. 234.

9.123. O. Hanske-Petitpierre, Y. Yacoby, J. Mustre de Leon, E.A. Stern, and J.J. Rehr, *Phys. Rev. B*, **44**, 6700–6707 (1992); B. Ravel, E. A. Stern, *Physica B*, **208** and **209**, 316 (1995).

9.124. N. Sicron, B. Ravel, Y. Yacoby, E. A. Stern, F. Dogan, and J. J. Rehr, *Phys. Rev. B*, **50**, 13168 (1994–II).

9.125. G. Burns and F. Dacol, *Ferroelectrics*, **37**, 661 (1981).

9.126. F. Gervais, *Ferroelectrics*, **53**, 91 (1984).

9.127. K. H. Ehses, H. Bock, and K. Fischer, *Ferroelectrics*, **37**, 507 (1981).

9.128. K. Itoh, L. Z. Zeng, E. Nakamura, and N. Mishima, *Ferroelectrics*, **63**, 29 (1985).

9.129. T. P. Dougherty, G. P. Wiederrecht, K. A. Nelson, M. H. Garrett, H. P. Jensen, and C. Warde, *Science*, **258**, 770 (1992).

9.130. D. Reinen and C. Friebel, *Struct. Bonding*, **37**, 1–60 (1979).

9.131. M. Lambert and R. Comes, *Solid State Commun.*, **7**, 305 (1969).

9.132. I. B. Bersuker and B. G. Vekhter, *Izv. Akad. Nauk. SSSR Ser. Fiz.*, **33**, 199 (1969).

9.133. I. B. Bersuker, *Teor, Eksp. Khim.*, **27**, 262–269 (1991).

9.134. W. J. A. Maaskant and W. G. Haye, *J. Phys. C*, **19**, 5295–5308 (1986).

9.135. W. J. A. Maaskant, *Struct. Bonding*, **83**, 55–87 (1995).

9.136. C. P. J. M. Van der Vorst and W. J. A. Maaskant, *J. Solid State Chem.*, **34**, 301 (1980).

9.137. W. J. A. Maaskant, *New J. Chem.*, **17**, 97 (1993).

10 Electron Transfer, Redox Properties, and Electron-Conformational Effects

Chemical interaction begins with electronic charge redistribution that initiates chemical transformations.

In chemical processes with transition metal participation charge transfer by coordination, intra- and intermolecular electron transfer, and electron-conformational effects are most important. Examples of these phenomena as controlled by the electronic structure are discussed briefly in this chapter.

10.1. ELECTRON TRANSFER AND CHARGE TRANSFER BY COORDINATION

Intramolecular Charge Transfer and Intermolecular Electron Transfer

As stated in Section 1.2, the chemical bond is an electronic phenomenon and the bonding is based mostly on collectivization of the electrons of interacting atoms or atomic groups. Any chemical interaction of atoms is thus associated with electronic charge redistribution, which generates further transformations. Charge transfers are especially important in coordination compounds because of the heterogeneity introduced by d and f states in the otherwise almost homogeneous organic and main-group systems, in the sense that they contain only $nsnp$-electron atoms.

In Section 5.2, methods of quantum-chemical calculation of charge redistribution and integral and orbital charge transfer are discussed, and Sections 6.3 and 6.4 provide many examples of such calculations. Here we try to give a more general understanding of the origin of electron transfer and charge transfer by coordination and to correlate these quantities with the electronic structure of the coordinating subunits.

The problem is to find *a measure of ability of the molecular system to transmit or to accept electronic charge*, based on the electronic structure of the system. Many attempts have been made to solve this problem, beginning with that based on Pauling's *electronegativity* [10.1]. Its explicit formulation states that the ability to transmit electronic charge is related directly to the ionization energy E_I, while the acceptance of electrons can be linked to the electron affinity E_A. The possibility of charge transfer by coordination is characterized by some combinations of E_I and E_A of the interacting atomic groups. A more flexible treatment of this phenomenon is based on the empirical conception of *hard and soft acids and bases* [10.2], discussed below.

The charge transfer between two strongly interacting molecular species within the stable coordination system is an *intramolecular charge transfer*. Also, there is much interest in *intermolecular electron transfer*. The latter takes place when there are two weakly interacting (almost independent) systems between which an electron transfers instantly from one to another. Such electron transitions are most important for *electron transfer reactions* and for various *electron-conformational effects* (Section 10.3). The possibility of intermolecular electron transfer and its rate are related directly to the electronic structure of the interacting molecules.

Consider two weakly interacting systems that can conventionally be called the donor (D) and the acceptor (A). Figure 10.1 shows a simplified scheme of the highest occupied and lowest unoccupied molecular orbitals (HOMO and LUMO; Section 6.2) of D and A. If there is no significant bonding between D and A, the electron transfer $D \rightarrow A$ (the reaction $D + A \rightarrow D^+ + A^-$) is energetically convenient in case I (Fig. 10.1a) when the ionization energy E_I of D is smaller than the electron affinity E_A of A, $E_I < E_A$. In case II the electron transfer is formally not convenient (Fig. 10.1b) since $E_I > E_A$, but the electron transfer can still take place when promoted by some excitation process, as shown in Fig. 10.1c.

This picture of electron transfer between a noninteracting DA pair is rather oversimplified in many aspects. First, the absolute value of the ionization energy is taken to be equal to the HOMO energy; this is in accordance with the Koopmans theorem, which has a limited applicability and is unacceptable in a number of cases (Sections 6.2 and 6.4). The accurate value E_I can be calculated as the difference between the total ground-state energies of D and D^+, and accordingly, E_A is the difference between the total energies of A^- and A. Hence the schemes of Fig. 10.1 remain valid if one substitutes total energies $E(D))$, $E(D^*)$, $E(A)$, and $E(A^-)$, for $\varepsilon(HOMO)$ and $\varepsilon(LUMO)$ of the donor and acceptor respectively (D^* denotes the corresponding excited state).

The HOMO–LUMO energies or, more precisely, the total energies of D, D^*, A, and A^- are related directly to the *standard electron potentials* E_{ox}^0 *and* E_{red}^0 and/or to the (approximately the same) *half-width potentials* $E_{1/2}^{ox}$ *and* $E_{1/2}^{red}$ in polarography (for details, see [10.3]). Thus the electronic structure parameters above (sometimes roughly the HOMO and LUMO energies) characterize the redox properties of the systems under consideration.

Electron transfer in DA pairs in which D and A are coordination systems was studied in many cases by experimental methods. In coordination DA pairs, D is generally an anion and A is a cation. An example of a well-studied series of ion pairs is $M(CN)_6^{4-}$ anions (M = Fe, Ru, Os) with $Ru(NH_3)_5L^{3+}$ cations (L = pyridine or substituted pyridine) [10.4]. The rate of electron transfer

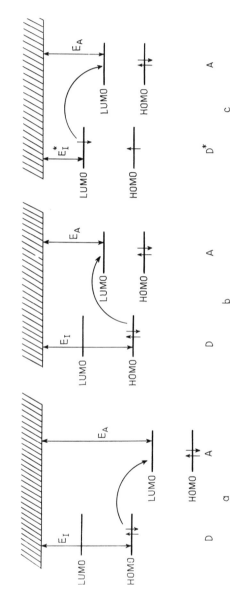

Figure 10.1. Simple scheme of HOMO–LUMO energy levels of a donor–acceptor pair DA: (*a*) the ionization energy E_I of D is larger, than the electron affinity E_A; (*b*) $E_I < E_A$; (*c*) for the excited state D^* of the case (*b*) $E_I^* > E_A$ (note that all E_I and E_A are negative).

depends on the energy barrier between the DA pair, which in turn is determined by the orbitals involved in the D–A orbital overlap during the instant interaction realizing the electron tunneling and associated vibronic coupling. Reference [10.5] provides an example of *ab initio* calculations of the electron transfer process

$$Co(NH_3)_6{}^{2+} + Co(NH_3)_6{}^{3+} \rightarrow Co(NH_3)_6{}^{3+} + Co(NH_3)_6{}^{2+} \quad (10.1)$$

which can be presented electronically as the transition (Section 6.2)

$$[t_{2g}^5 e_g^2(^4T_{1g})] + [t_{2g}^6(^1A_{1g})] \rightarrow [t_{2g}^6(^1A_{1g})] + [t_{2g}^5 e_g^2(^4T_{1g})] \quad (10.2)$$

From the electronic transition scheme (10.2), it is seen that in this electron transfer each of the two complexes undergoes spin transition ($^4T_{1g} \rightarrow {}^1A_{1g}$ and $^1A_{1g} \rightarrow {}^4T_{1g}$, respectively). As in the spin-crossover phenomenon described in Section 8.4, these spin transitions are accompanied by changes in the ligand configuration via vibronic coupling, and they involve excited d states.

The electron transfer reaction (10.1) is symmetrical ($D^+ \equiv A$, $A^- \equiv D$): The direct and inverse reactions have the same probability. Electron transfer also takes place between asymmetrical pairs of coordination systems, including those mentioned above. For them the products of electronic transfer D^+ and/or A^- may be unstable, initiating secondary reactions. For example, from the two cases,

$$Co^{III}(NH_3)_6{}^{3+} + Ru^{II}(CN)_6{}^{4-} \rightarrow Co^{II}(NH_3)_6{}^{2+} + Ru^{III}(CN)_6{}^{3-} \quad (I)$$

$$Co^{III}(sep)^3] + Ru^{II}(CN)_6{}^{4-} \rightarrow Co^{II}(sep)^{2+} + Ru^{III}(CN)_6{}^{3-} \quad (II)$$

only case I produces a secondary reaction of dissociation into Co, NH_3, and $Ru^{III}(CN)_6{}^{3-}$. As distinct from $Co^{II}(NH_3)_6{}^{2+}$, $Co^{II}(sep)^{2+}$ (sep = sepulchrate) with cage-type ligands is rather stable: It does not decompose easily and the decomposition rate cannot compete with the back electron transfer process.

If the bonding between coordinated groups that participate in the electron transfer is sufficiently strong, so that they can be considered as one molecule, the electron transfer is of the aforementioned intramolecular nature. However, in large systems there may be electron transfers between molecular sites which are not bonded directly. Figure 10.2 illustrates the absorption spectrum of electron transfer in the complex *cis*-$(NH_3)_5Ru^{II}NCRu^{II}(bpy)_2CNRu^{III}(NH_3)_5{}^{7+}$ [10.6], which is in fact a mixed-valence compound (Section 10.2). Owing to the large distances (and hence weak coupling) between different sites of this system, the spectrum looks as if it resulted from electron transfer reactions. The authors (in [10.6]) succeeded in decomposing the integral absorption curve into components, each corresponding to a certain site-to-site electron transfer, shown in Fig. 10.2 by arrows.

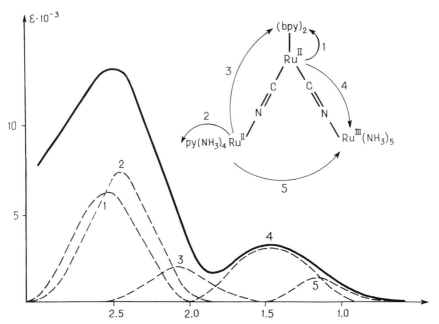

Figure 10.2. Absorption spectrum of the shown tricenter mixed-valence complex approximately resolved into component bands corresponding to the various site-to-site optical electron transfer transitions. (From [10.6].)

A large subject of *outer-sphere electron transfer* between metal complexes in solution is not considered here in detail. In these cases the rate of electron transfer is determined by the nonadiabatic process in which the sluggish motion of solvent structure and/or counterions is important [10.7]. Also relevant to electron transfer is the more general topic of *tunneling phenomena in chemistry*, which include in addition to *electron tunneling* of different origin, the tunneling of heavy particles, atoms, and atomic groups [10.8].

Redox Capacitance

We begin the discussion of intramolecular charge transfer with a simple picture of charge transfer by coordination. Consider two coordinating systems, 1 and 2, and denote their HOMO and LUMO energy levels by ε_1^0, ε_2^0, ε_1^u, and ε_2^u, respectively. As mentioned above, in the Koopmans approximation (Sections 2.2, 6.2, and 6.4) the energies of the HOMO are equal to the corresponding ionization potentials taken with the opposite sign. The LUMO energies ε_1^u and ε_2^u under the same assumptions could be attributed to the electron affinities E_{Ai}, but this is a rough approximation: The addition of an electron to or removal from the system changes the ε^u and ε^0 levels significantly and may even change the sign of ε^u (since the latter is nearer to zero than ε^0).

As discussed above, if $\varepsilon_1^0 > \varepsilon_2^u$, the electronic charge from the HOMO of 1 transfers to the LUMO of 2, following the principle of minimum energy, and if $\varepsilon_2^0 > \varepsilon_1^u$, charge transfer takes place in the opposite direction. How large will the charge transfer Δq be in these cases? To answer this question a characteristic of the system which gives a measure of its *charge capacity* is needed [10.9, 10.10]. It is well known that MO energy levels are dependent on the total charge in the system, $\varepsilon = \varepsilon(q)$, and these functions are used in many semiempirical versions of self-consistent calculations of electronic structure of coordination compounds (Section 5.5). In particular, following the iterative extended Hückel method, the function $\varepsilon(q)$ can be approximated by a three-term polynomial [cf. Eq. (5.107)]:

$$\varepsilon(q) = aq^2 + bq + c \qquad (10.3)$$

Thus when the charge of the system changes by Δq the MO energy level changes by $\Delta\varepsilon = 2aq\,\Delta q + b\,\Delta q$. The derivative

$$C = \frac{dq}{d\varepsilon} \qquad (10.4)$$

which is equal to *the amount of charge that increases the HOMO energy level of the system by a unity*, can be called the redox capacitance [10.9]. As seen from Eq. (10.3), at $q = 0$, $C = (dq/d\varepsilon)_0 = 1/b$.

Provided that the redox capacitances of the two coordinating systems are known, the charge transfer can be calculated directly. Indeed, by coordination, the MOs of the two systems are collectivized and the coordinated system acquires a common HOMO energy level ε^0 (another requirement is that the chemical potentials of the two coordinated system equalize; see below). The charge transfers for the two subsystems are equal in absolute value and opposite in sign (we assume that $\varepsilon_1^0 > \varepsilon_2^u$):

$$-\Delta q_1 = \Delta q_2 = \Delta q \qquad (10.5)$$

Hence

$$\varepsilon_1^0(-\Delta q) = \varepsilon_2^u(\Delta q) = \varepsilon^0 \qquad (10.6)$$

Then in the linear approximation with respect to Δq, which is valid either when Δq is sufficiently small or when C is constant (or both), we have (Fig. 10.3)

$$\Delta q = C_1(\varepsilon_1^0 - \varepsilon^0)$$

$$-\Delta q = C_2(\varepsilon_2^u - \varepsilon^0)$$

$$\Delta q = \frac{\varepsilon_1^0 - \varepsilon_2^u}{C_1^{-1} + C_2^{-1}} \qquad (10.7)$$

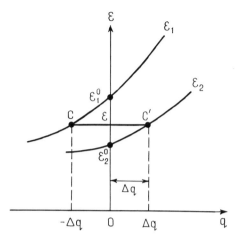

Figure 10.3. Graphic solution of the equation of charge transfer in the linear approximation. In the coordinated system $\varepsilon_1^0(-\Delta q) = \varepsilon_2^u(\Delta q) = \varepsilon^0$. The curves $\varepsilon_1^0(q)$ and $\varepsilon_2^u(q)$ after (10.3) are taken arbitrary.

$$\varepsilon_1^0 - \varepsilon^0 = \frac{(\varepsilon_1^0 - \varepsilon_2^u)}{1 + C_1/C_2} \tag{10.8}$$

$$\varepsilon_2^u - \varepsilon^0 = \frac{\varepsilon_1^0 - \varepsilon_2^u}{1 + C_2/C_1} \tag{10.9}$$

$$\frac{\varepsilon_1^0 - \varepsilon^0}{\varepsilon_2^u - \varepsilon^0} = -\frac{C_2}{C_1} \tag{10.10}$$

If $\varepsilon_2^u > \varepsilon_1^0$, the sign of Δq changes to the opposite. There may also be other combinations of energy spectra, leading to similar formulas [10.10]; a formula for charge transfer, somewhat similar to (10.7), was derived by Huheey [10.11].

Equations (10.7) to (10.10) show that the energy-level difference and redox capacitance determine the main important features of the charge transfer by coordination. In particular, if the capacitance of one of the systems is much larger than that of the other, $C_2 \gg C_1$, then

$$\Delta q \approx C_1(\varepsilon_1^0 - \varepsilon_2^u) \tag{10.11}$$

$$\varepsilon_2^u \approx \varepsilon^0 \tag{10.12}$$

and the charge transfer is determined by the relative value of the energy level and capacitance of system 1. This is often the case with catalysts (see below).

Consider in more detail the physical meaning of redox capacitance. By definition (10.4) C is a measure of ability to accept or transmit charge without changing the energy levels significantly. It is clear that this property depends on the electronic structure, in particular, on the possibility of the excess charge to delocalize over larger regions of space, thus reducing the interelectron repulsion. The larger the delocalization, the higher the capacitance C. The latter becomes especially large for conjugated systems for which the addition of an electron results in the population of the next energy level in the valence band. In this case the redox capacitance can be estimated using approximate expressions for the energy levels calculated in the model of a potential box [10.12]:

$$\varepsilon_n \approx \frac{n^2\pi^2}{2mN^2l^2} \tag{10.13}$$

where N is the number of chain members, l their linear dimension, and m the mass of the electron. Population of the first $N/2$ levels (for simplicity, N is even) by N electrons (by one electron from each chain member) yields the ground state with an energy gap $\varepsilon_{(N/2)+1} - \varepsilon_{N/2} = \Delta\varepsilon$ to the first excited state (the $(N/2) + 1$ state). To occupy the latter a two-electron charge, $\Delta q = 2e$, should be added. Consequently,

$$C \sim \frac{4mel^2}{\pi^2\hbar^2} \frac{N^2}{N+1} \tag{10.14}$$

or approximately, for $N \gg 1$,

$$C \sim aN \tag{10.15}$$

with $a = 4mel^2/\pi^2\hbar^2$.

It is seen that the redox capacitance is directly proportional to the number of chain members N in the conjugated system. Note that the same energy gap $\Delta\varepsilon$ to the excited states determines the polarizability of the system, and hence larger capacitance means larger polarizabilities. On the other hand, $\Delta\varepsilon$ equals the quantum of the first long-wave optical transition with the wavelength λ_m, $\Delta\varepsilon = hc/\lambda_m$ (c is the speed of light), and hence $C = 2e\lambda_m/hc$. Thus the redox capacitance for conjugated molecules can also be estimated using empirical data on light absorption.

By way of illustration, consider the charge transfer by coordination of hydrogen peroxide (system 1) to iron porphyrins (system 2). Semiempirical calculations of H_2O_2 in different charge states allow us to estimate the curve $\varepsilon_1(q)$, from which the approximate value of C_1 follows directly: $C_1 = 0.02\ e/$Kk (1Kk $= 10^3$ cm^{-1}), while for iron porphyrin $C_2 = 0.6\ e/$Kk [10.9]. This means that $C_2 \gg C_1$, and the simplified formulas (10.11) and (10.12) can be applied. For the HOMO–LUMO energy-level difference, we have $\varepsilon_2^0 - \varepsilon_1^u = 55$ Kk. Hence the charge transfer from the porphyrin to hydrogen peroxide is $\Delta q \approx C_1(\varepsilon_2^0 - \varepsilon_1^u) \approx 1.1e$. This transfer to the antibonding orbital of the H_2O_2 molecule strongly activates its O—O bond.

The notion of redox capacitance is important, first, to problems of redox catalysis. It also contributes significantly to an understanding of the origin of high-oxidation states (e.g., phthallocyanins can accept several electrons simultaneously [10.13]). This property is of special interest to ferredoxins, which play an important role in a series of biological processes, where they serve for both electron transfer and electron storage, the latter allowing for synchrony multielectron processes. In accordance with the results discussed above, the ferredoxins realize these functions owing to their high redox capacitance. This is confirmed experimentally. For instance, for the cluster $Fe_4(\eta^5\text{-}C_5H_5)_4(\mu_3\text{-}S)_4^n$, four inverse stages of one-electron charge transfers corresponding to the cluster states with $n = -1, 0, 1, 2$, and 3 have been revealed by means of the cyclic voltamperimetric method [10.14]. For calculation of redox capacitance for such systems, see [10.15].

Hard and Soft Acids and Bases

The necessity to introduce a property of molecular systems that characterizes their ability to act as an acid or base (i.e., as above, the ability to transmit or accept electronic charge) was recognized by chemists long ago based on chemical intuition. Pearson [10.2] introduced the term *hard and soft acids and bases* (HSAB) for this reason.

According to the definition of HSAB, molecular systems are divided into four groups [10.2]: "(1) soft acid — the acceptor atom is of low positive charge, large size and has polarizable outer electrons; (2) hard acid — the acceptor atom is of high positive charge, small size and has not easily polarized outer electrons; (3) soft base — the donor atom is of low electronegativity, easily oxidized, highly polarizable and with low-lying empty orbitals; (4) hard base — the donor atom is of high electronegativity, hard to oxidize, of low polarizability and with only high energy empty orbitals". Based on this definition, an empirical rule was formulated: Hard acids prefer to coordinate with hard bases and soft acids prefer to coordinate with soft bases. Hard–hard coordination represents a more ionic bonding, while the soft–soft one is more covalent.

This rather qualitative formulation contains many uncertainties that have been subjected to criticism. More quantitative formulations based on the electronic structure have recently been reached [10.16]. Based on the density functional theory (Section 5.2), the authors [10.16] defined the *chemical potential* μ as a derivative of the total energy of the system E with respect to the electronic charge produced by N electrons, while the nuclear charge Z is fixed:

$$\mu = \left(\frac{\partial E}{\partial N}\right)_z = \left(\frac{\partial E}{\partial q}\right)_z \tag{10.16}$$

In the finite difference approximation the μ value can be derived from the curve $E = E(\mu)$ and expressed roughly by known parameters as follows:

$$-\mu \approx \frac{I + A}{2} = \chi_\mu \tag{10.17}$$

where I is the ionization potential, A the electron affinity, and χ_μ the Mulliken electronegativity of the molecular system.

The quantity μ is, in fact, of significant importance to different topics in chemistry. In particular, if two molecular systems that have different chemical potentials μ_1 and μ_2 coordinate, the joint system acquires a common value of

$$\mu = \mu_1(-\Delta q) = \mu_2(\Delta q) \tag{10.18}$$

Now, the *chemical hardness* of an acid or base denoted by η (the softness is $\sigma = 1/\eta$) is defined as follows [10.2, 10.16]:

$$\eta = \frac{1}{2} \frac{\partial \mu}{\partial q} = \frac{1}{2} \frac{\partial^2 E}{\partial q^2} \tag{10.19}$$

Similar to (10.17), in the same finite difference approximation

$$\eta = \frac{I - A}{2} \tag{10.20}$$

It is seen from the definition (10.19) that the chemical hardness equals half the change of the chemical potential due to the increase in charge on the system by unity. In other words, chemical hardness characterizes the resistance of the chemical potential μ to the change in electronic charge of the system.

Consider two coordinated systems, 1 and 2, and denote their chemical potentials and hardness by μ_1 and μ_2, η_1 and η_2, respectively. In the linear approximation with respect to the charge transfer $\Delta q = \Delta q_1 = -\Delta q_2$

$$\mu_1 = \mu_1^0 + 2\eta_1 \Delta q$$
$$\mu_2 = \mu_2^0 - 2\eta_2 \Delta q$$

and

$$\Delta q = \frac{\mu_2^0 - \mu_1^0}{2(\eta_1 + \eta_2)}$$
$$= \frac{\chi_1^0 - \chi_2^0}{2(\eta_1 + \eta_2)} \tag{10.21}$$

Surprisingly, Eq. (10.21) for the charge transfer by coordination in the case of molecular systems with well-spaced energy levels, derived in [10.9, 10.10] based on the notion of redox capacitance, coincides exactly with Eq. (10.7). (Note the 16-year interval between publications [10.9] (or [10.10]) and [10.16].) Indeed, for systems with discrete energy levels ε_n the chemical potential μ is defined by the well-known statistical formula of the temperature dependence of the population $n(T)$ of the ε_n level:

$$n(T) = \left[1 + \exp\left(\frac{\varepsilon_n - \mu}{kT}\right)\right]^{-1} \tag{10.22}$$

At $T = 0$, $n(0) = 1$ for $\varepsilon_n < \mu$ and $n(0) = 0$ for $\varepsilon_n > \mu$. This means that for any system at $T = 0$, μ coincides with the highest occupied energy level ε^0, introduced above [see (10.6)]. The equality $\mu = \varepsilon^0$ also holds for all the much higher temperatures, for which the excited states are not significantly thermally populated. For the electronic energy levels under consideration this condition is obeyed, provided that there are no degenerate or pseudodegenerate electronic states.

With $\mu = \varepsilon^0$, Eq. (10.18) coincides with (10.6), the hardness η according to the definition (10.19) coincides with the inverse of redox capacitance (10.4) (with a factor $\frac{1}{2}$): $\eta = \frac{1}{2}C^{-1}$, and Δq (10.21) coincides exactly with Δq determined by Eq. (10.7).

The two coinciding formulas (10.21) and (10.7), as well as all others based on them, are certainly rather qualitative or semiquantitative and reflect the tendency of charge transfers by coordination, giving a general understanding of the origin and mechanism of this process. Taken as approximate relations aimed at qualitative understanding, it does not make very much sense to improve these relations at the expense of their further complication. For more exact calculations of charge transfers, approved quantum-chemical methods and computer programs should be employed (Section 6.4). Nevertheless, attempts to improve the expressions for the hardness η as a function of electronic structure parameters are being continued, and some of the results seem to be stimulating [10.2, 10.17].

It is clear that soft acids and soft bases, as well as hard acids and hard bases, fit each other better than soft acids with hard bases, or hard acids with soft bases. This statement can be confirmed using perturbation theory for chemical interactions (Section 11.1). In many cases the concept of HSAB correlates with the metal–ligand bonding energy ΔH. For illustration, the values of hardness η and charge transfers Δq in the coordination $M + CO$ for a series of transition metals M and their correlation with experimental ΔH values are shown in Table 10.1. The values χ^0 and η are estimated for the low-spin valence states of M (as they are expected to appear in the coordination with CO), and the Δq values are calculated by Eq. (10.22) (for CO $\chi^0 = 6.1\,eV$ and $\eta = 7.9\,eV$ [10.2]).

Table 10.1. Electronegativities χ^0 in the low-spin valence state of transition metals M, their hardness η, charge transfers Δq by coordination M + CO, and activation enthalpies ΔH for the dissociation of the first CO ligand from $M(CO)_n$

Transition Metal	χ^0 (eV)	η (V)	Δq (e)	$\Delta H^{\neq}$ (kcal)
V	2.24	1.24	0.211	V(CO) stable
Cr	2.47	1.58	0.196	40
Mn	3.10	2.19	0.149	37
Fe	2.55	1.55	0.188	42
CO	4.12	3.04	0.091	22
Ni	3.50	2.30	0.128	22
Cu	5.84	4.61	0.110	Very unstable
Mo	3.18	1.98	0.148	40
Ru	3.54	2.34	0.125	28
Pd	4.45	3.89	0.070	Unstable
Ag	6.87	5.57	−0.029	Very unstable
Pt	5.30	2.90	0.037	Unstable
Au	6.70	4.40	−0.026	Very unstable

Source: [10.2].

It is seen that the values Δq in Table 10.1 follow roughly the stability of appropriate complexes, the value ΔH of dissociation of the first ligand in $M(CO)_6$); for small and negative values of Δq the complexes are unstable and very unstable, respectively. Note that, in general, a full quantitative correlation between the total charge transfers Δq and bonding energies is not expected. Indeed, as stated elsewhere in this book (see, e.g., Sections 5.2, 6.3, and 11.2), the total charge transfer Δq results from orbital charge transfers Δq_i that may have opposite signs, and hence Δq can be small for large absolute values of Δq_i which determine the bonding.

10.2. ELECTRON TRANSFER IN MIXED-VALENCE COMPOUNDS

Mixed-Valence Compounds as Electronic Systems; a Two-Level Dimer

Mixed-valence (MV) compounds represent one of the most interesting examples of intramolecular electron transfer [10.18–10.20]. The increasing importance of these compounds is due to their applications as redox catalysts, models of metalloenzymes [10.21, 10.22], local centers in superconducting ceramics [10.23, 10.24], molecular electronic devices [10.25, 10.26], and so on.

In a larger definition [10.27], MV compounds contain ions of the same element in two different formal states of oxidation. These ions can be regarded as having the same oxidation state but with additional (excess) electrons that can either occupy one of the ions, changing its oxidation state and making it different from the others, or be delocalized over all (or a part of) the centers.

With respect to the excess electron delocalization, Robin and Day [10.27] suggested the following classification. All the MV compounds are divided into three classes, I, II, and III, with an additional division of class III into IIIA and IIIB. In class I the metal ions of the MV system are in ligand fields of different symmetry and/or strength (e.g., tetrahedral versus octahedral). In this case there are no active MOs formed by the AOs occupied by the excess electron at different centers because of the large difference in their energies and very small resonance integral w [see Section 5.1 and below, Eq. (10.26)]. Hence the excess electron is firmly trapped at one of the centers, making them essentially nonequivalent.

Class II contains metal ions in ligand fields of nearly identical symmetry differing from one another by distortions of only a few tenths Å. in these systems $w \neq 0$, and hence there are MOs in common formed by the AOs of the centers and occupied by the excess electron, but the valencies are still distinguishable with slight delocalization on the MO. In class IIIA the metal ions are grouped into polynuclear clusters within which they are equivalent (indistinguishable); inside the cluster w is maximal and the excess electron can be either localized or delocalized, depending on the bridging group. Finally, in class IIIB all the ions in the lattice are indistinguishable (complete delocalization over the cation sublattice).

Following this classification, coordination MV compounds belong mainly to class IIIA (polynuclear MV clusters) with a few exceptions which could be attributed to class II: for instance, when the small structural differences between the centers are of molecular or vibronic origin. The other cases are MV solids (mostly ionic solids); Table 10.2 illustrates some examples.

In what follows in this section, we consider only transition metal MV compounds. By definition they contain two or more equivalent coordination centers divided by some bridging atomic groups with one or more excess electrons, the number of which is smaller than the number of centers. If the excess electron is localized at one of the equivalent centers, they have different valence (oxidation) states, and this explains the origin of the name of these compounds.

The equivalency of the centers means that the excess electron can occupy any of the centers with equal probability. From the point of view of quantum mechanics, in the stationary state the excess electron should be uniformly distributed (delocalized) over all the centers. However, if the energy barrier for the electron transfer between the centers (created by the bridging groups) is sufficiently large, the excess electron can stay a relatively long time at one center, and its possible delocalization is controlled by the barrier height and temperature.

In these cases there is a lifetime of the electron at each center, and the observable properties of the MV compound depend on the method of obser-

Table 10.2. Examples of different classes of MV solids with the classification after Robin and Day [10.27]

Example	Classification	Remarks
Pb_3O_4	I	Red lead
Sb_2O_4	I	Mineral cervantite
$Fe_4[Fe(CN)_6]_3 \cdot 4H_2O$	II	Prussian blue
$Li_xNi_{1-x}O$	II	Hopping semiconductor
$La_{1-x}Sr_xMnO_3$	II	Ferromagnetism
$BaBi_{1-x}Pb_xO_3$	III	Superconductivity
$LiTi_2O_4$	III	Superconductivity
$K_2Pt(CN)_4Br_{0.30} \cdot 3H_2O$	III	Molecular metal; Paierls instability
Na_xWO_3	III	Bronze luster; metallic at high x
$M_xMo_6S_8$	III	Superconductivity
Fe_4S_4 (ferredoxins)	III	Enzymes

Source: [10.28].

vation, more precisely on the ratio of the time of measurement to the lifetime of electron localization (cf. the relativity rule concerning the means of observation in Section 9.2). Hence the notions of localized and delocalized electron may be to a certain extent conventional. *The localization–delocalization alternative is the main problem of MV compounds.*

To formulate the problem quantitatively, consider an MV dimer with two equivalent electronic closed-shell centers, 1 and 2, and one excess electron [10.29, 10.30]. Assume that the one-electron states at each center, φ_1 and φ_2, are nondegenerate. This means that when the excess electron is localized at one of the centers, it distorts its near-neighbor environment along the totally symmetric coordinate. The corresponding local distortion (breathing) coordinates can be denoted by Q_1 and Q_2, respectively. The coupling of the electronic state to these distortions is described by the vibronic constant F (Section 7.2):

$$F = \langle \varphi_1 | \left(\frac{\partial H}{\partial Q_1} \right)_0 | \varphi_1 \rangle = \langle \varphi_2 | \left(\frac{\partial H}{\partial Q_2} \right)_0 | \varphi_2 \rangle \qquad (10.23)$$

where H is the Hamiltonian.

For the system as a whole the presentations of the wavefunctions:

$$\Phi_\pm = (\tfrac{1}{2})^{1/2}(\varphi_1 \pm \varphi_2) \qquad (10.24)$$

and coordinates:

$$Q_\pm = (\tfrac{1}{2})^{1/2}(Q_1 \pm Q_2) \qquad (10.25)$$

are more convenient. Q_+ is a totally symmetric coordinate of the system describing the simultaneous (synchronous) breathing distortion of both centers, while Q_- gives the asymmetric distortion of the antiphase breathing of the two centers appropriate to the electron localization.

If there is an overlap between the two states φ_1 and φ_2, they form bonding Φ_+ and antibonding Φ_- states with an energy gap $2w$, where w is the resonance integral (Section 5.1):

$$w = \langle \varphi_1 | H | \varphi_2 \rangle \tag{10.26}$$

This parameter is most important for MV theory, characterizing the strength of the intercenter interaction and hence the electron transfer rate.

With the notations (10.23) to (10.26), the problem of the excess electron in the MV dimer under consideration can be formulated as a vibronic problem (Section 7.4). Indeed, the two states $\Phi_\pm$ at $Q_1 = Q_2 = 0$ ($Q_\pm = 0$) mix under the nuclear displacements Q_- and shift under Q_+. For the problems considered in this section the shift Q_+ is not important and can be excluded by an appropriate choice of the energy read-off (Section 7.4). The mixing of the two electronic states by the nuclear displacements Q_- taken as a perturbation results in the pseudo Jahn–Teller problem (7.67) with the solution—the adiabatic potential (7.69) (K_0 is the force constant of the Q_1 or Q_2 distortions):

$$\varepsilon_\pm(Q_-) = \tfrac{1}{2}K_0 Q_-^2 \pm (w^2 + F^2 Q_-^2)^{1/2} \tag{10.27}$$

These two curves are analyzed in Section 7.4. For $|w| > F^2/K_0$, both have a minimum at $Q_- = 0$. However, if [cf. (7.71)]

$$|w| < \frac{F^2}{K_0} \tag{10.28}$$

the lowest curve has a maximum at $Q_- = 0$ and two minima at $\pm Q^0$,

$$Q_-^{\ 0} = \left(\frac{F^2}{K_0^2} - \frac{w^2}{F^2}\right)^{1/2} \tag{10.29}$$

The curve $\varepsilon_-(Q)$ for this case is illustrated in Fig. 7.19 of the pseudo Jahn–Teller effect. It is seen that if the condition (10.28) is obeyed, that is, if the contribution to the energy due to the localization distortion Q_- is larger than that of the electron transfer w, the minimum energy (and the wavefunction) corresponds to the localization of the excess electron at one of the two centers. In the opposite case, when $|w| > F^2/K_0$, $Q_- = 0$ and there are no localization minima, the electron is uniformly delocalized over the two centers. Thus for MV dimers inequality (10.28) serves as the condition of localization of the excess electron.

One of the observables in MV systems is the *intervalence transition* (*IT*) *band of light absorption* produced by the transition from the minimum (localized) ground state to the excited (delocalized) state. An example of such an IT band is given above in Fig. 10.2. Both the frequency and probability (band shape) of the IT are dependent on the parameters of electron transfer w and vibronic coupling F. To calculate the band shape more accurately, one has to compute the vibronic states of the system with the potential (10.27), individual transitions between them, and the envelope band shape [10.20, 10.29, 10.30].

The simplest model of the MV dimer, discussed above, may be insufficient for describing real systems. The main complications may be caused by (1) more than one electronic state on each center that can be occupied by the excess electron, (2) influence of low-symmetry crystal fields and spin–orbital interactions, or (3) open-shell (nonzero spin) cores, resulting in magnetic exchange coupling between the centers [10.20]. Examples of the influence of these important factors on the electron transfer effects are discussed briefly below.

Several Electronic Levels: The Creutz–Taube Ion

The role of several possible states of the excess electron on each center can be elucidated using the Creutz–Taube (CT) ion as an example. The CT ion $[(NH_3)_5Ru(pyz)Ru(NH_3)_5]^{5+}$, where pyz = pyrazine, became a classical object of polynuclear MV system studies. Its relatively simple composition and structure make it possible to discriminate different factors influencing the intramolecular electron transfer. In the CT ion the two Ru^{3+} centers have $4d^5$ low-spin configuration with one excess electron which occupies the hole in the $(t_{2g})^6$ closed-shell configuration of each center or its corresponding MOs (see below).

The structure of the CT ion and the coordinate system chosen is shown in Fig. 10.4. Note that the X axis lies in between the nitrogen atoms; hence the

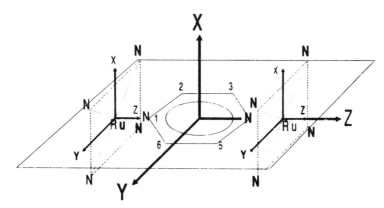

Figure 10.4. Structure and coordinate system of the Creutz–Taube ion.

three t_{2g} orbitals are d_{xz}, d_{yz}, and $d_{x^2-y^2}$ (the last one substitutes d_{xy} in the usual coordinate system, Section 4.2). Direct overlap between the d states of the two Ru centers is negligible. Therefore, their MO combinations of appropriate symmetry in the C_{2v} group of the system as a whole are as follows:

$$\varphi(a_{1g}) = \frac{1}{\sqrt{2}} (d^{(1)}_{x^2-y^2} + d^{(2)}_{x^2-y^2})$$

$$\varphi(b_{1u}) = \frac{1}{\sqrt{2}} (d^{(1)}_{x^2-y^2} - d^{(2)}_{x^2-y^2}) \qquad (10.30)$$

$$\varphi(b_{3g}) = \frac{1}{\sqrt{2}} (d^{(1)}_{yz} + d^{(2)}_{yz})$$

$$\varphi(b_{2u}) = \frac{1}{\sqrt{2}} (d^{(1)}_{yz} - d^{(2)}_{yz})$$

$$\varphi(b_{2g}) = \frac{1}{\sqrt{2}} (d^{(1)}_{xz} + d^{(2)}_{xz}) \qquad (10.31)$$

$$\varphi(b_{3u}) = \frac{1}{\sqrt{2}} (d^{(1)}_{xz} - d^{(2)}_{xz})$$

(the d orbitals are in fact MOs which they form with the ligands on each center). From these MOs only one, the b_{2g} type, overlaps the empty π^* orbital of the pyrazine bridge, forming the corresponding bonding and antibonding MOs of the whole system. The resulting MO energy scheme, including these states, is shown in Fig. 10.5. The relative positions of the energy level can be taken from electronic structure calculations performed in the X_α-DV approximation [10.31].

The population of six one-electron MOs by 11 electrons results in the ground-state spin doublet $^2B_{3u}$ and five excited doublets $^2B_{3g}$, $^2B_{2u}$, $^2B_{1u}$, ^2A_g, and $^2B_{2g}$, relatively close in energy, formed by the one-electron excitation from the corresponding one-electron MOs to the b_{3u} MO. These states are subject to further modification under spin–orbital interaction which is significantly strong in the Ru ion (the spin–orbital constant $\xi \approx 10^3$ cm^{-1}). The spin–orbital coupling mixes only the states that have the same parity, and hence the perturbation problem results in two secular equations of the third order. Their solution yields six new (Kramers) doublets, three even and three odd, that are linear combinations of the six energy terms above. These states can be used to solve the vibronic problem of a two-center MV system employing the ideas discussed above for a simpler case of a two-level system.

Since the electronic states above are orbitally nondegenerate, the interaction with the totally symmetric displacements on each center Q_1 and Q_2 only were taken into consideration. This means that one ignores the possible pseudo

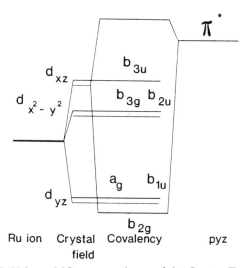

Figure 10.5. Valence MO energy scheme of the Creutz–Taube ion.

Jahn–Teller mixing of the electronic states by low-symmetry distortions, and this limitation remains open for further investigation (see below). Similar to (10.25), the combinations $Q_{\pm} = (1/2)(Q_1 \pm Q_2)$ characterizing distortions of the system as a whole are more convenient, but unlike the two-level case, in the CT six-level system the Q_{+} coordinate cannot be separated and neglected. This is due to the fact that there are three different electronic states that take part in the electron transfer (A_g, B_{2g}, and B_{3g} in the notation above), and they can be coupled in various ways to the totally symmetric (breathing) vibrations. Therefore, in general, three vibronic constants, $F_1(A_g)$, $F_2(B_{2g})$, and $F_3(B_{3g})$, should be introduced [10.20, 10.32] instead of one in the one-level one-center system discussed above. In [10.33] these three constants are taken equal to each other and to those for $Ru(NH_3)_6$; this approximation seems to be insufficient [10.32]. Indeed, while F_1 and F_3 characterize the coupling of "pure" d states with the breathing nuclear displacement, F_2 is the constant of interaction for an effective localized MO involving the pyrazine bridging group. It follows from the experimental data [10.34] that in $[Ru(NH_3)_5pyz)]^{3+}$ the distance Ru—N(Pyz) is smaller by 0.07 Å from that in $[Ru(NH_3)_5(pyz)]^{2+}$, whereas when passing from $Ru(NH_3)_6{}^{3+}$ to $Ru(NH_3)_6{}^{2+}$ the distance Ru—N increases by 0.04 Å. This means that the corresponding vibronic constant F_2 has opposite signs in these two cases.

Taking into account the linear terms of vibronic coupling of the type $F_i Q_{+}$ and $F_i Q_{-}$ as perturbations of the six states above, we come to the secular equation of the sixth order (see Eqs. (3.3) to (3.5) in [10.20]). Its numerical solution yields six sheets of the adiabatic potential surface in the space of the two coordinates Q_{+} and Q_{-}. The numerical values of the stretching frequency $\omega(Ru-N)$ needed for the computations was taken as $\omega \approx 500\ cm^{-1}$, and for an estimate of the vibronic constants F_1, F_2, and F_3, the data on g-factors in the ESR spectrum were used [10.32]. With the experimental data $g_x = 1.346$, $g_y = 2.799$, $g_z = 2.487$ [10.35], this yields

$$F_1 = 1.44$$
$$F_2 = -1.19 \tag{10.32}$$
$$F_3 = 0.76$$

Here we use dimensionless units which are related to the real magnitudes marked by an asterisk as follows: $Q = (M\omega/\hbar)^{-1/2}Q^*$, $F = (2M\omega^3\hbar)^{-1/2}F^*$, where ω is the frequency of the totally symmetric vibrations on each center, M is the effective mass of these vibrations, and all the energies, including w, are given in units of $\hbar\omega$.

The values of the vibronic constants (10.32) seem to be reasonable: $F_2 < 0$, as expected in view of the influence of the pyrazine, discussed above, and F_3 is smaller than F_1 because of the larger influence of the pyrazine ligand on the d_{yz} orbital than on $d_{x^2-y^2}$.

The six adiabatic potential surfaces were computed with the values F_i from Eq. (10.32) [10.32]. Their cross sections at $Q_- = 0$ are shown schematically in Fig. 10.6. The ground state 1 has a minimum at $Q_+ \neq 0$ $(Q_- = 0)$ which corresponds with electron delocalization. The optical IT band originates from the transition from the ground state to state 6. In the semiclassical approximation, the calculated maximum of the band is at 6600 cm^{-1}, in good agreement with the experimental value 6200 cm^{-1} [10.35].

One of the features of the IT band in the CT ion is that it is asymmetrical and slumps more slowly at higher frequencies; this also takes place at low temperatures down to 4.2 K. The asymmetry of the band can be explained by

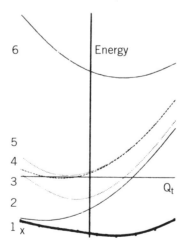

Figure 10.6. Extremal cross section $Q_- = 0$ of the AP of the Creutz–Taube ion. X is a saddle point between two minima at $Q_- \neq 0$.

the anharmonicity produced by both vibronic and spin–orbital interactions: At point X (Fig. 10.6) there is a saddle point on the ground-state surface with two minima at $Q_- \neq 0$. These minima do not influence the electron delocalization significantly because they are rather shallow, with a very small energy barrier between them. Note that this anharmonicity is due entirely to the multilevel structure and is not possible in the two-level case.

In the consideration above, the mixing of the electronic states by low-symmetry nuclear displacements, the pseudo Jahn–Teller effect on each center, is neglected. In the case of the CT ion this is justified by the fact that the excess electron is delocalized, and the alternative "localization–delocalization" is not very critical. In other cases the one-center pseudo Jahn–Teller effect may be more significant. In particular, it provides an additional channel of localization of the excess electron [10.20]: In addition to the totally symmetric distortions at each center caused by the localized electron (polaron effect), it produces low-symmetry distortions due to pseudo Jahn–Teller mixing with the excited states. Therefore, even when the usual polaron effect is weak (small vibronic coupling, rigid environment) and not enough for the excess electron localization ($|w| > F^2/K_0$), it can be localized by the pseudo Jahn–Teller distortion characterized by a different coupling constant F and other rigidness K_0 (Section 7.4).

On the other hand, the presence of an active excited state at each center which is coupled to the ground state via vibronic interactions means that the excess electron can delocalize through this excited state, which has another resonance integral w_e. If the latter is sufficiently large, the pseudo Jahn–Teller effect, admixing the excited state to the ground state, enhances the delocalization. The interplay of these factors causes new effects associated with the electron transfer.

In particular, it is shown [10.20] that if there is an electron localization in the presence of a pseudo Jahn–Teller effect, the latter is not symmetrical with respect to the two centers. In other words, the electron localization is accompanied by a low-symmetry distortion of one of the centers (an out-of-phase pseudo Jahn–Teller distortion) making them nonequivalent in this respect, too. Apparently, the different symmetries of the two centers $Mn^{3+}—Mn^{4+}$ observed in the MV system $[Mn_2O_2(bipy)_4]^{3+}$ with identical ligands [10.36] can be explained as due to the pseudo Jahn–Teller distortion that accompanies the electron localization. The Mn^{3+} ion in the high-spin state and C_2 symmetry of the ligands has a LUMO a_2 that is close to the HOMO a_1, and these two MOs, originating from the e_g MO in the octahedral symmetry, mix strongly under the A_2 nuclear displacements, producing the pseudo Jahn–Teller distortion.

Magnetic Implication

Many MV compounds have open-shell centers with nonzero spins besides that of the excess electron. In these cases additional magnetic interactions between this electron with each of the centers and between the centers themselves complicate the process of electron transfer [10.20]. The usual formulas of magnetic exchange coupling between the centers (Section 8.4) do not apply here because the migrating excess electron participates simultaneously in the formation of magnetic moments of both centers, making the Heitler–London approach, on which the HDVV model (8.80) is based, invalid. For open-shell (magnetic) centers with excess electrons, the concept of *double exchange*, developed for crystals, is very useful.

The idea of double exchange is as follows (see also [10.37]). In the ground state, spin S' of the excess electron localized at the magnetic center is parallel to spin S_1 of the latter because of the strong intraatomic exchange interaction J_0 (cf. Hund's rule, Section 2.2; for simplicity the electronic shell of each center is assumed to be filled in less than a half). Due to the equivalency of the centers and the electron transfer, the excess electron spin S' interacts similarly with the spin of the second center S_2. Therefore, the energy gain via exchange interaction is maximal when the spins of the two centers are parallel (the resonance interaction determining the electron transfer w is independent of the spin). Thus *electron transfer favors ferromagnetic ordering of the spin centers.* In application to MV compounds, the idea of double exchange, being in some aspects different from the crystal case, has been employed only recently [10.20].

Consider first a simple case of a three-electron problem in a MV dimer with $S_1 = S_2 = S = \frac{1}{2}$ and two nondegenerate one-electron states on each center φ_1, φ_1' and φ_2, φ_2', respectively [this case corresponds to the Ni(I)–Ni(II) dimer, for instance]. In the ground state the two orbitals φ_1 and φ_2 contain two electrons, one in each orbital, while the third (excess) electron can occupy orbital φ_1' or φ_2' with equal probability. If in addition to the intraatomic exchange integral J_0 mentioned above (an integral of the type (2.36) calculated by the functions φ_1 and φ_1' [10.20]), we introduce the intercenter exchange integral J (calculated by the functions φ_1 and φ_2, or φ_1' and φ_2, or φ_1 and φ_2'), the energy levels as a function of the total S are as follows [10.20]:

$$E(S = \tfrac{3}{2}) = -(J_0 + 2J) \pm w \tag{10.33}$$

$$E(S = \tfrac{1}{2}) = \pm[(J_0 - J)^2 + w^2 \pm (J_0 - J)w]^{1/2} \tag{10.34}$$

Equations (10.33) and (10.34) can be generalized to the case when the core spin is larger than 1/2 using the method suggested by Anderson and Hasegawa [10.38]. Denoting the total spin of each magnetic center without the excess electron by S_0 and $S_{max} = 2S_0 + 1/2$, we have [10.20]:

$$E(S < S_{max}) = \frac{J_0}{2} + J\left(S_0 + \frac{1}{2}\right) - JS(S + 1)$$

$$\pm\left[\left(S_0 + \frac{1}{2}\right)^2 (J_0 - J)^2 + w^2 \pm (J - J_0)w\left(S + \frac{1}{2}\right)\right]^{1/2} \tag{10.35}$$

$$E(S_{max}) = -J_0 S_0 + J(2S_0 + 1) - JS(S + 1) \pm w \tag{10.36}$$

For transition metal compounds the orders of magnitude of the parameters entering Eqs. (10.35) and (10.36) are as follows: $J_0 \sim 1$ to $10\,\text{eV}$, $w \leqslant 1\,\text{eV}$, $J \sim 10^{-1}$ to 10^{-3} eV, and the w values can be comparable in magnitude with J (see below).

It can be seen that Eq. (10.35) contains two pairs of states with an energy separation of about $2J_0$. The states of the upper pair are often called non-Hund states because they arise from the states with a local spin of $S_0 - \frac{1}{2}$, in contrast to the Hund rule. In most cases the non-Hund states can be neglected. Then, expanding Eq. (10.35) in powers of w/J_0 and keeping linear terms in w only, we obtain for the energies of the Hund states as functions of S:

$$E(S) = -JS(S + 1) \pm \frac{w(S + \frac{1}{2})}{2S_0 + 1} \tag{10.37}$$

Here the terms not containing S are omitted, and hence the energy read-off is changed compared with (10.35) and (10.36). This equation is also valid in the case of a more-than-half-filled electronic shell when there are no non-Hund states and no terms containing J_0.

Equation (10.37) emerges from other theories of energy spectra of MV dimers as well. Although approximate, the energy spectrum calculated by (10.35) agrees qualitatively with that obtained in quantum-chemical computations for specific MV dimers [e.g., Fe(II)–Fe(III) pairs in ferredoxin and in oxides]. It allows one to obtain a qualitative picture of the energy spectrum of the systems under consideration using the relation between w and J. For each value of the total spin there are two resonance states. The spacing of the centers of gravity of these doublets is determined by the intercenter exchange, while their splitting is linear dependent on the spin value (Fig. 10.7).

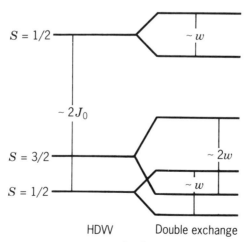

Figure 10.7. Electronic energy levels of a d^1–d^2 exchange-coupled MV dimer (the ratios between J_0, J, and ω are not obeyed numerically).

It follows that the double exchange favors the ferromagnetic spin ordering. However, unlike the HDVV ferromagnetic exchange interaction, the double exchange does not necessarily result in a ground state with a maximal spin when $J < 0$. The condition of such a ground state is rather difficult: $w > (2n + 1)(n + 1)J$, where n is the number of electrons in the ionic core. In general, it is clear that the migration of the electron results in an energy spectrum that is essentially different from that expected in the simple exchange scheme of the HDVV model (Section 8.4). In molecular crystals MV dimers are coupled by dipole–dipole interaction, which may produce phase transitions to charge-ordered states [10.39] in which the excess electron is localized.

Allowing for vibronic coupling with totally symmetric displacements on each center, one can assume that the constant of vibronic coupling F is the same for all spin states, provided that there is only one orbital state on each center (φ'_1 or φ'_2) which is occupied by the excess electron. In these cases each pair of states with the same spin mixes under the nuclear displacements Q_-, producing a pseudo Jahn–Teller adiabatic potential, quite similar to the one-electron case (10.27), with Q_- after (10.25). Following (10.37), the energy gap $2\Delta_S$ between the two mixing states with a given spin S is

$$\Delta_S = \frac{w(2S + 1)}{2(2S_0 + 1)} \tag{10.38}$$

and the adiabatic potential surfaces are

$$\varepsilon_S(Q_-) = \tfrac{1}{2}K_0 Q_-^2 \pm [\Delta_S^2 + F^2 Q_-^2]^{1/2} - JS(S + 1) \tag{10.39}$$

If the inequality $\Delta_S < F^2/K_0$ is satisfied, the lowest curve in (10.39) has two minima at $\pm Q_-^0$ (Fig. 7.19), each corresponding to the localization of the excess electron at one of the centers. For $\Delta_S > F^2/K_0$, $Q_-^0 = 0$, and the electron is delocalized over the two centers. Since ΔS is proportional to the total spin, a situation is possible when the excess electron is localized in states with small total spins and delocalized in states with larger spins.

More exact predictions of the magnetic behavior (and other properties) can be made based on detailed calculations of the vibronic energy states E_n of such systems with the potential (10.39). With the values E_n known, the magnetic moment μ is estimated by the following formula (Section 8.4):

$$\mu^2 = \frac{g^2\beta^2 \sum_{n,S} S(S + 1)(2S + 1) \exp[-E_n(S)/kT]}{\sum_{n,S} (2S + 1) \; exp[-E_n(S)/kT]} \tag{10.40}$$

Some results of such calculations are illustrated in Fig. 10.8 for a d^1–d^2 MV dimer with an antiferromagnetic exchange coupling $J < 0$. It is seen that in the absence of vibronic coupling ($F = 0$), the two centers, owing to the double exchange and large w, are ferromagnetically ordered with a magnetic moment of three unpaired electrons ($S = \tfrac{3}{2}$). The vibronic interaction changes the

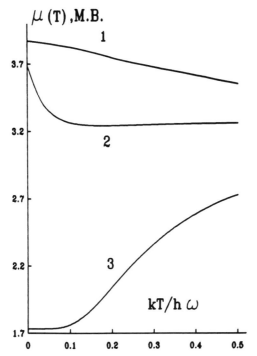

Figure 10.8. Temperature dependence of the magnetic moment μ of an MV d^1–d^2 dimer with $J = -0.2$, $w = 2$, and different vibronic constants: (1) $F = 0$; (2) $F = 1.8$; (3) $F = 3$ (in dimensionless units). (From [10.20].)

magnetic behavior, and for sufficiently large vibronic constants the ground state changes from $S = \frac{3}{2}$ to $S = \frac{1}{2}$. This phenomenon is formally similar to spin crossover (Section 8.4), where the ground-state spin changes as a function of the ligand field parameter.

The quantity of *electron transfer probability* P may also be important, especially for applications. It can be calculated directly from the vibronic energy spectrum of the system [10.20]: P is dependent on the spin S_0 and exchange parameters. In particular, for two systems with the same S, w, and $|J|$ but opposite signs of J, that is, for ferromagnetic P_F and antiferromagnetic P_{AF} systems (states),

$$\frac{P_F}{P_{AF}} = (2S_0 + 1)^2 \tag{10.41}$$

Mixed-Valence Trimers: Coexistence of Localized and Delocalized States

Regular three-center equivalent triangular clusters are most widespread, the series of carboxylates $[M_3O(RCOO)_6]L_3$ being a well-known example; in Fig. 10.9 the one with $M = Fe$, $R = CF_3$, and $L = H_2O$ is illustrated.

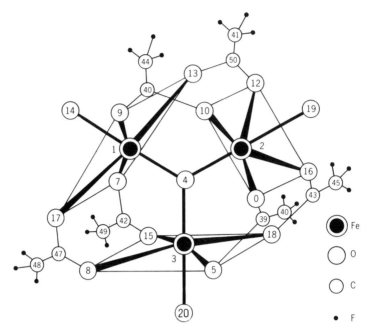

Figure 10.9. Structure of the MV trimer $M_3O(RCOO)_6L_3$ with M = Fe, R = CF_3, and three ligands L = H_2O marked by N 14, 19, 20, respectively.

Consider a trimer MV cluster and assume (as in the cases of simple dimers) that there is only one excess electron over the three metal centers with closed shells and only one nondegenerate state on each center φ_i, $i = 1, 2, 3$, to be occupied by the excess electron. In the triangular symmetry C_{3v} the three one-electron states φ_i form three MOs of A and E symmetry (Section 5.1):

$$\Psi_A = \frac{1}{\sqrt{3}}(\varphi_1 + \varphi_2 + \varphi_3)$$

$$\Psi_{E\theta} = \frac{1}{\sqrt{6}}(2\varphi_1 - \varphi_2 - \varphi_3) \qquad (10.42)$$

$$\Psi_{E\varepsilon} = \frac{1}{\sqrt{2}}(\varphi_2 - \varphi_3)$$

Denoting, as above, the intercenter resonance integral (the electron transfer parameter) by w [cf.(10.26)], we find easily that the splitting $\varepsilon(A) - \varepsilon(E) = 3w$.

As in the case of dimers, the excess electron, when localized at the center, distorts its environment, thus violating the C_{3v} symmetry of the system, and this distortion is very important for the electron transfer phenomenon. Since the φ_i are nondegenerate, only the totally symmetric local distortions Q_i ($i = 1$, 2, 3) are affected by the excess electron. Similar to the wavefunctions (10.42), it is more convenient to take these three local coordinates Q_1, Q_2, Q_3 in symmetrized combinations for the system as a whole (Section 7.1):

$$Q_A = \frac{1}{\sqrt{3}} (Q_1 + Q_2 + Q_3)$$

$$Q_\theta = \frac{1}{\sqrt{6}} (2Q_1 - Q_2 - Q_3) \tag{10.43}$$

$$Q_\varepsilon = \frac{1}{\sqrt{2}} (Q_2 - Q_3)$$

While Q_A is totally symmetric with respect to the C_{3v} symmetry, Q_θ and Q_ε form two components of the twofold-degenerate E displacement. The totally symmetric coordinate, similar to the dimer, can be separated by an appropriate choice of the energy read-off.

With these denotations, the electronic energy levels $\varepsilon_0(Q)$ as functions of the coordinates Q_θ and Q_ε can be found by solving the corresponding secular equation of the type (7.82), which in polar coordinates (7.37) $Q_\theta = \rho \cos \phi$, $Q_\varepsilon = \rho \sin \phi$ is [10.20]

$$\varepsilon_0^3 - \left(\frac{F^2\rho}{2} + 3w^2\right) \varepsilon_0 - 2w^3 - \left(\frac{2}{3}\right)^{1/2} F^3\rho^3 \cos 3\phi = 0 \tag{10.44}$$

where F is the constant of vibronic coupling with the Q_i displacements (10.23). The three roots of Eq. (10.44), ε_0^i ($i = 1$, 2, 3), together with the strain (deformation) energy $\frac{1}{2}K\rho^2$ form three sheets of the adiabatic potential surface:

$$\varepsilon_i(\rho, \phi) = \frac{1}{2}K\rho^2 + \varepsilon_0^i, \qquad i = 1, 2, 3. \tag{10.45}$$

For simplicity, we assume that the K values are the same for the two oxidation states of the center and, as above, we use dimensionless units for Q, F, and w.

It can be shown that ε_i are periodic functions of ϕ with a period of $2\pi/3$. The extremal points of the adiabatic potential (10.45) are at $\phi = \pi n/3$, $n = 0$, 1,..., 5. If $F < 0$, then for the lowest sheet the even values of n correspond to maxima along ϕ (and saddle points if other coordinates are included, see below), while the odd values give minima, and vice versa for $F > 0$.

Consider the radial dependence of the adiabatic potential in the extremal cross section $\phi = 0$ (or, equivalently, $Q_\varepsilon = 0$). The solutions of Eq. (10.45) in this case are

$$\varepsilon_1 = -\frac{F\rho}{\sqrt{6}} - w$$

$$\varepsilon_{2,3} = \frac{1}{2}\left[w - \frac{F\rho}{\sqrt{6}} \pm 3\left(\frac{F^2\rho^2}{6} - \frac{2w\rho}{3\sqrt{6}} + w^2\right)^{1/2}\right] \tag{10.46}$$

Further investigation of this expression requires knowledge of the sign of the parameter of intercenter electron transfer w. While for dimers the sign of w is not important, for trimers it is essential. If $w > 0$, the electronic doublet is lowest. Its adiabatic potential shape is then similar to that in the Jahn–Teller $E-e$ problem with the quadratic terms of the vibronic interaction included (Section 7.4).

In the case at hand the warped adiabatic potential (10.46) results from simultaneous Jahn–Teller and pseudo Jahn–Teller distortions in the linear approximation with respect to the vibronic interaction terms. The three minima of the adiabatic potential describe the three possibilities of localization of the excess electron at each of the three centers. Since there is no minimum at $Q_\theta = 0$ (where the three centers are equivalent), it follows that in trimer MV compounds with $w > 0$ fully delocalized electron distributions are not possible.

Systems with $w < 0$ seem to be more interesting. Indeed, they have singlet ground electronic terms, and the adiabatic potential shape, as in the case of the dimer, is completely determined by the parameter $|w|/F^2$. Substituting

$$\rho = \left(\frac{2}{3}\right)^{1/2} \frac{|w|}{F} (2\sqrt{2}\sinh t - 1) \tag{10.47}$$

into Eq. (10.46), we obtain the following transcendent equation for the extremal points of the adiabatic potential:

$$4\sqrt{2}\sinh 2t + \left[1 - \frac{|w|}{F^2}\right]\cosh t - 3\sinh t = 0 \tag{10.48}$$

Figure 10.10 illustrates the calculated shapes of the extremal cross sections of the adiabatic potential at $\phi = 0$ for different $|w|/F^2$ values. Its behavior at $\rho = 0$ can be investigated by means of the expansion of the potential function into a power series of ρ, keeping the terms up to ρ^2. If $|w|/F^2 < \frac{2}{9}$ (Fig. 10.10a), the point $\rho = 0$ is a local maximum, and the potential has three absolute minima M at $\rho \neq 0$. One more minimum at $\rho \neq 0$ in Fig. 10.10a at point S and two other equivalent at $\phi = 2\pi/3$, $4\pi/3$ are in fact saddle points. If $|w|/F^2 > \frac{2}{9}$, a minimum occurs at $\rho = 0$, but this is not necessarily accompanied by the disappearance of the minima at $\rho \neq 0$. In the interval $\frac{2}{9} < |w|F^2 < 0.255$, both types of minima coexist (Fig. 10.10b). Finally, when $|w|F^2 > 0.255$, the only minimum exists at $\rho = 0$ (Fig. 10.10c).

The minimum at $\rho = 0$ describes the state of the excess electron uniformly delocalized over the three centers, whereas the minima at $\rho \neq 0$ correspond to electron-localized (at one of the three centers) states. Thus we come to the conclusion that for three-center MV compounds in a certain interval of parameter values, *coexistence of localized and delocalized electron distributions is possible.* The region of parameter values required for this coexistence of two alternative electron distributions is relatively small. This region is expected to

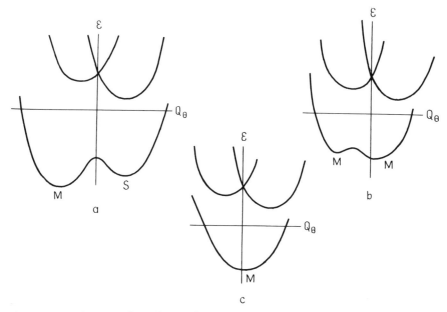

Figure 10.10. Cross sections $\phi = 0$ of the adiabatic potentials of an MV trimer as a function of $|w|/F^2$: (a) $|w|/F^2 < \frac{2}{9}$; (b) $\frac{2}{9} < |w|/F^2 < 0.255$; (c) $|w|/F^2 > 0.255$. M are minima, S is a saddle point.

increase when the difference in the frequencies of the local one-center totally symmetric vibration in two valence states is taken into account.

The coexistence of localized and delocalized electron distributions was observed first in a series of compounds of the type $[Fe(II)Fe_2(III)$-$(CH_3COO)_6L_3]$ (see [10.40] and references therein). In these works it is shown that the Mossbauer spectra, besides the lines corresponding to the Fe(II) and Fe(III) ions with the intensity ratio 1:2, also contain quadrupole doublets that are characteristic of iron ions in the intermediate oxidation state.

The MV systems under consideration form, as a rule, molecular crystals in which interaction between molecules depends on the intramolecular electron distribution and hence nuclear configuration distortions. It follows that the cooperative properties of MV compounds in the crystal state depend on their electron localization or delocalization state. Potential curves of the type considered above were used to analyze possible types of *phase transitions in crystals of MV trimers* [10.41].

The results above are based on the assumption that there is only one excess electron above the three zero-spin cores of the three centers, and the validity of these results for clusters containing Fe(II), Fe(III),... in the high-spin state (where for each value of the total spin there are several A and E energy levels) may be questioned. However, if one takes into account the exchange anisotropy (i.e., the difference in the exchange parameters in the pairs Fe(II)–Fe(III) and Fe(III)–Fe(III)), these levels are relatively displaced from each other, and it may be possible to consider each set containing the pair of A and E levels separately, as was done above.

For a more detailed evaluation of the trimer properties the vibronic energy levels and wavefunctions of the MV system with the adiabatic potentials (10.45) should be computed. Such computations have been performed for different values of the parameters F and w, and the results were used to evaluate the band shapes of the IT spectra. Figure 10.11 illustrates some results.

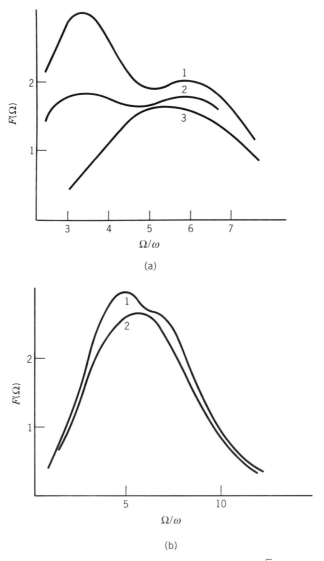

Figure 10.11. Shapes of intervalence transfer bands for $F = \sqrt{3}$ and (*a*) $w > 0$: (1) $w = 0.7$, (2) $w = 0.5$, (3) $w = 0.3$; (*b*) (1) $w = -0.7$, (2) $w = -0.3$.

One can see that the expected band shape is essentially dependent on both the vibronic coupling constant F and electron transfer parameter w, including its sign. If $w > 0$ (Fig. 10.11a), the orbital doublet E is lower than the singlet A, and the IT is $E \rightarrow A$, whereas for $w < 0$ (Fig. 10.11b) the transition is $A \rightarrow E$. In Section 8.2 optical transitions of this kind are discussed, but unlike them, the A and E states in the MV trimer originate from the same T term at $w = 0$ and their separation depends on the w value.

If the electronic coupling (electron transfer) between the centers is negligible ($w \approx 0$), the spectrum coincides with that expected from intersheet transitions in the Jahn–Teller $T-e$ problem (Sections 7.3 and 8.2; see also [10.42]). Here the adiabatic potential surfaces are presented by three intersecting (but not interacting) paraboloids (Fig. 7.17), while the wavefunction for each of them has a usual adiabatic form (they are mutually orthogonal). The theory of the band shapes due to transition between such multisheet surfaces is discussed briefly in Sections 8.1 and 8.2. When the three paraboloid minima are well displaced, which for MV trimers means strong localization of the excess electron, the band has a Gaussian shape, its parameters obeying the corresponding relationships for dimers [10.43].

In the opposite limit case of strong delocalization ($w \gg F^2$) pseudo Jahn–Teller mixing of the A and E states can be neglected. Then the shape of the intervalence transfer band coincides with that of the envelope of the $A \rightarrow E$ ($w < 0$) or $E \rightarrow A$ ($w > 0$) optical transition (Section 8.2). The band has either one or two maxima, depending on the magnitude of the vibronic constant F and the temperature. This result is quite different from that for dimers in the delocalized state, where the band is rather narrow and has no more than one maximum [10.30].

In Fig. 10.11 the shape of the intervalence transfer spectrum is shown for different w values. The dependence of the type of spectrum on the sign of w is seen explicitly. For negative w values, band intensity increases with the intercenter interaction. For a given F value, the double-humped structure of the band takes place for w values which are close to those of coexistence of localized and delocalized electron distributions. This does not mean that the two maxima of the band correspond to the transitions from two minima of the adiabatic surface: in this region the semiclassical approximation is invalid.

At $F = \sqrt{3}$ and very large negative w values (not shown in Fig. 10.11) the spectrum has one maximum, similar to the case of a corresponding Jahn–Teller system with weak vibronic coupling (Section 8.2). The double-humped structure of the band emerges with the increase in F. The IT spectra with two maxima are observed in some iron MV trimers [10.44]. For other examples of triangular metal clusters, see [10.45].

There are many MV trimers with equivalent centers that are not regular triangles; the linear system $trans$-$(NH_3)_5Ru(II)pyzRu(II)(NH_3)_4pyzRu(III)$-$(NH_3)_5{}^{7+}$ (1) [10.46] and the nonlinear system cis-$(NH_3)_5Ru(II)NCRu(II)$-$(bpy)_2CNRu(III)(NH_3)_5{}^{5+}$ (2) [10.47] may serve as examples. The inter-valence transfer bands for these systems have maxima at 0.59 μm^{-1} and 0.95

μm^{-1}, respectively. The IT absorption curve for a derivative of 2 in which one ammonia molecule is substituted by pyridine is given in Fig. 10.2.

Another interesting use of the MV trimer vibronic states is in the calculation of their *magnetic characteristics* [10.20].

Example: Ferredoxins

Delocalization of the excess electron in trimer MV systems, considered above for the case of zero-spin cores, raises the following question: To what extent are these results valid for systems with double exchange (nonzero-spin cores) that have a more complicated electronic energy spectrum? The trinuclear iron-sulfur cluster in the ferredoxin of the sulfate-reducing bacteria *Desulfovibrio gigas* [10.49], for which the presence of double exchange is revealed, seems to be an appropriate example to illustrate this problem.

In the reduced form this ferredoxin cluster has the valence composition Fe(III)–Fe(III)–Fe(II). The ESR and Mossbauer spectra show that the excess electron is delocalized over two of three centers [10.50]. Based on these data, a model was proposed that considers the double exchange between the pair Fe(II)–Fe(III) in addition to the HDVV exchange among all three ions [10.50]. This approach implies that one of the ions is strongly distinguished from the other two, so that its resonance interaction with the latter is negligible. However, x-ray structural data [10.51] show that any structural differences between the three ions are very small.

In [10.52], a model based on simultaneous magnetic and vibronic effects is suggested which explains the origin of the peculiar electron distribution in the ferredoxin cluster under consideration. The electronic configuration of the three centers is $d^5-d^5-d^6$ and all the iron ions are in the high-spin states. From the experimental data it is known that the ground state is a quintet $S = 2$, and the excited states with other $S \neq 2$ spins lie higher than the ground state by no less than 80 cm^{-1}. Therefore, approximately, for sufficiently low temperatures, only the quintet states can be considered. With the localized electron, there are five states of this kind.

Let us neglect the differences between the exchange integrals J_1 for the Fe(II)–Fe(III) pair and J_2 for Fe(III)–Fe(III). In other words, we assume that the system is exchange isotropic. Then the analytical expressions for the electronic energies (Eq. (10.46)) at $Q_\theta = Q_\varepsilon = 0$ are

$$\varepsilon(^5A^{(1)}) = \varepsilon(^5E^{(1)}) = \varepsilon(^5E^{(2)}) = -w$$

$$\varepsilon(^5A^{(2)}) = \varepsilon(^5A^{(3)}) = \varepsilon(^5E^{(3)}) = 0 \qquad (10.49)$$

$$\varepsilon(^5A^{(4)}) = \varepsilon(^5A^{(5)}) = \varepsilon(^5E^{(5)}) = \tfrac{5}{6}w$$

where the energies are read off from the position of nonsplit HDVV multiplet with $S = 2$. For an arbitrary relation between the exchange parameters J_1 and

J_2 and arbitrary values of vibrational coordinates, the electronic energies can be obtained by numerical calculations.

With the known electronic energies at fixed nuclei, one can determine the adiabatic potential of the system in the space of the active Q_ϑ and Q_ε or ρ and ϕ coordinates [see Eq. (10.44)]. In the case at hand, although the energy-level scheme depends on the sign of w, the shape of the lowest sheet of the adiabatic potential undergoes qualitatively the same transformations with the $|w|/F^2$ parameter for both positive and negative w values.

For small w values the minima of the lowest adiabatic potential are at $\phi = \pi/3$, π, $5\pi/3$, whereas at $\phi = 0$, $2\pi/3$, $4\pi/3$ there are saddle points. As a result, the extremal cross sections of the adiabatic potential have the same shape as in Fig. 10.10a. Similar to the case of a one-electron trimer, considered above, these minima correspond to the localization of the excess electron at one center. For some values of $|w|/F^2$, a pair of minima occur instead of one. The two minima in the pair occupy symmetrical position with respect to the directions $\phi = \pi/3$, π, $5\pi/3$, the minima in these directions thus becoming saddle points. The value $|w|/F^2$, for which the number of minima is doubling, depends on the relation between the exchange integrals; for the $|J_1 - J_2|$ values from 0.0 to 0.1, $0.7 < |w|/F^2 < 0.8$.

The possible existence of such six equivalent minima of the adiabatic potential in the space of E vibrations has been predicted from group-theoretical consideration [10.53]. Such a potential surface has been reported so far in the only case of hexafluorine d^0 transition metal complexes. The essential feature making these two otherwise physically different systems similar is the presence of more than two electronic states mixed by E-type vibronic interactions.

Upon further increase of the ratio $|w|/F^2$, the six minima merge in pairs into three minima at $\phi = 0$, $2\pi/3$, $4\pi/3$; these points thus become minima instead of saddle points, and the surface as a whole again becomes tri-minimal. But as opposed to previous cases, *the excess electron in these minima is delocalized over two centers of the trimer.*

If $|w|/F^2 \to \infty$, the energies of the minima and saddle points coincide and we obtain a continuous set of minima, a trough [10.42, 10.53]. The motion along the trough corresponds to a dynamical delocalization of the excess electron; this is not possible in other MV systems considered above, neither in dimers nor in one-electron trimers.

The observed charge distribution in the iron-sulfur ferredoxin can thus be explained as due to the joint influence of vibronic and interelectron (double exchange) interactions in a multilevel system with a special relation of parameter values, resulting in minima of the adiabatic potential that describe the partial (two-center) delocalization of the excess electron. X_α calculations [10.54] of once- and twice-reduced systems of the type $[Fe_3S_4(SH)_3]^{2-}$ with C_{3v} symmetry in which the competition between spin–spin coupling and delocalization is considered in detail confirms qualitatively the scheme of double exchange. The calculations show that the exchange constant J is larger in reduced form than in oxidized form, and w is negative (w/J varies from 1.5 to 3.0).

Obviously, one can expect that the number of novel effects that arise due to the combined vibronic and double exchange interactions will increase with the number of new MV trimers studied. More effects may be expected in four-center MV ferredoxins that have a delocalized electronic distribution and double exchange.

For a model four-center tetrahedral ferredoxin with one excess electron over the closed shell and one electronic state on each center, the following qualitative results are discussed briefly [10.20]. There are four symmetrized coordinates of the tetrahedron as a whole, Q_A, $Q_{T\xi}$, $Q_{T\eta}$, and $Q_{T\zeta}$ (Table 7.1), constructed from the local totally symmetric displacements on each center Q_i, $i = 1$, 2, 3, 4. From these coordinates the totally symmetric combination $Q = 1/\sqrt{3}(Q_{T\xi} + Q_{T\eta} + Q_{T\zeta})$ in the trigonal space is most informative in the problem under consideration. In the space of this coordinate the four energy levels as functions of Q (adiabatic potentials in the cross sections along Q) are [cf. Eqs.(10.44) and (10.45)]

$$\varepsilon_{1,2} = \frac{1}{2} Q^2 - w - \frac{FQ}{2\sqrt{3}} \tag{10.50}$$

$$\varepsilon_{3,4} = \frac{1}{2} Q^2 + w + \frac{FQ}{2\sqrt{3}} \pm 2 \left(w^2 - \frac{wQ}{2\sqrt{3}} + \frac{F^2Q^2}{12} \right)^{1/2} \tag{10.51}$$

By direct calculation one can easily make sure that if $w > 0$ there is always a minimum of the lowest curve at $Q \neq 0$ corresponding to the excess electron localized at one center. For negative w values, three possibilities similar to those for the trimers occur:

1. If $|w|/F^2 < 0.125$, the lowest curve has two minima, of which one is, in fact, a saddle point (in the extended space including other coordinates).
2. If $0.125 < |w|/F^2 < 0.152$, there are two minima at $Q = 0$ and $Q \neq 0$, respectively.
3. If $|w|/F^2 > 0.152$, the only minimum point occurs at $Q = 0$.

As for the electron charge distribution corresponding to these three cases, they are (1) an electron localized state, (2) coexistence of localized and delocalized states, and (3) electron delocalization.

Thus for the tetrahedral MV tetramers we come again to the possibility of *coexistence of states with localized and delocalized excess electron distributions.* Apparently, this effect has a more general meaning and can also be expected in clusters with larger numbers of MV centers. So far, to our knowledge, there has been no experimental confirmation of the coexistence effect in tetramers.

The shortest path between the adiabatic potential minima of the localized electron goes through the saddle points along the $Q_{T\xi}$, $Q_{T\eta}$, and $Q_{T\zeta}$ directions. The numerical calculation shows that the barrier height for the electron transfer through these saddle points in tetramers is always lower than the corresponding barrier in dimers for the same values of w and F. Perhaps this result explains why in the same conditions, that is, for the same structure of the near-neighbor environment of the iron ions and the same bridges between them, as well as approximately the same iron–iron distances, in iron-sulfur dimers the excess electron is rather localized, whereas in similar tetramers it is delocalized.

It is evident that other problems considered above for dimers and trimers by means of the vibronic theory can be treated similarly for tetramers and other MV systems. However, for MV systems with the number of centers $n > 3$, some new problems may occur. In particular, two and more excess electrons migrating among the MV centers may be important in these systems. Note that in trimers two excess electrons are equivalent to one excess hole. For two or more excess electrons, the electron distribution and dynamics are determined by the competition of vibronic, intra-atomic, and intercenter interactions. Examples of MV systems with two excess electrons are known among cubic [10.55] and quadratic tetramers [10.56], six-nuclear clusters [10.57], and others.

The coexistence of localized and delocalized electronic distributions in polynuclear clusters considered in this section is somewhat analogous to the coexistence of localized and delocalized excitonic states in crystals [10.58, 10.59].

10.3. ELECTRON-CONFORMATIONAL EFFECTS IN BIOLOGICAL SYSTEMS

Distortions Produced by Excess Electronic Charge; Special Features of Metalloenzymes

Conformational changes produced by excess electronic charge [*electron-conformational effects*, ECEs)] are of widespread interest, especially in biological processes. In essence, the problem of ECE is an extension of that formulated in Section 7.4 as electronic control of configuration instability. Similarly, formulation of the problem of ECE is as follows: Is it possible to predict conformational changes in molecular system induced by the addition or removal of electronic charge?

Solution of this problem is facilitated by the vibronic theory (Sections 7.2 and 7.4). In Section 7.2 the notion of vibronic constants is introduced [see Eq. (7.22)]: $F_{\Gamma*}^{(\Gamma,\Gamma')} = \langle \Gamma(\partial V/\partial Q_{\Gamma*})_0 | \Gamma' \rangle$, where Γ and Γ' are two electronic states (for the diagonal constant $F_{\Gamma*}^{(\Gamma)}$ $\Gamma = \Gamma'$), V is the electron–nuclear plus nuclear–nuclear interaction in the Hamiltonian of the system, and $Q_{\Gamma*}$ is the symmetrized coordinate of nuclear displacements that belongs to the representation Γ^*. The vibronic constant has the dimensionality of a force, and the diagonal constant $F_{\Gamma*}^{(\Gamma)}$ also has the physical meaning of a force (Section 7.2):

$F_{\Gamma*}^{(\Gamma)}$ means the force with which the electronic cloud in the state Γ acts on the nuclear configuration in the direction of the symmetrized coordinate $Q_{\Gamma*}$.

It is clear that in the equilibrium configuration $F_{\Gamma*}^{(\Gamma)} = 0$, but any excess electronic charge (electron addition) or positive charge (electron removal) violates the equilibrium of forces and makes $F_{\Gamma*}^{(\Gamma)} \neq 0$. *The nonzero force distorts the nuclear configuration.*

How can one determine the direction $Q_{\Gamma*}$ of this distortion and its magnitude as a function of the excess charge and electronic state Γ? If the excess charge is an integer (one or more electrons added or removed), the problem can be solved directly by computing the wavefunction $|\Gamma'\rangle$ of the new state Γ' with the additional charge (not to be confused with the excited state) and the matrix element for the $Q_{\Gamma*}$ coordinates, for which F is nonzero. But this is very difficult, especially for large (biological) molecular systems, and it cannot be carried out, in principle, when the excess charge is fractional (which is often the case).

A more realistic possibility for solving this problem is provided by the semiempirical method based on the notion of *orbital vibronic constants* f introduced in Section 7.2: $f_{\Gamma*}^{(i,j)} = \langle i|(\partial V'/\partial Q_{\Gamma*})_0|j\rangle$, where $|i\rangle$ are molecular orbitals and V' is the one-electron operator of electron–nuclear and nuclear–nuclear Coulomb interaction (7.24'). *The diagonal orbital vibronic constant* $f_{\Gamma*}^{(i)}$ *has the physical meaning of the force with which the electron on the ith MO acts on the nuclear configuration in the direction* $Q_{\Gamma*}$. It is very important that, owing to the additive electron–nuclear interactions, the diagonal integral vibronic constant according to Eq. (7.26) equals the sum of the orbital constants multiplied by their electron occupation numbers (in the state Γ) q_i^Γ.

In fact, Eq. (7.26) solves the foregoing problem approximately. Indeed, for a stable system in the equilibrium configuration,

$$F_{\Gamma*}^{(\Gamma)} = \sum_i q_i^\Gamma f_{\Gamma*}^{(i)} = 0 \qquad (10.52)$$

With the excess charge the population of the MO changes: $q_i^{\Gamma'} \rightarrow q_i^\Gamma + \Delta q_i$, and

$$F_{\Gamma*}^{(\Gamma')} = \sum_i q_i^{(\Gamma')} f_{\Gamma*}^{(i)} = \sum_i \Delta q_i f_{\Gamma*}^{(i)} \qquad (10.53)$$

This equation is approximate because it implies that the excess charge does not change the orbital vibronic constants. This is true roughly to the second order of the charge alteration $(\Delta q/N)^2$, where N is the total number of electrons, provided that the charge transfer Δq is sufficiently small.

Following Eq. (10.53), the distortion force $F_{\Gamma*}^{(\Gamma')}$ produced by the excess charge Δq_i can easily be found, provided that the orbital vibronic constants $f_{\Gamma*}^{(i)}$ are known. In particular, for one electron added to the system $F_{\Gamma*}^{(\Gamma')} = f_{\Gamma*}^{(i)}$, where i is the MO occupied by the excess electron. In many cases the numerical value of the constants $f_{\Gamma*}^{(i)}$ can be estimated using empirical data obtained from spectroscopic and diffraction experiments; examples are given in Sections 7.2 and 11.2. But the sign of f and the symmetry of $Q_{\Gamma*}$ (i.e., the direction of distortion) can be established from general considerations without detailed calculations.

As stated in Section 7.2, $f_{\Gamma*}^{(i)} > 0$ for bonding MO (in the $Q_{\Gamma*}$ direction), $f_{\Gamma*}^{(i)} < 0$ for antibonding MO, $f_{\Gamma*}^{(i)} \approx 0$ for nonbonding MO, and the absolute value of $f_{\Gamma*}^{(i)}$ follows quantitatively the measure of the MO bonding. The symmetry of $Q_{\Gamma*}$ can be obtained directly from Eq. (7.25) using the selection rules (Section 3.4). Following these rules, the diagonal matrix element (7.25) is nonzero if the symmetrical square of the symmetry type Γ_i (to which the ith MO belongs) contains the symmetry Γ^* of $Q_{\Gamma*}$: $\Gamma^* \in [\Gamma_i \times \Gamma_i]$. If Γ_i is nondegenerate, $\Gamma_i \times \Gamma_i$ contains only the totally symmetric representation A_1. Thus addition of electronic charge on nondegenerate MOs produces totally symmetric distortions which do not change the symmetry of the system. If Γ_i is degenerate, the product $[\Gamma_i \times \Gamma_i]$ also contains nontotally symmetric representations, and the charges on such MOs produce deformation of the system. These deformations are exactly the same as in the Jahn–Teller effect (Section 7.3): Doubly degenerate E MOs produce E deformations with two components, Q_θ and Q_ε (Fig. 7.1), or (for particular cases) B_1 and B_2 deformations (Fig. 9.27); orbital triplets T produce E and T_2 deformations; and so on. Each of these deformations is characterized by its own orbital vibronic constant $f_{\Gamma*}^{(i)}$.

If several MOs are populated, the total distortion equals the sum of the distortions produced by each MO. They can be found approximately using Eq. (7.41):

$$Q_{\Gamma*}^0 = \frac{F_{\Gamma*}}{K_\Gamma} \qquad (10.54)$$

where $K_{\Gamma*}$ is the force constant of the $Q_{\Gamma*}$ distortion [a more exact equation for $Q_{\Gamma*}^0$ is given in Section 11.2, Eq. (11.24)]. Substituting (7.26) into (10.54), we have

$$Q_{\Gamma*}^0 = \frac{\Sigma_i \Delta q_i f_{\Gamma*}^{(i)}}{K_{\Gamma*}} \qquad (10.55)$$

The symmetrized distortions $Q_{\Gamma*}$ determine the Cartesian displacement of the nuclei, as shown in Table 7.1; if several types of $Q_{\Gamma*}$ (for several types of Γ^*) are nonzero, the corresponding Cartesian displacements should be summed as vectors.

As seen from Eq. (10.55), the nuclear configuration distortion is directly proportional to the charge transfers Δq_i and orbital vibronic couplings $f_{\Gamma*}^{(i)}$ of the corresponding MO, and inversely proportional to the rigidity of the system $K_{\Gamma*}$ in the $Q_{\Gamma*}$ direction. While $f_{\Gamma*}^{(i)}$ characterizes quantitatively the participation of the excess charge in the alteration of bonding (and hence the occurrence of the distortion force) along Q_{Γ^*}, K_{Γ^*} characterizes the resistance of the system to deformations in the Q_{Γ^*} direction.

There are cases when the electronic charge added to the system formally does not change its MO occupation numbers but changes the MO itself. Such charge effects take place, for instance, by protonation or deprotonation, coordination of small molecules (e.g., oxygenation), and so on. Additional charges emerging from coordination or protonation produce important ECE since they violate the initial charge equilibrium conditions. In the vibronic scheme, the alteration of the MO changes the orbital vibronic constant $f_{\Gamma^*}^{(i)}$ and violates Eq. (10.52), making the force $F_{\Gamma^*}^{(i)}$ and hence the distortion nonzero, even when the MO population numbers q_i^{Γ} remain the same. The change in $f_{\Gamma^*}^{(i)}$ means that the bonding nature of the corresponding MO is changed, and hence the condition of pseudo Jahn–Teller instability (7.71) changes. A detailed treatment of this source of electron-conformational effects has not yet been performed.

In general, many biological systems are very susceptible to vibronic instabilities, mainly for two reasons:

1. Conformational flexibility, that is, the presence of *soft modes* (small K_0 values)
2. The presence of prosthetic groups with HOMO–LUMO levels close in energy (small Δ values)

Both sources of vibronic deformation, (10.55) and (7.71), are strongly dependent on K_{Γ^*}, the rigidity of the system with respect to the Q_{Γ^*} distortion. Biological systems are specific in possessing low-rigidity K_{Γ^*} in certain directions Γ^* (soft modes). This circumstance favors large distortions.

Examples of the second special feature in biological systems, the presence of close energy levels, ground and excited, are provided by metalloenzymes. The latter have a group of such active states due to the metal d orbitals (organic conjugated prosthetic groups may also be very similar to this type). *The prosthetic group is a site triggering (initiating) electron-conformational transitions, due its close-in-energy electronic levels and the soft modes of the biological environment.*

Example: Trigger Mechanism of Hemoglobin Oxygenation; Comparison with Peroxidase

One of the most important properties of hemoglobin is its ability to oxygenate (to absorb oxygen) and to deoxygenate (to release oxygen) with a special S-type kinetics. In other words, hemoglobin absorbs oxygen sharply (much more sharply than do most absorbers) in an oxygen environment, and releases oxygen sharply in a medium where there is no oxygen. Perutz was the first to show that this kinetics of oxygenation is due to a special mechanism based on $T \rightarrow R$ conformational transitions and consequent cooperative effects induced by out-of-plane and in-plane displacements of the iron atom in the metal–

porphyrin active center by oxygenation (trigger mechanism) [10.60] (see also [10.61, 10.62]).

Hemoglobin has four similar active iron centers, each containing an iron porphyrin group. The four centers are linked by protein chains. Figure 10.12 shows schematically the structure of the iron porphyrin center with the imidazole moiety as the fifth ligand (cf. Fig. 9.12). The iron centers are linked via this ligand, which is a part of the protein chain. In the Perutz mechanism [10.60] (see also [10.62]) the position of the iron atom in the deoxy form of the hemoglobin centers is out of the plane of the porphyrin ring, and in this position only two of the four centers are open to oxygen coordination. When the oxygen molecule occupies the sixth coordination position (Fig. 10.12), the iron metal goes back to the porphyrin plane. This displacement pulls the imidazole moiety of the proximal histidine, which being linked to the protein chain, produces the conformational transition $T \to R$; this, in turn, opens the other two iron centers for oxygen coordination, sharply increasing the curve of oxygenation.

In the absence of oxygen the release of one oxygen molecule initiates an inverse process of the iron out-of-plane displacement and $R \to T$ transition with corresponding consequences for the desorption curve. Thus the local changes in the iron atom position with respect to the porphyrin ring as a result of oxygen coordination initiate conformational transitions in the protein (trigger mechanism) important to the biologic function.

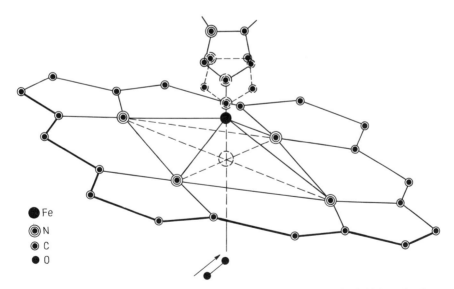

Figure 10.12. Structure of the active center of hemoglobin (mioglobin)—the iron porphyrin center with the imidazol moiety of the proximal histidine. The in-plane position of the iron atom is shown by dashed lines. The approaching oxygen molecule is also indicated.

In Section 9.2 it is shown that the in-plane position of the iron atom in the porphyrin ring is very soft or even unstable with respect to its out-of-plane displacement. In the planar metalloporphyrin symmetry D_{4h} this displacement has A_{2u} symmetry, and the instability is due to the sufficiently strong pseudo Jahn–Teller effect (Section 7.4), the mixing of the MO a_{2u} of mainly the porphyrin ring with the metal d_{z^2} orbital of a_{1g} symmetry under A_{2u} displacements ($A_{1g} \times A_{2u} = A_{2u}$). As seen in Fig. 10.13, the energy-level separation $2\Delta = \varepsilon(a_{1g}) - \varepsilon(a_{2u})$ is relatively small, making the inequality (7.71) quite real. Under the influence of the imidazole moiety of the proximal histidine residue, these MO energy levels of iron porphyrin, as shown in Fig. 10.13, do not change very much; at least the Δ value remains sufficiently small to soften the iron atom out-of-plane displacement.

The picture changes when the oxygen molecule is coordinated to the sixth position. The repulsion of the iron antibonding d_{z^2} orbital by the oxygen negative charge raises this iron energy level significantly and makes the energy gap $2\Delta \approx 3\,\text{eV}$ (see Fig. 10.13; for CO as the sixth ligand $2\Delta \approx 4\,\text{eV}$). With this Δ value the inequality (7.71) does not hold, and there is no more softness or instability of the in-plane position of the iron atom: it returns to the plane of the porphyrin ring.

An important feature of iron out-of-plane and back in-plane displacements is that they change the spin of the ground state of the system (cf. spin crossover, Section 8.4). In Fig. 10.13 the real ground-state spins are indicated. In the out-of-plan position of the iron atom in deoxyhemoglobin, the spin is higher than in the oxygenated in-plane position. In accordance with the results of Sections 4.3 and 6.2, the electron distribution over the AOs or MOs depends on the pairing energy Π and its ratio to the energy gap Δ between the MO levels occupied by the electrons in the unpaired (high-spin) configuration (Eqs. (4.48) and (4.49)]. Therefore, it is not surprising that in the configuration with larger Δ values the ground state is a low-spin one.

Note that the iron–porphyrin stereochemistry and the spin states are controlled by the same MO energy-level arrangements. It is worth emphasizing this fact because the earlier explanations of Hoard [10.63] and Perutz [10.60] (see also [10.61, 10.64]) are based on the fact that the atomic radii in the high-spin configuration are larger (the antibonding d orbitals of the high-spin configuration are more extended) than in the low-spin case. Hence the iron atom (ion) cannot fit the hole in the porphyrin ring when it is in the high-spin state, while it fits this hole in the low-spin state. In Section 9.2 we criticise this rough ball-hole fitting.

There are many examples with different metals [10.61] when the atom (ion) is in the in-plane position for smaller holes and it is out-of-plane for larger holes. A detailed analysis of out-of-plane displacements of the metal atom in metalloporphyrins and spin state/geometry relationship is given in the review article [10.61], and it is shown that the vibronic approach discussed above explains qualitatively the features of the phenomenon observed.

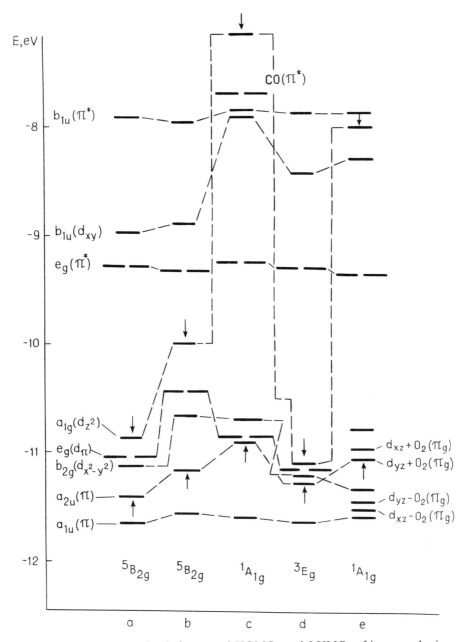

Figure 10.13. MO energy levels for several HOMOs and LUMOs of iron porphyrin with indication of the ground term: (*a*) the out-of-plane displacement of the iron atom is $\Delta R = 0.49\,\text{Å}$; (*b*) deoxy form of hemiglobin (Hb), $\Delta R = 0.62\,\text{Å}$; (*c*) Hb—CO; (*d*) planar iron porphyrin $\Delta R = 0$; (*e*) Hb—O_2. The energy gap 2Δ between the two states that mix under the iron out-of-plane displacement are shown by arrows.

In real metalloenzymes the metal porphyrin active center is subject to the influence of the protein environment, which controls a variety of their functions [10.65, 10.66]. For instance, the same iron–porphyrin active center with different proximal ligands (and other differences in the next coordination spheres) has quite different activities with respect to interaction with oxygen: In hemoglobin the O_2 molecule initiates $T \to R$ conformational transitions and it is released without being strongly activated, whereas in cytochrome P-450 coordinated oxygen is strongly activated and performs hydroxylation of saturated hydrocarbons (catalase and peroxidase act in a similar way). Chemical activation by coordination (also to cytochrome P-450) is considered in Section 11.2; here we simply compare, by way of example, hemoglobin and peroxidase, to illustrate the role of protein environment.

Figure 10.14 shows schematically the configuration of the iron porphyrin active center in the two metalloenzymes, hemoglobin and peroxidase, without (Fig. 10.14, I) and with (Fig. 10.14, II) coordinated oxygen (in peroxidase the coordinated molecule is H_2O_2, which is modeled by O_2). The essential difference in these two cases is in the structure of the protein not shown in the figure. In both cases the iron atom is out-of-plane without oxygen and goes back into the plane by oxygenation.

In hemoglobin the iron displacement toward the porphyrin plane triggers the conformational $T \to R$ transition, discussed above, without strong activation of the O—O bond, whereas in peroxidase, where the protein environment is different from that of hemoglobin, the move of the iron atom toward in-plane position deprotonates the imidazole residue, with the result that the distal nitrogen acquires an excess electron. This process, initiated by oxygen coordination, has an important influence on the coordinated molecule O_2.

Figure 10.15 shows the MO energy levels of the oxygenated active center FeP (P is porphyrin) of peroxidase with neutral imidazole, $Fe(ImH)O_2$, and deprotonated imidazole, $Fe(Im^-)O_2$, computed [10.67] in the semiempirical MO LCAO IEH (SCCC) approximation (Section 5.5). It is seen that due to the excess electron, the energy level of the (mainly) lone-pair hybridized sp^3 orbital of the distal nitrogen rose up to the LUMO, which is (mainly) a strongly antibonding π^* MO of the oxygen molecule; the energy gap between the HOMO and LUMO Δ becomes smaller than the pairing energy π [Eqs. (4.48) and (4.49)]. This results in separation of the lone pair and population of the antibonding π^* MO of the O_2 molecule and its activation (Section 11.2).

Interestingly, the enzyme as a whole acts in the regime of feedback: The oxygen molecule coordinates to the active center of the enzyme and the latter, depending on its function, activates the O—O bond (peroxidase, cytochrome P-450, catalase), or not (hemoglobin). In the case of cytochrome P-450, the mechanism of population of the antibonding π^* MO of the coordinated oxygen is similar to that of peroxidase. Calculations of the MO energy levels of the oxygenated active center of cytochrome P-450 [10.68] in the same approximation as above show that in the deprotonated form the energy of the lone

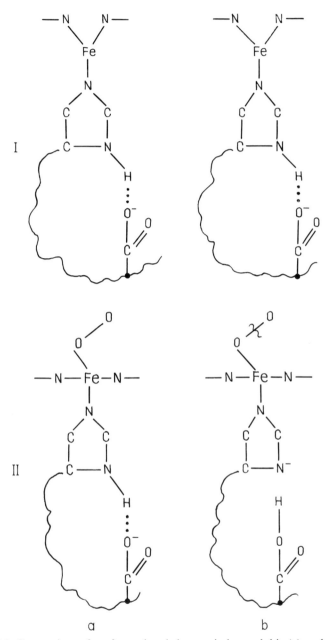

Figure 10.14. Comparison of conformational changes in hemoglobin (*a*) and peroxidase (*b*) by oxygenation. The iron atom from the out-of-plane position (I) returns back into the plane (II) with the effect of a T—R conformational transition in hemoglobin (not shown in the Figure) and deprotonation of peroxidase that leaves an excess electron on the distal nitrogen atom.

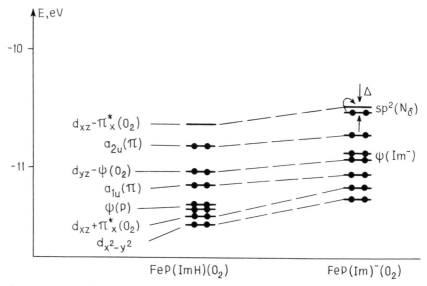

Figure 10.15. MO energy levels of the active center of oxygenated peroxidase with imidazol, neutral $FeP(ImH)O_2$, and deprotonated $FeP(Im^-)O_2$. In the last case one electron from the distal nitrogen lone pair populates the strongly antibonding π^* MO of the coordinated oxygen molecule [$\psi(O_2)$ and $\psi(Im)$ are the corresponding MOs].

pair of the sulfur atom goes up, comes close to, and populates the antibonding π^* MO of O_2, significantly activating the coordinated oxygen molecule. An additional electron acts in the same direction, reducing the O—O bonding up to its cleavage [10.69] (Section 11. 2).

In fact, the problem of electron transfer in biological systems is of general importance. For an instant view of some topics related to this problem, see, for example, [10.65, 10.66, 10.70–10.72].

REFERENCES

10.1. L. Pauling, *The Nature of the Chemical Bond*, 3rd ed., Cornell University Press, Ithaca, N. Y., 1960.

10.2. R. Pearson, *Coord. Chem. Rev.*, **100**, 403–425 (1990); *Inorg. Chim. Acta*, **198–200**, 781–786 (1992).

10.3. M. Chanon, M. D. Hawley, and M. A. Fox, in: *Photoinduced Electron Transfer*, Part A, ed. M. A. Fox and M. Chanon, Elsevier, Amsterdam, 1988, pp. 1–59.

10.4. J. C. Curtis and T. J. Meyer, *Inorg. Chem.*, **21**, 1562 (1982).

10.5. M. D. Newton, In: *The Challenge of d and f Electrons: Theory and Computation*, ACS Series 394, American Chemical Society, Washington, D.C., 1989, pp. 378–392.

10.6. V. Balzani and F. Scandola, in: *Photoinduced Electron Transfer*, Part D, ed. M. A. Fox and M. Chanon, Elsevier, Amsterdam, 1988, pp. 148–178.

10.7. J. Jortner and M. Bixon, J. *Chem. Phys.*, **88**, 167–170 (1988); I. Rips and J. Jortner, *J. Chem. Phys.*, **87**, 2090–2104 (1987); J. T. Hynes, *J. Phys. Chem.*, **90**, 3701–3706 (1986); A. M. Kjaer and J. Ulstrup, *Inorg. Chem.*, **25**, 644–651 (1986); A. Haim, *Comments Inorg. Chem.*, **4**, 113–119 (1985).

10.8. J. Jortner and B. Pullman, Eds., *Tunnelling*, D. Reidel, Dordrecht, The Netherlands, 1986; V. I. Gol'danskii, L. I. Trakhtenberg, and V. N. Fleurov, *Tunnelling Phenomena in Chemical Physics*, Gordon and Breach, New York, 1989.

10.9. I. B. Bersuker, *Biofizika*, **12**, 732–735 (1967); *Structure and Properties of Coordination Compounds* (Russ), Khimia, Leningrad, 1971.

10.10. I. B. Bersuker and S. S. Budnikov, *Teor. Eksp. Khim.*, **3**, 799–810 (1967); in: *Proceedings of the 4th International Congress on Catalysis*, Vol. 1, Nauka, Moscow, 1970, pp. 82–94.

10.11. J. E. Huheey, *J. Phys. Chem.*, **69**, 3284–3291 (1965).

10.12. W. Kauzmann, *Quantum Chemistry: An Introduction*, Academic Press, New York, 1957.

10.13. R. Taube, *Chem. Zvesti*, **19**, 215 (1965).

10.14. Trinh–Toan, Boan Keng Teo, and J. A. Ferguson, *J. Am. Chem. Soc.*, **99**, 408–416 (1977).

10.15. A. F. Shestakov, in: *Redox Metalloenzymes and Their Models: Theoretical and Methodological Aspects* (Russ.), Part 2, ed. G. I. Likhtenshtein, Akad. Nauk SSSR, Chernogolovka, 1982, pp. 19–34

10.16. R. G. Parr and R. G. Pearson, *J. Am. Chem. Soc.*, **105**, 7512 (1983).

10.17. R. F. Nalewajski, *Acta Phys. Polonica*, **A77**, 817–832 (1990).

10.18. D. B. Brown, Ed., *Mixed–Valence Compounds*, D. Reidel, Dordrecht, The Netherlands, 1980.

10.19. K. Prassides, Ed., *Mixed–Valency Systems: Applications in Chemistry, Physics and Biology*, NATO ASI Series C, Vol. 343, Kluwer, Dordrecht, The Netherlands, 1991, 451 pp.

10.20. I. B. Bersuker and S. A. Borshch, *Adv. Chem. Phys.*, **81**, 703–782 (1992)

10.21. S. J. Lippard, in: Ref. [10.18], p. 427.

10.22. N. S. Hush, *Coord. Chem. Rev.*, **64**, 135 (1985).

10.23. A. R. Bishop, R. L. Martin, K. A. Muller, and Z. Tesanovic, *Z. Phys. B*, **76**, 17 (1989).

10.24. L. J. de Jongh, in: Ref. [10.19], pp. 223–246.

10.25. J. P. Lannay, in: *Molecular Electronic Devices II*, ed. F. L. Carter, Marcel Dekker, New York, 1987, p. 39; C. Jioachim and J. P. Lannay, *J. Mol. Electron.*, **6**, 37–50 (1990).

10.26. I. B. Bersuker, S. A. Borshch, and L. F. Chibotaru, *Chem. Phys.*, **136**, 379–384 (1989).

10.27. M. B. Robin and P. Day, *Adv. Inorg. Chem. Radiochem.*, **10**, 247 (1967).

10.28. C. N. R. Rao, in: *Theoretical and Experimental Aspects of Valence Fluctuation in Heavy Fermions*, Plenum Press, New York, 1987, pp. 235–242.

10.29. P. N. Schatz, in: Ref. [10.19], p. 115.

10.30. K. Y. Wong and P. N. Shatz, *Prog. Inorg. Chem.*, **28**, 369 (1981) .

10.31. L. T. Zhang, J. Ko, and M. J. Ondrechen, *J. Am. Chem. Soc.*, **109**, 1666–1671 (1987) .

10.32. S. A. Borshch and I. N. Kotov, *Zh. Strukt. Khim.*, **32**, 44–48 (1991).

10.33. K. Neuenschwander, S. Piepho, and P. N. Schatz, *J. Am. Chem. Soc.*, **107**, 7862–7869 (1985).

10.34. M. E. Grees, C. Creutz, and C. O. Quicksoll, *Inorg. Chem.*, **20**, 1522 (1981).

10.35. U. Furholz, H. B. Burgi, P. E. Wagner, A. Stebler, J. H. Ammeter, E. Krausz, R. J. H. Clark, M. Stead, and A. J. Ludi, *J. Am. Chem. Soc.*, **106**, 121 (1984).

10.36. P. M. Plaksin, R. C. Stoufer, M. Methew, and G. J. Palenik, *J. Am. Chem. Soc.*, **94**, 2121 (1972).

10.37. G. Blondin, S. Borshch, and J. -J. Jirerd, *Comments Inorg. Chem.*, **12**, 315–340 (1992).

10.38. P. V. Anderson and H. Hasegawa, *Phys. Rev.*, **100**, 675 (1955).

10.39. S. I. Klokishner and B. S. Tsukerblat, *Chem. Phys.*, **125**, 11 (1988); S. I. Klokishner, B. S. Tsukerblat, and B. L. Kushkulei, *New J. Chem.*, **17**, 43 (1993).

10.40. H. G. Jang, S. J. Geib, Y. Kaneko, M. Nakano, M. Sorai, A. L. Pheingold, B. Monter, and D. V. Hendrickson, *J. Am. Chem. Soc.*, **111**, 173 (1989).

10.41. T. Cambara, D. N. Hendrickson, M. Sorai, and S. Oh, *J. Chem. Phys.*, **85**, 2895 (1986); R. M. Stratt and S. H. Adachi, *J. Chem. Phys.*, **86**, 7156 (1987).

10.42. I. B. Bersuker, *The Jahn–Teller Effect and Vibronic Interactions in Modern Chemistry*, Plenum Press, New York, 1984.

10.43. N. S. Hush, in: *Progress in Inorganic Chemistry*, ed. F. A. Cotton, Vol. 8, Wiley, New York, 1967.

10.44. R. D. Cannon, unpublished results.

10.45. R. D. Cannon and R. P. White, *Prog. Inorg. Chem.*, **36**, 195–298 (1988).

10.46. A. von Kameke, G. M. Tom, and H. Taube, *Inorg. Chem.*, **17**, 1790 (1978).

10.47. C. A. Bignozzi, S. Roffia, and F. Scandola, *J. Am. Chem. Soc.*, **107**, 1644 (1985).

10.48. M. I. Belinski, *Mol. Phys.*, **60**, 793 (1987).

10.49. B. H. Huynh, J. J. C. Moura, I. Moura, T. A. Kent, J. Le Gall, A. X. Xavier, and E. Munck, *J. Biol. Chem.*, **255**, 3242 (1980).

10.50. V. Papaefthymiou, J. J. Girerd, I. Moura, J. J. Moura, and E. Munck, *J. Am. Chem. Soc.*, **109**, 4703 (1987).

10.51. C. P. Kissinger, *J. Am. Chem. Soc.*, **110**, 8721 (1988).

10.52. S. A. Borshch and L. F. Chibotaru, *Chem. Phys.*, **135**, 375 (1989).

10.53. I. B. Bersuker and V. Z. Polinger, *Vibronic Interactions in Molecules and Crystals*, Springer-Verlag, New York, 1989, 420 pp.

10.54. S. F. Sontum, L. Noodleman and D. A. Case, in: *The Challenge of d and f Electrons: Theory and Computation*, ed. D. S. Salahub and M. C. Zerner, ACS Symposium Series 394, American Chemical Society, Washington, D.C., 1989, pp. 366–377.

10.55. T. G. Spiro, Ed., *Iron–Sulphur Proteins*, Wiley-Interscience, New York, 1982.

10.56. J. J. Girerd and J. P. Launay, *Chem. Phys.*, **74**, 217 (1983).

10.57. D. Coucovanis, M. G. Kanatzidis, A. Salifogloa, W. R. Dunham, A. Simopoulus, J. R. Sams, V. Papaegthymion, A. Kostikas, and C. F. Strouse, *J. Am. Chem. Soc.*, **109**, 6863 (1987).

10.58. V. V. Hizhnyakov and A. V. Sherman, *Phys. Status. Solidi* (*b*), **92**, 77 (1979).

10.59. K. Prassides and P. N. Schatz, *J. Phys. Chem.*, **93**, 83 (1989).

10.60. M. F. Perutz, *Br. Med. Bull.*, **32**, 195 (1976); *Nature*, **219**, 29 (1968).

10.61. I. B. Bersuker and S. S. Stavrov, *Coord. Chem. Rev.*, **88**, 1–68 (1988).

10.62. M. Weissbluth, *Haemoglobin: Cooperativity and Electronic Properties*, Chapman & Hall, London 1974, 175 pp.

10.63. J. L. Hoard, in: *Porphyrins and Metalloporphyrins*, Ed. K. M. Smith, Elsevier, Amsterdam, 1975, p. 317.

10.64. W. R. Sheidt and C. Reed, *Chem. Rev.*, **81**, 543 (1981).

10.65. S. J. Lippard and J. M. Berg, *Principles of Bioinorganic Chemistry*, University Science Books, Mill Valley, California, 1994.

10.66. W. Kaim and B. Schwederski, *Bioinorganic Chemistry: Inorganic Elements in the Chemistry of Life. An Introduction Guide*, Wiley, New York, 1994.

10.67. S. S. Stavrov, I. P. Decusar, and I. B. Bersuker, *Molecul. Biol.*, **22**, 837–843 (1988).

10.68. S. S. Stavrov, I. P. Decusar, and I. B. Bersuker, *Molecul. Biol.* (Russ), **21**, 338–346 (1987).

10.69. S. S. Stavrov, I. P. Decusar, and I. B. Bersuker, *New J. Chem.*, **17**, 71 (1993).

10.70. M. K. Johnson, R. B. King, D. M. Kurtz, Jr., Ch. Kutal, M. L. Norton, and R. A. Scott, Eds., *Electron Transfer in Biology and Solid State: Inorganic Compounds with Unusual Properties*, ACS Advances in Chemistry Series 226, American Chemical Society, Washington, D.C., 1990, 412 pp.

10.71. R. A. Sheldon, Ed., *Metalloporphyrins in Catalytic Oxydation*, Marcel Dekker, New York, 1994.

10.72. G. Berthon, Ed., *Handbook of Metal–Ligand Interaction in Biological Fluids*, Bioinorganic Chemistry, Vols. I and II, Marcel Dekker, New York, 1995.

11 Reactivity and Catalytic Action

Chemical transformations form the heart of all of chemistry; the problem is to establish how the chemical reaction, its rate and mechanism, depend on the structure and properties of the reactants, and to be able to control this phenomenon.

11.1. ELECTRONIC FACTORS IN REACTIVITY

Chemical Reactivity and Activated Complexes

The main quantitative characteristic of a chemical reaction is the *reaction rate* [11.1]. It depends on *kinetic factors* including temperature and reactant concentrations, as well as on the mechanisms of *elementary acts*. While kinetic factors have similar features in many reactions and can be controlled relatively easily, the elementary act is the most important stage determining the specificity and huge diversity of chemical reactions.

The general presentation of the energetics of elementary chemical reactions is usually given by means of the notion of the *energy barrier of the reaction*. Consider an elementary reaction starting with stable reactants and ending with stable products. To transform the reactant molecules to products, some of the chemical bonds should break down, while others are formed. Usually, a breakdown or activation of some bonds is required first and then the activated bonds undergo transformations. It means that in the process of the chemical reaction the energy of the system should first increase (activation), then decrease, due to new bond formation.

The energy of this process is lower when the displacements of atoms realizing the breakdown of some bonds and formation of others are correlated (concerted). The generalized coordinate of these atomic displacements is called *the reaction coordinate* or *reaction pathway* (shortly, reaction path). The latter is thus the generalized coordinate of correlated motions of the atoms of the reactants toward the products, along which the activation energy is minimal. The reaction pathway follows in detail (in intermediate points) the bond breakdowns and new bond formation determining the mechanism of the elementary act of the reaction, the *reaction mechanism*.

With respect to the reaction coordinate Q, the AP energy of the system (Section 7.1) appears as shown in Fig. 11.1. The configuration Q_D of the point

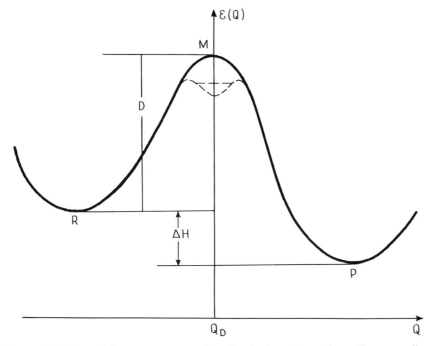

Figure 11.1. Potential energy curve of a chemical reaction along the generalized coordinate Q that describes the reaction pathway from the reagents (R) to the products (P) via the activated state (M). D and ΔH denote the activation energy and heat of the reaction, respectively. The dashed lines show the possible relative stability of the activated state.

M for which the energy of the interacting system is maximal is called the *activated state of the reaction* or *activated complex*. Sometimes the activated state is relatively stable (this possibility is shown in Fig. 11.1 by the dashed line). The energy D is called the *activation energy* of the chemical reactions, and ΔH is the *heat of the reaction*. For direct and inverse reactions D may be different, while ΔH changes sign from exothermic to endothermic reaction, or vice versa. There are reactions with zero activation energy, $D = 0$; their rate is determined by kinetic factors only. By definition, the activation energy along Q is minimal; hence along other coordinates the value of D is higher. This means that in the extended space the point M is a saddle point: a maximum along Q and a minimum along other coordinates orthogonal to Q.

The rate of the elementary act of the reaction is determined by the rate of overcoming the energy barrier between the reactants and products, or tunneling through the barrier, or both. To overcome the barrier, the higher vibrational states in the well of the reagents should be populated. Following the Boltzmann distribution we come to the *Arrhenius* exponential dependence of the reaction rate P on the activation energy D:

$$P = P_0 \exp\left(-\frac{D}{kT}\right) \tag{11.1}$$

where P_0 is the maximal rate (formally at $T = \infty$). Tunneling reactions are effective at low temperatures when the energy barrier cannot be overcome, and their rate is independent of temperature [11.2, 11.3].

The energy barrier of the reaction is directly dependent on the structure of the reactants and the mechanism of the elementary act. For a specific mechanism, the elementary reaction rate depends on the structure of the reactants. The notion of *chemical reactivity* characterizes the relative ability of a molecular system to interact with other molecules determining the reaction rate.

Consider a simple case of the substitution reaction $AB + C \rightarrow A + BC$. Assume that the activation of this reaction is due to the activation of the A—B bond (described by the increasing portion of the reaction curve of Fig. 11.1). When a specific activation is reached, binding with the C atom begins and the energy lowers; this is described by the decreasing portion of the reaction curve. For a more rigorous description of this process it is necessary to introduce two reaction coordinates: the two interatomic distances $R_1(A—B)$ and $R_2(B—C)$. The potential energy surface of the reaction in terms of the coordinates R_1 and R_2 (more precisely, the changes of these coordinates ΔR_1 and ΔR_2 from their equilibrium positions) is shown schematically in Fig. 11.2.

It is evident that the more active (reactive) atom C, the earlier the turning point M is reached and the lower the activation energy. This is illustrated in Fig. 11.3, which shows the projection of the reaction trajectory on the plane perpendicular to Fig. 11.2. The activation energy D varies for different C atoms and reaches the value of the dissociation energy for monomolecular reactions.

The activation energy D may be taken as a quantitative measure of the reactivity of the molecule in the reaction under consideration. The activated state (or the activated complex) is most important in the chemical reaction because it determines both the activation energy (the rate of the reaction) and the reaction mechanism. Chemical literature reports much work in which the activated states are treated as real states and the problem of their possible experimental observation and investigation is discussed. However, based on the first principles of quantum mechanics, it can easily be shown that the activated state of a chemical reaction, understood literally (the activated complex in its ground unstable state), cannot in principle be observed experimentally as a stationary state. Indeed, according to quantum mechanics, the Schrödinger equation (1.5) for a saddle-point potential at the point M of Fig. 11.1 has no stationary (localized) solutions. It follows that spectroscopic observation of an activated state that corresponds to this point is impossible. In the case of an additional minimum at the top of the reaction curve (dashed line in Fig. 11.1)

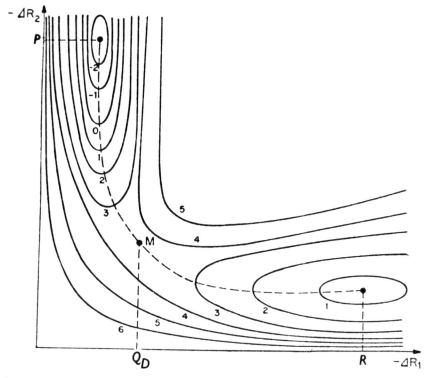

Figure 11.2. Potential energy surface of the $AB + C = A + BC$ reaction shown by equipotential curves in ΔR_1 and ΔR_2 coordinates. The energy increases with the number of the curve. The reaction trajectory is indicated by dashed line. M is the turning point at the barrier maximum of Fig. 11.1.

the state in this well may be observed, provided that its depth is larger than the kinetic energy of the corresponding nuclear motions; in this case state M can be regarded as an isomeric state of the system.

Some additional possibilities for studying the dynamics of nuclear motions along the classical adiabatic potential surfaces are provided by the modern high-resolution frequency- and time-resolved spectroscopy (femtosecond spectroscopy) [11.4], but these techniques do not allow us to see the transition state of the chemical reaction as a definitive molecular configuration either.

However, the impossibility of observing the activated complex in its ground state at point M does not mean that it cannot be observed in excited states. One of the recent results of the vibronic theory of instability is just the proof of the existence of stable excited states of unstable ground-state configurations. In Section 7.4 the theorem of uniqueness of the vibronic origin of molecular dynamic instability is discussed, and it is shown that the instability of the ground-state of any molecular group is due to vibronic mixing with stable excited states. In application to activated complexes of chemical reactions, the

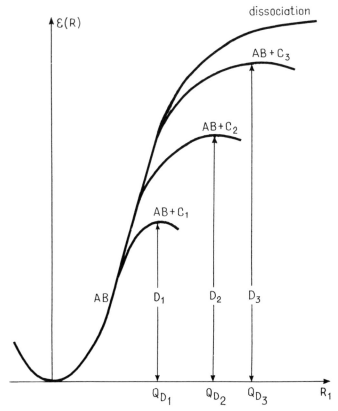

Figure 11.3. Increasing portion of the reaction potential energy curve along the reaction pathway in Fig. 11.2 characterizing the activation of the A—B bond. The turning point and the reaction barrier height D depend on the reactivity of the C atom.

theorem thus predicts the existence of stable excited states in the configuration Q_D.

In the two-level approximation, the curvatures of the adiabatic potential energy curves for the two states, ground 1 and excited 2, which are mixed by vibronic coupling, are given by Eq. (7.94), from which it follows that if at the point Q_D, the ground state is unstable in the Q direction, $K_1 < 0$, $K_0 < F^2/\Delta$, the excited state is stable in this direction, $K_2 > 0$ [11.5].

The stable excited state 2 that causes the instability of the ground state 1 is subject to selection rules which emerge from the condition that $F \neq 0$. Since $F = \langle 1|(\partial V/\partial Q)_{Q^*}|2\rangle$, F is nonzero if the product of symmetry representations of states 1 and 2 contain the representation of Q [Eq. (3.33)].

Figure 11.4 shows the adiabatic potential curves of a chemical reaction in the two-level case, including the reactants, activated state, and the excited state of the activated complex at $Q = Q_D$. The problem is to link the observable parameters of the stable excited state of the activated complex with that of the

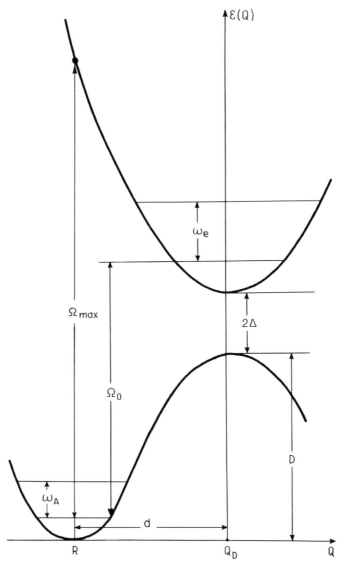

Figure 11.4. Potential energy curve of a chemical reaction including the stable excited state of the activated complex which causes the instability of the ground state. Parameters linking the two states are also shown.

nonobservable ground state. The spectroscopic characteristics of the excited state contain at least the information about the configuration Q_D; many of them contain the excited-state frequency ω_e of Q vibrations. The energy level of the stable excited state can, in principle, be estimated from the resonances in *molecular beam experiments* (scattering of the reactants as a function of their kinetic energy).

By way of illustration, we show here one of the possibilities of estimating some parameters of the activated state of the chemical reaction using spectroscopic data of the broad band of light absorption due to the transitions from the initial state of the reactants R to the stable excited state of the activated complex, provided that this absorption can be observed experimentally. Denoting the frequencies of the band maximum and pure electronic transition by Ω_{max} and Ω_0, respectively, and using the formulas of Section 8.1 and the obvious relations between the parameters in Fig. 11.4, we have

$$D \cong \hbar[\Omega_0 + \tfrac{1}{2}(\omega_R - \omega_e)] - 2\Delta \tag{11.2}$$

$$\frac{1}{2}K_2d^2 = \hbar\left(\Omega_{max} - \Omega_0 + \frac{\omega_e}{2}\right) \tag{11.3}$$

where ω_R and ω_e are the Q vibrational frequencies of the reactants and excited activated complex, respectively, $K_2 = M\omega_e^2$, D is the barrier height, and d is the distance between the coordinates of the minimum R and activated complex maximum, determining the barrier width.

In fact, the vibrational frequencies are much smaller than the electron transition frequencies, and in accordance with the discussion above, Δ is also much smaller than Ω_0. Under these assumptions formulas (11.2) and (11.3) may be simplified:

$$D \cong \hbar\Omega_0 - 2\Delta \approx \hbar\Omega_0 \tag{11.4}$$

$$\tfrac{1}{2}K_2d^2 = \hbar(\Omega_{max} - \Omega_0) \tag{11.5}$$

and in the often used approximation $K_0^{(1)} \approx K_0^{(2)}$ [cf. (7.94)],

$$K_1 = K_2 - \frac{2F^2}{\Delta} \tag{11.6}$$

Thus, by investigating the light absorption band shape observed in the process of the chemical reaction, we can estimate the barrier height D and its width d, and (using additional estimates for the vibronic constant F and energy separation Δ) the curvature of the barrier near its maximum. These parameters, together with the nuclear configuration at the point M, almost completely characterize the activated state of the chemical reaction.

Frontier Orbitals and Perturbation Theory

An important problem is to relate the reactivity and reaction mechanism to the electronic structure of the interacting systems. As mentioned in the introduction to Chapter 7, any molecular interaction begins with changes in the much less inertial electron distribution, which then impels the nuclei (via vibronic coupling) to rearrange. Therefore, the beginning of the chemical reaction should be sought for in the electronic structure of the reactants. In Section 10.1 the interaction between two atomic groups is considered from the point of view of their possible bonding; here a similar interaction is studied with respect to the chemical reaction.

The outer electrons enter the interaction first. The electronic charge distribution decreases exponentially with distance to the nuclei, and the exponential power is proportional to the electron energy (Section 2.1). Hence the outer electrons have the highest energy, and they form the HOMO. However, the interaction of the HOMO of one system with the HOMO of the other one (Fig. 11.5a) does not change significantly the electron distribution and bonding because the HOMO–HOMO interaction results in equal population of the bonding and antibonding orbitals (Section 5.2). The inclusion of the LUMO makes a number of bonding electrons uncompensated (Fig. 11.5b). The HOMO–LUMO states form *frontier orbitals*. The theory of molecular interactions in the approximation of frontier orbitals has been developed by Fukui [11.6].

In many cases, particularly with transition metal participation, not one but several HOMOs and LUMOs, as well as single-electron occupied MOs (SOMOs), are active in intermolecular interaction. In a more rigorous treatment all the orbitals of corresponding symmetry may be involved in the interaction, but in a qualitative treatment some of these MOs which are energetically close to the frontier orbitals may be discriminated as giving the major contribution. This group of active MOs is sometimes called *generalized frontier orbitals.*

With respect to the frontier orbitals, chemical reactions can be divided in three types, illustrated schematically in Fig. 11.6. In the first type (Fig. 11.6a) the reactant participates in the interaction with its HOMO, while the reagent provides its LUMO (*electrophilic reagent*). This interaction has a donor–acceptor nature (Sections 6.1 and 10.1). In Fig. 11.6b the second type of

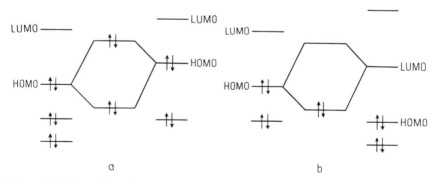

Figure 11.5. Frontier orbital interactions: HOMO–HOMO interaction (*a*) does not result in MO bonding (the bonding and antibonding MO compensate each other), while HOMO–LUMO (*b*) does.

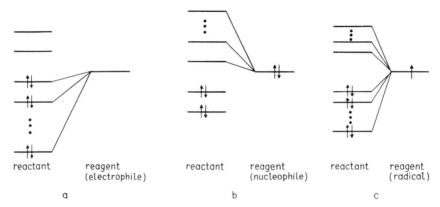

Figure 11.6. HOMO and LUMO in the interaction of a reactant with an electrophile (a), nucleophile (b), and radical (c) reagent. (From [11.6].)

reaction, with a *nucleophilic reagent*, is shown. Finally, the reagent shown in Fig. 11.6c has an unpaired electron on the HOMO (*radical reagent*) that takes part in the interaction as both a HOMO and a LUMO; the reactant in this condition also takes part with both HOMO and LUMO (*exchange reactions*).

In Section 10.1 it is shown how the interaction between two molecular groups results in electron charge transfer from one of them to another, but the energetics of this interaction which determines the activation energy and reaction path is not considered. At a relatively large distance between the interacting groups, their interaction energy can be estimated using perturbation theory [11.6, 11.7]. The idea is to consider the frontier orbitals of the reactant perturbed by the formation of MOs with the reagent. The situation is quite similar to that of weak covalency considered in Section 5.2. Using formulas similar to (5.30) and (5.31) and omitting the intermediate transformations, we come to the following expression for the energy of two interacting atoms s and t of the molecular groups which includes also their electrostatic interaction:

$$\Delta E = -\frac{q_s q_t}{R_{st}\varepsilon} + 2 \sum_m \sum_n \frac{(c_s^m c_t^n \beta_{st}^{mn})^2}{E_m^s - E_n^t} \tag{11.7}$$

In this equation q_s and q_t are the atomic charges of the s and t atoms, R_{st} the distance between them, ε the effective dielectric constant of the medium (solvent), c_s^m and c_t^n the LCAO coefficients of the MO m and n, respectively, in which the atoms s of the reactant and t of the reagent take part, β_{st}^{mn} is the corresponding resonance integral between the two interacting states, and E_m and E_n are the MO energies. The summation in (11.7) is carried out over the

HOMOs m in the reactant and LUMOs n in the reagent in donor–acceptor (electrophilic) reactions (Fig. 11.6a), and vice versa in nucleophilic reaction (Fig. 11.6b). In exchange reactions with radicals, both m and n contain HOMOs and LUMOs, but in each term one of the indices belongs to HOMOs and the other to LUMOs (otherwise, the corresponding contribution to the bonding, as shown in Fig. 11.5, is near zero). If there is more than one pair of interacting atoms, Eq. (11.7) should be summed over all of them.

The first term in Eq. (11.7) stands for the electrostatic interaction between atoms s and t, while the second term gives the covalency contribution of orbital overlap and formation of MOs. The latter demands nonzero overlap between HOMOs of the reactant with LUMOs of the reagent, or vice versa, and small energy separation between them. Depending on the electronic structure of the interacting groups, the two terms in Eq. (11.7), electrostatic and covalent, can be of different orders of magnitude. Therefore, it is convenient to consider two cases:

1. *Charge-controlled reactions.* The electrostatic term is predominant and the covalent contribution can be neglected.
2. *Orbital-controlled reactions.* The electrostatic term can be neglected as compared with the covalent contribution.

Certainly, there are reactions that may be classified as both charge and orbital controlled.

The difference between charge- and orbital-controlled reactions may be very significant. Indeed, electrostatic interactions are scalar and do not require specific orientations of the interacting species, whereas MO formation is possible only between appropriately oriented orbitals that give nonzero overlap. Hence, unlike charge-controlled reactions, *orbital-controlled reactions are stereoselective* (cf. below, "Orbital Symmetry Rules in Reaction Mechanisms").

Provided that the mutual orientation of the interacting groups is not restricted, there are always such positions when the overlap of the appropriate orbitals is nonzero. But this does not mean that the reaction is necessarily orbital controlled (even when the charges are small), since the energy difference $E_m - E_n$ between the overlapping orbitals may be large, making the corresponding terms in (11.7) small. Orbital-controlled reactions require small energy gaps between the overlapping HOMOs of the reactant and the LUMOs of the reagent (Fig. 11.6). In particular, when the HOMOs and LUMOs in question are almost degenerate, $E_n \approx E_n$, the perturbation problem should be solved for the degenerate state; then the covalent contribution to the interaction energy ΔE is due mainly to this degenerate interaction:

$$\Delta E \approx 2c_s^m c_t^n \beta_{st}^{mn} \qquad (11.8)$$

In practical use, the quantity called *superdelocalizability* S may be convenient [11.6]. Following Eq. (11.7), it is defined as follows:

For an electrophilic reaction:

$$S_t(E) = 2 \sum_n^{occ} \frac{(c_t^n)^2}{X_{st}^n} \tag{11.9}$$

where X_{st}^n is a dimensionless energy parameter,

$$\frac{1}{X_{st}^n} = \sum_m^{uno} \frac{(c_s^m)^2 \beta_{st}^{mn}}{E_n^s - E_m^t} \tag{11.10}$$

and "occ" and "uno" means that the summation is performed over occupied
and unoccupied states, respectively;

For nucleophilic reagents:

$$S_t(N) = 2 \sum_n^{uno} \frac{(c_t^n)^2}{X_{st}^n} \tag{11.11}$$

$$\frac{1}{X_{st}^n} = \sum_m^{occ} \frac{(c_s^m)^2 \beta_{st}^{mn}}{E_n^s - E_m^t} \tag{11.12}$$

The role of orbital overlap in the interaction of reactant and reagent allowed
Fukui [11.6] to formulate a *general orientation principle: "A majority of
chemical reactions are liable to take place at the position and in the direction
where the overlapping of HOMO and LUMO of the respective reactants is
maximum; in the electron-donating species, HOMO predominates in the overlapping interaction, whereas LUMO does so in an electron-accepting reactant; in the
reacting species which have SOMOs these play the part of HOMO or LUMO,
or both."*

The theory of frontier orbitals in reactivity was created basically for organic
compounds [11.6, 11.7], but its basic features, especially the perturbation
theory formulas (11.7) and (11.8), can also be applied to coordination compounds with metal atom participation. In these cases, the number of interacting
HOMOs and LUMOs is usually larger than for organic compounds. For this
reason, and due to the specificity of d orbitals, the exploitation of reactivity
indices (frontier electron density, delocalizability, superdelocalizability, etc.
[11.6]) is less important for transition metal coordination compounds.

The electrostatic term in (11.7) may be more important for transition metal
compounds than for organic compounds. Even when the interacting atomic
groups are initially neutral, they may become charged as a result of the
interaction due to charge transfer. These charges may be small, and hence the
covalent term in the interaction can be predominant, but the electrostatic term
cannot a priori be neglected.

To illustrate the application of the frontier orbitals approach to reactions with coordination compounds, we consider the mechanism of formation of polyuretans,

$$R'—OH + O{=}C{=}N—R \rightarrow R'—O—C—N—R$$
$$\underset{\displaystyle O \quad H}{\overset{\displaystyle \| \quad |}{}}$$

(11.13)

catalyzed by square–planar complexes of dioximines of Ni(II), Pd(II), and Pt(II) [11.8]. Experimentally, the reaction rate of polymerization depends very weakly on the nature of the metal. On the other hand, electronic structure calculations of the catalyst complexes and the reagents (the isocyanates and alcohols) in the semiempirical iterative extended Hückel (SCCC) approximation (Section 5.5) show that the HOMOs of the complexes which comprise the metal AOs are far (in energy) from the LUMOs of the reagents, while the HOMOs of the latter are far from LUMOs with participation by metal AOs.

The strongest interactions after Eq. (11.7) are expected between the outer atoms O, N, and H of the dioximine ligand of the complex with the oxygen atom of the alcohol and N and O atoms of the isocyanate. For them, the overlapping HOMOs and LUMOs are close in energy, and hence the interaction energies can be compared by the relative magnitude of the numerator of the second term in (11.7), that is, after (11.8): $\gamma_{st} = (c_s^m c_t^n \beta_{st}^{mn})^2$. The electrostatic term in (11.7) proved to be by one or two orders of magnitude smaller than the covalent contributions.

The estimations show that the interaction is the strongest between the oxygen of the alcohol and the hydrogen of the complex, by more than an order of magnitude stronger than between other pairs of atoms. For the isocyanate, the interactions $N_s–N_t$ and $O_s–N_t$ (s and t stand for the complex and reagent, respectively) are the strongest and not very much different from each other.

These estimates allow us to suggest the intermediate state of the reaction above as shown in Fig. 11.7. This arrangement of the two reagents provides the strongest interactions between the oxygen and hydrogen atoms of the catalyst with, respectively, the nitrogen atom of the isocyanate and the oxygen of the alcohol, and allows for direct contact between the hydrogen atom of the alcohol and the nitrogen atom of the isocyanate, promoting reaction (11.13). Thus the frontier orbital perturbation approach predicts a quite unusual mechanism of the catalytic reaction in this case including coordination of the reagents to the outer ligand atoms and not to the metal atoms as in most cases.

Orbital Symmetry Rules in Reaction Mechanisms

Orbital symmetry rules in the mechanisms of chemical reactions follow directly from the treatment of intermolecular interactions that precede the reaction described above. The covalent contribution to the intermolecular interaction is determined by the overlap of the orbitals of the interacting atoms, which

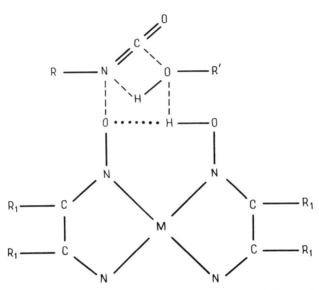

Figure 11.7. Suggested intermediate state (transition complex) of the catalytic reaction of formation of polyuretans based on estimations of molecular interactions in the approximation of frontier orbital perturbation theory. (From [11.8].)

depends strongly on their mutual orientation. Therefore, for orbital-controlled reactions, the energy of the activated complex (the energy barrier of the reaction) depends strongly on the mutual orientation of the interacting species. The overlap between atomic orbitals is nonzero if they possess the same symmetry properties (Section 2.1) in the molecular configuration of the activated complex. Hence the latter should be chosen to satisfy the condition of the same symmetry of the close-in-energy interacting HOMOs and LUMOs of the reactants and reagents. In other words, in order to have a low-energy barrier for the chemical reaction, the orbital overlaps that promote the formation of new bonds should be sufficiently large to compensate for breakdown of the old bonds in the process of the reaction.

Orbital symmetry rules have been suggested and developed for widespread use by Woodward and Hoffmann [11.9]. In Fig. 11.8 a visual treatment of these rules is reproduced for the formation of cyclobutane from two ethylene molecules through a rectangular activated complex. It is seen that the formation of σ bonds between the carbon atoms of the two molecules (before the break of the π bonds in each molecule) can take place only by involving excited MOs that have the same symmetry as the ground state. If the corresponding excited state, as in ethylene, is too high in energy, the formation of σ bonds does not compensate for the breakdown of the π bonds. This is seen from the one-electron MO correlation diagram (Fig. 11.8), in which the energy levels of the newly formed MOs and the old MOs that are broken in the process of the reaction are connected by straight lines. As a result of the high position of the

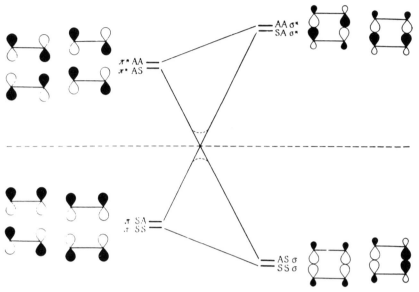

Figure 11.8. MO energy-level correlation diagram that illustrates the Woodward–Hoffmann orbital symmetry rule in formation of cyclobutane from two ethylene molecules. S and A indicate the symmetry properties, symmetric and antisymmetric, respectively, with respect to reflections in the two symmetry planes of the rectangular transition state. (From [11.9].)

appropriate excited MO, the corresponding lines in Fig. 11.8 are going up steeply, and in the intermediate area corresponding to the activated complex the energy is rather high — the reaction barriers assume high values. The latter can be estimated approximately if taking into account that the excited state above corresponds to a transition of the bonding electrons to the antibonding MO in both molecules, resulting in an excitation energy of about 5 eV (115 kcal/mol).

A more detailed energy-level diagram for this system [11.9] shows that along with the excited states mentioned, there are others much lower in energy but of different symmetry. In particular, when starting from an appropriate excited state, the two ethylene molecules can form the cyclobutane molecule (also in its excited state) without any activation barrier. Consequently, this reaction is allowed photochemically.

Further developments and other formulations of the orbital symmetry rules were suggested by Pearson based on the work of Bader [11.10] and others (see [11.11] and references therein). We demonstrate here a treatment that employs the pseudo Jahn–Teller effect and vibronic interactions [11.5], which seems to be more appropriate to transition metal coordination compounds under consideration. From the results obtained in Section 7.4, we know that if there is only one active excited state Γ' which is sufficiently close in energy to the ground state of the activated complex, the nuclear configuration of the latter softens in a particular direction $Q_{\Gamma*}$, and this coordinate is determined by the condition that the vibronic constant $F^{(\Gamma\Gamma')} = \langle \Gamma | (\partial V / \partial Q_{\Gamma*})_0 | \Gamma' \rangle$ is nonzero. The symmetry Γ^* of this $Q_{\Gamma*}$ displacement is determined by the symmetries of the wavefunctions of states Γ and Γ': $\Gamma^* = \Gamma \times \Gamma'$. Therefore, if the symmetries of the ground and corresponding excited states of the activated complex are known, the direction of its labilization (and hence a specific mechanism of the reaction) can be predicted.

Consider two approaching molecular closed-shell systems A and B, each of them being stable separately. The interaction forms a united system leading to the activated complex AB, with its own symmetry and own energy spacing Δ. What are the directions of the lowest (or even negative) curvature of the adiabatic curve of the AB system? To answer this question using the conclusions, given above, the symmetry of the wavefunctions of the ground and low-lying excited states must be known.

If the reaction mechanism is considered by approaching the two molecules A and B without changing the symmetry of the united system AB, the coordinate of the reaction $Q_{\Gamma*}$ is totally symmetric with respect to the activated complex AB. Therefore, the reaction mechanism under consideration is allowed if and only if the wavefunctions of the HOMO of A ψ_1^A and LUMO of B ψ_2^B (or ψ_1^B and ψ_2^A) have the same symmetry (a nonzero overlap integral) in the AB configuration. Otherwise, the reaction mechanism is forbidden. This explains the origin of the term *preservation of orbital symmetry in chemical reactions*. This term is also related to the general preservation rules [11.12]. If there are several close-in-energy HOMOs and/or LUMOs, the contribution of each pair of them should be examined and the results summarized (strictly speaking, the multilevel problem must be solved).

By comparison with the Woodward–Hoffmann formulation, one can see that the vibronic approach contains all the features of the phenomenon. For two ethylene molecules approaching in parallel and forming a rectangular activated state, the curvature of the adiabatic potential K after (7.94) in the direction Q^* of these motions does not decrease: The energy distance Δ to the excited state for which $F \neq 0$ is very large. When moving along Q^*, the energy of the ground (bonding) state is increasing while the energy of the excited (antibonding) one is decreasing. As a result, the Δ value decreases with consequent decrease in the curvature $K = K_0 - F^2/\Delta$. For smaller Δ values, K becomes negative, and at a certain point the energy reaches its maximum, as shown in Fig. 11.8 by the dashed line. When passing on this point, K_1 increases due to the formation of new bonds. If the initial Δ value is large, a great increase in energy is required before Δ becomes small enough and the negative vibronic contribution to the curvature F^2/Δ turns down the AP. As a result, the reaction barrier is high.

It is seen that the vibronic approach repeats the main point of the Woodward–Hoffmann orbital symmetry rules and gives them an additional foundation based on the first principles of quantum mechanics stemming from

Eq. (7.71) (or more general equations in Section 7.4). Also of great importance is the fact that unlike the qualitative Woodward–Hoffmann approach, *the vibronic approach allows one to deduce quantitative criteria for the favorable mechanism of the chemical reaction.* Indeed, for a given reaction, one can look over all the low-lying excited states of the activated complex and determine the Q_{Γ^*} coordinate, for which the vibronic contribution F^2/Δ (more precisely, its summary value) is maximal.

The orbital symmetry rules in their qualitative version are widely used in the study and prediction of chemical reaction mechanisms. Examples of these applications can be found in monographs, reviews, and original papers (see in [11.9, 11.11]). A forbidden reaction between two molecular system may become allowed by interaction with a third molecular system (the catalyst). Consider an example. As a forbidden reaction we again take the formation of cyclobutane from two ethylene molecules via a rectangular activated complex. Its MO energy-level correlation diagram is given in Fig. 11.8. Following Mango and Schachtschneider [11.13] (who were the first to consider this case), we assume that both molecules are *cis*-coordinated to the transition metal atom M with the two C=C bonds perpendicular to the plane of the metal–ligand bonds. The two coordinated parallel ethylene molecules, by moving along the Q^* coordinate, which preserves their rectangular arrangement, produce the coordinated cyclobutane molecule (the atom M is beyond the plane):

$$
\begin{array}{c}
\text{C} \qquad \text{C} \\
\| \cdots \text{M} \cdots \| \\
\text{C} \qquad \text{C}
\end{array}
\quad \xrightarrow{\ Q^*\ } \quad
\begin{array}{c}
\text{C} \cdots \text{C} \\
\diagdown \diagup \\
\cdot \ \text{M} \ \cdot \\
\diagup \diagdown \\
\text{C} \cdots \text{C}
\end{array}
\qquad (11.14)
$$

One can see that with the participation of the metal atom, the symmetry of the system and the symmetry of the reaction coordinate Q_* are the same as for the reaction without catalyst participation. In particular, the two planes of symmetry are preserved with respect to reflections, and hence classification of the MO states remains as shown in Fig. 11.8. However, *the positions of the MO energy levels under the influence of the metal atom vary considerably, and this is the main effect of the catalyst in the scheme under consideration.*

In Fig. 11.9 the MO correlation diagram for this case with a transition metal M that has the electronic configuration d^8 is given; only the π bonds of the ethylene molecules (transforming to σ bonds in cyclobutane) and the d states of metal atom are shown, the s and p states of the metal being omitted. These energy levels must be populated by six electrons (two electrons from the d^8 metal and two electrons from each of the two ethylene molecules). Compared with the MO correlation diagram without metal participation (Fig. 11.8), the energy levels of the cyclobutane molecule vary substantially: for the antibonding σ MO ($\sigma_1^* + \sigma_2^*$) with SA symmetry the energy lowers, while the AS

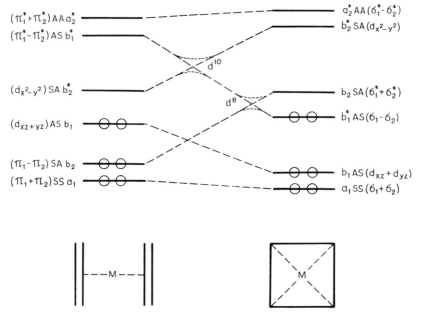

Figure 11.9. Correlation diagram for cyclobutane formation from two ethylene molecules with catalyst (transiton metal M) participation. The crossings d^8 and d^{10} illustrate the reaction barrier formation for the corresponding metal d^n configurations.

$(\sigma_1 - \sigma_2)$ MO energy increases. As a result, the energy gap Δ between the ground and excited states decreases and the vibronic reduction of the AP curvature in the Q^* direction increases; the reaction barrier (shown in Fig. 11.9 as the d^8 crossing) becomes essentially smaller. For d^{10} metals these (or somewhat changed) energy levels should be occupied by eight electrons, resulting in the population of the $b_2^*(d_{x^2-y^2})$ and $b_2(\sigma_1^* + \sigma_2^*)$ levels (Fig. 11.9). It does not increase the reaction barrier significantly (shown in Fig. 11.9 as d^{10} crossing). These results agree well with the empirical data on the d^8 and d^{10} metal activity as catalyst in the reactions in question [11.13]. Quantitatively, the catalyst influence depends critically on the magnitude of the energy level shifts produced by coordination and the corresponding vibronic constant F in Eq. (7.71).

11.2. ELECTRONIC CONTROL OF CHEMICAL ACTIVATION VIA VIBRONIC COUPLING

Chemical Activation by Electron Rearrangement

As stated in Section 11.1, the dependence of the rate of the elementary act of the chemical reaction on the activation energy D is exponential [Eq. (11.1)], and hence even a small change of D results in a considerable change in the reaction rate; D can be taken as a measure of reactivity of one of the reactants in the reaction with the other one. To lower the D value from D_0 to D means to activate the molecule; therefore, the change $-\Delta D = D - D_0$ can be called *chemical activation*. How can the chemical activation be controlled?

To answer this question, consider the rising portion of the adiabatic curve of the chemical reaction from the minimum R to the maximum M (Fig. 11.1). For simplicity, let us present this curve by a cubic polynomial:

$$\varepsilon(Q) = a + bQ + cQ^2 + dQ^3 \tag{11.15}$$

The constants in this polynomial have a clear physical meaning. By choosing the energy read off $\varepsilon(0) = 0$ we get $a = 0$; $(\partial\varepsilon/\partial Q)_0 = F$ is the force acting upon the nuclear framework in the Q direction at $Q = 0$, that is, it coincides with the definition of the linear vibronic constant F (Section 7.2), hence $b = F$; $(1/2)(\partial^2\varepsilon/\partial Q^2)_0 = K$ is the curvature at the minimum (or the force constant), hence $c = (1/2)K$; and d is the cubic anharmonicity that is convenient to denote by $d = -\gamma$. If the point $Q = 0$ is taken at the minimum, then the system is in equilibrium at this point, $F = 0$, and the reaction curve looks as follows (the subscript zero at ε, K and γ means initial values):

$$\varepsilon_0(Q) = \tfrac{1}{2}K_0Q^2 - \gamma_0Q^3 \tag{11.16}$$

By differentiating, one can easily make sure that in these notations (Q_{0D} is the coordinate of the maximum of the energy barrier)

$$D_0 = \frac{K_0^3}{54\gamma_0^2} \tag{11.17}$$

$$Q_{0D} = \frac{K_0}{3\gamma_0} = \left(\frac{6D_0}{K_0}\right)^{1/2} \tag{11.18}$$

It is seen that D_0 is determined by only two parameters, K_0 and γ_0, if $F = 0$, and by all three parameters F, K_0, and γ_0, if $F \neq 0$. Hence to change the D_0 value, one must change some of these parameters or all of them *without changing the reactants*. The only way to do this is to change their electronic state, *to rearrange the electronic structure*. There are several possibilities to rearrange the electronic structure of the reactants, including excitation, oxidation, reduction, ionization, and chemical perturbation by means of another molecular system. The latter thus acts as a *catalyst*.

To study the influence of electron rearrangements on chemical activation, the results of vibronic coupling theory (Sections 7.2 and 7.4), especially the use of orbital vibronic constants (Section 7.2), may be efficient [11.14–11.17]. Below the orbital vibronic constants $f_{\Gamma*}^{(ij)}$ and orbital contribution to the force constant $k_{\Gamma*}^i$ as defined by (7.25) and (7.29), respectively, are employed. For simplicity the indication of the representation Γ^* of the reaction coordinate Q is sometimes omitted.

As formulated in Section 7.2, the diagonal linear orbital vibronic constant (OVC) $f_{\Gamma^*}^i$ equals the force with which the electron of the ith MO distorts the nuclear framework in the direction of the symmetrized displacements Q_{Γ^*} minus the proportion of the nuclear repulsion force in this direction per electron. Consequently, the total force distorting the molecule in this direction (the integral vibronic constant $F_{\Gamma^*}^\Gamma$) is given by Eq. (7.26): $F_{\Gamma^*}^\Gamma = \Sigma_i q_i^\Gamma f_{\Gamma^*}^i$ (the "addition rule"); q_i^Γ is the electron occupation number for the ith MO in the electronic state Γ under consideration. If the system is (statically) stable with respect to the Q_{Γ^*} displacement,

$$F_{\Gamma^*}^\Gamma = \sum_i q_i^\Gamma f_{\Gamma^*}^i = 0 \qquad (11.19)$$

The OVC are different for different orbitals: The nuclear repulsion per electron is independent of the MO, whereas the electron distribution changes considerably from one MO to another. In particular, in diatomics the bonding influence of the electron of the bonding MO is stronger than the nuclear repulsion per electron, $f_R^i > 0$ (R is the interatomic distance), whereas for the antibonding orbitals the opposite is true: $f_R^i < 0$. The OVC are thus a measure of the MO bonding. At the point of stability these different values of OVC compensate each other exactly and Eq. (11.19) holds.

Similar relationships can be obtained for the off-diagonal OVC and orbital force constants given by Eqs. (7.28) and (7.29). It is seen from Eq. (7.29) that for the orbital contributions to the force constants there is also an addition rule: The curvature of the reaction curve in a given direction $K_{\Gamma^*}^\Gamma$ equals the sum of orbital contribution $k_{\Gamma^*}^i$. However, unlike the OVC, the orbital coefficient $k_{\Gamma^*}^i$ is determined not only by the ith MO, but also by the contribution of vibronic mixing with other MOs. From the latter, only the free and not fully occupied MOs should be taken into account: The contribution of the mixing of any two fully occupied MOs vanishes because it enters the sum for $K_{\Gamma^*}^\Gamma$ twice with opposite signs.

As indicated in Section 7.2, the introduction of orbital vibronic and force constants considerably amplifies the description of the electronic structure of molecules by means of the one-electron MO scheme. The energy levels and wavefunctions of each MO describe the static electronic structure and properties of the molecular system in the given one-electron state. Supplemented with OVC and orbital contributions to force constants and anharmonicities, these MOs become *vibronic molecular orbitals* (VMOs) that give much more complete information, adding to the static electronic structure new parameters of dynamic behavior (Fig. 7.4).

The supplementary parameters in the VMO characterize the molecular structure in more detail. They give a quantitative estimate of the bonding or antibonding nature of the MO even in polyatomic systems, where, owing to the complicated genealogy of the orbitals, examination of the bonding properties of individual MOs with respect to the many AOs may be difficult or even impossible. The orbital vibronic and force constants supplement the energy characteristics (bonding, antibonding, nonbonding) with information about the direction and magnitude of the distortion and/or softening (or hardening) of the nuclear framework by the MO electron.

The additional orbital vibronic parameters of molecular structure are of special importance in investigation of the origin of molecular transformations, in particular, in analyzing the origin of activation of a given molecule by another. As mentioned in Section 11.1, any influence of a given molecular system upon another begins with a change in the mobile electron distribution, which, in turn, may change the much more inertial nuclear configuration. If sufficiently small, the variation in electronic structure in the first approximation can be described by changes in the MO electronic occupation numbers — orbital charge transfers Δq_i (Section 5.2). In Section 10.1 it is shown how the total charge transfer Δq can be estimated by the parameters of the interacting molecules, and in Section 6.3 orbital charge transfers Δq_i are calculated. If these Δq_i values are not very large (on the order of one electron), one can use them, following the Koopmans approximation, as characteristics of the intermolecular influence by coordination.

In fact, the MO wavefunctions are also changed by coordination, but this may be neglected in the first approximation as a second-order effect, provided that the intermolecular interaction is sufficiently small. The second-order terms can be taken into account in a more sophisticated consideration. Although not all kinds of intermolecular interactions can be considered accurately by just orbital charge transfers, the main effects of coordination and chemical activation in catalysis are described qualitatively well by this approach in many cases.

The electronic redistribution is thus presented by the new orbital occupation numbers $q_i^\Gamma + \Delta q_i$. If the initial system is stable, $F_{\Gamma*} = 0$ and $K_{\Gamma*} > 0$ (the superscript Γ of the electronic state of the system as a whole is omitted), the substitution $q_i \rightarrow q_i + \Delta q_i$ in Eqs. (11.19) and (7.29) lead to the following relationships:

$$F_{\Gamma*} = \sum_i \Delta q_i f_{\Gamma*}^i \tag{11.20}$$

$$\Delta K_{\Gamma*} = K'_{\Gamma*} - K_{\Gamma*} = \sum \Delta q_i k_{\Gamma*}^i + \sum_{i \neq j} \frac{q_i \Delta q_j |f_{\Gamma*}^{(ij)}|^2}{\Delta_{ji}} \tag{11.21}$$

Hence the electronic rearrangement, taken into account by the changes in orbital occupancies Δq_i, results in a nonzero distorting force $F_{\Gamma*} \neq 0$ and a change of the force constant $\Delta K_{\Gamma*}$ in the direction $Q_{\Gamma*}$, for which $f_{\Gamma*}^i \neq 0$ (i is the index of the MO for which $\Delta q_i \neq 0$). The direction of the distorting force

$Q_{\Gamma*}$ depends on the symmetry Γ_i of the ith MO. As mentioned in Section 7.2, $f^i_{\Gamma*}$ is nonzero if the symmetric product $[\Gamma_i \times \Gamma_i]$ contains Γ^* (Section 3.4 and Table A1.14). If Γ_i is nondegenerate, $\Gamma = A_1$ is totally symmetric: Electrons of nondegenerate MOs distort the molecule in the direction of totally symmetric displacements A_1 which do not change the symmetry of the system. If Γ_i is degenerate, Γ^* can be nontotally symmetric, but it should be Jahn–Teller active.

As for the change of the force constant $\Delta K_{\Gamma*}$, Γ^* can be of any symmetry allowed in the appropriate point group. This is seen directly from the second term in Eq. (7.29), in which $f^{(ij)}_{\Gamma*}$ is nonzero if $\Gamma^* = \Gamma_i \times \Gamma_j$, while Γ_i and Γ_j for the ground and excited states, respectively, may belong to any symmetry representations. Similar expressions can be obtained for the change of the anharmonicity constants $\Delta\gamma$.

The occurrence of distorting forces and changes in the force constants and anharmonicities due to electronic rearrangements directly explain the change in the reactivity of the molecule—its chemical activation. With the new constants of the rearranged electronic structure F, $K = K_0 + \Delta K$, and $\gamma = \gamma_0 + \Delta\gamma$, the rising portion of the reaction curve becomes as follows (Fig. 11.10):

$$\varepsilon(Q) = FQ + \tfrac{1}{2}KQ^2 - \gamma Q^3 \tag{11.22}$$

This curve is different from (11.16) and yields different activation energy of the reactions D:

$$D = \frac{(K - 6\gamma Q_0)^3}{54\gamma^2} \tag{11.23}$$

where Q_0 is the new equilibrium position of the reactants:

$$Q_0 = \frac{K}{6\gamma}\left[1 - \left(1 + \frac{12\gamma F}{K^2}\right)^{1/2}\right] \tag{11.24}$$

The new position of the maximum of the energy barriers is

$$Q_D = \frac{K}{3\gamma} - Q_0 \tag{11.25}$$

Compared with the nonperturbed values, we have

$$\frac{D}{D_0} = \frac{(K - 6\gamma Q_0)^3 \gamma_0^2}{K_0^3 \gamma^2} \tag{11.26}$$

All the parameters in the right-hand side of (11.26) can be, in principle, estimated from empirical data. K and K_0 are directly related to IR spectra:

$K_0 = M\omega_0^2$ and $K = M\omega^2$, where ω_0 and ω are the corresponding vibrational frequencies of the reactants in the initial and electronically rearranged states, respectively, Q_0 is the new equilibrium position (distortion) in the rearranged state. The coefficients in the cubic terms γ_0 and γ can be expressed by spectroscopic anharmonicity correction ωx [11.18]: $\gamma_0/\gamma = \beta^{1/2}(K_0/K)^{3/2}(\hbar\omega/\hbar\omega_0) = \beta^{1/2}(K_0/K)$, where $\beta = \omega_0 x_0/\omega x$ is the ratio of the anharmonicity corrections in the initial and rearranged system, respectively (usually, for small rearrangements β does not differ very much from a unity).

For chemical activation, as defined at the beginning of this section, we have

$$-\Delta D = D - D_0 = D_0\left[1 - \left(\frac{K}{K_0}\right)^3\left(\frac{\gamma}{\gamma_0}\right)^2\right] + \frac{K^2 Q_0}{3\gamma} - 2KQ_0^2 + 4\gamma Q_0^3 \qquad (11.27)$$

The last term proportional to Q_0^3 is very small (Q_0 is on the order of 10^{-1} Å) and can be neglected. The remaining expression can be presented by two terms:

$$-\Delta D = D_0\left(1 - \beta\frac{K}{K_0}\right) + KQ_0(Q_D - Q_0) \qquad (11.28)$$

where $Q_D - Q_0 = Q'_D$ is the coordinate distance between the minimum R and maximum M in the rearranged system (Fig. 11.10), and in the same notation $Q_D = (6\beta D_0/K_0)^{1/2} - Q_0$. The following formula may be convenient:

$$\frac{D}{D_0} = \frac{K}{\beta^{1/2}K_0}\left(\frac{Q'_D}{Q_{0D}}\right)^3 \qquad (11.29)$$

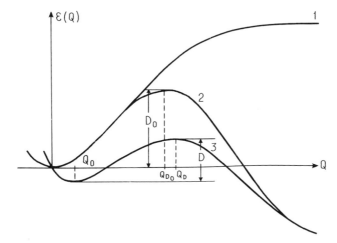

Figure 11.10. Section of the adiabatic potential of a molecular system in the direction of the reaction path Q: (1) free molecule (*dissociation curve*); (2) influence of another reactant (*reaction curve*); (3) influence of the catalyst (*activation curve*).

For activation of diatomics the presentation of $\varepsilon(Q)$ by the *Morse potential* [11.1] may be useful. For the unperturbed molecule

$$\varepsilon_0(Q) = D_0\{1 - \exp[-\alpha_0(R - R_0)]\}^2 \tag{11.30}$$

while for the activated molecule

$$\varepsilon(Q) = D\{1 - \exp\{-\alpha(R - R_0')\}\}^2 \tag{11.31}$$

By differentiating, one can get the distorting force $F = d\varepsilon/dR$, force constant $K = d^2\varepsilon/dR^2$, and anharmonicity $\Gamma = d^3\varepsilon/dR^3$. For the unperturbed molecule, from Eq. (11.30) at $R = R_0$:

$$F_0(R_0) = 0$$
$$K_0(R_0) = 2D_0\alpha_0^2 \tag{11.32}$$
$$\gamma_0(R_0) = -6D_0\alpha_0^3$$

In the activated state $R_0' = R_0 + \Delta R$, and (see [11.17])

$$F(R_0) = F_0(R_0) + \Delta F(R_0)$$
$$K(R_0) = K_0(R_0) + \Delta K(R_0) \tag{11.33}$$
$$\gamma(R_0) = \gamma_0(R_0) + \Delta\gamma(R_0)$$

$$\Delta R = \frac{1}{\alpha}\ln y \tag{11.34}$$

$$K = \frac{K(R_0)}{y(2y - 1)}$$

where

$$\alpha = \frac{K(R_0)(1 - y)}{F(R_0)}(2y - 1)$$

$$y = \frac{5a + 4}{8a + 1} + \left\{\left[\frac{5a + 4}{8(a + 1)}\right]^2 - \frac{1}{4}\right\}^{1/2}$$

$$a = -\frac{[K(R_0)]^2}{F(R_0)\gamma(R_0)}$$

For the activation (dissociation) energy in the activated state D we get, from Eq. (11.34), $D = K/2\alpha^2$.

The first term in the expression (11.28) of the chemical activation gives the contribution of the softening (hardening) by the electronic rearrangement, while the second term stands for a similar contribution of the distortion force (it equals the work of the force $F = K_0 Q$ along the way, Q'_D).

In both terms the anharmonicity is essential since it forms the barrier itself (without anharmonicity there is no maximum of the reaction curve), although the influence of the change of anharmonicity (the ratio of anharmonicities) in the first term presented by the parameter β can be small. With the parameters F and ΔK calculated after Eqs. (11.20) and (11.21), and $\Delta \gamma$ after a similar equation, one can estimate K and γ (K_0 and γ_0 are assumed to be known) and the chemical activation $-\Delta D$. A much easier way to use Eq. (11.26) or (11.27) is to estimate the parameters from empirical data.

For a polyatomic molecule with many degrees of freedom the consideration and conclusions above apply to each normal coordinate, in particular, to the reaction path (which may be a linear combination of normal coordinates). In these multidimensional cases the expressions obtained above can also be used for determining the change in molecular geometry by electronic rearrangement (Section 9.2).

Equation (11.28) is approximate and it is valid only for small changes in the electronic structure, compared with the initial structure. Such electronic rearrangement occurs in the above-mentioned processes of oxidation, reduction, excitation, ionization, and coordination of one molecular system to another molecule (in the process of chemical reaction) or to a coordination center as a ligand, on a solid surface (chemical adsorption), and so on.

The last cases can be joined together under the common title *activation by coordination*. This topic is of special importance. The changes of the MO occupancies due to the charge transfers to the coordination center and back are fractional. While cases of integer charge transfers (oxidation, reduction, ionization, excitation, etc.) can in principle be treated by other methods (e.g., by quantum-chemical calculations of the electronic structure of the initial and final systems), the approximate analysis of the properties of molecular systems with fractional charges can apparently be carried out only by the approach described above involving the OVC (although calculations using modified MO LCAO versions are not excluded).

The assumption of weak changes in the electronic structure by coordination is valid in many systems. It is confirmed by spectroscopic investigations indicating that the coordinated molecule preserves its main individual structural features slightly changed by coordination. Certainly, this assumption is not always true. For example, the hydrogen molecule may dissociate by coordination, thus changing its structural features. Even in these cases, if the process is evaluated according to the foregoing approach at an early stage when the coordination is still sufficiently weak, the direction of the reaction, as well as some other features, may be predicted qualitatively.

Activation of Small Molecules by Coordination: Semiempirical Approach

Chemical activation due to electron rearrangements induced by interaction with (or coordination to) another molecular system are of special interest: They model the activation mechanism in catalysis. To consider this activation in the scheme above, the orbital charge transfers Δq_i, orbital vibronic constants f^i, orbital contributions to the force constant k^i, and anharmonicities for each MO of the activated molecule are needed [see Eqs. (11.20) and (11.21)].

First we note the importance of the number of MOs involved in the coordination. Mono-, di-, and multiorbital bonds with ligands are considered in Section 6.3. The number of active MOs is important, in particular, because it influences the orbital charge transfers Δq_i. As mentioned in Section 6.3, the total transfer by coordination $\Delta q = \Sigma_i \Delta q_i$ cannot be very large, owing to thermodynamic restrictions (cf. the electroneutrality principle proposed by Pauling [11.19]). Therefore, the Δq_i values may be large, in principle, only if more than one orbital is involved in the bonding and if different Δq_i values have different signs. When only two orbitals are involved in the bonding—the HOMO and LUMO (diorbital bonds; Section 6.3)—the two Δq_i values often have opposite signs: $\Delta q_1 < 0$ and $\Delta q_2 > 0$. If the HOMO is bonding (i.e., $f^{(1)} > 0$) and the LUMO is antibonding ($f^{(2)} < 0$), the resulting distorting force, according to Eq. (11.20), is

$$F = \Delta q_1 f^{(1)} + \Delta q_2 f^{(2)} = -(|\Delta q_1||f^{(1)}| + |\Delta q_2||f^{(2)}|) \qquad (11.35)$$

In other words, the contribution of the two orbitals to the distorting force is determined by the sum of their absolute values. Similar conclusions are valid for the contributions to the change of the force constant and anharmonicities. Consequently, the greater the absolute values of the two charge transfers of opposite sign, the greater the mutual vibronic influence of the interacting molecular systems. If there are more than two orbitals active in the bond formation, the possibility of a favorable combination of charge transfers that would increase the vibronic influence (but preserve the required small total charge transfer Δq) increases. Hence we arrive at a conclusion concerning *the role of multiorbital bonding in the vibronic influence* of one molecular system upon another as, for example, in catalysis. *The special role of transition and rare earth elements and their compounds in catalysis is due to the ability to form multiorbital bonds with various molecular groups.*

For concrete calculations values of OVC and orbital charge transfers Δq_i are required. The latter are determined by the electronic structure of both the coordinated molecule and the coordination center, as well as by the geometry of coordination. The mode of coordination determines which orbitals of the coordinated molecule overlapping those of the coordination center have the largest changes Δq_i. Examples in Section 6.3 illustrate how these values can be calculated. Following are some semiempirical schemes and illustrative examples.

Consider the HOMO–LUMO two-orbital approach. Neglecting the changes in the MO occupation numbers for the inner orbitals, one can essentially simplify Eqs. (11.20) and (11.21), which can be written in the following form (1 and 2 refer to HOMO and LUMO, respectively):

$$F = f^{(1)} \Delta q_1 + f^{(2)} \Delta q_2$$
$$\Delta K = k^{(1)} \Delta q_1 + k^{(2)} \Delta q_2 \qquad (11.36)$$
$$\beta - 1 = C^{(1)} \Delta q_1 + C^{(2)} \Delta q_2$$

where $k^{(1)}$ and $k^{(2)}$ are the above-mentioned force constant coefficients, which are combinations of the second- and first-order vibronic constants and show how the force constant changes by adding one electron on the corresponding MO (Section 7.2). The last equation in (11.36) describes the linear dependence of the ratio of anharmonicity constants β on the charge transfers Δq_i, the coefficients $C^{(1)}$ and $C^{(2)}$ being complicated combinations of cubic and lower-order OVC. For anharmonicity changes, another presentation is also possible:

$$\Delta \gamma = \gamma - \gamma_0 = \gamma^{(1)} \Delta q_1 + \gamma^{(2)} \Delta q_2 \qquad (11.37)$$

All six coefficients $f^{(1)}, f^{(2)}, k^{(1)}, k^{(2)}, C^{(1)}$, and $C^{(2)}$ (or $\gamma^{(1)}$ and $\gamma^{(2)}$ instead of $C^{(1)}$ and $C^{(2)}$) can be determined easily if the values of F, ΔK, and β (or $\Delta \gamma$) are known for any two independent processes of electronic rearrangements (for two pairs of values of Δq_1 and Δq_2 excluding the trivial values $\Delta q_1 = \Delta q_2 = 0$). These processes may be ionization ($\Delta q_1 = -1$, $\Delta q_2 = 0$), reduction ($\Delta q_1 = 0$, $\Delta q_2 = 1$), or excitation ($\Delta q_1 = -1$, $\Delta q_2 = 1$), provided that the empirical parameters above are available for them.

In the empirical scheme the distorting force and force constant are determined from empirical data measured at the equilibrium position of the reactants with the rearranged electronic structure. Therefore, the corresponding quantities should be determined at $Q = Q_0$ and not at $Q = 0$. In particular, by direct differentiation of $\varepsilon(Q)$ from (11.22), one can make sure that

$$F(Q = Q_0) = \left(\frac{\partial \varepsilon}{\partial Q}\right)_{Q_0} = -KQ_0 + 3\gamma Q_0^2 \qquad (11.38)$$

$$K(Q = Q_0) = \left(\frac{\partial^2 \varepsilon}{\partial Q^2}\right)_{Q_0} = K - 6\gamma Q_0 \qquad (11.39)$$

and $\Delta K = K - K_0 - 6\gamma Q_0$. On the other hand, γ_0 can be obtained from Eq. (11.17):

$$\gamma_0 = \left(\frac{K_0^3}{54D_0}\right)^{1/2} \qquad (11.40)$$

where for a free molecule D_0 is the dissociation energy (in the Q direction).

As mentioned above, the force constants K and K_0 can be obtained from the appropriate vibrational frequencies. The magnitude of distortion Q_0 in the Q direction is in general available from x-ray (or other diffraction method) analysis of the structure of the electronically rearranged system as compared with the unperturbed system. For simple electronic rearrangements, such as ionization or excitation, Q_0 can be obtained from spectroscopic data. For some systems, especially for simple molecules, Q_0 can be also determined from the known K value using empirical relations between K and Q_0; then the theory becomes one-parametrical [11.20].

If the coefficients in Eqs. (11.36) are known, the distorting force F and the alteration of the force constant ΔK and anharmonicity constant β (or $\Delta\gamma$) can be estimated for given charge transfers Δq_1 and Δq_2. Consequently, the change in the activation energy $-\Delta D$ can be evaluated from Eq. (11.28) as a function of D_0. On the other hand, if F, ΔK, and $\Delta\gamma$ (or β) are estimated from Eqs. (11.38) to (11.40), Δq_1 and Δq_2 can be obtained for the HOMO and LUMO:

$$\Delta q_1 = \frac{\Delta K f^{(2)} - k^{(2)} F}{k^{(1)} f^{(2)} - k^{(2)} f^{(1)}}$$

$$\Delta q_2 = \frac{\Delta K f^{(1)} - k^{(1)} F}{k^{(1)} f^{(2)} - k^{(2)} f^{(1)}}$$

(11.41)

This empirical parametrization may be inconvenient if the vibronic constants are very sensitive to the processes employed for their estimation. If this is the case, the processes closest to the expected charge transfers must be used in the empirical scheme. The calculated OVC and force constant coefficients are devoid of this fault. The Q_0 value, which is a small difference in the interatomic distances of the two states of the molecule, free and activated by coordination, is difficult to obtain from experimental data: direct x-ray measurements may lack the high accuracy needed, and Q_0 in the crystal structure may not coincide with that in the free system. This difficulty can be overcome if one uses empirical correlations between interatomic distances and frequencies of vibrations. These correlations make it possible to obtain Q_0 values that are more appropriate to the interatomic distances in the activated molecule than that obtained by x-ray measurements.

Equation (11.28) can also be used as an empirical relation between ΔD and D_0:

$$-\Delta D = aD_0 + bD_0^{1/2} + c$$

(11.42)

where the coefficients a, b, and c [not to be confused with a, b, c, and d in Eq. (11.15)] are functions of the empirical parameters K, K_0, Q_0, and β (or γ_0). The D_0 value, the activation energy for the reaction under consideration in the absence of the catalyst influence, is often unknown. In these cases the curves $-\Delta D = f(D_0)$ can be plotted for each coordination center. By comparison of these curves, the relative efficiency of different coordination centers in lowering the activation energy can be established for each value of D_0.

Following are several examples of chemical activation considered by the vibronic theory in its semiempirical versions.

Activation of Carbon Monoxide. The electronic structure and spectroscopic parameters of the CO molecule and its ions are well studied. This makes it possible to obtain estimates of its main orbital vibronic constants and force constant coefficients. For CO the HOMO is 5σ and the LUMO is 2π (Fig. 6.9). Using the empirical data [11.21–11.23] for the force constant K_0 of the free molecule, and K, Q_0, and β for the CO^+ ion ($\Delta q_1 = -1$, $\Delta q_2 = 0$) and the CO excited state $5\sigma^2 2\pi^0 \rightarrow 5\sigma^1 2\pi^1$ ($\Delta q_1 = -1$, $\Delta q_2 = 1$), we easily find from Eqs. (11.38) to (11.40):

$$f^{(1)} = -4.54 \cdot 10^{-4} \text{ dyn} \qquad f^{(2)} = -12.1 \cdot 10^{-4} \text{ dyn}$$

$$k^{(1)} = -0.080 \cdot 10^6 \text{ dyn/cm} \qquad k^{(2)} = -0.83 \cdot 10^6 \text{ dyn/cm} \qquad (11.43)$$

$$C^{(1)} = 0.134 \qquad C^{(2)} = 0.0104$$

Here for the $5\sigma^1 2\pi^1$ configuration, the values averaged over the two excited states $A^1\Pi$ and $a^3\Pi$ were taken, and the approximate value $\Delta K = K - K_0$ is taken instead of Eq. (11.39).

It is evident from both the OVC and force constant coefficients that the two orbitals 5σ and 2π are both antibonding, the latter being much more antibonding than the former. In addition to the qualitative statement about the nature of the corresponding MO (bonding or antibonding), which can often (but not always) be made without using the vibronic approach, the latter also gives the quantitative degree of MO participation in the chemical bonding, and separately in the distorting force, force constant, anharmonicity correction, and so on.

With knowledge of the constants (11.43), one can analyze the behavior of the CO molecule in various situations, in particular, when it is coordinated to another molecular system or solid surface. If the charge transfers Δq_1 and Δq_2 are known, F, ΔK, β, and hence $-\Delta D$ (for given D_0) can be evaluated, and vice versa. It is seen from these constants that the greater the positive values of Δq_1 and Δq_2 (i.e., the greater the increase in the occupancy of the HOMO and LUMO), the larger the negative force F (which acts as an antibonding factor to lower the activation energy) and the negative ΔK (acting in the same way). However, since the HOMO is fully occupied by electrons in the free molecule, Δq_1 can be only negative or zero. Consequently, the activation of the CO molecule is greater the larger the positive value of Δq_2 and the smaller the negative value of Δq_1, the former being much more important. Hence in reactions in which the activation energy is determined by the activation of the CO molecule when it is linearly coordinated to a catalyst, the latter is more efficient the greater its π-donor properties and the lower its σ-acceptor ability.

Consider the linear coordination of carbon monoxide to the Ni atom on an NiO surface. The value $K = 1.710 \cdot 10^6$ dyn/cm is known from empirical data [11.24]. Using the above-mentioned relationship between K and Q_0 for the CO molecule [11.21],

$$K = (1.7957 - 8.2926 Q_0 \, [\text{Å}]) \cdot 10^6 \text{ dyn/cm} \tag{11.44}$$

one can obtain $Q_0 = 0.0126 \text{ Å}$, $F = -2.119 \cdot 10^{-4}$ dyn, and $\Delta K = K - K_0 = -0.192 \cdot 10^6$ dyn/cm, and consequently estimate $\Delta q_1 = -0.32$, $\Delta q_2 = 0.26$ according to (11.45) and $\beta = 0.93$ according to Eq. (11.40). These data provide in detail the mechanism of activation of the CO molecule when coordinated to the NiO surface: As a result of coordination there is about a 0.3 electron charge transfer to the metal from the weak antibonding 5σ orbital and about a 0.3 electron charge transfer from the metal to the strong antibonding 2π orbital of CO.

Using the relationship (11.44) between K and Q_0, a general correlation between the CO vibrational frequency and the charge transfers Δq_1 and Δq_2 from the HOMO 5σ ($-\Delta q_1$) and to the LUMO 2π (Δq_2) by coordination has been derived. It is shown in Fig. 11.11 in the form of a correlation diagram. Note, however, that the empirical function $K = f(Q_0)$ (11.44) is valid in the region where the coordination is sufficiently strong and does not apply to very small charge transfers and CO frequency changes.

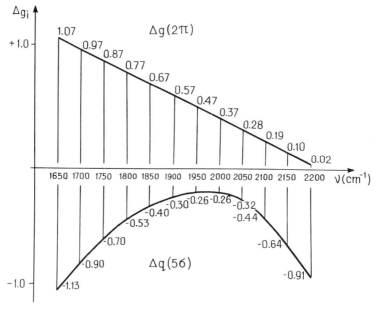

Figure 11.11. Correlation diagram between CO vibration frequencies v (in cm^{-1}) and orbital charge transfers from the HOMO 5σ ($-\Delta q_1$) to the coordination center, and from the latter to the LUMO 2π (Δq_2) in linear end-on coordination.

The change in CO reactivity when coordinated to the NiO surface can be estimated from Eq. (11.42) (D_0 and ΔD in kcal/mol):

$$-\Delta D = 0.17D_0 + 1.40D_0^{1/2} - 0.78 \qquad (11.45)$$

This curve $-\Delta D = f(D_0)$, together with similar curves for CO activation in a series of coordination systems, including metal carbonyls and metal surfaces, is shown in Fig. 11.12. Similar results for activation of CO by coordination to polynuclear clusters are illustrated in Figs. 11.13 and 11.14 and Table 11.1 [11.20]. It is seen that simultaneous (bridged) coordination to two or three centers of the cluster (denoted by μ_2 and μ_3, respectively) results in much stronger activation, $-\Delta D$, than for one-center coordination (μ_1). In $Rh_6(CO)_{16}$ and $Co_2(CO)_8$ the effect of bridged coordination is seen explicitly (Table 11.1 and Fig. 11.14). In the rhodium polynuclear complex triple-bridged coordination to three centers simultaneously results in activation of the CO molecule, which in reactions with $D_0 = 100$ kcal is more than five times larger than in simple one-center coordination to the same system.

The charge transfers in Table 11.1 explain in more detail the origin of this bridged multicenter coordination effect: The orbital charge transfer from the complex to the strongly antibonding MO 2π of CO is much larger in the case of triple bridged coordination.

To get numerical estimation of the activation $-\Delta D$, knowledge of D_0 is needed. By way of illustration, if we assume that the activation energy of the reaction $CO + O_2 \rightarrow CO_2 + O$ in flame is $D_0 = 48$ kcal/mol [11.25], and this is associated with the activation of the CO bond only, then by linear

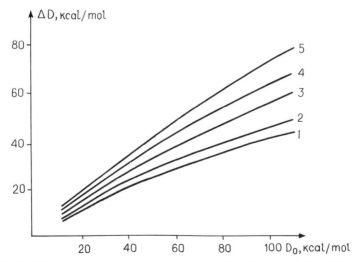

Figure 11.12. Vibronic activation of carbon monoxide. Curves $-\Delta D = f(D_0)$ for CO coordination to surfaces PdO (1) and NiO (2), and in carbonyls $Fe(CO)_5$ (3), $V(CO)_5$ (4), $Mn(CO)_5$ (5).

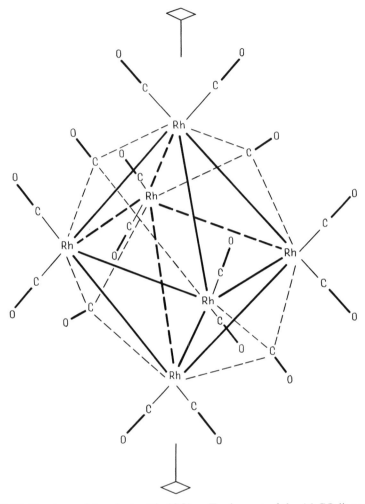

Figure 11.13. Structure of the cluster $Rh_6(CO)_{16}$. Twelve out of the 16 CO ligands are one-center (μ_1) coordinated, while four ligands are tricenter (μ_3) coordinated (shown by dashed lines).

coordination to the NiO surface, the activation energy, in accordance with Eq. (11.45), is lowered by $\Delta D = -17$ kcal/mol and becomes equal to $D = D_0 + \Delta D = 31$ kcal/mol. The experimental value for this reaction on the NiO catalyst (in excess CO) is $D_{expt} = 25.4$ kcal/mol [11.26].

Activation of Dinitrogen. In the nitrogen molecule N_2 the HOMO is 5σ and the LUMO is 2π (Fig. 6.7), as in CO. Using the data for N_2^+ and the excited states $a^1\Pi_g$ and $B^3\Pi_g$ emerging from the $(5\sigma)^2 \rightarrow (5\sigma)^1(2\pi)^1$ excitation

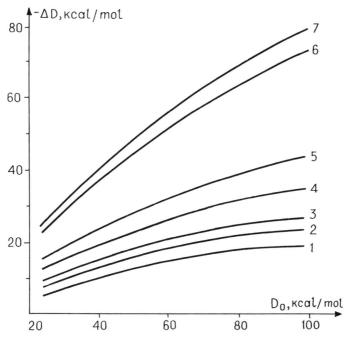

Figure 11.14. Activation of CO by coordination to polynuclear clusters: curves $-\Delta D = f(D_0)$ for $Rh_6(CO)_{16}$-μ_1 (1), $Rh_4(CO)_{12}$-μ_1 (2), $Co_2(CO)_8$-μ_1 (3), $Mo(CO)_6$-μ_1 (4), $Mn_2(CO)_{10}$-μ_1 (5), $Co_2(CO)_8$-μ_2 (6), $Rh_4(CO)_{12}$-μ_2 (7), $Rh_6(CO)_{16}$-μ_3 (8); μ_1, μ_2, and μ_3 indicate one-center, two-center, and tricenter coordinations, respectively, as illustrated in Fig. 11.13.

Table 11.1. Charge transfers to the CO molecule $\Delta q_1(5\sigma)$ and $\Delta q_2(2\pi)$ and chemical activation $-\Delta D$ (for $D_0 = 100$ kcal/mol) in some polynuclear clusters

Complex[a]	$\Delta q_1(5\sigma)$	$\Delta q_2(2\pi)$	$-\Delta D$ for $D_0 = 100$ (kcal/mol)
$Mo(CO)_6$	-0.31	0.32	30
μ_1-$Mn_2(CO)_{10}$	-0.27	0.36	36
μ_1-$Os_3(CO)_{12}$	-0.28	0.31	28
μ_1-$Co_2(CO)_8$	-0.33	0.27	22
μ_2-$Co_2(CO)_8$	-0.42	0.67	74
μ_1-$Rh_6(CO)_{16}$	-0.39	0.22	15
μ_3-$Rh_6(CO)_{16}$	-0.53	0.77	78

[a] μ_1, μ_2, and μ_3 stand for simultaneous coordination to one, two, and three centers, respectively.

($\Delta q_1 = -1$, $\Delta q_2 = 1$) and Eqs. (11.36), we get

$$f^{(1)} = 3.51 \cdot 10^{-4} \text{ dyn} \qquad f^{(2)} = -8.18 \cdot 10^{-4} \text{ dyn}$$

$$k^{(1)} = 0.286 \cdot 10^6 \text{ dyn/cm} \qquad k^{(2)} = -0.785 \cdot 10^6 \text{ dyn/cm} \qquad (11.46)$$

$$C^{(1)} = 0.112 \qquad C^{(2)} = 0.127$$

It is seen that in contrast to the CO molecule, the HOMO 5σ is bonding, while the LUMO 2π is antibonding, like CO. Therefore, an activation center with high π-donor and high σ-acceptor properties is needed for the activation of linearly coordinated dinitrogen. If the coordination center is a π acceptor, a charge transfer may take place from either the inner bonding π_u orbital when the coordination is linear end-on (longitudinal) or from the antibonding σ_u orbital for transversal (side-on) coordination. It follows that the activation depends on the geometry of coordination.

The constants (11.46) and empirical data pertaining to the coordinated nitrogen molecule can be employed to estimate the charge transfers Δq_1 and Δq_2 and the reduction in activation energy $-\Delta D$ as a function of D_0 using Eqs. (11.36). Table 11.2 presents some examples of such calculations [11.5, 11.16]. The value Q_0 is estimated from the empirical formula (the N_2 stretching vibrational frequency in the coordinated state v is given in 10^3 cm^{-1}):

$$Q_0[\text{Å}] = 0.9482 - 0.7105v + 0.1302v^2 \qquad (11.47)$$

It is seen from Table 11.2 that, as above, the major contribution to the $-\Delta D$ value comes from the linear effect, the distorting force (b and c terms); the contributions from the softening (second-order) effect is smaller.

Activation of NO. As distinct from the two examples above, the HOMO and LUMO in the free NO molecules are realized in the same antibonding 2π orbital occupied by one electron. Calculations [11.27] show that in addition to this orbital, the lower fully occupied 5σ orbital takes part in the charge transfers by coordination (the remaining inner orbitals are practically not involved). The coefficients a, b, and c determining the chemical activation $-\Delta D$ [Eq. (11.37)] are given in Table 11.3 for several coordination systems. In contrast to previous examples, the iron complex produces deactivation of the coordinated NO molecule ($-\Delta D < 0$), thus being an anticatalyst for corresponding reactions with NO.

Activation of Localized Double Bonds (C_2H_4 and C_3H_6). The semiempirical scheme described above yields [11.16]:

$$f^{(\pi)} = 4.5 \cdot 10^{-4} \text{ dyn} \qquad f^{(\pi^*)} = -0.46 \cdot 10^{-4} \text{ dyn}$$

$$k^{(\pi)} = 0.40 \cdot 10^6 \text{ dyn/cm} \qquad k^{(\pi^*)} = -0.39 \cdot 10^6 \text{ dyn/cm} \qquad (11.48)$$

Table 11.2. Semiempirically calculated charge transfers Δq_1 and Δq_2 and vibronic reduction of the activation energies $-\Delta D$ by coordination of the n_2 molecule to different complexes (D and $-\Delta D$ in kcal/mol)

Coordination System[a]	ν' (cm^{-1})	K' (10^6 dyn/cm)	$\Delta K = K' - K$ (10^6 dyn/cm)	Q_0 (Å)	F (10^{-4} dyn)	Δq_1	Δq_2	$-\Delta D = aD + bD^{1/2} + c$		
								a	b	c
[RuH$_2$(N$_2$)(PPh$_3$)$_3$]	2147	1.900	−0.394	0.0229	−4.19	−0.16	0.44	0.14	2.72	−2.87
[{Ru(NH$_3$)$_5$}$_2$N$_2$]$^{4+}$	2100	1.821	−0.479	0.0303	−5.25	−0.57	0.41	0.21	3.37	−4.81
[Os(NH$_3$)$_5$N$_2$]Br$_2$	2028	1.695	−0.599	0.0428	−6.77	−1.00	0.40	0.31	4.32	−8.94
[(tol)(PPh$_3$)$_2$Mo-N$_2$-Fe(C$_5$H$_5$)(dmpe)]$^+$	1930	1.536	−0.758	0.0619	−8.54	−1.21	0.52	0.37	5.62	−16.9

[a]tol, toluol; dmpe, Me$_2$PCH$_2$CH$_2$PMe$_2$.

Table 11.3. a, b, and c coefficients in the relation $-\Delta D = aD + bD^{1/2} + c$ for NO activated by coordination in different complexes

Coordination System	a	b	c
$Fe(CN)_5NO^{2-}$	-0.03	-0.78	-0.19
$Mn(CN)_5NO^{2-}$	0.02	0.23	-0.02
$Mn(CN)_5NO^{3-}$	0.18	2.6	-2.83
$Cr(CN)_5NO^{3-}$	0.27	3.56	-5.77
$Cr(CN)_5NO^{4-}$	0.37	4.26	-9.58
$V(CN)_5NO^{3-}$	0.35	4.19	-9.07

and $\beta \approx 1$. Together with the relationship between the length of the C=C bond R and the change of the vibrational frequency by coordination $\Delta v = v_0 - v$ [11.28] ($v_0 = 1640$ cm^{-1}, Δv in 10^3 cm^{-1}):

$$R(\Delta v) = 0.56855(\Delta v)^2 - 0.0298906\Delta v + 1.334 \qquad (11.49)$$

the parameters (11.48) allow one to estimate the charge transfers to the HOMO π and LUMO π^* and the activation $-\Delta D$. Examples are given in Table 11.4 (empirical data from [11.29]). It is seen that the activation $-\Delta D$ is much smaller here than that for CO (Table 11.1) and N_2 (Table 11.2) (see also Section 6.3 and [11.30]).

Activation of Hydrogen. Due to the small number of electrons in H_2, even small charge transfers strongly activate the H—H bond. Therefore, in most cases H_2 decomposes in the coordinated state, although there are examples when hydrogen is coordinated as a molecule. In the latter cases the vibrational frequencies observed are about $v = 2600$ to 3100 cm^{-1} (in the free molecule, $v_0 = 4401$ cm^{-1}) with the H—H distance $R \sim 0.75$ to 0.86 Å ($R_0 = 0.74$ Å). To

Table 11.4. Semiempirical calculated charge transfers to HOMO (Δq_1) and LUMO (Δq_2) and chemical activation $-\Delta D$ of the localized double bond in C_2H_4 and C_3H_6 coordinated to oxide catalyst CuO/γ-Al$_2$O$_3$ ($-\Delta D$ and D_0 in kcal/mol)

System	v (cm^{-1})	Q_0 (10^{-3} Å)	Δq_1	Δq_2	$-\Delta D = aD_0 + bD_0^{1/2} + c$		
					a	b	c
Cu—C_2H_4	1545	5.8	-0.11	0.15	0.094	0.18	-0.01
Cu—C_3H_6	1550	5.2	-0.10	0.14	0.088	0.16	-0.01

Table 11.5. Orbital charge transfers to the bonding (Δq_1) and antibonding (Δq_2) MOs and chemical activation $-\Delta D$ of coordinated molecular hydrogen as a function of the frequency of stretching vibration $v(H\!-\!H)$ in the coordinated state for $D_0 = 103$ kcal/mol

v (cm^{-1})	Δq_1	Δq_2	$-\Delta D$ (kcal/mol)
4200	-0.06	0.06	30.5
3800	-0.17	0.16	53.7
3400	-0.27	0.25	70.3
3000	-0.35	0.34	81.7
2600	-0.43	0.41	89.4
2200	-0.49	0.38	94.8
2000	-0.52	0.50	96.1

employ the one-parametric version of the vibronic theory of chemical activation, the following relationship between the interatomic distance R and vibrational frequency change $\Delta v = v - v_0$ can be suggested (Δv in 10^3 cm^{-1}) [11.25]:

$$R(\text{Å}) = 0.7412 + 0.5034\,\Delta v - 0.1541\,(\Delta v)^2 \tag{11.50}$$

The orbital vibronic constants for the two active MOs, bonding 1 and antibonding 2, are [11.31]

$$\begin{aligned} f^{(1)} &= 1.58 \cdot 10^{-4}\,\text{dyn} & k^{(1)} &= 0.418 \cdot 10^6\ \text{dyn/cm} \\ f^{(2)} &= -0.41 \cdot 10^{-4}\,\text{dyn} & k^{(2)} &= -0.0348 \cdot 10^6\,\text{dyn/cm} \end{aligned} \tag{11.51}$$

and the anharmonicity coefficient $\beta \approx 0.5$ is much smaller than in the examples above.

With these data, using Eqs. (11.41) and (11.45), one can estimate the charge transfers Δq_1 and Δq_2 to the bonding and antibonding MOs of H_2, and the chemical activation $-\Delta D = D_0 - D$ for different Δv values. Some results are given in Table 11.5 ($-\Delta D$ is given for model reactions with $D_0 = 103$ kcal/mol).

The main conclusion drawn from these data is that the homolytic cleavage of the H_2 molecule,

$$L_n M \!-\! \begin{array}{c} H \\ | \\ H \end{array} \ \Longleftrightarrow\ L_n M \begin{array}{c} \diagup H \\ \diagdown H \end{array} \tag{11.52}$$

takes place roughly at $v = 2400$ to 2100 cm^{-1} and $R = 1.0$ to 1.2 Å.

Activation of Oxygen by Hemoproteins. For oxygen coordination the semiempirical scheme used above should be completed because, as shown by calculations, three MOs are active in the orbital charge transfers (oxygen forms a three-orbital bond with the active site of the coordination system). Activation of oxygen by metalloenzymes is of special interest to biology. In Section 10.3 the electronic structure and some electron-conformational effects in these systems are discussed. In this section we consider in more detail the mechanism of activation of oxygen by several hemoproteins [11.17].

One of the versions of the vibronic theory of chemical activation is *to calculate the orbital charge transfers* Δq_i to and from the coordinated molecule and, using the equations above, to estimate the expected distortion, changes of vibrational frequencies, anharmonicities, and activation by coordination. This version has been employed in the treatment of oxygen activation in question. It gives additional possibilities to verify the applicability of the vibronic approach by comparison of the estimated frequencies of vibrations with those observed experimentally.

The active site of several metallobiochemical systems considered below can be presented by the general formula $Fe(P)(L)$, where P is porphyrin and L is the aminoacid moiety of the protein: hemoglobin ($L = ImH$, Im denotes imidazol), horseradish peroxidase (HRP, $L = Im^-$), cytochrome P-420 ($L = SHCH_3$), and cytochrome P-450 ($L = SCH_3^-$); as well as $L = C_6F_4HS^-$; except the latter, the same systems with an additional electron were also tried. Their electronic structures with the coordinated oxygen were calculated in the semiempirical MO LCAO IEH (SCCC) approximation (Section 5.5). The three active MOs participating in bonding with the active side of the enzyme are $3\sigma_g$, $1\pi_u$, and $1\pi_g{}^*$. The corresponding three orbital vibronic constants $f^{(i)}$, force constant $k^{(i)}$, and anharmonicity $\gamma^{(i)}$ coefficients were calculated by Eqs. (11.36) using spectroscopic data for oxygen and its ions [11.17]. The constants were calculated for the coordination of the superoxide ion $O_2{}^-$, which is more appropriate to the state of coordinated oxygen in hemoproteins. Unlike the examples above, the Morse potential formulas (11.30–11.34) were employed in calculation of the vibrational frequencies and interatomic distances in the coordinated state. The $3\sigma_g$ and $1\pi_u$ MOs of oxygen are bonding, whereas $1\pi_g^*$ is antibonding, but with respect to the superoxide ion their contribution to the force constant has an opposite sign.

Some results for the calculated charge transfers, interatomic distances, vibrational frequencies, and activation of coordinated oxygen are illustrated in Table 11.6. It is seen that the activation is due mainly to the contribution of the charge transfer $\Delta q(1\pi_g^*)$ to the antibonding $1\pi_g^*$ MO of oxygen, which in the case of cytochrome P-450 + e- is about 0.91, and in the approximation when the sulfur lone pair is split and contributes to this MO (Section 10.3) it is even larger. However, the value of ~ 1.4 electron in Table 11.6 is overestimated because of the use of the "frozen orbital" approximation.

The resulting activation of the O—O bond is significant: In the coordinated state D is almost half of the dissociation energy D_0, but it is still larger than

Table 11.6. Parameters of O_2 activation by hemoprotein models Fe(P)(L) without and with an additional electron[a]

Parameter	ImH		Im$^-$		SHCH$_3$		SCH$_3^-$		C$_6$F$_4$HS$^-$
	Hb	Hb+e	HRP	HRP+e	P-420	P-420+e	P-450	P-450+e	
$\Delta q(1\pi_u)$	−0.04	−0.04	−0.06	−0.04	−0.05	−0.06	−0.05	−0.05	−0.04
$\Delta q(3\sigma_g)$	−0.11	−0.13	−0.13	−0.13	−0.11	−0.13	−0.11	−0.12	−0.10
$\Delta q(1\pi_g)$	0.56	0.86	0.60	0.88	0.55	0.86	0.74	0.91	0.64
			1.30	1.58			1.21	1.38	
K, mdyn/Å	6.28	5.40	6.04	5.37	6.27	5.35	5.73	5.31	6.07
			4.53	4.36			4.79	4.56	
$\omega^{(1)}$, cm^{-1}	1115	1067	1132	1067	1154	1066	1102	1061	1134
			979	962			1008	989	
$\omega^{(2)}$, cm^{-1}	1193	1112	1111	1110	1198	1115	1150	1111	1174
ω_{exp}, cm^{-1}	1160						1140		1140
R_0, Å	1.30	1.35	1.32	1.35	1.31	1.35	1.33	1.36	1.32
			1.41	1.45			1.42	1.42	
R_{exp}, Å	1.15–1.22								1.21–1.28
D, eV	4.62	4.08	4.45	4.05	4.69	4.04	4.34	3.97	
			2.63				3.36	2.98	

[a] The data obtained from electron-lone-pair splitting by deprotonation are given in second rows.

the activation energies of the reaction of hydroxylation by cytochrome P-450 (ca 1 eV). One of the reasons of this discrepancy is that D_0 in Eq. (11.32) is not the dissociation energy, but the activation energy in the corresponding associative reactions without the catalyst participation, which is smaller than the dissociation energy (Fig. 11.3). The calculated frequencies of O_2 stretching vibrations are in good agreement with the experimental data available.

Jahn–Teller (pseudo Jahn–Teller) Induced Chemical Activation

The influence of the electronic rearrangement on the reactivity of a molecular system, discussed above in this section, is attributed to the occurrence of a distorting force F, the changes of the force constant ΔK, and anharmonicity $\Delta \gamma$ in this rearrangement [Eq. (11.22)]. The change in the rising portion of the reaction curve as a function of the reaction pathway and the consequent change of the activation energy $-\Delta D = D_0 - D$ are shown in Fig. 11.10. By comparison, one can see that near the origin $Q = 0$ these changes are similar with those produced by the Jahn–Teller or pseudo Jahn–Teller effect (Sections 7.3 and 7.4 and Fig. 7.5). In the Jahn–Teller effect (degenerate electronic states) a nonzero force distorts the high-symmetry nuclear configuration in the direction of active coordinates Q; if the quadratic terms of vibronic interactions are taken into account, the force constant (adiabatic potential curvature) K_0 is also changed. In the pseudo Jahn–Teller effect (mixing of the ground and excited electronic state by the nuclear displacements) the force constant K_0 changes.

If a Jahn–Teller system enters a chemical reaction, the behavior of its AP curves without and with the Jahn–Teller effect as a function of the Jahn–Teller active coordinate Q [$\varepsilon_0(Q)$ and $\varepsilon(Q)$, respectively] exactly follow Eqs. (11.16) and (11.22), respectively, as illustrated in Fig. 11.10. It means that the Jahn–Teller effect produces the same chemical activation $-\Delta D$ as a corresponding electronic rearrangement induced by, say, a catalyst. The chemical activation $-\Delta D$ is determined here by the same Eq. (11.26), with Q_0 and $\Delta K = K - K_0$ taken from the Jahn–Teller effect formulas (Section 7.3).

If the second-order vibronic interaction terms in the Jahn–Teller effect are neglected (linear vibronic coupling), $K = K_0$, $\gamma = \gamma_0$, and the expression (11.28) is simplified:

$$-\Delta D = (12 D_0 E_{JT})^{1/2} - 4E_{JT} \tag{11.53}$$

where E_{JT} is the Jahn–Teller stabilization energy (Section 7.3). For instance, for a system with a doubly degenerate electronic E term $E_{JT} = F^2/2K_0$ [see Eq. (7.41)] and

$$-\Delta D = F \left(\frac{6D_0}{K_0} \right)^{1/2} - \frac{2F^2}{K_0} \tag{11.54}$$

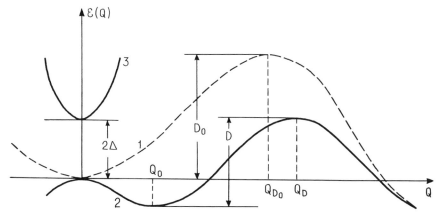

Figure 11.15. AP curves along the reaction path Q without (1) and with (2) the strong pseudo Jahn–Teller effect that mixes the states 1 and 3.

Calculation of the chemical activation $-\Delta D$ for a pseudo Jahn–Teller system becomes more complicated since an additional important parameter emerges: the energy gap 2Δ between the ground state and the excited states whose admixing produces the instability of the ground state of the reactant. Fig. 11.15 shows the two reaction curves without and with the pseudo Jahn–Teller effect in the direction of the reaction path Q. The curve modified by the pseudo Jahn–Teller effect differs from the Jahn–Teller curve (Fig. 11.10) by the behavior near the origin $Q = 0$.

If the condition of the strong pseudo Jahn–Teller effect (7.71) ($K_0 < F^2/\Delta$) holds, the ground state is unstable at $Q = 0$ and the adiabatic potential curve (7.69) has two minima at the points (7.72) with the curvature at the minima points given by Eq. (7.73). Taking into account the anharmonicity produced by the interacting reactants, we can present the reaction curve by adding a negative cubic anharmonicity term to Eq. (7.69), which unlike (11.16) or (11.22), should be of sixth order to preserve the symmetry conditions:

$$\varepsilon(Q) = \tfrac{1}{2}K_0 Q^2 - (\Delta^2 + F^2 Q^2)^{1/2} - \gamma Q^6 \tag{11.55}$$

This curve has not been studied in detail as yet. Approximately, by expansion of the square root, we get

$$\varepsilon(Q) = \frac{1}{2}KQ^2 + \frac{F^4}{4\Delta^3}Q^4 - \gamma Q^6$$

$$K = K_0 - \frac{F^2}{\Delta} \tag{11.56}$$

and for the points of the minimum Q_0 (stable reactants) and maximum Q_D (activated complex), we have

$$Q_{0,D} = \left\{ \frac{F^4}{12\Delta^3\gamma} \left[1 \pm \left(1 + \frac{24\gamma K\Delta^6}{F^8} \right)^{1/2} \right] \right\}^{1/2} \tag{11.57}$$

with the minus sign for Q_0 and plus sign for Q_D. Note that due to (7.71), $K < 0$, and hence for real roots of Eq. (11.57) the condition $|24\gamma K| < F^8/\Delta^6$ must be obeyed. If this inequality is sufficiently strong, we have approximately

$$Q_0 \approx \frac{\Delta}{F} \left(1 - \frac{K_0\Delta}{F^2} \right)^{1/2} \tag{11.58}$$

$$Q_D \approx \left(\frac{F^8 + 6\Delta^6 K\gamma}{6\gamma\Delta^3 F^4} \right)^{1/2} \tag{11.59}$$

By substituting the parameters (11.57) or (11.58) and (11.59) into (11.55), one can estimate the activation energy modified by the pseudo Jahn–Teller effect,

$$D = \varepsilon(Q_D) - \varepsilon(Q_0) \tag{11.60}$$

and the chemical activation $-\Delta D = D_0 - D$ [with D_0 according to (11.17)].

Thus Jahn–Teller and pseudo Jahn–Teller induced chemical activation is similar to the effect of electronic rearrangement discussed above. However, along with this important similarity there are essential differences between the two sources of chemical activation. First, they are concerned with the direction Q, in which the activation energy is lowered, that is, with the allowed mechanism of the reaction. In systems with the Jahn–Teller and pseudo Jahn–Teller effects, this direction is predetermined by the electronic structure of the molecule itself (by Jahn–Teller and pseudo Jahn–Teller active displacements), whereas in the case of catalyst influence the direction of chemical activation depends on the nature of the catalyst: on the electronic rearrangement it produces in the reagents. Therefore, different catalysts in principle can cause different dominating mechanisms. In particular, the catalyst can change the mechanism of the reaction that takes place without the catalyst. On the contrary, for vibronically induced reactions of free molecules the mechanism is known a priori, and this allows us to predict the reaction course and its products.

By way of example, consider the reaction of substitution in, or monomolecular decomposition of, an octahedral transition metal complex in an orbitally twofold degenerate electronic E state [11.5, 11.32], for example, for bivalent copper known as having a strong Jahn–Teller effect. The adiabatic potential of this system has three equivalent minima; in each of them the octahedron is elongated along the tetragonal axis, two ligands being farther away from the central atom than the other four (Sections 7.3 and 9.2). In Q_θ and Q_ε coordinates (Figs. 7.1 and 9.27) the distortion takes place along the Q_θ direction; Q_θ is thus the coordinate along which the activation energy of the reaction is lowered (the system is chemically activated) compared with other directions *ceteris paribus*.

As a result, the reaction above should proceed in the direction of the Q_θ displacements, for which the two axial ligands move away from (while the other four approach) the CA (Fig. 9.27). For the monomolecular reaction this implies dissociation of the ligands in the *trans* positions, while for the substitution reaction it means the formation of *trans*-substituted complexes. Visually, this conclusion corresponds to the following picture: When in the pulsating motions during a specific lifetime (Section 9.2) elongated octahedra are formed, the two ligands in *trans* positions are weaker bonded and reached more easily by the attacking reagents. Thermodynamic considerations (Section 9.3) also support the conclusion about the formation of *trans*-substituted complexes of E-term metals.

This result is in complete agreement with the well-known empirical data on the behavior of octahedral complexes of Cu(II) in solution or gas phase. It is known that these complexes lose two ligands in the *trans* position, forming square-planar structures, while substitution reactions produce only *trans*-substituted complexes (except for bidentate ligands, which can be *cis*-coordinated only). On the other hand, if the Jahn–Teller effect in such a system is quenched (say, by the significant differences between the ligands), the potential barrier essentially increases and the reaction rate decreases. This conclusion also agrees with the experimental data [11.33], in which the substitution reaction rate decreases about 3000 times when passing from $[Cu(H_2O)_6^{2+}]$ to $[Cu(tren)H_2O]^{2+}$, where tren = 2,2′,2″-triaminotriethylamine. According to the predictions of the theory, similar effects are expected in octahedral complexes of high-spin Cr(II) and Mn(III), Ag(II),... that have E ground terms.

The expected vibronically induced mechanisms of reactions are different for other nonoctahedral molecular configurations with E terms and for other terms. They can be determined qualitatively from known Jahn–Teller and pseudo Jahn–Teller distortions. There are other suggested applications of the vibronic effects in determining the chemical reaction rate and mechanisms [11.34–11.39].

11.3. DIRECT COMPUTATION OF ENERGY BARRIERS OF CHEMICAL REACTIONS

Methods of numerical computation of the electronic structure of coordination compounds involving computer and supercomputer calculations (Chapters 5 and 6) allow one, in principle, to obtain the AP surface for any chemical reaction, including catalytic reactions, and to get the numerical value of the energy barrier of the reaction. Practically, the solution of this problem may be very laborious and expensive. However, the continuous improvement of methods and algorithms, as well as computer power, makes it possible to solve more and more complicated problems; the number of papers devoted to these topics increases rapidly.

In this section we discuss several examples that illustrate the state of the art in direct numerical calculations of chemical reactions with coordination compounds, including the cases when the latter act as catalysts.

Substitution Reactions: The *trans* Effect

In Section 9.3 devoted to mutual influence of ligands in coordination compounds, the phenomena of *trans* and *cis* influences in stereochemistry problems are discussed. It was emphasized there that one should distinguish between the thermodynamic aspect of the problem (*trans influence*), in which the mutual influence is considered from the static (energetic) point of view, and the kinetic aspect (*trans-effect*), in which the influence of mutual ligand positions on the reaction rate of substitution or decomposition reactions is discussed. In this section we evaluate *mutual ligand effects* — the influence of a given ligand on the energy barrier of the reactions of substitution (or decomposition) of another ligand in the *trans* or *cis* position.

There is an obvious correlation between the mutual influence and mutual effect of ligands as they are defined above. For instance, as shown in Section 9.3, the *trans* influence results in a weakening of the metal–ligand bond in the *trans* position to the influencing ligand. It is clear that the weaker bond is subject to easier cleavage or substitution. Hence the *trans* influence also promotes the *trans* effect. On the other hand, for a bimolecular associative S_N2 reaction, the formation of the reaction barrier may be much more dependent on the details of the electronic structure of both reactants (especially the entering, leaving, and *trans*-directing ligands) than the stereochemistry of the initial complex, where mainly one ligand is specified.

Square-planar complexes of Pt(II) were suggested first [11.40] and are best studied (see the reviews [11.41–11.44]) as demonstrating the *trans* effect in the most explicit way. Usually, the series of ligands arranged in decreasing *trans*-effect power (the *trans*-effect series) is given as follows [11.45]:

$$C_2H_4, CO > CN > NO_2^- > SCN^- > I^- > Br^- > Cl^- > NH_3 > OH^-$$

$$(11.61)$$

In Section 9.3, many examples are given (see Tables 9.8 to 9.11) and the role of some major electronic features of the ligand and the metal in realizing the *trans* influence is discussed. It is shown that the σ donor properties of the ligand enhanced by its π acceptor properties are most important in the *trans* influence. Basically, these ideas are also valid in the *trans* effect, with the

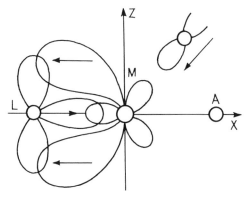

Figure 11.16. $\sigma-\pi$ interaction in the *trans* effect: The π-acceptor properties of the *trans*-influencing ligand L enhanced by its σ-donor properties lower the electron density along the *trans* bond M—A, thus favoring substitution of the *trans* ligand A.

distinction that in the latter the π-acceptor properties are more important. The qualitative reason of this π effect was discussed first by Chatt et al. [11.46] and by Orgel [11.47] and is illustrated in Fig. 11.16.

If the ligand L in the square-planar complex of Pt(II) has the strongest π-acceptor properties (low-energy empty π orbitals), there is a back donation to this ligand from the d_{xz} orbital of the metal M [which is occupied in the d^8 configuration of Pt(II)]. This back donation results in a significant rearrangement of the electron distribution in which the electron density along the *trans* bond M—A is decreased. In turn, this lowers the repulsion from the entering group in the *trans* position, thus lowering the energy of the transition state of the reaction and hence the activation energy. Hence the *trans* influence, which weakens and elongates the M—A bond, also contributes to lowering the activation energy by making the approach of the attacking group and the removal of the leaving group easier. Note that if the d_{xz} orbital is not occupied, this π effect disappears.

As discussed in Sections 6.3, 9.3, and 11.2, the σ-donor and π-acceptor properties of a diorbital bonded ligand are interdependent due to the electron interaction effects. Indeed, the electron repulsion does not allow for high local concentration of electronic charge, which means that the total charge transfers to and from the ligand should be relatively small. This restriction is not concerned directly with orbital charge transfers. In σ-donor and π-acceptor effects, the corresponding orbital charge transfers have opposite signs, and hence their sum may be smaller than each of them. Therefore, in the presence of strong π donation to the ligand its σ-donor property increases, and vice versa, strong σ donation increases the π-acceptor ability.

This mutual enhancing of σ-donor and π-acceptor properties is important in understanding both the *trans* influence (Section 9.3) and *trans* effect. For instance, H^- has empty π states $(2p_z)$, but in the free ion they are rather high in energy and hence H^- is a poor π acceptor. However, in the coordinated states with strong σ donation, its $2p_z$ orbital energy decreases significantly, and H^- becomes a better π acceptor.

It was also shown [11.48] that the competition of the ligands in the σ donation to the metal results in a characteristic *cis effect* — the influence on the reaction rate of ligand substitution in the *cis* position. The *cis* effect is opposite to the *trans* effect: *the larger the trans effect of a given ligand, the smaller its cis effect*. This means that the reaction rate for the substitution of A is determined, first, by the *trans*-effect power of the *trans*-ligand L, and second, by the *cis*-effect power (opposite to *trans*-effect power) in the *cis* positions of L and A. The higher the *trans* effect of the *trans*-ligand and the lower the *trans* effect of the *cis*-ligands, the higher the reaction rate for the given ligand.

These general qualitative considerations were recently confirmed by direct numerical calculations of transition states and energy barriers of substitution reactions with some square-planar Pt(II) and Rh(I) complexes [11.49]. To reveal the *trans*-effect in its details, four reaction systems were computed; for each of them the reactions of substitution of two ligands in *trans* positions to other two ligands were studied to compare their *trans* effect:

$$\text{cis-Rh(CO)H(NH}_3)_2 + \text{NH}_3 \qquad \text{comparing CO and H}^- \qquad (11.62)$$

$$\text{cis-PtH(Cl)(NH}_3)_2 + \text{NH}_3 \qquad \text{comparing H}^- \text{ and Cl}^- \qquad (11.63)$$

$$\text{cis-Pt(CH}_3)\text{Cl(NH}_3)_2 + \text{NH}_3 \qquad \text{comparing CH}_3^- \text{ and Cl}^- \quad (11.64)$$

$$\text{cis-Rh(Cl)NH}_3(\text{H}_2\text{O})_2 + \text{H}_2\text{O} \qquad \text{comparing Cl}^- \text{ and NH}_3 \quad (11.65)$$

In all these reactions the entering and leaving groups are identical [NH_3 for the first three cases and H_2O for (11.65)]. It simplifies the calculations since it allows one to use additional symmetry considerations. In particular, the identical entering and leaving groups should occupy equivalent positions, provided that there are no additional local minima (intermediates) on the reaction curve. For simplicity all the groups are chosen neutral.

The computation scheme is as follows. For each substitution reaction the geometry (bond lengths and angles) of the five-coordinated complex (associative S_N2 reaction) is optimized, yielding the energy of the transition state. With respect to the energies of the free reactants, this gives the activation energy. In all eight reactions (11.62) to (11.65) the transition state is a trigonal bipyramid with a mirror plane for the two equivalent groups, entering and leaving. Then allowing one of these groups to move away and calculating the optimized geometry and energy for several fixed positions of this leaving group, one gets the reaction curve along the reaction coordinate. This curve is illustrated in Fig. 11.17 for the reaction $\text{cis-PtHCl(NH}_3)_2 + \text{NH}_3$ with the substitution *trans* to H. The optimized ligand positions along the reaction path are shown in Fig. 11.18.

All the calculations were carried out by the *ab initio* MO LCAO RHF (restricted Hartree–Fock) method with effective core potentials (Sections 5.3

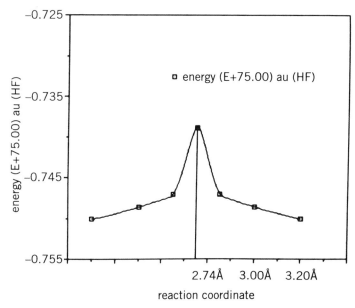

Figure 11.17. AP energy curve along the reaction coordinate for the substitution reaction *cis*-PtHCl(NH$_3$)$_2$ + NH$_3$ with H$^-$ as a *trans*-directing ligand. (After [11.49].)

and 5.4) (calculation details are described well in [11.49]). The validity of the results, especially the role of electron correlations, was examined by comparison with more sophisticated valence approaches. Total energies and activation energy differences (relative *trans* effects) are given in Table 11.7.

We begin the discussion of these results with Fig. 11.18, showing the optimized geometries for several fixed positions of the entering group NH$_3$ in reaction (11.63) with H$^-$ as the *trans*-directing ligand. To understand the origin of these geometries, let us also employ the qualitative σ MO energy-level scheme for the initial square-planar 16-electron complex M(d^8)L$_4$ shown in Fig. 11.19. It is seen that the LUMO is the pure p_z orbital of M, which is perpendicular to the plane of the complex, while the three (d_{xy}, d_{xz}, d_{yz}) orbitals form three occupied lone pairs. When π bonding is taken into account, these lone pairs form the corresponding MOs which have asymmetric distribution shifted toward the most *trans*-directing ligand (Fig. 11.16).

At large distances $R = 3.2$ Å (Fig. 11.18a) when there is no significant orbital overlap and bonding, the entering group NH$_3$ forms an angle $\alpha = 78.5°$ with the plane; it is smaller than 90°, due to the repulsion from the corresponding d_ε lone pair which has an asymmetrical distribution, mentioned above (Fig. 11.16). At smaller distances the NH$_3$ lone pair overlaps with the empty p_z orbital to form a square-pyramidal structure and the angle α increases (Fig. 11.18b and c). At equilibrium distances ($R = 2.73$ Å) the optimized transition state is reached that has a pseudotrigonal–bipyramidal structure (Fig. 11.18d); here a compromise between the repulsion of the two NH$_3$ ligands, their

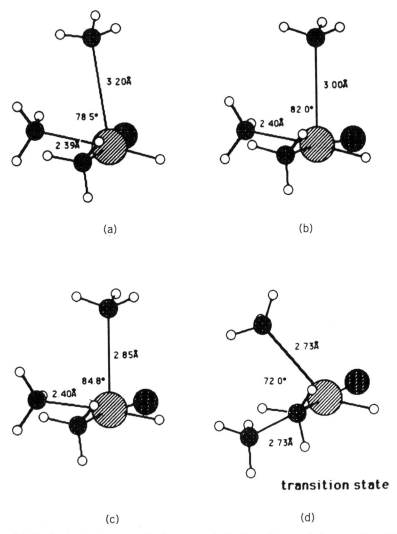

Figure 11.18. Optimized geometries for several fixed positions of the entering NH_3 group in reaction (11.63) with H^- as the *trans*-directing ligand. (After [11.49].)

repulsion from the d_{xz} lone pair and bonding to the rehybridized p-d metal orbitals (including p_z) is reached.

The less repulsion from the lone pairs (directly dependent on the π-acceptor properties of the *trans*-directing ligand) and the weaker the metal-leaving and metal-entering bonds [dependent on the σ-donor properties of the *trans*-directing ligand; Eq. (9.30)], the lower the energy of the transition state and hence the activation energy. These features are effective when the metal M has

Table 11.7. Total energies of transition states and activation energy differences for different *trans*-directing ligands (relative *trans* effect) in several substitution reactions

System	*Trans*-Directing Ligand	Total Energy of Transition State (a.u.)	Activation Energy Difference (kcal/mol)
$Rh(CO)H(NH_3)_3$	CO	−78.0154	10.92
	H^-	−78.0239	
$PtHCl(NH_3)_3$	H^-	−75.7388	1.57
	Cl^-	−75.7363	
$Pt(CH_3)Cl(NH_3)_3$	CH_3-	−83.3640	0.38
	Cl-	−82.3634	
$RhCl(NH_3)(H_2O)_3$	Cl^-	−97.8882	7.28
	NH_3	−97.8766	

Source: [11.49].

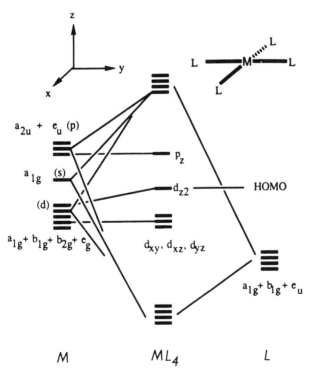

Figure 11.19. σ MO energy levels of a 16-electron square planar d^8 transition metal complex ML_4.

occupied d_ε orbitals and they have a sufficient diffusiveness to overlap strongly with (and to be easily shifted to) the ligands. Second- and third-row d^8 transition metals are good examples. Thus direct numerical calculations confirm the qualitative statements above: *The trans-effect is produced by an appropriate combination of π-acceptor and σ-donor properties of the trans-directing ligand and occupied diffuse d_ε orbitals of the central atom.* Numerical data of Table 11.9 show that among the considered ligands the *trans* effect follows the series $CO > H^- > CH_3^- > Cl^-$, in agreement with the experimental order.

In [11.49], many other details of the *trans*-effect mechanism are elucidated. In particular, it is shown that the monomolecular mechanism of substitution reactions (via dissociation) have a considerably higher activation energy than in the associative mechanism.

Ligand Coupling and Cleavage Processes

Transition metal systems are often involved in chemical transformations as catalysts. In principle, catalytic reactions are similar to noncatalytic reactions with the distinction that in the former the yield of the reaction contains one of the reactants unchanged. In Sections 6.3 and 11.2, the specific features of transition metals important to catalytic action are emphasized: Due to the active *d*-electron configurations, *transition metals and their complexes form multiorbital bonds with ligands, with the effect of strong activation of the latter.*

Another important feature is that due to high coordination numbers, coordination compounds may serve as a *center of organization* of the reactants promoting specific mechanisms of the reaction. This organizational aspect of coordination is of special importance in *supramolecular chemistry*, where the coordination system, via *recognition, orientation, and termination*, serves as a center of self-organization, self-assembling, and *templating* [11.50, 11.50a].

A relatively simple example of catalytic action of coordination compounds is provided by the bonding of two coordinated carbon ligands CR_2 in *cis* positions forming an olefin C_2R_4, and the inverse process of cleavage of a coordinated olefin into a *bis*-carbene complex:

$$
\begin{array}{ccc}
\underset{\overset{|}{R}}{\overset{\overset{|}{R}}{\underset{/\,|\,\backslash}{\overset{\backslash\,|\,/}{M}}}}\!\!\begin{array}{c}C\!-\!R\\[2pt]\\[2pt]C\!-\!R\end{array}
& \rightleftharpoons &
\underset{/\,|}{\overset{\backslash\,|}{M}}\!-\!\begin{array}{c}\overset{|}{R}\\ C\!-\!R\\ C\!-\!R\\ \underset{|}{R}\end{array}
& (11.66)
\end{array}
$$

The first attempt to model this reaction was made for the "naked metal" $M = W$ and $R = H$ [11.51]. Figure 11.20 illustrates the MO energy-level diagram as a function of the angle between the two carbon ligands obtained by the semiempirical extended Hückel method (Section 5.5). It is seen that for

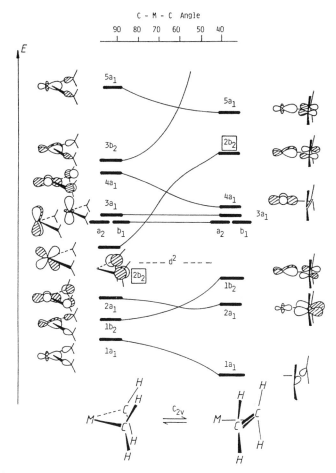

Figure 11.20. MO energy-level correlation diagram for a C_{2v} least motion coupling/decoupling of two methylene groups coordinated to a unligated metal center. Symmetries of the AOs that form the corresponding MOs and the d^2 configuration limit are also shown. (After [11.51 and 11.52].)

a metal M with more than two electrons (in W they are six) the least motion process of direct bonding of two methylene groups CH_2 to form C_2H_4 is forbidden by symmetry considerations (Section 11.1) since it involves the highly excited state $2b_2$ of C_2H_4.

A more complete analysis, using the same approach and the same method of calculation, was carried out for the reactions [11.52]

$$(CO)_4Cr(CH_2)_2 \rightarrow (CO)_4Cr(C_2H_4) \tag{11.67}$$

$$(CO)_4Cr(CD_2)_2 \rightarrow (CO)_4Cr(C_2D_4) \tag{11.68}$$

where D is a π-donor substituent (D = NR_2).

For reaction (11.67) the MO energy-level diagram [11.52] shows that as compared with the unligated metal (Fig. 11.20), the ordering of the important MO levels under the influence of the four π-acceptor CO groups changes significantly, and the $2b_2$ MO is no longer involved in the reaction. Instead, the $1b_2$ level in the olefin side is pushed up and unoccupied, and hence the reaction of direct least motion coupling is again formally forbidden. It means that somewhat less symmetric pathways, although formally allowed by symmetry considerations, may still encounter energy barriers.

The qualitative MO energy-level diagrams give no definite information about the relative thermodynamic stability of the two systems and the height of the energy barrier between them. To answer these questions, calculations of the potential energy surface for the least-motion coupling (uncoupling) of the two methyl groups in the reaction (11.67) were carried out [11.52]. Two coordinates, the angles θ (between the two M—C bonds, Fig. 11.21) and Φ (between the CH_2 plane and the M—C bond) were chosen to describe the reaction pathway. Their variation with the fixed position of the $Cr(CO)_4$ group results in the potential energy surface presented schematically in Fig. 11.22. The surface has two minima for the olefin and bis-carbene complexes, respectively. Interestingly, in this model the olefin complex, which is a 16-electron system, is more stable than the 18-electron bis-carbene complex; the allowance for a more relaxed configuration with optimization of the CO ligand positions is expected to lower the energy of the olefine complex. The energy barrier between these two minima is rather low, about 10 kcal/mol, and will be probably lower for a less symmetric reaction pathway.

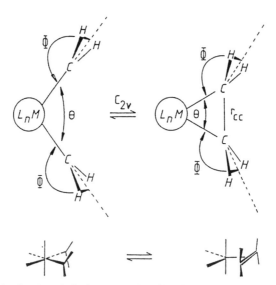

Figure 11.21. Angles θ and ϕ chosen to describe the least-motion convergence of two CH_2 groups and their planes bending in the reaction (11.67). Instead of θ, the interatomic r_{CC} distance can be equally useful. (From [11.52].)

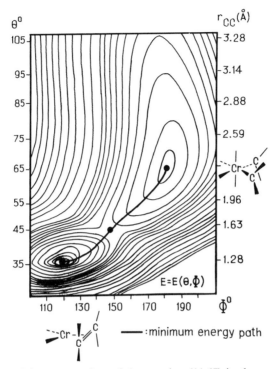

Figure 11.22. Potential energy surface of the reaction (11.67) in the space of the angles θ and ϕ characterizing the two methyl group least-motion convergence and their planes bending (Fig. 11.21). Contour lines at 0.1 eV separation. (From [11.52].)

With good π-donor substituents in the methyl groups CD_2, the situation changes drastically. The experimental data show that in this instance the bis-carbene d^6 complexes $M(CO)_4(CD_2)_2$ are stable compounds and can be obtained from the corresponding olefine. The calculations by the approach above confirm these results and reveal their origin. The corresponding potential energy surface in (θ, Φ) coordinates for

is illustrated in Fig. 11.23. It is seen that here there is only one minimum for the bis-carbene complex and no minimum for the olefin complex. The MO energy-level diagram [11.52] shows that the reaction under consideration is symmetry allowed and may proceed without energy barrier, the olefin complex

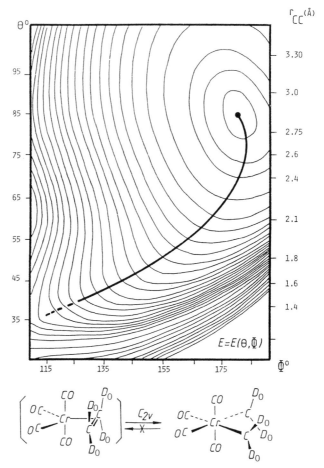

Figure 11.23. Potential energy surface for the reaction (11.68). The coordinates θ and ϕ (or r_{CC}) are given in Fig. 11.21. (After [11.52].)

being thermodynamically less stable (all the MO energy levels are lower in the bis-carbene complex). The reason for this result can be revealed from analysis of the MO energy-level reordering produced by the donor substitution.

CO Reductive Hydrogenation and Carbonyl Insertion Reactions

Insertion reactions of the type

$$
\begin{array}{c}
\text{L} \\
| \\
\text{R}-\text{MA}_n
\end{array}
\longrightarrow
\begin{array}{c}
\text{R}-\text{L} \\
| \\
\text{MA}_n
\end{array}
\qquad (11.69)
$$

where R is an unsaturated ligand and M—L is a metal–hydrogen or metal–carbon bond, are most important in modern organometallic chemistry as the key processes in many catalytic cycles. In this subsection we consider the examples of CO insertion into M—H and M—CH$_3$ bonds and a related reaction of coordinated CO hydrogenation [11.53, 11.54]. The next subsection is devoted to olefin insertion reactions.

Homogeneously catalyzed carbon monoxide hydrogenation underlies many processes, including conversion of the CO/H$_2$ mixture into liquid fuels and production of methanol, ethanol, ethylene glycol, and so on. The process can be presented as either a nucleophilic addition of a hydride ion,

$$L_nM—CO + H^- \rightarrow L_nM—C\begin{smallmatrix} O^- \\ \diagup \\ \diagdown \\ H \end{smallmatrix} \tag{11.70}$$

or as an insertion of a carbonyl ligand into the metal–hydride bond:

$$L_nM—CO \rightarrow L_nM—C\begin{smallmatrix} H \\ \diagup \\ \diagdown \\ O \end{smallmatrix} \underset{-L}{\overset{+L}{\rightleftarrows}} L_{n+1}M—C\begin{smallmatrix} H \\ \diagup \\ \diagdown \\ O \end{smallmatrix} \tag{11.71}$$

Reactions (11.70) and (11.71), despite an apparent similarity, are significantly different: There are many examples known of the first reaction, whereas examples of the second type, (11.71), are generally lacking, at least when resulting in a monohaptoformyl ligand. The hydride affinity of coordinated carbonyls is large (ranging from 45 to 55 kcal/mol) as compared with that of bare carbon monoxide (a few kcal/mol only), and this indicates the significant activation of the CO molecule by coordination to transition metals (Section 11.2).

To reveal the reaction mechanism, calculations of the AP energy surfaces in the space of possible reaction pathway for prototype reactions have been performed [11.53]. The calculations carried out in the *ab initio* SCF-MO-LCAO approximation (Section 5.3) with extended basis sets and other computational details are discussed in [11.53]. The role of electron correlation effects was also elucidated by means of additional CI calculations (Section 5.3).

For the hydrogenation reaction, the reaction of addition of H$^-$ to Fe(CO)$_5$ has been considered:

$$H^- + Fe(CO)_5 \rightarrow Fe(CO)_4(CHO)^- \tag{11.72}$$

The approach of the hydride is restricted by the plane containing the two axial and one equatorial ligand. The numerical results are illustrated in Fig. 11.24,

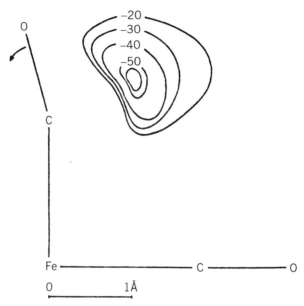

Figure 11.24. Contour lines of the AP energy surface for the approach of the hydride ion (energies in kcal/mol with respect to separated reactants). The inner contour is -52 kcal/mol. (After [11.53].)

and the optimized geometry is shown in Fig. 11.25. In the geometry calculations the axial Fe–C–O angle is allowed to relax, while the Fe—C and C—O distances are kept fixed at 1.81 Å and 1.22 Å, respectively.

Figure 11.24 shows explicitly that the reaction is highly exothermic (ca. 52 kcal/mol) and proceeds without any energy barrier. This is in agreement with the corresponding gas-phase studies [11.55]. The calculation details also explain the origin of the high affinity of the activated CO ligand to the H^- ion and the role of the metal d_{z^2} orbital in this process [11.53].

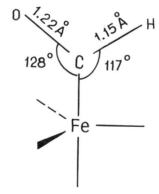

Figure 11.25. Optimized geometry of the hydrogenated system $Fe(CO)_4(CHO)^-$.

For the CO ligand insertion reaction (11.69) the prototype example is taken as follows:

$$HMn(CO)_5 \rightarrow Mn(CO_4)(CHO) \tag{11.73}$$

In principle, there are two main pathways for this reaction to proceed: (1) proper insertion of the CO molecule into the Mn—H bond, or (2) migration of the hydrogen atom to the CO ligand to form the CHO ligand. To reveal the true process, a two-dimensional AP energy surface as a function of two critical angles, α and β, has been calculated [11.53]. The two angles are chosen as shown in Fig. 11.26, while the two out-of-plane (perpendicular to the reaction plane) carbonyl ligands are kept frozen; the Mn—C and C—O bond lengths of the nonreacting in-plane CO ligands are also kept fixed at their experimental value. For each point of the surface (α, β) the Mn—H bond length (d), the angles γ and δ, and the C—O bond length were successively optimized, and then the Mn—C bond length (R) was optimized for representative points of the minimum energy path.

The results are illustrated in Fig. 11.27, and the successive geometries of the system along the resulting reaction path are given in Fig. 11.28. It is seen that the process is best described as a hydride migration and not CO insertion, in agreement with experimental data and other theoretical investigations.

Figure 11.28 shows that the CO groups almost do not change their geometry during the reaction. The geometry of the transition state D is near to that of a five-coordinated formyl intermediate F. The energy profile as a function of the H–Mn–C angle η for configurations of Fig. 11.28A to F is shown in Fig. 11.29. In the SCF approximation the energy barrier of the reactions is about 15 kcal/mol, with a small difference between D and F (about 4 kcal/mol). However, CI calculations with electron correlation effects carried out for the three key structures A, D, and F show a large increase in the reaction barrier of the forward reaction and its decrease (almost to zero) for the inverse reaction (Fig. 11.29). The analysis of computational details [11.53] show that the reason of this result is in the poor description of the difference

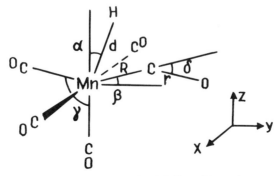

Figure 11.26. Variation coordinates for the CO ligand insertion reaction (11.73).

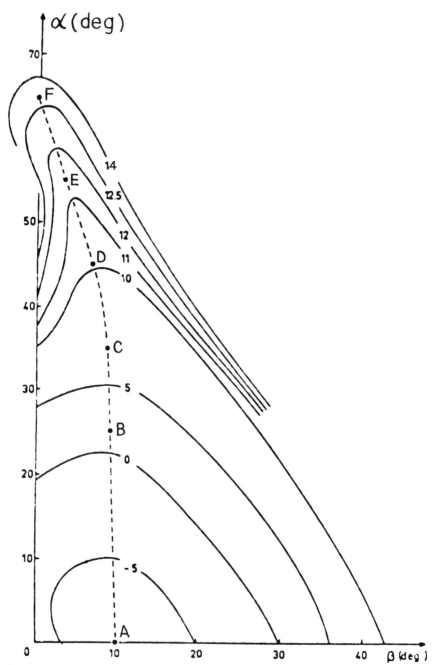

Figure 11.27. Contour lines of the AP energy surface for the reaction (11.73) as a function of the angles α and β specified in Fig. 11.26. The distance $R = 1.885\,\text{Å}$ is kept constant and the energies are read off the point $\alpha = \beta = 0$. (From [11.53].)

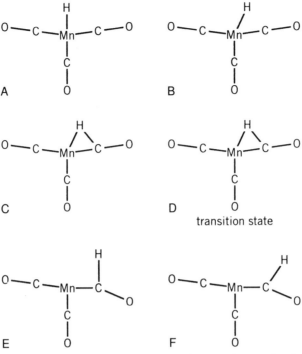

Figure 11.28. In-plane geometry changes along the reaction path of the hydride migration in the HMn(CO)₅ system (the two CO ligands perpendicular to the plane are not shown). (After [11.53].)

in the π back-donation to CO and CHO without CI. This feature is also inherent in other processes in which similar ligand changes occur. For the whole process HMn(CO)₅ + CO → Mn(CO)₅CHO in which the number of CO ligands is not changed, the calculated endothermicity 3.7 kcal/mol agrees satisfactorily with the experimental value of 5 kcal/mol.

Thus the failure to observe CO insertion into a metal hydride bond, resulting in a monohaptoformyl ligand, is well accounted for by the high-energy barrier and the close proximity of the transition state with the five-coordinated intermediate.

For a similar reaction of CO ligand insertion into a M—CH₃ bond,

$$\underset{\substack{\displaystyle | \\ \displaystyle \text{C} \\ \displaystyle | \\ \displaystyle \text{PH}_3}}{\overset{\substack{\displaystyle \text{O}}}{\text{CH}_3 - \text{M} - \text{H}}} \longrightarrow \underset{\substack{\displaystyle | \\ \displaystyle \text{PH}_3}}{\overset{\substack{\displaystyle \text{CH}_3 \diagdown \quad \diagup \text{O} \\ \displaystyle \text{C}}}{\text{M} - \text{H}}} \tag{11.74}$$

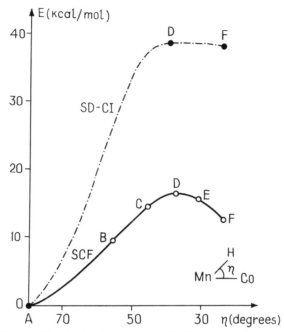

Figure 11.29. SCF and CI calculated energy profiles for the hydride migration process (Fig. 11.28) as a function of the H—Mn—C angle η.

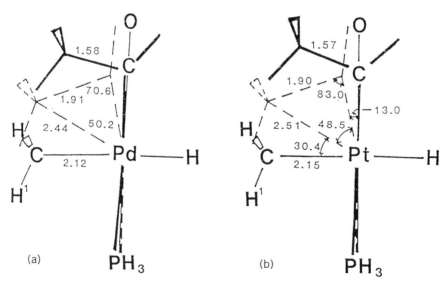

Figure 11.30. Optimized geometries (in Å and degrees) of $M(CH_3)H(CO)PH_3$ and $M(COCH_3)H(PH_3)$ (heavy solid lines), M = Pd, Pt, and the transition states (dashed lines) for the CO insertion (CH_3 migration) and inverse reaction. (After [11.54].)

with M = Pd, Pt, calculations were carried out [11.54] for electronic structure, energy, and geometry of the initial complex, activated state, and the product of the reaction. The authors use the *ab initio* SCF MO LCAO restricted Hartree–Fock (RHF) approximation with gradient technique for geometry optimization and perturbation theory (MP2) for electron correlation effects (Section 5.3), as well as relativistic effective core potentials for the metal atoms (Section 5.4).

The optimized geometries are illustrated in Fig. 11.30, while relative energies are given in Table 11.8. It is shown that, similar to the previous case of CO insertion into the M—H bond, where in fact there is no insertion but H migration, there is no CO insertion into the M—CH_3 bond either: The reaction proceeds as a CH_3 migration toward the CO ligand in accordance with experimental data. The energy of a similar transition state with CO migration instead of CH_3 [a configuration in which the angles (CO)–Pd–(H) and (CH_3)–Pd–(PH_3) of the true transition state in Fig. 11.30 are exchanged] is about 7 kcal/mol higher than for the true transition state.

The reason for this reaction pathway is in the strong repulsion of the carbonyl lone pair from the occupied d_{xy} orbital of the metal in the case of CO migration (Fig. 11.31a). This repulsion is significantly lower when the CH_3 group migrates (Fig. 11.31b) due to the lower electron density in the hybridized sp^3 orbital pointed toward the lobe of the metal d_{xy} orbital (approximately one electron instead of two electrons in the CO lone pair) and the longer distance M—CH_3.

Although the structural changes in the two reactions above with Pd and Pt are almost the same, the energetics is quite different (Table 11.8): The activation energy in the Pt complex is considerably higher than in Pd (at the MP2 level 21.8 and 13.5 kcal/mol, respectively). This is due to the differences in the M—CH_3 and M—CO bond energies, which are, respectively, by 7 and 11 kcal/mol larger in the Pt complex than in the Pd one, while the Pt—$COCH_3$

Table 11.8. Relative energies (in kcal/mol) of the transition states and products of the reaction (11.82), $M(CH_3)H(CO)(PH_3) \rightarrow M(COCH_3)H(PH_3)$ with M = Pd,Pt calculated without (RHF) and with (MP2) electron correlation effects

M	Method	Reactant	Transition State	Product
Pd	RHF	0.0	25.8	19.1
	MP2	0.0	13.5	8.8
Pt	RHF	0.0	31.3	23.0
	MP2	0.0	21.8	17.6

Source: [11.54].

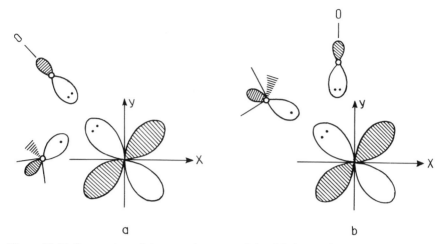

Figure 11.31. Interaction of the two electrons of the CO lone pair and one electron of the CH_3 sp^3 orbital with the metal occupied d_{xy} orbital in the CO migration (a), and in the true transition state of CH_3 migration (b).

Table 11.9. Relative energies (in kcal/mol) for the CO insertion (R migration) reaction
$Pd(R)H(CO)(PH_3) \rightarrow Pd(COR)H(PH_3)$

R	Reactant	Transition State	Product
CHF_2	0.0	40.9	31.2
C_2H_5	0.0	23.1	14.6

bond is stronger by 9 kcal/mol than Pd—$COCH_3$. Therefore, the Pd reaction is more endothermic by $7 + 11 - 9 = 9$ kcal/mol, and hence it has a higher energy barrier.

The reaction was also studied for CH_3-substituted ligands [11.54]. Some results for the energetics of these reactions obtained for the fixed ligand geometries (Fig. 11.30) are given in Table 11.9.

Olefin Insertion Reactions and Polymerization with Ziegler–Natta Catalysts

Olefin insertion reactions are of great importance in organometallic chemistry and chemical industry as basic models of the corresponding polymerization processes. Formally, olefin insertion is similar to the CO insertion considered above. For ethylene and its derivatives $CX_2{=}CH_2$ insertion into the M—H bond (and inversely, β-elimination), the following four reactions were investigated [11.54]:

$$
\begin{array}{ccc}
\begin{array}{c} CX_2{=}CH_2 \\ | \\ H{-}M{-}PH_3 \\ | \\ H \end{array}
&
\rightleftharpoons
&
\begin{array}{c} \quad\;\; CX_2{-}CH_2 \\ \diagup \quad\quad | \\ H \qquad\quad M{-}PH_3 \\ \qquad\quad | \\ \qquad\quad H \end{array}
\end{array}
\qquad (11.75)
$$

$$M = Ni, \quad X = H \qquad\qquad (11.76)$$

$$M = Pd, \quad X = H \qquad\qquad (11.77)$$

$$M = Pt, \quad X = H \qquad\qquad (11.78)$$

$$M = Pd, \quad X = F \qquad\qquad (11.79)$$

The method of numerical computation is the same RHF method with MP2 corrections mentioned above for the CO insertion reactions (for details see [11.54]). By means of this method the optimized geometries and energies of the transition states and products of the reaction were calculated. The results are illustrated in Table 11.10 and Fig. 11.32. It is seen that the reaction proceeds as a hydrogen migration via a four-center transition state; its optimized geometry is shown for the Pt reaction (11.78). For other reactions the transition state is similar, with the distinction that for Ni and Pd the transition state is achieved much earlier with smaller distortions of the M—H and C=C bonds (Ni produces smaller distortions than Pd) and corresponding structural differences in the products (see below). This is also seen from the energy profile of the reaction given by the energies in Table 11.10: The activation barrier is much lower for the Ni reaction (11.76) and increases for similar reactions (11.77) and (11.78) with Pd and Pt. On the other hand, the reaction for the Ni complex is also most exothermic.

These effects are due, first, to the strength of the $M{-}(C_2H_4)$ bond, which is very weak for Ni and increases for Pd and Pt. Therefore, the Pt–ethylene complex is more stable than the ethyl complex. The stability of the ethyl complex increases in the order Pt < Pd < Ni, while the barrier height for the

Table 11.10. Relative energies of the reaction
$M(H)_2(CH_2CX_2)(PH_3) \rightarrow M(CH_2CHX_2)(H)(PH_3)$ **(in kcal/mol)**

M	X	Ethylene Complex	Transition State	Ethyl Complex
Pd	H	0.0	8.0	−3.0
		0.0	5.1[a]	3.0[a]
	F	0.0	7.3	−11.1
Ni	H	0.0	0.6	−31.5
Pt	H	0.0	12.5	4.2

[a] Calculated including correlation effects in the MP2 approximation with RHF geometries.

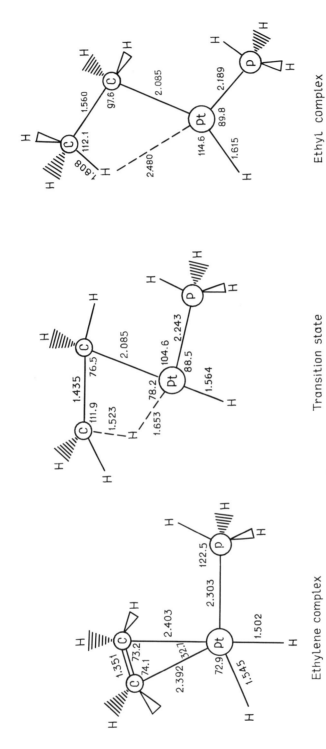

Ethylene complex   Transition state   Ethyl complex

Figure 11.32. Optimized geometries (distances in Å and angles in degrees) of ethylene and ethyl complexes and the transition state of the reaction $Pt(CH_2{=}CH_2)(H)_2PH_3 \rightarrow Pt(CH_3{-}CH_2)(H)PH_3$ (for simplicity, some of C—H bond lengths and angles are not indicated; see [11.54]).

migration reaction decreases in the order $Pt > Pd > Ni$. In general, the metal–ethylene bonds are relatively weak, indicating a strong *trans* influence of the hydride ligand.

As for the inverse reaction of β elimination, its activation energy depends mainly on two effects: (1) the remaining interaction of the migrating hydrogen H^β with the metal atom M via σ donation from the CH^β bonding orbital to the empty $d + p$ orbital of the free coordination place (the *inverse inter-action*), and (2) the back donation from the occupied metal d_{xy} orbital to the antibonding σ^* orbital of CH^β that weakens the bonds in the latter. The calculations [11.54] show that the inverse interaction is stronger in the Pd reaction (11.77) and decreases toward Pt and Ni. The back donation, on the contrary, is stronger in the Pt complex because of its more diffuse d_{xy} orbital and very poor in Ni, for a similar reason. The activation energy is thus a result of interplay of these two reasons, illustrated qualitatively in Table 11.11.

As mentioned above, insertion reactions underlie many catalytic processes. In particular, ethylene insertion into the $M\!-\!CH_3$ bond is the basic elementary act in the well-known Ziegler–Natta catalytic ethylene polymerization process:

$$TiCl_4 \cdot Al(CH_3)_3 + C_2H_4 \rightarrow TiCl_4Al(CH_3)_2C_3H_7 \qquad (11.80)$$

Cossee [11.56] first considered this process as ethylene coordination to Ti followed by a C_2H_4 insertion reaction into the $Ti\!-\!CH_3$ bond in the *cis* position, and brought up ligand field and semiempirical MO LCAO consider-ations in support of this reaction mechanism. A more detailed calculation of the energies along the reaction pathway and the activation energy with a detailed analysis of the electronic origin of the phenomenon was carried out in [11.57].

The schematic representation of the catalyst–ethylene complex and the notation of atoms are shown in Fig. 11.33. The reaction mechanism is assumed, in principle, similar to that suggested earlier by Cossee [11.56]; 12 steps, which

Table 11.11. Comparison of activation energies of the β-elimination reactions $M(CH_2CHX_2)(H)(PH_3) \rightarrow M(H)_2(CH_2CX_2)PH_3$ with a qualitative estimate of inverse and back-donative interaction

M	X	Activation Energy	Inverse Interaction	Back Donation
Pd	H	11.0	Excellent	Good
	F	18.4	Fair	Good
Ni	H	32.1	Good	Poor
Pt	H	8.3	Good	Excellent

Source: [11.54].

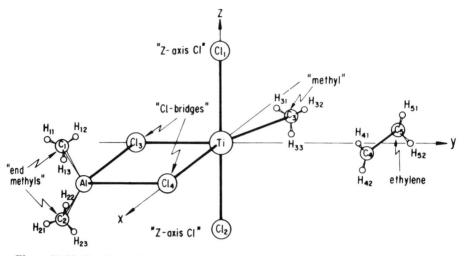

Figure 11.33. Catalyst–ethylene complex in the Ziegler–Natta catalytic polymerization of ethylene and the denotation of the atoms. (After [11.57].)

include ethylene bonding and CH_3 migration, are taken as the reaction pathway. Calculations were carried out by means of the *ab initio* all-electron SCF MO LCAO (RHF) approximation (Section 5.3). The parameters for the 12 steps and the calculated energies are given in Table 11.12, while the denotation of the corresponding angles are shown in Fig. 11.34. For each Ti–C_2H_4 distance, optimization of the other parameters is performed. The visual presentation of the reaction by means of molecular models is given in Fig. 11.35.

Figure 11.36 shows the evolution of the total energy of the system as a function of the reaction path presented by the 12 reaction steps above. It is seen that at the beginning of the ethylene approach to the complex, the energy lowers slightly, due to the metal–ethylene bond formation with a bonding energy of more than 4 kcal/mol. At step 6 with the Ti—C_2H_4 distance of 2.4 Å, the energy increases, and this increase is attributed mainly to the ethylene's π-bond breaking. The highest activation of the C=C bond is reached at step 7 [$R(\text{Ti}—C_4) = 2.24$ Å, $R(\text{Ti}—C_5) = 2.39$ Å], after which the bonding to the CH_3 group increases and the energy decreases; at step 8 the π bond finally breaks. This and many other details are seen explicitly from the MO populations and charge distribution, discussed thoroughly in [11.57].

Photochemical Reactions of Organometallics

From the computational point of view, photochemical reactions are more complicated because they involve excited states for which quantum-chemical calculations are in general more difficult. On the other hand, photochemical

Table 11.12. **Geometries and total energies for a sequence of configurations describing the pathway of reaction (11.80)**[a]

Reaction Step	$R(Ti-C_4)$ (Å)	$R(Ti-C_5)$ (Å)	$R(Ti-C_3)$ (Å)	$R(C_3-C_5)$ (Å)	α_R (°)	α_0 (°)	α_1 (°)	r (Å)	$\Delta E = E_i - E_1$ (kcal/mol)
1	7.03	7.03	2.15	5.55	45	0	117	1.337	0
2	5.04	5.04	2.15	3.58	45	0	117	1.337	−0.3
3	4.06	4.06	2.15	3.23	28	0	117	1.337	−1.9
4	3.08	3.08	2.15	2.71	13	8	115	1.383	−4.2
5	2.59	2.59	2.15	2.70	8	16	115	1.383	−3.0
6	2.40	2.43	2.15	2.69	0	19	114	1.406	+5.3
7	2.24	2.39	2.26	2.48	−2.3	23	113.25	1.428	15.0
8	2.18	2.49	2.36	2.31	−4.6	28	112.49	1.451	10.9
9	2.14	2.60	2.48	2.14	−6.9	34	111.74	1.473	1.1
10	2.124	2.73	2.59	1.95	−9.2	40.9	110.98	1.495	−14.2
11	2.126	2.88	2.65	1.75	−11.5	47.8	110.23	1.518	−34.5
12	2.15	3.03	2.81	1.54	−13.8	54.7	109.47	1.54	−40.3

[a] The energy is read off the first step with $E_1 = -3112.1681$ a.u. The geometrical parameters are defined as shown in Fig. 11.34 [11.57].

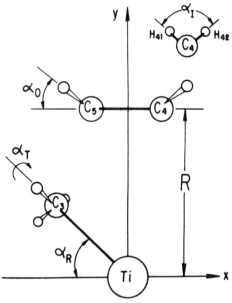

Figure 11.34. Geometrical variation parameters describing the 12 steps of Table 11.12 in the pathway of reaction (11.80).

reactions are rather widespread in chemistry. By way of examples demonstrating theoretical possibilities, the following reactions are considered [11.58]:

$$Fe(CO)_5 \xrightarrow{h\nu} Fe(CO)_4 + CO \tag{11.81}$$

$$HCo(CO)_4 \xrightarrow{h\nu} Co(CO)_4 + H \tag{11.82}$$

$$Mo(CO)_5L + L' \xrightarrow{h\nu} Mo(CO)_4LL' + CO \quad L = NH_3, PMe_3, C(OMe)H, C_2H_4 \tag{11.83}$$

The calculations were carried out by the *ab initio* SCF MO LCAO method with CI (Section 5.3). For all these reactions the AP energy curves of the ground and excited electronic states as a function of the corresponding coordinate of the ligand elimination were calculated and analyzed. Strictly speaking, these curves are cross sections of the multidimensional potential energy surface, with some additional symmetry and bond-length restrictions.

For reaction (11.81), the reaction path of equatorial CO ligand elimination is taken as shown in Fig. 11.37a maintaining C_{2v} symmetry of the product. For reaction (11.82), both the reactant and product $Co(CO)_4$ belong to C_{3v} symmetry (in their ground states), and therefore the reaction coordinate is taken as shown in Fig. 11.37b.

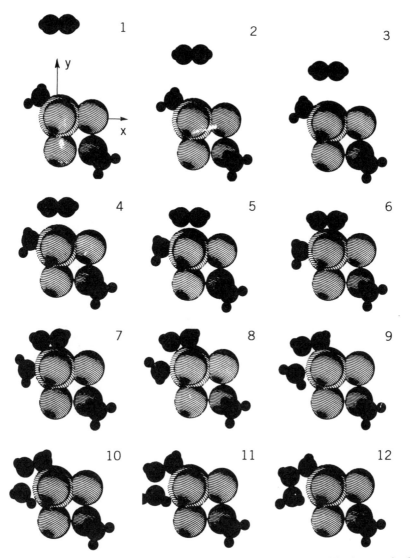

Figure 11.35. Molecular models for the 12 steps of the reaction (11.80) characterized in Table 11.12. (After [11.57].)

The results of the computations for these two reactions are illustrated in Figs. 11.38 and 11.39. It is seen that in both cases there are excited states that lead to ligand elimination without any energy barrier. However, not all the excited states evolve to the dissociation products in their ground state. A special feature of the two reactions is that their products have a triplet-spin ground state, $^3B_2$ and $^3A_1$ for reactions (11.81) and (11.82), respectively. This means that since the ground states are singlets, only singlet–triplet excitation

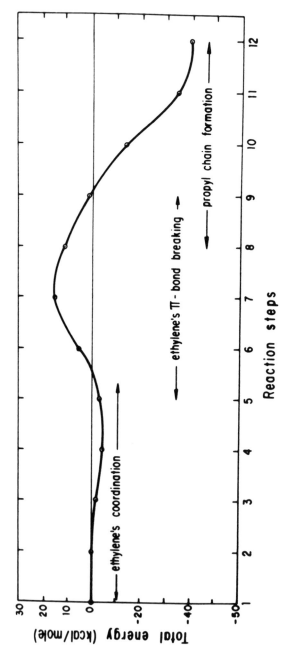

Figure 11.36. Evolution of the total energy (the reaction curve) in the Ziegler–Natta polymerization process (11.80) along a given pathway specified by the 12 steps in Table 11.12. (After [11.57].)

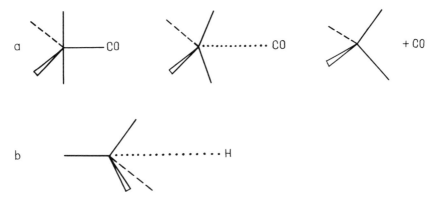

Figure 11.37. Reaction pathway for CO elimination from Fe(CO)$_5$ (a) and hydrogen dissociation from HCo(CO)$_4$ (b).

may lead to direct photolysis of the ligands above: $^1A_1 \rightarrow {}^3E'$ in CO photo-elimination from Fe(CO)$_5$, and $^1A_1 \rightarrow {}^3A_1$ in hydrogen elimination from HCo(CO)$_4$.

These electronic singlet → triplet transitions are spin forbidden (Section 8.1), and therefore they are difficult to be realized experimentally, especially in the presence of singlet–singlet transitions which are close in frequency and much more intensive (e.g., $^1A_1' \rightarrow {}^1E'$ in Fig. 11.38). Therefore, another mechanism of photolysis has been suggested [11.58]. It is illustrated in Fig. 11.40: The reactant is photoexcited to a singlet state and then, by way of radiationless transition (via spin–orbital coupling and thermal effects), the corresponding triplet state is populated. From the latter the molecule, according to Figs. 11.38

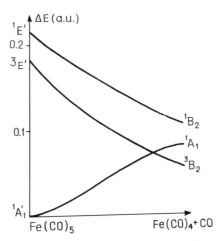

Figure 11.38. Potential energy curves for the ground and two excited states in the dissociation process Fe(CO)$_4$ + CO under the C_{2v} constraint. (From [11.58].)

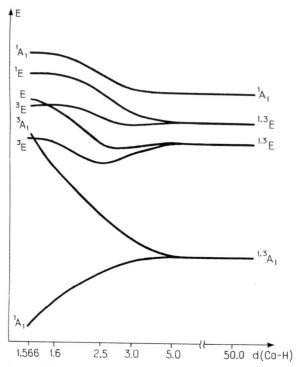

Figure 11.39. Potential energy curves (ground and excited) for dissociation of the Co—H bond in HCo(CO)$_4$ under C_{3v} constraint. (From [11.58].)

and 11.39, can dissociate to the products in the ground state without any energy barrier.

In fact, this mechanism of photolysis is valid (qualitatively) in many other photochemical reactions where homolysis of a σ bond takes place with the reactant in a singlet ground state and the products in a triplet state. In [11.58] some examples are listed: dissociation of hydrogen from metal hydrides HMn(CO)$_5$, HReCp$_2$, HW(CO)$_3$Cp, photochemical cleavage of the metal–carbon bond in RMn(CO)$_5$ with R = CH$_3$, C$_6$H$_5$, metal–silicon bond in R$_3$SiCo(CO)$_4$, R = Et, Ph, metal–metal bond in Mn$_2$(CO)$_{10}$, and so on.

The photosubstitution reactions (11.83) proceeds via a primary photodissociation of a CO ligand:

$$M(CO)_5L \rightarrow M(CO)_4L + CO \tag{11.84}$$

and subsequent thermal reaction of the unsaturated species M(CO)$_4$L with L':

$$M(CO)_4L + L' \rightarrow M(CO)_4LL' \tag{11.85}$$

One of the important features of this photoreaction is its stereospecificity: Its product M(CO)$_4$LL' is a *cis*-disubstituted derivative. Therefore, one of the

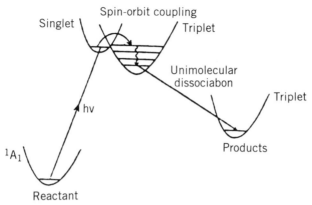

Figure 11.40. Mechanism of photochemical dissociation starting from a singlet ground state to triplet-state products.

goals of the calculations [11.58] is to explain this characteristic *cis* stereospecificity.

The photoelimination reaction (11.84) is more complicated than (11.81) or (11.82), and therefore the calculations have been restricted to a number of remarkable points of the potential energy surface. Figure 11.41 illustrates some

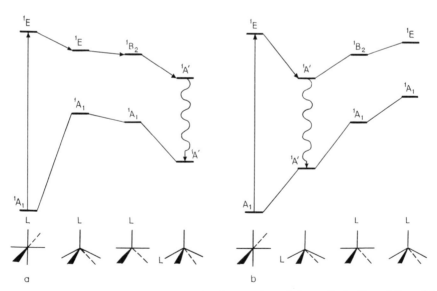

Figure 11.41. State correlation diagram for the reaction of photoelimination of the axial (*a*) or equatorial (*b*) carbonyl ligand from $Mo(CO)_5L$ and consequent internal rotation (the eliminated ligand is not shown in the products). The straight arrows show the evolution of the excited state to the configuration of minimum energy, while the waved arrows show the radiationless transitions to the ground state.

results of such calculations, showing the evolution of two important singlet states of the photoexcitation in the reaction and subsequent transformations [the final product $M(CO)_4LL'$, as a d^6 hexacoordinated complex, is assumed to have a singlet ground state].

The mechanism of the photoreaction of ligand substitution (11.83) emerging from these computations is as follows. By photoexcitation of the complex $M(CO)_5L$, the reaction of CO elimination takes place, resulting in a square-pyramidal structure with C_{4v} symmetry and the axial ligand eliminated (second scheme in Fig. 11.41a), or a pseudo-square-pyramidal structure if the equatorial ligand is eliminated (Fig. 11.41b).

In the axial elimination, the excited state in the C_{4v} configuration has no minimum energy (Fig. 11.41a): By an internal rotation (pseudorotation, or Berry rotation; Section 9.2) via an intermediate trigonal–bipyramidal structure, it transforms itself to the pseudo-square-pyramidal structure shown on the right-hand side of Fig. 11.41a. From this metastable state, a radiationless transition to the ground state takes place. In the case of equatorial ligand elimination the system happens to this minimum-energy pseudo-square-pyramidal structure directly, without additional transformations (Fig. 11.41b). It is seen that beginning with this configuration, the subsequent reaction (11.85) proceeds as an addition of L' to the only free coordination place of the photoinduced complex, resulting in a cis-bisubstituted system $M(CO)_4LL'$.

For another example of ab $initio$ calculated photochemical reaction, the photochemical cleavage of the metal–hydrogen bond in aluminum porphyrins, see [11.59].

REFERENCES

11.1. H. Eyring, J. Walter, and G. E. Kimball, *Quantum Chemistry*, Wiley, New York, 1947.

11.2. V. I. Goldanskii, V. A. Benderskii, and L. I. Trakhtenberg, *Adv. Chem. Phys.*, **75**, 349 (1989); Yu. I. Dakhnovskii and V. V. Nefedova, *Phys. Lett. A*, **157** 301 (1991).

11.3. J. Jortner and B. Pullman, Eds., *Tunnelling*, D. Reidel, Dordrecht, The Netherlands, 1986.

11.4. A. H. Zewail, *Femtochemistry. Ultrafast Dynamics of the Chemical Bond*, Vols. 1 and 2, World Scientific, Singapore, 1994; M. Gruebele and A. H. Zewail, *Phys. Today*, **43**, 24 (1990); P. R. Brooks, *Chem. Rev.*, **88**, 407 (1988).

11.5. I. B. Bersuker, *The Jahn–Teller Effect and Vibronic Interactions in Modern Chemistry*, Plenum Press, New York, 1984.

11.6. K. Fukui, *Theory of Orientation and Stereoselection*, Springer-Verlag, Berlin, 1975, 134 pp.

11.7. G. Clopman, in: *Chemical Reactivity and Reaction Paths*, ed. G. Clopman, Wiley, New York, 1974, Chap. 4.

11.8. D. G. Batyr, F. A. Spataru, S. S. Budnikov, and I. B. Bersuker, *Koord. Khim.*, **7**, 1632–1637 (1981).

11.9. R. B. Woodward and R. Hoffmann, *The Conservation of Orbital Symmetry*, Verlag Chemie, Weinheim, Germany, 1970.

11.10. R. F. W. Bader, *Can. J. Chem.*, **40**, 1164 (1962).

11.11. R. G. Pearson, *Symmetry Rules for Chemical Reactions*, Wiley-Interscience, New York, 1976.

11.12. T. F. George and J. Ross, *J. Chem. Phys.*, **55**, 3851 (1971).

11.13. F. D. Mango and J. H. Schachtschneider, *J. Am. Chem. Soc.*, **89**, 2484–2486 (1967); **93**, 1123–1130 (1971); F. D. Mango, *Adv. Catal.*, **20**, 291 (1969).

11.14. I. B. Bersuker, *Kinet. Katal.*, **18**, 1268–1282 (1977); *Chem. Phys.*, **31** , 85–93 (1978).

11.15. I. B. Bersuker, *(IUPAC) Coordination Chemistry*, Vol. 20, ed. D. Banerjea, Pergamon Press, Oxford, 1980, pp. 201–218

11.16. I. B. Bersuker and S. S. Budnikov, in: *Fundamental Research in Homogeneous Catalysis*, ed. A. E. Shilov, Gordon and Breach, London, 1986, Vol. 2, pp. 557–568.

11.17. S. S. Stavrov, I. P. Decusar, and I. B. Bersuker, *New J. Chem.*, **17**, 71 (1993).

11.18. L. D. Landau and E. M. Liphshitz, *Quantum Mechanics: Nonrelativistic Theory*, Nauka, Moscow, 1974, 752 pp.

11.19. L. Pauling, *J. Chem. Soc.*, 1461 (1948); in: *Proceedings of the Symposium on Coordination Chemistry*, Copenhagen, 1953, p. 25.

11.20. A. P. Svitin, S. S. Budnikov, I. B. Bersuker, and D. V. Korol'kov, *Teor. Eksp. Khim.*, **18**, 694–699 (1982); V. I. Bykov and A. P. Svitin, *Methods of Calculation of Parameters of Molecular Activation* (Russ.), Nauka, Novosibirsk, 1988, 211 pp.

11.21. Yu. Ya. Kharitonov and O. V. Bazileva, *Zh. Neorg. Khim.*, **23**, 867 (1978).

11.22. K. S. Krasnov, V. S. Timoshinin, T. G. Danilova, and S. V. Khandozhko, *Handbook of Molecular Constants for Inorganic Compounds* Israel Program for Scientific Translations, Jerusalem, 1970.

11.23. A. A. Radsig and V. M. Smirnov, *Handbook on Atomic and Molecular Data* (Russ.), Atomizdat, Moscow, 1980.

11.24. V. I. Yakerson, L. I. Lafer, and A. M. Rubinshtein, in: *Problemi Kinetiki i Kataliza*, Vol. 16, Nauka, Moscow, 1975, p. 49.

11.25. V. N. Kondratiev, *Constants for Gas Phase Reactions* (Russ.), Nauka, Moscow, 1970. .

11.26. A. Belyanskii and G. Dereni, in: *Electronic Effects in Adsorption and Catalysis* (Russ.), ed. F. F. Vol'kenstein, Mir, Moscow, 1969, pp. 227–251.

11.27. R. F. Fenske and K. L. LeCock, *Inorg. Chem.*, **11**, 437 (1972).

11.28. R. Rericha, *Coll. Czech. Chem. Commun.*, **40**, 2577–2591 (1975).

11.29. A. A. Efremov, U. A. Zokhov, and A. A. Davidov, *Kinet. Katal.*, **22**, 702–705 (1981).

11.30. R. Rericha, *J. Mol. Struct.* (*Theochem.*), **227**, 305–310 (1991).

11.31. S. S. Budnikov, *Teor. Eksp. Khim.*, **28**, 34–38 (1992).

11.32. I. B. Bersuker, *Teor. Eksp. Khim.*, **1**, 5 (1965).

11.33. G. Canley, D. Cross, and P. Knowless, *J. Chem Soc. Chem. Commun.*, **20**, 837 (1976).

11.34. I. B. Bersuker, Ed., *The Jahn–Teller Effect: A Bibliographic Review*, IFI/Plenum, New York, 1984, 589 pp.

11.35. M. J. C. Dewar, S. Kirschner, H. W. Kollmar, and L. E. Wade, *J. Am. Chem. Soc.*, **96**, 5242 (1974).

11.36. M. S. Gordon and J. W. Caldwell, *J. Chem. Phys.*, **70**, 5503 (1979).

11.37. M. D. Glick, J. M. Kusray, and J. E. Edincott, *J. Am. Chem. Soc.*, **95**, 5097 (1973).

11.38. B. G. Vekhter and M. L. Rafalovich, *Chem. Phys.*, **21**, 21 (1977).

11.39. D. B. Moss, Lin Chin-tung, and D. B. Rorabacher, *J. Am. Chem. Soc.*, **95**, 5179 (1973).

11.40. I. I. Chernyaev, in: *Selected Works: Complex Compounds of Platinum* (Russ.), Nauka, Moscow, 1973.

11.41. F. R. Hartley, *Rev. Chem. Soc.*, **2**, 163–179 (1973).

11.42. E. M. Shustorovich, M. A. Porai-Koshitz, and Iu. A. Buslaev, *Coord. Chem. Rev.*, **17**, 1 (1975).

11.43. V. I. Nefedov and M. M. Hofman, *Mutual Influence of Ligands in Inorganic Compounds* (Russ.), in: *Itogi Nauki i Tekhniki, Ser.: Inorganic Chemistry*, Vol. 6, VINITI, Moscow, 1978, 192 pp.

11.44. J. D. Atwood, *Inorganic and Organometallic Reaction Mechanisms*, Brooks/Cole, Monterey, Calif., 1985.

11.45. F. A. Cotton and G. Wilkinson, *Advanced Inorganic Chemistry: A Comprehensive Text*, Wiley, New York, 1992.

11.46. J. Chatt, L. A. Duncanson, and L. M. Venanzi, *J. Chem. Soc.*, 4456–4469 (1955).

11.47. L. E. Orgel, *J. Inorg. Nucl. Chem.*, **2**, 137–140 (1956).

11.48. I. B. Bersuker, *Zh. Neorg. Khim.*, **9**, 36–41 (1964); *Zh. Strukt. Khim.*, **4**, 461 (1963).

11.49. Z. Lin and M. B. Hall, *Inorg. Chem.*, **30**, 646–651 (1991).

11.50. J. –M. Lehn, in: *Perspectives in Coordination Chemistry*, ed. A. F. Williams, C. Floriani, and A. E. Merbach, VCH, New York, 1992, pp. 447–462; G. Denti, S. Serroni, S. Campagna, A. Juris, M. Ciano, and V. Balzani, *ibid.*, pp. 153–164.

11.50a. D. H. Bush and N. A. Stephenson, *Coord. Chem. Rev.*, **100**, 119–154 (1990).

11.51. R. Hoffmann, C. N. Wilker, and O. Eisenstein, *J. Am. Chem. Soc.*, **104**, 632 (1982); C. N. Wilker, R. Hoffman, and O. Eisenstein, *New J. Chem.*, **7**, 535 (1983).

11.52. R. Hoffmann, in: *Quantum Chemistry: The Challenge of Transition Metals and Coordination Chemistry*, ed. A. Veillard, D. Reidel, Dordrecht, The Netherlands, 1986, pp. 253–276.

11.53. A. Dedieu and S. Nakamura, in: *Quantum Chemistry: The Challenge of Transition Metals and Coordination Chemistry*, ed. A. Veillard, D. Reidel, Dordrecht, The Netherlands, 1986, pp 277–299.

11.54. N. Koga and K. Morokuma, in: *Quantum Chemistry: The Challenge of Transition Metals and Coordination Chemistry*, ed. A. Veillard, D. Reidel, Dordrecht, The Netherlands, 1986, pp. 351–361; *New J. Chem.*, **15**, 749–755 (1991).

11.55. K. R. Lane, L. Sallons, and R. R. Squires, *J. Am. Chem. Soc.*, **107**, 5369 (1985).

11.56. P. Cossee, *J. Catal.*, **3**, 80–88 (1964).

11.57. O. Novaro, E. Blaisten-Barojas, E. Clementi, G. Giunchi, and M. E. Ruiz-Vizcaya, *J. Chem. Phys.*, **68**, 2337–2351 (1978).

11.58. C. Daniel and A. Veillard, in: *Quantum Chemistry: The Challenge of Transition Metal and Coordination Chemistry*, ed. A. Veillard, D. Reidel, Dordrecht, The Netherlands, 1986, pp. 363–375.

11.59. M. M. Rohmer and A. Veillard, *New J. Chem.*, **15**, 795–799 (1991)

Appendixes

Appendix 1: Tables of Characters of Irreducible Representations of the Most Usable Symmetry Point Groups (Chapter 3) and Direct Products of Some Representations

The Cartesian coordinates x, y, and z and some of their combinations, as well as rotations around the axes R_x, R_y, and R_z which belong to the corresponding representation, are also indicated; for degenerate representations the corresponding degenerate combinations are shown in parentheses.

Table A1.1. Point groups C_s, C_2, and C_i

C_i		C_2		C_s		E	I
						E	C_{2z}
						E	σ_z
A_g	$R_x, R_y, R_z,$ $x^2, y^2, z^2,$ xy, xz, yz	A	$z, R_z, x^2, y^2, z^2, xy$	A'	x, y, R_z, z^2, xy	1	1
A_u	x, y, z	B	$x, y, R_x, R_y,$ xz, yz	A''	z, R_x, R_y	1	-1

Table A1.2. Point groups C_{2h} and C_{2v}

C_{2h}		C_{2v}		E	C_2	σ_h	I
				E	C_2	σ_v	σ_v'
A_g	R_z, x^2, y^2, z^2, xy	A_1	z, x^2, y^2, z^2	1	1	1	1
B_g	R_x, R_y, xz, yz	B_2	y, R_x, yz	1	-1	-1	1
A_u	z	A_2	R_z, xy	1	1	-1	-1
B_u	x, y	B_1	x, R_y, xz	1	-1	1	-1

Table A1.3. Point groups C_{3v} and D_3

| C_{3v} | | D_3 | | E | $2C_3$ | $3\sigma_u$ |
				E	$2C_3$	$3C_2$
A_1	z, x^2+y^2, z^2	A_1	x^2+y^2, z^2	1	1	1
A_2	R_z	A_2	z, R_z	1	1	-1
E	$(x, y), (R_x, R_y),$ $(x^2-y^2, xy),$ (xz, yz)	E	$(x, y), (R_x, R_y)$	2	-1	0

Table A1.4. Point group D_{3d}

D_{3d}		E	$2C_3$	$3C_2$	I	$2S_6$	$3\sigma_d$
A_{1g}	x^2+y^2, z^2	1	1	1	1	1	1
A_{1u}		1	1	1	-1	-1	-1
A_{2g}	R_z	1	1	-1	1	1	-1
A_{2u}	z	1	1	-1	-1	-1	1
E_g	(R_x, R_y)	2	-1	0	2	-1	0
E_u	(x, y)	2	-1	0	-2	1	0

Table A1.5. Point groups C_{6v} and D_{3h}

| C_{6v} | | D_{3h} | | E | C_2 | $2C_3$ | $2C_6$ | $3\sigma_v$ | $3\sigma_v'$ |
				E	σ_h	$2C_3$	$2S_3$	$3C_2$	$3\sigma_v$
A_1	z, x^2+y^2, z^2	A_1'	x^2+y^2, z^2	1	1	1	1	1	1
A_2	R_z	A_2'	R_z	1	1	1	1	1	1
B_2		A_1''		1	-1	1	-1	1	-1
B_1		A_2''	z	1	-1	1	-1	-1	1
E_2	(x^2-y^2, xy)	E'	$(x, y), (x^2-y^2, xy)$	2	2	-1	-1	0	0
E_1	$(x, y), (R_x, R_y),$ (xz, yz)	E''	$(R_x, R_y), (xz, yz)$	2	-2	-1	1	0	0

Table A1.6. Point group D_{2h}

D_{2h}		E	C_{2z}	C_{2y}	C_{2x}	I	σ_z	σ_y	σ_x
A_g	x^2, y^2, z^2	1	1	1	1	1	1	1	1
A_u	xyz	1	1	1	1	-1	-1	-1	-1
B_{1g}	R_z, xy	1	1	-1	-1	1	1	-1	-1
B_{1u}	z	1	1	-1	-1	-1	-1	1	1
B_{2g}	R_y, xz	1	-1	1	-1	1	-1	1	-1
B_{2u}	y	1	-1	1	-1	-1	1	-1	1
B_{3g}	R_x, yz	1	-1	-1	1	1	-1	-1	1
B_{3u}	x	1	-1	-1	1	-1	1	1	-1

Table A1.7. Point groups C_{4v} and D_{2d}

C_{4v}					E	C_2	$2C_4$	$2\sigma_v$	$2\sigma'_v$
		D_{2d}			E	C_2	$2S_4$	$2C'_2$	$2\sigma_d$
A_1	$z, x^2 + y^2, z^2$	A_1	$x^2 + y^2, z^2$		1	1	1	1	1
A_2	R_z	A_2	R_z		1	1	1	-1	-1
B_1	$x^2 - y^2$	B_1	$x^2 - y^2$		1	1	-1	1	-1
B_2	xy	B_2	z, xy		1	1	-1	-1	1
E	$(x, y), (R_x, R_y),$ (xz, yz)	E	$(x, y), (R_x, R_y),$ (xz, yz)		2	-2	0	0	0

Table A1.8. Point group D_{4h}

D_{4h}		E	$2C_4$	C_2	$2C'_2$	$2C''_2$	I	$2S_4$	σ_z	$2\sigma_v$	$2\sigma_d$
A_{1g}	$x^2 + y^2, z^2$	1	1	1	1	1	1	1	1	1	1
A_{1u}		1	1	1	1	1	-1	-1	-1	-1	-1
A_{2g}	R_z	1	1	1	-1	-1	1	1	1	-1	-1
A_{2u}	z	1	1	1	-1	-1	-1	-1	-1	1	1
B_{1g}	$x^2 - y^2$	1	-1	1	1	-1	1	-1	1	1	-1
B_{1u}		1	-1	1	1	-1	-1	1	-1	-1	1
B_{2g}	xy	1	-1	1	-1	1	1	-1	1	-1	1
B_{2u}		1	-1	1	-1	1	-1	1	-1	1	-1
E_g	$(R_x, R_y), (xz, yz)$	2	0	-2	0	0	2	0	-2	0	0
E_u	(x, y)	2	0	-2	0	0	-2	0	2	0	0

Table A1.9. Point group $C_{\infty v}$

$C_{\infty v}$		E	$2C_\infty(\phi)^a$	$\cdots$	$\infty \sigma_v$
$A_1 \equiv \Sigma^+$	$z, x^2 + y^2, z^2$	1	1	$\cdots$	1
$A_2 \equiv \Sigma^-$	R_z	1	1	$\cdots$	-1
$E_1 = \Pi$	$(x, y), (R_x, R_y), (xz, yz)$	2	$2\cos\phi$	$\cdots$	0
$E_2 = \Delta$	$(x^2 - y^2, xy)$	2	$2\cos 2\phi$	$\cdots$	0
$E_3 = \Phi$		2	$2\cos 3\phi$	$\cdots$	0
$\cdots$		$\cdots$	$\cdots$	$\cdots$	$\cdots$

Table A1.10. Tetrahedral point group T_d

T_d		E	$8C_3$	$3C_2$	$6S_4$	$6\sigma_d$
A_1	$x^2 + y^2 + z$	1	1	1	1	1
A_2		1	1	1	-1	-1
E	$(2z^2 - x^2 - y^2, x^2 - y^2)$	2	-1	2	0	0
T_1	(R_x, R_y, R_z)	3	0	-1	1	-1
T_2	$(x, y, z), (xy, xz, yz)$	3	0	-1	-1	1

Table A1.11. Octahedral point group O_h

O_h		E	$8C_3$	$3C_2(=C_4^2)$	$6C_4$	$6C_2$	I	$8S_6$	$3\sigma_h$	$6S_4$	$6\sigma_d$
A_{1g}	$x^2+y^2+z^2$	1	1	1	1	1	1	1	1	1	1
A_{1u}		1	1	1	1	1	-1	-1	-1	-1	-1
A_{2g}		1	1	1	-1	-1	1	1	1	-1	-1
A_{2u}		1	1	1	-1	-1	-1	-1	-1	1	1
E_g	$(2z^2-x^2-y^2, x^2-y^2)$	2	-1	2	0	0	2	-1	2	0	0
E_u		2	-1	2	0	0	-2	1	-2	0	0
T_{1g}	(R_x, R_y, R_z)	3	0	-1	1	-1	3	0	-1	1	-1
T_{1u}	(x, y, z)	3	0	-1	1	-1	-3	0	1	-1	1
T_{2g}	(xy, xz, yz)	3	0	-1	-1	1	3	0	-1	-1	1
T_{2u}		3	0	-1	-1	1	-3	0	1	1	-1

Table A1.12. Icosahedral group I_h

I_h		E	$12C_5$	$12C_5^2$	$20C_3$	$15C_2$	I	$12S_{10}$	$12S_{10}^3$	$20S_6$	15σ
A_g	$x^2+y^2+z^2$	1	1	1	1	1	1	1	1	1	1
A_u		1	1	1	1	1	-1	-1	-1	-1	-1
T_{1g}	(R_x, R_y, R_z)	3	$(1+\sqrt{5})/2$	$(1-\sqrt{5})/2$	0	-1	3	$(1-\sqrt{5})/2$	$(1+\sqrt{5})/2$	0	-1
T_{1u}	(x, y, z)	3	$(1+\sqrt{5})/2$	$(1-\sqrt{5})/2$	0	-1	-3	$-(1-\sqrt{5})/2$	$-(1+\sqrt{5})/2$	0	1
T_{2g}		3	$(1-\sqrt{5})/2$	$(1+\sqrt{5})/2$	0	-1	3	$(1+\sqrt{5})/2$	$(1-\sqrt{5})/2$	0	-1
T_{2u}		3	$(1-\sqrt{5})/2$	$(1+\sqrt{5})/2$	0	-1	-3	$-(1+\sqrt{5})/2$	$-(1-\sqrt{5})/2$	0	1
G_g		4	-1	-1	1	0	4	-1	-1	1	0
G_u		4	-1	-1	1	0	-4	1	1	-1	0
H_g	$(2z^2-x^2-y^2, x^2-y^2,$	5	0	0	-1	1	5	0	0	-1	1
	$xy, xz, yz)$										
H_u		5	0	0	-1	1	-5	0	0	1	-1

Table A1.13. Point group O' (octahedral double group)

Mulliken Notation	Bethe Notation	E	Q	$4C_3$ $4C_3^2Q$	$4C_3^2$ $4C_3Q$	$3C_4^2$ $3C_4^2Q$	$3C_4$ $3C_4^3Q$	$3C_4^3$ $3C_4Q$	$6C_2$ $6C_2Q$
A_1'	Γ_1	1	1	1	1	1	1	1	1
A_2'	Γ_2	1	1	1	1	1	-1	-1	-1
E'	Γ_3	2	2	-1	-1	2	0	0	0
T_1'	Γ_4	3	3	0	0	-1	1	1	-1
T_2'	Γ_5	3	3	0	0	-1	-1	-1	-1
E_2'	Γ_6	2	-2	1	-1	0	$\sqrt{2}$	$-\sqrt{2}$	0
E_3'	Γ_7	2	-2	1	-1	0	$-\sqrt{2}$	$\sqrt{2}$	0
G'	Γ_8	4	-4	-1	1	0	0	0	0

Table A1.14. Direct products of irreducible representations of simple groups $\Gamma_i \times \Gamma_j$ (I), simple with double groups $\Gamma_i \times \Gamma_\alpha$ (II), and double with double groups $\Gamma_\alpha \times \Gamma_\beta$ (III) presented as a sum of Γ_i

I. $\Gamma_i \times \Gamma_j$

$$\Gamma_1 \times \Gamma_i = \Gamma_i, \qquad \Gamma_2 \times \Gamma_2 = \Gamma_1$$
$$\Gamma_2 \times \Gamma_3 = \Gamma_3, \qquad \Gamma_2 \times \Gamma_4 = \Gamma_5$$
$$\Gamma_2 \times \Gamma_5 = \Gamma_4, \qquad \Gamma_3 \times \Gamma_3 = \Gamma_1 + \Gamma_2 + \Gamma_3$$
$$\Gamma_3 \times \Gamma_4 = \Gamma_4 + \Gamma_5, \qquad \Gamma_3 \times \Gamma_5 = \Gamma_4 + \Gamma_5$$
$$\Gamma_4 \times \Gamma_4 = \Gamma_1 + \Gamma_3 + \Gamma_4 + \Gamma_5$$
$$\Gamma_4 \times \Gamma_5 = \Gamma_2 + \Gamma_3 + \Gamma_4 + \Gamma_5$$
$$\Gamma_5 \times \Gamma_5 = \Gamma_1 + \Gamma_3 + \Gamma_4 + \Gamma_5$$

II. $\Gamma_i \times \Gamma_\alpha$

$$\Gamma_1 \times \Gamma_6 = \Gamma_2 \times \Gamma_7 = \Gamma_6, \qquad \Gamma_2 \times \Gamma_6 = \Gamma_1 \times \Gamma_7 = \Gamma_7$$
$$\Gamma_3 \times \Gamma_6 = \Gamma_3 \times \Gamma_7 = \Gamma_8, \qquad \Gamma_3 \times \Gamma_8 = \Gamma_6 + \Gamma_7 + \Gamma_8$$
$$\Gamma_4 \times \Gamma_6 = \Gamma_5 \times \Gamma_7 = \Gamma_6 + \Gamma_8$$
$$\Gamma_5 \times \Gamma_6 = \Gamma_4 \times \Gamma_7 = \Gamma_7 + \Gamma_8$$
$$\Gamma_4 \times \Gamma_8 = \Gamma_5 \times \Gamma_8 = \Gamma_6 + \Gamma_7 + 2\Gamma_8$$

III. $\Gamma_\alpha \times \Gamma_\beta$

$$\Gamma_6 \times \Gamma_6 = \Gamma_7 \times \Gamma_7 = \Gamma_1 + \Gamma_4$$
$$\Gamma_6 \times \Gamma_7 = \Gamma_2 + \Gamma_5$$
$$\Gamma_6 \times \Gamma_8 = \Gamma_7 \times \Gamma_8 = \Gamma_3 + \Gamma_4 + \Gamma_5$$
$$\Gamma_8 \times \Gamma_8 = \Gamma_1 + \Gamma_2 + \Gamma_3 + 2\Gamma_4 + 2\Gamma_5$$

Appendix 2: General Expressions for the Matrix Element $V_{mm'}$ of Perturbation of the States of One d Electron in Crystal Fields of Arbitrary Symmetry

For the calculation of the matrix element $V_{mm'}$ after (4.8) we use the following expansion (2.37):

$$\frac{1}{|\mathbf{r} - \mathbf{R}_i|} = \sum_{k=0}^{\infty} K_k(r, R_i) P_k(\cos \gamma_i)$$

where $K_k(r, R_i)$ is given by Eq. (2.38), and $\cos \gamma_i = \cos \vartheta \cos \vartheta_i + \sin \vartheta \sin \vartheta_i \cos(\phi - \phi_i)$. Then

$$V_{mm'} = \int \Psi_m^* V \Psi_{m'} \, d\tau = \sum_{i=0}^{N} eq_i \sum_{k=0}^{\infty} F_k(R_i) \theta_k^{mm'}(\vartheta_i, \phi_i) \tag{A2.1}$$

where

$$F_k(R) = R^{-(k+1)} \int_0^R r^k R_{n2}^2(r) r^2 \, dr + R^k \int_R^{\infty} r^{-(k+1)} R_{n2}^2(r) r^2 \, dr \tag{A2.2}$$

$$\theta_k^{mm'} = \int_0^{\pi} \int_0^{2\pi} Y_2^m(\vartheta, \phi) Y_2^{m'}(\vartheta, \phi) P_k(\cos \gamma_i) \sin \vartheta \, d\vartheta \, d\phi \tag{A2.3}$$

To simplify the last equation the expansion of the product of spherical functions $Y_2^m Y_2^{m'}$ in spherical functions Y_1^m can be employed [A2.1, p. 60]:

$$Y_2^m(\vartheta, \phi) Y_2^{m'*}(\vartheta, \phi) = \frac{(-1)^{m'}}{(4\pi)^{1/2}} \left[\tfrac{5}{3} C_{-m'm}^{224} C_{00}^{220} Y_4^{m-m'}(\vartheta, \phi) \right.$$

$$\left. + \sqrt{5}\, C_{-m'm}^{222} C_{00}^{222} Y_2^{m-m'}(\vartheta, \phi) + 5 C_{-m'm}^{220} C_{00}^{220} Y_0^{m-m'}(\vartheta, \phi) \right] \tag{A2.4}$$

where $C_{\alpha\beta}^{abc}$ denotes the Clebsh–Gordan coefficients which are tabulated [A2.2]. By substituting Eq. (A2.4) into (A2.3), we obtain integrals that are calculable directly by means of the following formula of addition of spherical functions:

$$\int Y_l(\vartheta, \phi) P_n(\cos \gamma_i) \sin \vartheta \, d\vartheta = \frac{4\pi}{2n + 1} Y_l(\vartheta_i, \phi_i) \delta_{ln} \tag{A2.5}$$

In so doing, we obtain

$$\theta_n^{mm'}(\vartheta_i, \phi_i) = A_{mm'} Y_4^{m-m'}(\vartheta_i, \phi_i) \delta_{n4} + B_{mm'} Y_2^{m-m'}(\vartheta_i, \phi_i) \delta_{n2} + D_{mm'} Y_0^{m-m'}(\vartheta_i, \phi_i) \delta_n \tag{A2.6}$$

where

$$A_{mm'} = (-1)^{m'} \frac{10\sqrt{\pi}}{27} C_{-m'm}^{224} C_{00}^{224}$$

$$B_{mm'} = (-1)^{m'} \frac{2\sqrt{\pi}}{5} C_{-m'm}^{222} C_{00}^{222} \tag{A2.7}$$

$$D_{mm'} = (-1)^{m'} (10\sqrt{\pi})\, C_{-m'm}^{220} C_{00}^{220}$$

Owing to the δ symbols, only three terms in the infinite sum (A2.1) are nonzero, and only the summation over the N nuclei remains:

$$V_{mm'} = \sum_{i=1}^{N} eq_i [A_{mm'} F_4(R_i) Y_4^{m-m'}(\vartheta_i, \phi_i) + B_{mm'} F_2(R_i) Y_2^{m-m'}(\vartheta_i, \phi_i)$$

$$+ D_{mm'} F_0(R_i) Y_0^{m-m'}(\vartheta_i, \phi_i) \qquad (A2.8)$$

This is the general formula for the matrix element of the perturbation of d states (2.1) by crystal fields.

The functions $F_k(R)$ after (A2.2) depend on the radial part of the d functions. For instance, taking the $3d$ function as a hydrogenlike one with an effective screening parameter α (Section 2.1),

$$R_{32}(r) = \frac{4\sqrt{10}}{3} \alpha^{7/2} r^2 e^{-\alpha r} \qquad (A2.9)$$

we obtain

$$F_k^{3d}(R) = \frac{\alpha}{360} \left\{ \left[\frac{(k+6)!}{(2x)^{k+1}} \right] - (2x)^6 [A_{k+6}(2x) - A_{5-k}(2x) \right\} \qquad (A2.10)$$

where $x = \alpha R$ and

$$A_n(y) = \int_1^{\infty} \exp(-yr) r^n \, dr \qquad (A2.11)$$

are auxiliary function often used in quantum-chemical calculations (tables of $A_n(y)$ see in [A2.3]). If n is an integer, the integrals (A2.11) can be calculated directly, yielding analytical expressions for the functions $F_k(R)$. In particular, for F_0, F_2, and F_4 in (A2.8), we have

$$F_0 = \alpha \left\{ x^{-1} - e^{-2x} \left[x^{-1} + \frac{5}{3} + \frac{4x}{3} + \frac{2x^2}{3} + \frac{2x^3}{9} + \frac{2x^4}{45} \right] \right\} \qquad (A2.12)$$

$$F_2 = \alpha \left[\frac{14}{x^3} - e^{-2x} \left(\frac{14}{x^3} + \frac{28}{x^2} + \frac{28}{x} + \frac{56}{3} + \frac{28x}{3} + \frac{11x^2}{3} + \frac{10x^3}{9} + \frac{2x^4}{9} \right) \right]$$

$$(A2.13)$$

$$F_4 = \alpha \left[\frac{315}{x^5} - e^{-2x} \left(\frac{315}{x^5} + \frac{630}{x^4} + \frac{630}{x^3} + \frac{420}{x^2} + \frac{210}{x} + 84 + 28x \right. \right.$$

$$\left. \left. + 8x^2 + 2x^3 + \frac{2x^4}{5} \right) \right] \qquad (A2.14)$$

The values $F_k(R)$ are tabulated in [A2.4].

REFERENCES

A2.1. G. Ya. Liubarskii, *The Theory of Groups and Its Application in Physics*, Gostek-hteorizdat, Moscow, 1957, 354 pp.

A2.2. A. Simon, in: *Deformation of Atomic Nuclei*, Oak Ridge National Laboratory, Oak Ridge, Tenn., 1957, pp. 353–379.

A2.3. M. Kotani, A. Amemiya, E. Ishiguro et al., *Tables of Molecular Integrals*, Maruzen, Tokyo, 1955, 230 pp; H. Preuss, *Integraltafeln zur Quantumchemie*, Bd. 1, Springer-Verlag, Berlin, 1956, 162 S; Bd. 2, 1957, 143 S; Bd. 3, 1961, 363 S; Bd. 4, 1960, 145 S.

A2.4. C. J. Ballhausen and E. M. Ancmon, *Mat. Fys. Medd. Dan. Vid. Selsk.*, **31**, 1–38 (1958).

Appendix 3: Calculation of the Destabilization and Splitting of the States of One d Electron in Crystal Fields of Different Symmetries

The calculations of the d-electron energy-level splitting in the field of ligand–point charges performed in Section 4.2 can be extended to the case of ligand–dipoles. Consider the dipole system as two charges $-q$ and $+q$ situated at a distance ΔR from each other and assume that in the complex the dipoles of the ligands are oriented along the metal–ligand bonds with the negative pole toward the CA. Performing the calculation of the perturbation of the d states twice, first in the field of six negative charges $-q$ at distances R from the CA and then in the field of similar positive charges at a larger distance $R + \Delta R$, and summing up the results, we have [cf. (4.14)]

$$\varepsilon_1 = \varepsilon_2 = eq\{[6F_0(R) + F_4(R)] - [6F_0(R + \Delta R) + F_4(R + \Delta R)]\}$$

$$\varepsilon_1 = \varepsilon_4 = \varepsilon_5 = eq\{[6F_0(R) - \tfrac{2}{3}F_4(R)] - [6F_0(R + \Delta R) - \tfrac{2}{3}F_4(R + \Delta R)]\}$$

$$(A3.1)$$

For $\Delta = \varepsilon_1 - \varepsilon_2$ we get

$$\Delta = eq\tfrac{5}{3}[F_4(R) - F_4(R + \Delta R)] \qquad (A3.2)$$

These equations can be simplified if we assume that the ligand–dipoles can be considered as point dipoles; that is, the dimensions of the dipoles ΔR are much smaller than the metal–ligand distance R, $\Delta R \ll R$. Under this assumption

$$\frac{1}{\Delta R}[F_k(R + \Delta R) - F_k(R) \approx \frac{dF_k(R)}{dR} = F'_k(R)$$

Taking into account that $q\,\Delta R = \mu$, we obtain easily Eq. (4.30)

$$\Delta = -\tfrac{5}{3}e\mu F'_4(R) \qquad (A3.3)$$

Tetragonally Distorted Octahedron and Planar Square. If the six ligand–point charges q form tetragonal symmetry D_{4h}, then

$$R_1 = R_4 \qquad R_2 = R_3 = R_5 = R_6 \qquad (R_1 \neq R_2)$$

$$\vartheta_1 = 0 \qquad \vartheta_2 = \vartheta_3 = \vartheta_5 = \vartheta_6 = \frac{\pi}{2} \qquad \vartheta_4 = \pi \qquad (A3.4)$$

$$\phi_2 = 0 \qquad \phi_3 = \frac{\pi}{2} \qquad \phi_5 = \pi \qquad \phi_6 = \frac{3\pi}{2}$$

With these relations the nonzero matrix elements $V_{mm'}$ are the same as for the octahedron, and the roots ε are the same, as in Eqs. (4.13). However, the expressions for $V_{mm'}$ do not coincide with (4.12):

$$V_{22} = eq\{2F_0(R_1) + 4F_0(R_2) - \tfrac{4}{7}[F_2(R_1) - F_2(R_2)] + \tfrac{2}{21}F_4(R_1) + \tfrac{1}{14}F_4(R_2)\}$$
$$V_{11} = eq\{2F_0(R_1) + 4F_0(R_2) + \tfrac{2}{7}[F_2(R_1) - F_2(R_2)] - \tfrac{8}{21}F_4(R_1) - \tfrac{2}{7}F_4(R_2)]$$
$$V_{00} = eq\{2F_0(R_1) + 4F_0(R_2) + \tfrac{4}{7}[F_2(R_1) - F_2(R_2)] + \tfrac{4}{7}F_4(R_1) + \tfrac{3}{7}F_4(R_2)\}$$
$$V_{2-2} = \tfrac{5}{6}eqF_4(R_2) \qquad\qquad\qquad\qquad\qquad\qquad\qquad\qquad (A3.5)$$

For this reason all four roots (4.13) are different, and for the description of the splitting one should introduce, together with the cubic splitting parameter (4.15),

$$\Delta = 10Dq = \tfrac{5}{3}eqF_4(R) \qquad (A3.6)$$

two more parameters of the tetragonal splitting:

$$D_s = \tfrac{2}{7}eq[F_2(R_2) - F_2(R_1)]$$
$$D_t = \tfrac{2}{21}eq[F_4(R_2) - F_4(R_1)] \qquad (A3.7)$$

With these parameters the d electron energy levels are (the type of symmetry and d function are indicated in parentheses)

$$\varepsilon_1(A_{1g}; d_{z^2}) = E_0' + \tfrac{3}{5}\Delta - 2D_s - 6D_t$$
$$\varepsilon_2(B_{1g}; d_{x^2-y^2}) = E_0' + \tfrac{3}{5}\Delta + 2D_s - D_t$$
$$\varepsilon_3(B_{2g}; d_{xy}) = E_0' - \tfrac{2}{5}\Delta + 2D_s - D_t \qquad (A3.8)$$
$$\varepsilon_{4,5}(E_g; d_{xy}, d_{yz}) = E_0' - \tfrac{2}{5}\Delta - D_s + 4D_t$$

where $E_0' = 2F_0(R_1) + 4F_0(R_2)$.

In the limit of strong tetragonal distortion with two axial ligands at infinity, the octahedron transforms into a *planar-square configuration* with the CA in the center. The corresponding energy levels can be found by tending $R_1 \to \infty$. This results in $F_k(R_1) \to 0$ [see Eqs. (A2.12) to (A2.14)], and hence $D_s = \frac{2}{7}eqF_2(R_2)$, $D_t = \frac{2}{21}eqF_4(R_2)$. Note that in this case the three parameters Δ, D_s, and D_t are no longer independent: $D_t = \frac{2}{35}\Delta$, and the energy levels depend on only two parameters, D_s and Δ.

For $F_k(R_1) \to 0$ the nonzero matrix elements (A3.5) become as follows:

$$V_{22} = eq[4F_0(R) - \tfrac{4}{7}F_2(R) + \tfrac{1}{14}F_4(R)]$$

$$V_{11} = eq[4F_0(R) - \tfrac{2}{7}F_2(R) - \tfrac{2}{7}F_4(R)]$$

$$V_{00} = eq[4F_0(R) - \tfrac{4}{7}F_2(R) + \tfrac{3}{7}F_4(R)] \tag{A3.9}$$

$$V_{2-2} = eq\tfrac{5}{6}F_4(R)$$

This allows one to obtain the energy levels of d electrons in the square-planar crystal field of the ligands:

$$\varepsilon_1(A_{1g}; d_{z^2}) = E_0^S + \tfrac{9}{35}\Delta - 2D_S$$

$$\varepsilon_2(B_{1g}; d_{x^2-y^2}) = E_0^S + \tfrac{19}{35}\Delta + 2D_S$$

$$\varepsilon_3(B_{2g}; d_{xy}) = E_0^S - \tfrac{16}{35}\Delta + 2D_S \tag{A3.10}$$

$$\varepsilon_{4,5}(E_g; d_{xz}, d_{yz}) = E_0^S - \tfrac{6}{35}\Delta - D_S$$

where $E_0^S = 4F_0(R)$.

Tetrahedron and Cube. For four ligand–point charges forming a regular tetrahedron with the CA in the center, we have

$$q_i = q \qquad R_i = R \qquad i = 1, 2, 3, 4$$

$$\cos \vartheta_1 = \cos \vartheta_2 = -\cos \vartheta_3 = -\cos \vartheta_4 = \frac{1}{\sqrt{3}} \tag{A3.11}$$

$$\phi_1 = \frac{\pi}{4} \qquad \phi_2 = \frac{3\pi}{4} \qquad \phi_3 = \frac{5\pi}{4} \qquad \phi_4 = \frac{7\pi}{4}$$

For these coordinates the roots of the secular equation (4.7) are given by the same formulas (4.13) as for the octahedron, but with different expressions for the matrix elements $V_{mm'}$:

$$V_{22} = eq[4F_0(R) - \tfrac{2}{27}F_4(R)]$$

$$V_{11} = eq[4F_0(R) + \tfrac{8}{27}F_4(R)]$$

$$V_{00} = eq[4F_0(R) - \tfrac{4}{9}F_4(R)] \tag{A3.12}$$

$$V_{2-2} = eq\tfrac{10}{27}F_4(R)$$

Here, as in the octahedron (and any other cubic system),

$$V_{22} + V_{2-2} = V_{11}$$
$$V_{22} - V_{2-2} = V_{00}$$
(A3.13)

Therefore, among the five roots (4.13) there are only two different groups with two and three equal roots, respectively:

$$\varepsilon_1 = \varepsilon_2 = \varepsilon_3 = \varepsilon(T_2) = eq[4F_0(R) + \tfrac{8}{27}F_4(R)]$$
$$\varepsilon_4 = \varepsilon_5 = \varepsilon(E) = eq[4F_0(R) - \tfrac{4}{9}F_4(R)]$$
(A3.14)

Thus the splitting in the tetrahedral field is similar to that of the octahedral field (4.14), with the significant distinction that in the tetrahedron the E term is lower than T_2. Besides, the absolute value of the splitting in the tetrahedron $\Delta_T = \varepsilon(E) - \varepsilon(T_2)$,

$$\Delta_T = -eq\tfrac{20}{27}F_4(R) = -\tfrac{4}{9}\Delta$$
(A3.15)

is $\tfrac{4}{9}$ of the splitting in the octahedron, provided that the interatomic distances R and ligand–point charges q are the same in these two type of complexes.

Quite similar to the tetrahedral field, the cubic field created by eight ligand–point charges at the eight corners of a regular cube can be considered just by doubling the field strength. The resulting d electron energy terms and crystal field splitting Δ_k are

$$\varepsilon(T_2) = eq[8F_0(R) + \tfrac{16}{17}F_4(R)] = E_0^k + \tfrac{2}{5}\Delta_k$$
$$\varepsilon(E) = eq[8F_0(R) - \tfrac{8}{9}F_4(R)] = E_0^k - \tfrac{3}{5}\Delta_k$$
(A3.16)

$$\Delta_k = \tfrac{40}{27}eqF_4(R) = 2\Delta_T = \tfrac{8}{9}\Delta$$
(A3.17)

$$E_0^k = 8eqF_0(R) = 2E_0^T = \tfrac{4}{3}E_0$$
(A3.18)

Appendix 4: Matrix Elements of Crystal Field Perturbation of a Two-Electron Term $F(nd)^2$, V'_{ij}, $i,j = 1,2,\ldots,7$, After (4.34), Expressed by One-Electron Matrix Elements $V_{mm'}$ Given in Appendix 2

$$V'_{11} = V_{11} + V_{22}$$
$$V'_{12} = V_{10}$$
$$V'_{13} = (\tfrac{3}{5})^{1/2}V_{1-1} - (\tfrac{2}{5})^{1/2}V_{20}$$
$$V'_{14} = (\tfrac{1}{5})^{1/2}V_{1-2} - (\tfrac{4}{5})^{1/2}V_{2-1}$$
$$V'_{15} = -(\tfrac{3}{5})^{1/2}V_{2-2}$$
$$V'_{16} = 0$$

$$V'_{17} = 0$$

$$V'_{22} = V_{00} + V_{22}$$

$$V'_{23} = (\tfrac{3}{5})^{1/2} V_{0-1} + (\tfrac{2}{5})^{1/2} V_{21}$$

$$V'_{24} = (\tfrac{1}{5})^{1/2} V_{0-2}$$

$$V'_{25} = -(\tfrac{2}{5})^{1/2} V_{2-1}$$

$$V'_{26} = -V_{2-2} = (\tfrac{5}{3})^{1/2} V'_{15}$$

$$V'_{27} = 0$$

$$V'_{33} = \tfrac{3}{5}(V_{22} + V_{-1-1}) + \tfrac{2}{5}(V_{00} + V_{11})$$

$$V'_{34} = (\sqrt{3}/5)V_{-1-2} + (2\sqrt{3}/5)V_{21} + (2\sqrt{2}/5)V_{0-1}$$

$$V'_{35} = (\sqrt{6}/5)V_{20} + (\sqrt{6}/5)V_{0-2} - \tfrac{2}{5}V_{1-1}$$

$$V'_{36} = -(\tfrac{2}{5})^{1/2} V_{1-2}$$

$$V'_{37} = -(\tfrac{3}{5})^{1/2} V_{2-2}$$

$$V'_{44} = \tfrac{1}{5}(V_{22} + V_{-2-2}) + \tfrac{4}{5}(V_{11} + V_{-1-1})$$

$$V'_{45} = (\sqrt{3}/5)V_{21} + (2\sqrt{3}/5)V_{-1-2} + (2\sqrt{2}/5)V_{10}$$

$$V'_{46} = (\tfrac{1}{5})^{1/2} V_{20}$$

$$V'_{47} = (\tfrac{3}{5})^{1/2} V_{2-1} - (\tfrac{4}{5})^{1/2} V_{1-2}$$

$$V'_{55} = \tfrac{3}{5}(V_{11} - V_{-2-2}) + \tfrac{2}{5}(V_{00} + V_{-1-1})$$

$$V'_{56} = (\tfrac{3}{5})^{1/2} V_{10} + (\tfrac{2}{5})^{1/2} V_{-1-2}$$

$$V'_{57} = (\tfrac{3}{5})^{1/2} V_{1-1} - (\tfrac{2}{5})^{1/2} V_{0-2}$$

$$V'_{66} = V_{00} + V_{-2-2}$$

$$V'_{67} = V_{0-1}$$

$$V'_{77} = V_{1-1} + V_{-2-2}$$

Appendix 5: Matrix Elements of Crystal Field Perturbation of f Electron States (After [4.22])

The wavefunction of the f electron can be taken in the form

$$\Psi_{n3m} = R_{n3}(r)Y_{3m}(\vartheta, \phi) \qquad m = 0, \pm 1, \pm 2, \pm 3 \tag{A5.1}$$

Considering the ligand–point charge perturbation after Eq. (4.2), denoting the ligand coordinates by $\mathbf{R}(R_j, \vartheta_j, \phi_j)$ and the functions $F_n \equiv F_n(R_j)$ after (A2.2) with $R_{n3}(r)$ instead of $R_{n2}(r)$,

$$R_{43}(r) = (\tfrac{140}{9})^{1/2}\alpha^{9/2}r^3e^{-\alpha r} \tag{A5.2}$$

and following the computation procedure of Appendix 2, we get the following matrix elements of the ligand–point charge perturbation of the f states (the sum is taken over all the ligands j):

$$V_{00} = e^2 \sum_j [F_0 + \tfrac{2}{15}(3\cos^2\vartheta_j - 1)F_2 + \tfrac{1}{44}(35\cos^4\vartheta_j$$
$$- 30\cos^2\vartheta_j + 3)F_4 + \tfrac{25}{1716}(231\cos^6\vartheta_j - 315\cos^4\vartheta_j$$
$$+ 105\cos^2\vartheta_j - 5)F_6] \tag{A5.3}$$

$$V_{\pm 1 \pm 1} = e^2 \sum_j [F_0 + \tfrac{1}{10}(3\cos^2\vartheta_j - 1)F_2 + \tfrac{1}{264}(35\cos^4\vartheta_j - 30\cos^2\vartheta_j + 3)F_4$$
$$- \tfrac{25}{2288}(231\cos^6\vartheta_j - 315\cos^4\vartheta_j + 105\cos^2\vartheta_j - 5)F_6] \tag{A5.4}$$

$$V_{\pm 2 \pm 2} = e^2 \sum_j [F_0 - \tfrac{7}{264}(35\cos^4\vartheta_j - 30\cos^2\vartheta_j + 3)F_4$$
$$+ \tfrac{5}{1144}(231\cos^6\vartheta_j - 315\cos^4\vartheta_j + 105\cos^2\vartheta_j - 5)F_6] \tag{A5.5}$$

$$V_{\pm 3 \pm 3} = e^2 \sum_j [F_0 - \tfrac{1}{6}(3\cos^2\vartheta_j - 1)F_2 + \tfrac{1}{88}(35\cos^4\vartheta_j - 30\cos^2\vartheta_j + 3)F_4$$
$$- \tfrac{5}{6864}(231\cos^6\vartheta_j - 315\cos^4\vartheta_j + 105\cos^2\vartheta_j - 5)F_6] \tag{A5.6}$$

$$V_{32} = e^2 \sum_j [(\sqrt{6}/6)\exp(-i\phi_j)\sin\vartheta_j\cos\vartheta_j\{-F_2 + \tfrac{5}{22}(7\cos^2\vartheta_j - 3)F_4$$
$$- \tfrac{35}{1144}(33\cos^4\vartheta_j - 30\cos^2\vartheta_j - 30\cos^2\vartheta_j + 5)F_6] \tag{A5.7}$$

$$V_{31} = e^2 \sum_j (\sqrt{15}/2)\exp(-2i\phi_j)\sin^2\vartheta_j[-\tfrac{1}{15}F_2 + \tfrac{1}{22}(7\cos^2\vartheta_j - 1)F_4$$
$$- \tfrac{35}{3432}(33\cos^4\vartheta_j - 18\cos^2\vartheta_j + 1)F_6] \tag{A5.8}$$

$$V_{30} = e^2 \sum_j \sqrt{5}\tfrac{7}{44}\exp(-3i\phi_j)\sin^3\vartheta_j\cos\vartheta_j[F_4 - \tfrac{5}{26}(11\cos^2\vartheta_j - 3)F_6]$$
$$\tag{A5.9}$$

$$V_{3-1} = e^2 \sum_j \sqrt{15}\tfrac{7}{88}\exp(-4i\phi_j)\sin^4\vartheta_j[\tfrac{1}{3}F_4 - \tfrac{5}{26}(11\cos^2\vartheta_j - 1)F_6] \tag{A5.10}$$

$$V_{3-2} = -e^2 \sum_j \sqrt{6}\tfrac{385}{2288}\exp(-5i\phi_j)\sin^5\vartheta_j\cos\vartheta_j F_6 \tag{A5.11}$$

$$V_{3-3} = -e^2 \sum_j \tfrac{385}{2288}\exp(-6i\phi_j)\sin^6\vartheta_j F_6 \tag{A5.12}$$

$$V_{2-2} = e^2 \sum_j \tfrac{35}{88}\exp(-4i\phi_j)\sin^4\vartheta_j[\tfrac{1}{3}F_4 + \tfrac{3}{13}(11\cos^2\vartheta_j - 1)F_6] \tag{A5.13}$$

$$V_{1-1} = -e^2 \sum_j \exp(-2i\phi_j) \sin^2\vartheta_j [\tfrac{1}{5}F_2 + \tfrac{5}{66} (7\cos^2\vartheta_j - 1)F_4$$

$$+ \tfrac{175}{2288} (33\cos^4\vartheta_j - 18\cos^2\vartheta_j + 1)F_6] \tag{A5.14}$$

$$V_{10} = -e^2 \sum_j \sqrt{3} \exp(-i\phi_j) \sin\vartheta_j \cos\vartheta_j [\tfrac{1}{15}F_2 + \tfrac{5}{132} (7\cos^2\vartheta_j - 3)F_4$$

$$+ \tfrac{175}{3432} (33\cos^4\vartheta_j - 3\cos^2\vartheta_j + 5)F_6] \tag{A5.15}$$

$$V_{21} = e^2 \sum_j \sqrt{10} \exp(-i\phi_j) \sin\vartheta_j \cos\vartheta_j [-\tfrac{1}{10}F_2 - \tfrac{1}{33} (7\cos^2\vartheta_j - 3)F_4$$

$$+ \tfrac{35}{2288} (33\cos^4\vartheta_j - 30\cos^2\vartheta_j + 5)F_6] \tag{A5.16}$$

$$V_{20} = e^2 \sum_j (\sqrt{30}/6) \exp(-i\phi_j) \sin\vartheta_j [-\tfrac{1}{5}F_2 - \tfrac{1}{44} (7\cos\vartheta_j - 1)F_4$$

$$+ \tfrac{35}{572} (33\cos^4\vartheta_j - 18\cos^2\vartheta_j + 1)F_6] \tag{A5.17}$$

$$V_{2-1} = e^2 \sum_j \sqrt{10} \tfrac{7}{44} \exp(-3i\phi_j) \sin^3\vartheta_j \cos\vartheta_j [\tfrac{1}{3}F_4 + \tfrac{15}{52} (11\cos^2\vartheta_j - 3)F_6]$$

$$\tag{A5.18}$$

SUBJECT INDEX

FORMULA INDEX

Chemical formulas are listed by the coordination centers as they appear in the periodic table (separate ligands are not listed). Within each element entry the sequence is: complexes with simple ligands, polydentate coordination, (explicitly) crystal systems, and polynuclear compounds.